Principles of Phase Structures in Particle Physics

World Scientific Lecture Notes in Physics

World Scientific Lecture Notes in Physics – Vol. 77

Principles of Phase Structures in Particle Physics

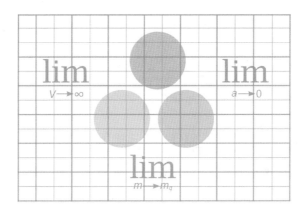

Hildegard Meyer-Ortmanns

International University Bremen, Germany

Thomas Reisz

Formerly of Universität Heidelberg, Germany

 **World Scientific**

NEW JERSEY · LONDON · SINGAPORE · BEIJING · SHANGHAI · HONG KONG · TAIPEI · CHENNAI

Published by

World Scientific Publishing Co. Pte. Ltd.

5 Toh Tuck Link, Singapore 596224

USA office: 27 Warren Street, Suite 401-402, Hackensack, NJ 07601

UK office: 57 Shelton Street, Covent Garden, London WC2H 9HE

1005 122791 T

British Library Cataloguing-in-Publication Data
A catalogue record for this book is available from the British Library.

PRINCIPLES OF PHASE STRUCTURES IN PARTICLE PHYSICS
World Scientific Lecture Notes in Physics — Vol. 77

ISBN-13 978-981-02-3441-6
ISBN-10 981-02-3441-4

Printed in Singapore by World Scientific Printers (S) Pte Ltd

To Lara Celine and Anne Sophie (H.M.-O.)

To My Parents (T.R.)

Preface

Although phase transitions of particle physics appear as exotic phenomena from the viewpoint of daily-life physics, their occurrence in the early universe has far reaching consequences for the status of the universe nowadays and makes these transitions conceptually interesting. "Little bangs" in heavy-ion collisions may reproduce these exotic states of matter in lab-experiments, since the available energies in these collisions are meanwhile sufficiently high to produce these states for a very short instant of time. The aim of the book is to derive the phase structure of particle physics by applying standard thermodynamic concepts to particle physics.

The book is based on a review about *Phase Transitions in Quantum Chromodynamics* by one of the authors (H. M.-O.), published in Rev. Mod. Phys. 68 (1996) 473, and on a series of lectures about *Phase Transitions and Critical Phenomena with Applications to Particle Physics* at the Universities of Wuppertal and Heidelberg. It was extended to include results on the phase structure of the electroweak part of the standard model, as well as further theoretical developments such as perturbation theory on the lattice at zero and finite temperature, dimensional reduction, flow equations, gap equations and various resummation techniques.

For a detailed understanding of the phase structure of quantum chromodynamics and the electroweak standard model, approximations in analytical and numerical calculations are unavoidable. The aim of the book is to explain these methods and to clearly specify the conditions under which the approximations are under control. The analytical part covers methods both in a continuum regularization and on a spacetime lattice. Since the lattice plays a distinguished role as a nonperturbative treatment of the phase structure of the standard model, it makes up the larger part of the book. In the chapter on numerical methods we focus on criteria for an appropriate interpretation of numerical data rather than on details of algorithms to produce them. Where possible we try to bridge the gap between different approaches to one and the same problem, for example between a phenomenological description of first-order phase transitions and rigorous results about their finite-size scaling analysis, or between lattice and continuum approaches, or between effective models and first-principle calculations. The presentation also contains more recent topics which cannot be found in other textbooks so far. To these belong generalized series expansions, dimensional reduction, resummation techniques and certain realizations of the renormalization group.

The book is addressed to graduate and Ph. D. students, to post-docs and interested colleagues, primarily from the field of particle physics. However, to a certain extent it should be of interest for colleagues from the field of statistical physics in general; some tools we are using are not standard ones from statistical physics, but were developed for solving problems in particle physics. Such tools have, for example, applications to the phase structure of spin glasses.

The style of the chapters varies from self-contained, detailed introductions, assuming basic knowledge from undergraduate lectures, to reviews of advanced research. In such reviews we explain the main ideas whose realizations are sometimes quite demanding in the techniques they use. To go there into detail would go beyond the scope of this book. Sections on perturbative approaches can be followed in detail if the reader has expertise in perturbative calculations. In general, we summarize the main ingredients in all chapters to make them self-contained, but the presentation is sometimes condensed so that a basic background knowledge in quantum field theory will facilitate the understanding. On the basis of this book it should be possible to follow the more recent literature on topics in this field.

The aim of the book is neither to give a complete or updated review on possible approaches nor a summary of the state of art nor a complete list of results for the phase structure of particle physics that are available today. The emphasis is on *principles*.

We are very much indebted to World Scientific for its patience with the submission of our manuscript. For one of us (H.M.-O.) it is also a pleasure to thank the International University Bremen, where it became possible to finalize the book during the last years, and in particular to her student Tiberiu Tesileanu for his support in providing a camera-ready manuscript.

Bremen, July 2006

Hildegard Meyer-Ortmanns

Contents

Chapter 1

Introduction

Phase transitions are found over a wide range of temperatures. They start near absolute zero with the Mott transitions [1] from superconductor to insulator, and are common between 10^2 [K] and 10^3 [K]. But have you ever asked yourself how much one should heat up "normal" condensed matter until it dissolves into its elementary constituents, quarks, gluons, electrons, Higgs-bosons, W-and Z-particles and the like? The answer is known. Nine orders of magnitude higher than the more familiar temperature range one finds the phase transitions of quantum chromodynamics (QCD). At a temperature typical of the QCD-scale ($\Lambda_{\overline{MS}} \sim 200$ [MeV]), nuclear matter melts during the QCD transition. At this temperature quarks and gluons cease to be confined inside hadrons, and begin moving freely. Three orders of magnitude higher is the electroweak phase transition, at which the electromagnetic and weak interactions are unified, and the particle spectrum of the electroweak part of the standard model changes according to the symmetry restoration that goes along with the electroweak phase transition.

Let us first give a short review about the phases of particle physics which the transitions transform into each other. In case of the electroweak model the transition is temperature-driven. It restores at high temperatures the full $SU(2) \times SU(2) \times U(1)$-symmetry that is spontaneously broken to $U(1)$ at low temperatures. The phases are characterized by different symmetries along with different particle spectra.

Above we have glibly mentioned the QCD transition as if it were unique. However, there are really two, and they take place if either the temperature or the density crosses some critical point. One expects, naively, a phase transition in QCD if the density rises to the point at which the hadrons

1

begin to overlap. This will be around 0.5 [GeV/fm^3], the energy density in a proton. Similarly, for temperatures above the QCD-scale parameter, $T > \Lambda_{\overline{MS}} = 200$ [MeV], one expects a temperature-driven phase transition. This book concentrates on the effects of temperature, as this is the area in which the lattice is strong and many partial, preliminary or even rigorous results are available, for which we outline the methods to achieve them, but there is no physical argument that would justify this restriction. Transitions at high baryon densities are as relevant as those at high temperatures.

Again, in most (although not all) approaches to the temperature-driven transition in QCD one assumes a spontaneous breaking of a symmetry as the characteristic feature of the transition. Here we have to distinguish between the chiral $SU(N_f) \times SU(N_f)$ symmetry for QCD with N_f massless flavors and the $Z(3)$ symmetry for QCD with infinitely heavy masses and three colors. ($Z(3)$ is the center of the gauge symmetry group $SU(3)$.) In the massless or so-called chiral limit the transition transforms the phases with broken-chiral symmetry at low temperature to the chiral-symmetric phase at high temperature, whereas the $Z(3)$ symmetry is spontaneously broken at high temperature, characterized by the deconfinement of quarks and gluons, and restored at low temperature in the confinement phase, in which quarks and gluons are confined to hadrons.

Obviously these limits of chiral and $Z(3)$ symmetries are idealizations of the physical case, in which only the up-and down quarks are light compared to the scale of the transition temperature, the charm-, bottom- and top-quark masses are heavy compared to this scale (therefore irrelevant for the thermodynamics), but the strange-quark mass is just of the order of T_c. Thus it is neither clear whether these transitions persist for physical quark masses nor whether they will occur together. If they take place at different temperatures, one expects the chiral transition to occur after the deconfinement transition ([4, 5]). Lattice results indicate, however, that both transitions coincide at the same critical point. One of them can then be regarded as driving the other, and the nature of the transition will be determined by the driving symmetry. From here on it is assumed that both happen simultaneously, unless stated otherwise. If we still talk about two transitions in the following, we mean "the" QCD transition and the electroweak transition.

These two transitions may seem exotic, but according to the big-bang theory they occurred at least once in the universe. The electroweak transition is predicted to have taken place at 10^{-10}[s], and the deconfinement transition at 10^{-6}[s] after the big bang. In the following let us focus on QCD. The deconfinement transition occurred, when the temperature of the universe dropped to the order of tera degrees, i.e. to a scale $T_c \sim (2.32 \pm 0.6) \cdot 10^{12}$[K] or 200 ± 50 [MeV]. (The conversion from natural energy units [MeV] to degrees Kelvin [K] is determined by the Boltzmann constant $k_B \sim 8.6 \cdot 10^{-5}$[eV/K] or 100 [MeV] $\sim 1.16 \cdot 10^{12}$[K].) These temperatures with time scales of 10^{-23}[s] and typical distances of 1 [fm] may now be achievable in the laboratory. For example, at the Brookhaven National Laboratory one nowadays reaches energy densities between $20 - 98$ [GeV/fm^3] in Au-Au (gold-on-gold)-collisions [2], and even higher initial energy densities are expected for nucleus-nucleus collisions at the Linear-Hadron-Collider (LHC) at CERN.

Thinking about experiments of particle physics at these labs, one usually associates particle collisions, measurements of cross sections or high-precision tests of the standard model. What are therefore the typical questions whose answers require a knowledge of the *phase structure of particle physics*? We list some of them in the following. For example, one may be curious about the interior of a very dense neutron star. What was the distribution of protons and neutrons before primordial nucleosynthesis set in? What was the particle spectrum when the electromagnetic and weak forces were still unified? What are the remnants of the transitions in the early universe? Are there specific remnants for the type of phase conversion at these early times? Are large-scale structures induced by these transitions? When mimicking the big bang by so-called little bangs at heavy-ion collisions, in which for very short instances of time very high energy densities are created, is it meaningful at all to apply thermodynamic concepts to the "fireballs", in particular those from equilibrium thermodynamics, and to call them "hot"? Is the observed particle yield in these collisions primordial? Are the signatures for the QCD transition unique and conclusive? Which heat-conducting processes lead to the cooling of the fireball, and is it rapid or slow as compared to the other involved time-scales? Independently of these realizations, what do we actually mean if we speak of "free" quarks and gluons if the gas made out of quarks and gluons is extremely viscous, a gas of pure "glue" would be characterized by a shear viscosity of the order of the Λ-QCD-scale, that is about 10^{16} times the viscosity of a classical gas under "normal" conditions (where it is 10^{-5}[kg/m s]) [3]. How are particles produced in a medium with long-range correlations (usually the correlations are assumed to fall off exponentially fast, which is justified unless the system produces particles at a second-order phase transition).

Last but not least, how should one measure the temperature if all standard thermometers have dissolved as well into their basic constituents?

Even after reading the more than six hundred pages of this book, most of these questions remain open to a certain extent, but you will understand why it is so inherently difficult to come up with definite answers.

After all, the tools needed for treating gases of elementary particles are not at all "elementary" themselves. The book is not about standard matter being composed of classical or quantum particles, with the only difference of an unusually high temperature. Our "particles" here are particle-like excitations of quantized fields with infinitely many degrees of freedom. Thus we have to deal with the *field theoretical* limit in addition to the *thermodynamic* limit. Let us assume that we use a lattice regularization of quantum field theory. It is then the continuum limit that we must control, in principle before we take the thermodynamic limit (thermodynamic limit in the sense that the volume in physical units (say in units of fermi) goes to infinity). As long as the lattice constant is finite we have a quantum mechanical system of a countable (finite or infinite) number of degrees of freedom. It may actually happen that a phase transition in such a quantum mechanical system does not "survive" the continuum limit. Lattice physicists call this phenomenon a bulk transition that is a mere lattice artifact from the point of view of continuum-quantum field theory. Later we give explicit examples from pure gauge theories. Because of the field theoretical aspects we spend a separate chapter on the very continuum limit, the conditions under which it is unique and methods of how to control it.

From critical phenomena one knows that many length- (or equivalently energy) scales play a role at a phase transition when fluctuations on all length scales appear simultaneously. This makes it impossible to find a suitable expansion parameter for a perturbative treatment. In addition, QCD is asymptotically free in the UV-limit, but has a coupling $g(T) \approx 1$ near the phase transition. So a nonperturbative approach must be used; the lattice regularization may be the only means of calculating the properties of the phase transition from first principles.

Away from the transition region, at extremely high temperatures, $T \gg \Lambda_{\overline{MS}}$ (where $\Lambda_{\overline{MS}}$ stands for the QCD-scale), with typical momenta $Q \approx T$, the coupling $Q^2/\Lambda_{\overline{MS}}^2$ becomes weak, and resummed finite-temperature perturbation theory [6, 7] may be applied. This high temperature defines a scale in space of the order of $1/T = l_p$, the so-called plasma length. A further complication as compared to "standard" condensed matter systems are two further length scales in the bulk of the order of $l_e = 1/(gT)$, the (color-) electric screening length, and $l_m = 1/(g^2 T)$, the (color-)magnetic screening length, so that we have $l_P < l_e < l_m$ [8]. While the first one gives the range over which perturbation theory may be applied, the second one corresponds to the scale, set by the chromoelectric mass, the third one to the scale, set by the chromomagnetic mass, for which not only naive perturbation theory fails, but also resummed perturbation theory, at least to date. The precise definition of these masses will be given later. Even the very gauge invariant definition of the chromomagnetic screening mass to all orders of perturbation theory is conceptually not clear. It is even less clear how to define it in a gauge invariant and nonperturbative way. What has been used so far is a gauge independent but variant definition by means of the polarization tensor of the gauge field ("gauge independent" refers to the independence of the gauge parameter within a certain class of gauges). According to the latter definition, the magnetic screening mass vanishes up to one loop order. The two-loop order is inherently ill defined because of intrinsic infrared problems of nonabelian massless gauge theories in (effectively) three dimensions. Therefore we spend an extended chapter on weak-coupling expansions at finite temperature in spite of the fact that the coupling is strong in the phase transition region. Although the corresponding long-range interactions are strong interactions that may lead to some kind of dynamical confinement in the plasma, the bulk properties, the equation of state etc., are still dominated by short-range interactions. So perturbation theory is valid at high temperatures for these quantities. More evidence that the chromomagnetic long-range sector of QCD has a complicated nonperturbative structure comes from the observed realization of dimensional reduction. In a $(3 + 1)$-dimensional Euclidean field theory the imaginary time dimension is proportional to the inverse temperature, $l_\tau \propto 1/T$. Thus the theory reduces, in the high-temperature limit, to a three-dimensional theory. This is plausible. But as it turns out, in case of QCD the three-dimensional theory must be treated nonperturbatively, it is given by the adjoint gauge-Higgs model without fermions, see section 4.6. Clearly it is important to clarify the validity range of perturbative expansions and to see how far perturbative results can be extrapolated towards the transition region. For example, in the framework of dimensional reduction applied to QCD, the screening mass can be calculated within perturbation theory for temperatures from high ones down to $2\,T_c$.

Let us assume for a moment that the challenges in the theoretical description of the phase structure can be mastered and the phase structure is predicted in a reliable way. Yet we are not done. To some extent the situation in particle physics is opposite to that in condensed matter physics. Phase transitions in condensed matter appear as changes of macroscopic properties as a function of external, tunable parameters like temperature or magnetic fields. Usually the task is the prediction and understanding of such macroscopic phenomena, at the transition point and its vicinity, in terms of a microscopic model that has to be found. In particle physics the microscopic (even fundamental) theories of strong and electroweak interactions are known, but the macroscopic manifestations of phase transitions in these theories must be derived. It is by far not obvious how to infer the QCD transition in the early universe from some relics nowadays, or from signatures in lab-experiments where the QCD transition can be reproduced in a so-called little bang of heavy-ions. (The energy density needed for reproducing the electroweak transition is out of the range of lab-experiments). In addition, apart from computer realizations or other theoretical treatments, a temperature of the order of more than one tera-degree is no longer an easily tunable parameter that allows for a perfectly controlled experimental reproduction of the transition as a function of temperature.

Since the predictions of observable remnants depend on the very phase structure, let us look in some more detail why this phase structure, in particular that of QCD with realistic values for the quark masses, has not fully settled in spite of several decades of computer simulations and analytical calculations.

A short view at predictions of the order of the QCD transition should illustrate how sensitively the results depend on the involved approximations. First let us briefly recall that the *order of a phase transition* is one of the basic thermodynamic classifications. A phase transition is said to be of *first order* if there is at least one finite gap in the first derivatives of a suitable thermodynamic potential in the thermodynamic limit. A finite latent heat goes often along with a gap in the order parameter. A transition is said to be of *second order* if there is an algebraic singularity in at least one of the second derivatives of the potential. If the thermodynamic potential is analytic over the whole temperature range (for temperature-driven transitions), the phase conversion is called a *crossover phenomenon*.

In general the order of the QCD transitions depends on the number of colors (N_c), the number of flavors (N_f), the current quark masses, and on more subtle effects related to the IR- and UV-cut-offs. The deconfinement transition for infinite quark masses is now believed to be of first order for three colors (see section 4.7.5 and references therein), and the equation of state is known in the limit of infinite quark masses. Similarly the chiral transition is believed to be of second order for two massless flavors and of first order for three or more massless flavors. These results do not contradict each other. They just indicate the sensitivity of the order to the involved approximations and should be taken as a warning not to jump to conclusions.

Ultimately we are, of course, interested in the physically relevant case of two light quarks (up and down), and one heavier quark (the strange quark). This case is more difficult to handle since the strange quark mass is of the same order as the transition temperature or QCD's intrinsic scale, so it can be neither treated as zero nor as infinite in comparison to QCD's scale (say $\Lambda_{\overline{MS}}$). It is therefore an open question as to whether the chiral and the deconfinement transitions persist for realistic quark masses. All indications so far point in the same direction: There are no QCD transitions at all, at least not in the thermodynamic sense of nonanalytic behavior in a thermodynamic potential like the free energy. Still the phases far below and far above the intrinsic scale of QCD may look quite differently, and a rapid crossover in the range of temperatures of the order of $\Lambda_{\overline{MS}}$ would offer the chance to see some specific signatures of the transition region. If we nevertheless keep the wording "transition" and "critical temperature", we either use it for the characteristic temperature scale of crossover phenomena or for true transitions in the artificial world of computer simulations with unrealistic quark masses.

The order of a phase transition has far-reaching phenomenological consequences. Phenomenological implications of the electroweak and the QCD transitions are supposed to have a strong impact even on the nowadays universe if the transitions were strongly first-order. For a first-order transition one expects metastabilities with latent heat, interfaces, supercooling and overheating etc.. The experimental consequences of a first-order transition make it relatively easy to see, especially if the plasma "explodes" into the hadronic phase [9]. A second-order transition, lacking a jump in the energy density, may be less easy to see experimentally. However, it has divergent correlation lengths (at least in the thermodynamic limit). These may also lead to observable consequences ([10–15]), in analogy to the well-known phenomenon of critical opalescence.

As indicated above, perhaps a smooth crossover is the most realistic scenario, as in the transition from a molecular or atomic gas to an electric plasma. In this case, one would have a smooth transition from a pion gas at low temperatures to a quark-gluon plasma at high temperatures, with a highly nontrivial mixture of excitations in the neighborhood of the crossover. A sharp crossover phenomenon with a rapid change in thermo-dynamic quantities over a small temperature interval (say of the order of 10 [MeV]) offers some chance for measurable effects in experiments. For example, double Φ-peaks in the dilepton-invariant mass spectrum are still predicted as a signature for the phase conversion, as long as the crossover phenomenon is rapid enough ([16, 17]).

Ultra-relativistic heavy-ion collisions have a finite volume expanding in time. Any finite volume smooths out the nonanalyticities in the free en-ergy. This is well known from computer simulations that are unavoidably performed in a finite volume. Moreover, the high-temperature matter cre-ated in the collisions may not reach equilibrium either, and has different processes dominating in different regions of phase space. It becomes a nontrivial task to find suitable "thermometers", to tune and measure the temperature of the transient hot plasma. There is thus no clear signal of the phase transition known, although a number of signals were proposed, and the quantities measured so far can often be modelled by both, a hot hadron gas and a quark-gluon plasma. Nevertheless there are meanwhile experimental signatures that the quarks in the initial state of the fireball are "decorrelated" in a sense that was made precise by Bialas [18]. The interpretation does not rely on any thermodynamic concept.

As mentioned before, relativistic heavy-ion collisions are sometimes called the *little bang* in contrast to the big bang in the early universe. The analogies between the little and the big bang go beyond the transient usually high-energy densities in both cases [19]. From pion-interferometry measurements it was possible to identify a transverse component in the primarily longitudinal expansion of the fireball, thus the expansion has analogous features to the Hubble expansion in the universe. Furthermore the counterpart of the decoupling of photons in the universe when it got transparent to photons is the decoupling of pions. Finally the role of the primordial nucleosynthesis, in the universe at a time scale of the order of three minutes after the big bang, has a pendant in the primordial hadrosyn-thesis in heavy-ion collisions, although in case of the hadrons one has to argue why the observed hadronic yield may be considered as primarily pri-mordial [19].

Pronounced differences between the little and the big bang come from the competing time and length scales in both cases. In the early universe we have extremely different competing scales. The Hubble expansion of the universe is rather slow in units of QCD. If a typical time scale of QCD is taken as $1/T_c$ ($\sim 1[\text{fm}/c] \sim 10^{-23}[\text{s}]$), the Hubble time is of the order of $10^{19}/T_c$. Thus there is enough time for equilibration before and after the transition in spite of the Hubble expansion. The spatial volume V of the universe, 10^{-6}s after the big bang, appears as almost infinite in units of QCD (in heavy-ion collisions it is of comparable size). In units of T_c, it is estimated as $V \cdot T_c^3 = 7.1 \times 10^{55} (\frac{200 \ [\text{MeV}]}{T_c})^3$, if the physical correlation length ξ is taken as $\xi(T_c) \cdot T_c = 1.38 \pm 0.24$ in the deconfinement phase [20]. This leads to a rather small upper bound on the amount of supercooling during a (hypothetical) first-order deconfinement transition ($\Delta T/T_c \sim 10^{-56}$). The effect of supercooling is an important ingredient in the reasoning why one should see remnants of the early QCD transition even nowadays. A first-order deconfinement transition might have led to inhomogeneities in the baryon-number density in the early universe. If these inhomogeneities had survived until the epoch of primordial nucleosynthesis, they had influenced the light-element abundances and led to deviations from values obtained in the standard scenario (without a phase transition) [21–23]. It is these deviations in the light element abundances which were predicted as a visible remnant nowadays. Meanwhile these predictions became rather questionable. Initial inhomogeneities which are compatible with the rather small supercooling are most likely insufficient to induce well separated proton- and neutron- rich regions. Pronounced inhomogeneities in the proton and neutron distributions are a precondition for a considerable change in the initial conditions of primordial nucleosynthesis. Therefore, we will no longer discuss hypothetical remnants of the QCD transition in the early universe, but later focus on some possible signatures in heavy-ion collisions.

The book is about *principles* of phase transitions in particle physics and does not intend to give a complete status report about the current state of the art. Such a status report would soon need for an update, since out of the numerous results about the phase structure of the standard model many results must be considered as preliminary. Our emphasis is on basic concepts. We demonstrate a variety of methods and use different regularization schemes to avoid the impression that there is one preferred and distinguished method which solves "everything". Whenever approximation schemes are difficult to control, it is quite useful to supplement numerical methods with analytical ones and vice versa. We also point on the various levels of reliability of the results (ranging from rigorous to conjectural), which were obtained so far, and mention conceptually open problems. In some examples we show how one should carefully translate results in different regularization schemes into each other. It is a dangerous attitude to treat lattice simulations as a black box which produces numbers, to forget about the context of their derivation and use them as an input in other regularization schemes. One of the reasons is that the couplings do depend on the scale and the regularization scheme which were used for their derivation.

The book is organized as follows. Chapter 2 gives the background from statistical physics, but our presentation goes beyond the standard repertoire of a lecture on statistical physics. After a phenomenological classification of phase transitions we start with the molecular-mean field approximation in the Ising model and its relation to variational estimates of the free energy. Its pendant in particle physics leads to reasonable estimates of the critical line in an SU(2) Higgs model for the electroweak transition. We then recall a mean-field approach in the spirit of Landau and Ginzburg, point out its intrinsic limitations and its range of applicability. Including the effect of fluctuations in a Landau Ginzburg model is no longer possible without approximations. A perturbative evaluation of the model about the exactly solvable Gaussian theory leads to infrared divergencies order by order in perturbation theory, a difficulty that was overcome by Wilson's renormalization group. In section 2.4 we therefore introduce the renormalization group for various applications. The very first step in the framework of the renormalization group is explained in detail to expose the conditions under which a truncation of the arising effective action is under control. The first two applications of the renormalization group are standard ones and concern critical phenomena. We derive critical exponents and the finite-size scaling behavior of thermodynamic quantities in the case of second-order transitions section (2.5). Further applications, considered later in the book (cf. section 4.7), refer to the derivation of low-energy effective actions in the framework of Polchinsky's flow equations [24]. To a certain extent we discuss some subtleties like "dangerous" irrelevant variables and the "large-field-problem" in the renormalization-group approach.

Section 2.6 deals with finite-size scaling analysis of first-order transitions. For a specific class of models we bridge the gap between the phenomenological description of finite-size effects in numerical simulations and a rigorous derivation of these effects (that is in general not possible). We list criteria how to distinguish first- and second-order transitions from the specific way they are anticipated in a finite volume and add a short outlook to linked cluster expansions as a tool for measuring finite-size effects. The chapter is concluded with a dictionary for thermodynamic potentials, order parameters, response functions and critical indices between condensed matter physics and particle physics.

In chapter 3 we summarize some basic knowledge on Euclidean quantum field theory at finite temperature. We identify the various limiting cases of QCD and the electroweak standard model which are of further interest for the following discussions. In particular we summarize the conjectures about the phase structure of QCD and the $SU(2)$ Higgs model, based on analogies to familiar spin systems of condensed matter physics. These analogies are quite useful as guideline for more rigorous studies, but the limitations of the analogies will also become clear so that more work is needed to support or disprove the conjectures. The primer to lattice gauge theory serves to fix the notation and to summarize the main ideas behind the lattice regularization: putting actions of scalar and gauge field theories on the lattice, translating lattice units into physical units, taking care on possible artifacts. It gives a derivation of the formal representation of the Polyakov-loop expectation value and its physical interpretation. The Polyakov loop is one of the basic observables in lattice QCD.

Chapter 4 deals with analytical methods on and "off" the lattice, where "off" means other continuum regularizations. First we compare asymptotic with convergent expansions. Examples for convergent expansions are polymer expansions, strong coupling expansions, linked and dynamical linked cluster expansions, whereas standard weak coupling expansions in a small coupling constant can be identified at best as asymptotic expansions (section 4.1). We spend a separate section on linked cluster expansions. Linked cluster expansions have a long tradition in statistical physics. In systems of statistical physics they were applied to fluids back in the thirties, later to magnetic systems in the fifties (see the references in [28]). In the eighties they were considered in connection with the triviality discussion of a Φ^4-theory in four dimensions at zero temperature [29]. An extension to linked cluster expansions at finite temperature and an increase in the accuracy [30] allowed a measurement of the tiny finite-temperature effects on the critical coupling in $O(N)$ models. The finite-temperature effects are so small that they are hidden in the background of zero-temperature contributions to the analytical expansions unless the expansions are performed to a very high order in the expansion parameter, for example to eighteenth order. But this order calls for a new technique to handle the graphs of which there are more than a billion contributing to this order. A further implementation of a finite rather than infinite volume was used to perform a finite-size scaling analysis within the framework of linked cluster expansions [31]. Since linked cluster expansions are formulated for interactions between fields on two lattice *sites*, they are not directly applicable to gauge theories, but to *effective scalar* field theories for particle physics. In principle, generalizations are possible to include fermions and (with considerably more effort) four-link interactions to develop a graphical expansion for the free energy of a lattice gauge theory. To our knowledge these generalizations are not worked out to date.

The next section (4.3) deals with weak-coupling expansions at zero temperature and the continuum limit of lattice gauge theories. Lattice results which do not "survive" the continuum limit have no physical meaning in the context of particle physics, therefore an understanding of performing this limit in the appropriate way belongs to the basics of understanding lattice gauge theory.

We then prepare the ingredients for the framework of dimensional reduction. These are weak-coupling expansions at finite temperature for massless theories (section 4.4). In particular we explain the magnetic-mass problem in three-dimensional gauge theories and indicate the reasons for why it is such a hard problem. Because of the inherent difficulties of perturbative expansions which are free of both UV-and IR-divergencies at finite temperature, we distinguish between their realization in an infinite and in a finite volume. In case of the infinite volume we refer to the work of Pisarski and Brateen ([6, 7]), in case of the finite volume to the work by one of us et al. ([25, 26]). Analytic estimates of the electric and magnetic screening masses are based on such perturbative estimates that were first derived by Linde [27]. Even the (gauge variant) very definitions of the electric and magnetic screening masses rely on these calculations. The version of a weak coupling expansion at finite temperature in a *finite* volume provides a systematic approach for calculating short-distance quantities in asymptotically free and massless field theories at finite temperature, in which even short-distance properties are usually not free of IR-problems. The expansion works at any order of perturbation theory and does not get stuck on the one-or two- loop level as other perturbative schemes in the infinite volume. Still it neither provides a solution of IR-problems at finite temperature in the *far infrared* region.

Section 4.5 on the constraint effective potential and gap equations [32] illustrates a resummation of graphs that neither solves the magnetic-mass problem, but leads to reasonable results for the phase structure of the $SU(2)$

Higgs model in connection with the electroweak transition when a certain assumption on the magnetic mass is made.

Since in the framework of Euclidean field theory a finite temperature is implemented as a finite extension in Euclidean time direction, one would naively expect a reduction of the theory in four dimensions at high temperature to an effective theory in three dimensions. The three-dimensional theory should be easier to handle because of one dimension less. The naive expectation is based on a classical geometrical argument: The extension in Euclidean time direction shrinks to zero at infinitely high temperature. On a quantum level one has to carefully specify the conditions under which dimensional reduction holds. In any case the effective three-dimensional theory will contain couplings which are induced from the original four-dimensional model. In an extended section on dimensional reduction (section 4.6) we first exemplify this approach in a scalar Φ^4-theory from four to three dimensions. As a second example we give an outline of dimensional reduction in QCD with dynamical fermions. It can be used for calculating the electric screening mass in the deconfinement phase. In the first two examples dimensional reduction is realized by means of perturbation theory in a small renormalized coupling. Alternatively dimensional reduction can be performed in a large-N expansion as we show for the reduction of the Gross-Neveu model from three to two dimensions. The spontaneous breaking of the parity symmetry in the Gross-Neveu model shares similarities with the chiral transition for two massless flavors in QCD. In a combination of a small-momentum and a large-N expansion we derive an effective two-dimensional scalar model for the Gross-Neveu model and identify the universality class of the effective scalar model by means of linked cluster expansions. The universality class of the chiral transition of two-flavor QCD was controversial for some time, so that the universality class of the corresponding transition in the Gross-Neveu model was of some interest too. Finally we summarize dimensional reduction of the electroweak standard model in four dimensions to an $SU(2)$ Higgs model in three dimensions. The phase structure of the effective three-dimensional model was frequently studied in Monte Carlo simulations. The results for the electroweak phase transition can be compared with Monte Carlo simulations of the original four-dimensional $SU(2)$ Higgs model.

Flow equations of Polchinsky [24] nowadays run under the name of the *exact* renormalization group. As other approaches, the exact renormalization group provides a framework for deriving an effective action at a low-energy scale starting from an effective action at a high-energy scale. The attribute "exact" refers to the one-loop nature of the set of equations that is not a result of a perturbative truncation, but holds exactly. However, one has to deal with an infinite set of coupled integro-differential equations for certain generating functionals. Unavoidably, their solutions involve truncations, for example in higher-order n-point functions. Examples in which these truncations are well under control are again $O(N)$ models (section 4.7).

In chapter 5 we turn to numerical methods in lattice field theories. We restrict our selection of topics to the main updating procedures which are used in Monte Carlo simulations of pure gauge theories and in gauge theories with fermions. Standard local updating procedures for pure gauge theories are the Metropolis and the heat bath algorithm, nowadays combined with over-relaxation techniques. For gauge theories with dynamical fermions one of the most prominent algorithms is the nonlocal Hybrid Monte Carlo algorithm. We explain its essential features. The purpose of this section is not to introduce into computational or technical details which are actually needed for writing the source code of the numerical simulations that get particularly demanding when fermions are included. The emphasis is on an appropriate interpretation of data and on criteria which are used for identifying the phase transitions from simulations in a finite volume, for finite quark masses and for a finite lattice constant. The main goal is to distinguish lattice artifacts from continuum physics. It will be shown how the scaling behavior can be controlled as a function of the lattice size, the lattice constant, and the quark mass. Bulk transitions are unwanted mere lattice artifacts for lattice physicists, as mentioned before, since they do not survive the continuum limit of a lattice regularized theory. Although they do not have any impact on continuum physics, they still may be significant for statistical systems on a physically realized lattice like a crystal with a finite physical lattice constant.

Calculations in lattice gauge theories for QCD or the electroweak part of the standard model are usually considered as "first-principle" calculations, although sometimes approximations creep in which shed some doubt on the pretension that the calculations deserve this attribute. In view of that it seems to be worthwhile to supplement models of lattice gauge theory by so-called effective models. In chapter 6 we consider such models for QCD. In the ideal case these models should be derived from the QCD Lagrangian, for example by integrating upon all field fluctuations up to a certain energy scale. The resulting effective action would then take a simple form in terms of new, effective degrees of freedom, for example in terms of mesons and glueballs. It would be valid for energies smaller or equal a certain value. This is easier said than done. Rather than deriving such actions they are usually postulated from certain guiding principles like symmetry postulates, based on experimental observations which considerably restrict the choice of the effective Lagrangian. The predictive power of such models is reduced as compared to full QCD, neither need these models be renormalizable so that more and more experimental input is needed the higher the order in the expansion parameter is. It finally becomes a quantitative question and difficult to decide whether the systematic errors due to the truncations of effective actions are larger than the systematic errors inherent in the lattice-approximation scheme. The answer depends on the possibility of controlling these errors.

Our final chapter 7 on phenomenological applications of QCD transitions deals with heavy-ion collisions. It is far from being complete or representative for the present state of the art of these experiments. The selected topics should only indicate places where these collisions are actually sensitive to the phase structure of QCD and which assumptions in the theoretical approaches may or may not be justified. An idealization of these collisions as an infinite system in equilibrium at zero baryon density and with adiabatic cooling from the plasma to the confinement phase is certainly inappropriate. The realization and interpretation of these experiments remain a challenge. Our purpose is just to indicate this challenge. After all, we may be curious about the observable signatures of this exotic state of matter that is produced for a very short instance of time, of the order of 10^{-23}s, in which the experimentalists try to mimic the big bang in a little bang.

Chapter 2

General Background from Statistical Physics

2.1 Generalities

2.1.1 *Phase transitions in statistical systems*

2.1.1.1 *First- and second-order transitions in the infinite volume limit*

The order of a phase transition is one of the basic thermodynamic classifications. It concerns the thermodynamic potential and its derivatives at the transition. The thermodynamic potential Ω is the free energy F in case of a ferromagnet or the Gibbs free energy $G = F + pV$ for a fluid. In Eqs. (2.1.1a-2.1.1d) we recall the basic thermodynamic formulas for a magnet.

$$Z(T, H, V) = \int \mathcal{D}U e^{-\beta \mathcal{H}(U)} \, , \qquad (2.1.1a)$$

$$F = -T \ln Z \, , \qquad (2.1.1b)$$

$$E = -\frac{\partial \ln Z}{\partial \beta} \, , \quad S = -\left(\frac{\partial F}{\partial T}\right)_H \, , \quad M = -\left(\frac{\partial F}{\partial H}\right)_T \, , \qquad (2.1.1c)$$

$$c_H = \left(\frac{\partial E}{\partial T}\right)_{H,V} \, , \quad c_{H,M} = T\left(\frac{\partial S}{\partial T}\right)_{H,M,V} \, , \quad \chi_T = \frac{1}{V}\left(\frac{\partial M}{\partial H}\right)_T \, . \qquad (2.1.1d)$$

The indices indicate the quantities that are kept fixed. Here $\mathcal{H}$ denotes the spin Hamiltonian, $\beta = T^{-1}$ is the inverse temperature, the Boltzmann constant k_B was set equal to 1 everywhere, H is an external magnetic field, V is the volume and $\int \mathcal{D}U$ stands for the sum over all spin configurations $\{U\}$, weighted with the Boltzmann factor $\exp\{-\beta\mathcal{H}\}$. Equation (2.1.1b) defines the free energy F, depending on T, H, and V, of a system in a finite volume V. In the *large*-volume limit the free energy density f is given as $f = \lim_{V \to \infty} F(V)/V$. The free energy is assumed to depend on the scaling fields T and H. In view of a finite-size scaling analysis we keep also the dependence on V. The parameters T and H are on an equal footing, both may drive a phase transition. One speaks of a *temperature* or a *field driven* transition. In QCD we focus our discussion on temperature driven transitions (although density- or "mass-driven" transitions may be considered as well). First derivatives of the free energy with respect to T or H lead to the internal energy E, the entropy S or the magnetization M according to Eq. (2.1.1c). The magnetization is the conjugate variable to the external field and plays the role of an order parameter in the case of a magnet. More generally it plays the role of an order parameter if the external field *explicitly* breaks the symmetry which may become spontaneously broken at a phase transition.

On the second level of derivatives (Eqs. (2.1.1d)) we have the specific heat c (at constant H and/or M, respectively) and the isothermal susceptibility χ_T. For a fluid the isothermal susceptibility would be replaced by the compressibility κ.

In the infinite volume limit a phase transition is signaled by a singularity (in the sense of nonanalyticity) in the thermodynamical potential Ω. If there is a finite discontinuity in at least one of the first derivatives of Ω, the transition is called *first-order*. In the case of a ferromagnet there is a jump in the magnetization, if one passes through the transition temperature from the phase of broken symmetry to the symmetric phase. This gives M the name of an *order parameter*, M indicates the order of spins. This way it tells us the phase in which the system would be encountered at a given temperature.

The remaining first derivatives of F with respect to T (the internal energy and entropy) usually also show a discontinuity at the transition point. A gap in the entropy is associated with a finite latent heat $\Delta Q = T_c \cdot \Delta S$, but there need not be such a gap. (Consider a transition in a ferromagnet between states of magnetization opposite in sign but equal in magnitude. The latent heat would vanish in this case, while the magnetization jumps between values of opposite sign.) Vice versa there may be a finite latent heat without a gap in the order parameter at the transition point.

The second derivatives of the thermodynamic potential at a first order transition are typically δ-function singularities (corresponding to the discontinuities in the first derivatives) or finite.

According to the original Ehrenfest classification of phase transitions, *n-th order transitions* are defined by the occurrence of discontinuities (rather than divergences) in the n-th order derivative of the appropriate thermodynamical potential. In M.E.Fisher's terminology one distinguishes between first-order and *continuous* (or higher order) transitions. In continuous transitions the first derivatives of Ω are continuous, whereas second derivatives are either discontinuous or *divergent*. In a *second-order transition* at least one of the second derivatives of Ω is divergent. (If there are at most finite discontinuities in the second derivatives, the transition is of higher than second order.) Hence the order parameter M will vanish continuously at the transition point.

The susceptibility χ_T and the specific heat c typically both diverge in a second-order transition. (Again it is not necessary that both of them diverge.) Here the divergences are power law singularities. They are characterized by *critical indices.*

Let us introduce the critical indices in a phenomenological way as indices that describe the singular behavior of certain response functions which is observed in experiments (say in a ferromagnet). If the experimental observations are extrapolated to the infinite volume, close to T_c the susceptibility scales according to

$$\chi \equiv \left(-\frac{\partial^2 f}{\partial h^2}\right)_T \simeq |1 - T/T_c|^{-\gamma} \qquad (2.1.2)$$

with an index γ and f as defined above. The specific heat scales according to

$$c \equiv -T\left(-\frac{\partial^2 f}{\partial T^2}\right)_h \simeq |1 - T/T_c|^{-\alpha} \qquad (2.1.3)$$

with critical index α, and the magnetization density $m(T)$ according to

$$m(T) \equiv \lim_{h\to 0_+} m(T,h) = \lim_{h\to 0_+}\left(-\frac{\partial f}{\partial h}\right)_T \simeq |1 - T/T_c|^{\beta} \qquad (2.1.4)$$

as $T \to T_c$ defining the exponent β, while on a critical isotherm one finds

$$m(T = T_c, h) \simeq_{h\to 0_+} h^{1/\delta} , \qquad (2.1.5)$$

defining the exponent δ. Finally, one can measure spin-spin two-point correlations $G(x, x')$ of the local magnetization $s(x)$

$$G(x, x') = <s(x)s(x')> - <s(x)><s(x')> . \qquad (2.1.6)$$

For large distances $|x - x'| \to \infty$ at $T \neq T_c$, G decays according to

$$G(x, x') \simeq \exp\{-|x - x'|/\xi(T)\}. \qquad (2.1.7)$$

Equation (2.1.7) gives a possible definition of the correlation length (for alternative definitions cf. section 2.3, Eq. (2.3.124 and the following discussions there). The scaling behavior of ξ close to T_c defines the exponent ν

$$\xi(T, h = 0) \simeq |1 - T/T_c|^{-\nu}. \qquad (2.1.8)$$

At the critical point $T = T_c$, $h = 0$, also $G(x, x')$ decays like a power rather than exponentially for large distances $|x - x'|$. The associated exponent η is called the anomalous dimension, since it gives an additive contribution to the naive dimension of the two-point function

$$G(x, x') \simeq |x - x'|^{-(D-2+\eta)}. \qquad (2.1.9)$$

For example, measurements in a ferromagnet in dimensions $D = 3$ yield $\alpha \simeq 0.1$, $\beta \simeq 0.32$, $\gamma \simeq 1.24$, $\delta \simeq 4.8$, $\nu \simeq 0.63$ and $\eta \approx 0.03$. As an experimental fact, the very same values are measured in systems that look quite differently from a ferromagnet on a microscopic level, for example in binary liquids. This experimentally observed phenomenon of universality will find its theoretical explanation in the renormalization-group approach applied to critical systems near second-order phase transitions. In the framework of the renormalization group we will derive the critical exponents in terms of scaling variables and the scaling-sum rules that hold between the exponents. Not all of the exponents are independent of each other. Here we only summarize the most familiar sum rules

$$\alpha + 2\beta + \gamma = 2 \qquad (2.1.10)$$

$$\beta(\delta - 1) = \gamma$$

$$\nu(2 - \eta) = \gamma$$

$$2 - D\nu = \alpha \, .$$

In section (2.4) we shall specify the assumptions under which Eqs. (2.1.10) hold.

2.1.1.2 *Landau's free energy*

One criterion for the order of the phase transition is given by Landau's theory [33–36]. It consists in an expansion of the free energy in powers of the order parameter. The allowed terms in this expansion are further selected by symmetry arguments. Phase transitions can be classified according to their transformation behavior of the order parameter under a symmetry transformation. In this introductory section we discuss only an order parameter described by a scalar field ϕ. In QCD applications the scalar ϕ will be replaced by an $O(N)$-vector with N components or an $SU(3)$ matrix parametrized by two independent directions of possible "ordering" (see section 3.1).

The ansatz of a free-energy functional for a scalar order parameter ϕ in d space-dimensions is given as

$$F\{\phi(x)\} = \int d^d x \left\{ a \left(\nabla\phi(x)\right)^2 + \frac{r}{2}\phi^2(x) + \frac{\lambda}{4!}\phi^4(x) - h\phi(x) \right\}. \quad (2.1.11)$$

For vanishing h this is the simplest form that admits spontaneous symmetry breaking. Although F is reflection invariant (if $h = 0$), the ground state need not be so. F may take its minimum for nonvanishing values $\pm\phi_0 \neq 0$ (later denoted as $\langle\phi\rangle$), depending on the values of a, r, and λ. The "couplings" a, r, λ and h should be considered as parameters, where a, $\lambda > 0$. In the example of a magnet the condition $\lambda > 0$ corresponds to the physical condition that the magnetization is bounded. The parameter r has the meaning of a mass squared, λ of a coupling strength of the interaction and h of an external field. In the vicinity of a second-order transition the order parameter is small (more generally, if fluctuations in the field are allowed, its average expectation value is small), hence one drops higher powers of ϕ. A further assumption is that ϕ is slowly varying in space (thus there are no higher derivative terms than $(\nabla\phi)^2$). A ϕ^3-term is missing, if a symmetry under sign inversion $\phi \to -\phi$ is required for vanishing h. For $h = 0$ it is easily verified that Eq. (2.1.11) predicts a second-order phase transition. For $r < 0$ two stable states are predicted with magnetizations $\pm\phi_0$. The condition $r = 0$ defines the critical temperature, thus one may write for r

$$r = \tilde{r}\,(T - T_c)\,. \quad (2.1.12)$$

Figs. 2.1.1 and 2.1.2 display the typical signatures of a second-order transition. The free energy F is plotted as a function of the constant average

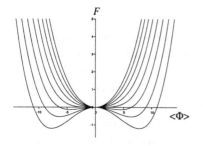

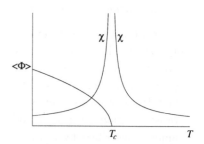

Fig. 2.1.1   Free energy F as a function
of $< \Phi >$ for temperatures above, and
below the transition temperature T_c in
case of a second-order transition.

Fig. 2.1.2   Typical behavior of an order parameter and a susceptibility for
a second-order transition.

of the order-parameter field $< \Phi >$. The nontrivial minima at $\pm\phi_0$ move
continuously inwards as the temperature is increased towards T_c, where the
$Z(2)$-symmetry is restored (Fig. 2.1.1). Fig. 2.1.2 illustrates the vanishing
of the order parameter as function of T and the power law divergence in
the susceptibility.

For fixed temperature $T < T_c$, a field-driven transition can be considered
as function of h. In this case the transition is of first order, ϕ_0 jumps from
$\sqrt{(-6r/\lambda)}$ to $-\sqrt{(-6r/\lambda)}$ as h changes sign.

Although the ansatz (2.1.11) has been originally proposed for second-
order transitions, temperature driven first-order transitions can be described as well. We mention two possibilities.

- $\lambda < 0$ in Eq. (2.1.11). The coupling λ can play the role of a
  renormalized coupling. In later applications of QCD, λ occurs as a
  renormalized coupling in an effective description of QCD. It varies
  as a function of the value which is chosen for the strange quark mass
  m_s. For certain values of m_s, λ becomes negative. If $\lambda < 0$, one
  has to include a term $\propto \phi^6$ with positive coefficient to stabilize the
  free-energy functional. Adding a term $(\kappa/6!) \cdot \phi^6(x)$ to Eq. (2.1.11)
  ($\kappa > 0$), F has now two local minima over a certain temperature
  interval $T_0 < T < T_1$, where T_0 and T_1 have the meaning of stability
  limits of the disordered phase in the ordered and the ordered in the
  disordered phase, respectively. At T_c the minima are equally deep,
  the order parameter jumps from $\phi_0 = \pm\sqrt{-24r/\lambda}$ to zero.
  In general λ may change its sign as function of an external parameter P. When λ changes its sign at some value P^*, a line

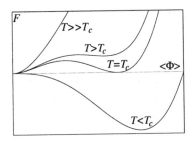

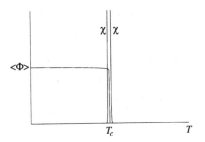

Fig. 2.1.3   (a) Free energy F as a function of $< \Phi >$ for temperatures above, and below the transition temperature T_c in case of a first-order transition.

Fig. 2.1.4   (b) Typical behavior of an order parameter and a susceptibility for a first-order transition.

$T_c(P)$ of second-order transitions ends at a so called *tricritical point* $T_t = T_c(P^*)$ and continues as a line of first-order transitions. Tricritical behavior is predicated to occur for QCD in certain limiting cases (see section 3.1).

- a cubic term in Eq. (2.1.11). If such a term is not suppressed by a symmetry argument, it admits a first-order temperature driven transition. For $\lambda > 0$, F has two minima at $\phi_0 = 0$ and at $\phi_0 \neq 0$ over a temperature interval $T_0 < T < T_1$. If the cubic term is written as $(\alpha/3!)\phi^3(x)$, the nontrivial minimum at T_c occurs for $\phi_0 = -6r/\alpha$. Fig.2.1.3 shows the typical shape of the free energy $F(< \Phi >)$ for a first-order transition with coexisting well separated minima at the transition temperature and for a certain range of temperatures above and below T_c. Fig.2.1.4 illustrates a typical behavior of the order parameter in the infinite volume limit, characterized by a jump in the order parameter value at T_c and a δ-function like singularity in the susceptibility.

The very existence of a cubic term in an effective potential for the electroweak phase transition has been under much debate in the last few years. In the simplest models for the electroweak transition the order parameter is an $O(N)$-vector field, r, λ and α are renormalized parameters.

A cubic term in the classical potential of Eq. (2.1.11) does not guarantee the first-order of the transition. If the transition is weakly first-order in the sense that the maximum between zero and the nontrivial minimum is not high, the transition may be washed out by fluctuations of the order-parameter field. Such fluctuations will be discussed below.

Landau's concept for the free energy and the criterion for the order of the transition is widely used in applications of particle physics. Two caveats should be mentioned at least. The first one concerns convexity properties, the second the validity range of the mean-field approximation.

• *Convexity properties.* It is known from general thermodynamic principles that thermodynamic potentials in thermal equilibrium are convex functions of their natural variables. The nonconvex shape of Figs. 2.1.3, 2.1.4 and the coexisting minima for a first-order transition are obviously at odds with the general expectation. Landau's free energy is a macroscopic concept. Its nonconvex and physical realization may be understood as a *coarse-grained free energy* arising in an intermediate step from a microscopic to a macroscopic scale.

Let us assume that we start with a microscopic spin action depending on spin variables s_i associated with sites i of a hyper-cubic lattice. As a first step short-wavelength fluctuations are eliminated by dividing the lattice into cells of linear dimensions L and introducing new variables $\Phi(x)$ in a so called block-spin transformation. This step will be explicitly performed in section 2.4. Such a block-spin transformation may be iterated. The general folklore then is (although it is extremely hard to prove it rigorously) that on a macroscopic scale, after a sufficient number of iterations, the effective action takes the form of Landau's free energy (cf. Eq. (2.1.11)) if one is close to a second-order phase transition. At such a transition the correlation length ξ diverges, and the condition $L << \xi$ on the linear cell size L is easily satisfied.

The bulk (coarse-grained) free energy F of a system governed by the effective action of the form of (2.1.11) is then obtained by integrating over all configurations $\{\phi\}$, remaining after the iterated block-spin transformations. It is given as

$$F = -\frac{1}{\beta} \ln Z = -\frac{1}{\beta} \ln \int \mathcal{D}\phi \exp\left\{-\beta S[\phi(x)]\right\} \qquad (2.1.13)$$

with S of the same form as F in (2.1.11).

If it is justified to evaluate Eq. (2.1.13) in a saddle-point approximation, i.e. to drop $\int \mathcal{D}\phi$ and take S at its minimum ϕ_0, the result is

$$F = S(\phi_0), \qquad (2.1.14)$$

the free energy coincides with its mean-field value, i.e. Landau's free energy, here termed S, because the conjecture is that Landau's ansatz (2.1.11) is obtained as an effective action S. Landau's free energy inherently corresponds to a mean-field approximation of the system. For a spin system, the order parameter is the magnetization. Landau's free energy is then an expansion in terms of mean values of spins. It does not include a summation over all spin configurations (here abbreviated as $\int \mathcal{D}s_i$) according to

$$F = -\frac{1}{\beta} \ln \int \mathcal{D}s_i \exp\left\{-\beta S[s_i]\right\}. \tag{2.1.15}$$

• *The validity of the mean-field approximation.* Next we address the question when one is allowed to ignore the fluctuations in the order-parameter field, i.e. to drop $\int \mathcal{D}\phi$ in Eq. (2.1.13) and take $S(\Phi_0)$ for F. On a microscopic level it amounts to a replacement of spin-spin interactions by some average background represented by $\phi \equiv M$. The validity of the mean-field approximation is guaranteed if the fluctuations $\delta\phi \equiv \phi(x) - \phi_0$ in the order-parameter field are small compared to the order parameter itself, i.e.

$$\left\langle (\delta\phi(x))^2 \right\rangle \ll \phi_0{}^2, \tag{2.1.16}$$

where the average $\langle \ldots \rangle$ should be taken over all cells of the coarse-grained lattice. Upon using the fluctuation-dissipation theorem, Eq. (2.1.16) translates to [37]

$$1 \ll R^D (1 - T/T_c)^{(4-D)/2}, \tag{2.1.17}$$

where R is the interaction range. Mean-field theory becomes exact if either the dimensionality, the range of interactions or the number N of order-parameter components becomes infinite. (In the last case the mean-field approximation amounts to a large-N expansion).

Mean-field theory often gives correct qualitative predictions for phase diagrams of three-dimensional systems. Three-dimensional systems play some role as effective models for high-temperature QCD. As the dimensionality increases, mean-field theory improves, while numerical calculations get harder. The critical dimension depends on the form of Landau's free-energy expansion. Symmetry considerations are essential for constructing Landau's free energy. As both finite-temperature transitions of QCD are supposed to be driven by symmetry breaking, it is natural to construct effective actions for QCD by using Landau's concept as a guideline.

If the condition (2.1.17) is violated, one should use the renormalization-group approach ([38–41]) to describe critical phenomena.

2.2   Generating functionals, n-point correlations and effective potentials

In the previous section we have introduced susceptibilities, the specific heat, critical exponents, Landau's free energy in terms of macroscopic thermodynamic quantities. In this section we give the associated microscopic description in terms of local field operators. This section also serves to fix the notation for the basic framework that we will use throughout this book.

We consider a hyper-cubic lattice Λ in D dimensions. In most of our applications, $D-1$ dimensions of extension L, $L \leq \infty$, are space-like, while one dimension with label 0 corresponds to Euclidean time with extension $L_0 \leq \infty$. Another case is a lattice in D space dimensions with volume $V = L^D$. For simplicity we choose scalar degrees of freedom $\Phi(x)$ which are associated with the sites of Λ_a, the lattice with lattice constant a. The partition function of the system $Z(J)$ in the presence of an external current $J(x)$ is then given as

$$Z(J) = \int_{-\infty}^{+\infty} \prod_{x \in \Lambda_a} d\Phi(x) \, e^{-S(\Phi) + a^D \sum_{x \in \Lambda_a} J(x)\Phi(x)} \tag{2.2.1}$$

with action $S(\Phi)$ and lattice constant a. As a standard example we consider the action of a scalar $\Phi^4 + \Phi^6$-theory

$$S(\Phi) = a^D \sum_{x \in \Lambda_{a=1}} \left\{ \frac{1}{2} \sum_{\mu=0}^{D-1} (\frac{1}{a}\widehat{\partial}_\mu \Phi(xa))^2 + \frac{m^2}{2}\Phi(xa)^2 \right.$$
$$\left. + \frac{\lambda}{4!}\Phi(xa)^4 + \frac{\sigma}{6!}\Phi(xa)^6 \right\} \tag{2.2.2}$$

with lattice-difference operator

$$\widehat{\partial}_\mu f(x) = f(x + \widehat{\mu}) - f(x), \tag{2.2.3}$$

$\widehat{\mu}$ is the unit vector in direction μ, and with bare mass parameter m and bare couplings λ and σ.

An alternative parametrization to (2.2.2) results for $S(\Phi)$ if we use the following reparametrization

$$\Phi(ax) = \sqrt{2\kappa}\, \Phi_0(x)/a^{\frac{D-2}{2}} \ , \ x \in \Lambda_{a=1} \equiv \Lambda \qquad (2.2.4)$$

$$m^2 = \frac{1}{\kappa}\,(1 - 2\lambda_0 + 3\sigma_0 - 2\kappa D)a^{-2} \qquad (2.2.5)$$

$$\lambda = \frac{3!}{\kappa^2}\,(\lambda_0 - 3\sigma_0) \qquad (2.2.6)$$

$$\sigma = \frac{90}{\kappa^3}\,\sigma_0\, a^{2D-6}\ , \qquad (2.2.7)$$

so that

$$S(\Phi_0) = -\frac{1}{2}\sum_{x \neq y\,\text{NN}} 2\,\kappa\,\Phi_0(x)\Phi_0(y) + \sum_{x \in \Lambda}\{\Phi_0(x)^2 \qquad (2.2.8)$$
$$+ \lambda_0\,(\Phi_0(x)^2 - 1)^2 + \sigma_0\,(\Phi_0(x)^2 - 1)^3\}\ .$$

NN denotes nearest-neighbor pairs. Equation (2.2.8) is an appropriate starting point for linked cluster expansions (cf. section (4.2)).

Expectation values of observables O with respect to the partition function $Z(J = 0)$ are defined by

$$< O >_Z = \frac{\int \prod_{x \in \Lambda_a} d\Phi(x)\, O(\Phi(x))\, e^{-S(\Phi)}}{\int \prod_{x \in \Lambda_a} d\Phi(x)\, e^{-S(\Phi)}}\ . \qquad (2.2.9)$$

Usually we omit the subscript Z. In a field-theoretical context $Z(J)$ plays the role of a generating functional of n-point Green functions $Z^{(n)}(x_1, \cdots x_n)$. It is represented (as formal power series) according to

$$Z(J) = \sum_{n \geq 0}\frac{1}{n!}(a^D)^n \sum_{x_1,\cdots,x_n \in \Lambda_a} Z^{(n)}(x_1, \cdots, x_n)\, J(x_1)\cdots J(x_n)\ . \qquad (2.2.10)$$

For $n = 0$, $Z^{(0)}(J) \equiv Z(J = 0)$. For $Z^{(n)}(x_1, \cdots x_n)$ we write

$$Z^{(n)}(x_1, \cdots, x_n) \equiv\, < \Phi(x_1)\cdots\Phi(x_n) >\, = \frac{\partial^n Z(J)}{\partial a^D J(x_1)\cdots\partial a^D J(x_n)}\bigg|_{J(x)\equiv 0}\ . \qquad (2.2.11)$$

Similarly, $W(J) \equiv \ln Z(J)$, proportional to the free-energy functional $F(J) = -T\ln Z(J))$, is the **generating functional** of the connected n-point Green functions $W^n(x_1, \cdots, x_n)$

$$W(J) = \sum_{n \geq 0}\frac{1}{n!}(a^D)^n \sum_{x_1,\cdots,x_n \in \Lambda_a} W^{(n)}(x_1, \cdots, x_n)\, J(x_1)\cdots J(x_n) \qquad (2.2.12)$$

with

$$W^{(n)}(x_1, \cdots, x_n) \equiv <\Phi(x_1) \cdots \Phi(x_n)>^c = \frac{\partial^n W(J)}{\partial a^D J(x_1) \cdots \partial a^D J(x_n)}\bigg|_{J(x)\equiv 0}.$$
(2.2.13)

The attribute "connected" comes from an associated graphical expansion of $W(J)$ in which the $W^n(x_1, \cdots, x_n)$ turn out to be connected in a graphical sense, while the $Z^n(x_1, \cdots, x_n)$ may be disconnected as well.

n-point susceptibilities are macroscopically defined as the n-th derivative of the magnetization with respect to an external field, their generic physical meaning is that of a response function at zero momentum on a stimulus in an external field, here the current J. They are obtained from (2.2.12) according to

$$\chi_n = \frac{1}{V}(a^D)^n \sum_{x_1, \cdots, x_n \in \Lambda_a} W^{(n)}(x_1, \cdots, x_n)$$
(2.2.14)

in a volume V, or, for translationally invariant systems and $V \leq \infty$

$$\chi_n = (a^D)^{n-1} \sum_{x_2, \cdots, x_n \in \Lambda_a} W^{(n)}(x_1, \cdots, x_n).$$
(2.2.15)

In particular, if we consider χ_2 defined by summing $W^{(2)}$ over x_2, Eq. (2.2.15) for $n = 2$ yields the well-known fluctuation-dissipation theorem.

The second moment μ_2 is defined as

$$\mu_2 = a^D \sum_{x \in \Lambda_a} x^2 \, W^{(2)}(0, x) \, .$$
(2.2.16)

Higher moments are defined accordingly.

Next we introduce the concept of the **effective action** Γ, defined via a Legendre transformation of $W(J)$ with respect to J

$$\Gamma(M) = W(J) - a^D \sum_{x \in \Lambda_a} J(x)M(x)$$
(2.2.17)

with

$$M(x) = \frac{\partial W(J)}{\partial a^D J(x)} \, .$$
(2.2.18)

As long as the definition of M via (2.2.18) is invertible to $J(M)$, $\Gamma(M)$ is uniquely defined via (2.2.17). In particular, in the symmetric phase $J = 0$

yields $M = 0$. Again, in a field theoretical context, Γ has the meaning of a generating functional. We write

$$\Gamma(M) = \sum_{n \geq 0} \frac{1}{n!} (a^D)^n \sum_{x_1, \cdots, x_n \in \Lambda_a} \Gamma^{(n)}(x_1, \cdots, x_n) \, M(x_1) \cdots M(x_n)$$

(2.2.19)

with

$$\Gamma^{(n)}(x_1, \cdots, x_n) \equiv <\Phi(x_1) \cdots \Phi(x_n)>^{1PI} = \frac{\partial^n \Gamma(M)}{\partial a^D M(x_1) \cdots \partial a^D M(x_n)}\bigg|_{M(x) \equiv 0}.$$

(2.2.20)

In a perturbative evaluation of $\Gamma(M)$ it is the set of so-called 1PI (one-particle irreducible) Feynman diagrams which contribute to $\Gamma^{(n)}$. This set is a subset of all connected diagrams contributing to $W^{(n)}$, but it is always possible to reconstruct the information that is contained in the set $\{W^{(m)} | m \leq n\}$ for given $n \in N$ from $\{\Gamma^{(m)} | m \leq n\}$, the set of 1PI diagrams.

For an N-component field Eq. (2.2.19) generalizes to

$$\Gamma(M) = \sum_{n \geq 0} \frac{1}{n!} \sum_{j_1, \cdots, j_n = 1}^{N} (a^D)^n \sum_{x_1, \cdots, x_n \in \Lambda_a} \Gamma^{(n)}_{j_1, \cdots, j_n}(x_1, \cdots, x_n)$$
$$\cdot M_{j_1}(x_1) \cdots M_{j_n}(x_n)$$

(2.2.21)

with

$$M_j(x) = \frac{\partial W}{\partial a^D J_j(x)}, \quad j = 1, \cdots, N.$$

(2.2.22)

Finally it is the coefficient $\Gamma^{(2)}$ that is used to define the renormalized mass m_R (in general different from the pole mass) and the wave function renormalization constant Z_R via the zero-momentum behavior of $\widetilde{\Gamma}^{(2)}$ in Fourier space

$$\widetilde{\Gamma}^{(2)}(p, -p) = -\frac{1}{Z_R}(m_R^2 + p^2 + O(p^4)) \text{ as } p \to 0,$$

(2.2.23)

where $\sim$ denotes the Fourier transform. Equation (2.2.23) implies that

$$m_R^2 = 2D\frac{\chi_2}{\mu_2}, \quad Z_R = 2D\frac{\chi_2^2}{\mu_2}.$$

(2.2.24)

The physical interpretation of $\Gamma(M)$ is that of an effective action related to the original action via a Legendre transformation of $W(J) = \ln\{\int \mathcal{D}\Phi \exp\{-S(\Phi) + \sum_x J(x)\Phi(x)\}\}$. Let V be the volume in physical units, i.e. $V = a^D|\Lambda_a|$, and $|\Lambda_a|$ the number of lattice sites. If we specialize $J(x)$ to constant $\overline{J}$ and define $\overline{M}$ as

$$\overline{M} = \frac{1}{V} \frac{dW(\overline{J})}{d\overline{J}}, \qquad (2.2.25)$$

$\Gamma(M)|_{M=\overline{M}}$ defines an **effective potential** V_{eff} according to

$$V V_{eff}(\overline{M}) = -\Gamma(\overline{M}). \qquad (2.2.26)$$

In general, the effective potential V_{eff} is not convex, neither in a finite nor in an infinite volume, but it has a convex hull. In the following we show that the convex hull of $V_{eff}(\overline{M})$ agrees with a convex potential $V_{stat}(\overline{M})$ that is convex by construction. It is the familiar definition in statistical physics. Again we start with $W(\overline{J})$, now for constant $J = \overline{J}$, and define

$$V_{stat}(\overline{M}) \equiv -\Gamma_{stat}(\overline{M}) = \sup_{\overline{J}}(\overline{J}M - w(\overline{J})) \qquad (2.2.27)$$

with $w(\overline{J}) = W(\overline{J})/V$, or, in a shorthand notation,

$$V_{stat}(\overline{M}) \equiv (Cw)(M), \qquad (2.2.28)$$

where C stands for "the convex hull of" and is defined by Eq. 2.2.27. This procedure usually runs under the name Legendre transformation in statistical physics. For the case of convex $w(\overline{J})$ both definitions agree, because (2.2.27) reduces to (2.2.26). Otherwise, only (2.2.27) gives a sensible definition of V_{stat}. The supremum in (2.2.27) makes the definition of Γ_{stat} unique even when the definition of J via the inversion of (2.2.25) is not (e.g., because of coexisting minima at a first-order phase transition). The definition (2.2.27) further ensures that V_{stat} as a function of M is convex. Therefore we introduce the short hand notation (2.2.28) for the combined operations on $W(J)$.

Next we prove that

$$C^2(V_{eff}(\overline{M})) = V_{stat}(\overline{M}). \qquad (2.2.29)$$

First we note that for any function $F(\Phi)$ that is convex for large $|\Phi|$

$$(C^2 F)(\Phi) = (\text{convex hull } F)(\Phi) \qquad (2.2.30)$$

or

$$(C^2 F)(\Phi) = \sup_h \, (h\Phi - CF(h)) \qquad (2.2.31)$$
$$= \sup_h \, (h\Phi + \inf_\Phi (F(\Phi) - h\Phi)) \ .$$

On the right hand side of (2.2.31) it is obvious that the operation C, twice applied to F, gives the convex hull of F. If F is convex $C^2 F(\Phi) = F(\Phi)$, i.e. C is invertible. Next we have to show that

$$C^2(V_{eff})(\overline{M}) \, = \, V_{stat}(\overline{M}) \, = \, (Cw)(\overline{M}) \qquad (2.2.32)$$

or

$$(C(V_{eff}))(I) \, = \, w(I) \, , \qquad (2.2.33)$$

since both $C(V_{eff})$ and w are convex functions of I. From the definition of V_{eff} and Γ_{stat}, (2.2.26) and (2.2.27), it follows

$$(C(V_{eff}))(I) \, = \, \sup_{\overline{M}} \left(I\overline{M} + w(\overline{J}) \, - \overline{J}\overline{M} \Big|_{\overline{J}=-\frac{d\Gamma(\overline{M})}{d\overline{M}}} \right) . \qquad (2.2.34)$$

Evaluating the supremum condition (2.2.34) leads to $\overline{J} = \overline{I}$ and therefore to

$$(C(V_{eff}))(I) \, = \, w(I) \, . \qquad (2.2.35)$$

Relation (2.2.35) tells us the following. Let us calculate V_{eff} in a field-theoretical framework, say in a perturbative expansion. What counts for a discussion of the phase structure is the convex hull of V_{eff}, because it agrees with V_{stat}. It is then a plateau in the convex hull that reflects the coexistence of nontrivial minima at a first-order phase transition, cf. Fig.(2.1.3).

A further relation is important for practical applications. It is the relation between V_{stat} and the so called **constraint effective potential** V_{cep} that is defined via a constraint which fixes a functional $O(\Phi)$ of the fields Φ

$$e^{-V \, V_{cep}(\overline{O},V)} \, = \, \int \mathcal{D} \, \Phi \, \delta(O[\Phi] - \overline{O}) \, e^{-S(\Phi)} \qquad (2.2.36)$$

so that

$$< O[\Phi] > = \, \frac{\int d\overline{O} \, \overline{O} \, e^{-V \, V_{cep}(\overline{O},V)}}{\int d\overline{O} \, e^{-V \, V_{cep}(\overline{O},V)}} \, . \qquad (2.2.37)$$

If $O(\Phi)$ is chosen as the order parameter

$$O[\Phi] = \frac{1}{V} \int d^D x \, \Phi(x) , \qquad (2.2.38)$$

it follows from the definition (2.2.36) and the translational invariance of the system that $\exp\{-VV_{cep}(\overline{\Phi})\}\Delta\Phi / \int d\overline{\Phi} \exp\{-VV_{cep}(\overline{\Phi})\}$ is the probability for finding a value for the scalar field at a point x, $\Phi(x)$, between $\overline{\Phi}$ and $\overline{\Phi} + \Delta\overline{\Phi}$.

In the infinite volume $V = \infty$ (and only there), the relation of $V_{cep}(\overline{\Phi})$ to $V_{stat}(\overline{\Phi})$, defined in (2.2.28), is given by

$$V_{stat}(\overline{\Phi}, V = \infty) = V_{cep}(\overline{\Phi}, V = \infty) . \qquad (2.2.39)$$

To prove (2.2.39) we first show that $C^2 V_{cep}(\overline{\Phi}, \infty) = V_{stat}(\overline{\Phi}, \infty)$. Thus V_{stat} is the convex hull of V_{cep}. If we can further show that $V_{cep}(\overline{\Phi}, V = \infty)$ is convex, it agrees with its convex hull and

$$V_{cep}(\overline{\Phi}, V = \infty) = V_{stat}(\overline{\Phi}, \infty) . \qquad (2.2.40)$$

For the first part of the proof we introduce on both sides of (2.2.36), with $O(\overline{\Phi})$ given by (2.2.37), an external current $\overline{J}$ and integrate over $\overline{\Phi}$

$$\int d\overline{\Phi} \, e^{-V \, V_{cep}(\overline{\Phi}, V)} \, e^{V \, \overline{J} \, \overline{\Phi}} = Z(\overline{J}, V) = e^{V \, w(\overline{J}, V)} . \qquad (2.2.41)$$

In the infinite-volume limit the saddle-point approximation to the integral in (2.2.41),

$$\int d\overline{\Phi} \, e^{-V \, (V_{cep}(\overline{\Phi}) - \overline{J} \, \overline{\Phi})} \simeq e^{-V \, \inf_{\overline{\Phi}} \, (V_{cep}(\overline{\Phi}) - J\overline{\Phi})} \qquad (2.2.42)$$

becomes exact. Then

$$w(\overline{J}, \infty) = \sup_{\overline{\Phi}} \, (\overline{J}\overline{\Phi} - V_{cep}(\overline{\Phi}, \infty) = \big(C(V_{cep})\big)(\overline{J}) . \qquad (2.2.43)$$

From (2.2.28) we know that $V_{stat}(\overline{\Phi}) = (Cw)(\overline{\Phi})$ leading to

$$V_{stat}(\overline{\Phi}) = \big(C^2 V_{cep}\big)(\overline{\Phi}) . \qquad (2.2.44)$$

For the proof that V_{cep} becomes convex in the infinite-volume limit, i.e. that

$$V_{cep}(\infty, \overline{\Phi}) \leq \mu \, V_{cep}(\infty, \overline{\Phi}_a) + (1 - \mu) \, V_{cep}(\infty, \overline{\Phi}_b) \qquad (2.2.45)$$

with

$$\overline{\Phi} = \mu\,\overline{\Phi}_a + (1-\mu)\,\overline{\Phi}_b\,, \quad 0 \le \mu \le 1 \qquad (2.2.46)$$

we refer to the original paper by [42]. Therefore $V_{cep}(\overline{\Phi}, \infty)$ agrees with its convex hull and relation (2.2.40) holds in the infinite-volume limit.

The constraint effective potential can be calculated analytically, but also numerically in Monte Carlo simulations. For sufficiently large volumes the convex shape is recovered when it is nonconvex for small volumes. We present a perturbative calculation in connection with gap equations in section (4.5).

2.3 The molecular-mean field approximation

Not only in our applications to particle physics, but also in systems of condensed matter physics it becomes inherently difficult to solve theories exactly in three or four dimensions. The more important it is to find approximations that can be controlled in a systematic way. The molecular field approximation is such an example. Originally it has been developed by P.Weiss as a self-consistent description of a ferromagnet. We briefly recapitulate the derivation of the equation of state of a ferromagnet. It then turns out that the same equation of state can be re-derived within a variational estimate for the free energy of an Ising model. This way the molecular field approximation achieves a similar status as the Hartree-Fock approximation for quantum mechanics. We will even go a step further and indicate for a scalar field theory, under which conditions the molecular field approximation may be regarded as the zeroth order in a systematic expansion scheme that later will be identified with the linked cluster expansion of a vertex functional.

2.3.1 *Self-consistent equation of state for a ferromagnet*

Let us consider a single spin $\sigma(x)$ at a site x in a ferromagnet made out of N spins. We look for an effective classical description of its energy, and its thermal expectation value. The spin is exposed to an effective classical local field $h_{eff}(x)$

$$h_{eff}(x) = \sum_{y \in \Lambda} J(x, y)\sigma(y) + h, \qquad (2.3.1)$$

consisting of two parts, a constant external magnetic field $h(x) = h \in \mathbf{R}$, and the field generated by all other spins $\sigma(y)$, $y \neq x$, $y \in \Lambda$, which interact with $\sigma(x)$. The interaction strength is parameterized by $J(x, y)$ with $J(x, x) = 0$. The Zeemann energy of the spins $\sigma(x)$ in the field $h_{eff}(x)$ is then given as

$$E(\sigma(x); h_{eff}(x)) = -h_{eff}(x)\sigma(x). \qquad (2.3.2)$$

For a given field $h_{eff}(x)$ the thermal expectation value of $\sigma(x)$ reads

$$< \sigma(x) > |_{h_{eff}(x)} = \frac{\sum_{\sigma(x)=\pm 1} \sigma(x)e^{-T^{-1}E(\sigma(x);h_{eff}(x))}}{\sum_{\sigma(x)=\pm 1} e^{-T^{-1}E(\sigma(x);h_{eff}(x))}}$$

$$= \tanh\left[T^{-1}\left(\sum_y J(x, y)\sigma(y) + h\right)\right] \qquad (2.3.3)$$

for x fixed. We set the Boltzmann constant k_B equal to 1 in the following. Summing now over all fluctuations of h_{eff}, we obtain for the full thermal expectation value of $\sigma(x)$

$$< \sigma(x) >= \sum_{\{\sigma\}} p(\{\sigma\}) \tanh\left[T^{-1}\left(\sum_y J(x, y)\sigma(y) + h\right)\right], \qquad (2.3.4)$$

where the sum runs over all spin configurations for fixed $\sigma(x)$ and

$$p(\{\sigma\}) = \frac{e^{+T^{-1}\left[\sum_{y \in \Lambda} J(x,y)\sigma(y)+h\right]} + e^{-T^{-1}\left[\sum_{y \in \Lambda} J(x,y)\sigma(y)+h\right]}}{\sum_{\{\sigma\}} \left(e^{+T^{-1}\left[\sum_{y \in \Lambda} J(x,y)\sigma(y)+h\right]} + e^{-T^{-1}\left[\sum_{y \in \Lambda} J(x,y)\sigma(y)+h\right]}\right)}$$

$$(2.3.5)$$

is the equilibrium distribution of all spins for fixed spin $\sigma(x)$. Since the argument of tanh does only depend on $h_{eff}(x)$, we may alternatively write

$$< \sigma(x) > = \int dh_{eff}(x) \, \varrho(h_{eff}(x)) \, \tanh(T^{-1}h_{eff}(x)) \qquad (2.3.6)$$

with

$$\varrho(h_{eff}(x)) = \sum_{\{\sigma\}} p(\{\sigma\}) \, \delta\left(h_{eff}(x) - \left(\sum_y J(x, y)\sigma(y) + h\right)\right). \qquad (2.3.7)$$

The essential simplification now comes from the assumption that the distribution of $h_{eff}(x)$ is peaked around an average value $< h_{eff}(x) >$ according to

$$\varrho(h_{eff}(x)) = \delta(h_{eff}(x) - < h_{eff}(x) >).$$ (2.3.8)

In this case we find

$$< \sigma(x) > = \tanh \left[T^{-1} < h_{eff}(x) > \right]$$
$$= \tanh \left[T^{-1} \left(\sum_y J(x,y) < \sigma(y) > +h \right) \right].$$ (2.3.9)

In a system which is invariant under translations, say for $J(x,y) = J$ for next neighbors and zero otherwise, $< \sigma(y) >$ is actually independent of y and equal to the average magnetization m. This finally leads to the equation of state $m = m(h, \beta)$ for an Ising ferromagnet in an external field h and at inverse temperature $\beta = T^{-1}$

$$m = \tanh \left(T^{-1}(J \, c \, m + h) \right).$$ (2.3.10)

Here c denotes the coordination number, i.e. the number of nearest neighbors. On a hyper-cubic lattice in D dimensions $c = 2D$. Equation (2.3.10) must be solved self-consistently. For simplicity let us consider the case of vanishing external field $h = 0$. A graphical representation of Eq. (2.3.10) then easily shows that $m = 0$ is the only solution of (2.3.10) for $\beta Jc < 1$, but $m \neq 0$ is the only solution for $\beta Jc > 1$, while the minimum for $m = 0$ becomes unstable. Thus $\beta_c Jc = 1$ must be identified as the transition point, or

$$T_c = J \cdot c.$$ (2.3.11)

Furthermore, by expanding Eq. (2.3.10) in terms of small m, the transition is seen to be continuous in the sense that $m(T)$ continuously vanishes for $T \to T_c$ from below.

2.3.1.1 *Critical exponents in the molecular-mean field approximation*

- Exponent β of the magnetization.
  We choose units so that $T_c = 1 = J \cdot c$. Expanding the equation of state for $h = 0$ about $m = 0$ yields

$$m = \tanh(T^{-1}m) \approx T^{-1}m - \frac{1}{3}(T^{-1}m)^3 + O(m^5).$$ (2.3.12)

For $m \neq 0$ it follows that

$$m^2 \approx 3(1 - T) , \qquad (2.3.13)$$

or

$$m \sim (T_c - T)^{1/2} , \qquad (2.3.14)$$

and

$$\beta = \frac{1}{2} . \qquad (2.3.15)$$

- Exponent δ at $T = T_c$, (i.e. for a critical isotherm). We consider the equation of state in the limit $h \to 0$ from above for $T = T_c = 1$. For small m, h we have

$$m = \tanh(m + h) = m + h - \frac{1}{3}(m + h)^3 + O(m + h)^5 , \quad (2.3.16)$$

i.e.

$$h \sim \frac{1}{3}m^3 , \qquad (2.3.17)$$

or

$$\delta = 3 . \qquad (2.3.18)$$

- Exponent γ, characterizing the singular behavior of the susceptibility.

  The order parameter susceptibility is defined as

$$\chi = \left. \frac{\partial m(h, T)}{\partial h} \right|_{h=0} \qquad (2.3.19)$$

with the derivative to be taken at constant temperature T. If we differentiate Eq. (2.3.10) and use the definition (2.3.19), we obtain

$$\chi \approx \frac{T^{-1}(1 + \chi)}{1 + \frac{m^2}{2T^2}} , \qquad (2.3.20)$$

or

$$\chi \approx \frac{1}{(1 + \frac{m^2}{2T^2})T - 1} . \qquad (2.3.21)$$

In the **symmetric phase** with $m = 0$ we find close to T_c

$$\chi \sim \frac{1}{T - T_c} , \qquad (2.3.22)$$

or, according to the definition of γ,

$$\chi \sim |T - T_c|^{-\gamma} \qquad (2.3.23)$$

so that

$$\gamma = 1 . \qquad (2.3.24)$$

In the **broken phase**, i.e. for $T \leq T_c = 1$, at $h = 0$ and in the vicinity of T_c, we find

$$m^2 \sim 3(1 - T),$$
$$\chi \approx \frac{1 - 3(1 - T)}{2(1 - T)} \sim (T_c - T)^{-1} , \qquad (2.3.25)$$

or,

$$\gamma' = 1. \qquad (2.3.26)$$

Here γ' denotes the exponent characterizing the singular behavior of χ when T_c is approached from the broken phase.

The equality of γ and γ' comes out correctly, as we shall see later, but whether the equality happens accidentally or has a deeper reason remains open within this approximation scheme. It is one of the merits of the renormalization group to explain the equality of critical exponents in the symmetric phase and in the broken phase on the basis of generic arguments, see section (2.4) below.

Let us summarize so far. In the molecular-mean field approximation the derivation of the equation of state and of the critical exponents turned out to be quite simple. The only place where we made an approximation was the assumption (2.3.8). Thus it is worthwhile to discuss under which conditions it is valid. First we notice that Eq. (2.3.10) neither depends on the geometry of the lattice nor on the dimension D. In particular the molecular-mean field approximation predicts a phase transition in $D = 1$ dimension. For $D = 1$ the prediction is known to be wrong. On the other hand, for $D \geq 4$, its predictions for the qualitative phase structure turn out to be correct. Moreover, the predicted values for T_c agree with the physically measured values the better the larger the coordination number c. For infinite coordination number the molecular-mean field approximation even becomes exact. This is what we show next.

The assumption that $h_{eff}(x)$ takes the form of a δ-distribution is nothing but a consequence of the central-limit theorem, applied to $\sigma(y)$ as statistically independent variables in the framework of the mean-field analysis. Let us see why. The spins σ are distributed according to

$$w_x(\sigma) = \frac{e^{T^{-1}\left[\sum_{y\in\Lambda} J(x,y)\sigma(x)\sigma(y)+h\sigma(x)\right]}}{\sum_{\{\sigma(y),y\in\Lambda\}} e^{T^{-1}\left[\sum_{y\in\Lambda} J(x,y)\sigma(x)\sigma(y)+h\sigma(x)\right]}}, \qquad (2.3.27)$$

so that $w_x(\sigma) \geq 0$ and $\sum_{\{\sigma\}} w_x(\sigma) = 1$. Let us define

$$a_x \equiv\; <\sigma(x)> = \sum_{\{\sigma\}} \sigma(x)\, w_x(\sigma)$$

$$b_x \equiv \sum_{\{\sigma\}} (\sigma(x) - a_x)^2\, w_x(\sigma)$$

$$A_n \equiv \sum_{x=1}^{n} a_x \;=\; O(n)$$

$$B_n \equiv \sum_{x=1}^{n} b_x \;=\; O(n). \qquad (2.3.28)$$

We consider the convolution $W_n(\tilde\sigma)$ of n probability densities $w_j(\sigma(j))$ of single spins $\sigma(j)$

$$W_n(\tilde\sigma) \equiv \sum_{\{\sigma(x=1)\}} \cdots \sum_{\{\sigma(x=n)\}} \left[\prod_{j=1}^{n} w_j(\sigma(x=j))\right] \delta\left(\tilde\sigma - \sum_{x=1}^{n} \sigma(x)\right). \qquad (2.3.29)$$

The asymptotic behavior of $W_n(\tilde\sigma)$ for large n is then predicted by the central limit theorem of probability theory, because the $w_x(\sigma)$ satisfy the assumptions of this theorem:

$$W_n(\tilde\sigma) \sim \frac{1}{\sqrt{2\pi B_n}} e^{-\frac{1}{2B_n}(\tilde\sigma - A_n)^2}\left(1 + O(\frac{1}{\sqrt{n}})\right). \qquad (2.3.30)$$

For a large coordination number c, and for $J(x,y) \equiv J/c$, independently of the sites x, y, we may therefore replace $\tilde\sigma \equiv \sum_{y\in U(x)} \sigma(y) = \sum_{y=1}^{c} \sigma(y)$ in (2.3.1) by $\sum_{y\in U(x)} <\sigma(y)> = c\cdot <\sigma>$, because of the definition of A_n. $U(x)$ denotes the set of sites $y \in \Lambda$ with $J(x,y) \neq 0$ and $\#U(x) = c$, or, since the right hand side is independent of x,

$$\rho(h_{eff}(x)) = \delta(h_{eff}(x)- <h_{eff}>) \qquad (2.3.31)$$

with

$$< h_{eff} >= J < \sigma > + h \equiv Jm + h, \qquad (2.3.32)$$

independent of x. Thus Weiss's approximation becomes exact in the limit of an infinite coordination number.

An example in which an infinite coordination number is realized is the **Curie-Weiss model for a spin glass** with couplings of infinite range. The Hamiltonian for a total number N of Ising spins in an external field $h(x)$ is given by

$$\mathcal{H}_N = -\frac{1}{2} \sum_{x,y \in \{1,...,N\},x \neq y} J(x,y)\, \sigma(x)\sigma(y) - h \cdot \sum_{x \in \{1...N\}} \sigma(x) \qquad (2.3.33)$$

with

$$\begin{matrix} J(x,y) = J/N \\ J(x,x) = 0 \end{matrix} \quad \text{for all} \quad x,y \in \{1...N\}. \qquad (2.3.34)$$

The coupling $J(x,y)$ must be scaled with N such that the thermodynamic limit $N \to \infty$ exists. Note that the only restriction for $J(x,y) \neq 0$ is $x \neq y$. It is also the thermodynamic limit in which the system becomes exactly solvable, as we show next.

Up to a constant the partition function of (2.3.33) reads

$$Z_N = \sum_{\{\sigma\}} \exp\left\{ \frac{\beta N}{2J} \left(\frac{1}{N/J} \sum_{x=1}^{N} \sigma(x) \right)^2 + \beta h \cdot \sum_{x=1}^{N} \sigma(x) \right\} \qquad (2.3.35)$$

with β the inverse temperature. A standard procedure then is a Gaussian linearization of the quadratic term in σ by means of an auxiliary field $\varphi \in R$. Thus we rewrite Z_N according to

$$Z_N = \sum_{\{\sigma\}} \int_{-\infty}^{+\infty} \frac{d\varphi}{\sqrt{\frac{2\pi J}{\beta N}}} \exp\left\{ -\frac{\beta N}{2J}\varphi^2 + \beta\varphi \sum_x \sigma(x) + \beta h \sum_x \sigma(x) \right\}$$

$$= \int_{-\infty}^{+\infty} \frac{d\varphi}{\sqrt{\frac{2\pi J}{\beta N}}} \exp\left\{ -\frac{\beta N}{2J}\varphi^2 \right\} \prod_{x=1}^{N} \left[\sum_{\sigma(x)=\pm 1} \exp\left\{ \beta(\varphi + h)\sigma(x) \right\} \right]. \qquad (2.3.36)$$

In the last step we have performed the sum over all configurations as the product over all sites x over the sum over all configurations at fixed site.

We obtain

$$Z_N = \int_{-\infty}^{+\infty} \frac{d\varphi}{\sqrt{\frac{2\pi J}{\beta N}}} \exp\left\{-N\left[\frac{\beta}{2J}\varphi^2 - \ln\left(2\cosh[\beta(\varphi + h)]\right)\right]\right\}. \quad (2.3.37)$$

Equation (2.3.37) now has an appropriate form for applying a saddle-point approximation in $1/N$. Note that the saddle-point approximation here corresponds to a large-N expansion with N the total number of spins. It becomes exact in the thermodynamic limit $N \to \infty$. For the free energy density we find

$$\begin{aligned} f(h, \beta) &= \lim_{N \to \infty} \left(-\frac{1}{\beta N} \ln Z_N\right) \\ &= \min_{\varphi} \left\{\frac{1}{2J}\varphi^2 - \frac{1}{\beta} \ln\left(2\cosh[\beta(\varphi + h)]\right)\right\} \\ &\equiv \min_{\varphi} \bar{f}(\varphi). \end{aligned} \quad (2.3.38)$$

The maximum condition for the integrand takes the same form as the equation of state for a ferromagnet as derived in (2.3.10)

$$\frac{1}{J}\varphi = \tanh(\beta(\varphi + h)). \quad (2.3.39)$$

The formal similarity suggests to identify the auxiliary field φ with the magnetization m if the units are chosen so that $J \cdot c = 1$. Actually φ and m must be identified, because

$$\begin{aligned} m &= \frac{1}{\beta} \frac{\partial(-\beta f(h, \beta))}{\partial h} = -\frac{\partial f}{\partial h} \\ &= \tanh(\beta(\varphi + h)) \\ &= \frac{1}{J}\varphi \ . \end{aligned} \quad (2.3.40)$$

Thus we have verified that for infinite-range couplings or infinite coordination number the equation of state of the molecular-mean field approximation becomes the exact equation of state of the Curie-Weiss model in the thermodynamic limit.

In applications to systems of particle physics couplings of infinite range may be generated in effective actions upon integration over a subset of degrees of freedom of some more fundamental local action. But usually one tries to argue why these non-local terms are negligible. Therefore the main interest here lies in clarifying the conditions under which the molecular-mean field approximation is reliable for systems with local actions and finite coordination number. In the next section we show how one can re-derive Eq. (2.3.39) within a variational estimate for the free energy. The main approximation, Eq. (2.3.31), will be replaced by an inequality that essentially follows from the convexity of the exponential function.

2.3.2 Variational estimates for the free energy of a spin system

For definiteness we consider a hyper-cubic lattice Λ in D dimensions with sites x. The lattice may be finite or infinite in all directions. The thermodynamic system of spins $\sigma(x)$ is described by an action (here synonymous for Hamiltonian) of the form

$$S[\sigma] = \overset{\circ}{S}[\sigma;\lambda] + S_{\text{int}}[\sigma;\lambda]. \qquad (2.3.41)$$

The splitting into the two terms on the right hand side of (2.3.41) is chosen in a way that expectation values w.r.t. $\overset{\circ}{S}$ are easily calculated while those with respect to S_{int} prevent an exact solution. The single parameter λ stands for a generic set of parameters which are introduced in a way that their sum drops out of the left hand side, λ will play the role of variational parameters. The partition function is first rewritten according to

$$\begin{aligned}
Z &\equiv \sum_{\{\sigma\}} \exp\{-\beta S[\sigma]\} \\
&= Z_0 \frac{\sum_{\{\sigma\}} \exp\{-\beta S_{\text{int}}[\sigma;\lambda]\} \exp\{-\beta \overset{\circ}{S}[\sigma;\lambda]\}}{Z_0} \\
&\equiv Z_0 \cdot < \exp\{-\beta S_{\text{int}}[\sigma;\lambda]\} >_0 .
\end{aligned} \qquad (2.3.42)$$

Here $< \cdots >_0$ denotes an average in the Gibbs ensemble that is defined via $\overset{\circ}{S}$, i.e. for an observable O

$$< O >_0 = \frac{\sum_{\{\sigma\}} O e^{-\beta \overset{\circ}{S}[\sigma;\lambda]}}{\sum_{\{\sigma\}} e^{-\beta \overset{\circ}{S}[\sigma;\lambda]}} . \qquad (2.3.43)$$

From the convexity of the exp-function it follows that

$$< \exp\{-\beta S\} >_0 \geq \exp < -\beta S >_0 \qquad (2.3.44)$$

for real valued S, because $< \cdots >_0$ corresponds to an average with a normalized probability distribution ρ (here $\rho = \rho(\sigma)$) so that $\sum_{\{\sigma\}} \rho(\sigma) =$

1). Thus we have

$$Z \geq Z_0 \exp\{-\beta < S_{\text{int}}[\sigma; \lambda] >_0\} , \qquad (2.3.45)$$

or, with $F = -k_B T \ln Z$,

$$F \leq \mathring{F} + < S_{\text{int}}[\sigma; \lambda] >_0 . \qquad (2.3.46)$$

If we express the free energy $\mathring{F}$ in terms of the internal energy $< \mathring{\mathcal{H}} > \equiv \mathring{S}$ and the entropy $\mathring{S}_E$ (to avoid confusion with the action $\mathring{S}$), we have

$$\mathring{F} = < \mathring{S} >_0 - T \cdot \mathring{S}_E , \qquad (2.3.47)$$

so that

$$F \leq < S >_0 - T \cdot \mathring{S}_E. \qquad (2.3.48)$$

$< \mathcal{H} >_0 \equiv < S >_0$ is the internal energy of the full system, S is as in (2.3.41), evaluated with respect to $< \cdots >_0$.

In the context of statistical physics the inequality (2.3.48) is called the Peierl's-Bogoliubov inequality. It yields an upper bound for the true free energy of the spin system. The quality of the upper bound depends on the choice of $\mathring{S}$ and S_{int} in (2.3.41). This is the place where the variational parameters enter the game. Within a given variational ansatz the parameters λ should be chosen such that the right hand side of Eq. (2.3.48) becomes minimal. The trial free energy then becomes

$$\overline{F} := \min_\lambda \left(< S[\sigma; \lambda] >_0 - T \cdot \mathring{S}_E(\lambda) \right). \qquad (2.3.49)$$

Within a given variational ansatz $\overline{F}$ gives the best approximation to the physical free energy F.

For an Ising ferromagnet with next neighbor interactions on a D dimensional lattice it is convenient to choose for $\mathring{S}$ a fully decoupled system of spins that is exposed to an effective auxiliary magnetic field λ, whose choice should be optimized. From a physical point of view this choice is natural for high temperatures and short range correlations. From a practical point of view it is an obvious choice, as it is the next-neighbor interactions that prevent a full factorization of the partition function. Thus we write for the action of the Ising ferromagnet in an external field h

$$\begin{aligned} S[\sigma; \lambda] &= -J \sum_{<xy>} \sigma(x)\sigma(y) - h \cdot \sum_{x \in \Lambda} \sigma(x) \\ &= \mathring{S}[\sigma; \lambda] + S_{\text{int}}[\sigma; \lambda] \end{aligned} \qquad (2.3.50)$$

with

$$\mathring{S}[\sigma; \lambda] = -\lambda \sum_{x \in \Lambda} \sigma(x) \tag{2.3.51}$$

and

$$S_{\text{int}}[\sigma; \lambda] = -J \sum_{<xy>} \sigma(x)\sigma(y) - (h - \lambda) \cdot \sum_{x \in \Lambda} \sigma(x). \tag{2.3.52}$$

$< xy >$ denote next-neighbor pairs of the lattice Λ. The trial free energy density $\overline{f}$ becomes

$$\overline{f} = \overline{F}/N = \frac{1}{N} \min_{\lambda} \{< S[\sigma; \lambda] >_0 - T \cdot \mathring{S}_E(\lambda)\}. \tag{2.3.53}$$

The internal energy is easily calculated as

$$\begin{aligned} < S[\sigma; \lambda] >_0 &= < -J \sum_{<xy>} \sigma(x)\sigma(y) - h \cdot \sum_x \sigma(x) >_0 \\ &= -J \cdot c \cdot N/2 < \sigma(x) >_0^2 - h \cdot N \cdot < \sigma(x) >_0 \\ &= -J \cdot c \cdot N/2 (\tanh(\beta\lambda))^2 - h \cdot N \cdot \tanh(\beta\lambda). \end{aligned} \tag{2.3.54}$$

Here we have used that there are $cN/2$ next-neighbor pairs. It should be noticed that the simplification comes from the factorization of $< \sigma(x)\sigma(y) >_0$ when it is evaluated with respect to $< \cdots >_0$.

The entropy density $\mathring{S}_E/N$ is calculated as

$$\mathring{S}_E = -\sum_{\{\sigma\}} p_0(\{\sigma\}) \ln p_0(\{\sigma\}), \tag{2.3.55}$$

where $p_0(\{\sigma\})$ denotes the Gibbs distribution generated from $\mathring{S}$, i.e.

$$p_0(\{\sigma\}) = \prod_{x \in \Lambda} \frac{\exp\{\beta\lambda\sigma(x)\}}{2\cosh\beta\lambda}. \tag{2.3.56}$$

Again we have set the Boltzmann constant $k_B = 1$. It follows that

$$\begin{aligned} \mathring{S}_E &= -\sum_{\{\sigma\}} p_0(\{\sigma\}) \left[\sum_{x \in \Lambda} (\beta\lambda\sigma(x) - \ln[2\cosh\beta\lambda]) \right] \\ &= -N \left[\beta\lambda\tanh(\beta\lambda) - \ln(2\cosh(\beta\lambda)) \right]. \end{aligned} \tag{2.3.57}$$

Here again we performed the sum over all configurations site by site. For the trial free energy density $\overline{f}(\lambda)$ we find

$$\overline{f}(\lambda) = -\frac{1}{2}Jc\tanh^2(\beta\lambda) + (\lambda - h)\tanh(\beta\lambda) - \beta^{-1}\ln(2\cosh(\beta\lambda)). \quad (2.3.58)$$

Minimization with respect to λ yields $f'(\lambda) = 0$, or

$$\beta\lambda = \beta[Jc\tanh(\beta\lambda) + h] \ . \qquad (2.3.59)$$

With $\tanh(\beta\lambda) = <\sigma(x)>_0 \equiv m$ (cf. also Eq. (2.3.3)), we have re-derived Eq. (2.3.10), i.e. the equation of state of the molecular-mean field approximation to the originally coupled spin system (2.3.52). Here the equation of state appears as the optimization condition within a variational estimate for the free energy. The auxiliary field λ determines the magnetization according to $\tanh\beta\lambda = m$.

Apparently there is some place for further optimizations by means of a sophisticated choice of $\overset{\circ}{S}$ and S_{int}. As the price for an optimized choice it gets in general harder to evaluate expectation values with respect to $< \cdots >_0$.

Two questions naturally arise.

- Is there any systematic way of improving on the molecular-mean field approximation so that the mean field level corresponds to the zeroth order in a systematic expansion scheme? In the case of the Ising model we have indicated already that the choice of $\overset{\circ}{S}$ in Eq. (2.3.51) is natural in particular from a physical point of view, because the Ising spins should behave as decoupled degrees of freedom for high temperatures. The linked cluster expansion, on the other hand, is a systematic expansion about a decoupled system. Thus it is not surprising that -under certain conditions- it is possible to identify the molecular-mean field approximation with the zeroth order of a linked cluster expansion for the vertex functional. Such a connection will be demonstrated for a scalar field theory in the next section.

- The second question concerns the applicability of the molecular-mean field approximation to models of particle physics.The variational estimates along the lines (2.3.45)-(2.3.48) obviously do not depend on the application to spin systems. The molecular-mean field approximation applied to an SU(2) gauge Higgs model in four dimensions leads to a value for the critical coupling which does differ from high precision Monte Carlo results only in the third digit [48]. In the next section we therefore illustrate, how the molecular-mean field approximation works for a scalar field theory, constructed in terms of an n-component order parameter field Φ, and next for the SU(2) Higgs model.

2.3.3 Molecular-mean field approximation for an N-component scalar field theory in D dimensions

Although the molecular-mean field approximation itself does not depend on the dimension D, the volume V, and the lattice topology, we want to be specific and choose Λ as a D-dimensional hyper-cubic lattice of volume $V = L^D$ with linear extension $L \leq \infty$. As an $O(N)$-invariant, Euclidean symmetric measure we have

$$\mathcal{D}_\Lambda \Phi = \prod_{x \in \Lambda} \mathrm{d}\mu(\Phi(x)) \qquad (2.3.60)$$

with translation invariant measure

$$\mathrm{d}\mu(\Phi(x)) = \mathrm{d}^N \Phi(x) e^{-\mathring{S}[\Phi(x)]} \qquad (2.3.61)$$

and

$$\int \mathrm{d}^N \Phi(x) = \prod_{a=1}^{N} \int_{-\infty}^{+\infty} \mathrm{d}\Phi^a(x) \qquad (2.3.62)$$

with $\Phi^a(x)$, $a = 1, \cdots, N$ an N-component scalar field, and $\mathring{S}(\Phi(x))$ the real-valued, $O(N)$-invariant, ultra-local part of the action depending on fields $\Phi(x)$ on single lattice sites x.

The partition function Z and the free energy density f are given as

$$\exp(-V \cdot f(h)) \equiv Z(h) = \int \mathcal{D}_\Lambda \Phi \, \exp\{-S_{\mathrm{int}}[\Phi] + \sum_{x \in \Lambda} h(x) \cdot \Phi(x)\} \quad (2.3.63)$$

with interaction part

$$S_{\mathrm{int}}[\Phi] = - \sum_{x,y \in \Lambda} v(x,y) \, \Phi(x) \cdot \Phi(y) \,, \qquad (2.3.64)$$

where we assume translational invariance

$$v(x,y) = v(x-y, 0) \qquad (2.3.65)$$

and the condition

$$0 < \sum_{x \in \Lambda} v(x) \equiv (2D) \cdot (2\kappa) < \infty. \tag{2.3.66}$$

Equation (2.3.66) excludes long-range couplings on an infinite lattice which do not decay sufficiently fast. This is to ensure the existence of the thermodynamic limit. Further details on v need not be specified now, since the molecular-mean field approximation is not sensitive to such details. To guarantee the very existence of the path integral of Z, in particular of a stable ground state, we further postulate that the total action $S(\Phi)$ satisfies

$$S[\Phi] = \sum_{x \in \Lambda} \mathring{S}[\Phi(x)] + S_{\text{int}}[\Phi] \geq \text{const} \sum_{x \in \Lambda} \Phi(x)^2, \tag{2.3.67}$$

with const > 0. An example for such an action is a $(\Phi^4 + \Phi^6)$-theory with N components

$$\mathring{S}[\Phi(x)] = \mathring{S}[\Phi(x); \lambda, \sigma] = \Phi^2(x) + \lambda(\Phi^2(x) - 1)^2 + \sigma(\Phi^2(x) - 1)^3 \tag{2.3.68}$$

with $\sigma > 0$ or $\sigma = 0$ and $\lambda \geq 0$, and

$$S_{\text{int}}[\Phi; \kappa] = -2\kappa \sum_{<xy>} \Phi(x) \cdot \Phi(y). \tag{2.3.69}$$

The case of $\lambda = 3\sigma$ corresponds to a pure Φ^6-theory and $\lambda < 3\sigma$ to negative quartic couplings. Already from a consideration of the classical potential one expects regions of first- and second-order phase transitions for the phase diagram of this theory.

Now we add and subtract a term to the action which depends on a set of variational parameters $\vec{H}$ that partly will be used to replace $S_{int}(\Phi)$ in the probability distribution of $< \cdots >_0$. The term is given by

$$S_H \equiv -H \cdot \sum_{x \in \Lambda} \Phi(x) \tag{2.3.70}$$

with H an x-independent N-component vector of auxiliary fields, explicitly breaking the $O(N)$ symmetry. Let us add a generic source-term $(\sum_{x \in \Lambda} \vec{h}(x) \cdot \Phi(x))$ to the action of the full partition function Z with $\vec{h}(x)$ an external scalar field with N components. In the following we write h and H for the vectors $\vec{h}$ and $\vec{H}$. We rewrite Z according to

$$Z(h) = \overline{Z}(H) \cdot < e^{A_h(H)} >_H \tag{2.3.71}$$

with the "molecular-mean field"-partition function $\overline{Z}(H)$ factorizing according to

$$\overline{Z}(H) := \prod_{x \in \Lambda} \overset{\circ}{Z}(H)$$

$$\overset{\circ}{Z}(H) = \overset{\circ}{\mathcal{N}} \int d\mu(\Phi) \, e^{H \cdot \Phi} \tag{2.3.72}$$

with normalization factor $\overset{\circ}{\mathcal{N}}$ chosen so that

$$\overset{\circ}{Z}(0) = 1 \, . \tag{2.3.73}$$

$< \cdots >_H$ is the expectation value w.r.t. $\overline{Z}(H)$ and

$$A_h(H) \equiv -S_{\text{int}}[\Phi] + \sum_{x \in \Lambda} \sum_{i=1}^{N} (h^i(x) - H^i) \, \Phi^i(x). \tag{2.3.74}$$

Again the positivity of the exp-function and the positivity of the measure imply an inequality similar to Eq. (2.3.44)

$$< e^{A_h(H)} >_H \geq \exp(< A_h(H) >_H), \tag{2.3.75}$$

or,

$$f(h) \leq \min_H \overline{f}(H, h) \tag{2.3.76}$$

with the true physical free energy density

$$f(h) = -\frac{1}{V} \ln Z(h) \tag{2.3.77}$$

and the trial free energy density

$$\overline{f}(H, h) \equiv - \left(\ln \overset{\circ}{Z}(H) + \frac{1}{V} < A_h(H) >_H \right)$$

$$= - \left[x(H) + 2D\kappa \sum_{i=1}^{N} (\partial_{H^i} x(H))^2 \right.$$

$$\left. + \sum_{i=1}^{N} (h^i - H^i) \, (\partial_{H^i} x(H)) \right] \tag{2.3.78}$$

with

$$x(H) \equiv \ln \overset{\circ}{Z}(H), \tag{2.3.79}$$

$h(x) = \text{const}$ and

$$\sum_{x \in \Lambda} v(x,0) = 2D \cdot 2\kappa \qquad (2.3.80)$$

in accordance with (2.3.64) and (2.3.66), which is in particular satisfied by

$$v(x,y) = \sum_{\mu=0}^{D-1} (2\kappa)(\delta_{x,y+\hat{\mu}} + \delta_{x,y-\hat{\mu}}) . \qquad (2.3.81)$$

Since (2.3.78) only depends on the projection of h on ∂_H, we may choose h parallel to H and specialize the set of variational parameters H to just one component $H \neq 0$. It follows that

$$\overline{f}(H,h) = -[x(H) + 2D\kappa(\partial_H x)^2 + (h-H)\partial_H x(H)]. \qquad (2.3.82)$$

The best approximation to the physical free energy is obtained for $H_0(h)$ which minimizes $\overline{f}(H(h),h)$. It should be noticed that in contrast to the true infinite-volume free energy, $\overline{f}(H_0(h),h)$ need not be convex as a function of h.

Because of the factorization properties of $\overline{Z}$, the molecular-mean field approximation will be independent on details of $v(x,y)$. It also does not depend on the topology of the lattice, in particular not on the finite temperature, implemented as a finite extension in Euclidean time.

The mean-field equation becomes

$$\partial_H \overline{f}\big|_{H_0(h)} = 0, \qquad (2.3.83)$$

or

$$4D\kappa(\partial_H x)\big|_{H_0(h)} + (h - H_0(h)) = 0 . \qquad (2.3.84)$$

The stability condition for the minimum at $H_0(h)$ is given as

$$(\partial_H^2 \overline{f}(H))\big|_{H_0(h)} > 0, \qquad (2.3.85)$$

or

$$\left[(4D\kappa\partial_H^2 x - 1)\partial_H^2 x + (4D\kappa\partial_H x + (h-H)) \cdot \partial_H^3 x\right]\big|_{H_0(h)} < 0. \qquad (2.3.86)$$

In view of an application to a scalar $(\Phi^4 + \Phi^6)$-theory with N components we are interested in an expansion of $\overline{f}(H_0(h))$ about its minimum at least up to sixth order, since we expect a tricritical line for which the order of the phase transition changes from first to second. For the couplings of the tricritical line the Taylor coefficients of the free energy vanish including terms of order H^4. An odd number of derivatives of $\overline{f}$ vanishes identically in the symmetric phase, thus we would need $\partial_H^6 \overline{f}$ for testing the stability of the solutions for the tricritical line. Furthermore we note that

$$\partial_H^n x(H)|_{H=0} = \begin{cases} \mathring{v}_n^c \, , n \text{ even} \\ 0 \, , n \text{ odd} \end{cases} . \tag{2.3.87}$$

Vanishing of the odd derivatives of x follows from the choice of $\mathring{S}$ that only depends only on even powers of Φ. Usually the coefficients $\mathring{v}_n^c$ appear in the generating functional $\ln \mathring{Z}(h)$ of ultra-local connected n-point functions

$$< \Phi^{a_1} \dots \Phi^{a_n} >^{\circ,c} = < \Phi^{a_1}(x_1) \dots \Phi^{a_n}(x_n) >^c_{x_1=\dots=x_n,v=0} , \tag{2.3.88}$$

$$\ln \mathring{Z}[h] = \sum_{n \geq 1} \frac{1}{n!} \sum_{a_1,\dots,a_n=1}^{N} h_{a_1} \dots h_{a_n} < \Phi^{a_1} \dots \Phi^{a_n} >^{\circ,c} ,$$
$$< \Phi^{a_1} \dots \Phi^{a_n} >^{\circ,c} \equiv \frac{K_n(a_1,\dots,a_n)}{(n-1)!!} \mathring{v}_n^c \tag{2.3.89}$$

with $K_n(a_1, \cdots, a_n)$ a total symmetric $O(N)$-tensor, accounting for the combinatorics of the internal $O(N)$-symmetry. For odd n, $K_n(a_1, \cdots, a_n) \equiv 0$, for even n $K_n(1, \cdots, 1) = (n-1)!!$. Since the variational parameter H enters the action in the same form as the external current h and since $\mathring{Z}(0) = 1$, $\ln \mathring{Z}(H)$ allows for an expansion in terms of H, which is formally identically with (2.3.89). For $n = 2$ we obtain

$$\partial_H^2 x = < (\Phi - < \Phi >_{\mathring{Z}})^2 >_{\mathring{Z}} \tag{2.3.90}$$

with $< \cdots >_{\mathring{Z}}$ denoting the expectation value with respect to the ultra-local partition function $\mathring{Z}$ of Eq. (2.3.72). For $H = 0$, $< \Phi >_{\mathring{Z}} = 0$, thus

$$\partial_H^2 x|_{H=0} = \mathring{v}_2^c = \mathring{v}_2 - \mathring{v}_1^2 = \mathring{v}_2 > 0, \tag{2.3.91}$$

since $d\mu(\Phi)$ is nondegenerate. The coefficients $\mathring{v}_n, n = 1, 2$, in (2.3.91) are part of the ultra-local n-point functions generated in an expansion of Z with respect to h (here equivalent to H)

$$\mathring{Z}[h] = \sum_{x \geq 1} \frac{1}{n!} \sum_{a_1,\dots,a_n=1}^{N} h_{a_1} \dots h_{a_n} \frac{K_n(a_1,\dots,a_n)}{(n-1)!!} \mathring{v}_n \tag{2.3.92}$$

with K_n as in (2.3.89). Explicitly the $\overset{\circ}{v}_n$ are given as

$$\overset{\circ}{v}_n = \frac{\int_{-\infty}^{+\infty} d^N\Phi \ (\Phi_1)^n \ \exp\{-\overset{\circ}{S}[\Phi]\}}{\int_{-\infty}^{+\infty} d^N\Phi \ \exp\{-\overset{\circ}{S}[\Phi]\}} \qquad (2.3.93)$$

independent of x, Φ_1 denotes the first component of $\Phi \in \mathbf{R}^N$. The combinatorics of the relation between the coefficient functions $\overset{\circ}{v}_n^c$ and $\overset{\circ}{v}_n$ is similar to that of connected and disconnected n-point functions ($W^{(n)}$ and $Z^{(n)}$) defined in section 2.2, Eqs. (2.2.10) and (2.2.13), respectively. For an explicit solution of Eqs. (2.3.83)-(2.3.85) we assume $\partial_H^2 x > 0$ for all H.

2.3.3.1 *Solutions of the mean-field equations*

Let us express d/dh in terms of derivatives with respect to H. We have

$$\frac{d}{dh} = \frac{\partial}{\partial h} + \frac{dH_0(h)}{dh} \frac{\partial}{\partial H} \qquad (2.3.94)$$

applied to the minimum condition for the free energy $\partial_H \overline{f}|_{H_0(h)} = 0$, or

$$4D\kappa \ (\partial_H x)_{H_0(h)} + (h - H_0(h)) = 0 \,, \qquad (2.3.95)$$

yields

$$\frac{dH_0(h)}{dh} = \frac{1}{1 - 4D\kappa \ \partial_H^2 x|_{H=H_0(h)}}. \qquad (2.3.96)$$

It follows for the trial free energy density

$$\overline{f}(H_0(h)) = 2D\kappa \ (\partial_H x)^2\big|_{H=H_0} - x(H_0(h)), \qquad (2.3.97)$$

and for the magnetization

$$\begin{aligned}
m &\equiv -\frac{d\overline{f}(H_0(h))}{dh} \\
&= -\frac{1}{(1 - 4D\kappa\partial_{H_0}^2 x)} \left\{ 4D\kappa(\partial_{H_0}x) \cdot (\partial_{H_0}^2 x) - \partial_{H_0}x \right\} \\
&= \partial_{H_0}x. \qquad (2.3.98)
\end{aligned}$$

It can be shown that (2.3.98) identifies the molecular-mean field approximation as the leading order of a linked cluster expansion of the vertex functional $R(\widehat{\Phi}, v)$, that is the generating functional of the one-particle-irreducible (or one-line-irreducible) susceptibilities, related to $\ln Z(h, v)$, Eq. (2.3.63), by a (nonstandard) Legendre transformation ($h \to \widehat{\Phi}$). (The origin of the "nonstandard" transformation is the difference in the expansion point: linked cluster expansions are expansions about a decoupled system, while "standard" perturbation theory is an expansion about an interaction-free theory.) Here we only indicate why this identification is allowed. On the left hand side of (2.3.98) we insert the molecular-mean field approximation for $\overline{f}(H_0(h))$. The result on the right hand side equals to $\partial_h \ln \mathring{Z}(h)$ because of the formal similarity between h and H. On the other hand, $\ln \mathring{Z}(h)$ gives the leading contribution to a linked cluster expansion of $\ln Z(h, v)$ about $v = 0$, cf. section (4.2) below. To leading order in v we furthermore have that

$$\partial_h \ln Z(h, v = 0) = \partial_{\widehat{\Phi}} R(\widehat{\Phi}, v = 0), \qquad (2.3.99)$$

where R is understood as above. The reason why the molecular-mean field approximation corresponds to the leading order of a linked cluster expansion of $R(\widehat{\Phi}, v)$ rather than $\ln Z(h, v)$ is not obvious from (2.3.98) and (2.3.99), but it is only seen in higher-order derivatives of $\overline{f}$ with respect to h. In the molecular-mean field approximation we find for

$$\chi_{n+1} \equiv \frac{d^n m}{dh^n} \qquad n = 2, 3, 4 \qquad (2.3.100a)$$

$$\chi_2 = (1 - 4D\kappa \partial_{H_0}^2 x)^{-1} \partial_{H_0}^2 x \qquad (2.3.100b)$$

$$\chi_3 = (1 - 4D\kappa \partial_{H_0}^2 x)^{-3} \partial_{H_0}^3 x \qquad (2.3.100c)$$

$$\chi_4 = (1 - 4D\kappa \partial_{H_0}^2 x)^{-4} \left(\partial_{H_0}^4 x + 3(\partial_{H_0}^3 x)^2 \frac{4D\kappa}{1 - 4D\kappa \partial_{H_0}^2 x} \right) \qquad (2.3.100d)$$

where the derivatives $\partial_{H_0}^n x$ should be understood as $\partial_H^n x$ evaluated at $H = H_0$. These expressions are not obtained if we express the χ_n to leading order in a power series in κ according to (2.2.14), following from the derivatives of $\ln Z(h, v)$. Instead they are obtained if we first express the χ_n in terms of the connected (1PI) n-point functions (generated by $R(\widehat{\Phi}, v)$ rather than $\ln Z(h, v)$) and next evaluate these (1PI) n-point functions to leading order. The leading order then gives exactly the terms $\partial_{H_0}^n x$ that we find on the right hand side of (2.3.100)in the molecular-mean field approximation .

For a calculation of the exponent η we further need to know the second moment μ_2 of the truncated two-point function. μ_2 is obtained as response of the magnetization to fluctuations in the external field, evaluated at zero momentum

$$\mu_2 \equiv \sum_y (z - y)^2 \left. \frac{dm(z)}{dh(y)} \right|_{h=\text{const}}. \qquad (2.3.101)$$

Here y, z are D-dimensional coordinates on the lattice. Thus μ_2 is sensitive to long-range fluctuations on a scale not smaller than p^{-2} if p denotes the momentum scale. Accordingly we consider small fluctuations of h about the average value

$$h = \frac{1}{V} \sum_{x \in \Lambda} h(x) = \text{const.} \tag{2.3.102}$$

Again we choose $h(y)$ parallel to $H(y)$. The trial free energy will then be given as the minimum of

$$\overline{F} = - \left(\sum_{y \in \Lambda} x(H(y)) + \frac{1}{2} \sum_{y,z} v(y, z)\, \partial_y x(H(y))\, \partial_z x(H(z)) \right.$$

$$\left. + \sum_{y \in \Lambda} (h(y) - H(y))\, \partial_y x(H(y)) \right) \tag{2.3.103}$$

with

$$v(y, z) = v(z, y)$$
$$x(H(y)) \equiv \ln \mathring{Z}(H(y)) \tag{2.3.104}$$

and

$$\partial_y \equiv \frac{\partial}{\partial H(y)}\,, \tag{2.3.105}$$

as before. For the mean-field equation it follows that

$$\partial_w \overline{F} = -\partial_w^2 x(w) \cdot \left(\sum_{y \in \Lambda} v(y, w)\partial_y x(H(y)) + (h(w) - H(w)) \right) = 0, \tag{2.3.106}$$

or, if we assume the nondegenerate case with $\partial_w^2 x(H(w)) > 0$,

$$\sum_{y \in \Lambda} v(y, w)\partial_y x(H(y)) + (h(w) - H(w)) = 0 \tag{2.3.107}$$

for all $w \in \Lambda$. On the subset of the solutions of the mean field approximation we have

$$\frac{d}{dh(z)} = \frac{\partial}{\partial h(z)} + \sum_{y \in \Lambda} \frac{dH(y)}{dh(z)} \frac{\partial}{\partial H(y)}, \tag{2.3.108}$$

applied to (2.3.107) it yields

$$0 = \delta_{wz} - \sum_{y \in \Lambda} \left(\delta_{wy} - v(w, y) \, \partial_y^2 x(H(y)) \right) \frac{dH(y)}{dh(z)}, \qquad (2.3.109)$$

or

$$\delta_{wz} = \sum_{y \in \Lambda} G^{-1}(w, y) \frac{dH(y)}{dh(z)}, \qquad (2.3.110)$$

where we have defined

$$\left(\delta_{wy} - v(w, y) \partial_y^2 x(H(y)) \right) \equiv G^{-1}(w, y). \qquad (2.3.111)$$

This leads to a magnetization

$$m(z) = -\frac{dF}{dh(z)} = -\frac{\partial \overline{F}}{\partial h(z)} + \sum_{y \in \Lambda} G(y, z) \, \partial_y \overline{F} \Big|_{H=H_0} \cdot \qquad (2.3.112)$$

The second term vanishes because of (2.3.106) and the first one yields

$$m(z) = \partial_z x(H(z)), \qquad (2.3.113)$$

so that from (2.3.110)

$$\frac{dm(z)}{dh(y)} = G(z, y) \, \partial_z^2 x(H(z)) \qquad (2.3.114)$$

and

$$\mu_2 = \sum_{y \in \Lambda} (z - y)^2 \left[G(z, y) \, \partial_z^2 x(H(z)) \right]_{h=\text{const}}, \qquad (2.3.115)$$

or, if we set $z = 0$ because of translational invariance,

$$\mu_2 = \left[\sum_{y \in \Lambda} y^2 G(0, y) \right] \cdot \partial_0^2 x \Big|_{h=\text{const}}, \qquad (2.3.116)$$

since $\partial_0^2 x|_{h=\text{const}}$ as well as $\partial_y^2 x(H(y))$ are independent of y. Next we express μ_2 via the Fourier transform of $G(0, y)$

$$(\partial_0^2 x) \sum_{y \in \Lambda} y^2 G(0, y) \Big|_{h=\text{const}} = - \left[\partial_0^2 x \cdot D \cdot \frac{\partial^2 \tilde{G}(k)}{\partial k_1^2} \Big|_{k=0} \right]_{h=\text{const}} \cdot \qquad (2.3.117)$$

Here we have assumed Euclidean invariance to write the D-dimensional Laplacian in terms of one-component k_1, k denotes the momentum vector in D dimensions. From the definition of $G^{-1}(z,y)$ it follows for the Fourier transform of $G(0,y)$ in y

$$\widetilde{G}(k)\Big|_{h=\text{const}} = \frac{1}{1 - \tilde{v}(k)\partial_y^2 x(H(y))}\Big|_{h=\text{const}}. \tag{2.3.118}$$

In particular we have for $\tilde{v}(k=0)$

$$\tilde{v}(0) = \sum_{x\in\Lambda} \int_{-\pi}^{+\pi} \frac{d^D k}{(2\pi)^D} e^{ikx}\tilde{v}(k) = \sum_{x\in\Lambda} v(x,0) = 4D\kappa, \tag{2.3.119}$$

cf. Eq. (2.3.66). For the second derivative of $\tilde{G}(k)$ we find

$$\frac{\partial^2 \tilde{G}(k)}{\partial k_1^2}\Big|_{h=\text{const},k=0} = \frac{\partial_y^2 x(y)\frac{\partial^2 \tilde{v}(k)}{\partial k_1^2}}{(1 - \tilde{v}(k)\partial_y^2 x(H(y)))^2}\Big|_{h=\text{const},k=0}, \tag{2.3.120}$$

where we have used that $\partial\tilde{v}(k)/\partial k_1 = 0$. Putting things together, we finally obtain for μ_2

$$\mu_2 = \frac{-\left[(\partial_0^2 x)^2 \cdot D \cdot \left(\frac{\partial^2 \tilde{v}(k)}{\partial k_1^2}\right)_{k=0}\right]_{h=\text{const}}}{(1 - \tilde{v}(k=0)\partial_0^2 x)^2}. \tag{2.3.121}$$

The correlation length ξ in terms of χ_2 and μ_2 is given as

$$\xi^2 = \frac{1}{2D}\frac{\mu_2}{\chi_2}. \tag{2.3.122}$$

This corresponds to the definition of the correlation length as the inverse renormalized mass m_R, if m_R^2 is defined via the zero momentum limit of the renormalized connected two-point function $\Gamma^{(2)}$

$$\widetilde{\Gamma}^{(2)} \stackrel{p\to 0}{=} -\frac{1}{Z_R}(m_R^2 + p^2) + O(p^4), \tag{2.3.123}$$

since

$$\xi^2 = \frac{1}{2D}\frac{\sum_{x\in\Lambda} x^2 < \Phi(x)\Phi(0)>^c}{\sum_{x\in\Lambda} < \Phi(x)\Phi(0)>^c}$$

$$= \frac{\frac{1}{2D}\frac{\partial^2}{\partial p^2}\widetilde{\Gamma}^{(2)}\Big|_{p=0}}{\Gamma^{(2)}\Big|_{p=0}}$$

$$= \frac{1}{m_R^2}. \tag{2.3.124}$$

(Here the attribute "renormalized" should not be seen in contrast to a "bare" mass as we later have to distinguish in the context of perturbation theory. It is the physical mass in natural units, while $m_R a$ is the physical mass in lattice units, that is a number.) In a Gaussian model (free field theory) or in the vicinity of a second-order phase transition the pole mass m_P goes to zero as well as m_R for vanishing momenta p. Therefore the definition of ξ according to Eq. (2.3.122) agrees with its definition as a screening length l_{scr} if the screening length is defined as the inverse pole mass m_P. For $m_R \simeq m_P$ we have

$$\left(\widetilde{\Gamma}^{(2)}\right)^{-1}_{p \to 0} = -\left(\frac{Z_R}{m_R^2 + p^2} + O(p^4)\right) \approx \frac{-Z_R}{m_p^2 + p^2} \,. \qquad (2.3.125)$$

A Fourier transformation of this negative inverse two-point function leads to the following decay for large distances $|x - y|$

$$\left(\widetilde{\Gamma}^{(2)}\right)^{-1}(x, y) \overset{|x-y| \to \infty}{\longrightarrow} |x - y|^{-(D-1)/2} \exp\{-m_P|x - y|\} \qquad (2.3.126)$$

for $D \geq 3$ with $m_P \equiv l_{scr}^{-1}$. From (2.3.100b) and (2.3.121) we then find for ξ

$$\xi^2 = \frac{1}{2} \frac{\partial^2_{H_0} x \left(-\frac{\partial^2 \tilde{v}}{\partial k_1^2}(0)\right)}{\left(1 - \partial^2_{H_0} x \cdot \tilde{v}(0)\right)} \,. \qquad (2.3.127)$$

2.3.3.2 *Critical exponents in the symmetric phase*

Since the molecular-mean field approximation is not sensitive to the volume, we may choose the external field $h = 0$ even if we expect a nontrivial minimum in the auxiliary field H because of a first-order phase transition (otherwise, for example in Monte Carlo simulations in a finite volume, one needs a nonvanishing external field as a trigger of the phase transition) .First note that for $h = 0$, $H_0(h) = 0$ is always a solution of (2.3.84). The stability condition then tells us when the local minimum at $H_0 = 0$ turns into a maximum. In case of a second-order phase transition this condition determines the critical coupling κ_c, at which the transition to the broken phase occurs. We find

$$\kappa_c = \frac{1}{4D\tilde{v}_2^c} \,. \qquad (2.3.128)$$

For a first-order transition this value of κ gives an upper bound within the mean field approximation on the true transition point at a higher temperature (or smaller value of κ). The true transition point is then defined by the degeneracy of the free energy for coexisting separated minima.

The symmetric phase is characterized by the following equations

$$f(H_0(h)) = 0 \qquad (2.3.129a)$$

$$m = \partial_{H_0} x = 0 , \qquad (2.3.129b)$$

because $\partial_H x(H) = O(H)$ so that it vanishes as $H \to 0$. Furthermore we have

$$\chi_2 = \mathring{v}_2^c \left(1 - 4D\kappa \mathring{v}_2^c\right)^{-1} \qquad (2.3.130a)$$

$$\chi_4 = \mathring{v}_4^c \left(1 - 4D\kappa \mathring{v}_2^c\right)^{-4} \qquad (2.3.130b)$$

$$\chi_6 = \left(\mathring{v}_6^c + \frac{10 \cdot \left(\mathring{v}_4^c\right)^2 4D\kappa}{\left(1 - 4D\kappa \mathring{v}_2^c\right)}\right) \left(1 - 4D\kappa \mathring{v}_2^c\right)^{-6} \qquad (2.3.130c)$$

$$\mu_2 = 4D\kappa\omega \left(\mathring{v}_2^c\right)^2 \left(1 - 4D\kappa \mathring{v}_2^c\right)^{-2} \qquad (2.3.130d)$$

with

$$4\kappa\omega \equiv -\frac{\partial^2}{\partial k_1^2} \, \widetilde{v}(0) , \qquad (2.3.131)$$

where ω has been introduced to explicitly exhibit the κ dependence.

A sufficient condition for the uniqueness of the solution $H_0(h = 0) = 0$ for $\kappa < \kappa_c$ of the mean-field equations and the minimum of the free energy is given by the following lemma.

Let

$$\partial_H^2 x(H) > 0 \qquad (2.3.132)$$

and

$$\partial_H^3 x(H) \lesseqgtr 0 \qquad (2.3.133)$$

for all $H \geq 0$ ($H \leq 0$), respectively. Note that (2.3.133) corresponds to the Lebowitz inequality which is always satisfied for an action of the form

$$\mathring{S}[\Phi] = \sigma\Phi^2 + \sum_{n=2}^{N} c_{2n}(\Phi^2)^n \qquad (2.3.134)$$

with coefficients $c_{2n} \geq 0$. Furthermore, let $\kappa < \kappa_c$ and $H_1 \neq 0$ with

$$\partial_H \overline{f}(H)\big|_{H_1} = 0 . \qquad (2.3.135)$$

The statement of the lemma then is that H_1 corresponds to a minimum again, that is

$$\partial_H^2 \overline{f}(H)\big|_{H_1} > 0 \,. \tag{2.3.136}$$

Thus, $H_0 = 0$ is the only solution of the mean-field equations for $h = 0$. For the proof we note that

$$\begin{aligned}
\partial_H^2 \overline{f}\big|_{H_1} &= -\left(4D\kappa \partial_{H_1}^2 x - 1\right)\partial_{H_1}^2 x \\
&\geq -\left(4D\kappa \mathring{v}_2^c - 1\right)\partial_{H_1}^2 x \\
&> -\left(4D\kappa_c \mathring{v}_2^c - 1\right)\partial_{H_1}^2 x = 0 \,.
\end{aligned} \tag{2.3.137}$$

The first inequality follows from the Lebowitz inequality, $\partial_{H_1}^2 x(H) > 0$ and $\partial_0^2 x = \mathring{v}_2^c$, the second from $\kappa < \kappa_c$. Therefore $\overline{f}$, as a continuous function of κ, cannot have a local maximum as it should have if there were two separated local minima at $H_0 = 0$ and $H_1 \neq 0$.

The standard definitions of the critical exponents in terms of κ rather than T are given by

$$\chi_2 \propto \mathcal{A}_{\chi_2}\left(1 - \frac{\kappa}{\kappa_c}\right)^{-\gamma} \tag{2.3.138a}$$

$$\xi \propto \mathcal{A}_\xi \left(1 - \frac{\kappa}{\kappa_c}\right)^{-\nu} \tag{2.3.138b}$$

for $\kappa \to \kappa_c$ from below or $T \to T_c$ from above. From (2.3.128) and (2.3.129c) it follows that

$$\gamma = 1 \,. \tag{2.3.139}$$

Equations (2.3.122), (2.3.130c) and (2.3.130f) imply for the correlation length ξ

$$\xi^2 = \frac{2\omega\kappa}{4D\kappa_c}\left(1 - \frac{\kappa}{\kappa_c}\right)^{-1} \,, \tag{2.3.140}$$

so that

$$\nu = \frac{1}{2} \,. \tag{2.3.141}$$

From the scaling relation

$$\eta = 2\nu - \gamma \tag{2.3.142}$$

it follows that

$$\eta = 0 \ . \tag{2.3.143}$$

The exponents (2.3.139), (2.3.141), (2.3.143) are called Gaussian exponents, because the same set of exponents is obtained as an exact result for a free field theory, and the action of a free field theory agrees with the action associated with a Gaussian fixed point. The stability condition for having such critical behavior is given by

$$\partial_H^4 \bar{f} > 0 \ , \tag{2.3.144}$$

or

$$4D\kappa_c \partial_H^4 x < 0, \tag{2.3.145}$$

since $\partial_H^3 \bar{f} = 0$ in the symmetric phase.

2.3.3.3   Critical exponents in the broken phase

Let us assume that κ_c is the critical coupling of a second-order phase transition, $\kappa < \kappa_c$ identifies the symmetric phase, characterized by the fact that for $h = 0$ and all $\kappa < \kappa_c$ the only stable solution of the mean-field equation $\partial_H \bar{f}|_{H_0} = 0$ is $H_0 = 0$. In the lemma of the last section we have given a sufficient condition for that.

For $h = 0$, the solution $H_0 = 0$ becomes unstable when $\partial_0^2 \bar{f} < 0$. Therefore $\kappa = \kappa_c$ is determined as $\partial_0^2 \bar{f} = 0$. The correlation length ξ diverges at $\kappa = \kappa_c$, and a second-order phase transition occurs to a spontaneously broken phase for which $\kappa > \kappa_c$. In the following we study the broken phase in the vicinity of the critical coupling κ_c, hence $\kappa - \kappa_c > 0$, but small.

Let us assume that $\partial_H^2 x > 0$ and that for all $\kappa < \kappa_c = (4D\mathring{v}_2^c)^{-1}$ $H_0 = 0$ is the only stable solution for $h = 0$ in the symmetric phase. Therefore for $|\kappa - \kappa_c|$ small we may expand $\partial_H x$ as a power series in H about $H = 0$. Let $\mathring{v}_4^c < 0$. As we prove next, for $h = 0$ in the broken phase, the mean-field equation (2.3.84) then has the O(N)-symmetric solution

$$H_0^2 = (\kappa - \kappa_c) \left(\frac{3!}{4D\kappa_c^2 (-\mathring{v}_4^c)} + O(\kappa - \kappa_c) \right) \tag{2.3.146}$$

as $\kappa \to \kappa_c^+$. The stability condition

$$4D\kappa \partial_{H_0}^2 x - 1 = -\frac{2}{\kappa_c}(\kappa - \kappa_c) + O\left((\kappa - \kappa_c)^2\right) < 0 \tag{2.3.147}$$

is satisfied as $\kappa \to \kappa_c^+$.

Proof: From an analogous expansion to (2.3.89) of $\ln \mathring{Z}[H]$ in H rather than h we find

$$\partial_H x = \mathring{v}_2^c H + \frac{\mathring{v}_4^c}{3!} H^3 + O(H^5) \,. \tag{2.3.148}$$

Therefore

$$
\begin{aligned}
0 &= 4D\kappa\partial_{H_0} x - H_0 \\
&= \frac{H_0}{\kappa_c}\left((\kappa - \kappa_c) + \frac{4D\kappa\kappa_c}{3!}\mathring{v}_4^c H_0^2 + O(H_0^4) \right)
\end{aligned} \tag{2.3.149}
$$

has for $\kappa > \kappa_c$ the nontrivial solution (2.3.146) with the right hand side of (2.3.146) > 0 for $\kappa > \kappa_c$. The stability condition (2.3.147) is satisfied as $\kappa \to \kappa_c^+$, because

$$
\begin{aligned}
4D\kappa\partial_{H_0}^2 x - 1 &= 4D\kappa\left(\mathring{v}_2^c + \frac{\mathring{v}_4^c}{2}H_0^2 + O(H_0^4) \right) - 1 \\
&= \frac{\kappa - \kappa_c}{\kappa_c} - \frac{3!}{2\kappa_c}(\kappa - \kappa_c) + O\left((\kappa - \kappa_c)^2\right) \tag{2.3.150} \\
&= -2\frac{\kappa - \kappa_c}{\kappa_c} + O\left((\kappa - \kappa_c)^2\right) \,. \tag{2.3.151}
\end{aligned}
$$

From (2.3.100b),(2.3.148) and (2.3.150) we find for the susceptibility

$$\chi_2 = \frac{\mathring{v}_2^c + O(\kappa - \kappa_c)}{\frac{2}{\kappa_c}(\kappa - \kappa_c) + O\left((\kappa - \kappa_c)^2\right)} \propto (\kappa - \kappa_c)^{-\gamma'} \tag{2.3.152}$$

as $\kappa \to \kappa_c^+$, thus

$$\gamma' = 1. \tag{2.3.153}$$

For the correlation length ξ we find from (2.3.127)

$$\xi^2 = \frac{1}{2}\frac{(\mathring{v}_2^c + O(\kappa - \kappa_c))\left(-\frac{\partial^2 \tilde{v}}{\partial k_1^2}(0)\right)}{\frac{2}{\kappa_c}(\kappa - \kappa_c) + O\left((\kappa - \kappa_c)^2\right)} \propto (\kappa - \kappa_c)^{-2\nu'} \tag{2.3.154}$$

as $\kappa \to \kappa_c^+$, or

$$\nu' = \frac{1}{2}. \tag{2.3.155}$$

For the magnetization it follows from (2.3.98) and (2.3.148) at $h = 0$

$$m = \partial_{H_0} x$$
$$= \mathring{v}_2^c H_0 + O(H_0^2)$$
$$\approx \left(\frac{3!}{(4D)^3 \kappa_c^4 (-\mathring{v}_4^c)} \right)^{1/2} (\kappa - \kappa_c)^\beta \qquad (2.3.156)$$

as $\kappa \to \kappa_c^+$, with

$$\beta = \frac{1}{2}. \qquad (2.3.157)$$

At $\kappa = \kappa_c$, for nonvanishing h, the mean-field equation

$$4D\kappa_c \partial_H x + (h - H) = 4D\kappa_c \left(\mathring{v}_2^c H + \frac{1}{3!} \mathring{v}_4^c H^3 + O(H^5) \right) + (h - H)$$
$$= \frac{4D\kappa_c}{3!} \mathring{v}_4^c H^3 + h + O(H^5)$$
$$= 0 \qquad (2.3.158)$$

has the solution

$$H(h) \approx \left(\frac{3!}{4D\kappa_c} \frac{1}{(-\mathring{v}_4^c)} \right)^{1/3} h^{1/3}, \qquad (2.3.159)$$

so that the magnetization behaves as

$$m = \partial_{H(h)} x = \mathring{v}_2^c H(h) + O(H^3)$$
$$\approx \left(\frac{3!}{(4D\kappa_c)^4} \frac{1}{(-\mathring{v}_4^c)} \right)^{1/3} h^{1/\delta}, \qquad (2.3.160)$$

as h goes to zero from above with

$$\delta = 3. \qquad (2.3.161)$$

Finally, we leave it as an exercise to calculate the exponents α in the symmetric phase and α' in the broken phase by means of the internal energy U and the specific heat c, where the derivatives with respect to the temperature have to be replaced by derivatives with respect to κ. The result is

$$\alpha = \alpha' = 0. \qquad (2.3.162)$$

In this derivation in the framework of the molecular-mean field approximation, the agreement of the exponents α, γ, ν, and η in the symmetric and in the broken phase appears as a coincidence. It will be explained in the framework of the renormalization group, cf. section (2.4).

2.3.3.4 First-order transitions within the molecular-mean field approximation

How can we identify a first-order transition in the molecular-mean field approximation? As mentioned above, κ_c, as defined by the stability condition (2.3.85), gives only an upper bound on the critical coupling in case of a first-order transition. The actual value of the transition coupling κ_1 is implicitly given by

$$\overline{f}\left(H_0(h=0)=0;\kappa_1\right)=\overline{f}\left(H_1(h=0)\neq 0;\kappa_1\right),\qquad (2.3.163)$$

where H_1 denotes a second nontrivial solution of (2.3.84) and (2.3.85). Clearly a jump in the magnetization M is only possible if the Lebowitz inequality is violated. It is then possible to prove the following lemma.

Let us assume that $\partial_H^2 x > 0$ and there exists a second local minimum $H_1 \neq 0$ for some $0 < \widetilde{\kappa} < \kappa$ so that $\partial_{H_1}\overline{f} = 0$ and $\partial_{H_1}^2\overline{f} > 0$. Then there exists a local neighborhood $U(\widetilde{\kappa})$ and a function $H_1(\kappa)$ so that for all $\kappa \in U(\widetilde{\kappa})$, $H_1(\kappa)$ is still a minimum of $\overline{f}$ and $d\overline{f}/d\kappa < 0$ for all $\kappa \in U(\widetilde{\kappa})$. For $\widetilde{\kappa}$ chosen as κ_1 such that $\overline{f}(H_1(\kappa_1)) = 0$ (same as $\overline{f}(H_0 = 0)$), it follows that $\overline{f}(H_1(\kappa)) > 0$ for $\kappa < \kappa_1$ and < 0 for $\kappa > \kappa_1$, respectively. So the main statement is that $\kappa < \kappa_1$ identifies the symmetric phase with $\overline{f}(0) < \overline{f}(H_1)$, $\kappa > \kappa_1$ the broken phase with $\overline{f}(0) > \overline{f}(H_1)$. Moreover the magnetization

$$m = \partial_{H_1} x \neq 0\,,\qquad (2.3.164)$$

as expected for the broken phase, here because of $H_1 \neq 0$ and $\partial_H^2 x > 0$.

2.3.3.5 Tricritical behavior

From a discussion of the classical potential of the action (2.3.68) -(2.3.69) one expects a tricritical point for fixed six-point coupling $\sigma > 0$. The tricritical point is characterized by three coexisting phases. It is specified by a coupling λ_{tric}, at which a line of first order transitions turns into a line of second-order transitions. A Taylor expansion of the free energy in fluctuations about the minimum solution H_0 starts with terms of order H^6. In the molecular-mean field approximation the condition for tricriticality is given as

$$\partial_{H=0}^4 x \;=\; 0 \;=\; \overset{\circ}{v}_4^c\,,\qquad (2.3.165)$$

and its stability as

$$\partial^6_{H=0} x < 0 \qquad (2.3.166)$$

or

$$\mathring{v}^c_6 < 0 \; . \qquad (2.3.167)$$

The existence of a broken phase and the critical exponents in both phases are derived in an analogous way as it was shown above with the only difference that the expansions of $x(H)$ and $m(H)$ must be performed to a higher order in H, adapted to the conditions (2.3.165) and (2.3.166).

The phase structure of the $(\Phi^4 + \Phi^6)$-theory can be re-derived in a linked cluster expansion of $\ln Z$ of this model. There it turns out that the molecular-mean field approximation at the *tricritical* point ("point" for fixed $\sigma > 0$, line for varying σ) becomes exact: in particular the (tri)critical exponents are predicted correctly. A deeper understanding of the reason why it becomes exact in the exceptional case of a tricritical point is provided by the Ginzburg-criterion, for which we refer to the literature [49].

2.3.4   *Variational estimates for the SU(2) Higgs model*

In this section we consider the $SU(2)$ Higgs model in four dimensions on a hyper-cubic lattice in a finite or infinite volume. We will argue in section (3.1.2.6) why the $SU(2)$ Higgs model is supposed to lead to an effective description of the electroweak phase transition.

The $SU(2)$ Higgs model is much more complex than the models we have considered so far. First of all we have two kind of fields, gauge fields $U \in SU(2)$, and scalar Higgs fields $\Phi(x)$, with Φ real multiples of $SU(2)$. The internal symmetry is now a gauge symmetry. The model is invariant under local $SU(2)$ gauge transformations, and the gauge group is nonabelian. The nonabelian nature has far reaching consequences, in particular it leads to several length scales as we have indicated in the introduction. Therefore the investigation of the phase structure in the presence of more than one "bulk" correlation length becomes a challenge also from the viewpoint of statistical physics. In spite of this complexity, variational estimates still provide a useful overview of the phase structure. They even give a reasonable quantitative estimate of the critical line of the Higgs transition, as we shall demonstrate in this section. Care must be taken, however, with respect to predictions of the order of the transition. Since the variational estimates can be made with very little computational effort, they should supplement Monte Carlo simulations, in particular in parameter ranges, where no independent results are available yet.

We consider a four-dimensional hyper-cubic temperature lattice Λ_0 of size $L_0 \times V_3$, with $L_0 = T^{-1}$ the inverse temperature in lattice units and V_3 the spatial volume. The lattice links are the set of nearest neighbor lattice sites, given by

$$\overline{\Lambda}_1 = \{(x;\mu) \mid x \in \Lambda_0, \mu = 0, \ldots, 3\}. \qquad (2.3.168)$$

The gauge field $U(x;\mu)$ is an $SU(2)$-valued field living on the lattice links $\overline{\Lambda}_1$. It is convenient to parameterize such an $SU(2)$ matrix according to

$$U = \phi_0(U)\, \mathbf{1}_2 + i\, \vec{\sigma} \cdot \vec{\phi}(U)$$
$$\equiv U_0 + i\, \vec{\sigma} \cdot \vec{U} \qquad (2.3.169)$$

with $U \simeq (U_0, \vec{U}) \in S_3$ and $\vec{\sigma} = (\sigma_1, \sigma_2, \sigma_3)$ the Pauli-matrices. The Higgs field Φ is assigned to the lattice sites Λ_0. Its values are real multiples of $SU(2)$,

$$\Phi(x) = \phi_0(\Phi(x))\, \mathbf{1}_2 + i\, \vec{\sigma} \cdot \vec{\phi}(\Phi(x))$$
$$\equiv \phi_0(x) + i\, \vec{\sigma} \cdot \vec{\phi}(x), \qquad (2.3.170)$$

with $\phi \simeq (\phi_0(x), \vec{\phi}(x)) \in \mathbf{R}^4$. We say that Φ is $cU(2)$ valued.
We introduce the action of the $SU(2)$ Higgs model via

$$Z = \int \mathcal{D}U \mathcal{D}\Phi \, \exp\left(-S(U,\Phi)\right),$$
$$S(U,\Phi) = S_W(U) + S_{hop}(U,\Phi), \qquad (2.3.171)$$
$$S_{hop}(U,\Phi) = -\sum_{x \in \Lambda_0} \left\{ \sum_{\mu=0}^{3} (2\kappa)\frac{1}{2}\mathrm{tr}\left(\Phi(x)^\dagger U(x;\mu)\Phi(x+\widehat{\mu})\right) \right\}$$

and

$$S_W(U) = \sum_{x \in \Lambda_0} \sum_{\mu<\nu=0}^{3} \overline{\beta}\left(1 - \frac{1}{2}\mathrm{Re}\,\mathrm{tr}\, U(x;\mu)U(x+\widehat{\mu};\nu)U(x+\widehat{\nu};\mu)^{-1}U(x;\nu)^{-1}\right).$$
$$(2.3.172)$$

S_{hop} is the hopping-parameter term in four dimensions. S_W is the gauge invariant Wilson action with gauge coupling $\overline{\beta} \equiv 4/g^2$ and $U \in SU(2)$. The Higgs self-interaction has been absorbed in the measure according to

$$\mathcal{D}\Phi = \prod_{x \in \Lambda_0} \exp\left(-\mathring{S}(\Phi(x))\right) d\nu(\Phi(x)), \qquad (2.3.173)$$

where

$$d\nu(\Phi) = d^4\phi,$$
$$\mathring{S}(\Phi) = \frac{1}{2}\mathrm{tr}\left(\Phi^\dagger\Phi\right) + \lambda\left(\frac{1}{2}\mathrm{tr}\left(\Phi^\dagger\Phi\right) - 1\right)^2 \qquad (2.3.174)$$
$$= \phi^2 + \lambda\left(\phi^2 - 1\right)^2.$$

Finally,

$$\mathcal{D}U = \prod_{x\in\Lambda_0}\prod_{\mu=0}^{3} d\mu_H(U(x;\mu)), \qquad (2.3.175)$$

with $d\mu_H(U)$ the normalized Haar measure on $SU(2)$,

$$d\mu_H(U(x;\mu)) = \frac{1}{2\pi^2}\sin^2\theta\sin\varphi_1 d\theta d\varphi_1 d\varphi_2 \qquad (2.3.176)$$

if $U(x;\mu) \in SU(2)$ is parameterized according to

$$U = \phi_0(U)\,\mathbf{1}_2 + i\,\vec{\sigma}\cdot\vec{\phi}(U)$$
$$\equiv U_0 + i\,\vec{\sigma}\cdot\vec{U} \qquad (2.3.177)$$

with $U \simeq (U_0, \vec{U}) \in S_3$, and

$$U_0 \equiv \phi_0(U) = \cos\theta$$
$$U_1 \equiv \phi_1(U) = \sin\theta\cos\varphi_1$$
$$U_2 \equiv \phi_2(U) = \sin\theta\sin\varphi_1\cos\varphi_2$$
$$U_3 \equiv \phi_3(U) = \sin\theta\sin\varphi_1\sin\varphi_2 \qquad (2.3.178)$$

$$0 \le \theta \le \pi$$
$$0 \le \varphi_1 \le \pi$$
$$-\pi \le \varphi_2 \le \pi,$$

so that

$$U = \cos\theta\mathbf{1}_2 + i\sin\theta\,\vec{\sigma}\cdot(\cos\varphi_1, \sin\varphi_1\cos\varphi_2, \sin\varphi_1\sin\varphi_2)$$
$$\mathrm{tr}\,U = 2\phi_0(U) = 2\cos\theta. \qquad (2.3.179)$$

In the presence of external currents $j_\mu(x)$, $(\mu = 1, \cdots, D = 4)$, $h(x)$,

$S(U, \Phi)$ of (2.3.171) should be replaced by

$$S_{\mathrm{GH}}(U, \Phi) = S_W(U) + S_{hop}(U, \Phi) - \sum_{x,\mu} \mathrm{tr}\,(j_\mu(x)^T U(x; \mu))$$

$$+ j_\mu^*(x) U(x; \mu)^{-1}) - \sum_x \mathrm{tr}\,(h(x)^T \phi(x) + \Phi^\dagger(x) h^\star(x))\,. \qquad (2.3.180)$$

The currents j_μ and h are complex 2×2 matrices. For the adjoint fields $U(x; \mu)^{-1}$ and $\Phi^\dagger$, the currents must be chosen accordingly to ensure that the partition function is real. Moreover one should gauge-average the external currents that break the local gauge symmetry explicitly and integrate over all gauge degrees of freedom. This means that for $\Lambda(x)$ and $T \in SU(2)$ the currents j_μ and h in (2.3.180) should be replaced by

$$j \to (j^\wedge)_\mu(x) = \Lambda(x)^T j_\mu(x) \Lambda(x + \widehat{\mu})^{-1\,\mathrm{T}}$$

$$h \to (h^\wedge)(x) = \Lambda(x)^T h(x) T^{-1\,\mathrm{T}}\,, \qquad (2.3.181)$$

and the integration over all gauge degrees of freedom should be performed

$$\int \prod_x \mathrm{d}\mu(\Lambda(x))\,\mathrm{d}\mu(T)\,, \qquad (2.3.182)$$

in addition to Haar measures $d\mu(\Lambda(x))$ and $d\mu(T)$. Direction (μ)-dependent currents j_μ are introduced for cases in which it is natural to distinguish space-like and time-like directions. In this section, however, we will not make use of this distinction. Now we proceed in an analogous way as before.

We replace Z, the full partition function of the $SU(2)$ Higgs model, by a partition function Z_{VE} that is related to Z by an inequality of the form of (2.3.45)

$$\exp(-Vf) \equiv Z \geq Z_{VE}\,\exp(< -(S - S_{VE}(\zeta)) >_{VE}) \equiv \exp(-V\overline{f}(\zeta))\,. \qquad (2.3.183)$$

We recall that Eq. (2.3.183) follows from the convexity of the exponential function and holds independently of the specific choice of Z and Z_{VE} if the measure is positive definite and normalized. V here is the four-dimensional volume, f denotes the true physical free-energy density defined via Z, $\overline{f}$ the trial free-energy density, defined via the second equality in (2.3.183). S refers to the original action in Z, in our case it is the SU(2) Higgs action, and S_{VE} to an ansatz for the action in Z_{VE}, depending on a generic set of variational parameters ζ.

Again, the most naive (but still sensible) ansatz for S_{VE} is an ansatz in the spirit of the molecular-mean field approximation, leading to a complete factorization of Z_{VE} over all sites and over all links according to

$$Z_{VE}(\zeta) = Z_{link}^{4L_0V_3}(\xi_1) \cdot Z_{site}^{L_0V_3}(\xi_2). \qquad (2.3.184)$$

The partition function depends on two variational parameters ξ_1 and ξ_2 and factorizes in a product over single-link (Z_{link}) and single-site (Z_{site}) partition functions, V_3 denotes the three-dimensional volume. Because of the ansatz of a factorization over time-like and space-like links and over all sites, the volume dependence, in particular the temperature dependence, is trivial by construction. The simplest choice towards this factorization is

$$S_{VE}(U, \Phi) = - \sum_{x \in \Lambda_0} \left[\sum_{\mu=0}^{D-1} \operatorname{tr} \left(J_\mu^T U(x; \mu) + U(x; \mu)^{-1} J_\mu^* \right) \right.$$

$$\left. + \operatorname{tr} \left(H^T \Phi(x) + \Phi(x)^\dagger H^* \right) \right] \qquad (2.3.185)$$

with translation-invariant (x-independent) variational parameters $\zeta = (J_\mu, H)$. From (2.3.180) and (2.3.185) it follows for

$$< (S_{GH} - S_{VE})(U, \Phi) >_{VE} = < S_W(U) >_J$$

$$-V \sum_{\mu=0}^{D-1} \operatorname{tr} \left(j_\mu - J_\mu \right)^T U_\mu + \overline{U}_\mu^1 (j_\mu - J_\mu)^\star \right)$$

$$+ < S_{hop}(U, \Phi) >_{VE} -V$$

$$\cdot \left(\operatorname{tr} \left((h - H)^T \overline{\Phi} + \overline{\Phi}^\dagger (h - H)^\star \right) \right), \qquad (2.3.186)$$

where we used the following shorthand notations

$$U_\mu(J) = \frac{\int d\mu(U) U \exp \left(\operatorname{tr} \left(J_\mu^T U + J_\mu^* U^{-1} \right) \right)}{\int d\mu(U) \exp \left(\operatorname{tr} \left(J_\mu^T U + J_\mu^* U^{-1} \right) \right)} \equiv \frac{\widehat{U}(J_\mu)}{\mathring{Z}(J_\mu)}, \qquad (2.3.187)$$

$\overline{U}_\mu^1(J)$ is given by the same formula as (2.3.187) with U replaced by U^{-1}. Furthermore

$$\overline{\Phi}(H) = \frac{\int d_0\Phi \ \Phi \ \exp \left(\operatorname{tr} \left(H^T \Phi + H^* \Phi^\dagger \right) \right)}{\int d_0\Phi \ \exp \left(\operatorname{tr} \left(H^T \Phi + H^* \Phi^\dagger \right) \right)} \equiv \frac{\widehat{\Phi}(H)}{Z^{SC}(H)}, \qquad (2.3.188)$$

where

$$d_0\Phi = d^4\phi \ e^{-\mathring{S}(\Phi)}, \qquad (2.3.189)$$

cf. Eq. (2.3.174), and $\overline{\Phi^\dagger}(H)$ with Φ replaced by $\Phi^\dagger$ in (2.3.188). The definitions of the second equality signs in (2.3.187) and (2.3.188) hold for the nominators and denominators, respectively. Moreover

$$
\begin{aligned}
< S_{\text{hop}}(U, \Phi) >_{\text{VE}} &= \sum_{x \in \Lambda_0} (-2\kappa) \sum_{\mu=0}^{D-1} \frac{1}{4} < \text{tr}\, \Phi(x)^\dagger U(x; \mu) \Phi(x + \widehat{\mu}) + \\
&\quad \text{tr}\, \Phi(x + \widehat{\mu})^\dagger U(x; \mu)^{-1} \Phi(x) >_{\text{VE}} \\
&= V(-2\kappa) \frac{1}{4} \sum_{\mu=0}^{D-1} \text{tr}\, \overline{\Phi^\dagger}(U_\mu + \overline{U}_\mu^1) \overline{\Phi} \,, \qquad (2.3.190)
\end{aligned}
$$

and

$$
< S_{\text{W}}(U) >_J = V \cdot \sum_{\mu < \nu} \overline{\beta} \left(1 - \frac{1}{2} \text{Re}\, \text{tr}\, \left[U_\mu(J) U_\nu(J) \overline{U}_\mu^1(J) \overline{U}_\nu^1(J) \right] \right) \,,
$$
$$(2.3.191)$$

since the product over scalar and gauge fields at different sites and links factorizes because of the definition of $< \cdots >_{VE}$ according to

$$
< \mathcal{O} >_{\text{VE}} \equiv \frac{\int \mathcal{D}U \mathcal{D}\Phi \mathcal{O} \exp\left(-S_{\text{VE}}(U, \Phi; J_\mu, H)\right)}{\int \mathcal{D}U \mathcal{D}\Phi \exp\left(-S_{\text{VE}}(U, \Phi; J_\mu, H)\right)} \,. \qquad (2.3.192)
$$

Putting everything together, we obtain for the trial free-energy density $\overline{f}(\zeta)$

$$
\begin{aligned}
\overline{f}(\zeta) = -\Bigg\{ &\sum_{\mu=0}^{D-1} \ln \mathring{Z}(J_\mu) + \ln Z^{\text{SC}}(H) \\
&- \sum_{\mu < \nu} \overline{\beta} \left(1 - \frac{1}{2} \text{Re}\, \text{tr}\, \left[U_\mu(J) U_\nu(J) \overline{U}_\mu^1(J) \overline{U}_\nu^1(J) \right] \right) \\
&+ 2\kappa \frac{1}{4} \sum_{\mu=0}^{D-1} \text{tr}\, \left(\overline{\Phi^\dagger}(H) \left(U_\mu(J) + \overline{U}_\mu^1(J) \right) \overline{\Phi}(H) \right) \\
&+ \sum_{\mu=0}^{D-1} \text{tr}\, \left((j_\mu - J_\mu)^{\text{T}} U_\mu(J) + (j_\mu^\dagger - J_\mu^\dagger)^{\text{T}} \overline{U}_\mu^1(J) \right) \\
&+ \text{tr}\, \left((h - H)^{\text{T}} \overline{\Phi}(H) + (h^\dagger - H^\dagger)^{\text{T}} \overline{\Phi^{\text{T}}}(H) \right) \Bigg\} \,. \qquad (2.3.193)
\end{aligned}
$$

To further evaluate (2.3.193), we restrict the current matrices j, J to four-vectors

$$j_\mu = j_\mu^0 \mathbf{1}_2 + i\vec{\sigma} \cdot \vec{j}_\mu,$$
$$J_\mu = J_\mu^0 \mathbf{1}_2 + i\vec{\sigma} \cdot \vec{J}_\mu, \tag{2.3.194}$$

with real valued j_μ^0, $\vec{j}_\mu$, J_μ^0, $\vec{J}_\mu$, so that j_μ, $J_\mu \in cU(2)$, and, for convenience,

$$h = \frac{1}{4}(h_0 - i\vec{\sigma}^{\mathrm{T}} \cdot \vec{h})$$
$$H = \frac{1}{4}(H_0 - i\vec{\sigma}^{\mathrm{T}} \cdot \vec{H}), \tag{2.3.195}$$

moreover, Φ according to (2.3.170), and U according to (2.3.169). It follows for Z^{SC} as defined in (2.3.188)

$$Z^{\mathrm{SC}}(H) = \exp\{W^{\mathrm{SC}}(H)\}$$
$$= \int \mathrm{d}^4\varphi \exp\left(-\overset{\circ}{S}(\varphi) + H \cdot \varphi\right)$$
$$W^{\mathrm{SC}}(H) = \sum_{n=1}^{\infty} \frac{1}{(2n)!} (H^2)^n \overset{\circ}{v}_{2n}^c$$
$$\widehat{\Phi}^{(\dagger)} = \left(\mathbf{1}_2 \frac{\partial}{\partial H_0} \overset{+}{(-)} i\vec{\sigma} \cdot \frac{\partial}{\partial \vec{H}}\right) Z^{\mathrm{SC}}(H)$$
$$\overline{\Phi}^{(\dagger)}(H) = \left(\mathbf{1}_2 \frac{\partial}{\partial H_0} \overset{+}{(-)} i\vec{\sigma} \cdot \frac{\partial}{\partial \vec{H}}\right) W^{\mathrm{SC}}(H)$$
$$= \partial_H^{\pm} W^{\mathrm{SC}}(H)$$
$$\mathrm{tr}\,(\partial_H^- W^{\mathrm{SC}})(\partial_H^+ W^{\mathrm{SC}}) = 2\left(\frac{\partial W^{\mathrm{SC}}}{\partial H}\right)^2$$
$$\mathrm{tr}\,(h^\dagger - H^\dagger)^{\mathrm{T}} \overline{\Phi}^\dagger(H) = \mathrm{tr}\,(h - H)^{\mathrm{T}} \overline{\Phi}(H)$$
$$= \frac{1}{2}\left((h - H)_0 \frac{\partial}{\partial H_0} + (\vec{h} - \vec{H}) \cdot \frac{\partial}{\partial \vec{H}}\right) W^{\mathrm{SC}}(H). \tag{2.3.196}$$

For the terms depending on the gauge fields we obtain

$$\mathrm{tr}\,(J_\mu^{\mathrm{T}} U + J_\mu^{\dagger\mathrm{T}} U^{-1}) = 4J_\mu^0 \phi_0(U). \tag{2.3.197}$$

In the following formulas we write J for J^0 and also suppress the dependence on μ

$$\overset{\circ}{Z}(J) = \int d\mu(U) \exp\left(\text{tr}\left(J^T U + J^{\dagger T} U^{-1}\right)\right)$$

$$= \frac{2}{\pi} \int_0^\pi d\theta \sin^2 \theta \exp\left(4J\cos\theta\right) .$$

$$= I_0(4J) - I_2(4J) \equiv F_0(J) \tag{2.3.198}$$

Here, I_n are the modified Bessel functions. Furthermore,

$$\widehat{U}(J) = \int d\mu(U)\, U\, \exp\left(\text{tr}\left(J^T U + J^{\dagger T} U^{-1}\right)\right)$$

$$\equiv \mathbf{1}_2 F_1(J) = \mathbf{1}_2 \cdot \frac{2}{\pi} \int_0^\pi d\theta \sin^2 \theta \cos\theta \exp\left(4J\cos\theta\right)$$

$$= \mathbf{1}_2 \frac{1}{2}\left(I_1(4J) - I_3(4J)\right) ,$$

$$\overline{\widehat{U}}^1(J) = \widehat{U}(J) ,$$

$$\frac{\partial}{\partial J} \ln \overset{\circ}{Z}(J) = 4\frac{F_1(J)}{\overset{\circ}{Z}(J)} . \tag{2.3.199}$$

It follows that

$$U_\mu(J) + \overline{U}_\mu^1(J) = 2U_\mu(J)$$

$$= 2\frac{\int d\mu(U)U \exp\left(\text{tr}\left(J_\mu^T U + J_\mu^{\dagger T} U^{-1}\right)\right)}{\int d\mu(U) \exp\left(\text{tr}\left(J_\mu^T U + J_\mu^{\dagger T} U^{-1}\right)\right)}$$

$$\equiv 2 <U>_J$$

$$= 2\frac{F_1(J)}{\overset{\circ}{Z}(J)}\mathbf{1}_2 = 2\cdot <\varphi_0(U)>_J \cdot \mathbf{1}_2 . \tag{2.3.200}$$

Hence the trial free-energy density $\overline{f}$ becomes

$$\overline{f}(\zeta) = -\left\{ \sum_{\mu=0}^{D-1} \ln \mathring{Z}(J_\mu) - \sum_{\mu<\nu} \overline{\beta} \left(1 - <\varphi_0(U)>^2_{J_\mu} < \varphi_0(U)>^2_{J_\nu} \right) \right.$$

$$+ 2\kappa \sum_{\mu=0}^{D-1} <\varphi_0(U)>_{J_\mu} \left(\frac{\partial W^{SC}}{\partial H} \right)^2$$

$$+ \left(1 + (h - H)_0 \frac{\partial}{\partial H_0} + (\vec{h} - \vec{H}) \cdot \frac{\partial}{\partial \vec{H}} \right) W^{SC}$$

$$\left. + \sum_{\mu=0}^{D-1} <\varphi_0(U)>_{J_\mu} 4(j_\mu - J_\mu) \right\} . \tag{2.3.201}$$

So far, H was a vector with four components. W^{SC} is $O(4)$-invariant and depends on $|H|$. Hence

$$\frac{\partial}{\partial H_i} W^{SC}(|H|) = \frac{H_i}{|H|} \frac{\partial}{\partial |H|} W^{SC}(|H|) . \tag{2.3.202}$$

Only the component of $\frac{\partial}{\partial H} W^{SC}(H)$ which is parallel to $h = (h_0, \vec{h})$ enters Eq. (2.3.201). Therefore we choose H to have just one component, the projection onto the h-direction, and treat H as a scalar in the following.

We now derive the mean-field equations for vanishing external currents j_μ and h and choose the variational parameters J_μ independent of the direction μ, $J_\mu = J$. Recall that J here stands for J^0, the zeroth component of the four-vector J of Eq. (2.3.194). We obtain

$$\frac{\partial \overline{f}}{\partial J} = -16D < \phi_0(U); \phi(U) >_J$$

$$\cdot \left[\beta^\star < \phi_0(u) >^3_J - J + (2\kappa) \frac{1}{4} \left(\frac{\partial W^{SC}}{\partial H} \right)^2 \right] \tag{2.3.203}$$

$$\frac{\partial \overline{f}}{\partial H} = -\left\{ \frac{\partial^2 W^{SC}}{\partial H^2} \left[(2\kappa)(2D) < \varphi_0(U) >_J \frac{\partial W^{SC}}{\partial H} - H \right] \right\} \tag{2.3.204}$$

with

$$\beta^\star = \frac{D-1}{2} \overline{\beta} \tag{2.3.205}$$

and

$$< \phi_0(U); \phi_0(U) >_J \equiv < \phi_0^2(U) >_J - (< \phi_0(U) >_J)^2 > 0 . \tag{2.3.206}$$

The last inequality follows upon rewriting

$$< \phi_0(U); \phi(U) >_J=$$

$$\frac{\frac{1}{2} \int d\mu(U) d\mu(U') \left[\phi_0(U) - \phi_0(U')\right]^2 \exp\left[4J\left(\phi_0(U) + \phi_0(U')\right)\right]}{\left(\int d\mu(U) \, \exp\left(4J\phi_0(U)\right)\right)^2},$$

$$(2.3.207)$$

because $\phi_0(U) = \phi_0(U')$ only on a subset of the group $G \times G$ of measure zero. If we further assume that $\frac{\partial^2 W^{SC}}{\partial H^2} > 0$ ($\frac{\partial^2 W^{SC}}{\partial H^2} \geq 0$ holds anyway), the mean-field equations simplify to

$$\beta^\star < \varphi_0(u) >_J^3 -J + (2\kappa)\frac{1}{4}\left(\frac{\partial W^{SC}}{\partial H}\right)^2 = 0 \qquad (2.3.208a)$$

$$(2D)(2\kappa) < \varphi_0(U) >_J \frac{\partial W^{SC}}{\partial H} - \widehat{H} = 0 \qquad (2.3.208b)$$

with solutions $H = \widehat{H}$ and $J = \widehat{J}$. For the second derivatives we obtain

$$\frac{\partial^2 \overline{f}}{\partial H \partial J} = -\frac{\partial^2 W^{SC}}{\partial H^2}(2D)(2\kappa)4 < \phi_0(U); \phi_0(U) >_J \frac{\partial W^{SC}}{\partial H} \qquad (2.3.209a)$$

$$\frac{\partial^2 \overline{f}}{\partial H^2} = -\frac{\partial^2 W^{SC}}{\partial H^2}\left[(2D)(2\kappa) < \phi_0(U) >_J \frac{\partial^2 W^{SC}}{\partial H^2} - 1\right] \qquad (2.3.209b)$$

$$\frac{\partial^2 \overline{f}}{\partial J^2} = -16D < \varphi_0(U); \varphi_0(U) >_J$$

$$\cdot \left[12\beta^\star < \phi(U) >_J^2 \cdot < \phi_0(U); \phi(U) >_J -1\right] . \qquad (2.3.209c)$$

Now we are ready to discuss the solutions $\widehat{H}$ and $\widehat{J}$ of the mean-field equations and the qualitative features of the phase structure of the $SU(2)$ Higgs model.

2.3.4.1 *Solutions of the mean-field equations of the $SU(2)$ Higgs model*

$\kappa = 0$: Pure SU(2) gauge theory

From Eq. (2.3.208b) it follows immediately that $\widehat{H} = 0$. (2.3.208a) leads to

$$\beta^\star < \phi(U) >_{\widehat{J}}^3 -\widehat{J} = 0 , \qquad (2.3.210)$$

or, in terms of modified Bessel functions,

$$\beta^\star \frac{1}{8} \left(I_1(x) - I_3(x)\right)^3 - \frac{x}{4} \left(I_0(x) - I_2(x)\right)^3 = 0 \qquad (2.3.211)$$

with $x \equiv 4J$. For given $\beta^\star$, Eq. (2.3.211) must be solved numerically. Let us call the solution $\widehat{J}(\beta^\star)$. $\widehat{J}(\beta^\star) = 0$ is always a stable solution in the sense that it leads to a local minimum of the trial free energy density $\overline{f}$. However, there may be nontrivial solutions $\widehat{J}(\beta^\star) \neq 0$ for large enough $\beta^\star$. The first-order transition point is then determined from the equality between the free energy density at $\widehat{J}(\beta^\star) = 0$ and $\widehat{J}(\beta^\star) \neq 0$. We have

$$\overline{f}(J) = -D\left\{ \ln \mathring{Z}(J) - \beta^\star \cdot \left(1 - <\varphi_0(U) >_J^4\right) - 4 <\varphi_0(U) >_J J \right\}$$
$$(2.3.212)$$

or

$$\overline{f}\left(\widehat{J}(\beta^\star)\right) = \overline{f}(0) - D\left(\ln \mathring{Z}\left(\widehat{J}(\beta^\star)\right) - 3\left(\frac{1}{\beta^\star}\right)^{1/3} \cdot \widehat{J}(\beta^\star)^{4/3} \right) \quad (2.3.213)$$

with $\overline{f}(0) = D \cdot \beta^\star$. Thus the critical gauge coupling $\beta_c^\star$ is determined by

$$\ln \mathring{Z}\left(\widehat{J}(\beta_c^\star)\right) - 3\left(\frac{1}{\beta_c^\star}\right)^{1/3} \cdot \widehat{J}(\beta_c^\star)^{4/3} = 0 \,. \qquad (2.3.214)$$

Qualitatively the results are as follows.

- For small $\beta^\star$, $\widehat{J}(\beta^\star) = 0$ is the only stable solution of (2.3.210).
- For larger $\beta^\star$, $\widehat{J}(\beta^\star) = 0$ and $\widehat{J}(\beta^\star) \neq 0$ both lead to local minima of the free energy, but $\overline{f}(\widehat{J} \neq 0) > \overline{f}(\widehat{J} = 0)$.
- For $\beta^\star > \beta_c^\star$, again $\widehat{J}(\beta^\star) = 0$ and $\widehat{J}(\beta^\star) \neq 0$ are stable solutions of (2.3.210), but now $\overline{f}(\widehat{J} \neq 0) < \overline{f}(\widehat{J} = 0)$.

Therefore a first-order phase transition at $\beta_c^\star$ from the confinement to the deconfinement phase is predicted, for which $\overline{f}(\widehat{J}(\beta_c^\star) \neq 0) = \overline{f}(0)$. At zero temperature this prediction of a first-order deconfinement phase transition was a first success of mean-field calculations in gauge theories [43]. At finite temperature we know, however, from Monte Carlo calculations that the deconfinement transition of the $SU(2)$ gauge theory in four dimensions is of second order. Hence the effective mean-field description fails in this aspect.

$\kappa \neq 0$, any β: Full $SU(2)$ Higgs model.

The solution $\widehat{J} = \widehat{J}(\beta^\star)$ of the pure gauge theory and $\widehat{H} = 0$ are always solutions of the mean-field equations (2.3.208) with

$$\frac{\partial^2 \overline{f}}{\partial \widehat{H} \partial \widehat{J}} = 0 \quad at \ \widehat{H} = 0 \,, \tag{2.3.215}$$

$\frac{\partial}{\partial \widehat{H}}$, $\frac{\partial}{\partial \widehat{J}}$ are shorthand notations for $\frac{\partial}{\partial H}\big|_{H=\widehat{H}}$, $\frac{\partial}{\partial J}\big|_{J=\widehat{J}}$,

$$\frac{\partial^2 \overline{f}}{\partial \widehat{J}^2} > 0 \,, \tag{2.3.216}$$

since the bracket on the right hand side of Eq. (2.3.209c) is smaller than zero, because $\widehat{J}(\beta^\star)$ is a stable solution of the pure gauge theory, and

$$\frac{\partial^2 \overline{f}}{\partial \widehat{H}^2} \gtrless 0 \tag{2.3.217}$$

for

$$(2\kappa)(2D) < \phi_0(U) >_{\widehat{J}(\beta^\star)} \frac{\partial^2 W^{\mathrm{SC}}}{\partial \widehat{H}^2} - 1 \lessgtr 0 \,. \tag{2.3.218}$$

Therefore, for $\beta^\star < \beta_c^\star$ we have

$$\widehat{J}(\beta^\star) = < \phi_0(U) >_{\widehat{J}(\beta^\star)} = 0, \tag{2.3.219}$$

so that $\widehat{J} = \widehat{H} = 0$ are stable solutions for all κ, and there is just a first-order transition at $\beta_c^\star$ driven by the gauge fields in which we are not interested here. On the other hand, for $\beta^\star > \beta_c^\star$, we find

$$\widehat{J}(\beta^\star) > 0, \quad < \phi_0(U) >_{\widehat{J}(\beta^\star)} > 0. \tag{2.3.220}$$

In this case $\widehat{H} = 0$ is a stable solution only for small κ. It becomes unstable when

$$-1 + 4D\kappa_c(\beta^\star, \lambda) < \varphi_0(U) >_{\widehat{J}(\beta^\star)} \mathring{v}_2^c(\lambda) = 0 \tag{2.3.221}$$

with

$$\mathring{v}_2^c(\lambda) = \left. \frac{\partial^2 W^{\mathrm{SC}}}{\partial H^2} \right|_{H=0}, \tag{2.3.222}$$

that is,

$$\kappa_c(\beta^\star, \lambda) = \left(4D\mathring{v}_2^c(\lambda) \left(\frac{\widehat{J}(\beta^\star)}{\beta^\star} \right)^{1/3} \right)^{-1} . \tag{2.3.223}$$

In case of a second-order phase transition, $\kappa_c(\beta^\star, \lambda)$ is the critical coupling at which the phase transition from the so-called Coulomb to the Higgs phase occurs. If the transition is of first order, $\kappa_c(\beta^\star, \lambda)$ gives an upper bound to the critical coupling, as we have argued above. The coefficient $\mathring{v}_2^c$ in (2.3.223) is given by

$$\mathring{v}_2^c = \mathring{v}_2 - \mathring{v}_1^2 = \mathring{v}_2 = \frac{\int_{-\infty}^{+\infty} \mathrm{d}^4\phi \; \phi_0^2 \; \exp\left(-\mathring{S}(\Phi(\phi)) \right)}{\int_{-\infty}^{+\infty} \mathrm{d}^4\phi \; \exp\left(-\mathring{S}(\Phi(\phi)) \right)} \tag{2.3.224}$$

with $\mathring{S}(\Phi)$ given by (2.3.174). $\widehat{J}(\beta^\star)$ is numerically determined as solution of (2.3.210). For $\lambda = 5 \cdot 10^{-4}$ and $\beta^\star = (3/2)\overline{\beta} = 12.0$ we obtain

$$\kappa_c(\beta^\star, \lambda) = 0.12973 . \tag{2.3.225}$$

The mean-field equations are rather sensitive to the value of κ_c. The estimate (2.3.225) is already quite reasonable, as a comparison with high precision Monte Carlo results shows. Monte Carlo simulations of the $SU(2)$ Higgs model in four dimensions for the same values of λ and $\overline{\beta}$ and on a lattice with $L_0 = 2$ and $V_3 = 32 \cdot 32 \cdot 256$ lead to $\kappa_c(\beta^\star, \lambda) = 0.12887(1)$ [45]. Here κ_c is determined as the critical coupling of a *first*-order transition. Therefore the meaning of (2.3.225) is that of an upper bound, derived from the stability condition, and obviously the upper bound is quite close to the actual transition coupling of the first-order transition.

2.3.5   *Improved variational estimates for the $SU(2)$ Higgs model*

As we mentioned in the introduction, in the variational ansatz of the molecular-mean field approach there is no space for implementing an asymmetry between temporal and spatial directions. Thus the results will be temperature independent by construction.

One can think of a variety of improvements of the molecular field ansatz, arguments can be given against or in favor of the various versions. For example, to implement a nontrivial temperature dependence, one can postulate a factorization of the partition function over all space-like links and over all sites of the $D - 1$-dimensional space-like hyperplane, i.e. over all time-like strings, but treat the system along these time-like strings more accurately, e.g. within an expansion scheme analogous to the linked cluster expansion.

Here we consider a different ansatz which is most plausible from a physical point of view and leads to the best agreement with Monte Carlo results for comparable sets of parameters. The ansatz treats the space-like degrees of freedom of the hopping term in three dimensions beyond the mean field level, but the time-like degrees of freedom within a mean field approach, so that the partition function factorizes over the space-like hyperplanes. The reason is that the spatial hopping term is supposed to contain the nonperturbative properties of the full model that drive the Higgs phase transition. While a mean field approach for all variables will be too rough to produce high quality results for κ_{crit}, it appears more reasonable for time-like variables, although a finite temperature effect on the Higgs transition gets lost this way. (We should remark that from results in a scalar Φ^4-theory in four dimensions, the finite-temperature effect on κ_{crit} is expected to be anyway quite small. In the Φ^4-theory, there is a tiny shift in the critical coupling κ_c from zero to finite temperature [46, 47] .)

Now we replace the variational ansatz S_{VE} of (2.3.185) by

$$
S_{VE1}(U, \Phi) = - \sum_{x \in \Lambda_0} \Bigg\{ 4\zeta_{link} \phi_0(U(x; 0)) + \sum_{\mu=1}^{3} 4\zeta_{cube} \phi_0(U(x; \mu))
$$
$$
+ \sum_{\mu=1}^{3} (2\kappa) \frac{1}{2} \text{tr} \left(\Phi(x)^\dagger U(x; \mu) \Phi(x + \widehat{\mu}) \right) \quad (2.3.226)
$$
$$
+ \xi \phi_0(\Phi(x)) \Bigg\},
$$

depending on variational parameters ζ_{link}, ζ_{cube} and $\xi \in \mathbf{R}$, and on the hopping parameter κ. The first two terms are non-gauge-invariant substitutes for the Wilson plaquette term of the gauge part S_W of the SU(2) Higgs model. We distinguish time-like from space-like directions. Time-like links are denoted as $(x; 0)$, space-like links as $(x; \mu)$, $\mu = 1, 2, 3$. The third term is a hopping term that couples Φ and U fields only in spatial directions. The last term depends on the third variational parameter ξ associated with the Higgs field Φ, while the remaining ultra-local action for Φ has been absorbed in the measure.

The partition function Z_{VE1} is then given as

$$Z_{VE1} = \int \mathcal{D}U\mathcal{D}\Phi \, \exp\left(-S_{VE1}(U,\Phi)\right) \qquad (2.3.227)$$

with $\mathcal{D}U$ as in (2.3.175) and $\mathcal{D}\Phi$ as in (2.3.173).

It is easily checked that the choice of S_{VE1} in Eq. (2.3.226) leads to a factorization of Z_{VE1} according to

$$Z_{VE1}(\zeta_{link},\zeta_{cube},\xi) = Z_{cube}(\zeta_{cube},\xi)^{L_0} \cdot Z_{link}(\zeta_{link})^{L_0 V_3}. \qquad (2.3.228)$$

L_0 equals the number of space-like hyper-surfaces, $L_0 \cdot V_3$ equals the number of time-like links. The partition function Z_{cube} of the space-like degrees of freedom reads

$$Z_{cube} = \int \prod_{x \in \Lambda_0^{(3)}} \left(d\nu(\Phi(x)) e^{-S^\circ(\Phi(x))} \prod_{\mu=1}^3 d\mu_H(U(x;\mu)) \right) \cdot \exp\left(-S_{cube}(U,\Phi)\right)$$

$$(2.3.229)$$

with $\Lambda_0^{(3)}$ the three-dimensional lattice and

$$S_{cube}(U,\Phi) = -\sum_{x \in \Lambda_0^{(3)}} \left\{ \sum_{\mu=1}^3 4\zeta_{cube}\phi_0(U(x;\mu)) + \xi\phi_0(\Phi(x)) \right.$$

$$\left. + \sum_{\mu=1}^3 (2\kappa)\frac{1}{2}\text{tr}\left(\Phi(x)^\dagger U(x;\mu)\Phi(x+\hat{\mu})\right) \right\}, \qquad (2.3.230)$$

while Z_{link} is an ultra-local one-link integral, of the same form as $\mathring{Z}(J)$ of Eq. (2.3.198),

$$Z_{link} = \int d\mu_H(U) \, \exp\left(-S_{link}(U)\right) \qquad (2.3.231)$$

with

$$S_{link}(U) = 4\zeta_{link}\phi_0(U). \qquad (2.3.232)$$

Expectation values $< O >_{cube}$, $< O >_{link}$, $< O >_{VE1}$ of observables O refer to Z_{cube}, Z_{link} and Z_{VE1}, respectively. Minimization of the trial free energy density $\overline{f}(\zeta_{link},\zeta_{cube},\xi)$ in terms of these expectation values leads to three equations. The first one $\left(\partial\overline{f}/\partial\xi\right) = 0$ is solved by $\xi = 0$ in the symmetric phase. The remaining two equations are given by

$$\frac{\bar{\beta}}{4} \frac{\partial \widetilde{W}^{1,2}_{cube}}{\partial \zeta_{cube}} + \frac{\partial}{\partial \zeta_{cube}} < \phi_0(U) >_{cube}$$

$$\cdot \left[\bar{\beta}(< \phi_0(U) >_{cube})^3 + \frac{\bar{\beta}}{2}(< \phi_0(U) >_{link})^2 < \phi_0(U) >_{cube} -\zeta_{cube} \right]$$

$$= 0 \qquad (2.3.233)$$

and

$$\left[\frac{3\bar{\beta}}{2}(< \phi_0(U) >_{cube})^2 < \phi_0(U) >_{link} -\zeta_{link} \right]$$

$$\cdot \frac{\partial}{\partial \zeta_{link}} < \phi_0(U) >_{link} = 0. \qquad (2.3.234)$$

Here we have used the shorthand notation

$$< \phi_0(U) >_{cube} \equiv < \phi_0(U(x;1)) >_{cube},$$

$$\widetilde{W}^{1,2}_{cube} \equiv \frac{1}{2} < \mathrm{tr}\, U(x;1)U(x+\widehat{1};2)U(x+\widehat{2};1)^{-1}U(x;2)^{-1} >_{cube}$$

$$- (< \phi_0(U(x;1)) >_{cube})^4 \qquad (2.3.235)$$

with $x \in \Lambda_0^{(3)}$. The link in $< \phi_0(U) >_{link}$ is time-like. In (2.3.235) we used the lattice symmetries. Equations (2.3.233), (2.3.234) should be solved for ζ_{link}, ζ_{cube}, and ξ as series in κ. The stability condition for the symmetric minimum at $\xi = 0$ is given as

$$3\bar{\beta}\, \frac{\partial^2 \widetilde{W}^{1,2}_{cube}}{\partial \xi^2}\bigg|_{\xi=0}$$

$$+12 \left[\bar{\beta}(< \phi_0(U) >_{cube})^3 + \frac{\bar{\beta}}{2} < \phi_0(U) >_{cube} (< \phi_0(U) >_{link})^2 - \zeta_{cube} \right]$$

$$\cdot \frac{\partial^2}{\partial \xi^2} < \phi_0(U) >_{cube}\bigg|_{\xi=0} \qquad (2.3.236)$$

$$+ \left[(2\kappa)\, 2 < \phi_0(U) >_{link} \frac{\partial}{\partial \xi} < \phi_0(\Phi) >_{cube}\bigg|_{\xi=0} -1 \right]$$

$$\cdot \frac{\partial}{\partial \xi} < \phi_0(\Phi) >_{cube}\bigg|_{\xi=0} < 0$$

with

$$< \phi_0(\Phi) >_{cube} \equiv < \phi_0(\Phi(x)) >_{cube}, \qquad (2.3.237)$$

$x \in \Lambda_0^{(3)}$. For an equality sign, (2.3.236) determines κ_{crit} order by order in the expansion.

Note the two type of expectation values $< \cdot >_{link}$, $< \cdot >_{cube}$, entering Eqs. (2.3.233), (2.3.234), (2.3.236). While the single-link expectation values $< \cdot >_{link}$ can be evaluated exactly, i.e. in terms of modified Bessel functions, the $< \cdot >_{cube}$-expectation values must be approximated. The derivatives of $\widetilde{W}_{cube}^{1,2}$ w.r.t. ζ_{cube} or twice with respect to ξ induce up to five-point functions in U and six-point functions in four Us and two Φs. Because of the bad signal/noise ratio, Monte Carlo calculations of such connected correlations would not be feasible. This is the place where one either gets stuck or has to invent a new expansion technique. We developed so-called dynamical linked cluster expansions (DLCEs) to evaluate expectation values of the type $< U \cdots U \Phi \cdots \Phi >_{cube}$ (all internal and configuration space indices suppressed) order by order in κ [50]. We explain dynamical linked cluster expansions in a separate section, once we have introduced linked cluster expansions. Here we just want to point out the specific features of an action of the form (2.3.226).

Note that the scalar fields Φ and the gauge fields U are coupled by the hopping-parameter term, but both fields, Φ as well as U do have their own dynamics, specified by the remaining terms in the action (2.3.226) and by the measures $\mathcal{D}\Phi$ and $\mathcal{D}U$. This feature strongly reminds us to an action of spin glasses with the following replacements. Substitute spin variables σ for the scalar fields Φ and site-dependent spin couplings $J(x, y)$ for the gauge fields $U(x; \mu)$. Both σ and J have their own dynamics (fast and slow, respectively) and are coupled via a term of the form of the hopping term in (2.3.226).

In case of our equations (2.3.233), (2.3.234), (2.3.236) it is the U-dependence of the hopping parameter term that prevents us from evaluating expectation values $< U \cdots U \Phi \cdots \Phi >_{cube}$ by means of ordinary linked cluster expansions. Clearly, the price for improving the accuracy in the third or fourth digit of $\kappa_c(\beta^\star, \lambda)$ is anyway high. In Monte Carlo simulations one has to spend of the order of months of CPU-time. In DLCEs, the number of connected DLCE graphs contributing, for example, to order κ^4 to a four-point function of two Us and two Φs is about 100. Not all correlations in (2.3.233),(2.3.234),(2.3.236) appear in the form of susceptibilities. In a product of 4 Us, for instance, the configuration space indices are fixed to the boundary of a plaquette. Such features must be noticed for calculating the graphical weights.

These remarks may indicate the complexity of the actual evaluation of Eqs. (2.3.233), (2.3.234), (2.3.236) and illustrate the consequences that an improved variational ansatz can have. Here we omit any further details and state the result for $\kappa_c(\beta^*, \lambda)$ obtained from (2.3.236) with an equality sign. For the same values of β^* and λ as before ($\beta^* = 12.0, \lambda = 5 \cdot 10^{-4}$), and a lattice of size $L_0 = 4$ (that drops out of the variational equations), and $V_3 = \infty^3$, we find [48]

$$\kappa_c(\beta^*, \lambda) = 0.1282(1). \qquad (2.3.238)$$

It improves the variational ansatz of the molecular-mean field approach in the third digit, including the third digit it also agrees with the corresponding Monte Carlo result. This agreement is quite satisfactory, because the variational estimates combined with DLCEs amount to a rather different methodical approach as compared to Monte Carlo simulations. They thus provide an independent check of the numerical results.

2.3.6 Summary

In summary, the molecular-mean field approximation is applicable to spin systems as well as to gauge theories. It leads to a reasonable first estimate of the location of a phase transition and of the overall-phase structure. The main shortage is the independence of the molecular-mean field approximation on the dimension D and on the topology of the lattice, by construction. It turns out that its predictions are the better the larger D or the larger the number of nearest neighbors, as we have illustrated for the Curie-Weiss model. In cases in which the finite temperature is expected to have only little impact on a phase transition that occurs also at zero temperature, it is reasonable to neglect the temperature dependence of the transition in this first approximation. The results for the critical line, corresponding to the electroweak phase transition in the $SU(2)$ Higgs model, obtained from variational estimates in the molecular-mean field approximation, are really encouraging. On a quantitative level they are in good agreement with high-precision Monte Carlo results.

2.4   Renormalization group

2.4.1   *Generalities*

The renormalization group is neither a group nor a universal procedure calculating a set of renormalized parameters from a set of starting values. It is more a generic framework with possibly very different realizations. Our applications of the renormalization group have in common the idea of deriving a set of new (renormalized) parameters, characteristic for a different scale, from a first set of parameters, while keeping some long distance physics unchanged. In general it need not be the long distance physics, but may be any other physical aspect of interest. The degrees of freedom are partitioned into disjoint subsets so that to each subset belong at least two. Specific for the renormalization group is the partitioning according to their length scale, or equivalently, according to their high- and low-momentum modes. Otherwise the partitioning is a commonly used practice in quantum field theory that is convenient for integrating upon the subsets in sequence (cf. the section on dimensional reduction). Successive integrations of modes according to their frequency, momentum or length scale lead to what is called the renormalization group. One computes a sequence of effective models defined by a sequence of effective actions S_0, S_1, ..., in which S_0 denotes the original action. These models have the property that for any observable $\mathcal{O}$ there is a corresponding sequence of effective observables

$$\mathcal{O}_0 = \mathcal{O}, \ \mathcal{O}_1, \ \ldots, \tag{2.4.1}$$

such that they have identical expectation values

$$< \mathcal{O}_0 >_{S_0} = < \mathcal{O}_1 >_{S_1} = \ldots . \tag{2.4.2}$$

The choice of the subsets of degrees of freedom is arbitrary to a large extent, but an appropriate partition is crucial for making the integration feasible.

Since the change in scale goes along with a reduction in the number of degrees of freedom, the iterated procedure should lead to a simpler description of the very aspect of interest, say the long-distance physics. An example for such a simplified description are critical phenomena in a $(3 + 1)$-dimensional $SU(2)$ gauge theory . The general conjecture is that integrating out the short-range fluctuations leads to an Ising model in three dimensions which shares the universality class with the $SU(2)$ gauge theory at the transition point of the deconfinement transition.

The renormalization group is not restricted to critical phenomena, although critical phenomena are one of the most important applications in our context. In a separate chapter we discuss the flow equations of Polchinsky (section 4.7). The renormalization-group flow there will be used to derive an effective action for discussing the phase structure of a scalar field theory with O(4) symmetry, but a very promising application of Polchinsky's flow equations lies in alternative proofs of the very *renormalizability* of a theory. Further applications concern the derivation of low-energy properties from some fundamental theory like QCD. More concretely, a (so far hypothetical) example is an effective Lagrangian in terms of mesonic fields, derived via the renormalization group from the very QCD Lagrangian in terms of quark and gluonic degrees of freedom.

Last but not least a different realization of the renormalization group may be used for investigating the very *UV-limit*-or if we use a lattice cutoff- the *continuum limit* of a quantum field theory.

Specific for the application of the renormalization group to critical phenomena is an implementation of the experimental observation that properties of a system remain unchanged under a change of scale. In the theoretical description this feature will be reflected by the existence of so-called fixed points of the renormalization-group transformation. First we will derive the behavior of the renormalization group in the vicinity of these fixed points. The linearization leads to the important result that thermodynamic quantities can be represented in a scaling form. Several manifestations of critical phenomena follow then automatically, which are otherwise understood only on a phenomenological level (e.g. in the molecular mean field approximation, series expansions or Monte Carlo simulations). To list some of these:

- Second-order phase transitions fall in so-called universality classes.
- For a fixed universality class there is an upper critical dimension above which the exponents take on their mean field values.
- Scaling relations hold as equalities, although pure thermodynamic considerations predict them only as inequalities.
- Critical exponents take the same values independently of whether the transition parameter (e.g. the critical temperature) is approached from above or below.

Only in exceptionally simple cases such as the one- and two-dimensional Ising model fixed points and critical exponents can be calculated exactly.

In general it is inherently difficult to control the involved approximations that go along with the truncation of the effective actions. Below we will show in more detail where the difficulties come from.

In the following section we consider the renormalization group as a computational tool. Again, there are different ways of implementing the change of scale. For definiteness and simplicity we choose the framework of block-spin transformations.

2.4.2   Block-spin transformations

To explain the basic ideas we focus on the theory of a single spin-zero bosonic field for simplicity. We consider a field theory on a D-dimensional hyper-cubic lattice $\Lambda = (a\mathbf{Z})^\mathbf{D}$ with lattice spacing a. The scalar fields make up a vector space $\mathcal{F}_\Lambda$ of real valued fields $\Phi : \Lambda \to \mathbf{R}$. The action S_Λ is a functional $S_\Lambda : \mathcal{F}_\Lambda \to \mathbf{R}$, and the partition function is given by $Z_\Lambda(0)$, with

$$Z_\Lambda(J) = \int \mathcal{D}\mu_\Lambda(\Phi) \, \exp\left(-S_\Lambda(\Phi) + (J, \Phi)_\Lambda\right), \qquad (2.4.3)$$

where $J \in \mathcal{F}_\Lambda$ and

$$\mathcal{D}\mu_\Lambda(\Phi) = \prod_{x \in \Lambda} d\Phi(x),$$
$$(J, \Phi)_\Lambda = a^D \sum_{x \in \Lambda} J(x)\Phi(x). \qquad (2.4.4)$$

Z_Λ is the generating functional of full (in contrast to truncated) correlation functions.

Expectation values of observables $\mathcal{O}(\Phi)$ are defined by

$$< \mathcal{O} >_{S_\Lambda} = \frac{1}{Z_\Lambda(0)} \int \mathcal{D}\mu_\Lambda(\Phi) \, \mathcal{O}(\Phi) \exp\left(-S_\Lambda(\Phi)\right). \qquad (2.4.5)$$

More precisely, the generating functional (2.4.3) and the expectation value (2.4.5) are first defined on a lattice with finite volume, $|\Lambda| < \infty$, the infinite volume limit will then be taken in an appropriate way. For the moment we just assume that the limit exists and that this $Z_{\Lambda=\infty}$ is a smooth function of the external source J, at least for sufficiently small J.

Complete or full correlation functions on the lattice Λ are defined by

$$
\begin{aligned}
G_n(x_1,\ldots,x_n)_\Lambda &\equiv\; <\Phi(x_1)\cdots\Phi(x_n)>_\Lambda \\
&= \frac{\int \mathcal{D}\mu_\Lambda(\Phi)\;\exp(-S_\Lambda(\Phi))\Phi(x_1)\cdots\Phi(x_n)}{\int \mathcal{D}\mu_\Lambda(\Phi)\;\exp(-S_\Lambda(\Phi))} \\
&= \frac{\partial^n Z_\Lambda(J)}{\partial J(x_1)\cdots\partial J(x_n)}\Bigg|_{J=0}.
\end{aligned}
\tag{2.4.6}
$$

In the very end, in our applications, the external source J is set to zero or to a constant value after differentiation. The generating functional of connected correlations is defined via

$$
W_\Lambda(J) \equiv \ln Z_\Lambda(J) \equiv -F_\Lambda(J) \equiv -|\Lambda|f_\Lambda(J),
\tag{2.4.7}
$$

with f_Λ the free-energy density in the presence of an external current, and connected correlation functions are defined by

$$
G_n^c(x_1,\ldots,x_n)_\Lambda = \frac{\partial^n}{\partial J(x_1)\cdots\partial J(x_n)}W_\Lambda(J)\Bigg|_{J=0}.
\tag{2.4.8}
$$

Note that full and connected correlation functions are related according to

$$
G_n(x_1,\ldots,x_n)_\Lambda = \sum_{\mathcal{P}\in\mathcal{P}(x_1,\ldots,x_n)}\;\prod_{P\in\mathcal{P}} G_n^c(P).
\tag{2.4.9}
$$

$\mathcal{P}(x_1,\ldots,x_n)$ denotes the set of partitions of the mutually distinct labels $x_1,\ldots,x_n$ into mutually disjoint subsets.

We define the block lattice Λ_L for integer $l \in \mathbf{N}$ by decomposing Λ into disjoint cubic blocks. Each block consists of l^D sites of Λ. The center points of these blocks constitute the hyper-cubic lattice $\Lambda_L = (la\mathbf{Z})^D$. Λ_L is a lattice with lattice spacing $a' = la$. For $x \in \Lambda_L$ we set $\underline{x}$ as the set of all $y \in \Lambda$ that belong to the block with center x. There are l^D such $y \in \Lambda$. For $x \in \Lambda$, we denote by $\overline{x} \in \Lambda_L$ the unique lattice site of the block lattice such that $x \in \underline{\overline{x}}$.

Analogous notions on Λ like the partition function, correlations, observables generalize to Λ_L in an obvious way. For $\Phi, \Psi \in \mathcal{F}_{\Lambda_L}$,

$$
(\Phi, \Psi)_{\Lambda_L} = (la)^D \sum_{x\in\Lambda_L} \Phi(x)\Psi(x).
\tag{2.4.10}
$$

We further need a rescaling operation to the original lattice Λ, that is a lattice with lattice spacing a again. Only then we shall be able to define contractions and fixed points and implement the semi-group property of renormalization-group transformations, as it will become clear below. For $\Psi \in \mathcal{F}_\Lambda$ and $l \in \mathbf{N}$ we define $s_l \Psi \in \mathcal{F}_{\Lambda_L}$ by

$$(s_l \Psi)(x) = \Psi(\frac{x}{l}), \qquad (2.4.11)$$

for all $x \in \Lambda_L$. Obviously,

$$(\Phi, \Psi)_\Lambda = l^{-D} (s_l \Phi, s_l \Psi)_{\Lambda_L}. \qquad (2.4.12)$$

Next we introduce the notions of scale-dependent effective actions and observables. The (renormalization) transformation of the action S_Λ is defined by the Boltzmann factor

$$\exp\left(-S'_\Lambda(\Psi)\right) \equiv \exp\left(-(R_l S_\Lambda)(\Psi)\right)$$
$$= \int \mathcal{D}\mu_\Lambda(\Phi)\, P(\Phi, s_l \Psi)\, \exp\left(-S_\Lambda(\Phi)\right). \qquad (2.4.13)$$

Effective observables are defined by

$$\mathcal{O}'(\Psi)\, \exp\left(-S'_\Lambda(\Psi)\right) = \int \mathcal{D}\mu_\Lambda(\Phi)\, P(\Phi, s_l \Psi)\, \mathcal{O}(\Phi)\, \exp\left(-S_\Lambda(\Phi)\right). \qquad (2.4.14)$$

Usually P is a distribution on $\mathcal{F}_\Lambda \times \mathcal{F}_{\Lambda_L}$. It specifies which degrees of freedom are integrated out and to which extent the integration is performed. For the case of a block-spin transformation a commonly used choice is

$$P : \mathcal{F}_\Lambda \times \mathcal{F}_{\Lambda_L} \to \mathcal{D}(\mathcal{F}_{\Lambda_L})$$
$$P(\Phi, \overline{\Phi}) = \prod_{x \in \Lambda_L} \delta(\overline{\Phi}(x) - \frac{1}{b}(C\Phi)(x)). \qquad (2.4.15)$$

Here the block-averaging operator $C : \mathcal{F}_\Lambda \to \mathcal{F}_{\Lambda_L}$ is defined by

$$(C\Phi)(x) = \frac{1}{l^D} \sum_{y \in \underline{x}} \Phi(y). \qquad (2.4.16)$$

The parameter b is a positive number that describes the rescaling of the block spin. For the moment it is just a constant in the definition of P. As we will see below its value in the vicinity of fixed points is actually related to the wave-function renormalization constant of the model and to the critical exponent η , which is usually called the anomalous dimension.

Short-wave length fluctuations with a wave length λ satisfying $a < \lambda < la$ are integrated out by the block spin integration. The effective action S'_Λ describes then the fluctuations with wavelength $\lambda > la$. We emphasize that the renormalization transformation as defined by (2.4.13) is not restricted to block spins (2.4.15), in which the high frequency modes are integrated out completely, completely in the sense of a δ-constraint. Also other convenient choices for P exist. P works as a kind of weight function which specifies the degree to which the high frequency modes are integrated out. Different choices lead to different types of renormalization-group transformations as we will see later. For example, the block-spin condition is often relaxed to

$$P(\Phi, \overline{\Phi}) = \mathcal{N} \exp\left(-\frac{\alpha}{2l^D}(\overline{\Phi} - \frac{1}{b}C\Phi, \; \overline{\Phi} - \frac{1}{b}C\Phi)_{\Lambda_L}\right), \qquad (2.4.17)$$

with $\alpha, b > 0$. For finite α the block spin integration is smoothed, the former definition is recovered in the limit $\alpha \to \infty$.

In order to guarantee the invariance of the partition function under a block-spin transformation, P is subject to the constraint

$$\int \mathcal{D}\mu_{\Lambda_L}(\overline{\Phi}) \; P(\Phi, \overline{\Phi}) = 1, \qquad (2.4.18)$$

so that for variables Ψ, defined again on the original lattice Λ, we have

$$\int \mathcal{D}\mu_\Lambda(\Psi) \; P(\Phi, s_l \Psi) = 1 \qquad (2.4.19)$$

with $\overline{\Phi} = s_l \Psi$. Obviously the block spin function (2.4.15) satisfies the constraint (2.4.18). It follows that

$$\int \mathcal{D}\mu_\Lambda(\Psi) \; \exp\left(-S'_\Lambda(\Psi)\right) = \int \mathcal{D}\mu_\Lambda(\Phi) \; \exp\left(-S(\Phi)\right) \qquad (2.4.20)$$

and

$$\int \mathcal{D}\mu_\Lambda(\Psi) \; \mathcal{O}'(\Psi) \; \exp\left(-S'_\Lambda(\Psi)\right) = \int \mathcal{D}\mu_\Lambda(\Phi) \; \mathcal{O}(\Phi) \; \exp\left(-S(\Phi)\right). \qquad (2.4.21)$$

Therefore the partition functions of the original and of the effective theory agree as they should, because we want to leave the physical system unchanged, and the expectation values of the effective observables reproduce the values in the original theory.

For $J \in \mathcal{F}_\Lambda$ we define the effective partition function as a function of the external current J, that is the partition function for the effective model after the renormalization transformation,

$$Z'_\Lambda(J) \equiv (R_l Z_\Lambda)(J) \qquad (2.4.22)$$

$$= \int \mathcal{D}\mu_\Lambda(\Psi) \; \exp\left(-S'_\Lambda(\Psi) + (J, \Psi)_\Lambda\right). \qquad (2.4.23)$$

The effective free energy is given by

$$W'_\Lambda(J) \equiv \ln Z'_\Lambda(J) \equiv -F'_\Lambda(J) \equiv -l^{-D}|\Lambda|f'_\Lambda(J). \tag{2.4.24}$$

The factor l^D results from the change of the length scale. With these definitions we obtain for the block-spin transformation (2.4.15) the identities

$$Z'_\Lambda(J) = Z_\Lambda(\widehat{J}),$$
$$W'_\Lambda(J) = W_\Lambda(\widehat{J}), \tag{2.4.25}$$

where for all $x \in \Lambda$

$$\widehat{J}(x) = \frac{1}{bl^D}J(\frac{\overline{x}}{l}). \tag{2.4.26}$$

For a block-spin transformation (2.4.17) the right hand side of (2.4.25) contains correction terms in $1/\alpha$ that vanish for $\alpha \to \infty$ or $J = 0$. For $J = 0$ the invariance of Z_Λ and W_Λ are guaranteed anyway, as we have seen in (2.4.20). The identities (2.4.25) lead to relations that are often referred to as exact renormalization-group equations. They are used to derive the flow equations for effective actions, generating functionals and correlation functions (4.7).

To prove (2.4.25) we write

$$Z'_\Lambda(J) = \int \mathcal{D}\mu_\Lambda(\Psi) \exp\left(-S'_\Lambda(\Psi) + (J,\Psi)_\Lambda\right)$$
$$= \int \mathcal{D}\mu_\Lambda(\Phi)\mathcal{D}\mu_\Lambda(\Psi) P(\Phi, s_l\Psi) \exp\left(-S_\Lambda(\Phi) + (J,\Psi)_\Lambda\right). \tag{2.4.27}$$

Under the integral sign we have

$$(J,\Psi)_\Lambda = \frac{1}{l^D}(s_l J, s_l \Psi)_{\Lambda_L}$$
$$= a^D \sum_{x\in\Lambda_L}(s_l J)(x)\frac{1}{bl^D}\sum_{y\in\underline{x}}\Phi(y)$$
$$= a^D \sum_{y\in\Lambda}\Phi(y)\left(\frac{1}{bl^D}(s_l J)(\overline{y})\right),$$
$$= (\Phi,\widehat{J})_\Lambda.$$

First integrating upon the Ψ-field and then applying (2.4.19), we obtain (2.4.25).

The behavior of the free-energy density and of the correlation functions under the renormalization-group transformation is easily obtained from (2.4.25). The free-energy density transforms as

$$f'(J) = l^D f(\hat{J}).$$ (2.4.28)

For the full Green functions

$$G'_n(x_1, \ldots, x_n)_\Lambda = \frac{\int \mathcal{D}\mu_\Lambda(\Psi)\Psi(x_1)\cdots\Psi(x_n)\exp\left(-S'_\Lambda(\Psi)\right)}{\int \mathcal{D}\mu_\Lambda(\Psi)\exp\left(-S'_\Lambda(\Psi)\right)}$$

$$= \frac{\partial^n Z'_\Lambda(J)}{\partial J(x_1)\cdots\partial J(x_n)}\bigg|_{J=0}$$

we have

$$G'_n(x_1, \ldots, x_n)_\Lambda = \frac{1}{(bl^D)^n} \prod_{i=1}^{n} \sum_{y_i \in l\underline{x}_i} G_n(y_1, \ldots, y_n)_\Lambda.$$ (2.4.29)

Note that on both sides of Eq. (2.4.29) the sites belong to the original lattice Λ. On the left hand side this is because of the rescaling operator that is involved in the block-spin transformation. Similarly, for the connected Green functions

$$G^{c\,\prime}_n(x_1, \ldots, x_n)_\Lambda = \frac{\partial^n W'_\Lambda(J)}{\partial J(x_1)\cdots\partial J(x_n)}\bigg|_{J=0}$$

we obtain the scaling relations

$$G^{c\,\prime}_n(x_1, \ldots, x_n)_\Lambda = \frac{1}{(bl^D)^n}\Big(\prod_{i=1}^{n} \sum_{y_i \in l\underline{x}_i}\Big) G^c_n(y_1, \ldots, y_n)_\Lambda.$$ (2.4.30)

So far we have partially integrated upon the degrees of freedom in a single block-spin transformation. Is there any advantage of this rewriting? The specific choice of block spin amounts to an integration upon the high-frequency modes. These are the degrees of freedom which fluctuate on the scale between a and the block size la. It is plausible but not necessary that this system, specified by the constrained integration according to the block-spin transformation, has a short correlation length of the order of la. (Below we specify more precisely the conditions under which this conjecture is valid.) The advantage of a short correlation length is at hand: the integration then amounts to solving a noncritical theory as long as the block length $l = a'/a$ is not too large. For noncritical systems a variety of methods can be employed: analytic methods such as convergent expansions if the couplings are sufficiently small or Monte Carlo simulations without critical slowing down.

Let us consider an almost critical system so that the correlation length of the total system on the original lattice is large but finite, say $\xi < \infty$. The correlation length on the block lattice ξ' becomes smaller by a factor of $1/l$ if l is the block length. $\xi' = \xi/l$ is either measured in units of the block lattice a' or, after rescaling (2.4.11) of the block lattice back to the original lattice, in units of the lattice spacing a. Such an almost critical system with large correlation length is approached by a sequence of block-spin transformations each of which treats noncritical systems with short-range correlations. Let us call the result of a large number n of such iterations $Z_\Lambda^{(n)}(J)$. Critical systems with infinite correlation length are described by appropriate limits of noncritical systems: The above procedure should be repeated starting from a system with larger correlation length $\widetilde{\xi} > \xi$ and ending up with $\widetilde{Z}_\Lambda^{(n)}(J)$ and so on.

Let us come back to the noncritical subsystems we have to deal with on the way of deriving critical behavior. The above conjecture that the subsystems have a correlation length of the order of la further suggests that it is sufficient to include only local interactions in the effective action $S_\Lambda'(\Psi)$. In many practical applications, $S_\Lambda'(\Psi)$ is actually truncated because the truncation considerably simplifies the iteration of the renormalization-group transformation and the investigation of critical behavior. Necessary conditions for truncating $S_\Lambda'(\Psi)$ are that the system that is described by the partition function $\exp(-S_\Lambda'(\Psi))$, (2.4.13), has a correlation length of the order of la (as discussed above) and that the so-called large-field domains can be ignored. Roughly speaking, large-field domains are field configurations Ψ for which the action $S_\Lambda'(\Psi)$ takes large values. It is too naive to argue that large-field domains are suppressed by small Boltzmann factors. Depending on the amount of entropy many small Boltzmann factors may sum up to a finite or even divergent contribution. For such field configurations any truncation of the action can no longer be justified. This sheds some doubts on the very concept of using one effective action $S_\Lambda'(\Psi)$ for all field configurations simultaneously.

We devote the remainder of this section to a discussion under which conditions the auxiliary system with partition function $\exp(-S_\Lambda'(\Psi))$ (2.4.13) actually has a correlation length $\widehat{\xi}$ of the order of $a' = la$ and indicate the possible problems related to the large-field domains. To this end, let us separate the quadratic part of the action,

$$S_\Lambda(\Phi) = \frac{1}{2} \left(\Phi, v^{-1}\Phi \right)_\Lambda + V(\Phi), \qquad (2.4.31)$$

with v the free propagator and V the "interaction" part. It is convenient to parameterize the field Φ in a way that is adapted to the block spin integration

$$\Phi = A\overline{\Phi} + \rho, \tag{2.4.32}$$

with block spin $\overline{\Phi} = s_l\Psi$, and with the map $A : \mathcal{F}_{\Lambda_l} \to \mathcal{F}_\Lambda$ chosen such that it satisfies the following conditions. Let C be the averaging operator defined in (2.4.16).

(1)

$$\frac{1}{b}\,CA = 1_{\mathcal{F}_{\Lambda_l}}, \tag{2.4.33}$$

so that $\overline{\Phi} - b^{-1}C\Phi = -b^{-1}C\rho$. That is, ρ is a field with block average zero if the block spin constraint is chosen according to (2.4.15).

(2)

$$\left(\rho, v^{-1}A\overline{\Phi}\right)_\Lambda = 0, \tag{2.4.34}$$

so that

$$\left(\Phi, v^{-1}\Phi\right)_\Lambda = \left(\rho, v^{-1}\rho\right)_\Lambda + \left(A\overline{\Phi}, v^{-1}A\overline{\Phi}\right)_\Lambda, \tag{2.4.35}$$

i.e. there is no mixing term between the fluctuation field ρ and the block spin field $\overline{\Phi}$ in the quadratic part of the action. Equivalently, $v^{-1}A\overline{\Phi}$ is constant inside the blocks.

These conditions uniquely determine the operator A. To see this let us first introduce the adjoint operator $C^\dagger$ of C. From the definition of C it follows that

$$C^\dagger : \mathcal{F}_{\Lambda_l} \to \mathcal{F}_\Lambda, \quad (C^\dagger\Theta)(x) = \Theta(\overline{x}) \text{ for } x \in \Lambda, \tag{2.4.36}$$

so that $(\psi, C\phi)_{\Lambda_l} = (C^\dagger\psi, \phi)_\Lambda$. The second condition then implies that $A = vC^\dagger B$ with some map $B : \mathcal{F}_{\Lambda_l} \to \mathcal{F}_{\Lambda_l}$, and from the first condition we get $\frac{1}{b}CA = \frac{1}{b}(CvC^\dagger)B = 1_{\mathcal{F}_{\Lambda_l}}$. Hence

$$A = bvC^\dagger(CvC^\dagger)^{-1}. \tag{2.4.37}$$

Now the δ-function of the block spin may be regarded as a limit of a Gaussian exponential,

$$\prod_{x\in\Lambda_l} \delta((C\rho)(x)) \simeq \mathcal{N}_\alpha \exp\left(-\frac{\alpha}{2}(\rho, C^\dagger C\rho)_\Lambda\right) \tag{2.4.38}$$

as $\alpha \to \infty$. Inserting the decomposition (2.4.32) and implementing the algebraic constraints (2.4.33) and (2.4.34) via (2.4.37) according to $A^\dagger v^{-1} A = b^2 (CvC^\dagger)^{-1}$, the block spin integration can be rewritten as

$$\exp\left(-S'_\Lambda(\Psi)\right) \equiv \exp\left(-\frac{1}{2} b^2 (s_l \Psi, (CvC^\dagger)^{-1} s_l \Psi)\right)_{\Lambda_l}$$

$$\cdot \mathcal{N} \int \mathcal{D}\mu_\Lambda(\rho) \, \exp\left(-\frac{1}{2}(\rho, \Gamma^{-1}\rho) - V(\rho + A s_l \Psi)\right), \qquad (2.4.39)$$

with some constant $\mathcal{N}$. The propagator Γ of the fluctuation field ρ is given by

$$\Gamma = \lim_{\alpha \to \infty} (v^{-1} + \alpha C^\dagger C)^{-1}. \qquad (2.4.40)$$

It can be shown that the fluctuation field propagator Γ and the operator A exponentially decay according to

$$|\Gamma(x, y)| \le c_1 \exp\left(-\frac{K_1 |x - y|}{la}\right), \qquad x, y \in \Lambda,$$

$$|A(x, y)| \le c_2 \exp\left(-\frac{K_2 |x - y|}{la}\right), \qquad x \in \Lambda, y \in \Lambda_l. \qquad (2.4.41)$$

Recall that we are interested in the locality properties of $S'_\Lambda(\psi)$. In perturbation theory the locality of $S'_\Lambda(\psi)$ easily follows from (2.4.41) as the integrand in (2.4.39) is expanded as a formal series in powers of the interaction V. Since the path integrals reduce to Gaussians, the measure does not cause any trouble, and the logarithm of both sides directly leads to statements about the effective action.

Outside the range of perturbation theory the exponential decay (2.4.41) does no longer imply the same conclusions for arbitrary block spin configurations $\overline{\Phi} = s_l \Psi$. The trouble is caused by the so-called large-field configurations. Here we only indicate the large-field problem. For any details we refer the interested reader to the literature ([54, 55]).

Let us define effective actions $S'_Y(\Psi)$ for arbitrary subsets $Y \subseteq \Lambda$ by a formula similar to (2.4.13). $S'_Y(\Psi)$ can then be uniquely represented by a Möbius transformation

$$S'_Y(\Psi) = \sum_{X \subseteq Y} z(\Psi; X)$$

$$= \sum_{x \in Y} z_1(\Psi(x)) + \sum_{x_1, x_2 \in Y} z_2(\Psi(x_1), \Psi(x_2)) + \cdots \qquad (2.4.42)$$

in which the $z(\Psi; X)$ depend only on $\Psi(x)$ with $x \in X$. Now, S' is said to have good locality properties if $z(\Psi; X)$ is small for large X, if the size of X is measured by the length of the shortest tree containing all sites of X as its vertices. Let an energy density be defined by

$$\mathcal{E}(\Psi; x) = \sum_{X \ni x} \frac{z(\Psi; X)}{|X|}, \qquad (2.4.43)$$

with $|X|$ the number of lattice sites of X. Now $x \in \Lambda$ belongs to the "large field" domain with respect to the configuration Ψ, together with the sites in an appropriate neighborhood of x, if $\mathcal{E}(\Psi; x) > \zeta$ for some appropriately chosen $\zeta > 0$ such that $e^{-\zeta}$ is a small number. The remainder of the lattice is the "small field" region. Within the small field regions, the action has good locality properties. These domains are thus the perturbative regions. The truly nonperturbative regions are the large-field domains. Within the large-field domains the interactions of the fluctuation fields can be highly nonlocal, imposing long-range fluctuations on the $\rho(x)$. At least in weakly coupled models these large-field domains are very rare, so that their contributions to the effective action are small.

In the more general case, in particular for strongly coupled systems, the truncation of the effective action is subtle and potentially dangerous. The problem can be overcome by use of convergent expansions of the effective Boltzmann factor $\exp(-S'_\Lambda(\Psi))$, rather than expanding the effective action $S'_\Lambda(\Psi)$ itself. The general framework is provided by the theory of polymer systems. In this way it becomes possible to control the entropy factor coming from the measure, since "large fields" associated with small Boltzmann factors may sum up to nonnegligible contributions. A sophisticated machinery - the rigorous renormalization group - has been developed which is beyond the scope of this book. Its application to QCD and the standard model, however, is very tough and only partially worked out, see e.g. [57].

Finally we mention the so-called multi-grid approach of [54]. A multi-grid is a multilayer lattice consisting of the sequence of block lattices, supplemented by an appropriate topology which provides the notion of distance of lattice sites on different block lattices. A Euclidean field theory can then be written as an auxiliary statistical mechanical system on the multi-grid. In the multi-grid approach there is no need for truncating effective actions. Large volume and continuum limits are recovered as particular thermodynamic limits on the multi-grid. Again the interest in this approach lies in the possibility of avoiding uncontrolled truncations of effective actions.

2.4.3   *Iteration of the block-spin transformation*

For many interesting models the partition function (2.4.3) and the expectation values (2.4.5) for a large class of operators $\mathcal{O}$ can be shown to exist in the thermodynamic limit, at least if correlation length and coupling constants are sufficiently small. Both analytical and numerical efficient methods are then available for an accurate quantitative computation of the expectation value $< \mathcal{O} >_{S_\Lambda}$. On the other hand we are interested in the case in which the correlation length ξ is large in units of the lattice spacing a, because a diverging correlation length ξ is in common to both the continuum limit and the thermodynamic limit of critical phenomena. In the continuum limit the lattice spacing a is sent to zero and ξ is fixed in physical units, but diverges in units of a. The limit can only exist if the coupling constants of S_Λ are adjusted as functions of renormalized, physical coupling constants, defined on the scale of the correlation length ξ. Similarly, in the thermodynamic limit of critical phenomena the correlation length diverges in physical units. The lattice spacing there is kept finite and has a physical meaning, but the lattice volume is sent to infinity. For both cases the one step procedures of the last section are in general not sufficient to project on the long-range properties in which we are interested here.

In the last section we have discussed a single block-spin transformation, defined by the integration upon the high frequency modes. Now we assume that the auxiliary statistical mechanical system with partition function given by the effective Boltzmannian (2.4.13) has a correlation length of the order of the block length la, and the above mentioned techniques apply if l is not too large. The effective model described by the effective action S'_Λ has a correlation length that is already reduced by a factor of l^{-1}, but $\xi' = \xi/l$ is still large if ξ is.

The idea now is to iterate the renormalization transformation $a \to a'$. The procedure of the last section easily generalizes to more than one block-spin integration, leading to a sequence of lattices with lattice spacings

$$a \to a' = la \to a'' = l^2 a \to \cdots \to a^{(n)} = l^n a \qquad (2.4.44)$$

and, combined with a rescaling to the original lattice units, defines a sequence of effective actions

$$S_\Lambda \to S'_\Lambda = R_l S_\Lambda \to S''_\Lambda = R_l^2 S_\Lambda \to \cdots \to S_\Lambda^{(n)} = R_l^n S_\Lambda , \qquad (2.4.45)$$

all of which are associated with the same lattice Λ once the rescaling operator has been applied.

The iteration of block-spin transformations leads to a flow of effective actions or effective Boltzmann factors on various length scales. Most important is the scaling property of these transformations. For block lengths l_1, l_2,

$$R_{l_1 l_2} = R_{l_1} \circ R_{l_2} \ , \ R_1 = \mathrm{Id}, \qquad (2.4.46)$$

where $\circ$ denotes the composition and Id the identity map. For instance, applied to the action S_Λ or the partition Z_Λ, we have

$$R_{l_1 l_2} S_\Lambda = R_{l_1}(R_{l_2} S_\Lambda), \qquad (2.4.47)$$

or

$$R_{l_1 l_2} Z_\Lambda = R_{l_1}(R_{l_2} Z_\Lambda). \qquad (2.4.48)$$

The maps R_l together with this decomposition law provide a semi-group, usually called the renormalization group (RG). Many results obtained by renormalization-group studies are based on this property. Note that the inverse transformation of R_l need not exist and in general may not even exist so that the semi-group property is the best one can have. The reason is explained below: microscopically different actions should be mapped to the same action on a macroscopic scale if they are supposed to describe a second-order phase transition.

The rescaling to the original lattice that was introduced above allows for a definition of fixed points of the map R_l in the space it acts upon. R_l acts on an infinite-dimensional function space. As mentioned in the introduction, the fixed points correspond to scale invariant actions. The idea then is to iterate the renormalization-group transformations and to study the resulting flow in the vicinity of these fixed points. Under a renormalization-group transformation, the correlation length ξ changes according to $\xi' = \xi/l$. At a fixed point of R_l, $\xi' = \xi$, which implies that $\xi = 0$ or $\xi = \infty$. Fixed points that correspond to critical behavior have $\xi = \infty$. They belong to the critical manifold. The flows close to such fixed points thus describe the critical behavior of the model and explain universality as different actions are mapped on the same fixed-point action in course of the renormalization-group flow.

In practice, the effective actions are truncated to an appropriate finite subset of operators. This truncation is a delicate problem and requires a careful analysis as outlined in the last section. In the following we assume that such an analysis has been made with the result that the renormalization-group flow may be considered in a finite space of coupling constants.

First of all let us see that the very existence of a fixed point relates the rescaling parameter b to the anomalous dimension η. Away from the critical point, truncated correlation functions (of a partition function that corresponds to a pure state) behave in the large-distance region, where $|x_i - x_j| >> 1$, according to

$$G_n^c(x_1, \ldots, x_n) \simeq P(x_1, \ldots, x_n) \exp\left(-\frac{1}{\xi}T(x_1, \ldots, x_n)\right), \qquad (2.4.49)$$

with $T(x_1, \ldots, x_n)$ the length of the shortest tree with $x_1, \ldots, x_n$ as its vertices, defined according to

$$T(x_1, \ldots, x_n) = \min_{\pi \in \Pi_n} \sum_{i=1}^{n-1} |x_{\pi(i)} - x_{\pi(i+1)}|, \qquad (2.4.50)$$

where Π_n is the set of all permutations of numbers $1, 2, \ldots, n$. $P(x_1, \ldots, x_n)$ is some function of the $x_i - x_j$ that is polynomially bounded. In particular, for the two-point function we obtain

$$G_2^c(x, 0) \simeq \frac{\text{const}}{|x|^{2d_\Phi}} \exp\left(-\frac{1}{\xi}|x|\right) \qquad (2.4.51)$$

as $|x| \to \infty$. (2.4.51) defines the canonical infrared (IR-) dimension d_Φ of the field. For bosonic fields in a D dimensional space of D unbounded directions $d_\Phi = (D-2)/2$. (This also agrees with the engineering dimension of Φ defined in a generic D-dimensional field theory of bosonic fields.) On the other hand, at the critical point, the correlation length diverges, i.e. $\xi = \infty$, and the two-point function decays only with a power of the distance,

$$G_2^c(x, 0) \simeq \frac{c}{|x|^{2d_\Phi + \eta}} \qquad (2.4.52)$$

for large $|x|$. The large-distance behavior of the two-point function (2.4.52) defines the critical exponent η, which is called the anomalous dimension.

Let us consider the renormalization-group recursion (2.4.30) in the large-distance region where all $|x_i - x_j| >> 1$. In this region we have

$$G_n^{c\prime}(x_1, \ldots, x_n)_\Lambda \simeq \frac{1}{b^n} G_n^c(lx_1, \ldots, lx_n). \qquad (2.4.53)$$

In particular,

$$G_2^{c\prime}(x, 0)_\Lambda \simeq \frac{1}{b^2} G_2^c(lx, 0). \qquad (2.4.54)$$

At a fixed point, $G_2^{c\prime}(x,0) = G_2(x,0)$, so that together with (2.4.52) we obtain

$$\frac{c}{|x|^{2d_\Phi+\eta}} = \frac{1}{b^2}\frac{c}{|lx|^{2d_\Phi+\eta}}, \qquad (2.4.55)$$

which implies that

$$b = l^{-d_\Phi-\frac{\eta}{2}}. \qquad (2.4.56)$$

The result is amazing. On the one hand, the block-spin parameter b appeared as a free input parameter of the block-spin transformation with block size parameterized by l, cf.(2.4.15). On the other hand, Eq. (2.4.56), derived for a renormalization-group flow close to the fixed point, reveals b as a fixed function of the block size and the anomalous dimension η. Therefore the rescaling parameter b must be fine-tuned so that the very fixed point exists.

2.4.4 Field renormalization

A set of independent random numbers
To solve this paradox we consider very simple examples, in which the necessity of fine-tuning b becomes evident also on the level of the action. The first example is a set of independent random numbers x_i, $i = 1, \cdots n$, distributed according to

$$\int_{-\infty}^{+\infty} dx_1 \cdots \int_{-\infty}^{+\infty} dx_n\, f(x_1)\cdots f(x_n) \equiv \int \mathcal{D}_n x \equiv 1. \qquad (2.4.57)$$

What is the distribution $F(\overline{x})$ of the "naive" average

$$\overline{x} = \frac{c_n}{n}\sum_{i=1}^{n} x_i \qquad (2.4.58)$$

with

$$F(\overline{x}) = \int \mathcal{D}_n x\, \delta\left(\overline{x} - \frac{c_n}{n}\sum_{i=1}^{n} x_i\right)? \qquad (2.4.59)$$

For $c_n = n$ the answer is included as a special case of the central limit theorem of probability theory, stating that, independently on details of the single distributions $f(x_i), i = 1, \cdots n$, for large n the limit distribution $F(\overline{x})$

is given by

$$F_n(\overline{x}) \simeq \frac{1}{\sqrt{2\pi n v}} \exp\left(-\frac{1}{2nv}(\overline{x} - n < x >)^2\right), \qquad (2.4.60)$$

where $< x >$ is defined according to

$$< x > = \int_{-\infty}^{+\infty} dx \, x \, f(x) \qquad (2.4.61)$$

and

$$v = \int_{-\infty}^{\infty} dx \, (x - < x >)^2 \, f(x) . \qquad (2.4.62)$$

For $c_n = 1$ (corresponding to the naive ansatz for the block spin) an explicit calculation with $f(x)$ chosen as a simple Gaussian, say

$$f(x) = \sqrt{\frac{1}{2\pi v}} \, \exp\left(-\frac{1}{2v}(x - w)^2\right), \qquad (2.4.63)$$

leads to $< x >= w$ and

$$F_n(\overline{x}) = \sqrt{\frac{n}{2\pi v}} \, \exp\left(-\frac{n}{2v}(\overline{x} - w)^2\right), \qquad (2.4.64)$$

as one could read off also from (2.4.59),(2.4.60) by rescaling the δ-function in the definition of $F(\overline{x})$. Neither (2.4.60) nor (2.4.64) are self-reproducing with increasing n.

However, if we set $c_n = \sqrt{n}$ and, in addition, shift x_i by the average $< x >$, i.e.

$$\overline{x} = \frac{1}{\sqrt{n}} \sum_{i=1}^{n} (x_i - < x >), \qquad (2.4.65)$$

we find for $f(x)$ given by (2.4.63) that

$$F(\overline{x}) = \int \mathcal{D}_n x \, \delta\left(\overline{x} - \frac{1}{\sqrt{n}} \left(\sum_i (x_i - < x >)\right)\right)$$

$$= \frac{1}{\sqrt{2\pi v}} \, \exp\left(-\frac{1}{2v}\overline{x}^2\right) \qquad (2.4.66)$$

with n-independent width and height, thus a limit distribution with finite "fixed-point" exponent and amplitude. Equation (2.4.66) shows that we have to adjust the average both in an additive and multiplicative way.

If we translate our system of n independent random numbers to a field theoretical language, we may interpret the system as having n independent scalar fields Φ on a lattice distributed according to

$$f(\Phi_i) = \exp\left(-\frac{m^2}{2}\Phi_i^2\right), \; i = 1, \ldots, n, \qquad (2.4.67)$$

and apply one block-spin transformation, in which the block size n is sent to infinity. The block field is defined according to

$$\overline{\Phi} = \frac{1}{\sqrt{n}} \sum_{i=1}^{n} \Phi_i. \qquad (2.4.68)$$

The result corresponding to (2.4.66) then means that the one-point action $(m^2/2)\Phi_i^2$, $i = 1, \cdots, n$ gets only reproduced on the block level in terms of $\overline{\Phi}$ if we choose b^{-1} of (2.4.15) or c_n in (2.4.58) as $\sqrt{n}$. Since the fields Φ_i are chosen as independent random numbers, one block transformation with block size $n \to \infty$ is equivalent to an iterated infinite number of block transformations, each of finite size.

A set of dependent random numbers
So far our field theoretical system is a rather trivial one, since the fields are completely decoupled. The simplest generalization is to a free Gaussian-field theory by introducing a kinetic term. In the notation of (2.4.66) it leads to consider

$$F(\overline{\Phi}) = \left(\prod_{i\in\Lambda} \int_{-\infty}^{\infty} d\Phi_i\right) \exp\left(-\frac{1}{2}\sum_{i}\left[\sum_{\mu=0}^{D-1}(\Phi_{i+\hat{\mu}} - \Phi_i)^2 + m^2\Phi_i^2\right]\right)$$

$$\cdot \, \delta\left(\overline{\Phi} - \frac{c_n}{n}\sum_{i=1}^{n}\Phi_i\right). \qquad (2.4.69)$$

$F(\overline{\Phi})$ may be interpreted as a block-spin transformation in which the whole lattice consists of one block. The generalization to block-spin transformations of the form (2.4.13), (2.4.15) leads to

$$\exp\left(-S'_\Lambda(\Psi)\right) \equiv \int \left(\prod_{z\in\Lambda} d\Phi(z)\right) \exp\left(-S_\Lambda(\Psi)\right)$$

$$\cdot \prod_{x\in\Lambda_l} \delta\left(\overline{\Phi}(x) - \frac{1}{b}\frac{1}{l^D}\sum_{y\in\underline{x}}\Phi(y)\right), \qquad (2.4.70)$$

with $s_l \, \Psi(x) = \overline{\Phi}(x)$, $b = 1/c_n$ and

$$S_\Lambda(\Phi) \; = \; \frac{1}{2} a^D \sum_{x \in \Lambda} \left[\sum_{\mu=0}^{D-1} \left(\frac{1}{a} \left[\Phi(x + a\widehat{\mu}) - \Phi(x) \right] \right)^2 + m^2 \Phi(x)^2 \right] . \quad (2.4.71)$$

The question here is under which conditions $\exp\left(-S_\Lambda'(\Psi)\right)$ converges to a limit distribution with a fixed-point action. Note that in terms of random numbers the kinetic term leads to a generalization to dependent random numbers. Thus from a mathematical point of view our main task can be posed as *the attempt of finding a limit distribution in case of an infinite set of dependent random numbers.*

Let us fix b from the postulate that the coefficient of the kinetic term (proportional to p^2, with p the momentum) in (2.4.39) stays normalized to one after the block-spin transformation in coordinate space. From (2.4.39) we know that for $V = 0$ the integral upon the fluctuation field ρ (for one block-spin transformation) gives a constant contribution, while the b-dependent part of the effective action is given by

$$\frac{1}{2} b^2 (s_l \psi, O s_l \psi)_{\Lambda_l} \quad (2.4.72)$$

with $s_l \, \Psi(x) = \overline{\Phi}(x)$, $x \in \Lambda_l$, and

$$O \; = \; (CvC^\dagger)^{-1} . \quad (2.4.73)$$

More explicitly we have

$$\frac{1}{2} b^2 (la)^{2D} \sum_{x,y \in \Lambda_l} \Psi(\frac{x}{l}) O(x,y) \Psi(\frac{y}{l})$$

$$= \; \frac{1}{2} b^2 (la)^{2D} \sum_{x,y \in \Lambda} \Psi(x) O(lx, ly) \Psi(y)$$

$$= \; \frac{1}{2} b^2 l^D \int_{-\pi/a}^{\frac{\pi}{a}} \frac{d^D p}{(4\pi)^D} \, |\widetilde{\Psi}(p)|^2 \, \widetilde{O}(\frac{p}{l}) . \quad (2.4.74)$$

The first equality sign holds in the infinite volume. For the second equality we inserted the Fourier transforms

$$\psi(x) = \int_{-\frac{\pi}{a}}^{\frac{\pi}{a}} \frac{d^D p}{(4\pi)^D} \, e^{ipx} \, \widetilde{\psi}(p)$$

$$O(lx, ly) = \int_{-\frac{\pi}{a}}^{\frac{\pi}{a}} \frac{d^D p}{(4\pi)^D} \, e^{ipl(x-y)} \, \widetilde{O}(p) . \quad (2.4.75)$$

To fix b we do not need the explicit form of the small-momentum expansion of $\widetilde{O}(p)$, but just consider the quadratic term of the form $c(\frac{p}{l})^2$. In a fixed-point action the quadratic term should transform into the same form after a block-spin transformation. By (2.4.74) this is guaranteed if

$$cb^2 l^{D-2} = 1 \,, \tag{2.4.76}$$

or

$$b = l^{-(D-2)/2}\, c^{-1/2}. \tag{2.4.77}$$

((2.4.77)should be compared with (2.4.56), with $\eta = 0$ and $d_\Phi = (D-2)/2$.) The conclusion is that for a given renormalization-group transformation we have to fine-tune the field renormalization to "hit" the fixed point distribution in case there is one. But the concrete realization of this fine-tuning procedure may look quite different as we shall illustrate now.

Once again we consider a free scalar field theory in D dimensions, but now in the continuum, and perform a renormalization-group transformation in momentum space rather than a block-spin transformation in coordinate space. We introduce a sharp ultraviolet momentum cutoff right from the beginning. The action is then given by

$$S(\widetilde{\Phi}) = \int_{k^2 \leq \Lambda^2} \frac{d^D k}{(2\pi)^D}\, \widetilde{\Phi}(k)^2\, (k^2 + m^2) \,. \tag{2.4.78}$$

Integration upon the high frequency modes with $\Lambda^2/l^2 \leq k^2 \leq \Lambda^2$ (l standing for an infrared cutoff in coordinate space) yields the effective action S', defined via

$$
\begin{aligned}
\exp(-S') &= \int_{\widetilde{\Phi}(k),\Lambda^2 l^{-2} \leq k^2 \leq \Lambda^2} \mathcal{D}\widetilde{\Phi}\, e^{-S(\Phi)} \\
&= \exp\left[- \int_{k^2 \leq \Lambda^2 l^{-2}} \frac{d^D k}{(2\pi)^D}\, |\widetilde{\Phi}(k)|^2\, (k^2 + m^2) \right] \\
&\quad \cdot \int \mathcal{D}\Phi\, \exp\left[- \int_{\Lambda^2 l^{-2} \leq k^2 \leq \Lambda^2} \frac{d^D k}{(2\pi)^D}\, |\widetilde{\Phi}(k)|^2\, (k^2 + m^2) \right].
\end{aligned}
\tag{2.4.79}
$$

The last integral contributes an additive constant to S'. The quadratic term in $\widetilde{\Phi}(k)$ reads

$$\int_{k^2 \leq \Lambda^2 l^{-2}} \frac{d^D k}{(2\pi)^D}\, |\widetilde{\Phi}(k)|^2\, (k^2 + m^2) \,, \tag{2.4.80}$$

or, in terms of momenta $k' \equiv k \cdot l$ with $|k| \leq \Lambda l^{-1}$

$$S'(\widetilde{\Phi}) = \int_{|k'|^2 \leq \Lambda^2} \frac{d^D k'}{(2\pi)^D} \frac{1}{l^D} |\widetilde{\Phi}(\frac{k'}{l})|^2 \left(\frac{k'^2}{l^2} + m^2 \right) . \qquad (2.4.81)$$

In analogy to coordinate space we still have to rescale the momenta in the arguments of the fields $\widetilde{\Phi}$ to ensure the semi-group property of the renormalization transformation and to allow for the possibility of having a fixed point in coupling constant space of the actions. Let

$$(\widetilde{s}_l \psi)(\frac{k'}{l}) = \psi(k') . \qquad (2.4.82)$$

If we now define $\widetilde{\Phi}'$, the analogue of the block-spin field $\overline{\Phi}$ before, according to

$$\widetilde{\Phi}'(k') = \frac{1}{l^{(D+2)/2}} \widetilde{\Phi}(\frac{k'}{l}) = \left(\widetilde{s}_l \widetilde{\Phi} \right) \left(\frac{k'}{l} \right) \qquad (2.4.83)$$

and

$$m^2 l^2 = m'^2 , \qquad (2.4.84)$$

the effective action $S'(\widetilde{\Phi}')$ takes the same form as (2.4.80)

$$S'(\widetilde{\Phi}') = \int_{k'^2 \leq \Lambda^2} \frac{d^D k'}{(2\pi)^D} |\widetilde{\Phi}'(k')|^2 (k'^2 + m'^2) . \qquad (2.4.85)$$

Thus the factor of the field renormalization is fixed as $1/l^{(D+2)/2}$.

Note that with $\Lambda = \infty$ the condition on the momenta upon integration of the high momentum modes would be empty.

Equation (2.4.84) reveals the mass as a relevant scaling variable, cf. the next section. Thus we have to adjust it to its critical value $m^2 = 0$ right from the beginning if the action shall converge to the fixed point action.

We remark that this simple renormalization group procedure characterized by a sharp momentum cutoff in a continuum formulation does only work out quite well, because the field theoretical model was most simple. As soon as interactions are included, the above procedure may no longer be well defined, while the block spin procedure, indicated above, still provides a well defined framework.

Clearly the necessity of an appropriate field renormalization remains in a generic field theoretical model with "random numbers" coupled via interaction terms. The right field renormalization is then a delicate issue. In the worst case the fixed point action may be missed by a naive choice of the prefactor. Fortunately, in four dimensions, $\eta = 0$ for a scalar Φ^4-theory, the same theory in three dimensions has $\eta \sim 0.05$, still an order of magnitude smaller than in two dimensions. Scalar models are supposed to share the universality class with QCD in certain limiting cases, in which the chiral or the deconfinement transition are of second order, cf. section 2.5.2. Under all these assumptions a wrong choice of the normalization factor of the block spin may have less dramatic consequences.

2.4.5 *Linearized renormalization-group transformation and universality*

In principle, the renormalization-group transformations R_l act in an infinite-dimensional function space. In practice, the effective actions are truncated to a finite subset of operators. As we have seen above, the truncation is a delicate problem and requires a careful analysis. For the following let us assume that such an analysis was performed so that the truncation is justified. Thus we consider the renormalization-group flow in a finite space of coupling constants.

Let us choose a linear parametrization in the space of actions according to

$$S = \sum_{i=1}^{n} g_i \mathcal{O}_i \qquad (2.4.86)$$

with real coupling constants g_i. The $\mathcal{O}_i$ are operators which are linearly independent in field space. Suppose that the renormalization-group transformation is considered as a smooth map in the space of actions parameterized by the set of couplings $\{g\}$

$$g' = R_l(g),$$
$$g_i' = (R_l)_i(g) \ , \ i = 1,\ldots,n. \qquad (2.4.87)$$

The renormalization-group property (2.4.47) implies that

$$R_{l_1 l_2}(g) = R_{l_1}(R_{l_2}(g)). \qquad (2.4.88)$$

Let g^* be a fixed point of R, and let R be a smooth map in the vicinity of g^*,

$$g^* = R_l(g^*). \qquad (2.4.89)$$

We consider the flow in the vicinity of g^*. With $g = g^* + \delta g$,

$$g' = R_l(g) = R_l(g^* + \delta g) = R_l(g^*) + (DR_l)(g^*)\delta g + O((\delta g)^2), \quad (2.4.90)$$

with $(DR_l)(g^*)$ the Jacobian of R_l at g^*,

$$(DR_l)(g^*)_{ij} = \frac{\partial(R_l)_i}{\partial g_j}(g^*). \quad (2.4.91)$$

For $\delta g' = g' - g^*$ we thus obtain

$$\delta g' = (DR_l)(g^*)\delta g + O((\delta g)^2). \quad (2.4.92)$$

Ignoring the correction term, we obtain the so-called linearized renormalization-group transformation.

Let us assume that $(DR_l)(g^*)$ is diagonalizable over the real numbers with eigenvalues real and positive. The eigenvectors v_i, $i = 1, \ldots, n$, then provide a basis of $\mathbf{R}^n$, the truncated space of coupling constants, with

$$(DR_l)(g^*)v_i = \lambda_{l,i} v_i , \quad i = 1, \ldots, n. \quad (2.4.93)$$

The eigenvalues $\lambda_{l,i}$ are real and positive by assumption. We write $\delta g = \sum_{i=1}^{n} \delta \bar{g}_i v_i$ with new coupling constants $\delta \bar{g}_i$ which transform under the linearized renormalization-group transformation according to

$$\delta \bar{g}_i' = \lambda_{l,i} \delta \bar{g}_i. \quad (2.4.94)$$

Depending on their eigenvalues $\lambda_{l,i}$, we distinguish three cases of coupling constants $\delta \bar{g}_i$.

- Relevant coupling constants $\delta \bar{g}_i$ are couplings with $\lambda_{l,i} > 1$. Under multiple application of the diagonalized renormalization-group transformation $(DR_l)(g^*)$

$$(\lambda_{l,i})^m \delta \bar{g}_i \to \infty \quad \text{as } m \to \infty. \quad (2.4.95)$$

Unless $\delta \bar{g}_i$ is zero from the very beginning, the distance from the fixed point increases with the number of renormalization-group transformations.

- Irrelevant coupling constants $\delta \bar{g}_i$ are couplings with $0 < \lambda_{l,i} < 1$. We have

$$(\lambda_{l,i})^m \delta \bar{g}_i \to 0 \quad \text{as } m \to \infty. \quad (2.4.96)$$

Thus irrelevant couplings flow towards the fixed point and die out with an iteration of the renormalization-group transformation.

- Marginal coupling constants $\delta\bar{g}_i$ are couplings with $\lambda_{l,i} = 1$. In this case a more detailed analysis beyond the linearized renormalization group is required to determine whether these couplings stay marginal or change their nature beyond the linearized level.

When iterating the renormalization-group transformation in the vicinity of the fixed point, the relevant parameters must be adjusted (to $\delta\bar{g}_i = 0$ as long as the linearized renormalization-group transformation is considered) to achieve convergence to the fixed point, while irrelevant parameters drop out under iteration. We emphasize that the irrelevant couplings parameterize the critical manifold only in the vicinity of the fixed point, when they were identified within the linearized renormalization group.

If marginal parameters occur, a more detailed analysis is required to determine their properties. For instance, the quartic coupling constant g_4 of a $g_4\Phi^4$-theory is marginal in $D = 4$ dimensions (the upper critical dimension). However, a detailed analysis (by means of the rigorous renormalization group) reveals that this coupling actually is irrelevant in the infrared and scales logarithmically according to

$$g_4 \sim \frac{1}{\log m} \quad \text{as } m \to \infty, \qquad (2.4.97)$$

if m denotes the number of iterations. Hence g_4 turns out to be an irrelevant parameter. Marginal parameters may also imply that there is no longer an isolated fixed point that describes the critical system, but a complete submanifold of fixed points. In this case critical behavior is not universal.

Let us assume the case that there are actually no marginal parameters, then we have found an explanation of universality. For a finite number of relevant parameters, models which differ only by irrelevant couplings are attracted to the same fixed point. Their critical exponents agree, as we shall see below.

There may be more than one isolated fixed point in coupling-constant space. Every fixed point has its own domain of attraction. In this case there is in general more than one universality class. Universality still holds in the sense that all models have the same critical behavior whose coupling constants belong to the domain of attraction of the same fixed point.

We conclude this section with the list of assumptions which were made in conjunction with the linearized renormalization group. The renormalization-group map R_l may be considered as a map in the space of effective actions, $R_l : S_\Lambda \to S'_\Lambda$, (and need not be considered between

Boltzmann factors). Furthermore it may be constrained to a space of a finite number of coupling constants g. Linearized about the assumed fixed point g^*, the Jacobian $(DR_l)(g^*)$ is diagonalizable with real and positive eigenvalues. In many interesting cases these assumptions are justified, as it was verified by means of the rigorous renormalization group. Counterexamples are provided in situations with limit cycles (rather than fixed points) in coupling constant space (in connection with chaotic behavior), or, when the effective action does not even exist, a phenomenon that is related to the large-field problem in strongly coupled models or in connection with discontinuity-fixed points.

2.4.6   Scaling-sum rules

The semi-group property (2.4.47) or (2.4.88) implies for the linearized renormalization-group transformation that

$$(DR_{l_1 l_2})(g^*) = (DR_{l_1})(g^*) \cdot (DR_{l_2})(g^*)$$
$$= (DR_{l_2})(g^*) \cdot (DR_{l_1})(g^*), \qquad (2.4.98)$$
$$(DR_1)(g^*) = 1.$$

It follows that the $(DR_l)(g^*)$ are simultaneously diagonalizable for all l, and that

$$\lambda_{l_1 l_2, i} = \lambda_{l_1, i} \lambda_{l_2, i} \quad , \quad \lambda_{1, i} = 1 \qquad (2.4.99)$$

for all i. So far we have discussed block-spin transformations on hypercubic lattices, with integer l as the side length of a symmetric block in units of the lattice spacing a. It is sometimes convenient to interpolate the transformation laws to real $l \geq 1$ or to use another cutoff than a lattice, also preserving the semi-group property (2.4.47) or (2.4.88). In that case, (2.4.99) implies that $\lambda_{l,i}$ may be rewritten according to

$$\lambda_{l,i} = l^{\nu_i},$$
$$\nu_i = \left. \frac{d\lambda_{l,i}}{dl} \right|_{l=1}, \qquad (2.4.100)$$

because (2.4.100) solves

$$l \frac{d\lambda_{l,i}}{dl} = \left. \frac{d\lambda_{l,i}}{dl} \right|_{l=1} \cdot \lambda_{l,i}, \qquad (2.4.101)$$

which follows from (2.4.99) by taking the derivative with respect to l_2, say. This way $\lambda_{l,i}$ and ν_i are determined as solutions of (2.4.101).

Relevant couplings have $\nu_i > 0$, irrelevant ones $\nu_i < 0$ and marginal $\nu_i = 0$. Although the particular form (2.4.100) is not necessary for the following derivation of the scaling-sum rules, it considerably simplifies the notation below.

We derive the scaling-sum rules or the so-called standard scaling relations between critical exponents for models with two relevant coupling constants. All other couplings are assumed to be irrelevant. In this case only two exponents turn out to be independent. The two relevant couplings are denoted by τ and h. In many applications, τ may be identified with the reduced temperature $\tau = T - T_c$, if T_c denotes the critical temperature. $\tau < 0$ holds then in the low-temperature phase, sometimes related to the breakdown of a global symmetry, $\tau > 0$ in the (symmetric) high-temperature phase. In the background of critical phenomena in ferromagnets, h is an external field, in connection with QCD it stands for a current quark mass. Whatever the physical meaning of h is, in the action h is linearly coupled to the fluctuation field Φ

$$S_\Lambda^{ext}(\Phi; h) = h \, a^D \sum_{x \in \Lambda} \Phi(x). \qquad (2.4.102)$$

In Landau's mean field theory Φ plays the role of the order-parameter field. We also keep the largest irrelevant eigenvalue, denoted by λ, (as representative for all other irrelevant couplings) to keep an eye on their scaling behavior. Under a renormalization-group transformation

$$\tau \to l^{x_\tau} \tau,$$
$$h \to l^{x_h} h, \qquad (2.4.103)$$
$$\lambda \to l^{-x_\lambda} \lambda$$

with exponents $x_\tau, x_h, x_\lambda > 0$. Since the eigenvalues λ of the renormalization-group transformation (defined in (2.4.93)) will furthermore be written in the form (2.4.100), no confusion should arise when we call the irrelevant couplings λ, too, in the following. We should remark that in general the action need not be parameterized in terms of τ and h, if τ and h stand for the relevant directions in coupling-parameter space. The first step is then to identify the relevant directions by diagonalizing the matrix (2.4.91). Here we assume that this step has been done, and τ and h are the relevant parameters.

The exponent x_h of h is easily related to the critical exponent η. Under a block-spin transformation (2.4.13-2.4.15), we obtain

$$h\, a^D \sum_{x \in \Lambda} \Phi(x) = ha^D \sum_{y \in \Lambda_l} \sum_{x \in \underline{y}} \Phi(x) = ha^D \sum_{y \in \Lambda_l} (bl^D)\Psi(\tfrac{y}{l})$$

$$= h' a^D \sum_{x \in \Lambda} \Psi(x), \tag{2.4.104}$$

where $h' = bl^D h$. (The last equality holds only in the infinite volume.) With b chosen according to (2.4.56) we obtain

$$x_h = D - d_\Phi - \frac{\eta}{2}. \tag{2.4.105}$$

From (2.4.28) and the linearized renormalization-group transformation (2.4.103), close to the fixed point the free-energy density behaves as

$$f(l^{x_\tau}\tau, l^{x_h}h, l^{-x_\lambda}\lambda) = l^D f(\tau, h, \lambda). \tag{2.4.106}$$

Furthermore, for large spatial separations $|x_i - x_j| \gg 1$, the scaling relations (2.4.30) for the connected correlation functions become

$$G_n^c(x_1, \ldots, x_n | l^{x_\tau}\tau, l^{x_h}h, l^{-x_\lambda}\lambda)_\Lambda = l^{n(D - x_h)} G_n^c(lx_1, \ldots, lx_n | \tau, h, \lambda)_\Lambda. \tag{2.4.107}$$

We have explicitly indicated the dependence on the coupling constants τ, h and λ.

For the derivation of the sum rules below we assume that the free-energy density and derived quantities are analytic functions of τ, h and λ away from the critical manifold. In particular $f(\tau, h, \lambda)$ is analytic whenever $\tau \neq 0$ or $h \neq 0$.

2.4.6.1 *Anomalous dimension*

In (2.4.105) we have already obtained a relation for the anomalous dimension

$$\eta = 2(D - d_\Phi - x_h). \tag{2.4.108}$$

2.4.6.2 *Correlation length*

A suitable definition of the correlation length is as the inverse pole mass. As a physical length scale it should behave under a renormalization transformation according to

$$\xi(\tau, h, \lambda) \simeq l\xi(l^{x_\tau}\tau, l^{x_h}h, l^{-x_\lambda}\lambda). \tag{2.4.109}$$

Let us check the scaling behavior if ξ is calculated via

$$\xi(\tau, h, \lambda)^2 = \frac{\sum_{x \in \Lambda} x^2 G_2^c(x, 0)}{\sum_{x \in \Lambda} G_2^c(x, 0)}(\tau, h, \lambda), \tag{2.4.110}$$

i.e. via the two-point function of momentum zero. (Close to criticality, the definition of ξ according to (2.4.110) is equivalent to its definition by the inverse pole mass.) The dominant contribution comes from the region where $x \sim \xi$. Using (2.4.107) for $n = 2$ we obtain

$$\xi(l^{x_\tau}\tau, l^{x_h}h, l^{-x_\lambda}\lambda)^2 = \frac{\sum_{x \in \Lambda} x^2 G_2^c(x, 0)}{\sum_{x \in \Lambda} G_2^c(x, 0)}(l^{x_\tau}\tau, l^{x_h}h, l^{-x_\lambda}\lambda)$$

$$\simeq \frac{\sum_{x \in \Lambda} x^2 G_2^c(lx, 0)}{\sum_{x \in \Lambda} G_2^c(lx, 0)}(\tau, h, \lambda) = \frac{1}{l^2}\frac{\sum_{x \in \Lambda}(lx)^2 G_2^c(lx, 0)}{\sum_{x \in \Lambda} G_2^c(lx, 0)}(\tau, h, \lambda)$$

$$\simeq \frac{1}{l^2}\xi(\tau, h, \lambda)^2. \tag{2.4.111}$$

Hence we confirm (2.4.109) as expected. Let us set $h = 0$ and $l^{x_\tau}\tau = \pm\epsilon$ with some small but positive number ϵ so that the τ dependence of the relevant couplings is fully contained in the amplitude of ξ. $\tau \to 0\pm$ then implies $l \to \infty$. Together with (2.4.111) we get

$$\xi(\tau, 0, \lambda) \simeq \left(\epsilon|\tau|^{-1}\right)^{1/x_\tau}\xi(\pm\epsilon, 0, (\frac{|\tau|}{\epsilon})^{x_\lambda/x_\tau}\lambda)$$

$$\simeq \text{const } |\tau|^{-1/x_\tau} \quad , \quad \text{as } \tau \to 0\pm. \tag{2.4.112}$$

It follows that

$$\nu = \nu' = \frac{1}{x_\tau}. \tag{2.4.113}$$

2.4.6.3 *Two-point susceptibility*

For the other quantities we proceed in a similar way as for the correlation length. For the two-point susceptibility we get from (2.4.107)

$$\chi_2(l^{x_\tau}\tau, l^{x_h}h, l^{-x_\lambda}\lambda) = \sum_{x \in \Lambda} G_2^c(x, 0 | l^{x_\tau}\tau, l^{x_h}h, l^{-x_\lambda}\lambda)$$

$$\simeq l^{2(D-x_h)} \sum_{x \in \Lambda} G_2^c(lx, 0 | \tau, h, \lambda) \simeq l^{D-2x_h} \sum_{x \in \Lambda} G_2^c(x, 0 | \tau, h, \lambda)$$

$$= l^{D-2x_h}\chi_2(\tau, h, \lambda). \tag{2.4.114}$$

Thus

$$\chi_2(\tau, h, \lambda) \simeq l^{2x_h - D}\chi_2(l^{x_\tau}\tau, l^{x_h}h, l^{-x_\lambda}\lambda). \tag{2.4.115}$$

As before we set $h = 0$ and $l^{x_\tau}\tau = \pm\epsilon$ and obtain

$$\chi_2(\tau, 0, \lambda) \simeq \left(\frac{|\tau|}{\epsilon}\right)^{-(2x_h - D)/x_\tau} \chi_2(\pm\epsilon, 0, (\frac{|\tau|}{\epsilon})^{x_\lambda/x_\tau}\lambda)$$

$$\simeq \text{const } |\tau|^{-(2x_h - D)/x_\tau} \quad , \quad \text{as } \tau \to 0\pm, \tag{2.4.116}$$

so that

$$\gamma = \gamma' = \frac{2x_h - D}{x_\tau}. \tag{2.4.117}$$

2.4.6.4 *Vacuum expectation value*

The one-point function yields the vacuum-expectation value of the field Φ (usually identified with an order parameter, in particular for ferromagnets it is the magnetization). From (2.4.107) it follows for the scaling relation of the one-point function

$$M(l^{x_\tau}\tau, l^{x_h}h, l^{-x_\lambda}\lambda) \equiv G_1^c(0 | l^{x_\tau}\tau, l^{x_h}h, l^{-x_\lambda}\lambda)$$

$$= l^{D-x_h}G_1^c(0 | \tau, h, \lambda) \equiv l^{D-x_h}M(\tau, h, \lambda), \tag{2.4.118}$$

or

$$M(\tau, h, \lambda) = l^{x_h - D}M(l^{x_\tau}\tau, l^{x_h}h, l^{-x_\lambda}\lambda). \tag{2.4.119}$$

For $h = 0$ and $l^{x_\tau}\tau = -\epsilon$ we get

$$M(\tau, 0, \lambda) = \left(\frac{|\tau|}{\epsilon}\right)^{(D-x_h)/x_\tau} M(-\epsilon, 0, (\frac{|\tau|}{\epsilon})^{x_\lambda/x_\tau}\lambda)$$

$$\simeq \text{const } |\tau|^{(D-x_h)/x_\tau} \quad , \quad \text{as } \tau \to 0-, \tag{2.4.120}$$

that is

$$\beta = \frac{D - x_h}{x_\tau}. \tag{2.4.121}$$

On the other hand, setting $\tau = 0$, but choosing the other relevant coupling h so that $l^{x_h} h = \pm\epsilon$ takes a noncritical value, we get

$$M(0, h, \lambda) = \left(\frac{|h|}{\epsilon}\right)^{(D - x_h)/x_h} M(0, \pm\epsilon, (\frac{|h|}{\epsilon})^{x_\lambda/x_h}\lambda)$$

$$\simeq \text{const } |h|^{(D - x_h)/x_h} \quad, \quad \text{as } h \to 0\pm, \tag{2.4.122}$$

hence

$$\delta = \frac{x_h}{D - x_h}. \tag{2.4.123}$$

2.4.6.5 *Specific heat*

The specific heat $C = \partial^2 f/\partial\tau^2$ scales under a renormalization-group transformation according to

$$C(l^{x_\tau}\tau, l^{x_h}h, l^{-x_\lambda}\lambda) = \frac{\partial^2 f(l^{x_\tau}\tau, l^{x_h}h, l^{-x_\lambda}\lambda)}{\partial(l^{x_\tau}\tau)^2}$$

$$= l^{D - 2x_\tau}\frac{\partial^2}{\partial\tau^2}f(\tau, h, \lambda) = l^{D - 2x_\tau}C(\tau, h, \lambda), \tag{2.4.124}$$

or

$$C(\tau, h, \lambda) \simeq l^{2x_\tau - D}C(l^{x_\tau}\tau, l^{x_h}h, l^{-x_\lambda}\lambda). \tag{2.4.125}$$

For $h = 0$ and $l^{x_\tau}\tau = \pm\epsilon$ we get

$$C(\tau, 0, \lambda) \simeq \left(\frac{|\tau|}{\epsilon}\right)^{(D - 2x_\tau)/x_\tau} C(\pm\epsilon, 0, (\frac{|\tau|}{\epsilon})^{x_\lambda/x_\tau}\lambda)$$

$$\simeq \text{const } |\tau|^{-(2x_\tau - D)/x_\tau} \quad, \quad \text{as } \tau \to 0\pm, \tag{2.4.126}$$

thus

$$\alpha = \alpha' = \frac{2x_\tau - D}{x_\tau}. \tag{2.4.127}$$

2.4.6.6 *Summary*

We have considered the linearized renormalization-group transformation for the case of two relevant coupling constants, while all other couplings were irrelevant by assumption. The critical exponents were obtained in terms of two exponents related to the two relevant eigenvalues. Hence only two of the critical exponents are independent. If we choose ν and η as independent exponents, the scaling-sum rules become

$$x_\tau = \frac{1}{\nu} \quad \text{and} \quad x_h = D - d_\Phi - \frac{\eta}{2}, \qquad (2.4.128)$$

and for the other critical exponents

$$\nu' = \nu,$$
$$\alpha' = \alpha = 2 - D\nu,$$
$$\beta = \nu(d_\Phi + \frac{\eta}{2}), \qquad (2.4.129)$$
$$\gamma' = \gamma = \nu(D - 2d_\Phi - \eta),$$
$$\delta = \frac{2(D - d_\Phi) - \eta}{2d_\Phi + \eta}.$$

Here, D is the infrared dimension of space, which is equal to the canonical dimension if all directions are unbounded (otherwise the bounded directions are excluded from D), and d_Φ denotes the infrared dimension of the fields defined in (2.4.51). For conventional magnetic systems in D dimensions we have $d_\Phi = (D - 2)/2$. In this case the above relations take the familiar form

$$\nu' = \nu,$$
$$\alpha' = \alpha = 2 - D\nu,$$
$$\beta = \frac{\nu}{2}(D - 2 + \eta), \qquad (2.4.130)$$
$$\gamma' = \gamma = \nu(2 - \eta),$$
$$\delta = \frac{D + 2 - \eta}{D - 2 + \eta}.$$

2.4.7 *Violation of scaling-sum rules for critical exponents*

Violation of scaling-sum rules can be caused in many ways. To trace back their origin we must check the assumptions, under which the relations were derived. Some of them were listed at the end of section (2.4.5). In addition, the free-energy density f and its derived quantities were assumed to be analytic away from the critical manifold, at least in the vicinity of the fixed point.

For instance, the critical exponents β and δ were related to x_τ and x_h by (2.4.120) and (2.4.122), assuming that the limits

$$\lim_{\lambda \to 0} M(-\epsilon, 0, \lambda) = M(-\epsilon, 0, 0) \tag{2.4.131}$$

and

$$\lim_{\lambda \to 0} M(0, \pm\epsilon, \lambda) = M(0, \pm\epsilon, 0) \tag{2.4.132}$$

both exist for nonvanishing ϵ. This means that irrelevant variables can be directly set to zero as long as the relevant variables are nonvanishing, so that we stay away from the critical surface.

However, we have to be aware that this assumption is sometimes false! Irrelevant variables that violate smoothness assumptions like (2.4.131) or (2.4.132) are called *dangerous irrelevant variables*. A frequent situation is a coupling constant which appears to be irrelevant under a renormalization-group transformation, but its nonvanishing value is essential for ensuring the very existence of the partition function.

We give a simple example of a dangerous irrelevant variable, which is related to the observation that the scaling sum rules are violated above the upper critical dimension D_c. For the Φ^4 theory, $D_c = 4$, and for dimension $D > 4$, the critical infrared behavior is described by mean field theory with critical exponents $\eta = \alpha = 0$, $\beta = \nu = 1/2$ and $\delta = 3$. Let us write the partition function of the model in the form

$$Z(\mu, h, \lambda) = \int \prod_{x \in \lambda} d\Phi(x) \, \exp\left(-S(\Phi)\right),$$

$$S(\Phi) = \frac{1}{2}(\Phi, -\widehat{\Delta}\Phi) + \sum_{x \in \Lambda} \left(-\frac{\mu^2}{2}\Phi(x)^2 + \frac{\lambda}{4!}\Phi(x)^4 - h\Phi(x)\right). \tag{2.4.133}$$

Tree-level evaluation of the first renormalization-group step yields as renormalized coefficients

$$\begin{aligned} \mu'^2 &= l^2(\mu^2 + O(\lambda)), \\ h' &= l^{(D+2)/2}h, \\ \lambda' &= l^{4-D}(\lambda + O(\lambda^2)). \end{aligned} \tag{2.4.134}$$

With $\tau = -\mu^2$, the fixed point is given by $\tau = h = \lambda = 0$. τ and h are the two relevant variables, λ is irrelevant. The corresponding critical exponents as defined by (2.4.103) take the values

$$x_\tau = 2 \ , \ x_h = (D+2)/2 \quad \text{and} \quad x_\lambda = D - 4. \tag{2.4.135}$$

Let us discuss the expectation value $M = <\Phi(x)>$ in the broken phase with $\tau = -\mu^2 < 0$, as the transition is approached. According to (2.4.120), for small $\epsilon > 0$,

$$M(\tau, 0, \lambda) \simeq \left(\frac{|\tau|}{\epsilon} \right)^{(D-x_h)/x_\tau} M(-\epsilon, 0, (\frac{|\tau|}{\epsilon})^{x_\lambda/x_\tau} \lambda)$$

$$\simeq \left(\frac{|\tau|}{\epsilon} \right)^{(D-2)/4} M(-\epsilon, 0, (\frac{|\tau|}{\epsilon})^{(D-4)/2} \lambda) \tag{2.4.136}$$

as $\tau \to 0-$ and λ finite. Under the assumption that M on the right hand side stays finite, we get the critical exponent $\beta = (D-2)/4$.

To understand how $M(-\epsilon, 0, \lambda)$ actually behaves as $\lambda \to 0$ we make a saddle-point expansion of the partition function Z for small h with the result

$$W(\mu, h, \lambda) = \frac{\ln Z(\mu, h, \lambda)}{V}$$

$$\simeq \frac{3}{2} \frac{\mu^4}{\lambda} \left(1 + \left(\frac{8}{3} \right)^{1/2} (\lambda^{1/2} \frac{h}{\mu^3}) + \frac{1}{6} (\lambda^{1/2} \frac{h}{\mu^3})^2 \right) + \frac{1}{2} \ln(2\pi)$$

$$- \frac{1}{2} \int_{-\pi}^{\pi} \frac{d^D p}{(2\pi)^D} \ln \left(\widehat{p}^2 + 2\mu^2 (1 + \left(\frac{3}{8} \right)^{1/2} (\lambda^{1/2} \frac{h}{\mu^3}) - \frac{1}{8} (\lambda^{1/2} \frac{h}{\mu^3})^2) \right),$$

with volume V and lattice momentum $\widehat{p}^2 = \sum_{i=0}^{D-1} (2 \sin \frac{p_i}{2})^2$. We set the lattice spacing $a = 1$ for simplicity. Taking the derivative once with respect to h at $h = 0$, we obtain for small $\lambda/|\tau|$

$$M(\tau, 0, \lambda) \simeq \left(\frac{3! |\tau|}{\lambda} \right)^{1/2} \left(1 - \frac{\lambda}{4|\tau|} \int_{-\pi}^{\pi} \frac{d^D p}{(2\pi)^D} \frac{1}{\widehat{p}^2 + 2|\tau|} \right). \tag{2.4.137}$$

This means that even for finite τ with $\lambda \to 0$, $M(\tau, 0, \lambda)$ diverges as $\lambda^{-1/2}$. Thus we have identified λ as a dangerous irrelevant variable. Using this singular behavior of M on the right hand side of (2.4.136), we convert the behavior as $\lambda \to 0$ into the behavior as $\tau \to 0$. As $\tau \to 0$, $M(\tau, 0, \lambda)$ then actually scales according to

$$M(\tau, 0, \lambda) \simeq \left(\frac{3! |\tau|}{\lambda} \right)^{1/2}. \tag{2.4.138}$$

This is either seen directly from (2.4.137), or by using (2.4.137) on the right hand side of (2.4.136). Hence, in dimensions $D > 4$, the critical exponent β becomes $\beta = 1/2$, in agreement with the mean-field analysis.

In complete analogy, the other critical exponents are obtained from the scaling relations (2.4.106) of the free-energy density and the derived quantities above. The correct singular behavior must be taken into account as the irrelevant coupling constant λ approaches 0. This solves the apparent discrepancy between the standard scaling-sum rules of the critical exponents and the mean-field relations above the upper critical dimension in a satisfactory way.

2.5 Finite-size scaling analysis for second-order phase transitions

The qualitative effect of a finite volume $V = L^D$ in D dimensions will be a *rounding* of singularities in thermodynamic functions. The rounding effects differ for first- and second-order phase transitions in characteristic features. These features allow to anticipate in a finite volume what kind of phase transition in the thermodynamic limit will occur. By means of the linearized renormalization group we will derive the *scaling of the peak* in thermodynamic functions as a function of L, the linear volume size, the *shift of the critical parameter* and the *width of the critical region* if the phase transition is of second order. In the derivation we will make use of the fact that in the thermodynamic limit the correlation length ξ diverges as the critical parameter is approached ($T \to T_c$ or $\tau \to 0$ after $L \to \infty$). Therefore the same kind of analysis does not apply to first-order phase transitions.

The important point about the step of $L = \infty$ to finite L is the reduction of the total number N of degrees of freedom. In our notation this is achieved by decreasing the volume from an infinite to a finite size while keeping the lattice spacing a of the fundamental lattice fixed. Alternatively N may be reduced by keeping the linear extensions L fixed, but increasing the fundamental lattice spacing from zero to a finite value. For our applications the first realization is more convenient.

Let us start with the central scaling relation for the free-energy density which replaces (2.4.106), valid in the infinite volume, here in a finite volume

$$f(l^{x_\tau}\tau, l^{x_h}h, l \cdot L^{-1}, l^{-x_\lambda}\lambda) = l^D \cdot f(\tau, h, L^{-1}, \lambda) \tag{2.5.1}$$

with L the linear extension of the system. For simplicity we choose the same extension in all directions.

Let us see under which assumptions the scaling relation (2.5.1) may be postulated.

- Obviously the inverse volume is put on an equal footing with other relevant scaling fields like temperature, external magnetic field or mass with eigenvalue $x_{L^{-1}} = 1$. Under a block spin transformation L transforms to L/l or $L^{-1} \to l\, L^{-1}$.

  Both formally and physically L^{-1} plays a similar role as a relevant mass parameter or an external magnetic field. It cuts off a diverging correlation length to a finite value unless it takes its "critical" value $L^{-1} = 0$.

- In (2.5.1) it is further assumed that the very existence of a fixed point, the spectrum of the scaling operators and the associated eigenvalues y_τ, y_h, y_λ are not affected by the finite volume, so that y_τ, y_h, and y_λ themselves are independent of L. Clearly this assumption is only justified if L is sufficiently large. It must be so large that it allows an implementation of the finite subset of couplings and their associated (less local) interaction terms to which the effective action is truncated after a number of renormalization-group steps.

- The renormalization-group transformations can only have a fixed point for $L = \infty$. Hence the critical value of L^{-1} is zero. The linearization of the renormalization-group transformations is only valid in the vicinity of the fixed point. Therefore L must be sufficiently large also in order to justify the linearization of the renormalization group.

- As in the infinite volume case we omit the contribution to the free energy which stays regular in the infinite-volume limit by assuming that it is independent of L, more precisely that the analytic dependence on L does not contribute to the leading singular behavior as $L \to \infty$.

- Finally, as in the infinite volume case, we expressed the free-energy density directly in terms of τ, h, λ, and L^{-1}. Again τ and h stand for two *generic* relevant variables that need not have the physical meaning of a reduced temperature or an external magnetic field.

In (2.4.111) of the last section we checked that the correlation length in the infinite volume scales as a usual length scale with l^1 if the relevant parameters are close to their critical values. An analogous derivation of the scaling behavior in a finite volume leads to

$$\xi(\tau, h, L^{-1}, \lambda) \approx l \cdot \xi(l^{x_\tau}\tau, l^{x_h}h, l^1 L^{-1}, l^{-x_\lambda}\lambda). \qquad (2.5.2)$$

In the infinite volume we have focused on two limits: $\tau \to 0$ for $h = 0$ and $h \to 0$ for $\tau = 0$. In the finite volume we consider $\tau \to 0$ if the finite volume is sent to infinity first, i.e. in the limit $L = \infty$ (and $h = 0$ for simplicity). In this case we should recover well-known formulas for the scaling behavior in the thermodynamic limit. The other order of limits is $L \to \infty$ for $\tau = 0$ (and $h = 0$ for simplicity). This limit is of main interest in this section.

For the first case we choose a renormalization-group transformation so that

$$l^{x_\tau}|\tau| = \epsilon. \qquad (2.5.3)$$

For simplicity we choose $\epsilon = 1$. The expected nonanalytic behavior in τ then drops out of the argument of f so that

$$f(\tau, h, L^{-1}, \lambda) = |\tau|^{D/x_\tau} \cdot \tilde{f}(|\tau|^{-x_h/x_\tau} \cdot h, \, |\tau|^{-1/x_\tau} L^{-1}, \, |\tau|^{x_\lambda/x_\tau} \lambda). \qquad (2.5.4)$$

If we express τ in terms of the correlation length ξ_∞ in the infinite volume limit

$$|\tau| \propto \xi_\infty^{-1/\nu}, \qquad (2.5.5)$$

we get

$$f(\tau, h, L^{-1}, \lambda) = (\xi_\infty)^{-D/(\nu \cdot x_\tau)} \tilde{f}(\xi_\infty^{x_h/(\nu x_\tau)} \cdot h, \xi_\infty^{1/(\nu x_\tau)} L^{-1}, \xi_\infty^{-x_\lambda/(\nu x_\tau)} \cdot \lambda), \qquad (2.5.6)$$

or, with $x_\tau = 1/\nu$ by (2.4.128)

$$f(\tau, h, L^{-1}, \lambda) = \xi_\infty^{-D} \tilde{f}(\xi_\infty^{x_h} \cdot h, \xi_\infty/L, \xi_\infty^{-x_\lambda} \cdot \lambda). \qquad (2.5.7)$$

In the thermodynamic limit, $\xi_\infty/L \to 0$, since $L \to \infty$, but $\xi_\infty < \infty$ as long as $\tau \neq 0$.

As one could have naively expected, the free-energy density becomes a function of the ratio ξ_∞/L. In a second-order phase transition ξ_∞ sets an intrinsic scale. Whether L is large or small must be compared with ξ_∞.

For the other order of limits we choose a renormalization-group transformation so that

$$l \cdot L^{-1} = 1, \qquad (2.5.8)$$

here to eliminate the nonanalytic behavior in L from the argument of f. Together with (2.5.5) the scaling relation (2.5.1) becomes

$$f(\tau, h, L^{-1}, \lambda) \;=\; L^{-D} \cdot \hat{f}\left((L/\xi_\infty)^{1/\nu}, L^{x_h} \cdot h, L^{-x_\lambda} \cdot \lambda\right) . \qquad (2.5.9)$$

Note that now the limit of interest corresponds to $L/\xi_\infty \to 0$ since first the (pseudo)critical parameters are adjusted (in particular $\tau \to 0$ at finite L first) and next $L \to \infty$. (In a finite volume the "critical" parameters are called pseudo-critical, because they differ from their infinite-volume values, and f is not really singular.) This is the limit in which finite-size effects should be pronounced as soon as $\xi_\infty \geq O(L)$, or, if the correlation length in the finite volume becomes of the order of the linear size L.

Let us see what the scaling relations (2.5.7) and (2.5.9) imply for their derivatives. From now on we set the relevant parameter $h = 0$ as well as the irrelevant parameters $\lambda = 0$ to ensure that $\tau = 0$ corresponds to the fixed point couplings for $L = \infty$. First we check that (2.5.7) implies the correct thermodynamic limit for the two-point susceptibility. We have (with $2x_h - D = \gamma/\nu$, cf. (2.4.128) and (2.4.129))

$$\begin{aligned}
\chi_2(\tau, L^{-1}) &\propto \xi_\infty^{2x_h - D} \frac{\partial^2}{\partial\left(\xi_\infty^{x_h} h\right)^2} \widetilde{f}\left(\xi_\infty^{x_h} h, \xi_\infty/L, 0\right)\big|_{h=0} \\
&= \xi_\infty^{\gamma/\nu} \mathcal{A}_{\chi_2}^\pm \quad (\text{as } L \to \infty) \\
&\propto |\tau|^{-\gamma} \mathcal{A}_{\chi_2}^\pm \quad (\text{as } \tau \to 0).
\end{aligned} \qquad (2.5.10)$$

For the second relation we used (2.4.117) and defined the proportionality factor, the second derivative of $\widetilde{f}$ at $h = 0$ in the limit of $\xi_\infty/L = 0$, as amplitude $\mathcal{A}_{\chi_2}^\pm$. $\mathcal{A}_{\chi_2}^\pm$ obviously does depend on χ_2, but also on the sign of τ which determines whether the critical region is approached from above $(+)$ or below $(-)$. The reason is that for $L = \infty$, $\widetilde{f}$ on the right hand side of (2.5.4) does depend on the sign of τ (although we have suppressed this dependence in the notation). In the thermodynamic limit we just obtain the familiar result.

More generally, let us consider the approach of a critical point for a thermodynamic quantity P. Let the singular behavior of P be characterized by an exponent μ so that

$$P(\tau, L = \infty) \;\propto\; \mathcal{A}_P^\pm \cdot |\tau|^{-\mu} \qquad (2.5.11)$$

with finite amplitude $\mathcal{A}_P^\pm$ in the infinite volume. The $\pm$- sign also here stands for the approach of the critical point from above $(+)$ or below $(-)$, the index P shall indicate that the amplitude is not universal, it does depend on P. Because of the above considerations which led to (2.5.10) we expect for the analogous relation to (2.5.11) in a finite volume

$$P(\tau, L < \infty) \; \propto \; |\tau|^{-\mu} Q_P^{\pm} \left(\xi_\infty(\tau)/L \right) \tag{2.5.12}$$

with $Q_P^{\pm} \left(\xi_\infty(\tau)/L \right)$ the amplitude of P in a finite volume if L is sufficiently large and τ sufficiently small. In the thermodynamic limit $(\xi_\infty/L \to 0)$ we must have

$$\lim_{x \to 0} Q_P^{\pm}(x) = \mathcal{A}_P^{\pm} \tag{2.5.13}$$

for $x \equiv \xi_\infty/L$. (2.5.13) has been verified for χ_2 in (2.5.10). To derive the singular behavior of P in the opposite limit $(L/\xi_\infty(\tau) \to 0)$, we have two possibilities. Either we study the appropriate derivative of f. For $P = \chi_2$ we obtain this way from (2.5.9)

$$\chi_2(\tau, L^{-1}) \; = \; L^{-D} \frac{\partial^2}{\partial h^2} \widehat{f} \left((L/\xi_\infty)^{1/\nu}, L^{x_h} \cdot h, L^{-x_\lambda} \cdot 0 \right) \Big|_{h=0}$$

$$= \; L^{2x_h - D} \frac{\partial^2}{\partial h^2} \widehat{f} \left((L/\xi_\infty)^{1/\nu}, h, 0 \right) \Big|_{h=0}, \tag{2.5.14}$$

or, with (2.4.128) and (2.4.129)

$$\chi_2(\tau = 0, L^{-1}) \; \propto \; L^{\gamma/\nu}. \tag{2.5.15}$$

Alternatively, the singular behavior of P in the limit of $L/\xi_\infty(\tau) \to 0$ follows from the very fact that for finite L $\lim_{\tau \to 0} P(\tau, L, h, \lambda)$ must be finite with $\tau = T - T_c(L = \infty)$. A system in a finite volume can never have a true thermodynamic singularity, not even as the parameters take on their critical values corresponding to the critical point in the infinite volume limit (here $\tau \to 0$). In addition, P will not even take its finite maximum at $\tau = 0$, but at some shifted value $T_c(L)$ that differs from $T_c(L = \infty)$, as we show below. Thus the nonanalyticity of $P(\tau, L = \infty)$ that is proportional to $|\tau|^{-\mu}$ as $\tau \to 0$ at $L = \infty$ must be compensated by the scaling behavior of $Q_P^{\pm}(\xi_\infty(\tau)/L)$ as $\tau \to 0$ for finite L. Let $y \equiv L/\xi_\infty(\tau)$. We must have

$$Q_P^{\pm} \left(y^{-1} \right) \; \propto \; |\tau|^{\mu} \; \propto \; y^{\mu/\nu} \quad \text{as } y \to 0, \tag{2.5.16}$$

since $\tau \propto (y/L)^{1/\nu}$. It follows

$$P(\tau = 0, L^{-1}) \propto \lim_{\tau \to 0} |\tau|^{-\mu} Q_P^{\pm} \left(y^{-1} \right)$$

$$\propto \lim_{\tau \to 0} |\tau|^{-\mu} y^{\mu/\nu}$$

$$\propto L^{\mu/\nu} < \infty \qquad \text{for } L < \infty. \qquad (2.5.17)$$

In particular, if $P = \chi_2$ and $\mu = \gamma$, we recover Eq. (2.5.15).

For $P = c$, the specific heat, and $\mu = \alpha$ we find

$$c(\tau = 0, L^{-1}) \propto L^{\alpha/\nu}, \qquad (2.5.18)$$

and for $P = \xi$, the correlation length, and $\mu = \nu$

$$\xi(\tau = 0, L^{-1}) \propto |\tau|^{-\nu} Q_\xi^{\pm} \left((\xi_\infty(\tau)/L \right)$$

$$\propto |\xi_\infty| \cdot y$$

$$\propto L. \qquad (2.5.19)$$

At the critical point the correlation length diverges linearly in L. The linear divergence is also easily seen as follows. For $L/\xi_\infty \to 0$ we set $l \cdot L^{-1} = \epsilon = 1$ in (2.5.2) and $\tau \propto \xi^{-\nu}$. It follows that

$$\xi(\tau, L^{-1}) \propto L \cdot \xi \left((L/\xi_\infty)^{1/\nu} \right), \qquad (2.5.20)$$

or again the result (2.5.19).

We remark that the finite-size scaling analysis predicts the scaling behavior of thermodynamic functions at a location where they have their singularity in the *infinite* volume (for which $\tau = T - T_c(L = \infty) = 0$). Actually this is not quite the maximum in the finite volume. Therefore, more precisely, the statement about the scaling behavior is that $P(L^{-1}, T_c(L), h, \lambda)$ grows at least as $L^{\mu/\nu}$ at its finite-volume maximum. Further notice that the value $T_c(L)$, for which P takes its maximum in the finite volume, depends on P itself. Therefore the pseudo-critical temperature, or, more generally the pseudo-critical parameter, is a function of both L and P.

$$T_c|_{L<\infty} = T_c(L, P). \qquad (2.5.21)$$

2.5.1 Shift in the pseudo-critical parameter

As we have seen in (2.5.17), P depends on the ratio $L/\xi_\infty(\tau) \propto L \cdot \tau^\nu$. For fixed L we choose $\tau = \tau_{max}$ so that P becomes a maximum at $u_{max} = L \cdot \tau_{max}^\nu$. If we change L to L' but want to stay at the maximum of P at

$u = u_{max}$, we have to choose $\tau'_{max} = (L/L')^{1/\nu}\tau_{max}$, or, in other words, τ_{max} scales according to

$$\tau_{max} \propto L^{-1/\nu} \qquad (2.5.22)$$

with $\tau_{max} = T_c(L, P) - T_c(L = \infty)$ and $T_c(L, P)$ determined from the maximum of P as $T \to T_c(L, P)$. Thus the shift of the pseudo-critical parameter scales with $L^{-1/\nu}$.

The shift of T_c also gives a measure for the width σ of the critical region by saying that all parameters T belong to the critical region for which $T \in [T_c(L = \infty), T_c(L, P)]$. Hence σ scales like τ_{max}

$$\sigma(\tau, L^{-1}) \propto L^{-1/\nu}. \qquad (2.5.23)$$

In identifying the critical region as the interval $[T_c(L = \infty), T_c(L, P)]$ we have assumed that $T_c(L, P) > T_c(L = \infty)$. Such a shift towards larger values of T is found in systems with periodic boundary conditions. Because of their constraint on momenta to multiples of $2\pi/L$, periodic boundary conditions tend to suppress fluctuations in the system and therefore to increase $T_c(L, P)$. For free boundary conditions a shift in the opposite direction is often observed.

We conclude with some remarks. To see true critical behavior all relevant parameters must take their critical values. In our choice this means that we have to choose $\tau = h = L^{-1} = 0$. Monte Carlo calculations are unavoidably performed in a finite volume. The finite-size scaling behavior which is predicted by the renormalization group analysis will be observed if

- the volume is sufficiently large
- τ and/or h are sufficiently small

so that the scaling ansatz (2.5.1) for the free-energy density is justified.

If we start in a finite volume and then consider the usual thermodynamic limit in the sense of $\xi_\infty(\tau)/L \to 0$, we rediscover the familiar formulas of the infinite volume. Finite-size effects are negligible in this limit. In the opposite limit, $L/\xi_\infty(\tau) \to 0$, finite-size effects are pronounced. Their form was predicted from (2.5.9) to (2.5.17). The "crossover region" consists of values of L and $\tau(h)$ for which $L/\xi_\infty(\tau) \approx O(1)$. This case may be realized either because for given large L, τ is so small that the correlation length becomes of the order of the linear size of the system. In this case the observed behavior will be predicted by (2.5.17), (2.5.22) and (2.5.23).

However, $L/\xi_\infty(\tau) \approx O(1)$ may be realized just because L is not sufficiently large. In practice this case often occurs in numerical simulations when it is rather time consuming to run the simulations in a large volume. If L^{-1} and τ are not small but such that $L/\xi_\infty(\tau) \approx O(1)$, finite-size effects are still pronounced in this case, but their quantitative rounding is no longer predicted by the formulas we derived above. Thus one should keep in mind the very conditions under which the finite-size scaling formulas have been derived to interpret possible deviations from (2.5.17), (2.5.22), (2.5.23) in numerical simulations.

The same kind of renormalization-group analysis will not apply to first-order transitions. Replacing τ by $\xi_\infty^{-1/\nu}$ we made use of a diverging correlation length ξ_∞ in the thermodynamic limit. Even more, the very ansatz (2.5.1) for the scaling behavior of the free energy assumed the existence of a fixed point with a scale invariant fixed point action. Scale invariance is a typical manifestation of second-order transitions, but not of first-order ones.

2.5.2   *τ-like and h-like scaling fields*

In our discussions so far we considered two scaling fields τ and h, corresponding to two relevant directions in coupling-parameter space in the renormalization group. As we mentioned above, τ and h need not have the physical meaning of a reduced temperature or a reduced magnetic field. Just for simplicity the standard notation is adapted to an Ising ferromagnet in an external field h at temperature T with action

$$S = -T^{-1} \sum_{<xy>} \sigma(x)\sigma(y) - h \cdot \sum_{x \in \Lambda} \sigma(x) \qquad (2.5.24)$$

and the phase diagram of Fig.(2.5.1). In the (T, h) plane there is a line

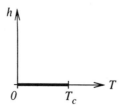

Fig. 2.5.1   Line of temperature driven first-order phase transitions in a ferromagnet at zero magnetic field ending in a second-order transition at T_c

of first-order phase transitions for $h = 0$, terminating in a critical point $(h_c = 0, T_c)$, at which the transition is of second order. $(h_c = 0, T_c)$ is called a critical endpoint. It is a peculiarity of the Ising model that the action (2.5.24) is directly parameterized in terms of scaling fields T and h that correspond to independent fluctuations of the scaling operators. Let us write a generic action in lattice units in the form

$$S = \sum_{i=1}^{n} g_i \sum_{x \in \Lambda} O_i(x) \qquad (2.5.25)$$

with scaling fields g_i and operator densities $O_i(x)$. The fluctuation matrix M in a finite volume is then defined as

$$M_{ij} = <(O_i - <O_i>) \cdot (O_j - <O_j>)> = \frac{1}{V} \cdot \frac{\partial^2 \ln Z(V, g_i)}{\partial g_i \partial g_j} \qquad (2.5.26)$$

with

$$<O_i> \equiv \frac{\int \mathcal{D} \Phi \, O_i \, e^{-S[\Phi]}}{\int \mathcal{D} \Phi \, e^{-S[\Phi]}} \qquad (2.5.27)$$

and the shorthand notation

$$O_i = \sum_{x \in \Lambda} O_i(x) \qquad (2.5.28)$$

if we consider one sort of fields Φ. For the Ising model with $\Phi = \sigma$ and for $h = 0$, M_{ij} is already diagonal, in particular at the critical endpoint, because of the $h \to -h$-symmetry of the action along the line of first-order phase transitions. The fluctuations in τ and h directions are independent of each other. It is these directions that enter the critical exponents describing the singular behavior.

In the generic case the action is neither parameterized in terms of τ and h, nor is the fluctuation matrix (2.5.26) diagonal. Let us focus on a typical question which arises in connection with identifying a universality class. We assume that we have determined a line of first-order phase transitions and localized the critical endpoint $g_i^\star$ at which the phase transition becomes of second order. Examples are liquid-vapor systems at their critical endpoint, the U(1)-Higgs model at the critical point at which the Higgs and confinement phases merge and, last but not least, the SU(2) Higgs model at the critical Higgs mass for which the electroweak phase transition is of second order and turns into a crossover phenomenon for larger Higgs masses. For a scalar order parameter a possible conjecture for the universality class at the critical endpoint is the Ising universality. How should one proceed to prove such a conjecture? First one should find τ-like and h-like directions for which the fluctuation matrix in a finite volume becomes diagonal at the critical point. Otherwise a mismatch with the Ising exponents is not conclusive. Let us consider the most simple case with an action consisting of two terms with couplings g_1 and g_2. After diagonalizing the action in (g_1, g_2) we obtain

$$
M = \left. \begin{pmatrix} < (O_\tau - < O_\tau >)^2 > & 0 \\ 0 & < (O_h - < O_h >)^2 > \end{pmatrix} \right|_{g_\tau^\star(L), g_h^\star(L)},
$$
(2.5.29)

where O_τ, O_h, and τ, h are linear combinations of O_{g_1}, O_{g_2} and g_1, g_2, respectively. $(g_\tau^\star, g_h^\star)$ denote the new coordinates of the critical endpoint in a finite volume. As a function of the volume the eigenvalues should scale according to

$$
< (O_\tau - < O_\tau >)^2 >_{g_\tau^\star(L), g_h^\star(L)} \equiv \chi_\tau \propto L^{\alpha/\nu}
$$
$$
< (O_h - < O_h >)^2 >_{g_\tau^\star(L), g_h^\star(L)} \equiv \chi_h \propto L^{\gamma/\nu}
$$
(2.5.30)

with α, ν, γ taking Ising values if the conjecture about sharing the Ising universality was right.

Other manifestations of sharing the Ising universality show up in a similar shape of the two-dimensional *joint probability distributions* of the operators O_τ and O_h, $P(O_\tau, O_h)$, and of the operators of the Ising action $P(\sum_{<ij>} \sigma_i \sigma_j, \sum_i \sigma_i)$. For the connection of these distributions P with the constraint effective potential we refer to (2.6.55), (2.6.57), (2.6.61) of section 2.6. A quantitative description of this similarity leads to consider higher moments of $P(O_\tau, O_h)$ of which the second moments are just the elements of the fluctuation matrix. For details on the practical realization of finding the τ-like and h-like directions also in an extended space of couplings g_i, $i > 2$, we refer to the literature [60–62].

2.6   Finite-size scaling analysis for first-order phase transitions

A finite-size scaling analysis for first-order transitions is inherently more difficult than for second-order transitions. The correlation length stays finite even in the infinite volume limit as $T \to T_c$ after $L \to \infty$, so ξ_∞/L is no longer a sensible scaling variable as in Eq. (2.5.7). We have

$$\lim_{T \to T_c} \lim_{L \to \infty} \xi(L) < \infty \tag{2.6.1}$$

for a first-order transition. In the same order of limits the correlation length diverges with an exponent $-1/\nu$ for a second-order transition. There it is the thermal eigenvalue $1/\nu$ of the renormalization-group transformation that controls rounding and shifting of the algebraic singularities, both being of the order $L^{-1/\nu}$. Note that in the reversed order of limits its $\lim_{L \to \infty} \lim_{T \to T_c(L)}$ the correlation length diverges linearly in L both for first- and second-order transitions. Thus the right order of limits is important to correctly identify a first-order transition.

The goal of finite-size scaling analysis for first-order transitions is to predict the rounding and shifting of δ-function singularities in the second derivative of a thermodynamic potential due to the finite volume (e.g., in the specific heat due to a latent heat, or in the susceptibility due to a jump in the order parameter).

Before we go into the subtleties of refined criteria, which are based on the precise form of the rounding and shifting as a function of L, we summarize some qualitative signatures of a first-order transition in a typical Monte Carlo simulation. (Qualitative signatures are essentially the large- or infinite-volume signatures that we call "naive" criteria for inferring the order of the transition.)

- Some thermodynamic quantities (e.g., the internal energy) are almost discontinuous at the transition.
- A starting configuration that is half-ordered, half-disordered relaxes to very different equilibrium states on both sides of T_c (rather than frequently tunneling between both sides). In a second-order transition the system relaxes to an equilibrium configuration independently of the initial condition.
- At infinite volume, tunneling between both phases is completely suppressed; at small volumes tunneling may be mixed with fluctuations of statistical origin. For large volumes tunneling events are clearly visible in a Monte Carlo simulation. The system is in the ordered phase with a probability proportional to $\exp\{-L^D f_o(\beta)\}$, and in the disordered phase with a probability proportional to $\exp\{-L^D f_d(\beta)\}$, where f_o and f_d denote the free-energy densities in the ordered phase for $T < T_c$ and in the disordered phase for $T > T_c$, respectively. On the other hand, f_o in the disordered phase and f_d in the ordered phase represent

some kind of metastable free-energy density in the vicinity of T_c. If the time history is followed over a number of Monte Carlo iterations, it shows flip-flops between states of different "magnetizations". The frequency of flip-flops decreases with increasing volume.

- Another manifestation of tunneling shows up in the probability distribution P_e for the internal energy density $e(\beta)$. For large volumes it is sharply peaked at the energy values of the ordered (e_o) and disordered (e_d) phases. The deep valley between these peaks reflects the rare number of tunneling events. If the initial condition is an ordered start, the probability is large to find e_o for e. For a disordered start, it is large to find e_d. This is nothing but a sign of *metastability*.

- *Hysteresis effects* are observed even away from the transition temperature.

The qualitative signature of a pronounced double-peak structure can be made more quantitative when it is analyzed as a function of the lattice size. In this way we are led to *refined criteria*.

A precise form for finite-size scaling was sought following different approaches. We focus on systems for which the free energy has several distinct local minima in the transition region. This includes Ising-like systems with two realizations of the ordered phase and a disordered phase, or q-state Potts models with q realizations of the ordered phase and a disordered phase. First we derive relations between the partition function, the constraint effective potential and the distribution function $P(z)$ of some parameter z that is defined in terms of the field variable Φ. In particular z is the averaged order-parameter field

$$z = \frac{1}{\beta V} \int_x \Phi(x) \, ,$$

or the internal energy e per lattice degree of freedom. As usual let us define the expectation value of an observable O according to

$$< O >= \frac{1}{Z} \int \mathcal{D}\Phi \, O[\Phi] e^{-S[\Phi]} \tag{2.6.2}$$

in a finite volume βV with partition function

$$Z = \int \mathcal{D}\Phi \exp\{-S(\Phi)\} \equiv \exp\{-\beta V f\}$$

with f the free-energy density and Φ chosen as a scalar field for simplicity. We assume translation invariance so that expectation values like $< \Phi(x) >$

do not depend on x. Here and in the following V denotes the spatial volume $V = L^d$ with d the spatial dimension. In a field theoretical context the total volume is given by βV, because $\beta = T^{-1}$ is the extension in Euclidean time direction.

Let $\mathcal{F}(\Phi(x))$ denote a function of $\Phi(x)$, for example any monomial in Φ, in particular the internal energy per lattice site or the one-plaquette action in case of a pure gauge theory. If we introduce a constraint $\delta(y - \frac{1}{V} \int_x \mathcal{F}(\Phi(x))), y \in \mathbf{R}$, into the partition function, we probe the distribution of the density $\mathcal{F}(\Phi(x))$ (cf. Eq. (2.6.4) below) and are naturally led to the definition of a constraint effective potential in terms of y. We write for an observable O

$$
\begin{aligned}
< O > &= \frac{1}{Z} \int dy \int \mathcal{D}\Phi \, \delta \left(y - \frac{1}{V} \int_x \mathcal{F}(\Phi(x)) \right) O(\Phi) e^{-S[\Phi]} \\
&= \frac{1}{Z} \int dy \, \widetilde{O}(y, V) e^{-V \cdot V_{cep}(y, V)} \\
&= \int dy \, \widetilde{O}(y, V) P(y, V) \,.
\end{aligned}
\tag{2.6.3}
$$

The constraint effective potential $V_{cep}(y, V)$ is defined by

$$
e^{-V \cdot V_{cep}(y, V)} \equiv \int \mathcal{D}\Phi \, \delta \left(y - \frac{1}{V} \int_x \mathcal{F}(\Phi(x)) \right) e^{-S[\Phi]} \,,
\tag{2.6.4}
$$

the distribution $P(y, V)$ by

$$
P(y, V) \equiv \frac{1}{Z} \cdot e^{-V \cdot V_{cep}(y, V)} \,,
\tag{2.6.5}
$$

so that

$$
\int dy \, P(y, V) = 1 \,.
\tag{2.6.6}
$$

Finally $\widetilde{O}(y, V)$ is a shorthand notation for

$$
\widetilde{O}(y, V) \equiv \frac{\int \mathcal{D}\Phi \, \delta \left(y - \frac{1}{V} \int_x \mathcal{F}(\Phi(x)) \right) O(\Phi) e^{-S[\Phi]}}{\int \mathcal{D}\Phi \, \delta \left(y - \frac{1}{V} \int_x \mathcal{F}(\Phi(x)) \right) e^{-S[\Phi]}} \,.
\tag{2.6.7}
$$

If $\Phi(x)$ has the meaning of an order-parameter field, $\mathcal{F}(\Phi(x))$ is chosen as $\Phi(x)$ and $O(\Phi) = \frac{1}{V} \int_x \Phi(x) \equiv m$, we have $\widetilde{O}(y, V) = y$ so that

$$
< m > = \int dy \, y P(y, V) \,.
\tag{2.6.8}
$$

From (2.6.8) we get an intuitive understanding of the constraint effective potential V_{cep} as it determines the associated distribution $P(y, V)$, that is, $\Delta y \cdot P(y, V)$ gives the probability of finding the order-parameter field $\Phi(x)$ (or equivalently m (because of translation invariance)) between y and $y + \Delta y$. In case of spontaneous symmetry breaking, in the broken phase, $P(y, V)$ will be peaked around nonvanishing values of y. V_{cep} itself has the meaning of a coarse-grained free energy that can be derived from a microscopic scale by integrating over short-range fluctuations (in principle, at least). Like any coarse-grained free energy, V_{cep} need not be convex in a finite volume.

As a trigger current let us introduce an external field h into the action that we need to test upon spontaneous symmetry breaking in a finite volume. The h-dependent term $-h \int_x \mathcal{F}(\Phi(x))$ is not absorbed in the definition of the constraint effective potential so that the formula for V_{cep} (2.6.4) remains unchanged, but

$$P(y, V, h) \equiv \frac{1}{Z(V, h)} e^{V(hy - V_{cep}(y,V))}$$

$$< F(\Phi(x)) >_h = \int dy\, y P(y, V, h) \qquad (2.6.9)$$

with

$$Z(V, h) = e^{-Vf(V,h)} = \int dy\, e^{V(hy - V_{cep}(y,V))} . \qquad (2.6.10)$$

It should be noticed that the partition function in (2.6.10) is represented as an ordinary integral over a real variable y.

As $V \to \infty$ only saddle point configurations contribute to Z and, accordingly, to the free-energy density $f(V, h)$. We have

$$\lim_{V \to \infty} f(V, h) = \lim_{V \to \infty} \left[-\sup_y (hy - V_{cep}(y, V)) \right] \qquad (2.6.11)$$

if the exponent in (2.6.10) in the presence of an external field has only one maximum. Otherwise the formula would generalize to a sum (integral) over all degenerated maxima. Here we will focus on systems with a finite number of suprema, for example with two suprema corresponding to the coexistence of the ordered and disordered phase at the critical temperature for zero external field. If there is a whole manifold invariant with the same value for the exponent in (2.6.11), formula (2.6.11) generalizes according to

$$\lim_{V \to \infty} f(V, h) = \lim_{V \to \infty} \left[\frac{-\ln \mathcal{M}(V)}{V} - \sup_y (hy - V_{cep}(y, V)) \right] \quad (2.6.12)$$

with $\mathcal{M} \equiv \int dy_{sp}$ denoting the measure of the supremum domain to account for this possible degeneracy.

For large volumes the h-dependent distribution of $\mathcal{F}(\Phi(x))$, in particular of the order-parameter field $\Phi(x)$, is given by

$$P(y, V, h) = \frac{1}{\mathcal{M}(V)} \exp \left\{ V \left[hy - V_{cep}(y, V) - \sup_y (hy - V_{cep}(y, V)) \right] \right\}$$
$$(2.6.13)$$

with $y = m$ if $\mathcal{F}(\Phi(x)) = \Phi(x)$. From (2.6.13) we see that in the thermodynamic limit $P(y, V, h)$ tends to zero except for the extrema.

In Fig. 2.6.1 we show how the extremum structure of $V_{cep}(y, V)$ translates to $P(y, V)$. For definiteness we interpret $P(y, V)$ as the distribution function of the magnetization $y = m$. In each figure the dashed lines correspond to the infinite-volume limit, the full lines to a measurement in a finite volume V. For a second-order transition and for $m \geq 0$ the characteristic features of the constraint effective potential are a single minimum of V_{cep}. It is located in the broken phase at $m \neq 0$, at the transition point at $m = 0$ and in the symmetric phase again at $m = 0$. The single peak of V_{cep} corresponds to a single peak in the order-parameter distribution at $m > 0$ in the broken phase and at $m = 0$ at the transition point and in the symmetric phase.

For first-order transitions and for $y = m \geq 0$, in an Ising-like case with two realizations of the ordered phase (of which we project only onto the positive magnetization for simplicity) and one realization of the disordered phase, the single minimum and single peak structures of second-order transitions are replaced by double-minimum and double-peak structures in an appropriate range of temperatures above and below T_c. As the volume increases, the peaks get shifted and sharpened. At the transition point, they finally approach two peaks of δ-distributions (dashed lines). The intermediate range between the peaks correspond to configurations in which the system tunnels between the ordered and disordered phases. In Monte Carlo simulations these tunneling events are manifest in so-called flip flops. The tunneling events become more and more rare the larger the volume is.

We should emphasize that a measurement for a *single* finite volume is not conclusive in general. The total number of saddle points and - in case

Second-order transitions

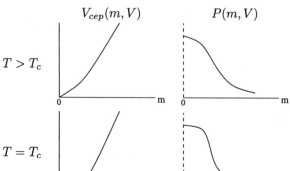

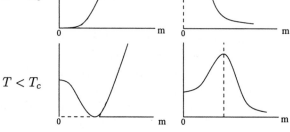

First-order transitions

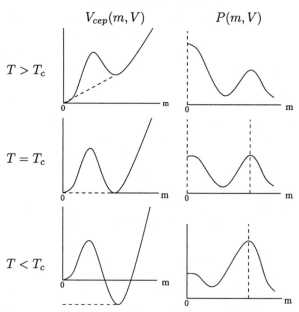

Fig. 2.6.1   Constraint effective potential $V_{cep}(m, V)$ versus order-parameter distribution $P(m, V)$ for $m \geq 0$, $m \equiv \frac{1}{V} \int_x \Phi(x)$, for second- and first-order transitions, above T_c, at T_c and below T_c. The dashed lines correspond to the thermodynamic limit.

of a continuous degeneracy - the saddle-point manifold with measure $\mathcal{M}$ do depend on V as indicated also in (2.6.12). Therefore, even if there is a finite number of minima, this number may change with V. For example, a double-peak structure for $y = m \geq 0$ turns out as an artifact of a finite and small volume if it disappears for larger volumes for which a single-peak structure remains. In this case a transition which is actually of second order would be misinterpreted as a first-order transition, unless the volume dependence of the order-parameter distribution is followed over several volumes.

So far our discussion refers to qualitative features of V_{cep} and P. Let us see how far we can get in a quantitative description of the scaling of the peak and its shift for a first-order transition if we now further restrict the class of systems which we consider.

2.6.1 Predictions for rounding and shifting of singularities in first-order transitions

In this section we consider a finite-size scaling analysis for classical statistical systems. Here the inverse temperature plays the role of an external parameter and $V = L^d$ is the total spatial volume. Ultimately we are interested in a finite-size scaling analysis of field theoretical models at finite temperature. In field theoretical models the inverse temperature is implemented as the finite extension in Euclidean time with appropriate boundary conditions on the field-degrees of freedom. The kind of "classical" finite-size scaling analysis we are going to derive is directly applicable to field theory in the following cases. If we consider coupling-driven phase transitions at zero temperature in $D = d + 1$ dimensions, the same kind of analysis goes through by replacing βV by L^D. If we are interested in a coupling driven transition at fixed inverse finite temperature β, the finite-size scaling behavior of the d-dimensional spatial volume should be described by the "classical" analysis of a system which becomes large in d directions and stays finite and constant in one direction. Furthermore, if it is possible to derive an effective, d-dimensional model from the field theory in D dimensions, again the classical analysis applies, but to the effective model. The analysis of a temperature-driven transition in a D-dimensional field theory gets more subtle. Ideally one should first take the continuum limit and next the thermodynamic limit, that is, send β to the pseudo-critical $\beta_c(L)$ in a finite physical volume, and then send the physical volume to infinity so that one could apply the standard finite-size scaling analysis to the continuum model at the pseudo-critical temperature. If instead the system is simulated at fixed extension N_τ in time direction, the bare coupling g_0 is next changed so that the given extension N_τ (translated to physical units) corresponds to the pseudo-critical temperature of a lattice system in a finite physical volume, if furthermore only the spatial volume is made larger and larger to perform a finite-size scaling analysis, and finally all the former steps are repeated for larger N_τ as a step towards the continuum limit, one has exchanged the order of limits, taking first the thermodynamic limit and next the continuum limit. It is not clear whether this exchange is justified.

We derive the scaling behavior of derived quantities of $\ln Z$ as a function of the finite volume. We focus on a class of models that is characterized by a partition function of the form

$$Z(\beta, V) = q e^{-\beta V f_o(\beta, h, V)} + e^{-\beta V f_d(\beta, h, V)} + o(e^{-V}). \qquad (2.6.14)$$

Equation (2.6.14) should be understood as the leading order of an expansion in the inverse volume V^{-1}, β the inverse temperature and V the d-dimensional spatial volume. f_o is the free-energy density of one of q versions of an ordered phase for $T < T_c$ and some kind of metastable free-energy density for $T > T_c$, f_d the corresponding quantity of the disordered phase. What we take here as an ansatz for Z was derived in [63] for sufficiently large q for models which allow a so-called contour representation. Included are the q-state Potts models and, in particular, the Ising model for $q = 2$ if $q = 2$ may be considered as sufficiently large in the sense that the derivation extends down to $q = 2$.

The precursor in the finite volume of the phase transition in the infinite volume occurs at an inverse pseudo-critical temperature $\beta_c(V)$. As we shall see, the definition of the pseudo-critical temperature in the finite volume is not unique. We start with the following choice. $\beta_c(V)$ is the temperature for which the Boltzmann weight of all q realizations of the ordered phase equals the weight of the disordered phase, that is

$$q \cdot e^{-\beta_c(V)V f_o(\beta_c(V), h=0, V)} = e^{-\beta_c(V)V f_d(\beta_c(V), h=0, V)}. \qquad (2.6.15)$$

The *equal-weight condition* (2.6.15) as a definition of the pseudo-critical temperature was first formulated in [66] for temperature-driven transitions. In connection with this definition, Challa, Landau and Binder in [66] and Landau and Binder in [65] proposed a phenomenological ansatz for the distribution function of the internal-energy density and the order parameter to which we come back below. The proposals differed in postulating that at $\beta_c(\infty)$ the peaks of the (Gaussian) distributions have either equal height [66] or equal weight [65]. From an intuitive point of view both postulates are plausible. What is finally the appropriate ansatz must be decided from its implications on the finite-size scaling behavior of the position of the susceptibility's maximum. This behavior can be calculated independently either analytically or with Monte Carlo simulations. As it turns out, for temperature-driven transitions the leading $1/V$-correction of the position of the maximum $\beta_c(\chi_{max})$ as compared to the infinite-volume value $\beta_c(\infty)$ is correctly reproduced by both ansätze (but not so for field-driven transitions).

In the following we set the external field $h = 0$ and consider q as independent of V and β (this condition may be violated if V is not sufficiently large as we mentioned above). We write

$$Z(\beta, V) = qZ_o(\beta, V) + Z_d(\beta, V) \qquad (2.6.16)$$

and

$$f_o(\beta, V) = -\frac{1}{\beta V} \ln Z_o(\beta, V)$$

$$f_o(\beta) \equiv - \lim_{V \to \infty} \left(\frac{1}{\beta V} \ln Z_o(\beta, V) \right), \qquad (2.6.17)$$

and accordingly for the quantities in the disordered phase, $f_d(\beta, V)$, $f_d(\beta)$. The critical temperature $\beta_c(\infty)$ in the infinite volume is defined according to

$$f_o(\beta_c(\infty)) = f_d(\beta_c(\infty)). \qquad (2.6.18)$$

In the following we assume that

$$q \, e^{-\beta V(f_o(\beta, V) - f_d(\beta, V))} \gtrless 1 \quad \text{for } \beta \lessgtr \beta_c(V). \qquad (2.6.19)$$

We may assume that $f_o(\beta, V)$ and $f_d(\beta, V)$ are analytic functions of β, also at $\beta_c(\infty)$, so that we are allowed to expand $f_{o(d)}$ about $\beta_c(V)$, but the total free-energy density $f(\beta, V = \infty) \neq f_o(\beta, \infty) + f_d(\beta, \infty)$ is nonanalytic at

$\beta_c(\infty)$. We write

$$-\beta V f(\beta, V) = \begin{cases} -\beta V f_d(\beta) + \ln\left[1 + q e^{-\beta V(f_o(\beta,V) - f_d(\beta,V))}\right] \\ \qquad\qquad\qquad\qquad \text{for } \beta \le \beta_c(V) \\ \\ -\beta V f_o(\beta) + \ln q + \ln\left[1 + \frac{1}{q} e^{-\beta V(f_d(\beta,V) - f_o(\beta,V))}\right] \\ \qquad\qquad\qquad\qquad \text{for } \beta \ge \beta_c(V) \end{cases}.$$
$$\tag{2.6.20}$$

It follows from (2.6.19) and (2.6.20) that

$$\beta f(\beta, V) = \beta f_d(\beta) + O(e^{-V}) \qquad \text{for } \beta \le \beta_c(V)$$
$$\beta f(\beta, V) = \beta f_o(\beta) + O(\frac{1}{V}, e^{-V}) \quad \text{for } \beta \ge \beta_c(V)\,, \qquad (2.6.21)$$

where

$$f_{o(d)}(\beta) \equiv \lim_{V \to \infty} f_{o(d)}(\beta, V) \tag{2.6.22}$$

as in (2.6.17).

Scaling of the shift in β_c

Next we derive the scaling of the shift of $\beta_c(V)$, defined via (2.6.15), compared to $\beta_c(\infty)$. Condition (2.6.15) implies

$$\ln q - V \cdot k(\beta_c(V)) = 0 \tag{2.6.23}$$

with

$$k(\beta_c(V)) \equiv \beta_c(V)\left[f_o(\beta_c(V), V) - f_d(\beta_c(V), V)\right]\,, \tag{2.6.24}$$

such that $k(\beta_c(\infty)) = 0$. Since $f_{o(d)}(\beta)$ are assumed as analytic functions of β, also at $\beta_c(\infty)$, we expand

$$k(\beta) = k(\beta_c(\infty)) + (\beta - \beta_c(\infty))\, k'(\beta_c(\infty)) + O(\beta - \beta_c(\infty))^2$$
$$k'(\beta_c(\infty)) = \beta_c(\infty) \cdot \left[\frac{\partial f_o(\beta)}{\partial \beta} - \frac{\partial f_d(\beta)}{\partial \beta}\right]_{\beta = \beta_c(\infty)}\,. \tag{2.6.25}$$

Since

$$df = \frac{s}{\beta^2}\, d\beta \tag{2.6.26}$$

with entropy density s, we find for $\beta < \beta_c(\infty)$

$$\frac{df}{d\beta} = \frac{df_d}{d\beta} = \frac{s_d}{\beta^2}\,, \tag{2.6.27}$$

for $\beta > \beta_c(\infty)$

$$\frac{\mathrm{d}f}{\mathrm{d}\beta} = \frac{\mathrm{d}f_o}{\mathrm{d}\beta} = \frac{s_o}{\beta^2}, \tag{2.6.28}$$

or

$$k'(\beta_c(\infty)) = \left.\frac{s_o(\beta) - s_d(\beta)}{\beta}\right|_{\beta_c(\infty)}$$
$$= e_o(\beta_c(\infty)) - e_d(\beta_c(\infty)) \tag{2.6.29}$$

with $e_{o(d)}$ the internal energy densities in the ordered (disordered) phase, respectively, if we use the standard thermodynamic relation $s = \beta \cdot (e - f)$. Equations (2.6.23), (2.6.25) and (2.6.29) now lead to

$$\beta_c(V) - \beta_c(\infty) = \frac{1}{V}\frac{\ln q}{[e_o(\beta_c(\infty)) - e_d(\beta_c(\infty))]}. \tag{2.6.30}$$

Thus the shift in the inverse critical temperature is proportional to V^{-1}, unless $q = 1$. Equation (2.6.30) is precisely the shift which is predicted in the phenomenological model mentioned above.

Next we want to derive

- the location of the maxima of the specific heat and the two-point susceptibility χ and their relation to $\beta_c(V)$ defined via (2.6.15)
- the height of the maxima
- the width of the critical region.

For $h = 0$ we expand up to second order in $(\beta - \beta_c(V))$

$$\beta f_{o(d)}(\beta, V) = g_{o(d)} + \epsilon g_{o(d)}^{(1)} + \frac{1}{2}\epsilon^2 g_{o(d)}^{(2)} + O(\epsilon^3), \tag{2.6.31}$$

with

$$\epsilon \equiv \beta - \beta_c(V)$$
$$g_{o(d)} \equiv \beta_c(V)f_{o(d)}(\beta_c(V), V)$$
$$g_{o(d)}^{(1)} \equiv \left.\frac{\partial}{\partial\beta}(\beta f_{o(d)}(\beta), V)\right|_{\beta=\beta_c(V)}$$
$$g_{o(d)}^{(2)} \equiv \left.\frac{\partial^2}{\partial\beta^2}(\beta f_{o(d)}(\beta), V)\right|_{\beta=\beta_c(V)}. \tag{2.6.32}$$

Using condition (2.6.15), the partition function takes the form

$$Z(\beta, V) = q \cdot e^{-V g_o} \cdot \left\{ e^{-V\left[\epsilon g_o^{(1)} + \frac{\epsilon^2}{2} g_o^{(2)}\right]} + e^{-V\left[\epsilon g_d^{(1)} + \frac{\epsilon^2}{2} g_d^{(2)}\right]} \right\}. \qquad (2.6.33)$$

For the specific heat $c_V(V)$ it follows that

$$
\begin{aligned}
c_V &= \beta^2 \frac{1}{V} \left(\frac{1}{Z} \frac{\partial^2 Z}{\partial \beta^2} - \left(\frac{1}{Z} \frac{\partial Z}{\partial \beta} \right)^2 \right) \\
&= -\beta^2 \, \frac{g_o^{(2)} e^{-V\epsilon\left[g_o^{(1)} + \frac{\epsilon}{2} g_o^{(2)}\right]} + g_d^{(2)} e^{-V\epsilon\left[g_d^{(1)} + \frac{\epsilon}{2} g_d^{(2)}\right]}}{e^{-V\epsilon\left[g_o^{(1)} + \frac{\epsilon}{2} g_o^{(2)}\right]} + e^{-V\epsilon\left[g_d^{(1)} + \frac{\epsilon}{2} g_d^{(2)}\right]}} \\
&\quad + \beta^2 \, V \cdot \frac{\left(\widetilde{g}^{(1)} + \epsilon \widetilde{g}^{(2)} \right)^2}{\left(2 \cosh \frac{V\epsilon}{2} \left[\widetilde{g}^{(1)} + \frac{\epsilon}{2} \widetilde{g}^{(2)} \right] \right)^2}
\end{aligned}
\qquad (2.6.34)
$$

with the shorthand notation

$$\widetilde{g}^{(j)} \equiv g_o^{(j)} - g_d^{(j)}, \quad j = 1, 2 \qquad (2.6.35)$$

for the first ($(j = 1)$) and second ($(j = 2)$) derivatives of $g_o - g_d$ with respect to β, respectively.

We distinguish two cases with respect to the volume V. For large V

- $\epsilon = O(1)$, but $\epsilon \neq O(\frac{1}{V})$, and
- $\epsilon = O(\frac{1}{V})$.

If $\epsilon = \beta - \beta_c(V) \simeq O(1)$, we are away from the phase-transition region. The second term in (2.6.34) is then exponentially suppressed in the volume compared to the first term that is of $O(1)$ if $g_o^{(2)} \simeq O(1)$. So far we have not made any assumption on the sign of ϵ, hence β^{-1} can be a temperature in the broken or in the symmetric phase.

If ϵ is of $O(\frac{1}{V})$, we are in the vicinity of the phase transition. For large V, the second term of (2.6.34) dominates the first one. Note that its maximum and therefore also the maximum of c_V as a function of β is not exactly at $\epsilon = 0$, i.e. at $\beta_c(V)$ defined via the equal weight condition (2.6.15), but slightly shifted by an amount of

$$\epsilon = \frac{\delta}{V^2} \qquad (2.6.36)$$

with $\delta = (4\widetilde{g}^{(2)})/(\widetilde{g}^{(1)})^3$ of $O(1)$, because $\partial c_V/\partial \epsilon = 0$ leads to

$$\frac{V}{2} \tanh \frac{V\epsilon}{2} \left(\widetilde{g}^{(1)} + \epsilon \widetilde{g}^{(2)}\right) = \frac{\widetilde{g}^{(2)} + O(\epsilon)}{\left(\widetilde{g}^{(1)} + \epsilon \widetilde{g}^{(2)}\right)^2} \qquad (2.6.37)$$

or

$$V^2 \epsilon \approx \frac{4\widetilde{g}^{(2)}}{\left(\widetilde{g}^{(1)}\right)^3} \equiv \delta, \qquad (2.6.38)$$

because the nominator of the second term in (2.6.34) is not symmetric about $\epsilon = 0$. If we denote the value of β, for which c_V takes its maximum, as $\beta_{max}(V, c_V)$, we conclude

$$\beta_{max}(V, c_V) = \beta_c(V) + O\left(\frac{1}{V^2}\right). \qquad (2.6.39)$$

Furthermore it is obvious from (2.6.34) that for all temperatures β^{-1} with $\epsilon \simeq O(\frac{1}{V})$, c_V increases proportional to V, in particular the maximum c_{max} at $\beta_c(V) + \frac{\delta}{V^2}$,

$$c_{max} \propto V, \qquad (2.6.40)$$

but also all points of the critical region, for which $\beta = \beta_{max}(V, c_V) + O(\frac{1}{V})$. Therefore the width σ of the critical region itself scales with V^{-1}

$$\sigma \propto \frac{1}{V}. \qquad (2.6.41)$$

It should be also noticed that δ in (2.6.38) and $\beta_{max}(V, c_V)$ are not universal. As indicated by the notation, the critical temperature $\beta_{max}^{-1}(V, c_V)$ defined via the maximum of the specific heat does depend on c_V as long as $V < \infty$. In the next step we will explicitly show that $\beta_{max}(V, \chi)$, defined via the maximum of the two-point susceptibility χ, differs from $\beta_{max}(V, c_V)$ by terms of $O(\frac{1}{V^2})$. Unless V is sufficiently large, such differences are actually visible in Monte Carlo simulations. (In our calculation it is in principle possible to explicitly calculate the shift of $\beta_{max}(V, c_V)$ with respect to $\beta_c(V)$ of $O(\frac{1}{V^2})$ if the expansion of $\beta f_{o(d)}(\beta, V)$ includes terms of $O(\epsilon^2)$.)

For the two-point susceptibility χ in a finite volume V

$$\chi = \frac{1}{V} \left(\frac{1}{Z} \frac{\partial^2 Z}{\partial h^2} - \left(\frac{1}{Z} \frac{\partial Z}{\partial h}\right)^2\right)\Bigg|_{h=0} \qquad (2.6.42)$$

we expand $f_{o(d)}(\beta, h, V)$ including terms of $O(\epsilon)$ and $O(h^2)$ according to

$$
\begin{aligned}
\beta f_j(\beta, h, V) = {}& g_{j0}(V) + h g_{j1}(V) + \frac{h^2}{2} g_{j2}(V) \\
& + \epsilon \left(g_{j0}^{(1)}(V) + h g_{j1}^{(1)}(V) + \frac{h^2}{2} g_{j2}^{(1)}(V) \right) \\
& + \mathcal{O}(\epsilon^2, h^3) ,
\end{aligned} \tag{2.6.43}
$$

where the first subscript j stands either for the ordered $((o))$ or the disordered $((d))$ phase and the second subscript for the first $((1))$ or second $((2))$ derivative with respect to h, the superscript $((1))$ stands for the first derivative with respect to β. All derivatives are to be taken at $\beta_c(V), h = 0$. Again we implement the equal-weight condition (2.6.15), now at $h = 0$, and write for Z

$$
\begin{aligned}
Z = q \cdot e^{-V g_{o0}} \Big\{ & e^{-V\left[h g_{o1} + \frac{h^2}{2} g_{o2} + \epsilon \left(g_{o0}^{(1)} + h g_{o1}^{(1)} + \frac{h^2}{2} g_{o2}^{(1)} \right) \right]} \\
& + e^{-V\left[h g_{d1} + \frac{h^2}{2} g_{d2} + \epsilon \left(g_{d0}^{(1)} + h g_{d1}^{(1)} + \frac{h^2}{2} g_{d2}^{(1)} \right) \right]} \Big\} .
\end{aligned} \tag{2.6.44}
$$

It follows for the two-point susceptibility χ

$$
\chi = - \frac{\left[g_{o2} + \epsilon g_{o2}^{(1)} \right] e^{-V\epsilon g_{o0}^{(1)}} + \left[g_{d2} + \epsilon g_{d2}^{(1)} \right] e^{-V\epsilon g_{d0}^{(1)}}}{e^{-V\epsilon g_{o0}^{(1)}} + e^{-V\epsilon g_{d0}^{(1)}}} + V \cdot \frac{\left[\widetilde{g}_1 + \epsilon \widetilde{g}_1^{(1)} \right]^2}{\left(2 \cosh \frac{V\epsilon}{2} \widetilde{g}_0^{(1)} \right)^2} \tag{2.6.45}
$$

with

$$
\begin{aligned}
\widetilde{g}_1 &\equiv g_{o1} - g_{d1} \\
\widetilde{g}_1^{(1)} &\equiv g_{o1}^{(1)} - g_{d1}^{(1)} \\
\widetilde{g}_0^{(1)} &\equiv g_{o0}^{(1)} - g_{d0}^{(1)} .
\end{aligned} \tag{2.6.46}
$$

Again, for large V and $\epsilon \simeq O(1)$, the first term gives a finite, nonvanishing contribution, while the second term vanishes exponentially fast as $V \to \infty$, but for large V and $\epsilon \simeq O(\frac{1}{V})$, the second term dominates. The maximum of χ occurs for

$$
\beta_{max}(V, \chi) = \beta_c(V) + \frac{\widetilde{\delta}}{V^2} \tag{2.6.47}
$$

with

$$\tilde{\delta} = \frac{4\tilde{g}_1^{(1)}}{\left(\tilde{g}_0^{(1)}\right)^2 \cdot \tilde{g}_1} \, . \tag{2.6.48}$$

A comparison of (2.6.48) with (2.6.38) explicitly shows the nonuniversality of δ and $\beta_{max}(V)$, but the maximum of χ scales again proportional to V, and the width of the critical region of the susceptibility in a large volume shrinks with $1/V$ as V increases.

In this section, the volume dependence of the height of the maxima of response functions, the width of the critical region, and the shift of the critical temperature were derived for a specific class of models characterized by (2.6.14). The observed absence of nontrivial critical exponents and the scaling with integer powers of V are supposed to hold as generic features of first-order phase transitions.

Two-Gaussian peak model
The equal-weight condition (2.6.15) was also an essential ingredient in the phenomenological two-Gaussian peak model for the probability distribution $P_L(e)$ of the internal energy density e, which was introduced in [65]. $P_L(e) \equiv P(e, L)$ is a basic quantity for temperature-driven transitions. For field-driven transitions e is replaced by the magnetization m. From (2.6.5) we know that $P_L(e)$ is proportional to a Boltzmann factor with an appropriate constraint effective potential in the exponent that has the meaning of some coarse-grained free energy. Let us assume that

$$P_L(e) \propto e^{-\beta V f(e)} \tag{2.6.49}$$

with $f(e)$ the free-energy density as a function of the internal-energy density e, and d the dimension of the spatial volume V. We expand $f(e)$ about its minimum in e

$$-\beta L^d e + L^d s_L(e) \approx -\beta L^d \cdot f_{min}(\beta, V) - \frac{1}{2} L^d \cdot \frac{\beta^2 \left(e - e_{min}(\beta, V)\right)^2}{c(\beta, V)} \tag{2.6.50}$$

with s_L the entropy density in the finite volume L^d, $f_{min}(\beta, V)$ the free-energy density

$$f_{min}(\beta, V) \equiv e_{min}(\beta, V) - s_L(e_{min}(\beta, V), V)/\beta \, , \tag{2.6.51}$$

and $e_{min}(\beta, V)$, the internal-energy density, a solution of

$$\left.\frac{\partial s_L}{\partial e}\right|_{e_{min}} = \beta \,, \tag{2.6.52}$$

$c(\beta, V)$ denotes the specific heat. For large volumes and away from the transition point, $P_L(e)$ becomes a simple Gaussian. The double-peak structure of $P_L(e)$ in the vicinity of a first-order transition is then easily understood as a consequence of Eq. (2.6.52) that has two solutions $e_o(\beta, V)$ and $e_d(\beta, V)$, corresponding to two coexisting phases. For large enough $V = L^d$ we can neglect the V-dependence of $e_{o(d)}(\beta, V)$ and $c(\beta, V)$ and expand $e_{min}(\beta)$ about $\beta_c(\infty)$ so that

$$e_{min}(\beta) = e_{o(d)} + c_{o(d)} \cdot \delta T \,, \tag{2.6.53}$$

with $e_{o(d)} \equiv e_{min}(\beta_c(\infty))$ the internal-energy density in the infinite volume at $\beta_c(\infty)$ in the ordered (o) and disordered (d) phase, and

$$\delta T \equiv \frac{\beta_c - \beta}{\beta_c^2} \tag{2.6.54}$$

with $\beta_c \equiv \beta_c(\infty)$. Accordingly, close to the phase transition, the energy distribution is written as a sum of two Gaussians

$$P_L(e) = A\left[\frac{a_o}{\sqrt{c_o}} \frac{e^{-V\beta_c^2[e - e_o - c_o \delta T]^2}}{2c_o} + \frac{a_d}{\sqrt{c_d}} \frac{e^{-V\beta_c^2[e - e_d - c_d \delta T]^2}}{2c_d}\right] . \tag{2.6.55}$$

The e-independent prefactors $a_o = q \cdot e^{-\Delta}$ and $a_d = e^{\Delta}$ can be written in a symmetrized form with

$$\Delta \equiv \frac{V}{2}\beta \cdot (f_o(\beta) - f_d(\beta))$$

$$\propto \frac{V}{2}(\beta - \beta_c)\left(e_o - e_d + \frac{1}{2}(c_o - c_d)\delta T\right) \tag{2.6.56}$$

and A a normalization factor. A is chosen such that $P_L(e)$ has the interpretation of a probability distribution with $\int de P_L(e) = 1$. The form of Eqs. (2.6.55-2.6.56) was proposed in [65]. By construction it leads to the right shift of $\beta_c(V)$ with respect to $\beta_c(\infty)$ if $\beta_c(V)$ is defined via the equal-weight condition.

Next let us see how the double-peak structure of $P_L(e)$ is compatible with the large-volume expansion of $Z(\beta, V)$ according to (2.6.14). If we choose $\mathcal{F}(\Phi(x))$ in (2.6.3) as $\int_x \mathcal{F}(\Phi(x)) = S(\Phi)$, i.e. as the action per site, and write $y \equiv e$, we define a constraint effective potential as a function of e, β and V by

$$e^{-\beta L^d V_{cep}(e,\beta,V)} \equiv \int \mathcal{D}\Phi \, e^{-\beta S[\Phi]} \delta \left(e - \frac{1}{L^d} S[\Phi] \right)$$

$$\equiv e^{-\beta L^d \cdot e} Q(e, \beta, V)$$

$$Q(e, \beta, V = L^d) \equiv \int \mathcal{D}\Phi \, \delta \left(e - \frac{1}{L^d} S[\Phi] \right). \tag{2.6.57}$$

We expand $V_{cep}(e, \beta, V)$ about its minima in e, say about e_o and e_d, respectively, including terms of $O(e - e_{o(d)})^2$. This leads to two additive contributions in a saddle-point evaluation of $\int de \exp\{-V \cdot V_{cep}(e, \beta, V)\}$, cf. Eq. (2.6.13). In view of this saddle-point expansion we replace $\int de \exp -V \cdot V_{cep}(e, \beta, V)$ according to

$$e^{-\beta L^d V_{cep}(e,\beta,V)} \simeq \alpha_o \cdot e^{-\frac{\gamma_0}{2} V (e-e_0)^2} + \alpha_d \cdot e^{-\frac{\gamma_d}{2} V (e-e_d)^2} \tag{2.6.58}$$

with appropriate e-independent coefficients $\alpha_{o(d)}(\beta, V)$, $\gamma_{o(d)}(\beta)$. By the same line of arguments as in (2.6.50)-(2.6.52), we end up with an expression for $\exp\{-V \cdot V_{cep}(e, \beta, V)\}$ that is proportional to $P_L(e)$ of Eq. (2.6.55). The proportionality factor must depend on β and V, since

$$\int de \, P_L(e) = 1, \tag{2.6.59}$$

but

$$\int de \, e^{-\beta V V_{cep}(e,\beta,V)} = Z(\beta, V = L^d). \tag{2.6.60}$$

From (2.6.3), (2.6.4), and (2.6.5) it follows that $Z(\beta, V)$ can be written as an ordinary integral according to

$$Z(\beta, V) = \int de \, e^{-\beta V e} Q(e, \beta, V) = \int de \, e^{-\beta V \cdot V_{cep}(e,\beta,V)}. \tag{2.6.61}$$

If we now evaluate $Z(\beta, V)$ in a saddle-point expansion about the minima of V_{cep} in e, we reproduce Z in the form of (2.6.14). In terms of Q according to (2.6.57), the two values of e_o and e_d correspond to two solutions of the δ-constraint in (2.6.57) for $e = e_{min}$ in the space of constant field configurations Φ. Writing $Q(e, \beta, V)$ as a sum of two δ-functions immediately gives Z of (2.6.14) from (2.6.61).

Moments and cumulants of $P_L(e)$

The peak structure of $P_L(e)$ was frequently studied in Monte Carlo simulations. But also certain moments and cumulants of the energy distribution were proved to suitably indicate the order of the transition in a finite volume. An important example is the so-called Binder cumulant proposed in [66]. It is defined according to

$$B_L(\beta) = \left(1 - \frac{< e^4 >_L (\beta)}{3 < e^2 >_L^2 (\beta)}\right)$$

$$< e^n >_L = \int de\, e^n P_L(e), \quad n \geq 0 \tag{2.6.62}$$

in a volume L^d with $P_L(e)$ given by (2.6.55). The large-volume limit of B_L is easily calculated from its definition and (2.6.55) with the following result. *Away from* T_c, for a first-order transition in a region of temperatures in which contributions of either the ordered or the disordered phase to $P_L(e)$ are negligible, B_L goes to a constant value

$$B_L \xrightarrow{V \to \infty} 2/3 \tag{2.6.63}$$

up to corrections in $1/L^d$. The same behavior is also found for second-order transitions *at* $\beta_c(V)$, whereas for first-order transitions *at* $\beta_c(V)$ we find

$$B_L(\beta_c(V)) = \frac{2}{3} - \frac{1}{3}(e_o - e_d)^2 \left(\frac{e_o + e_d}{e_o^2 + e_d^2}\right)^2 + \mathcal{O}\left(\frac{1}{L^d}\right)$$

$$\equiv B_{min}. \tag{2.6.64}$$

At $\beta = \beta_c(V)$, B_L takes its minimum value, therefore we refer to Binder's cumulant as B_{min}. The second factor of the first nontrivial term in B_{min} depends on an additive arbitrary constant in the energy e. Therefore Billoire et al. [64] proposed a different cumulant, called U_4 and defined according to

$$U_4 \equiv \frac{< (e- < e >)^4 >}{< (e- < e >)^2 >^2}, \tag{2.6.65}$$

of which the additive constant drops out. Their result is that $U_4 > 1$ apart from the first-order transition point.

The advantage of studying finite-size effects on quantities like $c_{max}(V)$, B_{min} or the minimum of U_4 are the power law corrections in $1/V$ in contrast to exponential corrections. Power-law corrections are easier identified in numerical simulations. Exponential corrections are found to bulk quantities like the average internal-energy density e, which are directly obtained as derivatives of Z. According to Eq. (2.6.14), Z itself has only exponential corrections from the finite volume.

2.6.2 Finite-size scaling with linked cluster expansions

So far we have derived the finite-size scaling (FSS) behavior of relations like (2.6.30), (2.6.40), (2.6.41) for a certain class of models, specified by a partition function which can be expanded according to (2.6.14). Their derivation was based on large-volume expansions and the equal-weight condition (2.6.15). For generic models similar relations are expected to hold, but this conjecture must be proved for any concrete case, either numerically or analytically. Linked cluster expansions (LCEs) provide an *analytic* tool to calculate the finite-size effects on n-point correlation functions and derived quantities. LCEs were elaborated to a high order in the expansion parameter, both in the infinite and in the finite volume for scalar models. Special cases are Ising models, Potts models, but also O(N)-symmetric scalar field theories for arbitrary N. Scalar O(N)-models with *continuous* O(N)-symmetries, i.e. for $N > 1$, are not covered by the analysis of Borgs and Kotecky [63] as they derive the representation (2.6.14) of the partition function which was the starting point of our finite-size scaling analysis so far.

In LCEs the action is split into a sum of an ultra-local part $\overset{\circ}{S}$ (depending on products of fields in which not only the fields themselves but also their products depend on single sites of the lattice) and a (next-)neighbor part S_{nn} with couplings between pairs of fields. The coupling is proportional to the generic expansion parameter κ. A Taylor expansion in κ of the logarithm of the partition function $\ln Z$ about the ultra-local contribution to $\ln Z$ finally leads to graphical expansions of n-point susceptibilities χ_n with coefficients $a_i^{(n)}$

$$\chi_n = \sum_i a_i^{(n)} (2\kappa)^i. \qquad (2.6.66)$$

For every n and order of the expansion i, $a_i^{(n)}$ is a sum over all connected graphs with i internal and n external lines each. For scalar models each graph adds as its weight a product of the inverse topological symmetry factor, an internal symmetry factor, a product of vertex contributions depending on the couplings involved in $\overset{\circ}{S}$, and, last but not least, a lattice embedding factor. For the precise form of this factorization and an introduction to LCEs we refer to section (4.2).

It is only the lattice embedding factor $I_{\Gamma_{n,i}}$ of a graph Γ with i internal and n external lines that is sensitive to the topology and size of the particular lattice that will change in passing from an infinite to a finite volume.

Apart from a trivial overall volume factor, $I_{\Gamma_{n,i}}$ counts the number of ways of embedding an abstract graph on a lattice of specified size and geometry, for example on a hyper-cubic lattice in $D = 4$ dimensions of size $L_0 \times L^3$. Extending LCEs to a finite volume requires a new calculation of the embedding factors for the various graphs. In section (4.2) we illustrate such calculations. Here we just assume that they were performed and summarize applications of LCEs in a finite volume.

Finite-size scaling analysis near the singularity at T_c

As we have seen above, a standard finite-size scaling analysis amounts to a calculation of the critical temperature $(\beta_c^{-1}(V))$ and various response functions such as the n-point susceptibilities $\chi_n(V)$ and the specific heat $c_V(V)$ in a finite volume V, along with a check of their scaling behavior according to relations like (2.6.30), (2.6.40), (2.6.41). As a peculiarity of LCEs in a finite volume, $\beta_c^{-1}(V)$ is an extrapolated quantity, because the series are truncated at some finite order. Thus χ_n or c_V can no longer be reliably evaluated at the critical temperature itself (proportional to the critical expansion parameter $\kappa_c(V)$), but only in its vicinity, say at $\approx 95\% - 99\%$ of T_c or κ_c. This feature is evident in particular in the infinite-volume limit: at $\kappa_c(V = \infty)$ one would work "on top" of the singularity. Therefore we proposed a so-called monotony criterion that avoids an evaluation of χ_n or c_V at $\kappa_c(V)$, but provides an alternative finite-size scaling analysis otherwise [67]. Although it was invented for LCEs, the criterion works for Monte Carlo simulations as well, as we shall explain in more detail below.

Effective potentials in a finite volume

A further application of LCEs in a finite volume is the calculation of an effective potential V_{eff} as it was introduced in (2.2.26) of section 2.2. The coefficients of V_{eff} as series in M are suitable sums over $a_1, ..., a_{2n}$ and $x_1, ..., x_{2n}$ of $\Gamma_{a_1,...,a_{2n}}^{(2n)}(x_1, ..., x_{2n})$, as can be read off from (2.2.21) (section 2.2). It is possible to re-express the coefficients in terms of quantities χ_n^{1PI} for which the graphical expansion of LCEs is directly available. The superscript $1PI$ stands for one-particle-irreducible. The result for a generic action which admits a linked cluster expansion (for instance (2.3.63)-(2.3.68) of section 2.3) is [70]

$$V_{eff}(M) = \frac{1 - 4D\kappa\chi_2^{1PI}}{\chi_2^{1PI}} \cdot \frac{M^2}{2} - \frac{\chi_4^{1PI}}{\left(\chi_2^{1PI}\right)^4} \cdot \frac{(M^2)^2}{4!}$$

$$- \frac{1}{\left(\chi_2^{1PI}\right)^6} \left(\chi_6^{1PI} - \frac{10\left(\chi_4^{1PI}\right)^2}{\chi_2^{1PI}}\right) \frac{(M^2)^3}{6!}$$

$$+ O(M^8). \tag{2.6.67}$$

The χ_{2n}^{1PI} are obtained from the 2n-point susceptibilities χ_{2n} when the graphical expansion of the χ_{2n} is restricted to 1PI-graphs. The reader may wonder why the coefficients are certain functions of χ_n^{1PI} and do not just appear in the form $\frac{1}{n!}\chi_n^{1PI}M^n$, as the $\Gamma_{j_1...j_n}^{(n)}(x_1...x_n)$ are vertex functions which amount to 1PI-graphical contributions to Feynman diagrams if they are evaluated in a perturbative expansion in the coupling. The reason is that the generating functional for the χ_n^{1PI} in a linked cluster expansion is some functional R and not Γ so that we have to follow the mapping $\Gamma \to \ln Z \to R$, where Γ and $\ln Z$ are related by a "standard" Legendre transformation, but $\ln Z$ and R by a different Legendre transformation, accounting for the fact that LCEs are expansions about the ultra-local part of the action and not about the purely kinetic part.

A nonconvex effective potential as a function of M can arise from the truncation of V_{eff} at finite order in M, and from metastable minima in a finite volume that anticipate a first-order phase transition in the symmetric phase. If we assume that the truncation at finite order in M is justified because higher order terms are negligible for values of M close to the trivial and nontrivial local minima, we may attribute nontrivial minima as signals for a first-order phase transition in a quantity that has the physical meaning of a coarse-grained free energy. Furthermore the nonconvex shape is only conclusive for a first-order phase transition if the volume is actually so large that it cannot result exclusively from the finite volume. Stated differently, even for second-order transitions a small volume may induce first-order signals which disappear when the volume gets larger and larger. Here we assume that the nonconvex shape is not misleading in this sense. Under these assumptions we may also estimate $\kappa_c(V)$ from the coexistence of different minima of V_{eff} leading to the same value of V_{eff}, or the vanishing of the coefficient of the quadratic term.

The monotony criterion

The monotony criterion is based on the following observations. For a certain interval of the scaling region response functions with a nonanalytic behavior in the infinite-volume limit show different monotony behavior for first- and second-order transitions. Examples for such functions are the specific heat and order-parameter susceptibilities that we investigated by means of a standard finite-size scaling analysis above. They are increasing in volume in a certain neighborhood of T_c for second-order transitions, and decreasing for first-order transitions for some range in the scaling region, which will be specified below.

For definiteness we fix the notation in terms of an order parameter susceptibility $\chi(T, L)$, considered as a function of the temperature T and the spatial size parameter L, so that a symmetric d-dimensional volume is of size L^d. By $T_c(L, \chi) \equiv T_c(L)$ we denote here the location of the maximum of $\chi(T, L)$ in the volume L^d associated with the true phase transition at $T_c \equiv T_c(\infty)$. We know already that in the infinite volume

$$\chi(t + T_c(\infty), L = \infty) < \infty \quad \text{as} \quad t \to 0 \qquad (2.6.68)$$

for a first-order transition at T_c, with a possible discontinuity at $t = 0$ in the associated order parameter, with t denoting the (not normalized) scaling field $t = T - T_c(\infty)$, whereas

$$\chi(t + T_c(\infty), L = \infty) \propto |t|^{-\gamma} \quad \text{as} \quad t \to 0 \qquad (2.6.69)$$

for a second-order transition with critical exponent $\gamma > 0$. On the other hand, in the opposite order of limits, $\chi(T_c(L), L)$ diverges in both cases as L approaches infinity. Stated differently, at $T_c(L)$, in the limit $L \to \infty$, χ has a δ-function singularity for a first-order transition and an algebraic singularity for a second-order transition, cf. (2.5.15 of section 2.5). It is this difference that is responsible for the different monotony properties for $t \neq 0$ in the finite volume.

The essential statement of the monotony criterion is the following. For a sufficiently large size $L_1 < \infty$, there is always a size L_2 with $L_2 > L_1$ and such that

$$\chi(\delta + T_c(L), L \geq L_2) < \chi(\delta + T_c(L_1), L_1) \quad \text{for first order,}$$
$$\chi(\delta + T_c(L), L \geq L_2) > \chi(\delta + T_c(L_1), L_1) \quad \text{for second order,} \qquad (2.6.70)$$

that is, $\chi(\delta + T_c(L), L)$ is decreasing or increasing in volume for a first- or second-order transition, respectively. The susceptibility is measured at fixed distance δ to the volume dependent maximum, constrained by

$$c_2 \cdot \sigma(L_2)^{1/(1+\epsilon)} < |\delta| < c_1 \cdot \sigma(L_1). \qquad (2.6.71)$$

Here $\sigma(L)$ is the width of the critical region in the volume L^d, $\sigma(L) \propto L^{-d}$ (cf. Eq. 2.6.41), or $\sigma(L) \propto L^{-1/\nu}$ (cf. Eq. (2.5.23 section 2.5), for a first- and second-order transition, respectively, with ν being the critical exponent of the correlation length. Moreover, c_1, c_2 and ϵ are positive constants, typically $\epsilon \approx O(1)$. Beyond the general constraint that both linear sizes L_1 and L_2 have to be sufficiently large, L_2 has to be sufficiently larger than L_1, so that $\sigma(L_2)$ is considerably smaller than $\sigma(L_1)$. The monotony criterion does *not* refer to values of L_2 close to L_1.

In our original reference [67] we made these statements more precise in order to show that the monotony behavior is neither a peculiarity of specific models nor an artifact of the series expansions which we used in applications of the criterion. It is a generic feature of models with first- and second-order transitions if the standard assumptions on their finite-size scaling behavior apply. Here we further comment on how to apply the criterion to series expansions and Monte Carlo simulations.

- Let λ denote a generic set of couplings parameterizing the transition surface in coupling-constant space. Choose two volumes $L_1^d \equiv V_1$ and $L_2^d \equiv V_2$ both sufficiently large in order to satisfy the standard assumptions of a finite-size scaling analysis. The regular contribution to χ and its induced analytic volume dependence are then negligible. In series expansions the larger volume is conveniently chosen to be infinite so that (2.6.70) and (2.6.71) with $V_2 = \infty$ and $\sigma(L_2) = 0$ give strong criteria in the whole scaling region. In Monte Carlo simulations both volumes V_1 and V_2 must be finite. The monotony criterion works equally well in this case, but some care is needed to ensure that the susceptibility is actually measured at a temperature that satisfies (2.6.71), i.e. out of a region where the difference in the volume dependence between first- and second-order transitions appears. In particular this concerns the lower bound on δ in (2.6.71), because for $L_2 < \infty$, also in case of a first-order transition, a small neighborhood around the peak at $T_c(L_2)$ exists, in which $\chi(T, L_2)$ is *increasing* (rather than decreasing) in volume because of the rounding of the δ-singularity. In summary

- V_2 should be chosen sufficiently larger than V_1 to ensure that the widths of the critical region behave like $\sigma(V_2) << \sigma(V_1)$.

- Choose T or κ, the points, at which the χ are evaluated, at the same distance δ from the volume-dependent position of the peak of χ.

- Choose δ according to (2.6.71) and finally
- define the ratio

$$r_{V_1,V_2} \equiv 1 - \frac{\chi_2(\delta + \kappa_c(V_1), V_1)}{\chi_2(\delta + \kappa_c(V_2), V_2)}. \qquad (2.6.72)$$

The monotony criterion then says that

$$r_{V_1,V_2} \begin{cases} > 0 & , \quad \text{2nd order} \\ < 0 & , \quad \text{1st order} \\ = 0 & , \quad \text{tricritical point for } \partial r/\partial \lambda \neq 0. \end{cases} \qquad (2.6.73)$$

If there is a crossover phenomenon rather than a true phase transition, r should vanish for a finite neighborhood of λ for sufficiently large V_1, V_2.

Deviations from these predictions in Monte Carlo simulations or series expansions may be caused by

- V_1, V_2 both too small
- V_2/V_1 too small
- $(\kappa - \kappa_c(V))/\kappa_c(V)$ too large. Contributions of the regular part cannot be neglected in this case.
- $(\kappa - \kappa_c(V))/\kappa_c(V)$ too small, so that $r > 0$ even for first order.

In particular the interval of allowed values of κ in Eq. (2.6.72) depends on the coupling λ. Note that the ratio of Eq. (2.6.72) does not include the generic volume dependence that is induced by the analytic part in χ. It is supposed to be negligible if $\kappa - \kappa_c(V)$ is sufficiently small. Assuming that one is lucky in simultaneously matching the various constraints, a result of $r_{V_1,V_2} > (<)0$ for all κ of the specified scaling region excludes (indicates) a first-order transition, respectively. A finite-size scaling analysis by means of the monotony criterion should be always useful when it is not possible to work directly at the (pseudo)critical temperature $T_c(V)$ or coupling $\kappa_c(V)$.

The monotony criterion was applied in the framework of convergent series expansions of $O(N)$-symmetric scalar models on the lattice, with Φ^4- and Φ^6-interactions, in three dimensions for $N = 1$ and $N = 4$ [68]. The expansions were performed to the twentieth order in the hopping parameter κ both in the infinite and in the finite volume. This class of models admits both first- and second-order phase transition regions. An application of the monotony criterion allowed a localization of the tricritical line in this model which was by two orders of magnitude more precise than an analysis in the

Table 2.6.1   Criteria for distinguishing between first- and second-order phase transitions in a large but finite volume $V \equiv L^d$ at the pseudo-critical inverse temperature $\beta_c(V)$, defined via the equal-weight condition.

Criterion	*First-order*	*Second-order*
$P_L(e)$	double peak	single peak
c_{max}	$\propto L^d$	$\propto L^{\alpha/\nu}$
χ_{max}	$\propto L^d$	$\propto L^{\gamma/\nu}$
$\beta_c(\infty) - \beta_c(V)$	$\propto L^{-d}$	$\propto L^{-1/\nu}$
B_{min}	$< \frac{2}{3}$	$\to \frac{2}{3}$

infinite volume. Other examples for successful applications of the monotony criterion are the Gross-Neveu model in three dimensions [71] and the $SU(2)$ Higgs model [70].

2.6.3   *Summary of criteria*

In Table 2.6.1 we summarize criteria for distinguishing first- and second-order transitions in a finite volume. The indicated volume dependence should be understood as the leading term in a large-volume expansion. Similarly the number 2/3 indicates a limit value, which is approached as $\beta \to \beta_c$. Recall that $\alpha/\nu \leq d$ and $\gamma/\nu \leq d$ for a second-order transition. Thus a typical test in a Monte Carlo simulation could be a calculation of $c_{max}(L^d)/L^d$. If this ratio as a function of L^d goes to a nonvanishing constant for large values of L, a first-order transition is signaled; if it approaches zero, the transition must be of second or higher order.

Let us take the size of the latent heat with respect to the internal energy as a measure for the strength of a first-order transition. Depending on this strength the criteria listed in Table 2.6.1 may be more or less practicable, in particular, if a weakly first-order transition shall be distinguished from a second-order transition. A suitable ground for testing these criteria with numerical methods are q-state Potts models. Potts models satisfy the ansatz (2.6.14) for the partition function that was the starting point of our finite-size scaling analysis. The Hamiltonian of a D-dimensional q-state Potts model is given as

$$\mathcal{H} = - \sum_{<ij>} \delta_{\sigma_i \sigma_j} \, . \tag{2.6.74}$$

The spin variables σ_i are associated with sites i of a D-dimensional hyper-cubic lattice. Each of them can take q different integer values. The sum extends over nearest neighbor pairs $< ij >$. The symmetry group leaving $\mathcal{H}$ invariant is the permutation group of q elements. Later we shall see that the $q = 3$-Potts model in three dimensions plays a distinguished role in QCD. It shares the global $Z(3)$-symmetry with the pure $SU(3)$ gauge theory. The restoration of the spontaneously broken $Z(3)$-symmetry from high to low temperatures is supposed to drive the deconfinement transition in the pure gauge theory. Beyond this application to QCD, other special cases are of particular interest. The strength of the first-order transition varies as a function of q. The $q = 5$-model in two dimensions, for example, has a weakly first-order transition accompanied by a tiny latent heat of $\Delta Q = E_d - E_o = 0.05292$ in lattice units and a large correlation length of 2000 in lattice units at the transition point [72]. In contrast, the transition in the $q = 10$-model is strongly first-order, $\Delta Q = 0.69605$ and $\xi \simeq 6$, both quantities again measured in lattice units [73]. Obviously the transition in the $q = 5$ Potts model is easily misinterpreted as being of second-order. The linear lattice size has to exceed the large value of $\xi \simeq 2000$.

Deviations from predictions according to the criteria of Table 2.6.1 may have different reasons: either q or L, the linear size of the system, are not sufficiently large, since (2.6.14) was derived in the large-q limit of Potts models, and for L out of the asymptotic regime.

Let us summarize so far. Computer simulations are always performed in a finite volume, whereas the order of a phase transition is usually formulated in the infinite-volume limit. Although this formulation is very convenient, it is not the only appropriate way to proceed. The good news comes from finite-size scaling analysis. From the information obtained in a finite volume one can predict whether the transition is going to be of first or second order in the thermodynamic limit. (Likewise a small mass can play the role of a scaling field. A finite-mass scaling analysis is suited for an extrapolation to the zero-mass limit as we shall see later.) From a practical point of view, a finite-size scaling analysis may be less useful, since it is derived for finite but large volumes. The original hope was to disentangle first- and second-order signatures even for moderate lattice sizes by using refined criteria. However, as it turns out, one has to have such large volumes to verify predictions of a finite-size scaling analysis that the naive criteria are equally applicable for inferring the order of the transition. This statement refers to weakly first-order transitions. Otherwise the asymptotic region of large volumes would be easily realized as $\xi/L \ll 1$ for moderate values of L when ξ is sufficiently small.

Typical volumes in Monte Carlo calculations for QCD are marginal in large lattice size. For QCD transitions it seems difficult ever to get into the asymptotic regime of large volumes. Right from the beginning one is restricted to much smaller volumes when simulating full QCD rather than spin models. The final efficiency of refined criteria is, however, a question of size, which has to be answered in the concrete model of interest. The value of the correlation length in a first-order transition depends on details of the dynamics. For QCD the largest correlation length is likely not to be small compared to the typical lattice size.

This concludes our overview of the concepts and results of statistical mechanics. As we turn now to applications in QCD and the electroweak standard model, we close this section with a dictionary, Table 1, in the end of this section. The dictionary gives the correspondence between thermodynamic quantities in statistical physics and their counterparts in QCD and the $SU(2)$ Higgs model as an effective description of the electroweak part of the standard model.

The second and third columns of Table 1 refer to transitions in liquid/gas or ferromagnetic systems. Two analogies may be seen between the magnet and the fluid systems. The first one is between the sets (T, p, V) of a fluid and $(T, H, -M)$ in a magnet. The second one is between (T, ρ, μ) in a fluid and (T, M, H) in a magnet. The external field H is the thermodynamic variable conjugate to the order parameter M. Likewise the chemical potential is conjugate to the density ρ, where ρ is the order parameter for a fluid/gas transition. The replacement of V by $-M$ and p by H transforms almost all equations for a fluid/gas system into the corresponding equations for a magnet [74].

In the last two rows various response functions are listed, describing the response of the system to a stimulus in the temperature or in an external field. They are second derivatives of a thermodynamic potential Ω, often chosen as the free energy F (where the subscripts in the Table refer to the various systems considered there), while the order parameter is obtained as a first derivative of Ω with respect to the conjugate field. The last column shows the associated critical exponents characterizing the singular behavior of thermodynamic functions in a second-order transition.

The QCD transitions refer to limiting cases of vanishing quark masses (chiral symmetry) or infinite quark masses ($Z(3)$-symmetry). If one is interested in the evolution of a QCD plasma fluid, T, p, V is an appropriate set of thermodynamic variables that enters the equation of state. A further column is devoted to the electroweak transition in the $SU(2)$ Higgs model.

Formally the current quark masses m_q will be shown to play the same role as external magnetic fields $h_i, i = 1, 2, 3$ in a ferromagnet. One may make use of this analogy to guess, from results in the massless or pure-gauge limits, the effect of finite quark masses on the order of the QCD transitions. There is no similar analogy on the classical level for the electroweak transition.

The most popular order parameters in QCD are the quark condensate $< \bar{q}q >$ with $q = u, d$, the up and down quark masses, respectively, (or both, $< \bar{q}q >$ and $< \bar{s}s >$ with $< \bar{s}s >$ the strange-quark condensate) for the chiral transition, and the expectation value of the Polyakov loop $< L >$ for the deconfinement transition. For the electroweak transition one choice of the order parameter is the famous vacuum-expectation value of the Higgs field $< \Phi >$, or rather than that $tr\Phi^\dagger\Phi$ (as displayed in the Table), which can be used also without gauge fixing.

The indices s and b at T_c refer to the approach of T_c from the symmetric (s) and the broken (b) phase. For the deconfinement transition, the phase of broken Z(3)-symmetry is realized *above* T_c, thus the —-sign here means from temperatures above T_c. The singular behavior of the "magnetization", i.e. the order parameter, is characterized by a thermal exponent β as $T \to T_c^b$ and by a "magnetic" exponent $1/\delta$ as the generic external field $H \to 0$ *at* T_c.

Table 1: Correspondencies between liquids, ferromagnets, QCD in the limiting cases of chiral and $Z(3)$-symmetries and the SU(2) Higgs model (as replacement for the electroweak standard model).

	Liquid	Ferromagnet	QCD, $m_q = 0$	QCD, $m_q \gg T_c$	SU(2) Higgs at m_H^c	critical behaviour
a. Basic quantities						
Set of thermodynamic variables	T, p, V	T, H, M	T, p, V or $T, m_q, <\bar{q}q>$	T, p, V or $T, m_q, <L>$	T, p, V or $T_{gen}, H_{gen}, M_{gen}$	
Phase transition from	liquid to gas	broken to symm. phase	chirally broken to chirally symm. phase	deconfinement to confinement	SU(2)×U(1) broken to symmetric phase	
Possible choice of order parameter	ρ	M	$<\bar{q}q>$	$<L>$	$<\text{tr}\,\Phi^\dagger\Phi>$	$\underset{T\to T_c^b}{\propto}(T-T_c)^\beta$
conjugate field H_{gen}	μ	H	$m_u, m_d, \ldots$	$m_u^{-1}, m_d^{-1}, \ldots$	H_{gen}	
b. Response to a stimulus in T						
specific heat c_{gen}	$-T\left(\frac{\partial^2 F_{liq}}{\partial T^2}\right)$	$-T\left(\frac{\partial^2 F_{mag}}{\partial T^2}\right)$	$-T\left(\frac{\partial^2 F_{ch}}{\partial T^2}\right)$	$-T\left(\frac{\partial^2 F_{gen}}{\partial T^2}\right)$	$-T\left(\frac{\partial^2 F_{SU(2)}}{\partial T^2}\right)$	$c_{gen} \sim (T-T_c)^{-\alpha}$
2-point correlation $G_{gen}(x,0) = <\sigma_{gen}(x)\sigma_{gen}(0)>_c$	$<\rho(x)\rho(0)>_c$	$<\sigma(x)\sigma(0)>_c$	$<\bar{q}(x)q(x)\bar{q}(0)q(0)>_c$	$<L(x)L^+(0)>_c$	$<\text{tr}\,\Phi(x)^\dagger\Phi(x)$ $\text{tr}\,\Phi(0)^\dagger\Phi(0)>_c$	$<\sigma_{gen}(x)\sigma_{gen}(0)>_c$ $\sim |x|^{-(D-2+\eta)}$
correlation length ξ_{gen}	$-\ln G_{gen}(r) \to r/\xi$ as $r\to\infty$					$\xi_{gen}\sim|T-T_c|^{-\nu}$ $T\to T_c^s$
c. Response to a stimulus in H_{gen}						
order parameter M_{gen} at $T=T_c$	$\rho\sim\mu^{1/\delta}$	$M\sim H^{1/\delta}$	$<\bar{q}q>\sim m_q^{1/\delta}$	$<L>\sim m_q^{-1/\delta}$	$<\text{tr}\,\Phi^\dagger\Phi>\sim H_{gen}^{1/\delta}$	$M_{gen\,(T=T_c)}\overset{\sim}{}H_{gen}^{1/\delta}$
susceptibility χ_{gen}	$-\frac{1}{V}\left(\frac{\partial V}{\partial p}\right)_{T,S}$	$\frac{1}{V}\left(\frac{\partial M}{\partial H}\right)_{T,S}$	$\left(\frac{\partial <\bar{q}q>}{\partial m_q}\right)_{T,S}$	$\left(\frac{\partial <L>}{\partial m_q^{-1}}\right)_{T,S}$	$\left(\frac{\partial <\text{tr}\,\Phi^\dagger\Phi>}{\partial H_{gen}}\right)_{T,S}$	$\chi_{gen}\sim(T-T_c)^{-\gamma}$

Chapter 3

Field Theoretical Framework for Models in Particle Physics

3.1 The standard model in limiting cases I: Spin models as a guideline for the phase structure of QCD

In a continuum notation the QCD partition function at zero temperature in four dimensions is given by

$$Z_{Mink} = \int \mathcal{D}A\mathcal{D}\psi\mathcal{D}\bar{\psi}\mathcal{D}c\mathcal{D}\bar{c} \; e^{i\int d^4x L_{Mink}(x)}, \qquad (3.1.1)$$

where the QCD-Lagrangian density is given as

$$L_{Mink} = L_g + L_m + L_{gf} + L_{gh}$$

with

$$L_g = -\frac{1}{4g^2} \sum_{\mu,\nu=0}^{3} \sum_{a=1}^{N_c^2-1} F_{\mu\nu}^a(x) F^{a\mu\nu}(x)$$

$$L_m = \bar{\psi}(x)(i\sum_{\mu}\gamma^\mu D_\mu - M)\psi(x)$$

and

$$F_{\mu\nu}^a(x) \equiv \partial_\mu A_\nu^a(x) - \partial_\nu A_\mu^a(x) + \sum_{b,c} f^{abc} A_\mu^b(x) A_\nu^c(x)$$

$$D_\mu \equiv \partial_\mu - i \sum_a \frac{\lambda_a}{2} A_\mu^a(x). \qquad (3.1.2)$$

Here A_μ^a are the gauge fields, $\psi_{\alpha,f,c}$, $\bar{\psi}_{\alpha,f,c}$ denote the quark fields, where α is a Dirac index, $f = 1,\ldots,N_f$ labels the flavors, $c = 1,\ldots,N_c$ labels the colors (N_f is the number of flavors and N_c the number of colors). c and $\bar{c}$ are ghost and antighost fields, respectively. The gauge-coupling constant is denoted as g. At this stage we do not distinguish between bare and renormalized coupling constants. The structure constants are denoted as f_{abc}, and λ_a are the hermitian generators of the Lie algebra (known as Gell-Mann matrices) of the fundamental representation of $SU(N_c)$, $a = 1 \ldots 8$ for $N_c = 3$. The quark mass matrix is denoted as M. The gauge fixing part L_{gf} and the ghost part L_{gh} are not further specified here.

Since we will usually use the framework of Euclidean quantum field theory, we also denote the corresponding QCD action in Euclidean space. Throughout the book we choose the sign convention of the Euclidean action such that the partition function reads

$$Z_{Eucl} = \int \mathcal{D}A\mathcal{D}\psi\mathcal{D}\overline{\psi}\mathcal{D}c\mathcal{D}\overline{c} \, e^{-S_{Eucl}}, \tag{3.1.3}$$

where the Euclidean action is given by

$$S_{Eucl} = \int d^4x \, L_{Eucl}(x) \tag{3.1.4}$$

with

$$L_{Eucl} = -\frac{1}{2g^2} \sum_{\mu,\nu=0}^{3} \operatorname{tr} F_{\mu\nu}(x) F_{\mu\nu}(x) + \overline{\psi}(x) \, (D + M) \, \psi(x), \tag{3.1.5}$$

where

$$F_{\mu\nu}(x) = \partial_\mu A_\nu(x) - \partial_\nu A_\mu(x) + [A_\mu(x), A_\nu(x)],$$

$$D = \sum_{\mu=0}^{3} \gamma_\mu \left(\partial_\mu + A_\mu \right). \tag{3.1.6}$$

The Lie-algebra $(su(N_c))$-valued gauge fields $A_\mu(x)$ are chosen as anti-hermitian. In Euclidean space, we choose the usual convention for the γ matrices according to

$$\{\gamma_\mu, \gamma_\nu\} = 2\delta_{\mu\nu} \cdot \mathbf{1}_s, \tag{3.1.7}$$

where $\mathbf{1}_s$ denotes the unit matrix in spinor space.

As a first step in the investigation of the phase structure one may consider certain limiting cases of Eq. (3.1.4). One limit is the pure gauge theory. This is the limit of infinitely heavy quark masses and will be discussed below. The topic of the following section is the limit of massless quarks, which is called the chiral limit. In the chiral limit the QCD Lagrangian is invariant under global $U(1)_V \times SU(N_f)_L \times U(1)_A \times SU(N_f)_R$

transformations. The $U(1)_V$-invariance corresponds to the baryon-number conservation. The invariance under axial $U(1)_A$-transformations is only classically preserved. On the quantum level it is broken via the axial anomaly. Even in the presence of an anomaly there remains an axial $Z_A(N_f)$ symmetry (Callan et al., 1976). At zero temperature the invariance under $Z_A(N_f) \times SU(N_f)_L \times SU(N_f)_R$ chiral transformations is assumed to be spontaneously broken by the QCD vacuum to the vector $SU(N_f)_V$-symmetry. For $N_f = 2$ this isospin symmetry is realized in the hadronic spectrum to a very good approximation, for $N_f = 3$ the realization of the $SU(3)_V$-symmetry is a little more questionable, nevertheless it is also frequently considered as approximate symmetry of QCD.

3.1.1 *Renormalization-group analysis in the chiral limit*

Rather than directly studying the QCD Lagrangian (3.1.2) one can analyze chiral symmetry breaking in an effective Lagrangian which shares the chiral symmetry properties of QCD. The Lagrangian is formulated in terms of a self-interacting $N_f \times N_f$ matrix field Φ. Chiral symmetry breaking is parameterized in terms of Φ_{ij} according to $\Phi_{ij} \sim \langle \bar{q}_i(1 + \gamma_5)q_j \rangle$, where Φ transforms under transformations of $G_f \equiv U_A(1) \times SU(N_f)_L \times SU(N_f)_R$ according to

$$\Phi \rightarrow \Phi' = \exp(i\alpha)\, U_+ \Phi U_- \,. \tag{3.1.8}$$

Here U_+ and U_- are arbitrary and independent $SU(N_f)$ matrices, α generates a $U_A(1)$ transformation. The expectation value $< \cdots >$ that defines Φ_{ij} stands for the normalized path integration over all gluonic and quark degrees of freedom under the constraint that Φ_{ij} equals to the quark condensate. More precisely, the partition function of QCD, Z_{QCD} is rewritten according to

$$\begin{aligned} Z_{QCD} &= \int \mathcal{D}A_\mu \, \mathcal{D}q \, \mathcal{D}\bar{q} \, \exp\left[-S(A_\mu, \bar{q}, q) \right] \\ &\equiv \int \mathrm{d}\Phi \, Z(\Phi) \equiv \int \mathrm{d}\Phi \, e^{-S_{eff}(\Phi)} \end{aligned} \tag{3.1.9}$$

with

$$\begin{aligned} Z(\Phi) &= \int \mathcal{D}A_\mu \, \mathcal{D}q \, \mathcal{D}\bar{q} \prod_{x,ij} \delta\left[\Phi_{ij}(x) - \bar{q}_i(x)(1 + \gamma_5)q_j(x) \right] \\ &\quad \cdot \exp\left\{ -S(A_\mu, \bar{q}, q) \right\} \end{aligned} \tag{3.1.10}$$

and $< O > (\Phi)$, in particular the above Φ_{ij}, is defined by

$$< O > (\Phi) \equiv \frac{1}{Z(\Phi)} \int \mathcal{D}A_\mu \, \mathcal{D}q \, \mathcal{D}\bar{q} \, O \prod_{x,ij} \delta \left[\Phi_{ij}(x) - \bar{q}_i(x)(1 + \gamma_5)q_j(x) \right]$$

$$\cdot \exp \left\{ -S(A_\mu, \bar{q}, q) \right\} \tag{3.1.11}$$

for an observable O. The flavor indices i, j run from $1 \cdots N_f$, color indices are suppressed, x labels the lattice sites if we choose a lattice regularization of the path integral, although at this stage the path integrals should be understood as a symbolic notation that will be defined later. In principle the effective action $S_{eff}(\Phi)$ in terms of scalar (mesonic) degrees of freedom should be derived from QCD according to (3.1.10). This is still a challenge for future work, because a rigorous derivation within the framework of the renormalization group is quite demanding.

The postulate of chiral symmetry is, however, quite restrictive concerning the possible *form* of the effective action for finite temperature QCD in four dimensions. A candidate for such an effective action is an $SU(N_f) \times SU(N_f)$-symmetric linear sigma model in three dimensions at zero temperature. Since the phase structure will be investigated by means of the ϵ-expansion about $\epsilon = 0$ in $d = 4 - \epsilon$ dimensions and $\epsilon = 1$ will be set in the very end, one has to start with the most general renormalizable Lagrangian density in $D = 4$ dimensions that is consistent with the chiral symmetry of QCD. In terms of scalar fields Φ this is the $SU(N_f) \times SU(N_f)$-symmetric linear sigma model in $D = 4$ dimensions. In Euclidean space it is given as

$$L = \sum_\mu \frac{1}{2} \operatorname{tr}(\partial_\mu \Phi^+(x))(\partial_\mu \Phi(x))$$

$$+ \frac{m^2}{2} \operatorname{tr} \Phi^+(x)\Phi(x) + \frac{\pi^2}{3} f_1 \left[\operatorname{tr} \Phi^+(x)\Phi(x) \right]^2$$

$$+ \frac{\pi^2}{3} f_2 \operatorname{tr} \left[\Phi^+(x)\Phi(x) \right]^2 + g(\det \Phi(x) + \det \Phi^+(x)) \tag{3.1.12}$$

(if we write e^{-S} for the Boltzmann factor with action $S = \int d^D x \, L$.) As a sufficient condition for stability at large values of Φ, f_2 and $(f_1 + f_2/N_f)$ have to be larger than zero for $N_f \leq 3$ "flavors". This is easily checked from the Lagrangian. For $N_f > 3$, because of the det-term, the stability analysis must be considered separately. The determinant-term accounts for the anomaly. It vanishes in the pure gauge case ($N_f = 0$) and in the limit of infinite colors $N_c = \infty$ (Witten, 1979). At zero temperature

Table 3.1.1   Conjectures about the order of the chiral transition as a function of the number of flavors(N_f) and the anomaly strength g.

N_f	$N_c = \infty$ $g = 0$	$g = \mathcal{O}(1)$ const	$g = g(T)$ $g(T) \propto d_{Inst}(T)$
1	second-order $\mathcal{O}(2)$ exponents	no transition	no transition
2	first-order	second-order $O(4)$ exponents	first-order
3	first-order	first-order	first-order
≥ 4	first-order	first-order	first-order

the vacuum expectation value $\langle\Phi\rangle$ is $SU(N_f)$-symmetric and different from zero, $m^2 < 0$. The spontaneous symmetry breaking is associated with an $SU(N_f)$ massless multiplet of Goldstone bosons and a massive flavor singlet called η'. Furthermore $N_f \leq 4$ is required for the applicability of the renormalization-group analysis within the framework of the ϵ-expansion.

In Table 3.1.1 we summarize results of a renormalization-group analysis of Pisarski and Wilczek (1984) about the order of chiral phase transitions. These transitions are driven by chiral-symmetry restoration, as the temperature is raised.

For each number of flavors three cases are distinguished: a vanishing anomaly corresponding to a vanishing number of flavors or an infinite number of colors, g=const of order 1, where 'const' refers to the assumed temperature independence of g, and $g = g(T)$. Here $g(T)$ is taken to be approximately equal to the instanton density d_{inst}, which is supposed to vanish at high temperatures. In the following we explain, how the conjectures of Table 3.1.2 arise in a renormalization-group analysis.

In section (2.4) we discussed the real-space renormalization-group approach to critical phenomena, in particular the tool of block-spin transformations. An alternative approach is the ϵ-expansion that is performed in momentum space. Rather than integrating over short length scales in the real-space approach, one integrates over large momenta. Normally a change in the scale of momenta induces intricate changes in the action and in derived quantities like correlation functions, with certain exceptions. Simple scaling behavior is recovered if the set of couplings reaches values such that any further change in the scale does not affect them. This is the point in coupling parameter space where the action approaches a fixed-point action. Since we are interested in the question whether the Lagrangian (3.1.12) admits a second- or first-order phase transition depending on the actual values of the couplings, we take the action of (3.1.12) as a candidate for a fixed-point action. Concerning the flow of couplings in infrared direction, the Lagrangian (3.1.12) contains only relevant terms. Irrelevant terms such as ϕ^6-terms are dropped from the beginning, because they would not influence the universality class in case of a second-order phase transition, and they are nonrenormalizable in the ultraviolet. (ϕ^6-terms are renormalizable in three dimensions, but nonrenormalizable in four dimensions, and $D = 4$ is the dimension that is the expansion point of the ϵ-expansion.) Therefore, up to irrelevant terms, a change in momentum scale will lead to an action of the same form as in (3.1.12), but with two running four-point couplings. What remains to be done is to calculate this actual change under a change of scale and to see whether there is a special set, so called fixed-point couplings, that is invariant under a change of momenta. In the present framework these fixed points are determined as zeros of the β-functions. β-functions give the change in the renormalized couplings under a change in momentum scale.

3.1.1.1   *Renormalization-group equation, scaling behavior of the Green functions and zeros of the β-function.*

At this point we go into more detail and recapitulate the derivation of the renormalization-group equations for Green functions, the generic form of their solutions and the relation between the scaling behavior of the Green functions and the zeros of the β-functions, although, by now, the derivation is quite standard and can be found in many textbooks. To simplify the combinatorics we calculate the β-function for an $O(N)$- symmetric scalar field theory rather than for the Lagrangian (3.1.12). The calculation is a double perturbative expansion in g and in $\epsilon = 4 - D$ as small expansion parameters. The coefficients of the β-functions depend analytically on the dimension D or on ϵ, whereas Green functions are only meromorphic functions of D or ϵ. The action of a critical $O(N)$-symmetric scalar field theory in terms of renormalized parameters in D dimensions is given by

$$S_R(\Phi) = \int d^D x \left\{ \frac{1}{2} (\partial_\mu \Phi(x))^2 + \frac{\mu^\epsilon g}{4!} \Phi(x)^2 + CT \right\} \qquad (3.1.13)$$

with counterterms

$$CT = \frac{1}{2}\left(Z_R - 1\right)(\partial_\mu \Phi)^2 + \frac{1}{2}\delta m^2 Z_R \, Phi^2$$
$$+ \frac{1}{4!}\mu^\epsilon g \left(Z_g Z_R^2 - 1\right)(\Phi(x)^2)^2 \qquad (3.1.14)$$

with wave-function renormalization constant Z_R and coupling- renormalization constant Z_g; g is the renormalized coupling. Because of the assumed criticality we have set the renormalized mass $m_R = 0$ from the beginning. Therefore the bare mass m_0 equals δm, the mass counter-term, and μ is an arbitrary normalization scale. The counter-term constants δm, Z_g, Z_R are fixed by the normalization conditions for given renormalized coupling g and $m_R^2 = 0$, namely

$$\widetilde{\Gamma}^{(2)}_{R11}(p = 0) = 0, \qquad (3.1.15)$$

the zero-mass condition, and

$$\frac{\partial^2}{\partial p^2}\widetilde{\Gamma}^{(2)}_{R11}(p)\bigg|_{p^2=\mu^2} = -1, \qquad (3.1.16a)$$

$$\widetilde{\Gamma}^{(4)}_{R1111}(p_1, p_2, p_3, p_4 = -(p_1 + p_2 + p_3))\bigg|_{sym} = -\mu^\epsilon g. \qquad (3.1.16b)$$

The index sym stands for the symmetric point, defined via

$$p_i p_j = \frac{1}{3}\mu^2(4\delta_{ij} - 1), \quad i,j = 1,\ldots,4. \qquad (3.1.17)$$

In this case μ is the normalization point, where the renormalized couplings are defined. The bare couplings are independent of the choice of μ. The relation between the bare (Γ_0) and the renormalized (Γ_R) n-point vertex functions is given as

$$\widetilde{\Gamma}^{(n)}_R(p; g, \mu, \Lambda) = Z_R^{n/2} \cdot \widetilde{\Gamma}^{(n)}_0(p; g_0, \delta m, \Lambda). \qquad (3.1.18)$$

Here and in the following we write p for $p_1, \cdots, p_{n-1}, p_n = -\sum_{i=1}^{n-1} p_i$. Since the bare parameters g_0 and m_0 are independent of the choice of μ, the renormalization-group equation is derived from the fact that

$$0 = \mu \left. \frac{\mathrm{d}}{\mathrm{d}\mu} \widetilde{\Gamma}_0^{(n)} \right|_{g_0,m_0,\Lambda}$$

$$= \mu \left. \frac{\mathrm{d}}{\mathrm{d}\mu} \left(Z_R^{-n/2} \widetilde{\Gamma}_R^{(n)}(p; g, \mu, \Lambda) \right) \right|_{g_0,m_0,\Lambda}$$

$$= \left[\mu \frac{\partial}{\partial \mu} + W\left(g, \frac{\mu}{\Lambda}\right) \frac{\partial}{\partial g} - \frac{n}{2} \eta\left(g, \frac{\mu}{\Lambda}\right) \right] \widetilde{\Gamma}_R^{(n)}(p; g, \mu, \Lambda) \quad (3.1.19)$$

with

$$W\left(g, \frac{\mu}{\Lambda}\right) \equiv \left. \mu \frac{\mathrm{d}}{\mathrm{d}\mu} g \right|_{g_0,m_0,\Lambda} \quad (3.1.20a)$$

$$\eta\left(g, \frac{\mu}{\Lambda}\right) = \left. \mu \frac{\mathrm{d}}{\mathrm{d}\mu} \ln Z_R \right|_{g_0,m_0,\Lambda} . \quad (3.1.20b)$$

We called the β-function W rather than β to later avoid confusion with the critical exponent, or the inverse temperature or the inverse gauge coupling, all of them termed β as well. W and the anomalous dimension η are dimensionless functions. This is the reason why they can only depend on the ratio of μ and the cutoff Λ. If the theory is renormalizable (as (3.1.12) is in $D = 4$ for $N_f \leq 4$ and superrenormalizable in $D = 3$), the infinite-cutoff limit

$$\lim_{\Lambda \to \infty} \left. \widetilde{\Gamma}_R^{(n)}(p; g, \mu, \Lambda) \right|_{g,\mu} \equiv \widetilde{\Gamma}_R^{(n)}(p; g, \mu) \quad (3.1.21)$$

exists to all orders in the renormalized coupling constant g. Therefore, in this limit, W and η become independent of μ/Λ

$$\lim_{\Lambda \to \infty} W\left(g, \frac{\mu}{\Lambda}\right) \equiv W(g)$$

$$\lim_{\Lambda \to \infty} \eta\left(g, \frac{\mu}{\Lambda}\right) \equiv \eta(g) , \quad (3.1.22)$$

and the renormalization group equation takes the form

$$\left[\mu \frac{\partial}{\partial \mu} + W(g) \frac{\partial}{\partial g} - \frac{n}{2} \eta(g) \right] \widetilde{\Gamma}_R^{(n)}(p; g, \mu) = 0 . \quad (3.1.23)$$

Equation (3.1.23) becomes homogeneous in the first derivatives if we make the ansatz

$$\widetilde{\Gamma}_R^{(n)}(p; g, \mu) = \exp\left\{ \frac{n}{2} \int_{\bar{g}}^{g} \frac{\eta(g')}{W(g')} \, \mathrm{d}g' \right\} \Psi_R^{(n)}(p; g, \mu) \quad (3.1.24)$$

with constant $\bar{g}$, leading to

$$\left(\mu\frac{\partial}{\partial\mu} + W(g)\frac{\partial}{\partial g}\right)\Psi_R^{(n)} = 0.\tag{3.1.25}$$

Therefore we can write

$$\Psi_R^{(n)}(p;g,\mu) \equiv F_R^{(n)}\left(p;\ln\mu - \int_{\hat{g}}^{g}\frac{dg'}{W(g')}\right),\tag{3.1.26}$$

or,

$$\widetilde{\Gamma}_R^{(n)}(p;g,\mu) = \exp\left\{\frac{n}{2}\int_{\bar{g}}^{g}\frac{\eta(g')}{W(g')}dg'\right\}F_R^{(n)}\left(p;\ln\mu - \int_{\hat{g}}^{g}\frac{dg'}{W(g')}\right).$$
$$\tag{3.1.27}$$

Since the renormalized n-point functions are homogeneous functions of degree $(D+n-\frac{Dn}{2})$ in p and μ, a change in the scale of momenta by a factor ρ leads to

$$\widetilde{\Gamma}_R^{(n)}(\rho\,p;g,\rho\cdot\mu) = \rho^{D+n-Dn/2}\,\widetilde{\Gamma}_R^{(n)}(p;g,\mu).\tag{3.1.28}$$

From (3.1.27) and (3.1.28) it follows that

$$\widetilde{\Gamma}_R^{(n)}(\rho\,p;g,\mu) = \rho^{D-(D-2)n/2}\exp\left\{\frac{n}{2}\int_{\bar{g}}^{g}\frac{\eta(g')}{W(g')}\,dg'\right\}$$
$$\cdot F_R^{(n)}\left(p;\ln(\mu/\rho) - \int_{\hat{g}}^{g}\frac{dg'}{W(g')}\right).\tag{3.1.29}$$

At this point it is convenient to define a running coupling $g(\rho)$ via

$$\ln\rho \equiv \int_{g}^{g(\rho)}\frac{dg'}{W(g')},\quad g(1)\equiv g,\tag{3.1.30}$$

leading to an equivalent definition of the β-function according to

$$W(g(\rho)) \equiv \rho\,\frac{dg(\rho)}{d\rho}.\tag{3.1.31}$$

By means of (3.1.28), (3.1.29), (3.1.30) we obtain

$$\widetilde{\Gamma}_R^{(n)}(\rho\,p;g,\mu) = \rho^{D-(D-2)n/2}\exp\left\{-\frac{n}{2}\int_{g}^{g(\rho)}\frac{\eta(g')}{W(g')}dg'\right\}$$
$$\cdot\widetilde{\Gamma}_R^{(n)}(p;g(\rho),\mu).\tag{3.1.32}$$

Equation (3.1.32) gives the generic nontrivial scaling behavior of the n-point Green functions under a change of momentum scale.

Next we investigate the infrared behavior of the Green functions of the *critical* $(m_R = 0)$-theory. The infrared limit of the Green functions is obtained as the limit $\rho \to 0$ or $\ln \rho \to -\infty$. In the limit $\rho \to 0$, for g and $g(\rho) < \infty$, $g(\rho)$ must approach a zero $g^\star$ of the β-function $W(g)$. Moreover, if we write in the vicinity of $g^\star$

$$W(g') \simeq (g' - g^\star) \cdot W'(g^\star) \qquad (3.1.33)$$

so that

$$\ln \rho = \int_g^{g(\rho)} \frac{dg'}{W'(g^\star)(g' - g^\star)}, \qquad (3.1.34)$$

we have for $g < g(\rho) < g^\star$ that $g' - g^\star < 0$, and only for $W'(g^\star) > 0$ the overall contribution of the integral is negative. For $g > g(\rho) > g^\star$, $g' - g^\star > 0$, but again the overall sign of the integral is only negative if $W'(g^\star) > 0$. Therefore, in this framework, the sufficient conditions for the occurrence of an IR-stable fixed point $g^\star$ are

$$\lim_{\rho \to 0} g(\rho) = g^\star \qquad (3.1.35)$$

with

$$W(g^\star) = 0, \qquad (3.1.36)$$

and

$$W'(g)|_{g^\star} > 0. \qquad (3.1.37)$$

Correspondingly, the sufficient conditions for the occurrence of an UV-stable fixed point are

$$\lim_{\rho \to \infty} g(\rho) = g^\star, \qquad (3.1.38)$$

with

$$W(g^\star) = 0 \qquad (3.1.39)$$

and

$$W'(g)|_{g^\star} < 0. \qquad (3.1.40)$$

3.1.1.2   *Computation of the β-function for an $O(N)$-symmetric scalar field theory*

Our next goal is to explicitly calculate the roots of the β-function of the $O(N)$ model (3.1.13), (3.1.14) according to (3.1.35)-(3.1.37). First we derive the relation between the bare and the renormalized parameters $g = g(\mu; g_0, m_0, \Lambda)$ from the normalization conditions (3.1.15)-(3.1.17). To one-loop order we have

$$\widetilde{\Gamma}_R^{(2)}(p) = \quad\underbrace{\qquad}\quad + \quad\bigcirc\quad + \quad\underbrace{\quad\square\quad}\quad + \mathcal{O}(g^2) \quad (3.1.41)$$

and

$$\widetilde{\Gamma}_R^{(4)}(p_1, p_2, p_3) = \quad\times\quad + \left(\,\times\!\!\bigcirc\!\!\times\quad + 2\,\text{perm}\right) \quad + \quad\times\!\!\otimes \quad + \mathcal{O}(g^3).\quad (3.1.42)$$

It follows that

$$\widetilde{\Gamma}_{R11}^{(2)}(p) = -p^2 + \frac{1}{2}(-\mu^\epsilon g)\frac{N+2}{3}\int \frac{\mathrm{d}^D k}{(2\pi)^D}\frac{1}{k^2}$$
$$-\delta m^2 - (Z_2 - 1)p^2 + \mathcal{O}(g^2)\,,$$

$$\frac{\partial\widetilde{\Gamma}_{R11}^{(2)}(p)}{\partial p^2} = -Z_R + \mathcal{O}(g^2) \quad\quad (3.1.43)$$

and

$$\widetilde{\Gamma}_{R1111}^{(4)}(p_1, p_2, p_3) = -\mu^\epsilon g + (-\mu^\epsilon g)^2 \frac{1}{2}\frac{N+8}{9}\int \frac{\mathrm{d}^D k}{(2\pi)^D}\frac{1}{k^2}$$
$$\cdot\left\{\frac{1}{(k+p_1+p_2)^2} + \frac{1}{(k+p_1+p_3)^2} + \frac{1}{(k-p_2-p_3)^2}\right\}$$
$$+(-\mu^\epsilon g)(Z_g Z_R^2 - 1) + \mathcal{O}(g^3)\,. \quad\quad (3.1.44)$$

From (3.1.16) we obtain

$$Z_R = 1 + \mathcal{O}(g^2) \quad\quad (3.1.45)$$

and from (3.1.15) and the fact that the integral in (3.1.43) vanishes in dimensional regularization that

$$\widetilde{\Gamma}_{R11}^{(2)}(p=0) = -\delta m^2 + \mathcal{O}(g^2) = 0 \quad\quad (3.1.46)$$

or

$$\delta m^2 = 0\,, \quad\quad (3.1.47)$$

that is, the bare mass vanishes to one loop order. This is an artifact of dimensional regularization and does not hold for other general UV-cutoffs. The relation between bare and renormalized n-point vertex functions (3.1.18) implies for the four-point function

$$\widetilde{\Gamma}_R^{(4)} = (1 + \mathcal{O}(g^2)) \, \widetilde{\Gamma}_0^{(4)} \,. \tag{3.1.48}$$

Therefore we can evaluate the condition (3.1.16b) with $\widetilde{\Gamma}_R^{(4)}$ on the left hand side replaced by $\widetilde{\Gamma}_0^{(4)}$ in terms of bare parameters

$$
\begin{aligned}
-\mu^\epsilon g &= \widetilde{\Gamma}_0^{(4)}(p; g_0, \delta m, \Lambda) \\
&= -g_0 + (-g_0)^2 \, \frac{1}{2} \, \frac{N+8}{9} \, \int \frac{d^D k}{(2\pi)^D} \, \frac{1}{k^2} \\
&\quad \cdot \left\{ \frac{1}{(k+p_1+p_2)^2} + \frac{1}{(k+p_1+p_3)^2} + \frac{1}{(k-p_2-p_3)^2} \right\}.
\end{aligned}
\tag{3.1.49}
$$

To calculate the integrals, we make use of the following formulas

$$\frac{1}{p^z} = \frac{1}{\Gamma(z)} \int_0^\infty dt \, e^{-pt} t^{z-1} \,; \quad \operatorname{Re} p > 0 \,, \ \operatorname{Re} z > 0 \,, \tag{3.1.50}$$

in particular for a propagator we have the Feynman parametrization

$$\frac{1}{k^2 + m^2} = \int_0^\infty dt \, e^{-t(k^2+m^2)} \,. \tag{3.1.51}$$

For a product of n propagators we need the "λ-trick"

$$\int_0^\infty dt_1 \ldots dt_n \, f(t) = \int_0^\infty dt_1 \ldots dt_n \, \delta\left(1 - \sum_{i=1}^n t_i\right) \cdot \int_0^\infty d\lambda \, \lambda^{n-1} f(\lambda t) \tag{3.1.52}$$

for appropriate function f such that the integrals exist.

Let $\operatorname{Re} a > 0$, $b \in \mathbf{C}^D$. For appropriate g let us define

$$I_D(g) = \int_{-\infty}^{+\infty} \frac{d^D k}{(2\pi)^D} \, g(k) \, e^{-a(k^2+2kb)} \,. \tag{3.1.53}$$

In particular for integer positive D we can easily check that

$$I_D(1) = \frac{1}{(4\pi a)^{D/2}} \, e^{ab^2} \,. \tag{3.1.54}$$

According to the usual rules of dimensional regularization we take the right hand side of (3.1.54) as a definition of the left hand side for noninteger D.

Let us define

$$\widetilde{I}_4(q) \equiv \int \frac{\mathrm{d}^D k}{(2\pi)^D} \frac{1}{k^2} \frac{1}{(k+q)^2} \,. \tag{3.1.55}$$

Equations (3.1.50), (3.1.52), (3.1.54) then imply

$$
\begin{aligned}
\widetilde{I}_4(q) &= \int_0^\infty \mathrm{d}t_1 \, \mathrm{d}t_2 \int \frac{\mathrm{d}^D k}{(2\pi)^D} \, e^{-t_1 k^2 - t_2 (k+q)^2} \\
&= \int_0^\infty \mathrm{d}t_1 \, \mathrm{d}t_2 \, \delta(1 - t_1 - t_2) \int_0^\infty \mathrm{d}\lambda \, \lambda \int \frac{\mathrm{d}^D k}{(2\pi)^D} e^{-\lambda [t_1 k^2 + t_2 (k+q)^2]} \\
&= \int_0^1 \mathrm{d}t \int_0^\infty \mathrm{d}\lambda \, \lambda \int \frac{\mathrm{d}^D k}{(2\pi)^D} e^{-\lambda [t k^2 + (1-t)(k+q)^2]} \\
&= \int_0^1 \mathrm{d}t \int_0^\infty \mathrm{d}\lambda \, \lambda e^{-\lambda(1-t)q^2} \frac{1}{(4\pi\lambda)^{D/2}} \, e^{\lambda(1-t)^2 q^2} \\
&= \int_0^1 \mathrm{d}t \int_0^\infty \mathrm{d}\lambda \, \frac{\lambda^{1-D/2}}{(4\pi)^{D/2}} \, e^{-\lambda q^2 (1-t)t} \\
&= \int_0^1 \frac{\mathrm{d}t}{(4\pi)^{D/2}} \frac{\Gamma(2 - D/2)}{[q^2 t(1-t)]^{2-D/2}} \,.
\end{aligned} \tag{3.1.56}
$$

With $D = 4 - \epsilon$ we finally obtain

$$
\begin{aligned}
\widetilde{I}_4(q) &= \int_0^1 \frac{\mathrm{d}t}{(4\pi)^2} \frac{2}{\epsilon} \left\{ 1 + \frac{\epsilon}{2} \left[\ln 4\pi - \ln(q^2 t(t-1)) + \Gamma'(1) \right] + \mathcal{O}(\epsilon^2) \right\} \\
&= \frac{1}{(4\pi)^2} \frac{2}{\epsilon} \left(1 + \frac{\epsilon}{2} \left[\ln 4\pi + 2 + \Gamma'(1) - \ln q^2 \right] + \mathcal{O}(\epsilon^2) \right) \,.
\end{aligned} \tag{3.1.57}
$$

Hence

$$g = \mu^{-\epsilon} g_0 \left(1 - \frac{g_0}{2} \frac{N+8}{9} \left\{ \widetilde{I}_4(p_1 + p_2) + \widetilde{I}_4(p_1 + p_3) + \widetilde{I}_4(p_2 + p_3) \right\} \right) \,. \tag{3.1.58}$$

At the symmetry point we have

$$(p_i + p_j)^2 = \frac{4}{3}\mu^2 \,. \tag{3.1.59}$$

It follows that

$$
\begin{aligned}
g = \mu^{-\epsilon} g_0 \Bigg(&1 - \frac{g_0}{2} \frac{N+8}{9} \cdot \frac{3}{(4\pi)^2} \frac{2}{\epsilon} \bigg\{ 1 + \\
&+ \frac{\epsilon}{2} \left[\ln 4\pi + 2 + \Gamma'(1) - \ln\left(\frac{4}{3}\mu^2\right) \right] + \mathcal{O}(\epsilon^2) \bigg\} \Bigg) \,.
\end{aligned} \tag{3.1.60}
$$

Now we are ready to calculate the β-function

$$W(g) \equiv \mu \frac{dg}{d\mu}\bigg|_{g_0,m_0,\Lambda}$$

$$= -\epsilon g_0 + \frac{2}{(4\pi)^2} \cdot \frac{(N+8)}{3} g_0^2 + \mathcal{O}\left(g_0^3, \epsilon g_0^2, \epsilon^2 g_0\right). \quad (3.1.61)$$

It remains to express g_0 as a function of g by inverting (3.1.60). The result is

$$W(g) = -\epsilon g + \frac{g^2}{(4\pi)^2} \cdot \frac{N+8}{3} + \mathcal{O}(g^3, \epsilon g^2, \epsilon^2 g). \quad (3.1.62)$$

First of all we see that $g^\star = 0$ is a zero of (3.1.62) with $W'(g)|_{g^\star=0} = -\epsilon < 0$, therefore for $\epsilon > 0$ $(g^\star = 0)$ is an UV-stable fixed point according to (3.1.38)-(3.1.40). Furthermore,

$$g^\star = \frac{(4\pi)^2 \cdot 3}{N+8} \cdot \epsilon \quad (3.1.63)$$

with $W'(g^\star) = \epsilon > 0$, therefore, for $\epsilon > 0$, $g^\star$ of (3.1.63) is an IR-stable fixed point according to (3.1.35)-(3.1.37). For $D = 3$ or $\epsilon = 1$ this result is known from other independent nonperturbative methods. The $O(N)$-symmetric scalar field theory in three dimensions has one nontrivial (non-Gaussian) infrared stable fixed point.

3.1.1.3   *Generalization to the $SU(N_f) \times SU(N_f)$-symmetric model*

Above we have derived the β-function for an $O(N)$-symmetric scalar field theory with one four-point coupling g. In case of the Lagrangian (3.1.12) we have two quartic couplings f_1 and f_2 and $SU(N_f) \times SU(N_f)$ as an internal symmetry. When $f_2 = 0$ for either $N_f = 2$ or $N_f \geq 3$, the chiral $SU(N_f) \times SU(N_f)$ linear-sigma model becomes the familiar $O(N)$-linear σ model that we have studied in the last section, with $N = 4$ or $N = 2 \cdot N_f^2$, respectively. Unless $f_2 = 0$, the combinatorics for calculating the internal symmetry factors become more complicated.

    In case of r $(r > 0)$ (relevant) couplings, Eqs. (3.1.35)-(3.1.36) are replaced by

$$\lim_{\rho \to 0} g_i(\rho) = g_i^\star, \quad i = 1, \ldots, r, \quad (3.1.64)$$

where

$$W_i(\{g\})|_{\{g^\star\}} = 0. \quad (3.1.65)$$

The stability criterion is now a condition on the matrix of derivatives

$$M_{il} \equiv \left.\frac{\partial W_i(\{g\})}{\partial g_l}\right|_{\{g^*\}}, \quad i, l = 1, \ldots, r \qquad (3.1.66)$$

to be positive definite, $M > 0$, i.e. all eigenvalues must be real and positive at $g_i = g_i^*$, $i \in \{1 \cdots r\}$. The sets $\{g\}$ and $\{g^*\}$ stand for the couplings close to and at the fixed point, respectively.

The β-functions for a chiral $SU(N_f) \times SU(N_f)$-linear σ model in the absence of an anomaly (case $g = 0$ in Table (3.1.1)) were derived within the same framework of an ϵ-expansion by Paterson (1981). Rather than directly calculating $W_i = \mu \frac{d}{d\mu} f_i$ for $i = 1, 2$ according to the definition (3.1.20) and (3.1.22) in the infinite cutoff limit, the result is obtained as a special case of the β-functions for the generic renormalized Lagrangian density that is quartic in the scalar fields

$$L = \frac{1}{2} \sum_{\mu,i} \partial_\mu \psi_i \partial_\mu \psi_i + \frac{\mu^\epsilon}{4!} \sum_{ijkl} g_{ijkl} \psi_i \psi_j \psi_k \psi_l + CT. \qquad (3.1.67)$$

Here $\psi_i, i = 1, \cdots, k$ denote k massless scalar fields, g_{ijkl} is a tensor of couplings and CT are the counterterms. In $D = 4 - \epsilon$ dimensions the tensor of β-functions to one loop order is given by

$$W_{ijkl} \equiv \mu \frac{dg_{ijkl}}{d\mu} = -\epsilon\, g_{ijkl} + \frac{1}{16\pi^2} \left[\sum_{p,q} g_{ijpq} g_{pqkl} + 2 \text{ perms} \right]. \qquad (3.1.68)$$

Now, by suitably specializing the coupling tensor g_{ijkl} to the linear combination (3.1.12), (3.1.12) can be recast into the form (3.1.67). Accordingly the elements of the tensor of β-functions become linear combinations of W_1 and W_2 with

$$W_1 \equiv \mu \frac{df_1}{d\mu} = -\epsilon f_1 + \frac{1}{8\pi^2} \left[\frac{1}{3}(N^2 + 4)f_1^2 + \frac{4}{3} N f_1 f_2 + f_2^2 \right]$$

$$W_2 \equiv \mu \frac{df_2}{d\mu} = -\epsilon f_2 + \frac{1}{8\pi^2} \left[2 f_1 f_2 + \frac{2}{3} N f_2^2 \right]. \qquad (3.1.69)$$

The derivation of these results can be found in the original literature.

As a next step the stability criterion (3.1.66) has to be applied to W_1 and W_2. The results are summarized in the first column of Table 3.1.1. For $0 \leq N_f < \sqrt{2}$, the IR-stable fixed point has $f_2^* = 0$ with $O(2N_f)$ critical exponents. No IR-stable fixed point occurs for $N_f > \sqrt{3}$ if f_1, f_2 are of order ϵ.

For some time it was taken for granted that the absence of an IR-stable fixed point implies a first-order phase transition [79]. The very absence or existence of an IR-stable fixed point is, however, less conclusive than originally supposed. The existence of an IR-stable fixed point does not exclude a first-order phase transition (there may be a region in coupling-parameter space, which is not attracted by that fixed point). Vice versa does the absence of an IR-stable fixed point not exclude a second-order phase transition. Instead of the fixed-point criterion precise conditions for the occurrence of a first-order phase transition have been specified by Iacobson and Amit (1981) [80]. The framework is again the renormalization-group approach, realized in the perturbative tool of the ϵ-expansion. Their prediction of first-order phase transitions applies to all multi-component ϕ^4-theories with more than one dimensionless coupling constant, i.e. Lagrangians of the form (3.1.67). As long as the quartic terms proportional to f_1 and f_2 in Eq. (3.1.12) are independent of each other (as they are e.g. for $N_f = 3$), the conditions of the criteria of Iacobson and Amit are satisfied by the linear-sigma model, and a first-order chiral transition is predicted for $N_f > 2$. Note that these criteria are still derived within a perturbative framework (f_1, f_2 both small), and with ϵ sufficiently small. Nonperturbative features are not attainable in this approach. It is quite common to extrapolate the results from four-to three-dimensional models by setting $\epsilon = 1$, since examples are known (Bak et al., 1976), where the results of the ϵ-expansion remain a good guide for $\epsilon = 1$ (although it is *a priori* not clear whether $\epsilon = 1$ may be considered as sufficiently small).

The second column of Table 3.1.1 subsums the suggestions for the order of the chiral transition in the presence of a temperature-independent strength of the anomaly, g is assumed to be of the order of the other couplings f_1 and f_2. For two flavors the det-term in Eq. (3.1.12) acts itself like a mass term. Thus it may change the order of the transition depending on the magnitude of g. For three flavors, the det-term is trilinear in the matrix elements of ϕ. A cubic term on the classical level is usually regarded as sufficient for inducing a first-order phase transition (although we know from the discussion above that a weakly first-order transition, predicted on the classical level, may be wiped out by fluctuations; hence the classical cubic term makes the first-order transition likely to occur, but does not guarantee it to happen).

For $N_f = 4$ Pisarski and Wilczek argue again with the absence of an IR-stable fixed point in favor of a first-order transition. (The $g \neq 0$ case is not covered by the analysis of Iacobson and Amit). As claimed by Paterson

(1981), the fixed point structure of the $SU(4) \times SU(4)$-linear sigma model is unchanged when a term $\propto g(\det \phi + \det \phi^+)$ is included. For $N_f > 4$ the det-term is an irrelevant operator and should not change the critical behavior, but the ϵ-expansion is no longer applicable because for $N_f > 4$ the four-dimensional theory which is the "expansion point" of the ϵ-expansion is no longer renormalizable.

The effective symmetry of the linear-sigma model (3.1.12) may change as function of temperature if the anomaly strength is temperature dependent and determined by the density of instantons $d_I(T)$. Since $d_I(T) \to 0$ as $T \to \infty$ (Gross et al., 1981), g could be small at the transition temperature compared to the $T = 0$-value. Thus the axial $U_A(1)$-symmetry would be partially restored. Predictions accounting for this partial symmetry restoration as result of decreasing g are listed in the third column of Table 3.1.1. For further details about consequences of an approximate $U_A(1)$-restoration we refer to the original references.

3.1.1.4 *Including nonzero bare masses*

The effective Lagrangian (3.1.12) can be extended to include nonzero bare (meson) masses. The most simple ansatz has the form (tr M ϕ), it is linear in the mass matrix M. Formally it acts as a background magnetic field. The formal analogy is evident if we recall that the linear sigma model can be rewritten as a multi-component ϕ^4-theory with an additional (multi-component) ϕ^3-term for $N_f = 3$ (the det-term) and now a symmetry breaking mass term, which is linear in ϕ. This way it takes the form of Landau's free energy functional Eq. (2.1.11), here for a multicomponent order-parameter field ϕ in an external field.

For later comparison we notice that the magnetic-field term is proportional to the mass and vanishes in the chiral limit. In the other extreme case of infinitely heavy quark masses, the effective magnetic field of pure gauge theories will be shown to be proportional to (e^{-m}), thus vanishing in the infinite-mass limit $(m \to \infty)$. We come back to this point at the end of the next section.

To summarize so far, conjectures about the order of the chiral transition as function of N_f are based on a perturbative renormalization-group analysis in momentum space. The analysis was performed for an effective $SU(N_f) \times SU(N_f)$ linear-sigma model sharing the chiral symmetry properties with QCD. The β-functions were derived within the framework of the ϵ-expansion. Conjectures are based on the absence of an IR-stable fixed point when a first-order transition is predicted, a criterion which should be taken with care.

From statistical physics it is known that in many cases a first-order transition remains first-order when a background field is introduced and the field is sufficiently weak. Otherwise the transition may be washed out completely. The relative size of the latent heat compared to the strength of the external field, i.e. the values of the quark masses, decides whether the $N_f = 3$-chiral transition is preserved under realistic QCD conditions or not. This question cannot be answered within a renormalization-group analysis, but only by detailed calculations in QCD. Nevertheless the renormalization-group approach may serve as a guideline and is a good place to start.

3.1.2   Limit of the pure $SU(N_c)$ gauge theory

In this section we deal with the *quenched limit* of QCD. The quenched limit is obtained as the number of flavors goes to zero or the quark masses are sent to infinity. The decoupling of heavy masses later will be discussed in connection with dimensional reduction. The theorem that predicts the decoupling of heavy fermion masses is the Ambjorn-Appelquist-Carazzone decoupling theorem. The gluonic vacuum in a background of infinitely heavy quarks may be probed by test quarks. In perturbation theory this means that virtual quark loops are suppressed.

In the quenched limit the theory has an extra global symmetry, which results from the periodicity of the gauge fields in the temperature direction. The partition function may be represented as functional integral over gauge fields, which are periodic in Euclidean time with period $1/T$. The periodicity condition arises as a consequence of the trace in the definition of the thermodynamical potential (and the fact that a principle $SU(N_c)$-fiber bundle over the torus is trivial). In a symbolic notation the partition function is given as

$$Z = \int_{A_\mu(1/T,\vec{x})=A_\mu(0,\vec{x})} \mathcal{D}A \, \exp\left\{-\frac{1}{g^2}\int_0^{(1/T)} dt \int d^3x \, \frac{1}{4}\mathrm{Tr}\sum_{\mu,\nu} F_{\mu\nu}^2\right\}. \quad (3.1.70)$$

(We still use a continuum notation for the path integral over all gauge fields A_μ with Yang Mills field strength $F_{\mu\nu}$.)
Ordinary gauge transformations $V(x) \in SU(N)$ act as

$$A_\mu'(x) = V(x)\left(A_\mu(x) - \partial_\mu\right)V(x)^{-1} \quad (3.1.71)$$

on gauge fields A_μ. The periodicity of gauge fields A'_μ is guaranteed if $V(x)$ are periodic themselves. However, as it turns out, these V are not the most general symmetry transformations that leave the measure and action invariant and keep the periodicity of the gauge fields. The transformations must be only periodic up to an element of the center of the gauge group,

$$V(0, \mathbf{x}) = c_N V(T^{-1}, \mathbf{x}) \quad \text{for all } \mathbf{x} \qquad (3.1.72)$$

with $c_N \in Z(N)$. The n-th element of $Z(N)$ is given as $\exp(2\pi i n / N)$. An explicit example for such a transformation is given by

$$V(x_0, \mathbf{x}) = \exp\left(i\frac{2\pi}{N}\nu x_0 T e_0\right), \qquad (3.1.73)$$

where

$$e_0 = \text{diag}(1, 1, \ldots, 1, -(N-1)) \qquad (3.1.74)$$

is an $N \times N$-hermitian diagonal matrix. It is easily verified that

$$V(x_0 + T^{-1}, \mathbf{x}) = \exp\left(i\frac{2\pi}{N}\nu\right) V(x_0, \mathbf{x}). \qquad (3.1.75)$$

It is also easily checked that these transformations leave all topologically trivial observables invariant. Examples are the topologically trivial Wilson loops such as space-like Wilson loops and loops appearing in the action, thus the action itself. In this sense the V act like local gauge transformations. An example for a topologically nontrivial observable is the Polyakov loop that will be discussed below. Although the Polyakov loop is a gauge invariant observable, it transforms nontrivially under the transformations V. Therefore V act locally but not globally as ordinary gauge transformations.

On the lattice, the action of this extra $Z(N)$-symmetry on the gauge fields U is given by

$$U'((\bar{x}_0, \mathbf{x}); 0) = c_N U((\bar{x}_0, \mathbf{x}); 0),$$
$$U'(x; \mu) = U(x; \mu) \quad \text{for all other links.} \qquad (3.1.76)$$

Here $\bar{x}_0$ is a fixed-time coordinate. Equation (3.1.76) replaces Eq. (3.1.71) and Eq. (3.1.73) in the continuum.

The issue of the transition from a confinement phase at low temperatures to the deconfinement phase at high temperatures can be related to the issue of whether the pure-glue vacuum of QCD is $Z(N)$-invariant like the action or not. As it turns out, the transition from the confinement- to the deconfinement phase may be explained as spontaneous breaking of this extra $Z(N)$-symmetry at finite temperature.

Qualitatively the quark-gluon plasma can be probed by a heavy test quark. The free energy of this test quark should be infinite in the confinement phase, but finite in the deconfinement phase. It may be computed as the expectation value of the Polyakov loop (We use "Polyakov loop" synonymously for "Wilson line" or "thermal Wilson loop"), cf. section (3.3.8). The Polyakov loop L is defined as the spatially local operator

$$L(\vec{x}) \equiv \text{Tr } P \exp \left(\int_{0}^{(1/T)} dt\, A_0(t, \vec{x}) \right) \qquad (3.1.77)$$

with (anti-hermitian) A_0, P stands for the path ordered product. In detail we will derive in a separate chapter (section 3.3.8) that $L(\vec{x})$ actually has the physical meaning of an isolated static quark. The Polyakov loop is a topologically nontrivial loop. It is closed due to the periodic boundary conditions and transforms nontrivially under the center transformations

$$L(\vec{x}) \longrightarrow c_N L(\vec{x}) . \qquad (3.1.78)$$

Therefore the expectation value of L

$$<L> \equiv \frac{1}{Z} \int \mathcal{D}A \; L(\vec{x}) \, e^{-S(A)} \qquad (3.1.79)$$

transforms according to

$$<L> = c_N <L>, \qquad (3.1.80)$$

where we have used (3.1.78) and the invariance of the Haar measure $\mathcal{D}A$ and the action $S(A)$ of (3.1.79) under (3.1.72). Since $c_N \neq 1$ in general, we conclude from (3.1.80) that $< L >= 0$ unless the $Z(N)$-symmetry is spontaneously broken. Therefore a nonvanishing value of $< L >$ may be taken as a signal for a spontaneous breakdown of $Z(N)$. This behavior qualifies $< L >$ as an order parameter of the confinement/deconfinement phases of QCD in the absence of dynamical quarks. We have

$$\langle L(\vec{x}) \rangle = e^{-T^{-1}F(\vec{x})} = \begin{cases} 0 & \text{confinement} \\ \neq 0 & \text{deconfinement} \end{cases} . \qquad (3.1.81)$$

With the interpretation that $F(\vec{x})$ equals the free energy of an isolated test quark, the identification of phases according to (3.1.81) is obvious.

It is natural to look for an effective action of the $SU(N)$ gauge theory in terms of the order-parameter field, this means in terms of L. Such an effective action could simplify the investigation of the phase structure of pure QCD if universality arguments may be used. As suitable candidate for such an effective action Svetitsky and Yaffe [82,83] proposed a $Z(N)$-spin theory. Their result of an integration over spatial gauge fields is an effective action in terms of Polyakov loops. The Polyakov loops (originally given in terms of elements of the $SU(N)$-Lie algebra) are replaced by $Z(N)$ variables as a reasonable approximation. By construction the effective action is invariant under $Z(N)$ transformations. The $Z(N)$-symmetry can be spontaneously broken at high temperatures, but is restored at low temperatures and ensures confinement. The action is argued to be short-ranged by the following reason. The high-temperature behavior of the original $(3 + 1)$-dimensional theory is determined by the dynamics of the three-dimensional $SU(N)$ gauge theory. Three-dimensional $SU(N)$ gauge theories show an area-law for Wilson loops [84], but with interactions that are supposed to stay short-ranged over the entire temperature interval.

The path integral representation of the original $SU(N)$ gauge theory for the partition function (3.1.70) is then replaced by

$$Z \simeq \sum_{\{\sigma(\vec{x})\}} e^{-g^{-2}(T)S_{eff}(\{\sigma(\vec{x})\})} \equiv e^{-T^{-1}Vf} . \tag{3.1.82}$$

Here $\sigma(\vec{x}) \in Z(N)$, the sum extends over all $Z(N)$-spin configurations. The free-energy density on the right hand side is supposed to be a good approximation to the free-energy density of the $SU(N)$ gauge theory, at least in reproducing the same qualitative features of the phase structure. $S_{eff}(\{\sigma(x)\})$ is the action of a $Z(N)$ spin model, its precise form will be given below. The number of states, N, equals the number of colors in the original $SU(N)$ gauge theory, $N \equiv N_c$.

In conclusion, it appears worthwhile to first study the phase structure of $Z(N)$ spin models in $d = D - 1$ dimensions rather than the phase structure of the original $SU(N)$ gauge theory in D dimensions.

3.1.2.1 *Effective versus physical temperatures*

In the original action of Eq. (3.1.70) the parameter g characterizes the interaction strength. When the time dependence of the field is dropped, the $\int_0^{(1/T)} dt$ can be performed to yield a prefactor of the remaining three-dimensional action $(-T^{-1}/g^2)$. The T^{-1}-factor can be absorbed by a rescaling of the fields such that the action in three dimensions takes the same form as the original action in four dimensions. Now the prefactor plays the role of an inverse temperature of the three-dimensional model. Thus g^2 is identified with T_{eff}, the effective temperature of a classical spin system described by the partition function of Eq. (3.1.82). For T_{eff} small, the "spin system" is in the ordered phase, we have $\langle L \rangle \neq 0$ corresponding to deconfinement. Deconfinement is realized at high physical temperatures T, for which g is small due to asymptotic freedom. A small value of g means a small value of T_{eff}, which is consistent with the initial assumption. Thus a low effective temperature T_{eff} corresponds to a high physical temperature T and vice versa. This explains why the order parameter vanishes in the low-temperature phase and signals "order" in the high-temperature phase.

The coupling g becomes manifestly temperature dependent, when high frequency contributions are integrated out in passing from the $(3 + 1)$- to the three-dimensional theory.

3.1.2.2   *The phase structure of $Z(N)$-spin models*

The phase structure of $Z(N)$-spin models has been studied in statistical physics for various values of N and space dimension $d = D - 1$. If a second order transition is predicted in the spin model, one could attempt to locate the renormalization-group fixed point and consider the simpler spin model as fixed-point theory in the universality class of the $SU(N)$-theory.

A table for various $Z(N)$-, $SU(N)$- and $U(1)$-models in dimensions $d = 2, 3$ and ≥ 4 can be found in [82,83]. Here we mention the cases of two and three colors.

The spin model, which is associated with a $(3 + 1)$-dimensional $SU(2)$ gauge theory at high temperatures, is the Ising model in three dimensions. The 3-d-Ising model is known to have a second-order phase transition.

The case of $N = N_c = 3$ in $d = 3$ is particular. In the space of three-dimensional $Z(3)$- symmetric theories no IR-stable renormalization-group fixed point is known. A specific realization of a $Z(3)$-symmetric spin theory is the three-state Potts model (cf. section 2.3), which is known to have a first-order transition. The famous cubic term on the classical level, driving the transition to first-order, is allowed by the $Z(3)$-symmetry. The potential of a $Z(3)$-symmetric theory of a single complex scalar field $L(\vec{x})$ may be written as a sum of a $U(1)$-symmetric term depending on $|L|^2$ and a term depending on Re $[L^3]$. In $d = 3$, a $U(1)$ gauge theory has a second-order transition (see e.g. Pfeuty and Toulouse, 1977). A term Re $[L(\vec{x})^3]$

explicitly breaks the $U(1)$-symmetry down to $Z(3)$. In the renormalization-group sense it is relevant enough to affect the critical behavior of the $U(1)$-theory.

Thus the conjecture of Svetitsky and Yaffe is that in the absence of an IR-stable fixed point the original $(3+1)$-dimensional $SU(3)$ gauge theory has a first-order finite temperature phase transition as well. For a discussion about how conclusive the absence of an IR stable fixed point actually is we refer to our discussion above in connection with the *chiral* phase transition of QCD. In this case, however, it turns out that the conclusion is not misleading but a good guideline. The conjecture was confirmed in early Monte Carlo calculations [86, 87], then first cast into doubt [88] and later reestablished [89]. Before we report on this controversy in more detail in section 4.7.5, let us consider the influence of dynamical quarks on the deconfinement transition.

3.1.2.3 *Influence of dynamical quarks*

From spin systems we know that often a first-order transition stays first-order if the magnetic field is sufficiently small. The gaps in thermodynamic quantities are only continuously deformed for a perturbatively small external field. The magnetization, however, ceases to be a good order parameter, as it is always finite due to the presence of an external background field.

Similarly the Polyakov loop may be expected to fail as an order parameter when dynamical quarks are included. The physical reason is easily understood. The free energy of an isolated test quark does no longer diverge ($\langle L \rangle \neq 0$ for all temperatures). When the flux tube between two test quarks is sufficiently stretched, a $q\bar{q}$-pair is popped out of the vacuum. The test quarks may always form finite-energy bound states with dynamical quarks.

Since the deconfinement transition was supposed to be driven by the spontaneous symmetry breaking of the global $Z(3)$-symmetry, we consider the transformation behavior of the QCD action under $Z(3)$-transformations in the presence of dynamical quarks. The generalized gauge transformations Eq. (3.1.72) are periodic only up to an element of the center of the gauge group. While the periodic boundary conditions on the gauge field are preserved under the generalized transformations, the anti-periodic boundary conditions on the quark fields are not. Imposing

$$\psi(\beta, \mathbf{x}) = -\psi(0, \mathbf{x}) \qquad (3.1.83)$$

on the quark fields as before, Eq. (3.1.83) transforms under a generalized gauge transformation with $V(\beta, \mathbf{x}) = c_N V(0, \mathbf{x})$ (i labeling the center elements) according to

$$\psi(\beta, \mathbf{x}) \rightarrow V(\beta, \mathbf{x})\psi(\beta, \mathbf{x}) = c_N\, V(0, \mathbf{x}\,\psi(\beta, \mathbf{x})$$
$$= -c_N\, V(0, \mathbf{x})\,\psi(0, \mathbf{x})\,. \qquad (3.1.84)$$

Equation (3.1.84) shows that the gauge-transformed fields $U \cdot \psi$ (under the generalized gauge transformation) do no longer satisfy anti-periodic boundary conditions. If we compare the path integral over all gauge field configurations in the absence and presence of dynamical quarks, satisfying Eq. (3.1.83), configurations differing by generalized gauge transformations get the same Boltzmann weight in the absence of quarks, but a different weight in their presence. Hence the effect of dynamical quarks is an explicit symmetry breaking of the $Z(N)$-symmetry. Its strength is determined by the values of the quark masses as it is easily seen on the level of an effective action, cf. the next section.

3.1.2.4   *Finite quark masses and external fields*

The analogy between an external field in a spin system and dynamical quarks in QCD becomes manifest when the quarks are integrated out. Integration over quark degrees of freedom induces an external field on the effective level of spin models [90, 91]. We refer to the work of [91] and anticipate results from lattice gauge theory.

We consider a four-dimensional $SU(3)$-lattice gauge theory with fermions at high temperatures. The original action consists of a gauge field part S_g and a fermionic part S_f. The fermionic part may be written as $S_f(U) = \sum \bar{\psi} M(U) \psi$ (dropping all indices, where U is the $SU(3)$ gauge field and M the fermion matrix, explicit expressions will be given later). When the Grassmann variables are integrated out, the effective action takes the form

$$S_{eff}(U) \;=\; S_g(U) + \mathrm{Tr}\,\ln M(U)\,. \qquad (3.1.85)$$

In the high-temperature- and strong-coupling limit of the $SU(3)$ gauge theory it can be shown that the gauge part S_g transforms into the action of a three-dimensional three-state-Potts model. (For the Potts model see section 2.3.) The fermionic part of the action (3.1.85) simplifies, when it is treated in a hopping-parameter expansion. The hopping-parameter κ may be related to the bare quark mass m of the original action according to

$$\kappa \sim \frac{1}{2} e^{-ma_0} , \qquad (3.1.86)$$

where $m \cdot a_0$ is the bare quark mass in lattice units, a_0 is the lattice spacing in time direction. (Relation (3.1.86) only holds at strong coupling and for small values of κ). The result for the fermionic term together with the simplified gauge part leads to the following effective action (for zero chemical potential)

$$S_{eff} = - \left[\frac{\beta a}{a_0} \sum_{n,\mu} \mathrm{Re}\ z_n^* z_{n+\hat{\mu}} + H(\kappa) \sum_n \mathrm{Re}\ z_n \right] . \qquad (3.1.87)$$

Here a is the lattice spacing in spatial directions. The sums go over all nearest neighbor pairs $(n, n + \hat{\mu})$, $\hat{\mu}$ is the unit vector in the positive μ-direction, and over all sites n, respectively, z_n are elements of $Z(3)$. The coupling H plays the role of the external field, which is a remnant of the dynamical fermions. The κ-dependence is explicitly known and approximately given as

$$H(\kappa) \sim 24\kappa . \qquad (3.1.88)$$

Together with Eq. (3.1.86) we see that H vanishes with κ for infinitely heavy quark masses, that is, in the limit of a pure gauge theory. The strength of the external field grows with decreasing quark mass. Hence sufficiently light quark masses will completely wash out the first-order transition in the three-dimensional three-state-Potts model. DeGrand and DeTar [91] obtain for the critical strength of the magnetic field $H_{cr} = \frac{2}{3}(\ln 2 - \frac{2}{3})$. The result was obtained in a mean-field analysis. It holds for a cubic lattice in three dimensions. At the critical field (mass) the line of first-order transitions terminates at a second-order critical point and disappears for larger fields (smaller masses).

It should be noticed that the effective action on the lattice Eq. (3.1.87) was derived (rather than postulated from generic principles), even though the derivation uses some approximations (strong-coupling and high-temperature expansions). A particular virtue of such kind of derivation is that it allows a calculation of the effective coupling in terms of parameters of the original action (here $H = H(\kappa)$). In this respect the effective Potts model differs from an effective Lagrangian of chiral perturbation theory, for example, where the Lagrangian parameters have to be fixed from an experimental input, cf.section 6.1. Here it remains to extrapolate the strong-coupling results on the lattice to continuum results (over a "long distance" in coupling-parameter space) and to show that the qualitative predictions of the effective $Z(N)$-Potts models survive the continuum limit.

3.1.2.5   *Summary*

In summary of this section, the renormalization-group analyses of Pisarski and Wilczek [77], and of Svetitsky and Yaffe [82] refer to the idealized limits of vanishing or infinite quark masses, respectively. They were performed in the $SU(N_f) \times SU(N_f)$ linear-sigma model and in $Z(N)$-spin-models rather than directly in QCD. Studies in these limiting cases reveal a dependence of the order of QCD transitions on the number of flavors and the number of colors. The case of physical interest is included for three colors and two or three light flavors. The outcome of the renormalization-group analysis is the message that one has to do hard work in the following sense. Model calculations in terms of a scalar field theory with an N-component order parameter field would suffice to model QCD, if the transition were also conjectured to be of second order in the case of three light flavors, or, more realistically, two light and one heavier flavor. One would be free to choose as simple a model as possible within the conjectured universality class. Since the deconfinement and the chiral transitions are expected to be of first order (for three colors and three massless flavors), the transition depends on details of the dynamics, in this case the dynamics of full QCD. Ultimately Monte Carlo simulations must be performed in *full QCD* for *physical* parameters. Such simulations should go along with an analysis of the finite-size scaling behavior, finite-mass scaling and finite-cutoff scaling behavior. They should be supplemented by analytical methods (cf. section 4.7.5 and 4.6, for example).

Compared to the scale of T_c, two quarks (up, down) are light, three are heavy (charm, bottom, top), but the strange quark happens to be just of the order of T_c. In the thermodynamics of QCD the quark masses play the same role as external magnetic fields in temperature-driven transitions of ferromagnets. From this analogy one must expect that the effect of finite masses depend on their actual values. Results in the idealized limits may even change qualitatively under their influence. Thus a challenge for further investigations of the phase structure of QCD is to find out which of the alternatives displayed in Fig. 3.1.1(a)-(b) is realized. Partly conjectural diagrams are shown in the (m, T)-plane, where m stands for a generic current-quark mass, and $N_f = 3 = N_c$ is assumed. For comparison, an

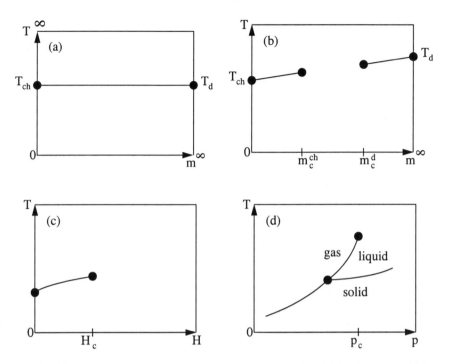

Fig. 3.1.1 Hypothetical phase diagrams of QCD in the (m, T) plane, where T is the temperature, and m stands for generic current-quark masses: (a) the transitions survive the external field and coincide, (b) both transitions are washed out for intermediate mass values; (c) a (H, T) diagram for a $Z(3)$-Potts model in three dimensions; and (d) a (p, T) diagram for a liquid/gas system. For further explanations see the text.

(H, T)-diagram for a $Z(3)$ Potts model in three dimensions and a (p, T)-diagram for a liquid/gas system are shown in Figs. 3.1.1(c)-(d). For $m = 0$, a first-order chiral transition is predicted at some temperature T_{ch}. The chiral transition continues to be of first order, but in a weakened form, as long as the mass is smaller than a critical value m_c^{ch} (corresponding to $H(m) < H_c$). For $m = \infty$ or $\kappa = 0$ the first-order deconfinement transition occurs at some temperature T_d. It will persist as long as $H(\kappa) \equiv H_d$ is smaller than some $\tilde{H}_c$ or $m > m_c^d$. Both transitions may "meet" and coincide for intermediate mass values (Fig. 3.1.1(a)) or they may not meet (Fig. 3.1.1(b)). In the latter case the discontinuities disappear completely. The dynamical quark masses are then too large for the chiral transition and too small for the deconfinement transition to persist. The chiral symmetric deconfining high-temperature region is then smoothly connected with the chiral symmetry-broken, confined, low-temperature world of daily life. To date lattice results suggest that Fig. 3.1.1(b) gives a more realistic description.

3.1.2.6   *The Columbia plot*

Although the analogy of the thermodynamics of QCD to the thermodynamics of a Potts model or a liquid/gas system is quite instructive, Figs. 3.1.1 may easily hide the complexity of the phase structure of QCD. Rather than one sort of fields (spins) we have nonabelian gauge fields and fermionic matter fields, the gauge fields described by $SU(N)$ matrices, the fermions by Grassmann variables. In addition we have two types of transitions, the chiral and the deconfinement transition with an inverted role of quark masses with respect to each other. As indicated in the discussion above, the order of the transitions depends on the number of flavors and the number of colors, and on the actual values of the quark masses, i.e. on the amount of *explicit* symmetry breaking. Therefore an analogy between one external magnetic field and "the" mass of QCD is too rough. In QCD one should distinguish at least between the masses $m_{u,d}$ of the up-and down-quarks (which both are of the same order of magnitude when compared to the scale of T_c), and the strange quark mass m_s which is of the order of T_c. From the viewpoint of statistical physics, QCD with *physical* parameters resembles more a spin-system with couplings to (at least) *two* external fields (with more than one bulk length scale).

Therefore we conclude this section with another representation of the expected phase structure of QCD. This is the so-called Columbia plot of Fig.3.1.2, first conjectured by [92,93]. Indicated are the presence or absence of the finite-temperature QCD transitions as a function of the quark masses $m_{u,d}[MeV]$ and $m_s[MeV]$. The solid lines indicate supposed second-order transitions, the shaded areas enclose mass values leading to first-order transitions. The concave/convex shape of the critical boundaries of second-order transitions is hypothetical. The dashed circle shall indicates the physical value of the ratio of strange- to up/down-quark masses. An error bar for the location of the physical mass point relative to the phase boundaries is omitted, due to the uncertainty in the location of the boundaries themselves. The mass point $(0,0)$ is the chiral limit of three flavors with an $SU(3) \times SU(3)$-symmetry in the continuum and a first-order chiral transition, as it was conjectured by the analysis of Pisarski and Wilczek [77]. (In a lattice regularization with Wilson or Kogut-Susskind fermions, the full

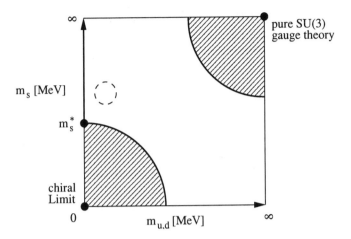

Fig. 3.1.2 Qualitative phase diagram, partly conjectural, for QCD with three colors and three flavors. Solid lines indicate supposed second-order transitions, shaded areas first-order transitions, including the $m_{u,d}$ and m_s axis that belong to the boundary. The lower left corner represents mass parameters for which the chiral transition is first-order, the upper right corner mass parameters for which the deconfinement transition is of first order. The dashed circle indicates a possible location of the physical mass point with $m_{u,d} \sim 10\,\text{MeV}$ and $m_s \sim 150\,\text{MeV}$. The solid circle at m_s^* refers to the tricritical value of the strange quark mass. On a quantitative level the extension of the shaded areas is merely conjectural as well as the concave/convex shape of the transition boundaries.

$SU(3) \times SU(3)$-symmetry is realized only in the continuum limit.) The pure gauge theory in the upper right corner of the diagram $m_{u,d} = \infty, m_s = \infty$ has an exact $Z(3)$-symmetry, both in the continuum and on the lattice for three colors. The transition is of first order as expected from the discussion of Svetisky and Yaffe [82, 83]. For sufficiently small (large, respectively) quark masses both transitions keep on being of first order and span up the shaded regions of Fig. 3.1.2. The precise extension of these regions is not known to date, because it is quite expensive to precisely locate the critical endpoints along the boundary of the first-order transition regions. The boundary lines consist of so-called critical quark masses. These are exceptional quark mass values for which the chiral (or the deconfinement) transition happens to be of second order in spite of having three quarks with finite masses.

The line $(m_{u,d} = 0, m_s)$ has an $SU(2) \times SU(2)$-(continuum) symmetry and a conjectured second-order transition above some tricritical value of the strange quark mass, where the line of first-order transition ends. The universality class of QCD may change along the second-order transition line. Several candidates for the universality class were proposed: the $O(4)$ model when $m_{u,d} = 0$ and m_s is large and a tricritical ϕ^6-model in the vicinity of the tricritical point [94,95], moreover an Ising model along the concave part when $m_{u,d} \neq 0$ [96]. As yet, the universality was identified only in single points in mass-parameter space along the phase boundary, cf. section 5.4.4. Also the location of the tricritical point on the $m_{u,d} = 0$ axis is not known to date. It would be of mere academic interest unless the physical mass point happens to be close to the tricritical point. Only in this case it would make sense to look for experimental manifestations of tricritical QCD.

In the area between the shaded regions, the confinement and deconfinement phases and the chiral-symmetric and chiral-symmetry broken phases are analytically connected. There is neither a chiral nor a deconfinement transition in the literal sense, but just a crossover between the different phases (which is more or less rapid, and hopefully rapid enough for the physical mass point to induce some remnants of the transition). As indicated in Fig. 3.1.2, it is rather likely that the physical mass point lies in this region.

In the original work of Brown et al. [92,93], the units of the axis of their plot were bare quark masses, i.e. quark masses in lattice units. The reason for using lattice units was obvious, because their plot contained results for sets of mass parameters that were actually used in numerical simulations. For example, Brown et al. found a first-order transition for $m_{u,d,s} \cdot a = 0.025$ (a the lattice constant) and in the infinite-mass limit $m_{u,d,s} \to \infty$, but no transition for $m_{u,d} \cdot a = 0.01$, $m_s \to \infty$ and $m_{u,d} \cdot a = 0.025$ and $m_s \to \infty$. Other parts of their diagram were conjectural as well. The diagram looked qualitatively the same as Fig. 3.1.2, although a precise quantitative transcription between physical units and lattice units is subtle. (The chiral limit always involves an extrapolation when it is performed on the lattice, and the physical quark-mass values depend on the scale and the renormalization scheme.) The location of the phase boundary of second-order transitions also depends on the fermionic formulation which is used on the lattice (Wilson or staggered fermions). Of course, in the continuum limit the location should become independent of such lattice artifacts, in particular the answer to the main question as to whether the physical mass point lies inside or outside the region of first-order phase transitions.

3.2 The standard model in limiting cases II: Phase transitions in the electroweak part of the standard model

In this section we review the phase structure of the electroweak standard model. We consider this model in certain limits that are supposed to contain the relevant degrees of freedom for the phase structure. The particle contents of the electroweak model with scalar fields, fermions, and gauge fields is richer than the particle structure of QCD. In particular we are interested in the role of the gauge fields. The question then arises as to whether we may replace the gauge fields by spin variables as we did in QCD and further neglect the fermions, so that the discussion would be reduced to the phase structure of certain spin models? Here the answer turns out to be negative, since the effect of the gauge fields is subtle and cannot be found on the classical level. Moreover we will focus on quantities which determine the order and the strength of the phase transition. We discuss suitable effective models from which we can guess the answer to these questions. From a phenomenological point of view one may ask why we should be interested at all in the electroweak transition, since there is no chance to reproduce this transition in nowadays lab-experiments; the needed high-energy densities are just not available in the near future. But it is the early universe, where this transition happened once. Therefore let us start with the electroweak transition in cosmology.

3.2.1 *The electroweak phase transition in cosmology*

The electroweak phase transition occurs at the order of $(\sqrt{G_F})^{-1}$ where G_F denotes the Fermi constant, $G_F = 1.16639 \times 10^{-5} GeV^{-2}$. The Fermi constant sets the scale of the electroweak interaction. At high temperatures (larger than $(\sqrt{G_F})^{-1}$) the $SU(2) \times U(1)_Y$-symmetry of this model is restored, where $U(1)_Y$ is the symmetry corresponding to the hyper-charge. At low temperatures (smaller than $(\sqrt{G_F})^{-1}$) the $SU(2) \times U(1)_Y$-symmetry is spontaneously broken via the Higgs mechanism, giving the gauge vector bosons W and $Z^{\pm}$ a finite mass, while the photon remains massless. The remaining symmetry is $U(1)_{em}$, the familiar electromagnetic $U(1)$. It then depends on the actual value of the Higgs mass (as we shall see) whether the transition between both phases actually goes along with a nonanalytic behavior in the free energy, or whether it proceeds smoothly as a crossover phenomenon. To date the experimental lower bound on the Higgs mass is $114.1 GeV$ at the 95% confidence level [97]. As numerous numerical and analytical calculations indicate, the phase conversion proceeds smoothly for such a large Higgs mass, but nevertheless, throughout this section, we speak of the electroweak phase "transition" as if there would be a true transition for all (hypothetical) Higgs masses. The pendant of the electroweak transition in the $SU(2)$ Higgs model that we have considered in section 2.3 is called the Higgs transition, sometimes also identified with the electroweak transition.

In contrast to the QCD transition, energy densities of the order of $100[GeV]$ cannot be produced in lab experiments, even not for very transient instants of time. However, the electroweak transition occurred once in the early universe, when the universe cooled down to a temperature of the order of $10^{15}K$, that was about 10^{-12}sec after the big bang. It is this very moment in the history of our universe where a gross feature of the nowadays world like the observed baryon asymmetry was fixed. The reason is that baryon-symmetry violating processes fall out of equilibrium at this temperature scale. If the asymmetry has been generated before or at the phase transition, this asymmetry gets fixed unless it is washed out by a sufficiently high sphaleron rate. Sacharov [98] formulated three necessary conditions for an explanation of the observed baryon asymmetry

- Obviously, baryon-symmetry-violating processes must occur.
- Secondly, CP-symmetry must be violated.
- Out-of-equilibrium processes must occur in the dynamical evolution.

The first two conditions are fulfilled within the minimal version of the standard model that we consider in this section. Baryon number-violating processes are not forbidden, but strongly suppressed at low temperatures. The suppression is determined by the sphaleron energy. Even the third condition could be realized in this model if the dynamics of the phase transition proceeds out of equilibrium. This possibility raised a strong interest in the order of the electroweak phase transition. Bubble nucleation, explosive phenomena or nonadiabatic phase conversion are characteristic features of a first-order transition if it is sufficiently strong. As quantitative measure for the strength of the transition serve a large latent heat and/or a large surface tension, large in units of the typical energy scale of the system, for example the transition temperature in natural units of energy.

In the following sections we describe various approaches (in the continuum in four dimensions, on the lattice in four dimensions and in three dimensions) to determine the order of the electroweak transition, more precisely its pendant in the corresponding simplified model. All these studies were motivated by the hope to explain the baryon asymmetry within the standard model.

3.2.2 Perturbative approach in four dimensions in the continuum

Naively one would expect that perturbation theory in four dimensions should be applicable for small gauge coupling g and scalar coupling λ, and also for small Higgs masses (because of the tree level relation $\lambda = m_H^2/2v^2$ with $v = 246 GeV$ the vacuum expectation value of the Higgs field) as long as one is in the broken phase. The gauge coupling stays even small in the phase transition region and above the transition in contrast to QCD where the gauge coupling is of order 1 in this region. However, even for small λ, g and small Higgs masses naive perturbation theory runs immediately into trouble. For example, if one tries to calculate the constraint effective potential V_{cep} that we have introduced in section 2.2 in order to read off the order of the phase transition, the breakdown of perturbation theory becomes manifest in imaginary tree level masses for small background fields, in spurious linear terms and in rational exponents of couplings (e.g. $\lambda^{3/2}$ if one goes to higher orders). One may try to cure these problems by resummations of classes of Feynman diagrams in analogy to pure scalar models (cf. section 4.4 and 4.5), but this procedure is only partially successful. The trouble is caused by the magnetic-mass problem of gauge theories, according to which the radiatively small gauge boson masses in the propagators prevent the loop expansion to be a systematic expansion in small couplings.

An attempt to circumvent these problems at some lower order of perturbation theory are gap equations, first formulated by Buchmüller, Helbig and Walliser [99] for scalar electrodynamics, and in detail discussed in this book for the electroweak model in section 4.5. Let us write

$$m_i^2 = m_{i,tree}^2 + \delta m_i^2 \qquad (3.2.1)$$

for the exact (physical) masses m_i, where the index i labels the various masses in the model, for example in the electroweak model the longitudinal and transverse gauge boson masses (l, t), respectively, or the Higgs and Goldstone boson masses (ϕ, χ), respectively. $m_{i,tree}$ denote the tree level masses and δm_i the mass counter-terms. The boson propagators are then used with the exact masses m_i^2 (rather than $m_{i,tree}^2$), but the radiative corrections δm_i^2 are treated as counter-terms in the action to compensate for the replacement in the propagators. The δm_i^2 must be determined self-consistently as solutions of the so-called gap equations at the corresponding loop order. In particular Buchmüller et al. [100] obtain for the transverse (magnetic) gauge boson mass the equation $m_T^2 = \frac{g^2 T}{3\pi} m_T$ at $\phi = 0$ with $m_T = \frac{g^2 T}{3\pi}$ as the only physically nontrivial solution. This value may have been expected from other estimates of the magnetic mass, but goes beyond the accuracy of the calculation which is otherwise only valid up to order g. Nevertheless let us see what Buchmüller et al. obtain for the constraint effective potential when they use this value for m_t^2 (up to an additional multiplicative factor γ to account for the overall uncertainty in calculating m_T^2 self-consistently). The authors start from the action

$$S = S_{gauge} + S_{Higgs} + S_{fermion} + S_{gaugefixing} + S_{ghost}. \quad (3.2.2)$$

Apart from the fermionic part $S_{fermion}$ all other terms are specified in detail in section 4.5, Eqs. (4.5.74)-(4.5.80). The fermionic part contains only the Yukawa coupling f_t of the top quark to the Higgs field, all other Yukawa couplings are much smaller and neglected. The connections between couplings and zero-temperature masses are given as

$$g = \frac{2m_W}{v}$$

$$f_t = \frac{\sqrt{2}m_t}{v}$$

$$\lambda = \frac{m_H^2}{2v^2}, \quad (3.2.3)$$

m_W is the physical vector boson mass of the W-boson.

The result for the constraint effective potential is then given as

$$\mathcal{V}_{\mathrm{CEP}}(\overline{\phi}) = \frac{\lambda}{4!}\overline{\phi}^4 + \left[\left(\frac{3g^2}{16} + \frac{\lambda}{12} + \frac{m_t^2}{2v^2}\right) T^2 - \frac{\lambda}{6}v^2\right]\frac{\overline{\phi}^2}{2}$$

$$- \frac{T}{12\pi}\left(m_\phi^3 + 3m_\chi^3 + 3m_L^3 + 6m_T^3\right), \quad (3.2.4)$$

where the masses are determined as solutions of the gap equations, cf. Eq. (4.5.107) in section 4.5. Also for a detailed discussion of this potential we refer to section 4.5. Here we only summarize the main features. The potential contains all terms of order g^3 or $\lambda^{3/2}$. The imaginary and the linear terms have disappeared to this order. The pure scalar part yields a second-order transition for all values of λ. The interaction with the gauge fields changes the order from second to first. As long as the transverse gauge boson mass vanishes, the transition would be of first order for all couplings $\lambda < \infty$. Inserting, however, a finite value for all plasma masses of vectors and scalars, in particular $m_T^2 = \gamma^2 \frac{g^4 T^2}{9\pi^2}$ for the magnetic mass squared, although the accuracy is inconsistent as indicated above, the first order changes to second order above some critical value of λ or m_H which depends on the uncertainty factor γ.

Furthermore an interesting relation between this critical Higgs mass and the vector boson mass is derived. According to this relation the magnetic mass determines the critical Higgs mass and vice versa. The effect of including the largest Yukawa coupling to the top quark on the effective potential is seen to be rather small, even on a quantitative level. Thus it seems to be a reasonable approximation to drop the fermions from the action already from the very beginning when the action is supposed to describe the phase structure of the electroweak model. Only for $m_{mag} \neq 0$ this framework predicts the existence of a critical endpoint that would never be seen in naive perturbation theory, although the gauge and scalar couplings are still quite small at this point.

Gap equations lead to reliable results for the phase structure up to Higgs masses of the order of $70[GeV]$ [100], and, as long as $m_{magn} \neq 0$, also in the vicinity of a vanishing vacuum expectation value of the Higgs field $< \Phi >= 0$. Still, as a manifestation of the magnetic mass problem, the magnetic mass is not self-consistently derived. Even a calculation to higher orders in the couplings (like g^4 in $\mathcal{V}_{\mathcal{CEP}}$) and a consistently derived magnetic mass of order g^2 would not solve the problem. No resummations are known that could remove the IR-singularities of perturbation theory in the infinite volume at these higher orders in g. For further discussions on this point we refer to section 4.4.

Therefore the lattice approach is needed not only for an independent check of continuum results in a common range of parameters. It is also necessary for extending the calculations to a larger range of parameters (in particular to larger Higgs masses) and studying the very existence of the critical endpoint without introducing the finite magnetic mass by hand.

3.2.3   The lattice approach in four dimensions

As the $U(1)$-symmetry remains almost unchanged at the transition (apart from the change of $U(1)_Y \longmapsto U(1)_{em}$) and, as a result of the gap equations, as also the top quark has little impact on the phase structure, it seems to be justified to consider the electroweak standard model without the $U(1)$-gauge bosons and the fermions. In this limit the standard model reduces to the $SU(2)$ Higgs model. On a Euclidean lattice, in four dimensions, its action reads

$$S[U, \Phi] = \beta \sum_p (1 - \frac{1}{2}\mathrm{tr}(U(\partial p))$$

$$+ \sum_x \{ \frac{1}{2}\mathrm{tr}\,(\Phi^\dagger(x)\Phi(x)) \,+\, \lambda_0 [\frac{1}{2}\mathrm{tr}(\Phi^\dagger(x)\Phi(x) - 1]^2$$

$$- \kappa \sum_{\mu=0}^{D-1} \mathrm{tr}\Phi^\dagger(x + \widehat{\mu})U(x,\mu)\Phi(x) \} \,. \tag{3.2.5}$$

We have encountered Eq. (3.2.5) in connection with variational estimates. Equation (3.2.5) has already a form that is suited for Monte Carlo simulations. We briefly recall that $U(x,\mu)$ are $SU(2)$ variables associated with links (x,μ) of the lattice. $U(\partial p)$ denotes the product of link variables along the boundary of a plaquette p. $\Phi(x) \in cU(2)$ are complex 2×2 matrices in isospin space describing the scalar Higgs field. The parameter β fixes the bare gauge coupling according to $\beta = 4/g^2$. The gauge part of the action has the Wilson form. The hopping parameter κ is related to the bare mass m_0 of the Higgs via

$$m_0^2 = (1 - 8\kappa - 2\lambda_0)/\kappa \tag{3.2.6}$$

with bare scalar coupling λ_0. A comparison with Eq. (2.2.6) of section 2.2 shows that λ_0 is related to the tree level coupling λ of the last section via $\lambda = 6/\kappa^2 \lambda_0$.

Tuning the bare Higgs mass therefore amounts to tuning λ_0 for fixed κ or κ for fixed λ_0. Let us assume that the spatial lattice extensions are infinite, but N_τ, the number of lattice sites in time direction, is finite. We write $1/N_\tau = T_{lat} = a \cdot T_{phys}$ so that the continuum limit is reached for $T_{lat} \mapsto 0$ or $N_\tau \mapsto \infty$ if the physical temperature T_{phys} is kept fixed. Before we focus on the Higgs transition as a function of κ for given N_τ, β and λ_0, we discuss the gross features of the phase structure in the full parameter range $N_\tau, \beta, \lambda_0, \kappa$, in particular to exclude an interference of the deconfinement transition with the Higgs transition.

3.2.3.1　Gross phase structure with deconfinement and Higgs transition

The gross features of the phase structure of the $SU(2)$ Higgs model in the full parameter space are deduced from Monte Carlo simulations starting from early simulations of [101, 102], later [103] and some analytic results (see e.g. [104]). For fixed scalar coupling λ_0 of order 1 consider the phase diagram of the $SU(2)$ Higgs model in the three-dimensional parameter space (T_{lat}, β, κ) of Figure 3.2.1 taken from [105].

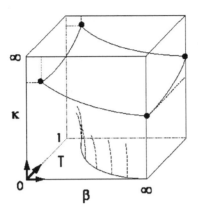

Fig. 3.2.1   Schematic phase diagram of the SU(2) Higgs model at finite temperature. For further explanations and the original reference see the text.

We distinguish three regions:

- $\beta < \beta_c$, $\kappa << 1$: confinement region
- $\beta > \beta_c$, $\kappa < \kappa_c$: deconfinement region
- $\beta >> 1$, $\kappa > \kappa_c$: Higgs region

Let us consider the different limiting cases. For $\kappa = 0$ we obtain the pure $SU(2)$ gauge theory which exhibits the deconfinement transition at a critical value β_c with $\beta_c < \infty$ as long as $N_\tau < \infty$ or $T_{lat} > 0$. The critical value β_c shifts towards larger values as T_{lat} decreases. In the presence of scalar fields, for $\kappa > 0$, but not too large, the deconfinement transition persists and extends into the cube. The results of [102], however, indicate that the deconfinement transition changes into a crossover phenomenon for sufficiently large κ. In the other extreme case the gauge fields are switched off by setting $\beta = \infty$, this is the limit in which the gauge fields are frozen to the pure gauge orbit. The remaining theory is an $O(4)$-symmetric scalar field theory with Φ a four-component vector, or, equivalently, a scalar theory with global $SU(2)_L \times SU(2)_R$-symmetry and Φ as in (3.2.5). This model was introduced in section 2.3. It plays a role as an effective model also for QCD with respect to mesonic degrees of freedom. As a function of

κ, for fixed T_{lat}, λ_0, this model has a second-order phase transition from an $O(4)$-symmetric to a broken phase with residual $O(3)$-symmetry. Varying T_{lat} generates the line $\kappa_c(T)$ on the right hand side of the cube.

Next we switch on the gauge fields by choosing $\beta < \infty$. The question concerning the effects of gauge fields on the Higgs transition is subtle, as we have seen already in connection with gap equations in the last section. It will be discussed in more detail below. Here we just state the result that for $\beta < \infty$ the transition line $\kappa_c(T_{lat})$ extends into the cube and spans a transition surface. As indicated in Figure 3.2.1, the transition surface may end before reaching the walls of the cube, because the transition turns into a crossover phenomenon for λ_0, T_{lat} sufficiently large and β sufficiently small.

In order to see in which part of the phase diagram the (substitute for) the electroweak phase transition takes place, we note that the weak $SU(2)$ tree level gauge coupling constant $g^2 = 0.5$ translates to a value of $\beta = 8$. According to the above classification $\beta \approx 8$ corresponds to the deconfinement region for $\kappa < \kappa_c$ and to the Higgs region for $\kappa > \kappa_c$. Choosing then $\beta > \beta_c$, the Higgs transition as a function of κ occurs between the Higgs phase and the deconfinement phase. When N_τ is increased towards the continuum limit, while keeping all physical parameters fixed, the lattice becomes more fine-grained, the coupling g correspondingly smaller and β larger, i.e. the Higgs transition shifts to even larger β-values. Therefore we can be sure that the Higgs transition always occurs between the Higgs and the deconfinement phase for the indicated choice of lattice parameters, we need not care about an interference with the deconfinement transition.

3.2.3.2   *The Higgs transition in four dimensions*

From now on we consider the electroweak phase transition in the four-dimensional $SU(2)$ Higgs model on the lattice (cf. Eq. (3.2.5) that was the starting point of Monte Carlo simulations of two groups [106, 107]. For fixed β, λ_0 and several values of N_τ the transition was determined as a function of κ. To convert the lattice results into continuum results, the vector boson mass $m_W[GeV] = 80 GeV$ was used as input in a lattice measurement of $(m_W \cdot a)(\beta, \lambda_0)$ at zero temperature (the zero temperature was realized in separate measurements by a lattice with the same extensions in all directions), so that the dimensionless number $(m_W \cdot a)$ and m_W in physical units yield $a(\beta, \lambda_0)$ as length scale in physical units.

Since the correlation length (in lattice units) associated with the Higgs mass at T_c is about ten times larger than the extension in Euclidean time direction ($\propto T_{(c,lat)}^{-1}$), one needs a large ratio of N_s/N_τ (N_s being the extension in spatial directions) on four-dimensional Euclidean lattices. This is the reason why Monte Carlo simulations in four dimensions used N_τ as small as 2 or 3 [107], although the continuum limit is obtained for $N_\tau \mapsto \infty$ when $T_{phys}[GeV]$ is kept fixed. Therefore one must be aware of strong scaling violations due to the rough discretization in time direction. Moreover it should be noticed that a finite temperature torus of two distinct sites and links in time direction does not provide an appropriate triangulation of a torus in the continuum, since a circle cannot be appropriately triangulated by two points.

Some control of scaling violations is provided by choosing two sets of lattice parameters (N_τ, g, λ_0), not independently of each other, but for a given first set chosen as $(N_{\tau 1}, g_1, \lambda_{01})$, the second set $(N_{\tau 2}, g_2, \lambda_{02})$ is chosen according to the perturbative renormalization group equations. These equations predict the change of couplings g and λ (the tree level continuum coupling (related to λ_0 via $\lambda = \lambda_0/4\kappa^2$) under a change of scale, here induced by passing from N_τ to $N_\tau + 1$, or, equivalently, $m_W \cdot a \equiv \exp(-\tau)$ to $m_W \cdot a' \equiv \exp(-\tau')$ because $N_\tau \propto (m_W \cdot a)^{-1}$. In terms of τ the renormalization-group equations then read

$$\frac{dg^2(\tau)}{d\tau} = -\frac{43}{48\pi^2} g^4 + O(\lambda^3, \lambda^2 g^2, \lambda g^4, g^6)$$
$$\frac{d\lambda(\tau)}{d\tau} = \frac{1}{16\pi^2} \left[96\lambda^2 + \frac{9}{32}g^4 - 9\lambda g^2\right] + O(\lambda^3, \lambda^2 g^2, \lambda g^4, g^6). \quad (3.2.7)$$

When the couplings for the second lattice with $N_\tau + 1$ time slices (and a corresponding value of τ') are chosen according to (3.2.7), ratios of physical observables such as mass ratios stay constant if one is close enough to the continuum limit. The reason is that lattice artifacts drop out of these ratios in this limit for an appropriate choice of observables. Thus a lattice measurement of such ratios with parameters tuned along these so-called lines of constant physics provides a check of how close one is to the continuum limit. It is also called a check of asymptotic scaling. (For readers who are not familiar with the lattice approach, we refer to section 3.3 for further explanations.)

For a given choice of parameters (g, λ_0, N_τ) the quantities of interest are the critical hopping parameter κ, and the order of the transition, measured for example via the probability distribution of an order parameter such as the expectation value of ρ^2 with ρ the absolute value of the Higgs field.

If the transition turns out to be of first order, one is further interested in characterizing its strength. (Recall that a strongly first-order transition goes along with explosive effects and gives the chance of a dynamics far out of equilibrium.) As a measure for the strength one often takes the interface tension between the coexisting phases, or, in the description of a coarse grained free energy, the barrier height between the coexisting minima.

Monte Carlo simulations in four dimensions certainly have the advantage of being from first principles: apart from the neglected fermions and $U(1)$ gauge fields no further, uncontrolled approximation enters the calculations. The error sources of finite size and finite lattice constant are well under control in principle, but less well in practice. Simulations on large asymmetric lattices that would be needed are beyond the present feasibility of numerical simulations. In view of this situation we have to look for alternatives. These alternatives could be a further reduction of the degrees of freedom or a reduction in the dimension from four to three.

3.2.3.3   *Effect of the gauge fields*

In analogy to QCD one may think of a further decimation of gauge degrees of freedom. As we have seen in section 3.1, both QCD in the chiral limit and in the limit of a pure gauge theory are well described by appropriate spin models, at least for an effective description of the phase structure. If we drop the gauge field variables in the $SU(2)$ Higgs model, we end up with an $O(4)$-symmetric scalar theory, as we have mentioned above, with a second-order phase transition for all couplings λ_0, including $\lambda_0 = \infty$. On the lattice it is now easy to show that one of the very effects of the gauge fields is a change of the phase transition to a crossover phenomenon before $\lambda_0 = \infty$ is reached so that the decimation of gauge fields is not allowed as a simplification. Let us see why. We start with the partition function of the $SU(2)$ Higgs model in four dimensions on the lattice

$$Z = \int \mathcal{D}U \mathcal{D}\Phi \; e^{-S(U,\Phi)}$$
$$S(U,\Phi) = S_W(U) \; + \; S_H(U,\Phi) \tag{3.2.8}$$

as in section 2.3 Eq. (2.3.171), and

$$\mathcal{D}U \; = \; \prod_{b \in \Lambda} dU(b) \tag{3.2.9}$$

$dU(b)$ being the Haar measure on $SU(2)$, and $\mathcal{D}\Phi$ as defined in section 2.3, Eqs. 2.3.173 and 2.3.174. Since $\Phi \in cU(2)$, we choose the parametrization

$$\Phi(x) = \rho(x) \cdot \Omega(x), \qquad \rho \in \mathbf{R}_+, \ \Omega(x) \in SU(2). \qquad (3.2.10)$$

The partition function becomes

$$Z = \int \mathcal{D}U \, \mathcal{D}\rho \, \mathcal{D}\Omega \, e^{-S(\rho,\Omega,\Phi)}$$

$$\int \mathcal{D}\rho = \int_0^\infty \prod_{x \in \Lambda} d\rho(x)$$

$$\int \mathcal{D}\Omega = \int \prod_{x \in \Lambda} d\Omega(x) , \qquad (3.2.11)$$

$d\Omega(x)$ being the Haar measure on $SU(2)$. Now we apply a gauge transformation $U \mapsto U'$ such that

$$U(x;\mu) = \Omega(x) \, U'(x;\mu) \, \Omega(x + \widehat{\mu})^{-1} , \qquad (3.2.12)$$

because the action and the measure are gauge invariant. It follows that

$$Z = \int \mathcal{D}U'\mathcal{D}\rho\mathcal{D}\Omega \, e^{-S_W(U')}$$
$$\cdot \exp\left(- \operatorname{tr}[\rho(x)\Omega(x)^\dagger \Omega(x)U'(x,\mu)\Omega(x + \widehat{\mu})^\dagger \rho(x + \widehat{\mu})\Omega(x + \widehat{\mu}))]\right)$$
$$\cdot \exp\left(-[\rho(x)^2 + \lambda(\rho(x)^2 - 1)])\right)^2$$
$$= \int \mathcal{D}U\mathcal{D}\rho \, \exp\left(-S_W(U)\right) \cdot \exp\left(- \operatorname{tr} \rho(x)U(x,\mu)\rho(x + \widehat{\mu})\right)$$
$$\cdot \exp -1/2 \operatorname{tr}(\rho(x)^2 \cdot 1) \cdot \exp\left(-\lambda(\rho(x)^2 - 1)^2)\right) . \qquad (3.2.13)$$

Hence the effect of the gauge transformation is to absorb the radial part of the Higgs field, but then, in the $\lambda \mapsto \infty$-limit, the Higgs field dynamics gets completely frozen, i.e. $\rho(x) = 1$ for all $x \in \Lambda_0$. Notice that not even an Ising dynamics is left, because of $\rho(x) \geq 0$. Therefore in the limit $\lambda \mapsto \infty$ we do not expect a Higgs transition to occur anymore. (The analogous argument in the continuum form of the action leads to $\rho^2 = 2v^2$ with v the Higgs vacuum expectation value. This gives the bound where the action becomes ill-defined unless some resummation is applied. The $\lambda \mapsto \infty$-limit is just not well defined in continuum perturbation theory.)

In addition, in comparison with the $O(4)$ model it is the very effect of the gauge fields to make the Higgs transition *fluctuation induced* first-order for sufficiently small Higgs masses, since the first-order is not obvious on the classical level. The classical potential contains no terms which are cubic or of power six in the fields; radial corrections, however, do generate cubic terms in several approximation schemes like re-summed perturbation theory (section 4.5) and also in a small momentum expansion after integrating upon the gauge field degrees of freedom [109].

As the decimation of gauge field degrees of freedom turned out to be not advisable, we have to look for further alternatives which facilitate the numerical treatment and circumvent the infrared problems of the analytic approach at the same time. Such an alternative is provided by the framework of dimensional reduction. Starting with the $SU(2)$ Higgs model in four dimensions in the continuum, one ends up again with an $SU(2)$ Higgs model, but in three dimensions which is then reformulated on the lattice. This model contains all "dangerous" perturbatively intractable infrared modes and is well suited for numerical simulations. In section 4.6.7 on dimensional reduction in the $SU(2)$ Higgs model we briefly mention the main intermediate steps on the way of deriving the dimensionally reduced action. Here we only state the result for the action that was the starting point for numerical calculations which are considered as conclusive in view of the order of the electroweak phase transition.

3.2.4 *The lattice approach in three dimensions*

The dimensional reduction from the four-dimensional $SU(2)$ Higgs model to a model in three dimensions proceeds in two steps. In a first step, nonstatic modes of the gauge fields (with nonzero Matsubara frequencies) are integrated over, while the temporal component A_0 of the gauge field is kept as adjoint Higgs field. The mass of this adjoint Higgs field is governed by chromo-electric screening, so it is heavier than the fundamental Higgs doublet and the three-dimensional gauge field, which gets its mass from chromo-magnetic screening. The parameters of this action are expressed in terms of the parameters of the original action in four dimensions. In a crude approximation one would just drop the A_0 field as it is heavier and should decouple.

Less crudely, the A_0 field is integrated over in the second step. The resulting action is again an $SU(2)$ Higgs model, in the continuum, but in three dimensions. It is given as

$$S = \int d^3x \left(\frac{1}{4}F_{\alpha,\beta}^b F_{\alpha,\beta}^b + (D_\alpha \Phi^\dagger)(D_\alpha \Phi) + m_3^2(\Phi^\dagger \Phi)(x) + \lambda_3(\Phi^\dagger \Phi)^2(x)\right)$$

$$\alpha, \beta, b = 1, 2, 3 \qquad (3.2.14)$$

with mass parameter $m_3 = m_3(\mu_3)$, renormalized at scale μ_3, and couplings λ_3, g_3. Via the parameters of the intermediate action (depending on A_0), the parameters m_3, λ_3, g_3 can be expressed in terms of the parameters of the original four-dimensional action. The relations can be found, for example, in [110]. The lattice version of (3.2.14) then reads

$$S = \beta_G \sum_p \left(1 - \frac{1}{2}\operatorname{tr} U(\partial p)\right) - \beta_H \sum_{x,\mu} \frac{1}{2}\operatorname{tr}(\Phi^\dagger(x)U(x,\mu)\Phi(x+\hat{\mu}))$$

$$+ \beta_R(\rho(x)^2 - 1)^2) \qquad (3.2.15)$$

with $U(x,\mu) \in SU(2)$, $\Phi(x) = \rho(x)\,\Omega(x)$, $\rho(x)^2 = \frac{1}{2}\operatorname{tr}(\Phi^\dagger(x)\Phi(x))$, $\Omega(x) \in SU(2)$, $U(\partial p)$ the product of U-variables along the boundary of a plaquette, and with couplings

$$\beta_G = \frac{4}{a g_3^2}$$

$$\beta_H = \frac{2(1 - 2\beta_R)}{6 + a^2 m_3^2}$$

$$\beta_R = \frac{\lambda_3}{g_3^2}\frac{\beta_H^2}{\beta_G}. \qquad (3.2.16)$$

This lattice model is studied numerically for various couplings β_G, hopping parameters β_H and λ_3/g_3^2. In typical Monte Carlo simulations the phase transition is then determined as a function of β_H for various gauge couplings (to check the continuum limit, cf. below) and for various effective Higgs masses $M_H^\star$, defined via the continuum couplings

$$\frac{\lambda_3}{g_3^2} =: \frac{1}{8}\left(\frac{M_H^\star}{80\,GeV}\right)^2. \qquad (3.2.17)$$

In principle, a number of bulk quantities can be used to study the order of the phase transition. In particular the square of the Higgs field condensate $\Phi^\dagger \Phi$ is well suited, as the scaling violations turn out to be moderate for this observable. Φ denotes the Higgs field of the $3d$-action in the continuum. As long as there is a finite difference $\Delta < \Phi^\dagger \Phi >$ between this expectation value in the symmetric and the broken phase at the phase transition, the transition is of first order. The difference is also proportional to the latent heat L_{heat}, i.e. to the release of energy during the transition

$$\frac{L_{heat}}{T_c^4} = \frac{M_H^{\star 2}}{T_c^3} \Delta < \Phi^\dagger \Phi > \qquad (3.2.18)$$

with $L_{heat} = \Delta \epsilon$ the difference in the internal energy densities and T_c the critical temperature in four dimensions. The precise form of the prefactor is not so obvious, because (3.2.18) makes use of the relation between 3d-observables and 4d-thermodynamics. The derivation of (3.2.18) can be found in [111]. Another suitable observable which may serve as order parameter for the electroweak transition and can be used for consistency checks is the link-expectation value E_{link} which is the volume average over $\text{tr}(\Phi^\dagger U \Phi)$. Finally, also a finite interface (or surface) tension between coexisting phases indicates the first-order and the strength of the transition.

Monte Carlo studies of [112, 113] have shown that the Higgs transition gets weaker as the Higgs mass is increased. In contrast to quark masses the Higgs mass does not enter the action in the form of an external field so that this effect would be quite simple to anticipate. The main interest therefore is the critical value of the Higgs mass for which the transition ceases to exist, and whether it lies above or below the lower experimental bound on the physical Higgs mass.

The precise localization of a critical endpoint is notoriously difficult, because weak first-order transitions are characterized by large correlation lengths and need even larger lattice extensions to keep the finite size errors under control. Let us summarize how one should proceed in principle. Simulating the Higgs transition for various Higgs masses, ideally one should take the continuum limit first, checking the scaling behavior for smaller lattice constants tuned via β_G while following lines of constant physics and using the indicated map between 3d-lattice parameters and 4d-continuum physics. (In practice this check may amount to a variation of β_G between two values only, because the simulations are quite expensive.) Next one should perform the thermodynamic limit, that is extrapolate the finite volume measurements to the infinite volume limit, infinite not only in lattice units, but also in physical units, because the physical correlation length will be diverging at the critical endpoint. Extrapolations enter therefore in two limits, giving rise to error sources which may not always be well under control, since the most simple extrapolations may not be the appropriate ones. In particular in view of very weak first-order transitions the result of the thermodynamic limit is not anticipated in a small physical volume, as it would be for strong first-order transitions with small correlation lengths. Yet it is encouraging that very different methods obtain the same conclusion, at least qualitatively, that the transition ends between $70 - 80$ $[GeV]$.

Since we have the tools at hand from sections 2.6 and 2.5, we focus now in some more detail on two issues:

- measurements of the surface tension as an indicator for a first-order electroweak transition
- localization of the critical endpoint via the vanishing of the gap in the condensate.

3.2.4.1 *Measurements of the surface tension*

In case of a first-order transition the high and low temperature phases of the electroweak transition would coexist and form interfaces. The additional contribution to the free energy due to the existence of such interfaces is called interface tension or, synonymously, surface tension. Here we review three methods to extract the surface tension from Monte Carlo measurements. The first method is called the two-coupling method, because an interface is enforced between the symmetric and the broken phase by simulating half of the lattice at coupling $\beta_{H\,c} + \Delta$, the other half by $\beta_{H\,c} - \Delta$ and extrapolating $\Delta \mapsto 0$ in the end. The extrapolation is done in different ways [114]. This method has been also used for measuring the surface tension at the deconfinement transition of QCD [115]. For the electroweak transition it was used by [116] in combination with a multi-histogram technique.

The second method is based on a two-state signal in the histogram of an order parameter like the magnetization in a ferromagnet, or the histogram of the internal energy, cf.section 2.6, or, as in the case of the electroweak transition, the modulus squared of the Higgs field $1/2\,\mathrm{tr}(\Phi^{\dagger})\Phi$ or the link $1/2\,\mathrm{tr}(\Phi^{\dagger}U\Phi)$, both averaged over the volume. The surface tension is estimated by comparing the minimum and the maxima of the doubly peaked histogram. The histogram at the minimum of the distribution can be obtained from the value at the maximum (corresponding to a pure phase) by a suppression factor with exponent proportional to the surface free energy costs of mixed configurations. This procedure leads to reliable estimates if the maxima are clearly separated by a broad minimum, if they are of equal height near phase coexistence and a minimal surface separates the pure phases from each other (it would be a two-dimensional disc in a three-dimensional cylinder geometry, for example). However, for the Higgs transition the height of the maxima and their widths are very different in both phases, along with the differences in the corresponding susceptibilities. In this sense the Higgs transition is asymmetric, and the interface between different phases is not minimal, but quite rough. Therefore this method is not so useful for measuring the surface tension at large Higgs masses where the typical features of the first-order are less pronounced.

In a third method one extracts the surface tension from the so-called tunneling correlation length ξ_L. Let us explain the idea and its physical meaning at the example of the three-dimensional Ising model, simulated slightly below the pseudo-critical T_c (i.e. at the critical temperature in a finite volume) in the broken phase. Roughly speaking, the correlation length determines the average length of magnetization domains in configurations, so that one has to register the number of flips of magnetization in space. These flips occur in interfaces perpendicular to the cylinder axis if one chooses a geometry with $L_3 \gg L_1, L_2$. The surface tension α determines this correlation length via the relation

$$\xi_L := E_{0a}^{-1} = C(T) \, e^{(L^2 \alpha(T))} . \qquad (3.2.19)$$

Equation (3.2.19) was derived in [117] in a semi-classical approximation. Here the dimension of α is understood as $1/area^2$, so that the surface energy is measured in units of temperature. The temperature-dependent amplitude $C(T)$ is assumed to be independent of the linear size L. The volume dependence of α is neglected, because it is exponentially suppressed. E_{0a} denotes the vacuum energy splitting in the broken phase that determines the tunneling amplitudes between states with positive and negative magnetization. These states become the degenerate ground state in the infinite volume limit (in which the correlation length diverges). This formula is only valid in a temperature region where tunneling dominates finite-size effects. It has been used to determine the surface tension in the three-dimensional Ising model in [118]. This value can be compared with experimental measurements in binary liquids.

The analytic form of the prefactor was derived for a scalar Φ^4 theory in three dimensions in a semi-classical approximation, including quadratic fluctuations around a kink solution. The model is in the universality class of the 3d-Ising model. The result is [119]

$$\xi_L := E_{0a}^{-1} = \widetilde{C} \left(\frac{\alpha a^2}{T_c} \right)^{-1/2} e^{L^2 a^2 \alpha / T_c} . \qquad (3.2.20)$$

Here α is the interface energy per interface area, and the temperature is written out explicitly. The prefactor contains fluctuations of the interface. The only L-dependence is contained in the exponent. The form (3.2.20) was used as ansatz for the $SU(2)$ Higgs model in four dimensions [120]. In [121] the prefactor was modified such that the L-dependence contained an additional fit parameter. Using this fit parameter the result fitted better to other measurements of the surface tension but is no longer derived as (3.2.20).

Also here the weak strength of the first order transition prevents a straightforward evaluation of the surface tension. Using a formula like (3.2.20) assumes that the tunneling correlation length is much larger than other typical bulk correlation lengths in the pure phases corresponding to the mass spectrum of the model. This may not be always the case [110]. Moreover the interfaces are quite rough. A typical result is $\alpha/T_c^3 = 70(26) \times 10^{-5}$ for $M_H^\star = 70 GeV$. Such a number may give an order of magnitude estimate for characterizing the weakness of the transition. It is not surprising when perturbative calculations of the surface tension via an effective potential deviate from numerical results. The reason is that the surface tension is sensitive to the actual shape of the effective potential. We have argued in detail in section 4.5 about the subtleties in distinguishing physically relevant and artificial shapes of the effective potentials.

3.2.4.2 *Localization of the critical endpoint*

We summarize the procedure of Gürtler et al. [121] who localized the critical endpoint by the breakdown of finite size scaling of Lee-Yang zeros and by the (infinite volume extrapolation of the) discontinuity $\Delta(\Phi\Phi)/g_3^2$ in the Higgs condensate. The model was simulated at gauge couplings $\beta_G = 12$ and 16 for (effective) Higgs masses $M_H^\star = 70, 74, 76$ and $80 [GeV]$ on lattices up to volume 96^3. The larger volume of β_G is a step towards the continuum limit, if simultaneously the tree level continuum couplings λ_3/g_3^2 and $m_3^2(\mu_3 = g_3^2)$ are fixed along the phase transition line as a line of constant physics, so that an increase of β_G amounts to a decrease of the lattice spacing a according to

$$\beta_G = \frac{4}{a g_3^2} \, . \tag{3.2.21}$$

As g_3^2 has the dimension of a mass, the quadratic scalar condensate is given in natural mass units. Comparing the quadratic terms in Φ, ρ in the actions

(3.2.14), (3.2.15) we find

$$\frac{\Delta < \Phi^\dagger \Phi >}{g_3^2} = \frac{1}{8}\beta_G\beta_H\Delta < \rho^2 > . \tag{3.2.22}$$

Here Φ denotes the Higgs field from the 3d-action. $\rho^2 = \frac{1}{L_3}\sum_x \rho(x)^2$, where $\rho(x)^2 = \frac{1}{2}\operatorname{tr}(\Phi^\dagger(x)\Phi(x))$ is the Higgs modulus squared and $<>$ denotes the average over the Monte Carlo measurements. $\Delta\rho^2 = < \rho_b^2 > - < \rho_s^2 >$ denotes the difference of the lattice condensate measured at the pseudo-critical hopping parameter between the broken ($< \rho_b^2 >$) and the symmetric ($< \rho_s^2 >$) phase. The critical endpoint is then located in the $(\beta_H, M_H^\star)$-plane, $M_H^\star$ as defined in Eq. (3.2.17) so that the result will be given in terms of $(\lambda_3/g_3^2)_c$. The pseudo-critical values of the hopping parameter β_H ("pseudo" because of the finite volume) are determined via the minimum of the Binder cumulant

$$B_{\rho^2}(L, \beta_H) = 1 - \frac{< (\rho^2)^4 >}{3 < (\rho^2)^2 >^2}, \tag{3.2.23}$$

cf. section 2.6, alternatively via the maximum of the susceptibility of ρ^2

$$C_{\rho^2}(L, \beta_H) = < (\rho^2)^2 > - < \rho^2 >^2, \tag{3.2.24}$$

or via a certain variant of the equal weight method, cf.section 2.6. Finiteness and shrinking of the Binder cumulant with increasing volume indicate the first-order nature of the transition at $M_H^\star = 70 GeV$. Also the maximum of the (interpolated) susceptibility increases almost linearly with the volume. The ρ^2-histograms at these pseudo-critical couplings show a decrease of the discontinuity with increasing $M_H^\star$ as the gap between the peaks gets more and more filled and the distance between the peaks gets smaller. The distance between the peaks determines the jump in $\Delta < \rho^2 >$. For a given Higgs mass $M_H^\star$ the thermodynamic limit of $\Delta < \rho^2 >$ is extrapolated from a fit according to

$$|\Delta < \Phi^\dagger \Phi >_\infty -\Delta < \Phi^\dagger \Phi >_L| \propto \frac{1}{L^2 a^2 g_3^2}. \tag{3.2.25}$$

The ansatz (3.2.25) is made in analogy to the derived finite size scaling behavior in the two-dimensional Potts model (cf.section 2.6). In case of very weak first-order transitions this scaling behavior sets in only for large lattice sizes, because of the long correlation lengths. Since the extrapolated values depend only weakly on β_G or on the lattice spacing a, the results are interpreted as "continuum" results. The gap in the condensates or the latent heat vanish at $M_H^\star = 76[GeV]$ if the results for the discontinuity from lattices with $L^3 \geq 64^3$ are used for $\beta_G = 12$ and 80^3 for $\beta_G = 16$. The result depends on the choice of lattice data which enter the fit. $M_H^\star = 74[GeV]$ marks the critical endpoint if only data for the two largest volumes are taken into account. Therefore the critical endpoint coupling is bounded from above according to

$$(\lambda_3/g_3^2)_{crit} < 0.107 , \tag{3.2.26}$$

corresponding to $M_H^\star = 74[GeV]$.

A different way to determine the critical endpoint proceeds via analyzing the position of the Lee-Yang zeros in the complex β_H-plane. For a first-order transition the zeros of the partition function move in the complex plane in a characteristic way as a function of the lattice volume. If the first zero of the partition function does no longer approach the real axis in the thermodynamic limit, it is taken as a signal for the change of a first-order transition into a crossover, leading to an estimate for $(\lambda_3/g_3^2)_{crit}$ to be 0.102(2), corresponding to $M_H^\star = 72.2(6)GeV$. For further details about Lee-Yang zeros as a tool for localizing the critical endpoint we refer to [121], and for an introduction to this method to [123].

3.2.4.3 *Universality class of the critical endpoint*

At the critical endpoint the electroweak transition is of second order as anticipated by the increasing correlation length and the decreasing latent heat with increasing Higgs mass. It is therefore challenging to determine the universality class of the critical endpoint although it is an academic question, since there is no electroweak phase transition for the physical Higgs mass. Rummukainen et al. [122] find strong similarities to the Ising universality class when they analyze the joint probability distributions of certain expectation values and apply a finite size scaling analysis to study the critical exponents. For example, the exponent ν can be determined from the slope of Binder's cumulant

$$B_L = 1 - \frac{1}{3} \frac{M^4}{<M^2>^2} \tag{3.2.27}$$

(cf. section 2.6, Eq. (2.6.62)) at the critical point, i.e.

$$\frac{\partial B_L}{\partial t}\Big|_{T_c} \propto L^{1/\nu} , \tag{3.2.28}$$

where L is the linear size, M the order parameter-like variable and t the distance from T_c. As we have outlined in the end of section 2.5, the three-dimensional $SU(2)$-Higgs action is written in terms of S_{hop} and $S_{(\Phi^2-1)^2}$, whose fluctuation matrix is not diagonal, i.e., fluctuations in the corresponding directions in coupling parameter space are not independent of each other. In contrast to that the Ising model is given in terms of $E \equiv -\sum \sigma(x)\sigma(y)$, $M \equiv \frac{h}{|h|}\sum \sigma(x)$, such that the fluctuation matrix is diagonal from the outset. The finite-size scaling analysis of the diagonal elements leads to the familiar critical exponents of the Ising model which characterize the singular behavior of the susceptibilities when approaching T_c. Coming back to the $SU(2)$ Higgs model, one first has to identify the E-like and M-like directions, the associated "E-terms" and "M-terms" are then linear combinations of the original terms in the action. Higher than second moments of the joint probability distribution are then studied to quantify deviations from Ising joint probability distributions. The authors claim that differences in higher order moments fade away with increasing volume and conclude that the electroweak transition at the critical endpoint belongs to the Ising-universality class.

3.2.5 *Summary of results and open questions*

The overall picture of the phase structure of the $SU(2)$ Higgs model is illustrated in Figure 3.2.2. Essentially it is based on four independent approaches:

- the effective potential derived from the gap equations [100]
- Monte Carlo simulations of the $SU(2)$ Higgs model in four dimensions [107, 108]
- Monte Carlo simulations of the $SU(2)$ Higgs model in three dimensions [121]
- an effective scalar potential in four dimensions derived from Monte Carlo integrations upon the gauge fields and a small momentum expansion about zero momentum. [109]

As a common qualitative result of these approaches the electroweak phase transition is of first order below the critical Higgs mass M_H^c, it gets weaker with increasing Higgs mass. The transition becomes of second order at the critical endpoint M_H and changes to a crossover phenomenon above M_H^c. The universality class of the critical endpoint is of academic

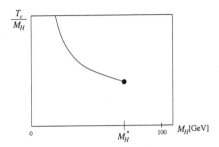

Fig. 3.2.2 Schematic phase diagram of the SU(2) Higgs model. M_H^* denotes the critical Higgs mass, which is of the order of 75 GeV. The electroweak transition ceases to exist for $M_H > M_H^*$.

interest from the present perspective, since the physical Higgs mass is certainly larger than the critical value. Still it is remarkable that only at the very critical endpoint a further decimation of the gauge degrees of freedom seems to be justified, because the universality class of this endpoint is claimed to be Ising-like [122]. The statement is based on a comparison of two-dimensional probability distributions of certain operators in the $SU(2)$ Higgs model and in the Ising model. This example of the electroweak standard model and say a liquid/gas system, both belonging to the Ising universality class, illustrate how different and intricate the microscopic structure can be, while the singularity structure at a second-order transition is the same, or at least quite similar.

Because of the lower bound on the experimental Higgs mass of $114[GeV]$, the physical Higgs mass is certainly much larger than the critical value, estimated as $72.2[GeV]$. Therefore the original hope seems to have gone that the third of Sacharov's conditions for baryogenesis can be satisfied, once the electroweak phase transition is strongly first-order.

Still it remains challenging to investigate the symmetric phase of the electroweak standard model, to study the particle spectrum at high temperatures and the subtle differences between the confinement phase in QCD and the symmetric phase of the electroweak model. The differences between both theories are induced by the scalar Higgs field components.

It is also the symmetric phase of the electroweak model, where the size of the (chromo)magnetic mass of the gauge bosons remains an open problem until the infrared problems are finally overcome by a suitable resummation technique to all orders of perturbation theory. The question of the size of the electric and magnetic screening masses in the electroweak standard model is of phenomenological interest. If the magnetic screening mass of the $SU(2)$ gauge part is actually as small as $O(g^2T)$ or less, it implies a longer ranged weak interaction in analogy to the Coulomb force in electrodynamics with possible cosmological implications. In QED it is well possible to determine the size of the electric and magnetic screening masses, because QED is asymptotically free in the infrared, hence perturbation theory works in a region which is just the region of physical interest. The long-range vacuum polarization tensor at momentum zero can be safely determined.

"Unfortunately", the weak interaction of a nonabelian gauge theory is asymptotically free in the ultraviolet. Perturbation theory works whenever the gauge coupling $g(R, T)$ is small. It is small for high temperatures and spatial distances $R < T^{-1}$, where the scale dependent renormalized coupling decays with a power law. But screening refers to an exponential decay at distances $R > T^{-1}$. Therefore it is an intricate and so far open question how perturbation theory can be transferred from the ultraviolet to the infrared region even when resummation techniques are applied.

Last but not least, the magnetic mass enters the sphaleron rate in a crucial way. It is the sphaleron rate that determines the fate of the baryon asymmetry after the phase transition once it was generated before, whatever mechanism led to its generation.

3.3   A primer to lattice gauge theory

3.3.1   *The QCD Lagarngian*

This section serves mainly to fix the notation of lattice gauge theory and to outline a few basic ideas. We list the essential definitions, summarize some tools which are relevant for investigating the phase structure, and point out sources of errors for misleading results. For pedagogical introductions we refer to the literature [129, 130]. The partition function of an $SU(N_c)$ gauge theory interacting with fermionic matter fields can be written in the Euclidean path integral formulation as

$$Z = \int \mathcal{D}A_\mu \mathcal{D}\psi \mathcal{D}\bar{\psi} \exp\left[-S(A_\mu, \bar{\psi}, \psi; g, m_i)\right], \qquad (3.3.1)$$

where the action depends on the gauge fields A_μ, $\mu = 0, 1, 2, 3$, quark fields $\bar{\psi}$, ψ, the gauge coupling g and the quark masses m_i. The action is given by

$$S = \int_0^\beta dx_0 \int d^3\mathbf{x} \; L(A_\mu, \bar{\psi}, \psi; g, m_i), \qquad (3.3.2)$$

where L is the QCD Lagrangian (Eq. (3.1.4)) for N_f flavors and N_c colors, i labels the quark flavors, $x = (x_0, \mathbf{x})$ denotes the points in configuration space. The chemical potential is set to zero. At finite temperature $T = \beta^{-1}$ the gauge- and matter fields have to satisfy the following boundary conditions

$$A_\mu(x_0, \mathbf{x}) = A_\mu(x_0 + \beta, \mathbf{x})$$
$$\psi(x_0, \mathbf{x}) = -\psi(x_0 + \beta, \mathbf{x}) \qquad \text{for all } x_0, \mathbf{x}, \mu \,. \qquad (3.3.3)$$
$$\bar{\psi}(x_0, \mathbf{x}) = -\bar{\psi}(x_0 + \beta, \mathbf{x})$$

Note that the finite temperature enters only via the boundary conditions, as we also shall see later in this section.

In finite-temperature physics a typical task is to evaluate the thermal expectation value of a physical observable O. The thermal expectation value, defined as

$$<O> \;= \frac{1}{Z} \text{ tr } e^{-\beta H} O, \qquad (3.3.4)$$

where tr denotes the trace in the physical Hilbert space, and H the Hamiltonian, takes the following form in the path integral formulation

$$<O> \;= \frac{\int_{pbc} \mathcal{D}A_\mu \mathcal{D}\psi \mathcal{D}\bar{\psi} O e^{-S}}{\int_{pbc} \mathcal{D} A_\mu \mathcal{D}\psi \mathcal{D}\bar{\psi} e^{-S}}. \qquad (3.3.5)$$

Here *pbc* stands for the periodic (antiperiodic) boundary conditions as specified in Eq. (3.3.3).

So far the discussion is rather formal. The functional integral (3.3.5) is by itself not well defined. Many regularization and renormalization schemes may be used to define the functional integration in an appropriate way. In a perturbative framework, a frequently used UV-regularization or cutoff scheme is dimensional regularization, in which the model (3.3.2) is considered in $D = 4 + \epsilon$ dimensions. Dimensional regularization preserves important properties such as locality and gauge invariance. Renormalization amounts to map the bare parameters (m_i and g in the notation used above) to renormalized ones in such a way that the limit $\epsilon \to 0$ exists. It is achieved by fixing the renormalized coupling constants to every finite order of the weak gauge coupling expansion. As the issue of renormalization is quite important both for zero and finite temperature quantum field theories, we devote a separate chapter to both.

3.3.2 Introducing the lattice cutoff

An alternative for an UV cutoff which does not refer to any perturbative expansion in the very definition is the lattice cutoff, where the lattice refers to the discretization of space-time. A lattice cutoff is particularly suited in view of the nonperturbative nature of the phase structure and the phase transitions themselves. Commonly the (3+1)-dimensional space-time continuum is discretized on a hypercubic lattice. The lattice provides a gauge invariant regularization scheme. Lattice results must be extrapolated to the continuum limit in the very end. Lattice artifacts which are specific for the discretization (so called $O(a)$ corrections) can be controlled via renormalization and the renormalization group equations. They should vanish in the continuum limit. In practice it may be quite subtle and intricate to disentangle lattice artifacts from continuum physics.

Next we introduce general notions of a space-time lattice and field theoretical models on the lattice, but only to the extent that is relevant for this book. In many of the later sections the notation is self-contained, but if it is not, we implicitly refer to the notation introduced now.

The four-dimensional hypercubic lattice is a set of sites denoted by

$$\Lambda_0 = a_0 \mathbb{Z} \times a_s \mathbb{Z}^3, \tag{3.3.6}$$

a_0 denotes the lattice spacing in Euclidean time direction and a_s the lattice spacing in the spatial directions (here assumed to be the same in all spatial directions). Λ_1 denotes the set of all links (or bonds) of the lattice,

$$\Lambda_1 = \{(x; \mu) \mid x \in \Lambda_0, \mu = 0, \ldots, 3\}. \tag{3.3.7}$$

In most cases we consider symmetric lattices with $a_0 = a_s$. In this case we simply write $a \equiv a_0 = a_s$. In some cases it is useful to consider asymmetric lattices with different a_0 and a_s as well (in connection with the derivation of thermodynamic relations, Hamiltonian and transfer matrix description etc.).

At finite temperature and/or for a finite spatial volume the lattice should have the topology of a torus in the corresponding directions. For instance, if L_0 denotes the extension of the lattice in 0-th direction, we define the finite temperature- $L_0 \times \infty^3$ lattice by

$$\Lambda_0^{L_0} = a_0 \left(\mathbb{Z}/L_0\right) \times a_s \mathbb{Z}^3, \tag{3.3.8}$$

with a_0 the lattice spacing in 0-th (temperature) direction and a_s the lattice spacing in spatial directions. The temperature is then given by $T = 1/(L_0 a_0)$. Furthermore $\Lambda_1^{L_0}$ denotes the set of all links of the lattice,

$$\Lambda_1^{L_0} = \{(x; \mu) \mid x \in \Lambda_0^{L_0}, \mu = 0, \ldots, 3\}. \tag{3.3.9}$$

In some cases it is convenient to generalize the notion of a lattice as a set of sites to a cell complex consisting of sites, links, plaquettes etc.

In the following we write just Λ for the lattice if no confusion can arise. It then depends on the context whether Λ stands for a symmetric or asymmetric, zero or finite temperature lattice or a pure spatial or space-time lattice. The baroque notation in (3.3.8) and (3.3.9) is convenient when various types of lattices are used simultaneously.

One of the first questions in lattice field theory is how to put a model on the lattice once it is defined in the space-time continuum. The question refers both to the framework of classical field theory, i.e. to the level of the classical action, and to quantum field theory. Most obviously, discretization of space and time implies that differentiation with respect to space and time must be replaced by a corresponding difference operation. Furthermore, symmetries as guiding principle for constructing many field theories should be either preserved or appropriately generalized.

3.3.3 Scalar field theories on the lattice

As usual let us first consider the example of a Φ^4-theory. The classical action in the continuum of a Φ^4-theory reads

$$S_c(\Phi) = \int d^4x \left(\frac{1}{2} \sum_{\mu=0}^{3} \left(\frac{\partial \Phi(x)}{\partial x_\mu}\right)^2 + \frac{1}{2} m_0^2 \Phi(x)^2 + \frac{\lambda_0}{4!} \Phi(x)^4\right). \tag{3.3.10}$$

On the lattice Λ with lattice spacing a, S_c naturally takes the following form of S_a

$$S_a(\Phi) = a^4 \sum_{x \in \Lambda} \left(\frac{1}{2} \sum_{\mu=0}^{3} \left(\frac{1}{a} \widehat{\partial}_\mu \Phi(x)\right)^2 + \frac{1}{2} m_0^2 \Phi(x)^2 + \frac{\lambda_0}{4!} \Phi(x)^4\right), \tag{3.3.11}$$

where $\Phi(x)$ has still engineering dimension -1 as in the continuum, and it is a scalar field, but it is only defined on a space-time lattice. The differentiation in (3.3.10) has been replaced in (3.3.11) by the difference operator

$$\widehat{\partial}_\mu \Phi(x) \;=\; \Phi(x + a\widehat{\mu}) - \Phi(x) \tag{3.3.12}$$

for $x \in \Lambda$. $\widehat{\mu}$ denotes the unit vector in the positive μ-direction. Although the form of (3.3.11) can be naively guessed, we have to ask as to whether the form of (3.3.11), or, more generally, the lattice-pendant of the continuum action is unique. It is not. The first requirement on a lattice action is that the classical action gets restored in the continuum limit $a \to 0$,

$$\lim_{a \to 0} S_a(\Phi) \;=\; S_c(\Phi). \tag{3.3.13}$$

The form of S_a is not uniquely fixed by the requirement (3.3.13). Further examples that also fulfill (3.3.13) are obtained by replacing the Φ^4-interaction according to

$$\frac{\lambda_0}{4!} \Phi(x)^4 \;+\; \frac{a\sigma_0}{6!} \Phi(x)^6, \tag{3.3.14}$$

$$\frac{\lambda_0}{4!} \Phi(x)^4 \;+\; \frac{a^2\sigma_1}{6!} \Phi(x)^6, \tag{3.3.15}$$

or

$$\frac{\lambda_0}{4!} \Phi(x)^4 \;+\; \frac{a^2\omega}{4} \sum_{\mu=0}^{3} \Phi(x)^2 \left(\frac{1}{a}\widehat{\partial}_\mu \Phi(x)\right)^2 \tag{3.3.16}$$

with σ_0 a coupling with dimension of a length, σ_1 dimensionless. On the classical level all these interactions are equivalent with respect to the continuum limit $a \to 0$. In quantum field theory, the classical level corresponds to the tree level of correlation functions for which no quantum fluctuations are taken into account. In order to investigate the small a behavior, it is clearly not sufficient to consider the action. Also the functional measure must be taken into account. On the lattice the partition function of the scalar model becomes

$$Z_s \;=\; \int \prod_{x \in \Lambda} d\Phi(x)\, e^{-S_a(\Phi)}, \tag{3.3.17}$$

and the correlation functions are given by

$$< \Phi(x_1) \cdots \Phi(x_n) > = \frac{1}{Z_s} \int \prod_{x \in \Lambda} d\Phi(x) \; \Phi(x_1) \cdots \Phi(x_n) \; e^{-S_a(\Phi)}.$$

(3.3.18)

It is the issue of renormalization theory to state sufficient conditions which guarantee the existence of the continuum limit of correlation functions, and to prove properties of the limit itself. This very limit is supposed to provide a definition of a continuum quantum field theory in a constructive way. In many cases the convergence to this limit is achieved by appropriately adjusting a finite number of bare parameters of the action and by a rescaling of the fields. The bare parameters become functions of corresponding renormalized coupling constants. The latter are defined by normalization conditions imposed on renormalized correlation functions at fixed Euclidean or on-shell momenta.

Details of the renormalization procedure are postponed to a separate section. For the moment let us just state a sufficient condition of renormalizability for the lattice theory (3.3.17), (3.3.18). We make two assumptions. Firstly, except for the mass m_0 of the bare mass term $(1/2)m_0^2\Phi(x)^2$, all coupling constants of the local lattice action $S_a(\Phi)$ are dimensionless. Secondly, the continuum limit of the lattice action is the action of a continuum field theory which is UV-finite by imposing a finite set of renormalization conditions (e.g. by the BPHZ finite part prescription). Under these assumptions the lattice theory is renormalizable. Furthermore, the limit is given by the renormalized continuum quantum field theory with the same coupling constants as obtained in the continuum within the same set of normalization conditions.

Let us apply these criteria to the lattice field theory described by (3.3.17), (3.3.18) with actions given by (3.3.11)-(3.3.16). We know that the four-dimensional Φ^4-theory in the continuum, described by the action (3.3.10), is UV finite by renormalizing the mass, the quartic coupling constant and the fields. The second condition of the above proposition on the action is then fulfilled because all actions converge as $a \to 0$ to the Φ^4 action. Whereas the first condition of the proposition is satisfied by (3.3.11), (3.3.15) and (3.3.16), it is obviously violated by (3.3.14) because σ_0 has the dimension of a length. This means that the continuum limit of the lattice field theories described by (3.3.11), (3.3.15) and (3.3.16) does exist and is the same in all cases, while the continuum limit of the theory with interaction (3.3.14) does not, at least not in perturbation theory. The agreement of the limit in the other cases reveals some kind of universality. The actions (3.3.11), (3.3.15) and (3.3.16) describe the same field theory in the continuum limit.

Change to dimensionless quantities

On the lattice it is often convenient to work with dimensionless quantities, in particular for numerical simulations. In lattice units with fixed lattice spacing $a = 1$ the task to construct the continuum limit then translates to identify the critical points where the correlation length in lattice units diverges. One is interested in properties of the lattice model in the vicinity of the critical point(s).

For example, the model (3.3.11) is conveniently parameterized in terms of dimensionless fields as follows. For the lattice sites $x \in \Lambda$ we write now $a\widehat{x}$ with $\widehat{x} \in \mathbf{Z}$. We identify x with $\widehat{x}$ by saying that x are the sites on the unit lattice (with lattice spacing set to unity). The dimensionless fields Φ_0 are defined by

$$\Phi(ax) = \frac{\sqrt{2\kappa}}{a} \, \Phi_0(x) \,,$$

$$m_0^2 = \frac{1}{a^2} \frac{1}{\kappa} \, (1 - 2\lambda - 8\kappa) \,, \tag{3.3.19}$$

and the coupling λ_0 by

$$\lambda_0 = \frac{6}{\kappa^2} \lambda \,. \tag{3.3.20}$$

In terms of these parameters the action becomes

$$S_a(\Phi) = \widehat{S}(\Phi_0) = -\frac{1}{2} \sum_{\substack{x,y\in\Lambda \\ \mathrm{NN}}} 2\kappa \, \Phi_0(x)\Phi_0(y)$$

$$+ \sum_{x\in\Lambda} \left(\Phi_0(x)^2 + \lambda(\Phi_0(x)^2 - 1)^2 \right) \,. \tag{3.3.21}$$

NN stands for nearest neighbors. Equation (3.3.21) is an appropriate starting point for linked cluster expansions and for Monte Carlo simulations. With

$$Z_0 = \int \prod_{x\in\Lambda} d\Phi_0(x) \, e^{-\widehat{S}(\Phi_0)}, \tag{3.3.22}$$

the correlation functions (3.3.18) are obtained as

$$< \Phi(ax_1) \cdots \Phi(ax_n) > = \left(\frac{\sqrt{2\kappa}}{a} \right)^n$$

$$\cdot \frac{1}{Z_0} \int \prod_{x \in \Lambda} d\Phi_0(x) \, \Phi_0(x_1) \cdots \Phi_0(x_n) \, e^{-\widehat{S}(\Phi_0)}. \qquad (3.3.23)$$

The form of the action in dimensionless quantities is also convenient in connection with condensed matter systems. In crystals, for example, the lattice constant has the physical meaning of a length scale rather than that of an inverse cutoff of large momenta. It is then kept fixed and set to unity. There one is interested in taking the thermodynamic limit rather than the continuum limit. In connection with constructing the continuum limit of a quantum field theory, however, there are various possibilities to construct the physical model with dimensionful quantities out of a given "bare" lattice model. They depend on the way in which the lattice spacing is (re-) introduced, in terms of masses or correlation length in lattice units, and on the appropriate rescaling of the fields in the vicinity of critical points.

As an example let us assume that we start from an action of the form

$$\widehat{S}(\Phi_0) = -\frac{1}{2} \sum_{\substack{x,y \in \Lambda_d \\ NN}} 2\kappa \, \Phi_0(x)\Phi_0(y) + \sum_{x \in \Lambda_d} \frac{\widehat{\sigma}_2}{2!} \, \Phi_0(x)^2 + \sum_{x \in \Lambda_d} \frac{\widehat{\sigma}_r}{r!} \, \Phi_0(x)^r \,,$$

$$(3.3.24)$$

typical for a model of statistical mechanics, in dimensionless units and try to find out whether we can construct a strictly renormalizable field theory out of it. Λ_d stands for a lattice in d dimensions, $r = 4, 6, 8$, $\widehat{\sigma}_r \geq 0$. Thus we have to find out for which values of r do we obtain a nontrivial (that is, non-free), strictly renormalizable field theory in the continuum. In order to answer this question we first rearrange the next-neighbor-term in the usual form of a kinetic term. After appropriate rescaling of Φ_0 we obtain

$$\widehat{S}(\Phi_0) = \frac{1}{2} \sum_{x \in \Lambda_d} \sum_{\mu=0}^{d-1} (\widehat{\partial}_\mu \Phi_0(x))^2 + \sum_{x \in \Lambda_d} \frac{1}{2} \, m_0^2 \, \Phi_0(x)^2 + \sum_{x \in \Lambda_d} \frac{\widehat{\sigma}_r}{r!} \, \Phi_0(x)^r$$

$$(3.3.25)$$

with $m_0 = m_0(d, \kappa, \widehat{\sigma}_2)$. To determine the engineering dimension of the continuum field $\Phi_c(a \cdot x)$, corresponding to the rescaled dimensionless Φ_0 on the lattice, we write

$$\Phi_0(x) = a^l \, \Phi_c(a \cdot x), \quad x \in \Lambda_d \qquad (3.3.26)$$

with l determined such that the kinetic term takes the familiar continuum form for $a \to 0$. This leads to

$$\frac{1}{2} \sum_{x \in \Lambda_d} \sum_{\mu=0}^{d-1} (\widehat{\partial}_\mu \, \Phi_0(x))^2 \longrightarrow \int d^d x \; \frac{1}{2} \sum_{\mu=0}^{d-1} \left(\frac{\partial \Phi(x)}{\partial x_\mu} \right)^2 \qquad (3.3.27)$$

with $\widehat{\partial}_\mu = a \cdot \partial_\mu$ the lattice derivative, resulting in the condition

$$2 + 2l = d \qquad (3.3.28)$$

or

$$l = \frac{d-2}{2}. \qquad (3.3.29)$$

Therefore the engineering dimension of the continuum field is fixed. Now we come back to the self-interaction part, proportional to $\Phi_0^r(x)$. We write

$$a^d \sum_{x \in \Lambda_d} \frac{1}{a^d} \frac{\widehat{\sigma}_r}{r!} \, a^{l \cdot r} \, \Phi_c^r(x) \longrightarrow \int d^d x \; \frac{\sigma_{rc}}{r!} \, \Phi_c^r(x) \qquad for \; a \to 0 \, ,$$

$$(3.3.30)$$

where the continuum coupling $\sigma_{rc} \equiv \widehat{\sigma}_r \, a^{l \cdot r - d}$ with $l = (d-2)/2$. We know that the condition for a strictly renormalizable field theory in $d > 2$ dimensions is a dimensionless coupling in the continuum, dimensionless as on the lattice. The coupling σ_{rc} becomes dimensionless if r is chosen such that $r \cdot (d-2)/2 - d = 0$, leading to $r = 6$ in $d = 3$, for example. (A positive exponent in a in front of the lattice coupling $\widehat{\sigma}_r$ would lead to a nonrenormalizable continuum theory, a negative exponent to a superrenormalizable one.)

So far we have sketched how to put scalar field theories on the lattice. We have indicated possible forms of the action and the field measure, the issues of renormalizability and universality of the continuum limit. It can be shown that many requirements generalize from scalar to more complex and realistic theories such as fermionic models, pure gauge theories and QCD.

3.3.4   *Gauge field theories on the lattice*

Next we turn to gauge theories. A gauge invariant lattice action is conveniently defined in terms of link variables $U(x; \mu)$, associated with a link $(x; \mu)$ leaving site x in positive direction μ. The link variables are elements of the gauge group , e.g. $SU(N_c)$, where N_c stands for the number of colors.

We denote the $SU(N_c)$ gauge field action on the lattice by $S_W(U)$. In the following we consider the Wilson action

$$S_W(U) = \beta \sum_{x \in \Lambda} \sum_{\mu < \nu = 0}^{3} \left(1 - \frac{1}{N_c} \mathrm{Re} \, \mathrm{tr} \, U(x; \mu) U(x + a\widehat{\mu}; \nu) \right.$$

$$\left. U(x + a\widehat{\nu}; \mu)^{-1} U(x; \nu)^{-1} \right) \tag{3.3.31}$$

with $\beta = (2N_c)/g_0^2$, g_0 denotes the bare gauge coupling constant. Each term of the sum contains a product of gauge field variables U along the boundary of an elementary plaquette of the lattice, tr denotes the trace in color space. In particular, $N_c = 3$ for $SU(3)$.

Apart from a constant, we sometimes use the shorthand notation of (3.3.31) according to

$$S_W(U) = -\frac{\beta}{N_c} \sum_{p} \mathrm{Re} \, \mathrm{tr} \, U(\partial p), \tag{3.3.32}$$

where the sum runs over all unoriented plaquettes of the lattice and $U(\partial p)$ denotes the product of the link variables U along the boundary ∂p of p.

On the asymmetric lattice, the Wilson action takes the following form

$$S_W = \beta_m \sum_{x \in \Lambda} \sum_{i < j = 1}^{3} P^{ij}(x) + \beta_e \sum_{x \in \Lambda} \sum_{i=1}^{3} P^{0i}(x), \tag{3.3.33}$$

where

$$P^{ij}(x) = 1 - \frac{1}{N_c} \mathrm{Re} \, \mathrm{tr} \, U(x; i) U(x + a_s\widehat{i}; j) U(x + a_s\widehat{j}; i)^{-1} U(x; i)^{-1},$$
$$1 \leq i < j \leq 3,$$

$$P^{0i}(x) = 1 - \frac{1}{N_c} \mathrm{Re} \, \mathrm{tr} \, U(x; i) U(x + a_s\widehat{i}; 0) U(x + a_0\widehat{0}; i)^{-1} U(x; 0)^{-1},$$
$$1 \leq i \leq 3, \tag{3.3.34}$$

and with

$$\beta_m = \frac{1}{\xi}\beta \,, \quad \beta_e = \xi\beta, \tag{3.3.35}$$

where

$$\xi = \frac{a_s}{a_0} \quad \text{and} \quad \beta = \frac{2N_c}{g_0^2}. \tag{3.3.36}$$

Upon parameterizing the link variables $U(x; \mu)$ in (3.3.31) according to

$$U(x; \mu) = \exp\left(a A_\mu(x)\right) \tag{3.3.37}$$

with $A_\mu(x)$ the Lie-algebra valued gauge fields, and expanding the action $S_W(U)$ in powers of the Lie-algebra valued fields A, we obtain for the Wilson action (3.3.31)

$$S_W(U) = -\frac{1}{2g_0^2} \int d^4x \sum_{\mu,\nu=0}^{3} \operatorname{tr} F_{\mu\nu}(x) F_{\mu\nu}(x) + O(a), \tag{3.3.38}$$

where

$$F_{\mu\nu}(x) = \partial_\mu A_\nu(x) - \partial_\nu A_\mu(x) + [A_\mu(x), A_\nu(x)]. \tag{3.3.39}$$

Therefore in the continuum limit $a \to 0$ the lattice action converges to the classical Yang-Mills action of a nonabelian gauge theory.

3.3.4.1 *Functional measure and the gauge orbit*

In most parts of this book we follow the convention that the Lie -algebra valued gauge fields $A_\mu(x)$ are anti-hermitian. The functional measure of the gauge field is defined as the product measure over all links,

$$\mathcal{D}U = \prod_{x,\mu} dU(x; \mu). \tag{3.3.40}$$

For every link $(x; \mu)$, $dU(x; \mu)$ denotes the invariant Haar measure on $SU(N_c)$, i.e.

$$d(gU) = d(Ug) = dU^{-1} = dU. \tag{3.3.41}$$

The partition function is given by

$$Z_g = \int \mathcal{D}U \, \exp\left(-S_W(U)\right). \tag{3.3.42}$$

Local gauge transformations are maps

$$\Lambda : \{U(x; \mu)\} \to \{U^\Lambda(x; \mu)\} \tag{3.3.43}$$

defined by

$$U^\Lambda(x; \mu) = \Lambda(x) U(x; \mu) \Lambda(x + a\widehat{\mu})^{-1}. \tag{3.3.44}$$

An observable $I(U)$ is called gauge invariant if $I(U^\Lambda) = I(U)$ for all gauge transformations Λ. In particular, the Wilson action $S_W(U)$ is a gauge invariant observable. The expectation value for every gauge invariant observable $I(U)$ is defined by

$$< I(U) >= \frac{\int \mathcal{D}U \ I(U) \ \exp{(-S_W(U))}}{\int \mathcal{D}U \ \exp{(-S_W(U))}} = \frac{1}{Z_g} \int \mathcal{D}U \ I(U) \ \exp{(-S_W(U))}.$$
(3.3.45)

Classical vacuum configurations are configurations that minimize the action. One can show that $S_W(U) \geq 0$ for all U. In the infinite volume the field configurations that satisfy $S_W(U) = 0$ so that $S_W(U)$ takes its minimal value are given by the pure gauge orbit

$$U(x; \mu) = \Lambda(x) \ \Lambda(x + a\widehat{\mu})^{-1}$$
(3.3.46)

with $\Lambda(x) \in SU(N_c)$.

In the finite volume, additional degeneracies can occur depending on the boundary conditions imposed. Field configurations that minimize the action in the finite volume are called toron configurations. They contain the pure gauge orbit as a submanifold, cf. section 4.4. In this section we consider, however, only the infinite volume case.

The vacuum manifold (3.3.46) is isomorphic to G/G_0, with

$$G = \bigotimes_{x \in \Lambda_0} SU(N_c)$$
(3.3.47)

as the group manifold of the gauge transformations and G_0 the submanifold of G that is defined by constant gauge transformations $\Lambda(x) \equiv \Lambda_0 \in SU(N_c)$. It is often useful to identify G/G_0 with the subset of G obtained by fixing the global gauge $\Lambda(x_0) = 1$ at some $x_0 \in \Lambda_0$,

$$G/G_0 = \{\Lambda \in G \mid \Lambda(x_0) = 1\}.$$
(3.3.48)

Let $I(U)$ be any gauge invariant observable. All field configurations that are related by a gauge transformation (3.3.44) are equivalent in the sense that $I(U^\Lambda) \equiv I(U)$, $S_W(U^\Lambda) \equiv S_W(U)$ and $\mathcal{D}U^\Lambda = \mathcal{D}U$. Equivalent field configurations contribute with equal weights to the partition function (3.3.42) and to the expectation values (3.3.45).

The perturbative evaluation of $< I(U) >$ amounts to an asymptotic power series expansion in the gauge coupling g_0. To obtain this series, it is sufficient to integrate only over a small neighborhood Ω of the pure gauge orbit (3.3.46). All other contributions are exponentially suppressed in g_0, because it is a saddle point expansion (about the minimum of the action). In order to lift the degeneracy with respect to gauge transformations we have to employ an appropriate parametrization of Ω. A gauge fixing procedure must be invoked that eliminates the gauge degrees of freedom. The basic idea of gauge fixing is to identify physical and gauge degrees of freedom first in a small neighborhood of $U(x;\mu) \equiv 1$. The small neighborhood is then extended to a finite neighborhood Ω of the gauge orbit by means of gauge transformations. One way of realizing this procedure is provided by the Faddeev-Popov trick. Because of gauge fixing $< I(U) >$ may be computed just from a neighborhood of $U(x;\mu) \equiv 1$.

The coefficients of the series with respect to g_0 are divergent in the continuum limit. The renormalization procedure must be invoked before the continuum limit can be taken. Renormalization converts the series in g_0 into a series with respect to a renormalized gauge coupling g. The corresponding series coefficients become UV finite, i.e. they are convergent as $a \to 0$. As it can be shown, this important property is ensured by gauge invariance and the property (3.3.38) of the lattice action, similarly as in the case of the scalar field theory, cf. the discussion above. For details we refer to the chapter on renormalization, perturbation theory and universality.

3.3.5   *Fermions on the lattice*

Fermions are represented by Grassman $\psi(x)$, $\overline{\psi}(x)$, satisfying the anti-commutation relations

$$\{\psi_\alpha(x), \psi_\beta(y)\} = \{\overline{\psi}_\alpha(x), \overline{\psi}_\beta(y)\} = \{\overline{\psi}_\alpha(x), \psi_\beta(y)\} = 0 , \quad (3.3.49)$$

(with α, β multiindices, representing color, flavor and spin degrees of freedom) and the integration rules

$$\int d\psi_\alpha(x)\, \psi_\alpha(x) = \int d\overline{\psi}_\alpha(x)\, \overline{\psi}_\alpha(x) = 1,$$
$$\int d\psi_\alpha(x) = \int d\overline{\psi}_\alpha(x) = 0 \qquad (3.3.50)$$

for all $x, y \in \Lambda_0$ (including $x = y$). (For Grassmann variables integration amounts to the same operation as left differentiation.)

The fermion action on the lattice takes the form

$$S_F(\psi, \overline{\psi}, U) = a^4 \sum_{x \in \Lambda} \overline{\psi}(x)\Big((D(U) + m_{F0})\, \psi \Big)(x), \qquad (3.3.51)$$

where $D(U)$ is a lattice regularization of the Dirac operator and m_{F0} denotes the bare fermion mass (matrix). Flavor, color and spinor indices are implicit in (3.3.51). The fermionic determinant is then given by

$$\det \left(a^4 \left(-D(U) - m_{F0} \right) \right) = \int \mathcal{D}\psi \mathcal{D}\overline{\psi} \, \exp\left(-S_F \right) \qquad (3.3.52)$$

with the fermion field measure

$$\mathcal{D}\psi \mathcal{D}\overline{\psi} = \prod_{x,\alpha} d\psi_\alpha(x) d\overline{\psi}_\alpha(x). \qquad (3.3.53)$$

Like the measure also the action (3.3.51) should be invariant under the gauge transformations (3.3.44) together with

$$\psi^\Lambda(x) = \Lambda\psi(x) \,, \ \overline{\psi}^\Lambda(x) = \overline{\psi}(x)\Lambda^{-1}. \qquad (3.3.54)$$

To find an appropriate form for $D(U)$ turns out to be more difficult and subtle. For the following discussion it is already instructive to consider a single free Dirac fermion . A naive translation and discretization of the continuum Dirac action leads to a lattice Dirac operator

$$D = \frac{1}{2a} \sum_{\mu=0}^{3} \gamma_\mu \left[\widehat{\partial}_\mu + \widehat{\partial}_\mu^* \right], \qquad (3.3.55)$$

with forward-difference operator $\widehat{\partial}_\mu$ given by (3.3.12) and backward-difference operator

$$\widehat{\partial}_\mu^* \psi(x) = \psi(x) - \psi(x - a\widehat{\mu}). \qquad (3.3.56)$$

This means that

$$D\psi(x) = \frac{1}{2a} \sum_{\mu=0}^{3} \gamma_\mu \left[\psi(x + a\widehat{\mu}) - \psi(x - a\widehat{\mu}) \right]. \qquad (3.3.57)$$

The Fourier transform $\widetilde{D}$ of D takes the form

$$\widetilde{D}(p) = i \sum_{\mu=0}^{3} \gamma_\mu \, \sin p_\mu a \qquad (3.3.58)$$

with p_μ taking continuous values as long as the volume is infinite. In addition to $\widetilde{D}(p = 0) = 0$, $\widetilde{D}$ has zeroes at the boundary of the Brillouin zone at $p_\mu = \pi/a$. Therefore the action (3.3.51) with the naive lattice Dirac operator (3.3.55) actually describes $2^4 = 16$ species of fermions in the naive continuum limit rather than the intended one species of the original continuum action.

The failure of the naive discretization of the Dirac operator actually turns out as a manifestation of a more fundamental problem of putting fermions on a lattice, in particular if one insists on an implementation of chiral symmetry. The failure is explained by the "No-Go" theorem of Nielsen and Ninomiya (1981a, 1981b). As usual let us assume that D is translation invariant. D then commutes with the momentum operator and has the same eigenfunctions e^{ipx}, i.e.

$$De^{ipx} = \tilde{D}(p) \cdot e^{ipx} \tag{3.3.59}$$

with some complex 4×4 matrix $\tilde{D}$ (the Fourier transform of D). The theorem of Nielsen and Ninomiya then states that the following properties cannot hold simultaneously:

(1) $\tilde{D}(p)$ is an entire $2\pi/a$-periodic function of the momenta p_μ.
(2) For small momentum p,

$$\tilde{D}(p) = i \sum_{\mu=0}^{3} \gamma_\mu p_\mu + O(ap^2) . \tag{3.3.60}$$

(3) For $p \neq 0 \mod 2\pi/a$, $\tilde{D}(p)$ is invertible.
(4) $\tilde{D}(p)$ anti-commutes with γ_5,

$$\{\tilde{D}(p), \gamma_5\} = 0. \tag{3.3.61}$$

Property (1) ensures that the lattice Dirac operator is "essentially" local in coordinate space, i.e. it couples only a finite number of neighbors, or more generally $D(x, y)$ decays exponentially fast for large $|x - y|$, with decay constant proportional to the inverse lattice spacing. Properties (2) and (3) ensure that the correct continuum limit is obtained for small a, with no doublers involved. Property (4) ensures that the same continuous chiral symmetry of the fermion action in the continuum is also realized on the lattice for $m_{F0} = 0$.

The "No-Go" theorem of Nielsen and Ninomiya [133] tells us that no lattice formulation of the fermionic action exists that is satisfactory in all aspects. In particular, if one insists on a local action, either continuous chiral symmetry is completely lost on the lattice, or too many flavors come out in the end.

We consider two popular choices, *Wilson fermions* and *staggered fermions* . The former choice gets rid of the species doubling at the expense of breaking all continuous chiral symmetries explicitly. The latter choice keeps a $U(1) \times U(1)$ chiral symmetry for all lattice couplings, a welcome feature in view of an investigation of chiral symmetry restoration. The price are too many flavors in the continuum limit (although the number is reduced with respect to the naive formulation) and a broken flavor symmetry on the lattice.

We briefly mention a realization of the Dirac operator, which was originally proposed in 1982, but was only recently rediscovered, namely Ginsparg-Wilson fermions [135]. The Ginsparg-Wilson-Dirac operator violates chiral symmetry as well, but in a form described by an irrelevant, local lattice operator.

$$\{\gamma_5, D\} = aD\gamma_5 D. \tag{3.3.62}$$

This leads to an *exact* chiral symmetry of the lattice action and the fermionic functional measure under a transformation, which deviates from its continuum version by an irrelevant term. It can be shown that, after renormalization, the continuum chiral symmetry is restored [136]. Unfortunately, solutions D to the Ginsparg-Wilson relation (3.3.62) are strongly anisotropic and slowly decaying over many lattice sites. It will still take some time to achieve sufficient computer capability and theoretical understanding until these fermions can be used in practice for finite temperature simulations of QCD.

For derivations and details of lattice fermions we refer to the textbooks (see e.g. [129] or reviews [130, 134]). Here we summarize only the results.

3.3.5.1 *The Wilson-fermion action and hopping parameters*

Wilson fermions get rid of the species doubling at the expense of breaking all continuous chiral symmetries explicitly. This is achieved by supplementing the naive lattice Dirac operator by an irrelevant mass-like term. The term is irrelevant in the sense that it vanishes in the classical continuum limit. The fermion action is given by (3.3.51), with lattice Dirac operator

$$D(U)\psi(x) = \frac{1}{2a} \sum_{\mu=0}^{3} \left[(\gamma_\mu - r)U(x; \mu)\psi(x + a\widehat{\mu}) \right.$$
$$\left. - (\gamma_\mu + r)U(x - a\widehat{\mu}; \mu)^{-1}\psi(x - a\widehat{\mu}) + 2r\psi(x) \right]. \tag{3.3.63}$$

Usually r is a positive real number. For free fermions, (3.3.63) becomes

$$D\psi(x) = \frac{1}{2a} \sum_{\mu=0}^{3} \gamma_\mu \left[\psi(x + a\widehat{\mu}) - \psi(x - a\widehat{\mu})\right]$$

$$- \frac{a}{2} r \sum_{\mu=0}^{3} \left(\frac{1}{a^2} \widehat{\partial}_\mu^* \widehat{\partial}_\mu\right) \psi(x). \qquad (3.3.64)$$

As for the case of scalar and gauge fields, it is often convenient to work with dimensionless fields and parameters. In analogy to (3.3.19), we write ax for the sites of the lattice with lattice spacing a so that $x \in \Lambda$ are the sites of the unit lattice. The dimensionless fields ψ_0, $\overline{\psi}_0$ are defined according to

$$\psi(ax) = \frac{\sqrt{2\kappa}}{a^{3/2}} \psi_0(x), \quad \overline{\psi}(ax) = \frac{\sqrt{2\kappa}}{a^{3/2}} \overline{\psi}_0(x), \qquad (3.3.65)$$

where the so called hopping parameter κ is given by

$$\kappa = (2m_{F0}a + 8r)^{-1}. \qquad (3.3.66)$$

The fermion action then becomes

$$S_F(\psi, \overline{\psi}, U) = \widehat{S}_F(\psi_0, \overline{\psi}_0, U) = \sum_{x \in \Lambda} \left\{\overline{\psi}_0(x)\psi_0(x) - \kappa \mathcal{P}(x)\right\}, \qquad (3.3.67)$$

with

$$\mathcal{P}(x) = \sum_{\mu=0}^{3} \overline{\psi}_0(x) \left[(r - \gamma_\mu)U(x; \mu)\psi_0(x + \widehat{\mu}) + (r + \gamma_\mu)U(x - \widehat{\mu}; \mu)^{-1}\psi_0(x - \widehat{\mu})\right].$$

$$(3.3.68)$$

For the free field case with $U = 1$, the critical value κ_c of the hopping parameter κ corresponding to $m_{F0}a = 0$ is given by $\kappa_c = 1/(8r)$. In many applications r is set to one so that in this case $\kappa_c = 1/8$.

While the flavor symmetry is well defined and conserved for all lattice spacings by the Wilson action, all chiral symmetries are explicitly broken even in the massless case ($m_{F0} = 0$ or $\kappa = \kappa_c$). 'All' chiral symmetries should be contrasted with remnants of the full invariance under certain subgroups. Recall our original intention to study the phase structure within the lattice approach. A chiral symmetry restoration at finite temperature in the massless limit should be signaled by a "melting" of the condensate $\langle \overline{\psi}\psi \rangle$. Such a melting is prejudiced right from the beginning, if an explicit chiral symmetry breaking is involved as it is in the Wilson formulation.

3.3.5.2 *Staggered fermions and flavor symmetries*

Staggered fermions are frequently used in studies of the chiral phase transition. A careful inspection of the origin of the species doubling in the naive formulation suggests the possibility of eliminating the unwanted fermions by doubling the effective lattice spacing. This amounts to a distribution of the fermionic degrees of freedom over the original lattice in such a way that the effective lattice spacing for each type of Grassmann variables is twice the fundamental lattice spacing. It turns out that in d dimensions $2^{d/2}$ fermion fields are necessary to place a different fermionic degree of freedom at each site of an elementary hypercube on the lattice. This indicates the origin of the integer multiple of four flavor degrees in the continuum for four dimensions, which is attainable by a description with staggered fermions.

A realization of these ideas is rather involved. The sites of the hypercube will be occupied by single-component spinors χ_f, $\bar{\chi}_f$, which may be multi-component in flavor space, $f = 1, \ldots, n$. (This integer number n should be distinguished from the desired number N_f of continuum flavors.) The χ's and $\bar{\chi}$'s are certain linear combinations of the original fields ψ, $\bar{\psi}$. In the end of a simulation in terms of $(\chi, \bar{\chi})$-fields, the results for the $(\psi, \bar{\psi})$-fields have to be reconstructed.

The action for staggered fermions in terms of $(\chi, \bar{\chi})$-fields on the asymmetric lattice is defined by

$$S_F \equiv a^4 \sum_{x,x' \in \Lambda} \sum_{f=1}^{n} \bar{\chi}_f(x) \, \mathcal{Q}_f(x, x') \, \chi_f(x') \tag{3.3.69}$$

with

$$\mathcal{Q}_f(x, x') = \sum_{\mu=0}^{3} D^{(\mu)}(x, x') + m_{f0} \, \delta_{x,x'}, \tag{3.3.70}$$

where

$$D^{(i)}(x, x') = \frac{1}{2a_s} \Gamma_i(x) \left[U(x; i)\delta_{x,x'-a_s\hat{i}} - U(x'; i)^{-1}\delta_{x,x'+a_s\hat{i}} \right],$$
$$\Gamma_i(x) = (-1)^{x_1 + \cdots + x_{i-1}}, \quad \text{for } i = 1, 2, 3, \tag{3.3.71}$$

and

$$D^{(0)}(x, x') = \frac{1}{2a_0} \Gamma_0(x) \left[U(x; 0)\delta_{x,x'-a_0\hat{0}} - U(x'; 0)^{-1}\delta_{x,x'+a_0\hat{0}} \right],$$
$$\Gamma_0(x) = (-1)^{x_1 + x_2 + x_3}. \tag{3.3.72}$$

The only remnant of the Dirac structure is hidden in the phases Γ_μ. The staggered fermion action depends on the bare quark masses m_{f0}. The index

f labels the species of 'staggered' flavors. Their number n is often set to 1, since n staggered species on the lattice correspond to $N_f = 4 \cdot n$ flavors in the continuum. The case $n = 1$ comes closest to $N_f = 3$ flavors, for which QCD is approximately chirally invariant. More precisely, for $n = 1$ and $d = 4$ dimensions one has 16 χ's in each elementary hypercube. It can be shown that these 16 χ's can be combined to define a quark field with four flavors in the continuum, more generally with $N_f = 4 \cdot n$ flavors. A nonzero lattice spacing breaks the flavor symmetry between the $4n$ flavors, and it breaks the chiral $U(4n) \times U(4n)$-symmetry of the continuum limit of the staggered fermion action for $m_{f0} = 0$ down to $U(n) \times U(n)$.

It is the $U(1) \times U(1)$ remnant of the full chiral symmetry (that is left for $n = 1$ and arbitrary lattice spacing) that is the desired feature for investigating chiral symmetry restoration.

Wilson and staggered fermions appear quadratically in the action. In general, fermion fields can only occur quadratically in a renormalizable four-dimensional field theory. Thus fermions are usually integrated out. This seems to be the only tractable way for numerical calculations, since the fermion fields are Grassmann variables. The integrals are performed using the well-known formulas of Berezin integration [132], cf. (3.3.49) and (3.3.50). The result for the full partition function of QCD is

$$
\begin{aligned}
Z &= \int \mathcal{D}U \mathcal{D}\psi \mathcal{D}\overline{\psi}\, e^{-S_W(U) - S_F(\psi, \overline{\psi}, U)} \\
&= \int \prod_{x,\mu} dU(x; \mu)\, e^{-S_W(U)} \det{}^l a^4(-\mathcal{Q}),
\end{aligned} \tag{3.3.73}
$$

where $\mathcal{Q}$ is given for Wilson fermions by

$$
\mathcal{Q} = D(U) + m_{F0}, \tag{3.3.74}
$$

with $D(U)$ from (3.3.63), and $\mathcal{Q}$ is given by Eq. (3.3.70) for staggered fermions with $m_{f0} = 0$. $\det \mathcal{Q}$ denotes the determinant of $\mathcal{Q}$ that extends over configuration space, spin and color indices. The integer power l equals to N_f, the number of continuum flavors, for Wilson fermions, and l equals n for staggered fermions (recall that $n = N_f/4$). The above result holds with the obvious generalizations if all flavor (and staggered flavor, respectively) masses are equal.

Both for Wilson and for staggered fermions, $\det \mathcal{Q}$ is nonnegative. In particular for Wilson fermions, $\det \mathcal{Q}(U)$ is positive for $m_{F0} > 0$ or, in the form (3.3.67), for values of the hopping parameter κ less than $1/(8r)$ [125].

We write in this case

$$Z = \int \prod_{x,\mu} dU(x;\mu) \, e^{-S_{\text{eff}}(U)} \, ,$$

$$S_{\text{eff}}(U) = S_W(U) - l \, \ln \det a^4 \mathcal{Q}, \qquad (3.3.75)$$

with real action $S_{\text{eff}}(U)$.

3.3.5.3 Sources of errors

For a positive definite $\mathcal{Q}$ one may use that $\det \mathcal{Q} = \exp(\text{tr} \ln \mathcal{Q})$, leading to

$$S_{\text{eff}}(U) = S_W(U) - \text{tr} \ln \mathcal{Q}(U). \qquad (3.3.76)$$

The symbol tr denotes the trace over space, spin and color indices. Here and in the following we omit the factor a^4 in front of $\mathcal{Q}$ for simplicity. Thus a simulation of Eq. (3.3.75) with $l = 1$ describes $N_f = 4$ − flavors in the continuum limit if staggered fermions are used for $\mathcal{Q}$. However, although $\det \mathcal{Q}$ is positive, the eigenvalues of $\mathcal{Q}$ are not always positive, so that $\mathcal{Q}$ is not positive (semi-) definite. This is an annoying feature in particular for a representation in terms of pseudofermions. Therefore, for staggered fermions, in actual simulations $(-l \cdot \text{tr} \ln \, \mathcal{Q}(U))$ in the effective action of Eq. (3.3.76) is replaced by $(-l \cdot \text{tr} \ln[\hat{\mathcal{Q}}\hat{\mathcal{Q}}^+(U)])$. Consider first $\hat{\mathcal{Q}} = \mathcal{Q}$. The term $\text{tr} \ln \mathcal{Q}\mathcal{Q}^+$ is proportional to $(\ln \det \mathcal{Q} \det \mathcal{Q}^+)$ or $\ln(\det \mathcal{Q})^2$, thus the replacement induces a further doubling of flavor degrees of freedom from n to $2n$. To compensate for this doubling, the number of degrees of freedom in $\mathcal{Q}$ is reduced by a factor of 2 by a further doubling of the effective lattice spacing. Let us call the corresponding fermion operator $\hat{\mathcal{Q}}$. Due to an even-odd symmetry of the determinant one has $\det \mathcal{Q} = \det \hat{\mathcal{Q}}_{\text{even}} \cdot \det \hat{\mathcal{Q}}_{\text{odd}}$. Thus $\text{tr} \ln \mathcal{Q}$ is actually simulated as $\text{tr} \ln \hat{\mathcal{Q}}\hat{\mathcal{Q}}^+$. This induces an error, which is not well under control. Another uncontrolled error entering the formulation for staggered fermions is due to flavor-exchanging currents. The flavor symmetry is violated for a finite lattice spacing. By claiming that their contribution vanishes in the continuum limit, one may induce an error similar to that one would make in claiming that the irrelevant additive terms in the Wilson action vanish in the continuum limit. It is well-known that the very irrelevant terms of the Wilson action are essential for reproducing the right axial anomaly in the continuum limit.

3.3.6   *Translating lattice results to continuum physics*

A basic step in understanding the lattice approach is the translation from lattice results into physical units . For illustration let us consider the measurement of a mass. Such a measurement on the lattice typically yields a dimensionless number of the order of 1. The dimensionless lattice mass m_{latt} is related to the physical mass m via the lattice spacing a. From simple dimensional arguments we have

$$m = m_{latt} \, a^{-1} \tag{3.3.77}$$

in units where $\hbar = c = 1$. (In these units $100 \, [\text{MeV}] \sim 0.5 \, [\text{fm}^{-1}]$.) In other words, the lattice mass m_{latt} is measured in units of a^{-1}. Other variables are obtained similarly, energy densities in units of a^{-4}, etc.

When a lattice mass is interpreted as $m \cdot a$ and m [MeV] is known from experiments, Eq. (3.3.77) gives "the" lattice spacing in physical units. More precisely it gives $a(g_0)$ [MeV^{-1}] at coupling g_0, if g_0 stands for the bare input parameter(s) of the lattice Lagrangian, which have been used in the measurement of m_{latt}. In a pure gauge theory g_0 is the bare gauge coupling. The lattice spacing obtained this way is unique, i.e., independent of the choice of the physical input m [MeV], only if one is in (or close to) the continuum limit.

Let us assume we have determined a [MeV^{-1}] or a[fm] from a first mass measurement as indicated above and measure a second mass $\tilde{m}_{latt}$ in lattice units. The physical value $\tilde{m}$ [MeV] is then predicted from the lattice simulation and can be compared with the experimental value. Furthermore, once $a(g_0)$ is known as function of g_0 in units of [MeV^{-1}] or [fm], it makes sense to associate strong couplings ($g_0 \gg 1$) with coarse grained lattices (say $a > 1$ [fm]), and weak couplings with a fine grain size close to the continuum description. The very existence of such a universal mapping between bare couplings and lattice spacings in physical units is based on the renormalization-group equation. In case of an $SU(N_c)$ gauge theory with N_c colors and N_f massless flavors the renormalization-group equation relates g_0^2 and a in the continuum limit (for perturbatively small couplings

$g_0^2 \to 0$) according to

$$a \, \Lambda_L = (b_0 g_0^2)^{-b_1/(2b_0^2)} \, e^{-1/(2b_0 g_0^2)},$$

$$b_0 = \frac{1}{16\pi^2} \left[11 \frac{N_c}{3} - \frac{2}{3} N_f \right] \tag{3.3.78}$$

$$b_1 = \left(\frac{1}{16\pi^2} \right)^2 \left[\frac{34}{3} N_c^2 - \left(\frac{10}{3} N_c + \frac{N_c^2 - 1}{N_c} \right) N_f \right],$$

where Λ_L denotes the scale parameter of QCD that represents the only parameter in QCD with N_f massless flavors, and has to be fixed from experiment. If lattice calculations are performed at sufficiently small values of g_0, an observable in lattice units should scale as a function of g_0 in a way that is determined by $a(g_0)$ according to Eq. (3.3.78). For a mass in lattice units this implies

$$(m \cdot a)(g_0) = m \cdot (a(g_0)) . \tag{3.3.79}$$

If such a scaling behavior is observed, the asymptotic scaling regime has been reached. The lattice spacing can then be replaced by the scale parameter Λ_L.

The physical volume V and temperature T are given in terms of the lattice spacing as

$$V = L_s^3 a_s^3, \quad T = \frac{1}{L_0 a_0}. \tag{3.3.80}$$

For $a_0 = a_s = a$, the volume and temperature may be converted in units of fm, if $a(g_0)$ is taken from Eq. (3.3.78).

For $a_0 \neq a_s$, the relations $a_0(\xi, g_0)$ and $a_s(\xi, g_0)$ can be determined in the case of a pure gauge theory (Eq. (3.3.33)) as follows. As in (3.3.35) and (3.3.36) we use the notations

$$\xi = \frac{a_s}{a_0}, \quad \beta = \frac{2N_c}{g_0^2} \tag{3.3.81}$$

and

$$\beta_m = \frac{1}{\xi} \beta, \quad \beta_e = \xi \beta. \tag{3.3.82}$$

β denotes the bare coupling on an isotropic Euclidean lattice. Here β is large when g_0 is small like the physical temperature. Thus one may also think of β as some kind of *effective temperature* (rather than effective *inverse* temperature as the notation suggests). The two relations which replace $g_0(a)$ of the isotropic case are written as

$$g_m^{-2}(a, \xi) = g_0^{-2}(a) + c_m(\xi) + O(g_0^2)$$
$$g_e^{-2}(a, \xi) = g_0^{-2}(a) + c_e(\xi) + O(g_0^2) \qquad (3.3.83)$$

with $g_m^2 \equiv 1/(\beta_m \, \xi)$, $g_e^2 \equiv \xi/\beta_e$. The coefficients c_m and c_e have been perturbatively determined [131]. With $a_s = a$, $a_0 = a/\xi$ and $a = a(g_0)$ from Eq. (3.3.78), Eq. (3.3.83) can be solved for a_s and a_0 e.g. in terms of g_m and g_e, the couplings of the pure gauge action, or, alternatively, in terms of ξ and g_0. The relations $a_s(\xi, g_0)$ and $a_0(\xi, g_0)$ replace $a(g_0)$ on an anisotropic lattice in thermodynamic calculations.

3.3.6.1 *The critical temperature T_c*

In QCD the critical temperature T_c depends on the number of colors (N_c), the number of flavors (N_f), the current quarks masses m_q, and the volume V. If we consider the pseudocritical temperature T_c, the finite volume induces a shift of T_c of the order of $1/V$, if the phase transition is of first order and V denotes the D-dimensional volume of the system. The finite light quark masses give an effect of a few percent compared to the value of T_c in the chiral limit. The strongest dependence comes from the number of flavors. In earlier simulations T_c varied about 100 [MeV] between $T_c \sim$ 150 [MeV] for two light flavors and $T_c(N_f = 0) \sim 260$ [MeV] for a pure gauge theory [126]. This is easily understood in a percolation picture. The transition is assumed to occur when a critical hadron density is reached where the hadrons start overlapping which we have briefly mentioned in the introduction. A much lower temperature is then needed for creating a critical hadron density out of light pions with a mass of the order of T_c than out of heavy glueball states. The lightest states in a pure gauge theory are glueballs with a mass of the order of $5T_c$.

For a smooth crossover phenomenon T_c is no longer defined. However, if there is a narrow temperature interval with rapid changes in thermodynamic quantities, it makes still some sense to associate a '*pseudocritical*' or "*crossover*" *temperature* T_c with the rapid crossover region. More precisely, the crossover temperature is defined as the temperature of maximum change in an observable (e.g. the chiral condensate). Alternatively it is the temperature of the peak in some susceptibility. Both pseudocritical values for T_c must only agree, if the susceptibility is the derivative of the particular observable whose maximal change defines T_c.

3.3.6.2   *Choosing the appropriate extension in time direction*

Usually one measures the critical coupling g_{0c} rather than the critical temperature T_c. Order parameters are plotted as functions of g_0 (or $\beta = 2N_c/g_0^2$) to show their behavior as a function of T. For simplicity let us consider an isotropic lattice, i.e. $a_s = a_0 = a$, and the limit of a pure gauge theory with bare coupling g_0. For a given number L_0 of lattices sites in the timelike direction the temperature may be varied by tuning a via g_0 according to Eq. (3.3.78). From a physical point of view it is not surprising that the temperature may be implicitly varied via the coupling, as both are related through asymptotic freedom. Starting in the strong coupling region $g_0 \gg 1$ and lowering g_0, the transition from the confinement to the deconfinement phase will be encountered at a certain coupling g_{0c}. Instead, g_0 can be kept fixed, but L_0 is varied from smaller to larger values (say from $L_0 = 2$ to $L_0 = 8$). This way the transition region is passed from the deconfinement to the confinement phase if g_0 and L_0 are in an appropriate range.

One would like to simulate the lattice system for large values of L_0 to keep the finite-size effects small. More precisely one should speak of finite-cutoff effects, because for small L_0, strong couplings are sufficient to reach the high-temperature phase, but at strong couplings one is far away from the continuum limit. Vice versa, for large L_0 rather small couplings are needed to reach the high-temperature phase. The larger L_0, the smaller is the critical coupling g_{0c}, which corresponds to the transition temperature T_c. In practice, the computer time rises rapidly if g_0 is small. This explains why typical extensions in time direction were initially limited to $L_0 = 6, 8$ or 12 when fermions were included.

The effect of fermions is a further reduction of the effective lattice spacing a at the same coupling g_0 compared to the pure gauge case. Even larger values of L_0 are needed to reach the transition region. Small values of L_0 require stronger couplings g_0 in the presence of fermions. One should keep in mind that one is far outside the asymptotic scaling region if dynamical fermions are simulated at small L_0.

3.3.6.3   *A test of asymptotic scaling*

A test of asymptotic scaling is the very first check of whether the measured critical coupling g_{0c} has some relevance for a critical temperature in the continuum limit. When the temporal size L_0 is increased, the transition should occur at a smaller lattice spacing $a(g_{0c})$ such that $T_c = 1/(L_0 a(g_{0c}))$

stays constant. For small couplings g_0, the scaling of $g_c(L_0)$ should be determined by

$$T_c/\Lambda_L = L_0^{-1}(b_0 g_{0c}^2)^{b_1/(2b_0^2)} e^{1/(2b_0 g_{0c}^2)} . \tag{3.3.84}$$

Equation (3.3.84) follows from Eqs. (3.3.78) and (3.3.80). Calculations for an $SU(2)$- and $SU(3)$ gauge theory have been performed. A measurement of g_{0c} shows strong violations of asymptotic scaling according to Eq. (3.3.84) unless L_0 is large enough.

3.3.6.4 *Translation to physical units*

A second check as to whether the critical temperature on the lattice has some relevance for the continuum limit is a translation to physical units. The outcome should be independent of the choice of the experimental input. In the pure $SU(3)$ gauge theory T_c is approximately independent of the physical input. For example, early measurements lead to 239(13) [MeV] from the string tension $\sqrt{\sigma} = 420 \pm 20[MeV]$, 239(23) [$MeV$] from the ρ-mass and 225(30) [MeV] from the nucleon mass [128].

For QCD with two light flavors in the staggered fermion formulation the *crossover temperature* was estimated as $140 - 160$ [MeV] [127]. The experimental input comes from the ρ-mass with 770 [MeV]. The critical temperature is then determined as

$$T_c \text{ [MeV]} = \frac{770 \text{ [MeV]}}{(m_\rho \cdot a)(6/g_{0c}^2, m_q a) \cdot L_0} . \tag{3.3.85}$$

In the ideal case the physical value of the ρ-mass (770[MeV]) just drops out of the ratio in (3.3.85) and T_c results in units of [fm^{-1}] as $1/(a(6/g_{0c}^2, m_q a) \cdot L_0$. What is measured, however, is the ρ-mass in lattice units $m_\rho \cdot a$ (depending on the bare gauge coupling g_{0c} and the bare quark masses m_q), calculated as a function of g_{0c} and $m_q \cdot a$ from a fit to several zero temperature simulations (see [127]) and references therein). Therefore the "drop out" may be slightly violated. Note also that in Eq. (3.3.85) the ρ-mass [MeV] is kept fixed at its physical value of 770 [MeV], while the bare quark masses $m_q \cdot a$ are allowed to vary to unphysical values, compatible with unphysical ρ-masses as well. One should further keep in mind how much the nucleon to rho mass ratio lies above its physical value over the range of lattice parameters, for which $m_\rho \cdot a$ is measured. The m_π/m_ρ ratio comes out easily twice to three times the experimental value once the bare quark masses are too large. Thus the ρ-mass is not (yet) a perfect candidate for the conversion from lattice units into physical units.

In earlier lattice simulations thermodynamic quantities have been plotted as function of $6/g_0^2$. Nowadays one finds plots vs T [MeV]. The translation should be taken with care as long as the mapping between $6/g_0^2$ and T is not (yet) unique due to violations of asymptotic scaling.

3.3.7   *Summary and outlook*

So far we have specified the measure, boundary conditions and lattice action for the path integral of Eq. (3.3.73). We have sketched how numbers from the lattice can be translated into physical units, and what a criterion for the relevance of lattice results for continuum physics looks like. Numerical and analytical methods can be utilized to attack the functional integrations of Eq. (3.3.73). Analytical methods are usually applicable in limits where small expansion parameters are available. For example, the *hopping-parameter* expansion is an expansion in small values of κ, around the limit of infinite bare quark masses. The *strong-coupling* expansion applies for large values of the bare gauge coupling g_0, so the inverse bare coupling provides a small expansion parameter, *linked-cluster expansions* apply for small inverse temperatures or small values of the nearest-neighbor couplings.

As mentioned above, it is intrinsically difficult to find a small expansion parameter in the transition region, where neither the bare coupling constant g_0 is strong nor the renormalized gauge coupling g is weak. Thus it is not surprising that many results on the phase structure in the vicinity of T_c are based on *numerical* simulations. The most important approach is the Monte Carlo method. In contrast to analytical methods, which will be extensively discussed in sections 4.1 - 4.7, we give only a summary of numerical methods in section 4.7.5. Their application to lattice gauge theories is illustrated with some examples from simulations in pure gauge theories and gauge theories with dynamical fermions. For further details we shall refer to the original literature.

Before we get in detail engaged in *Analytical Methods on the Lattice and in the Continuum* in the next chapter, we conclude this chapter with a section on the transfer-matrix approach and Polyakov loops. Throughout this book we use the Lagrangian formalism and the path integral formulation, as it is in general more convenient than the Hamiltonian formalism with a Schrödinger equation for the QCD-Hamiltonian, acting on wave functionals (rather than on wave functions). Yet it is instructive to translate one formulation into the other, illustrated with the example of the Polyakov-loop expectation value. We shall start from the Hamiltonian description at finite temperature that enters the partition function in the transfer matrix representation. We then consider the canonical partition function for an arbitrary color- charge configuration $\tau(x)$, in which $\tau(x)$ labels the irreducible representation, characterizing the color charge at point x. In this formulation the physical interpretation is obvious. The expression is then transformed into the path-integral expression for the expectation value of Polyakov loops. This expression is suited for numerical simulations, but its immediate physical interpretation would be less obvious than in the Hamiltonian description. Moreover, along the derivation we rederive a well-known result that we take otherwise for granted in this book: A finite physical temperature amounts to a finite extension in Euclidean time direction in the Euclidean path-integral formulation.

3.3.8   *Transfer matrix and Polyakov loops*

3.3.8.1   *Path integral formulation of finite temperature field theory*

The derivation of the path integral formulation of a field theory at finite temperature usually proceeds via the Hamiltonian formalism. A possible starting point is the field theory at zero temperature in its path integral description. Defining an appropriate Hilbert space and a transfer matrix, or, equivalently, a Hamiltonian operating in this space, the Schwinger functions coincide with the correlation functions of the corresponding field theory. Once the Hamiltonian description is obtained, in turn the partition function of the finite temperature equilibrium field theory is constructed in a standard way. We refer the reader who is interested in details to textbooks on finite-temperature field theory, although we rederive some basic relations also in this section.

The section is organized as follows. We start with the pure $SU(N)$ gauge theory at zero temperature in the Euclidean path integral formulation of lattice gauge theory and consider Euclidean N-point functions in the axial gauge. (The axial gauge at zero temperature is admissible because of the infinite extension in Euclidean time direction.) We then introduce the notions needed for the Hamiltonian formalism at zero temperature, that is the Hilbert space, projection operators on states transforming with respect to an irreducible representation of the (color) group $SU(N)$, the transfer matrix (without explicitly giving the form of the Hamiltonian) and end up with an expression for the Schwinger functions (see (3.3.124)) equivalent to the original expression (3.3.95). Not only for the N-point functions, but also for the partition function Z there is a correspondence between the Euclidean path integral formulation at $T = 0$ on the one side and the representation of Z, written as the trace over the transfer matrix $(\mathrm{tr}_{\mathcal{H}} \, \widehat{T}^m)$ in the limit $m \to \infty$, on the other side. The formulation via the transfer matrix is suited for introducing a finite temperature $T > 0$, such that Z can be written as $\mathrm{tr}_{\mathcal{H}} \, \widehat{T}^{L_0}$ with L_0 being the inverse temperature. Without derivation we state explicitly the form of $\widehat{T}$ for a pure gauge theory. Next we go backwards from the Hamiltonian to the Euclidean path-integral representation, but for $T > 0$. We start in section 3.3.8.4 from the path-integral expression of the partition function of the Polyakov loops in the irreducible representation τ of the gauge group G, an expression that is familiar from Monte Carlo simulations of Polyakov-loop expectation values. We then prove that this path-integral expression agrees with the partition function of a configuration of isolated (infinitely heavy) static quarks, located at sites x and transforming with respect to the representation $\tau(x)$ of $SU(N)$, formulated in terms of the transfer matrix (see (3.3.151)).

As an exercise it may be useful for the reader to recapitulate from quantum mechanics how the equivalence between the Schrödinger formalism and Feynman's path integral is shown (for example, for the very simple case of a free particle of mass m in three dimensions).

First we explore the relation between the path integral and the Hamiltonian description of a gauge theory. We use the equivalence of both descriptions to derive gauge invariant observables which describe localized color charges as irreducible representations of the gauge group. These gauge invariant observables are the well known Polyakov loops (or Wilson lines). Polyakov loops are used to describe isolated, static (infinitely heavy) quarks and their correlations. They are similar to the world lines of static quarks in a Wilson loop used as order parameter at zero temperature. Polyakov loops are closed due to the periodic boundary conditions in Euclidean time direction. They provide a criterion for confinement or deconfinement at finite temperature.

In order to have a well defined nonperturbative framework we use lattice gauge theory. We start with a lattice gauge theory with compact gauge group at zero temperature. For the current considerations it is useful to consider an asymmetric lattice, with different lattice spacings in spatial and in temporal direction. The four-dimensional lattice is denoted by

$$\overline{\Lambda}_0 = a_0 \mathbb{Z} \times a_s \mathbb{Z}^3 \,, \tag{3.3.86}$$

with a_0 the lattice spacing in Euclidean time direction and a_s the same lattice spacing in all spatial directions. For a more systematic discussion we should start with a lattice which is first finite in all directions, in which the spatial volume is sent to infinity later, but we ignore this complication for simplicity. $\overline{\Lambda}_1$ denotes the set of all links of the lattice,

$$\overline{\Lambda}_1 = \{(x; \mu) \mid x \in \overline{\Lambda}_0, \mu = 0, \ldots, 3\} \,. \tag{3.3.87}$$

The bars on $\overline{\Lambda}_1$ and $\overline{\Lambda}_0$ shall only indicate that the lattices are in four (rather than three) dimensions, and we shall also use lattices in three dimensions in this section. Let G denote any compact group, which plays the role of the gauge group. In the physical cases in which we are interested here, $G = SU(N)$, and we refer to this case for simplicity. We denote the invariant Haar measure on $SU(N)$ (usually written as $d\mu(U)$) simply by dU. The gauge fields $U(x; \mu)$ are $SU(N)$-valued functions associated with the links $\overline{\Lambda}_1$.

The Wilson action on the four-dimensional, asymmetric lattice was introduced in (3.3.33f), here we repeat it to make the section self-contained. It is given by

$$
\begin{aligned}
S_W(U) = \sum_{x \in \overline{\Lambda}_0} \Bigg\{ &\sum_{i \neq j = 1, 2, 3} \frac{\beta_m}{2} \Bigg[1 - \frac{1}{N} \\
&\cdot \operatorname{tr} U(x; i) U(x + a_s \widehat{i}; j) U(x + a_s \widehat{j}; i)^{-1} U(x; i)^{-1} \Bigg] \\
&+ \sum_{i = 1, 2, 3} \beta_e \Bigg[1 - \frac{1}{2N} \Big\{ \operatorname{tr} U(x; i) U(x + a_s \widehat{i}; 0) U(x + a_0 \widehat{0}; i)^{-1} U(x; 0)^{-1} \\
&+ \operatorname{tr} U(x; 0) U(x + a_0 \widehat{0}; i) U(x + a_s \widehat{i}; 0)^{-1} U(x; i)^{-1} \Big\} \Bigg] \Bigg\},
\end{aligned}
\tag{3.3.88}
$$

tr denotes the trace on $N \times N$-matrices. β_m and β_e are related to the bare gauge coupling constant g_0 by

$$\beta_m = \frac{a_0}{a_s} \frac{2N}{g_0^2} \,, \quad \beta_e = \frac{a_s}{a_0} \frac{2N}{g_0^2} \,. \tag{3.3.89}$$

Writing $U(x; \mu) = \exp A_\mu(x)$ and

$$A_0 = a_0 \sum_{d=1}^{N^2-1} \mathcal{A}_0^d T^d \,, A_i = a_s \sum_{d=1}^{N^2-1} \mathcal{A}_i^d T^d \,,$$

$$\mathrm{tr}\, T^d T^e = -\frac{1}{2} \delta_{de}, \tag{3.3.90}$$

where $\{T\}$ are a basis of the Lie-algebra of $SU(N)$. The action (3.3.88) reproduces the standard classical continuum action as a_0, $a_s \to 0$.

The dynamics of the gluon fields is expressed in terms of their correlation functions. With V denoting U or U^{-1},

$$< V(x^{(1)}; \mu^{(1)}) \cdots V(x^{(n)}; \mu^{(n)}) >$$
$$= \frac{1}{Z} \int \mathcal{D}U \, V(x^{(1)}; \mu^{(1)}) \cdots V(x^{(n)}; \mu^{(n)}) \, \exp\left(-S_W(U)\right), \tag{3.3.91}$$

where

$$Z = \int \mathcal{D}U \, \exp\left(-S_W(U)\right) \tag{3.3.92}$$

and

$$\mathcal{D}U \equiv \prod_{(x;\mu)\in\overline{\Lambda}_1} dU(x;\mu) \tag{3.3.93}$$

denotes the product Haar measure on $\overline{\Lambda}_1$.

3.3.8.2   *Quantum mechanics of gluons*

The next step is to construct the quantum mechanics of gluons We call it quantum mechanics, because on the lattice we have to deal with an infinite but countable set of degrees of freedom, treated now in the Hamiltonian formalism. One possible construction of the space of physical states of the Euclidean theory proceeds via the well-known Osterwalder-Schrader positivity condition of the above expectation value. On the lattice we have also a more natural procedure of constructing a physical Hilbert space, although at the price of fixing the gauge. For the following considerations it is very convenient to choose the axial gauge. In this gauge, all U-variables on the time-like links are set equal to unity, $U(x; 0) \equiv 1$. This is an admissible gauge, because for every gauge invariant observable $I(U)$, the expectation values with and without gauge fixing coincide,

$$< I(U) > = < I(U) >_{gf}, \tag{3.3.94}$$

where the subscript on the right hand side implies that all $U(x;0) \equiv 1$. The correlation functions for the gauge fixed theory become

$$< V(x^{(1)};i^{(1)}) \cdots V(x^{(n)};i^{(n)}) >_{gf} \qquad (3.3.95)$$

$$= \frac{1}{Z_{gf}} \int \prod_{(x;i)} dU(x;i) \; V(x^{(1)};i^{(1)}) \cdots V(x^{(n)};i^{(n)}) \; \cdot \; \exp\left(-S_W^{gf}(U)\right),$$

where

$$Z_{gf} = \int \prod_{(x;i)} dU(x;i) \; \cdot \; \exp\left(-S_W^{gf}(U)\right). \qquad (3.3.96)$$

The Latin indices i run over spatial components only, $i \in \{1,2,3\}$. The gauge fixed Wilson action reads

$$S_W^{gf}(U) = \sum_{x \in \overline{\Lambda}_0} \left\{ \sum_{i \neq j = 1,2,3} \frac{\beta_m}{2} \left[1 - \frac{1}{N} \right. \right. \qquad (3.3.97)$$

$$\left. \cdot \; \mathrm{tr}\, U(x;i)U(x + a_s\widehat{i};j)U(x + a_s\widehat{j};i)^{-1}U(x;i)^{-1} \right]$$

$$+ \sum_{i=1,2,3} \beta_e \left[1 - \frac{1}{2N} \mathrm{tr}\left\{ U(x;i)U(x + a_0\widehat{0};i)^{-1} + U(x + a_0\widehat{0};i)U(x;i)^{-1} \right\} \right] \right\}.$$

Towards the definition of the time zero Hilbert space $\mathcal{H}$ we define the spatial lattice by

$$\Lambda_0 = a_s \mathbb{Z}^3. \qquad (3.3.98)$$

Links of the spatial lattice are defined as Λ_1,

$$\Lambda_1 = \{(\mathbf{x};i) \mid \mathbf{x} \in \Lambda_0, i = 1,2,3\}. \qquad (3.3.99)$$

Furthermore we write

$$G_0 = \bigotimes_{\mathbf{x} \in \Lambda_0} G \qquad (3.3.100)$$

for the group of gauge transformations on the spatial lattice, cf. below, with corresponding product measure

$$d\mu_0(g) = \prod_{\mathbf{x} \in \Lambda_0} dg(\mathbf{x}), \qquad (3.3.101)$$

similarly for the corresponding product over the links

$$G_1 = \bigotimes_{(\mathbf{x};i)\in\Lambda_1} G,$$
(3.3.102)

and correspondingly

$$d\mu_1(U) = \prod_{(\mathbf{x};i)\in\Lambda_1} dU(\mathbf{x};i).$$
(3.3.103)

The Hilbert space $\mathcal{H}$ is then defined as the space of all complex valued, square integrable functions $\Psi(\{U(\mathbf{x};i)\})$ on G_1,

$$\mathcal{H} = L_2(G_1, d\mu_1).$$
(3.3.104)

For $\Phi, \Psi \in \mathcal{H}$, the scalar product of Φ and Ψ is given by

$$(\Phi, \Psi) = \int d\mu_1(U) \, \overline{\Phi}(U)\Psi(U),$$
(3.3.105)

where the bar denotes complex conjugation. The operators $\widehat{U}(\mathbf{x};i)$ act as multiplication operators on $\mathcal{H}$,

$$\left[\widehat{U}(\mathbf{x};i)_{\alpha\beta}\Psi\right](U) = U(\mathbf{x};i)_{\alpha\beta}\Psi(U),$$
(3.3.106)

and similarly for $\widehat{U}(\mathbf{x};i)^{-1}$. In many cases, an operator $\widehat{O}$ acting on $\mathcal{H}$ is represented by an integral kernel $\widetilde{O}(U, U')$,

$$(\widehat{O}\Psi)(U) = \int d\mu_1(U') \, \widetilde{O}(U, U') \, \Psi(U').$$
(3.3.107)

In particular the transfer matrix is represented by its integral kernel, as we shall see below.

The group G_0 as defined in (3.3.100) is the group of gauge transformations in $\mathcal{H}$. It corresponds to the group of time-independent gauge transformations of the Euclidean formulation which we are left with after the axial gauge fixing. For every $h = \{h(\mathbf{x}) , \mathbf{x} \in \Lambda_0\} \in G_0$, the gauge transformed U are defined by

$$\left(U^h\right)(\mathbf{x};i) = h(\mathbf{x})U(\mathbf{x};i)h(\mathbf{x} + a_s\widehat{i})^{-1}.$$
(3.3.108)

We notice that $(U^h)^l = U^{lh}$ for all $h, l \in G_0$. (3.3.108) induced a unitary (anti-) representation of G_0 in $\mathcal{H}$ according to

$$(\mathcal{U}(h)\Psi)(U) = \Psi(U^h),$$
(3.3.109)

for all $\Psi \in \mathcal{H}$, with $\mathcal{U}(h)\mathcal{U}(l) = \mathcal{U}(lh)$.

The Hilbert space $\mathcal{H}$ itself is not gauge invariant. But it can be decomposed according to irreducible representations $\{\tau(\mathbf{x}) , \mathbf{x} \in \Lambda_0\}$ of G_0,

$$\mathcal{H} = \bigoplus_{\{\tau(\mathbf{x})\}} \mathcal{H}_{\{\tau(\mathbf{x})\}} \tag{3.3.110}$$

with corresponding orthogonal projection operators

$$P_{\{\tau(\mathbf{x})\}} = \prod_{\mathbf{x} \in \Lambda_0} P_{\tau(\mathbf{x})}. \tag{3.3.111}$$

Explicitly they read according to the Peter-Weyl theorem

$$\left(P_{\{\tau(\mathbf{x})\}}\Psi\right)(U) = \left(\prod_{\mathbf{x} \in \Lambda_0} d_{\tau(\mathbf{x})} \int dh(\mathbf{x})\, \overline{\chi}_{\tau(\mathbf{x})}(h(\mathbf{x}))\right)\Psi(U^h), \tag{3.3.112}$$

where $d_{\tau(\mathbf{x})}$ is the dimension of the irreducible representation space of $\tau(\mathbf{x})$ of G and $\chi_{\tau(\mathbf{x})}$ its character, that is, for every $h \in G$

$$\chi_\tau(h) = \sum_{i=1}^{d_\tau} \tau(h)_{ii}. \tag{3.3.113}$$

Finally the bar in (3.3.112) denotes complex conjugation.

The dynamics of a quantum system is described by its transfer matrix. In general, the transfer matrix $\widehat{T}$ is a positive trace-class operator on $\mathcal{H}$. Hence $\widehat{T}$ is bounded and self-adjoint, and a Hamiltonian $\widehat{H}$ is defined by

$$\widehat{T} = \exp\left(-a_0\widehat{H}\right), \tag{3.3.114}$$

where $\widehat{H}$ is itself self-adjoint and bounded from below. This implies the existence of a lowest energy subspace. $\widehat{T}$ generates a translation in the positive, Euclidean time direction by one lattice spacing a_0.

Before we give the explicit expression of $\widehat{T}$, we define the Schwinger functions (the time-ordered Green functions) of the quantum system for a given $\widehat{T}$. They provide an alternative, equivalent description of the dynamics. For the appropriate transfer matrix $\widehat{T}$, the Schwinger functions will coincide with the corresponding Euclidean expectation values (3.3.95). This ensures the equivalence of the Euclidean and the Hamiltonian description.

We need the notion of the trace of an operator on $\mathcal{H}$. Let $\mathcal{O}$ be an operator and $\{\Psi_n\}$ be an arbitrary orthonormal basis of $\mathcal{H}$. The trace of

$\mathcal{O}$ is defined by

$$\text{tr}_{\mathcal{H}}\, \mathcal{O} = \sum_n (\Psi_n, \mathcal{O}\Psi_n). \tag{3.3.115}$$

It does not depend on the choice of the orthonormal basis.

Let $\widehat{\mathcal{P}}_j$ denote polynomials of time-zero fields $\widehat{U}(\mathbf{x};i)$ and $\widehat{U}(\mathbf{x};i)^{-1}$. The time translates of $\widehat{\mathcal{P}}_j$ are defined by

$$\widehat{\mathcal{P}}_{j,t} = \widehat{T}^t \widehat{\mathcal{P}}_j \widehat{T}^{-t}. \tag{3.3.116}$$

For integer $t_1 < t_2 < \cdots < t_n$, the Schwinger functions of a set $\widehat{\mathcal{P}}_{1,t_1}$, $\widehat{\mathcal{P}}_{2,t_2}$, $\ldots$, $\widehat{\mathcal{P}}_{n,t_n}$ are defined by

$$
\begin{aligned}
\mathcal{S}\left(\widehat{\mathcal{P}}_{1,t_1}, \widehat{\mathcal{P}}_{2,t_2}, \ldots, \widehat{\mathcal{P}}_{n,t_n}\right) &= \lim_{m \to \infty} \frac{\text{tr}_{\mathcal{H}}\, \widehat{T}^m \widehat{\mathcal{P}}_{1,t_1} \widehat{\mathcal{P}}_{2,t_2} \ldots \widehat{\mathcal{P}}_{n,t_n}}{\text{tr}_{\mathcal{H}}\, \widehat{T}^m} \\
&= \frac{\text{tr}_{\mathcal{H}}\, P_{vac} \widehat{\mathcal{P}}_{1,t_1} \widehat{\mathcal{P}}_{2,t_2} \ldots \widehat{\mathcal{P}}_{n,t_n}}{\text{tr}_{\mathcal{H}}\, P_{vac}},
\end{aligned}
\tag{3.3.117}
$$

where P_{vac} is the orthogonal projection operator onto the lowest energy subspace (or equivalently, the subspace of the largest eigenvalue of $\widehat{T}$). So far we have merely introduced the formalism.

We now give an explicit expression for the transfer matrix $\widehat{T}$ acting on $\mathcal{H}$, which is the transfer matrix of the Euclidean lattice gauge theory discussed above. $\widehat{T}$ is the product of three operators,

$$\widehat{T} = \widehat{\mathcal{M}}^\dagger \widehat{\mathcal{E}} \widehat{\mathcal{M}} \tag{3.3.118}$$

where $\widehat{\mathcal{M}}^\dagger$ denotes the Hermitian conjugate of $\widehat{\mathcal{M}}$. $\widehat{\mathcal{M}}$ and $\widehat{\mathcal{E}}$ are given as follows.

The "magnetic" part $\widehat{\mathcal{M}}$ amounts to a multiplication by a positive function

$$\left(\widehat{\mathcal{M}}\Psi\right)(U) = \mathcal{M}(U)\Psi(U) \tag{3.3.119}$$

with

$$
\mathcal{M}(U) = \exp\left\{ -\frac{1}{4}\beta_m \sum_{\mathbf{x} \in \Lambda_0} \sum_{i \neq j = 1,2,3} \left[1 - \frac{1}{N} \right.\right. \tag{3.3.120}
$$
$$
\left.\left. \cdot\, \text{tr}\, U(x;i) U(x + a_s\widehat{i}; j) U(x + a_s\widehat{j}; i)^{-1} U(x;i)^{-1} \right] \right\}.
$$

Furthermore, the "electric" part $\widehat{\mathcal{E}}$ is a positive convolution

$$(\widehat{\mathcal{E}}\Psi)(U) \;=\; \int d\mu_1(U')\,\widetilde{\mathcal{E}}(U,U')\,\Psi(U') \tag{3.3.121}$$

with

$$\widetilde{\mathcal{E}}(U,U') \;=\; \exp\Bigg\{-\beta_e \sum_{\mathbf{x}\in\Lambda_0}\sum_{i=1,2,3}\Bigg[1-\frac{1}{2N} \tag{3.3.122}$$
$$\mathrm{tr}\Big\{U(x;i)U(x;i)'^{-1}+U(x;i)'U(x;i)^{-1}\Big\}\Bigg]\Bigg\}.$$

As one can show, the operator $\widehat{T}$ just defined has all the required properties of a transfer matrix: It is strictly positive, bounded and thus selfadjoint in $\mathcal{H}$. These properties guarantee that $\widehat{T}$ may be written as (3.3.114) with $a_0 = 1$. Furthermore, it is gauge invariant in the sense that

$$\Big[P_{\{\tau(\mathbf{x})\}}\,,\,\widehat{T}\Big] \;=\; 0, \tag{3.3.123}$$

for all irreducible representations $\{\tau(\mathbf{x})\}$ of G_0.

Boundedness of $\widehat{T}$ and gauge invariance (3.3.123) are easily seen. The only less obvious statement is that $\widehat{T}$ is strictly positive. In the continuum limit, positivity follows from reflection positivity and time translation invariance, by using strong continuity of the time translation operator $S : t \to \widehat{T}^t$. On the lattice, an additional argument is required. The proof uses Fourier transform on G_0. For the proof we refer to the literature.

We can now formulate the reconstruction theorem, that is, the equality of the Schwinger functions (3.3.117) and the Euclidean correlation functions (3.3.95). Let P_j denote polynomials of the Euclidean time zero fields $U(\mathbf{x};i)$ (and $U(\mathbf{x};i)^{-1}$), $P_j = P_j(\{U(\mathbf{x};i)\})$, and let $\widehat{\mathcal{P}}_j$ be the corresponding polynomials $\mathcal{P}_j(\{\widehat{U}(\mathbf{x};i)\})$ of operators $\widehat{U}$ as defined in (3.3.106).

For $t_1 < t_2 < \cdots < t_n$, the Schwinger functions of a set $\widehat{\mathcal{P}}_{1,t_1}, \widehat{\mathcal{P}}_{2,t_2}, \ldots,$ $\widehat{\mathcal{P}}_{n,t_n}$ coincide with the corresponding Euclidean expectation values,

$$\mathcal{S}\left(\widehat{\mathcal{P}}_{1,t_1},\widehat{\mathcal{P}}_{2,t_2},\ldots,\widehat{\mathcal{P}}_{n,t_n}\right) \;=\; \Bigg\langle \mathcal{P}_1(\{U(\mathbf{x}+a_0 t_1;i)\}) \tag{3.3.124}$$
$$\cdot\,\mathcal{P}_2(\{U(\mathbf{x}+a_0 t_2;i)\})\cdots\mathcal{P}_n(\{U(\mathbf{x}+a_0 t_n;i)\})\Bigg\rangle_{gf}.$$

This means that all Euclidean correlation functions are reobtained as the appropriate Schwinger functions in the Hamiltonian description of the lattice gauge theory. We skip the proof in this context.

3.3.8.3 *Finite-temperature partition function*

Since we have now a Hamiltonian description of the Euclidean zero temperature lattice field theory at our disposal, we can easily specify the partition function for this system which is in equilibrium at some temperature T. With L_0 being any positive integer number, we interpret

$$\widehat{T}^{L_0} \; = \; \exp\left(-L_0 a_0 \widehat{H}\right) \tag{3.3.125}$$

as the Boltzmann operator for a lattice gauge theory in thermal equilibrium, with temperature $T = (L_0 a_0)^{-1}$. Let us recall that

$$P_{\{\tau(\mathbf{x})\}} \; = \; \prod_{\mathbf{x} \in \Lambda_0} P_{\tau(\mathbf{x})}, \tag{3.3.126}$$

as given by (3.3.112), projects onto the subspace of the Hilbert space $\mathcal{H}$ which is the irreducible subspace of the representation $\{\tau(\mathbf{x})\}$ of G_0. Hence

$$Z_{\{\tau(\mathbf{x})\}} \; = \; \mathrm{tr}_{\mathcal{H}} \, \widehat{T}^{L_0} P_{\{\tau(\mathbf{x})\}} \tag{3.3.127}$$

is the canonical partition function for the color charge configuration $\{\tau(\mathbf{x})\}$ at temperature $T = (L_0 a_0)^{-1}$. In particular, writing $\tau = e$ for the trivial representation of G and $P_0 = P_{\{e(\mathbf{x})\}}$, so that

$$(P_0 \Psi)(U) \; = \; \int d\mu_0(h) \, \Psi(U^h), \tag{3.3.128}$$

we get

$$Z \; = \; \mathrm{tr}_{\mathcal{H}} \, \widehat{T}^{L_0} P_0 \tag{3.3.129}$$

as the canonical partition function of the "vacuum", where at each lattice site there is no color charge. As a further example, let τ be an arbitrary irreducible representation of the gauge group G and $\mathbf{x}_0$ any fixed lattice site. With

$$P_{(\tau; \mathbf{x}_0)} \; = \; P_{\tau(\mathbf{x}_0)} \prod_{\mathbf{x} \in \Lambda_0} P_e, \tag{3.3.130}$$

we see that

$$Z_{(\tau; \mathbf{x})} \; = \; \mathrm{tr}_{\mathcal{H}} \, \widehat{T}^{L_0} P_{(\tau; \mathbf{x}_0)} \tag{3.3.131}$$

is the partition function of an isolated (infinitely heavy) colored quark with color charge τ (i.e. transforming according to the representation τ), located at $\mathbf{x}_0$ in the vacuum.

3.3.8.4   *Representation as path integral*

Our final step is to write the general canonical partition function (3.3.127) as path integral. As we shall see below, $Z_{\{\tau(\mathbf{x})\}}$ is equal to the expectation value of a gauge invariant quantity in a four-dimensional lattice gauge theory with one compact dimension, a "finite temperature" lattice. This gauge invariant quantity is nothing else but a product of Polyakov loops over all $\mathbf{x}$ in the corresponding representation.

We start with some definitions. Let L_0 be a positive integer. We define the finite temperature $L_0 \times \infty^3$ lattice by

$$\overline{\Lambda}_0^{L_0} = a_0 \mathbb{Z}/L_0 \times a_s \mathbb{Z}^3 \qquad (3.3.132)$$

with a_0 the lattice spacing in 0th (temperature) direction and a_s the lattice spacing in the spatial directions. $\overline{\Lambda}_1^{L_0}$ denotes the set of all links of the lattice,

$$\overline{\Lambda}_1^{L_0} = \{(x;\mu) \mid x \in \overline{\Lambda}_0^{L_0}, \mu = 0, \dots, 3\}. \qquad (3.3.133)$$

We impose periodic boundary conditions on the $U(x;\mu)$ in the temperature direction. Let us define for every spatial site $\mathbf{x}$ the Polyakov line

$$
\begin{aligned}
U_L(\mathbf{x}) &= \prod_{i_0=0}^{L_0-1} U((a_0 i_0, \mathbf{x}); 0) \\
&= U((0,\mathbf{x});0) U((1a_0,\mathbf{x});0) \cdots U(((L_0-1)a_0, \mathbf{x});0). \quad (3.3.134)
\end{aligned}
$$

The corresponding traces

$$\overline{\chi}_\tau(U_L(\mathbf{x})) \qquad (3.3.135)$$

are the Polyakov loops in the irreducible representation τ of the gauge group G. Let us define

$$Z'_{\{\tau(\mathbf{x})\}} = \int \mathcal{D}U \prod_{\mathbf{x}} \Big(d_{\tau(\mathbf{x})} \overline{\chi}_{\tau(\mathbf{x})}(U_L(\mathbf{x})) \Big) \exp\left(-S_W(U)\right), \qquad (3.3.136)$$

with measure

$$\mathcal{D}U = \prod_{(x;\mu)\in\overline{\Lambda}_1^{L_0}} dU(x;\mu), \tag{3.3.137}$$

where the product runs over all links, and the Wilson action is given by (3.3.88) but on the lattice $\overline{\Lambda}_0^{L_0}$ of (3.3.132).

The claim now is that (3.3.136) and (3.3.127) are identical, that is,

$$Z'_{\{\tau(\mathbf{x})\}} = Z_{\{\tau(\mathbf{x})\}}. \tag{3.3.138}$$

This means that (3.3.136) is the partition function of the lattice gauge theory with color charge distribution $\{\tau(\mathbf{x})\}$.

We prove (3.3.138) for the case of just one nontrivial representation at some $\mathbf{x}_0$ in order to simplify the notations, since the general case amounts to an obvious generalization. With the notation introduced in connection with (3.3.131), we will now show that

$$Z'_{(\tau;\mathbf{x}_0)} = Z_{(\tau;\mathbf{x}_0)} \tag{3.3.139}$$

Let us consider

$$Z'_{(\tau;\mathbf{x}_0)} = \int \mathcal{D}U \, d_{\tau(\mathbf{x}_0)} \chi_{\tau(\mathbf{x}_0)}(U_L(\mathbf{x}_0)) \exp\left(-S_W(U)\right). \tag{3.3.140}$$

Let m identify the mth time slice,

$$U^m(\mathbf{x};\mu) = U((ma_0,\mathbf{x});\mu), \ 0 \leq m \leq L_0 - 1, \tag{3.3.141}$$

moreover, for convenience

$$h^m(\mathbf{x}) = U^m(\mathbf{x};0) = U((ma_0,\mathbf{x});0). \tag{3.3.142}$$

We split the measure into a product of timelike and spacelike links for each time slice

$$\mathcal{D}U = \prod_{m=0}^{L_0-1} \left(\prod_{(\mathbf{x};i)\in\Lambda_1} dU^m(\mathbf{x};i) \cdot \prod_{\mathbf{x}\in\Lambda_0} dU^m(\mathbf{x};0) \right)$$
$$= \prod_{m=0}^{L_0-1} d\mu_1(U^m) d\mu_0(h^m). \tag{3.3.143}$$

Hence

$$Z'_{(\tau;\mathbf{x}_0)} = \int \prod_{m=0}^{L_0-1} \left(d\mu_1(U^m) d\mu_0(h^m) \right) \cdot$$ (3.3.144)

$$d_{\tau(\mathbf{x}_0)} \chi_{\tau(\mathbf{x}_0)} \left(h^0(\mathbf{x}_0) h^1(\mathbf{x}_0) \cdots h^{L_0-1}(\mathbf{x}_0) \right) \exp\left(-S_W(U, h)\right).$$

The next step is to introduce the following static-like gauge fixing. Under the integral sign we insert the identity

$$1 = \prod_{m=1}^{L_0-1} \int d\mu_0(k^m) \prod_{\mathbf{x}} \delta(k^m(\mathbf{x}) h^m(\mathbf{x}) k^{m+1}(\mathbf{x})^{-1} \mid 1_G),$$ (3.3.145)

where $\delta()$ denotes the Dirac delta function on G,

$$\int dg \, f(g) \delta(g \mid 1_G) = f(1_G)$$ (3.3.146)

for continuous f, and 1_G denotes the unit element of G. We emphasize that (3.3.145) is an admissible gauge fixing, in particular because the 0th time slice is not gauge fixed (the product over the slices starts at 1). We apply a gauge transformation

$$h^m(\mathbf{x}) \to k^m(\mathbf{x})^{-1} h^m(\mathbf{x}) k^{m+1}(\mathbf{x})$$
$$U^m(\mathbf{x}; i) \to k^m(\mathbf{x})^{-1} U^m(\mathbf{x}; i) k^m(\mathbf{x} + a_s\widehat{i}).$$ (3.3.147)

In this way, all h not associated with the 0-th time slice disappear, and we obtain

$$Z'_{(\tau;\mathbf{x}_0)} = \int \prod_{m=0}^{L_0-1} d\mu_1(U^m) \cdot d\mu_0(h) \cdot$$ (3.3.148)

$$d_{\tau(\mathbf{x}_0)} \chi_{\tau(\mathbf{x}_0)}(h) \exp\left(-S_W^{gf}(U, h)\right),$$

where

$$S_W^{gf}(U,h) = \sum_m \sum_{x \in \Lambda_0} \left\{ \sum_{i \neq j = 1,2,3} \frac{\beta_m}{2} \left[1 - \frac{1}{N} \right. \right. \tag{3.3.149}$$

$$\left. \cdot \operatorname{tr} U^m(x;i) U^m(x + a_s\widehat{i};j) U^m(x + a_s\widehat{j};i)^{-1} U^m(x;i)^{-1} \right]$$

$$+ (1 - \delta_{m,0}) \sum_{i=1,2,3} \beta_e \left[1 - \frac{1}{2N} \right.$$

$$\left. \cdot \operatorname{tr} \left\{ U^m(x;i) U^{m+1}(x;i)^{-1} + U^{m+1}(x;i) U^m(x;i)^{-1} \right\} \right]$$

$$+ \delta_{m,0} \sum_{i=1,2,3} \beta_e \left[1 - \frac{1}{2N} \operatorname{tr} \left\{ U^0(x;i) h(\mathbf{x} + a_s\widehat{i}) U^1(x;i)^{-1} h(\mathbf{x})^{-1} \right. \right.$$

$$\left. \left. \left. + h(\mathbf{x}) U^1(x;i) h(\mathbf{x} + a_s\widehat{i})^{-1} U^0(x;i)^{-1} \right\} \right] \right\},$$

with $U^{L_0} \equiv U^0$. Now the left hand side of $Z'_{\tau;\mathbf{x}}$ takes a form that can be interpreted in terms of integral kernels of the corresponding operators in the Hamiltonian formalism, namely upon inserting

$$\int d\mu_0(h) \, d_{\tau(\mathbf{x}_0)} \chi_{\tau(\mathbf{x}_0)}(h) \exp\left(-S_W^{gf}(U,h) \right) \tag{3.3.150}$$

$$= \prod_{m=1}^{L_0-1} \left(\mathcal{M}^\dagger \mathcal{E} \mathcal{M} \right) (U^{m+1}, U^m) \cdot \left(P_{(\tau,\mathbf{x}_0)} \mathcal{M}^\dagger \mathcal{E} \mathcal{M} \right) (U^1, U^0)$$

into (3.3.148), we get

$$Z'_{(\tau;\mathbf{x}_0)} = \int \prod_{m=0}^{L_0-1} d\mu_1(U^m) \cdot \prod_{m=1}^{L_0-1} \left(\mathcal{M}^\dagger \mathcal{E} \mathcal{M} \right) (U^{m+1}, U^m)$$

$$\cdot \left(P_{(\tau,\mathbf{x}_0)} \mathcal{M}^\dagger \mathcal{E} \mathcal{M} \right) (U^1, U^0) \tag{3.3.151}$$

$$= \operatorname{tr} \widehat{T}^{L_0} P_{(\tau;\mathbf{x}_0)}$$

$$= Z_{(\tau;\mathbf{x}_0)}.$$

This proves (3.3.138).

To summarize: we started from the path integral expression for the partition function of Polyakov loops in the irreducible representation τ of the gauge group G (Eq. 3.3.136). (3.3.136) was transformed to (3.3.151) via the Hamiltonian formalism. In (3.3.151) the physical interpretation can be read off. Actually in our proof we considered only one isolated infinitely heavy quark located at $\mathbf{x}_0$ in the vacuum background, just for simplicity. For the case that all static quarks transform with respect to the trivial representation we obtain the partition function of the vacuum. Reading Eqs. (3.3.148 -3.3.151) backwards, we have rederived a familiar result concerning the effect of a finite temperature in the path integral formulation: In order to interpret the right hand side of (3.3.151) as the finite temperature partition function of a Hamiltonian H, in (3.3.148) we have to start with a finite extension in Euclidean time direction with periodic boundary conditions on the gauge fields.

This book deals exclusively with stationary phenomena. In this section we have studied a static (chromo)electric charge (a quark) in a pure $SU(N)$ gauge theory and rederived the path integral representation of its partition function and its expectation value. The expectation value of the Polyakov loop plays the important role of an order parameter in QCD (strictly speaking without dynamical quarks). It is used to distinguish the confinement from the deconfinement phase. Correlations between these static chromo-electric charges are sensitive to screening effects. Therefore color-electric screening can be tested by measuring the distance dependence of correlation functions of two Polyakov loops which provides a gauge invariant definition of a color-electric screening mass. (We come back to such calculations in connection with dimensional reduction in QCD.) Note that such a definition of an electric screening mass does not rely on perturbation theory and may be considered as a definition from first principles.

At various places in this book we point toward the (chromo)magnetic-mass problem, related to its perturbative definition via the propagator in a specific gauge. One of the reasons are the (so far) intractable infrared problems. Therefore it would be rather worthwhile to derive a first-principle nonperturbative definition of a chromo-magnetic mass via the screening behavior of two chromo-magnetic currents. This magnetic mass would be defined in terms of observables that are easily accessible with Monte Carlo simulations of their path integral expressions. The very form of these observables is not yet known. It is tempting to proceed along analogous lines as (3.3.151) backwards to (3.3.136), i.e. to start from the partition function of chromo-magnetic currents in the Hamiltonian formalism and to construct the corresponding observables in the path integral formulation. However, to our knowledge such a derivation is still missing, in particular the very definition of the (chromo)magnetic currents in the Hamiltonian formalism.

Chapter 4

Analytic Methods on the Lattice and in the Continuum

4.1 Convergent versus asymptotic expansions

Among the various analytical methods we would like to contrast convergent versus asymptotic expansions. In common to both expansion schemes (and in contrast to Monte Carlo simulations) is the existence of a small expansion parameter. The physical meaning of the expansion parameter can be an inverse high temperature, a weak coupling constant, a small hopping parameter, an inverse large number of degrees of freedom, or the number of colors or flavors, for example. The difference in the convergence properties comes up in the following way.

To generate a *convergent* series, a first possibility is to present the partition function as a sum over disjoint unions of suitably defined polymers. The small expansion parameter, e.g. the inverse high temperature, is hidden in the so-called activities. A second possibility is to split the action into an ultra-local part $\mathring{S}$ of decoupled degrees of freedom and a next-neighbor part S_{nn}, multiplied with the hopping parameter κ

$$S(\Phi) = \sum_{x \in \Lambda} \mathring{S}(\Phi(x)) + \kappa \cdot S_{nn}(\Phi) \tag{4.1.1}$$

and to Taylor expand $\ln Z$ about a decoupled system in powers of κ. This way we are led to linked cluster expansions (see section 4.2 below). Convergent expansions lead to rigorous results about the existence of a mass gap or the existence of the thermodynamic limit of local observables. For convergent expansions it is worthwhile to reach a high order in the expansion parameter, because the accuracy of the result increases with increasing order of the expansion. If the expansion is a power series in a small parameter, a high order in the expansion parameter also allows an extrapolation of the radius of convergence of these series. Under certain conditions the radius of convergence contains information about the location and scaling behavior of physical singularities like those which characterize critical phenomena and occur at second-order phase transitions. This is the reason why we spend a separate chapter on linked cluster expansions.

On the other hand we know that the very proof of renormalizability of quantum field theories is based on a *perturbative* formulation, or, more precisely, of a split of the action into a free part S_{free} (quadratic in the fields) and an interacting part S_{int}, multiplied with a small bare or renormalized coupling λ, depending on whether the expansion is performed in bare or renormalized quantities

$$S(\Phi) = S_{free}(\Phi) + \lambda \cdot S_{int}(\Phi). \qquad (4.1.2)$$

The Feynman-graph expansion results from a perturbative (i.e. small λ-) expansion of $\exp(\lambda \cdot S_{int}(\Phi))$ in the integrand of the partition function and in derived quantities, and a term-wise evaluation of the Gaussian integrals.

Also the celebrated accuracy in experiments which serve as confirmation of quantum electrodynamics (QED) are based on a perturbative evaluation of QED. These are the $(g - 2)$-experiments for measuring the anomalous magnetic moment of the electron and the muon. For the electron [138]

$$(g - 2)/2 = 1159652187(4) \cdot 10^{-12} \qquad (4.1.3)$$

and for the muon [139]

$$(g - 2)/2 = 11659202(15) \cdot 10^{-10}. \qquad (4.1.4)$$

Nevertheless we can easily convince ourselves that standard perturbative expansions in a small coupling parameter cannot be convergent in the above sense, but are at best *asymptotic* expansions.

Asymptotic expansions can also arise if the total action is multiplied by a common large overall factor α such that it is natural to evaluate the path integral representation of the partition function in a saddle point approximation

$$Z = \int \mathcal{D}\Phi \, e^{-\alpha S(\Phi)}. \qquad (4.1.5)$$

Special cases of α are N with N the total number of degrees of freedom (cf. section 2.3) or $1/\hbar$ with $\hbar$ Planck's action quantum in connection with the semiclassical approximation to the Feynman propagator in quantum mechanics or in field theory.

Sometimes asymptotic expansions are Borel summable. It is then assumed that the Borel transforms contain some nonperturbative information about the original theory. An example for an application of Borel transformations are QCD sum rules.

Finally, if an expansion is neither convergent nor Borel summable nor an asymptotic expansion, or if the status of the expansion is just not known, we call the expansion *formal*. If the first few coefficients of a formal expansion are known, but no control exists about the higher order contributions, the first few coefficients may turn out to be meaningless when higher order coefficients are ill defined or divergent. In fact, the status of resummed perturbation theory in nonabelian gauge theories in four dimensions at finite temperature or in three dimensions at zero temperature is not clarified yet. As a manifestation of the magnetic mass problem, the terms beyond a certain order in the loops are not under control. For the vacuum polarization this order is the one-loop level.

Now we go into detail and start with the very definition of an asymptotic expansion.

4.1.1 *Asymptotic expansions*

Consider a function $f(\lambda)$ depending on a positive real parameter λ and a formal power series

$$S \equiv \sum_{n=0}^{\infty} a_n \lambda^n \; . \tag{4.1.6}$$

S is called asymptotic power series or expansion of $f(\lambda)$ about $\lambda = 0_+$, abbreviated according to

$$f(\lambda) \sim \sum_{n=0}^{\infty} a_n \lambda^n \,, \tag{4.1.7}$$

if for all $N \in \mathbf{N}$ positive real numbers K_N, λ_N exist such that the following inequality holds

$$\left| f(\lambda) - \sum_{n=0}^{N} a_n \lambda^n \right| \leq K_N \lambda^{N+1} \,, \tag{4.1.8}$$

for all $0 \le \lambda < \lambda_N$. Typically $\lambda_N \to 0$ or $K_N \to \infty$, and also typically the coefficients a_n are proportional to n-factorials $n!$.

Frequently asymptotic expansions result when an integration does not commute with an infinite summation of a convergent series. Consider for example the integral

$$\int_0^\infty dx \; \frac{e^{-x}}{1+x\lambda} = \int_0^\infty dx \; e^{-x} \sum_{\nu=0}^\infty (-1)^\nu \, x^\nu \, \lambda^\nu \; . \qquad (4.1.9)$$

Naively interchanging the summation and integration on the right hand side leads to

$$\sum_{\nu=0}^\infty (-1)^\nu \, \nu! \, \lambda^\nu \; , \qquad (4.1.10)$$

a series that is divergent. It is still an asymptotic expansion for the left hand side, as one can show.

Another example is provided by the very argument why a perturbative expansion in a small coupling parameter cannot be convergent. Rather than a full quantum field theoretical model let us consider just one degree of freedom, a Φ^4-theory at one (lattice) site x so that we have an ordinary integral

$$\int_{-\infty}^{+\infty} d\Phi(x) \; e^{-\frac{m^2}{2}\Phi(x)^2 - \frac{\lambda}{4!}\Phi(x)^4}$$

$$= \int_{-\infty}^{+\infty} d\Phi(x) \; e^{-\frac{m^2}{2}\Phi(x)^2} \cdot \sum_{\nu \ge 0} \frac{(-1)^\nu}{\nu!} \, (\frac{\lambda}{4!})^\nu \cdot \Phi(x)^{4\nu}$$

$$\sim \sum_{\nu \ge 0} a_\nu \, \lambda^\nu \qquad (4.1.11)$$

with a_ν appropriately defined. In the last step we have assumed that we may interchange the integral with the infinite sum. Now, if the right hand side would represent a convergent power series with radius of convergence $R > 0$, the series would be convergent for all couplings λ with $|\lambda| < R$, in particular for negative λ which is wrong as we know. Therefore the series on the right hand side can represent the integral on the left hand side at best in an asymptotic sense.

Similarly an asymptotic expansion can be generated when a *path* integral is evaluated term by term in a *saddle-point expansion* of the integrand.

Again, rather than (4.1.5) we consider a field theory in zero space dimensions at one point in time, i.e. an ordinary one-dimensional integral

$$I(\alpha) = \int_{-\infty}^{+\infty} dx \, e^{-\alpha f(x)} . \tag{4.1.12}$$

We sketch the proof that each saddle-point expansion generates an asymptotic expansion for the special case that

$$f(x) = x^2 + x^4 \tag{4.1.13}$$

in reminiscence to a Φ^4-theory. Thus we have the special case of only one global isolated minimum of $f(x)$ in the exponent at $x = x_0 = 0$. We shall see that $I(\alpha)$ for large α admits an asymptotic power series in $1/\alpha$.

- In the first step we split the integral $I(\alpha)$ for arbitrary $\epsilon > 0$ according to

$$I(\alpha) = \left[\int_{-\epsilon}^{+\epsilon} + \int_{-\infty}^{-\epsilon} + \int_{+\epsilon}^{+\infty} \right] dx \, e^{-\alpha f(x)} , \tag{4.1.14}$$

i.e. in a (small) neighborhood about the minimum at $x_0 = 0$ and the "rest" $R_1(x)$

$$I(\alpha) = \int_{-\epsilon}^{+\epsilon} dx \, e^{-\alpha f(x)} + R_1(x) \tag{4.1.15}$$

with constants c_1, c_2 such that

$$|R_1(x)| \leq c_1 \, e^{-c_2 \alpha} , \tag{4.1.16}$$

or

$$I(\alpha) = \int_{-\epsilon}^{+\epsilon} dx \, e^{-\alpha f(x)} + O(e^{-c_2 \alpha}) . \tag{4.1.17}$$

- In the next step we separate the quadratic term of f and Taylor expand the nonquadratic part up to *finite* order in N plus Taylor remainder R_2. In our example

$$e^{-\alpha f(x)} = e^{-\alpha x^2} \left[\sum_{\nu=0}^{N} \frac{1}{\nu!} (-\alpha \, x^4)^{\nu} + R_2(x) \right] . \tag{4.1.18}$$

An analysis of the Taylor remainder shows that a constant d_1 exists such that

$$\left| \int_{-\epsilon}^{+\epsilon} dx \, e^{-\alpha x^2} R_2(x) \right| \leq \frac{d_1}{\alpha^{N+3/2}} \, . \qquad (4.1.19)$$

Therefore we have

$$I(\alpha) = \left[\int_{-\epsilon}^{+\epsilon} dx \, e^{-\alpha x^2} \sum_{\nu=0}^{N} \frac{1}{\nu!} (-\alpha x^4)^\nu \right] + O(\alpha^{-N-3/2})$$

$$= \left[\sum_{\nu=0}^{N} \frac{1}{\nu!} (-\alpha)^\nu \int_{-\epsilon}^{+\epsilon} dx \, e^{-\alpha x^2} \, x^{4\nu} \right] + O(\alpha^{-N-3/2}). \qquad (4.1.20)$$

Notice that the interchange of $\sum$ and $\int$ in the last step was absolutely safe, because we have a *finite* sum of terms.

- Now we shift the α-dependence to the boundaries and obtain

$$I(\alpha) = \sum_{\nu=0}^{N} \frac{(-1)^\nu}{\nu!} \frac{1}{\alpha^{\nu+1/2}} \int_{-\epsilon\sqrt{\alpha}}^{+\epsilon\sqrt{\alpha}} dt \, e^{-t^2} \, t^{4\nu} + O(\alpha^{-N-3/2}) \, .$$
$$\qquad (4.1.21)$$

Finally we want to replace the integration boundaries by $\pm\infty$ to evaluate the integrals in a closed form. Since we are interested in the limit of large α, the error which is induced by this replacement is exponentially suppressed, since constants e_1 and e_2 exist such that

$$\left| \left(\int_{\epsilon\sqrt{\alpha}}^{\infty} + \int_{-\infty}^{-\epsilon\sqrt{\alpha}} \right) dt \, e^{-t^2} \, t^{4\nu} \right| \leq e_1 \, e^{-e_2\alpha} \, . \qquad (4.1.22)$$

It follows that

$$I(\alpha) = \sum_{\nu=1}^{N} \frac{(-1)^\nu}{\nu!} \alpha^{-\nu-1/2} \int_{-\infty}^{+\infty} dt \, e^{-t^2} \, t^{4\nu} + O(\alpha^{-N-3/2})$$

$$= \left[\sum_{\nu=1}^{N} \frac{(-1)^\nu}{\nu!} \Gamma(2\nu + \frac{1}{2}) \, \alpha^{-\nu} + O(\alpha^{-N-1}) \right] \alpha^{-1/2} \, . \qquad (4.1.23)$$

The last equality should be compared with (4.1.8). Since it holds for all positive integer N, we write

$$I(\alpha) \sim (\frac{1}{\alpha})^{1/2} \sum_{\nu=0}^{\infty} c_\nu \, (\frac{1}{\alpha})^\nu \qquad for \quad \frac{1}{\alpha} \to 0 \qquad (4.1.24)$$

with appropriately defined coefficients c_ν and call the right hand side for large α an asymptotic expansion of the left hand side.

Even in this very simple example of an "action" $f(x)$ with only one global isolated minimum we had to control neglected remainders at three places. Therefore it is obvious that the representation of an integral $I(\alpha)$ as an asymptotic power-series expansion in $1/\alpha$ gets much harder in the generic field theoretical case.

4.1.2   *Borel resummations*

The coefficients of asymptotic power series are often proportional to n-factorials. In these cases it is natural to apply a Borel transformation and to Borel-resum the series in the following steps. Again we consider a function $f(\lambda)$ and a formal series $S(\lambda) \equiv \sum_{n=0}^{\infty} a_n \lambda^n$ such that $S(\lambda)$ is an asymptotic series for $f(\lambda)$

$$f(\lambda) \sim \sum_{n=0}^{\infty} a_n \, \lambda^n \qquad (4.1.25)$$

in the sense of (4.1.7) and (4.1.8). As an infinite series $S(\lambda)$ is divergent for every $\lambda \neq 0$ if $a_n \propto n$. Next we try to enforce the convergence by defining

$$\widetilde{S}(\lambda) \equiv \sum_{n=0}^{\infty} \frac{a_n}{n!} \, \lambda^n \, . \qquad (4.1.26)$$

If the coefficients of $\widetilde{S}(\lambda)$ satisfy

$$R^{-1} \equiv \limsup_{\nu \to \infty} \left| \frac{a_\nu}{\nu!} \right|^{1/\nu} < \infty, \qquad (4.1.27)$$

it follows from the root criterion that $\widetilde{S}(\lambda)$ defines a convergent series for all λ with $|\lambda| < R$. We call this convergent series

$$\widetilde{f}(\lambda) = \sum_{n=0}^{\infty} \frac{a_n}{n!} \, \lambda^n. \qquad (4.1.28)$$

Now we apply some kind of inverse procedure to the first step and define a new function $\widehat{f}(\lambda)$ according to

$$\widehat{f}(\lambda) \equiv \int_0^{\infty} dt \, e^{-t} \, \widetilde{f}(\lambda t) \, . \qquad (4.1.29)$$

For the definition of $\widehat{f}(\lambda)$ to make sense we have to assume that $\widetilde{f}(\lambda t)$ has an analytic continuation in a strip along the positive real axis and has further properties such that the integral (4.1.29) exists. The idea behind the definition of $\widehat{f}(\lambda)$ becomes evident if we consider the very special case that $f(\lambda)$ has already a convergent series so that $\widetilde{f}$ is an entire analytic function. In this case the exchange of $\sum$ and $\int$ is allowed

$$\widehat{f}(\lambda) = \int_0^\infty dt\, e^{-t} \sum_{n=0}^\infty \frac{1}{n!}\, a_n\, (\lambda t)^n$$

$$= \sum_{n=0}^\infty \frac{a_n \lambda^n}{n!} \int_0^\infty dt\, e^{-t}\, t^n$$

$$= \sum_{n=0}^\infty \frac{a_n \lambda^n}{n!}\, \Gamma(n+1) \,=\, f(\lambda)\,. \qquad (4.1.30)$$

In general $\widehat{f}(\lambda)$ leads to a definition of a new function that reproduces the original $f(\lambda)$ for certain values of λ if additional conditions are satisfied. These conditions must be further specified. An example for such a specification is the Watson-Nevanlinna theorem [140].

Calan and Rivasseau [141] showed that the Φ^4-theory in four dimensions is locally Borel summable: the analogue of $\widetilde{f}(\lambda)$ exists, but because of the existence of renormalons, $\widetilde{f}(\lambda)$ cannot be analytically continued in a stripe along the positive real axis so that the analogue of $\widehat{f}(\lambda)$ is not defined.

In general the hope is that $\widehat{f}(\lambda)$ contains some nonperturbative information about $f(\lambda)$ when $f(\lambda)$ has been obtained as an asymptotic expansion in perturbation theory. This is also the reason why Borel resummations are applied to QCD sum rules.

Any contribution, however, which is exponentially suppressed in the inverse coupling λ with $e^{-1/\lambda}$ for small λ, is represented by zero in the asymptotic expansion. Therefore Borel transformations are not the right tool to gain information about such exponentially suppressed terms.

4.1.3 *Polymer expansions*

Polymer expansions originated in the work of Mayer and Montroll [142] and Kirkwood and Salsburg [143], but the notion of polymer expansions was generalized later (see e.g. [144]). Polymer expansions correspond to a generic framework that is much more general than the name may suggest. (Only in very special cases polymers are identified with a set of links on a lattice with the interpretation of chemical polymer chains.) It is convenient to separate the *algebraic* from the *analytic* aspects. The algebraic aspects refer to the combinatorics of the associated graphical expansion, the analytic aspects to convergence properties of the series.

Let us first focus on the algebraic aspects. Polymer expansions are an algebraic identity between a partition function Z and its logarithm $\ln Z$ in the sense that $Z = e^{\ln Z}$ for infinite (formal) series Z and $\ln Z$, where *both* Z and $\ln Z$ are series in suitably defined "activities". Therefore, roughly speaking, the existence of a polymer expansion is related to the commutation of $\ln$ and $\sum$ with appropriately defined terms in the sums. More precisely, the inverse function (exp) of the logarithm of a sum (here Z, which initially is not given as an exponential,) equals a sum over logarithms (namely $\exp \ln Z$). Thus the challenge is to define suitable notions of polymers, attributes of polymers of being disjoint or connected, and expansion coefficients, called polymer activities, in a way that a resummation of Z yields the desired series $e^{\ln Z}$. Before we can formulate a theorem about polymer expansions, we have to introduce a number of definitions.

We start with the set $\mathcal{P}$ of objects $\gamma_k, (k = 1, 2, ...)$ called polymers. To every pair $\gamma_i, \gamma_j \in \mathcal{P}$ we attribute the property of either being "connected" ($\gamma_i \cap \gamma_j \neq \emptyset$) or "disjoint" ($\gamma_i \cap \gamma_j = \emptyset$). Furthermore, to every polymer γ an activity $a(\gamma)$ is assigned. The activity $a(\gamma)$ is a function from $\mathcal{P}$ to $\mathbf{C}$. In particular, if activities are real-valued, they need not necessarily be positive. The support *supp* is a function from polymers γ to subsets of the cell-complex Λ, $\gamma \rightarrow supp(\gamma)$. In this context it is convenient to look at the lattice Λ as a cell-complex made of 0-cells (sites x), 1-cells (links b), 2-cells (plaquettes p), 3-cells (cubes c). The support of γ consists of those cells of Λ on which γ depends in a nontrivial way. The volume of a polymer is a function $\gamma \rightarrow |\gamma|$ from $\mathcal{P}$ to nonnegative real numbers. It is defined for polymers that are identified with their support. Therefore the polymers are subsets of $\mathbf{R}^D$ or $\mathbf{Z}^D$ for systems in D dimensions.

Next we introduce for integer k the notion of k-polymers $(\gamma_1, ..., \gamma_k) \equiv X$ with $|X| = k$. (In general k-polymers are not polymers themselves.) k-polymers are elements of the k-fold Cartesian product $\mathcal{P}_k \equiv \mathcal{P} \times \mathcal{P} \times \cdots \times \mathcal{P}$. For the partition function Z we need disjoint k-polymers, the set of disjoint k-polymers is called $\mathcal{D}_k \subset \mathcal{P}_k$, because the sum runs over nonempty collections of mutually disjoint polymers. For the logarithm $\ln Z$ we need connected polymers. Connected k-polymers are k-polymers of which the k polymers are connected in a sense which will be specified below. The set of connected polymers is called $\mathcal{C}_k \subset \mathcal{P}_k$ by an analogous reason.

The notion of activity extends to k-polymers $X \in \mathcal{P}_k$ by defining

$$a^X \equiv \prod_{\gamma \in X} a(\gamma) . \tag{4.1.31}$$

In particular a^X also factorizes over the polymers when X is a connected k-polymer. But notice that $X = (\gamma_1, ..., \gamma_k) \neq \gamma_1 \cup ... \cup \gamma_k$, hence the fact that a^X (4.1.31) is different from $a(\gamma_1 \cup ... \cup \gamma_k)$ for intersecting $\gamma_1, ..., \gamma_k$ is no contradiction.

Now it is convenient to represent a k-polymer X as a k-vertex graph $\Gamma^{(k)}(X)$. Each vertex represents one $\gamma_i \in X$. Two vertices γ_i, γ_j are connected by a line if $\gamma_i \cap \gamma_j \neq \emptyset$. By definition we set $\mathcal{C}_1 = \mathcal{D}_1 = \mathcal{P}$ and $\mathcal{D}_0 = \mathcal{P}_0$ as the empty set $\emptyset$. Therefore the connectivity properties of collections of polymers γ can be easily read off from the graphical representation. In particular connected k-polymers consist of X with $\Gamma^{(k)}(X)$ path-connected k-vertex graphs. Disjoint k-polymers of X are represented as totally disconnected graphs $\Gamma^{(k)}(X)$, these are graphs without lines. We denote the union of $\mathcal{D}_k$ over all $k = 0, 1, \ldots$, by $\mathcal{D}$ and the union of $\mathcal{C}_k$ over all $k = 1, 2, \ldots$ by $\mathcal{C}$,

$$\mathcal{D} = \bigcup_{k \geq 0} \mathcal{D}_k , \ \mathcal{C} = \bigcup_{k \geq 1} \mathcal{C}_k . \tag{4.1.32}$$

(As mentioned above, k-polymers need not (sometimes even may not) be polymers themselves.)

Finally we are ready to define for a given set of polymers $\mathcal{P}$ and an activity function $a(\gamma)$ a polymer-partition function Z as a power series in a

$$Z \equiv \sum_{k=0}^{\infty} \frac{1}{k!} \sum_{(\gamma_1, ..., \gamma_k) \in D_k} a(\gamma_1) \cdots a(\gamma_k) \equiv \sum_{X \in D} \frac{1}{|X|!} a^X . \tag{4.1.33}$$

The sum runs over all collections of pairwise *disjoint* polymers. What will be called polymer expansion of (the polymer-partition function) Z from now on is a resummation of (4.1.33) performed in a way that a power series results for Z in the form of Eq. (4.1.35), or, equivalently, for $\ln Z$ according to Eq. (4.1.36), again as a function of the activities $a(\gamma)$. So far we are not concerned about the convergence properties, but consider Z and $\ln Z$ as formal series. This resummation requires another important notion. It is the notion of an index $n(X)$ of a connected k-polymer X. Let us identify X with its associated connected graph $\Gamma^{(k)}(X) \equiv X$. For a graph $X \in C_k$, a subgraph G of X is either a single vertex of X or a subset of the lines of X

together with all vertices attached to them. Let $n_{\pm}(X)$ denote the number of connected subgraphs $G \subset X$ in C_k containing all vertices of X and which have an even (odd) number of lines, respectively. The index $n(X)$ is then defined according to

$$n(X) = n_+(X) - n_-(X) = \sum_{G \in X \cap C_k} (-1)^{l(G)} , \qquad (4.1.34)$$

where $l(G)$ denotes the number of lines in G. Now we are ready to state the theorem about the polymer expansion.

Theorem. As a formal power series in $a(\gamma)$ the partition function Z can be written as [144]

$$Z = \sum_{n=0}^{\infty} \frac{1}{n!} \prod_{j=1}^{n} \left[\sum_{k_j=1}^{\infty} \frac{1}{k_j!} \sum_{X^{(j)} \in C_{k_j}} \left(\sum_{G \in X^{(j)}} (-1)^{l(G)} \right) a^{X^{(j)}} \right] .$$
$$(4.1.35)$$

Equivalently to (4.1.35) we have

$$\ln Z = \sum_{k=1}^{\infty} \frac{1}{k!} \sum_{(\gamma_1,...,\gamma_k) \in C_k} n(\gamma_1, ..., \gamma_k) \, a(\gamma_1) \cdots a(\gamma_k) = \sum_{X \in C} \frac{n(X)}{|X|!} a^X$$
$$(4.1.36)$$

with the same activities as in the expansion of the partition function. In contrast to the series representation of Z in (4.1.33), the sum of $\ln Z$ runs over all *connected* graphs $X \in \bigcup_k C_k \equiv C$. The connected graphs X correspond to nonempty collections of not-necessarily distinct polymers, typically $X = (\gamma_1^{n_1}, ..., \gamma_N^{n_N})$, i.e. X may contain the polymers γ_i $n_i \geq 1$ times. The combinatorial factors $n(X)/|X|!$ have been defined above. The activities a^X are the product of activities over all polymers γ_i which belong to X according to the definition (4.1.31). (Actually it comes out of the proof of the theorem that the definition of a^X is convenient.) Below we give a short proof of the theorem.

Sometimes the sets X in (4.1.36) are called linked clusters [145]. This notion of linked clusters should not be confused with linked clusters occurring in linked cluster expansions (LCEs) of the next section. Linked clusters in LCEs refer to connected graphs which arise in a graphical representation of the Taylor expansion of $\ln Z$ about a decoupled theory. The small expansion parameter there is the hopping parameter or the inverse temperature in contrast to the activities here. Therefore the common feature between both notions is the representation of the expansion coefficients as connected graphs, but the graphical rules are quite different.

If the activities $a(X)$ are small enough, the polymer (linked cluster) expansion of the free energy (4.1.36) is convergent. Sufficient conditions for the convergence of the expansion are well known. They are derived by use of Kirkwood-Salsburg equations [145, 146]. However, even if the series are not convergent, but only asymptotic or formal series, the algebraic identities (4.1.35), (4.1.36) hold, because the proof relies on combinatorics. An important application of this case is the relation between the generating function for *connected* Feynman diagrams, $\ln Z$, and the generating function for *all* Feynman diagrams, Z, if Z denotes the partition function of a Euclidean field theory, and Z and $\ln Z$ are evaluated in perturbation theory. In this case the connected Feynman diagrams represent the truncated Green's functions, and the full class of diagrams the expectation values of generic n-point functions, cf. section 2.2.

We finally prove the representation (4.1.35) of Z or, equivalently, the polymer representation (4.1.36) of $\ln Z$. The starting point is the polymer representation (4.1.33) of Z. For every n-polymer $(\gamma_1, \ldots, \gamma_n) \in \mathcal{P}_n$ we define

$$\phi(\gamma_1, \ldots, \gamma_n) = \prod_{1 \leq i < j \leq n} (1 + \delta(\gamma_i, \gamma_j)), \qquad (4.1.37)$$

with

$$\delta(\gamma_i, \gamma_j) = \begin{cases} 0, & \text{for } \gamma_i \cap \gamma_j = \emptyset \\ -1, & \text{for } \gamma_i \cap \gamma_j \neq \emptyset \end{cases} \qquad (4.1.38)$$

We have $\phi(\gamma_1, \ldots, \gamma_n) = 1$ if $(\gamma_1, \ldots, \gamma_n) \in \mathcal{D}_n$ and $\phi(\gamma_1, \ldots, \gamma_n) = 0$ otherwise. With this definition, (4.1.33) is written as

$$Z = \sum_{n=0}^{\infty} \frac{1}{n!} \sum_{\gamma_1, \ldots, \gamma_n} \phi(\gamma_1, \ldots, \gamma_n) \cdot \prod_{i=1}^{n} a(\gamma_i). \qquad (4.1.39)$$

The summation over the polymers is not constrained. The next step is to rewrite $\phi(\gamma_1, \ldots, \gamma_n)$ appropriately as a sum over graphs. Let us write $\underline{n} = \{1, 2, \ldots, n\}$. A subgraph G of $\underline{n}$ is either a single vertex $j \in \underline{n}$ or a set of pairs = lines (i, j) with $1 \leq i < j \leq n$, together with all vertices $k \in \underline{n}$ that belong to at least one line of G. $\mathcal{G}(\underline{n})$ denotes the set of all subgraphs of $\underline{n}$. We then have

$$\phi(\gamma_1, \ldots, \gamma_n) = \sum_{G \in \mathcal{G}(\underline{n})} \prod_{(i,j) \in G} \delta(\gamma_i, \gamma_j). \qquad (4.1.40)$$

In order to rewrite $\phi(\gamma_1, \ldots, \gamma_n)$ in terms of *connected* subgraphs, we first sum over the elements of the set $\mathcal{P}(\underline{n})$ of all partitions of $\underline{n}$ into disjoint subsets

$$\phi(\gamma_1, \ldots, \gamma_n) = \sum_{P \in \mathcal{P}(\underline{n})} \prod_{P \in P} \left[\sum_{G \in \mathcal{G}_c(P)} \prod_{(i,j) \in G} \delta(\gamma_i, \gamma_j) \right], \qquad (4.1.41)$$

where $\mathcal{G}_c(P)$ denotes the set of *connected* subgraphs of P containing all vertices of P. The expression in brackets is precisely the index $n(\gamma_i | i \in P)$ of the $|P|$-polymer $(\gamma_i | i \in P)$ as defined in (4.1.34), with $|P|$ the number of elements of P.

We further reorganize the summations. Summing first over the number of sets of the partitions of $\underline{n}$ and then over the number of possible elements of the sets, we get

$$\phi(\gamma_1, \ldots, \gamma_n) = \sum_{m=1}^{\infty} \frac{1}{m!} \sum_{\substack{P_1, \ldots, P_m \\ (P_1, \ldots, P_m) \in \mathcal{P}(\underline{n})}} \prod_{i=1}^{m} n(\gamma_i | i \in P)$$

$$= \sum_{m=1}^{\infty} \frac{1}{m!} \sum_{\substack{n_1, \ldots, n_m \geq 1 \\ \sum n_i = n}} \sum_{\substack{P_1, \ldots, P_m, |P_i| = n_i \\ (P_1, \ldots, P_m) \in \mathcal{P}(\underline{n})}} \prod_{i=1}^{m} n(\gamma_i | i \in P). \qquad (4.1.42)$$

Inserting this expression for $\phi(\gamma_1, \ldots, \gamma_n)$ into the polymer representation (4.1.39) of Z yields

$$Z = 1 + \sum_{n=1}^{\infty} \frac{1}{n!} \sum_{m=1}^{\infty} \frac{1}{m!} \sum_{\substack{n_1, \ldots, n_m \geq 1 \\ \sum n_i = n}} \left(\sum_{\substack{P_1, \ldots, P_m, |P_i| = n_i \\ (P_1, \ldots, P_m) \in \mathcal{P}(\underline{n})}} \right)$$

$$\cdot \prod_{i=1}^{m} \sum_{\gamma_1, \ldots, \gamma_{n_i}} n(\gamma_1, \ldots, \gamma_{n_i}) \prod_{k=1}^{n_i} a(\gamma_k). \qquad (4.1.43)$$

The sum over the subsets $P_1, \ldots, P_m$ factorizes out because nothing depends on it any more. It gives a factor

$$\sum_{\substack{P_1,\ldots,P_m,|P_i|=n_i \\ (P_1,\ldots,P_m)\in\mathcal{P}(\underline{n})}} 1 = \frac{n!}{n_1!\cdots n_m!}. \tag{4.1.44}$$

Hence, (naively exchanging the infinite sums over n and m),

$$Z = 1 + \sum_{m=1}^{\infty} \frac{1}{m!} \prod_{i=1}^{m} \left(\sum_{n_i=1}^{\infty} \frac{1}{n_i!} \sum_{\gamma_1,\ldots,\gamma_{n_i}} n(\gamma_1,\ldots,\gamma_{n_i}) \prod_{k=1}^{n_i} a(\gamma_k) \right), \tag{4.1.45}$$

which is (4.1.35). This completes the proof.

Examples for polymer expansions in particle physics

Two examples for *convergent* expansions in particle physics are applications to

- pure SU(N)-Yang Mills theories on a Euclidean lattice and
- SU(2)-Yang Mills theories with quark or scalar matter fields on a Euclidean lattice.

The first example refers to the calculation of the string tension in an $SU(2)$ lattice gauge theory in four dimensions [147]. The partition function is represented as a sum over disjoint unions of polymers. The polymers consist of certain sets of plaquettes. The activity $a(\gamma)$ of a polymer γ is defined as

$$a(\gamma) = \int \prod_b dU(b) \prod_{p\in\gamma} f_p$$

$$f_p \equiv e^{-S_p} - 1 = \sum_{j\neq 0} c_j(\beta)\, \chi_j(U(\partial p))\,, \tag{4.1.46}$$

where S_p has the form of the Wilson action

$$-S = -\sum_{p\in\Lambda} S_p = \sum_{p\in\Lambda} \beta\, d_\chi^{-1}\, \mathrm{Re}\,\chi(U(\partial p))\,, \tag{4.1.47}$$

but χ refers to the character of *any* faithful representation of the gauge group G, in particular $G = SU(2)$ or $Z(2)$, including the fundamental representation which we have considered in the Wilson action above. The sum runs over all plaquettes p of the four-dimensional lattice Λ, $d_\chi = \chi(\mathbf{1})$ is the dimension of the faithful representation, Re denotes the real part and $\beta \equiv 2d_\chi g_0^{-2}$ with g_0 the strong gauge coupling. β is used as the small expansion parameter. The product in the measure of (4.1.46) runs over all links in the boundary ∂p of any plaquette p of γ, dU is the Haar measure on G. On the right hand side of (4.1.46) the single plaquette factors f_p are represented as character expansions, c_j are coefficients and $\chi_j(U(\partial p))$ denote the characters of U in the jth representation. This step is quite convenient in strong coupling expansions and will be explained in the next section in some detail.

Here the strong coupling expansion is combined with a polymer representation of the partition function, because the string tension is calculated via a linked cluster expansion for the free energy of vortices ("linked cluster" in the sense of (4.1.36)). Vortices are supposed to be the relevant gauge field configurations which drive the deconfinement transition in a pure gauge theory by their condensation. In these expansions it is shown that the vortex-free energy obeys an area law, and the coefficient of the area equals the string tension between static quarks.

The character expansion of f_p facilitates the concrete evaluation of activities that are group integrals of products of e^{-S_p}. For further details we refer to the original reference. The result of [147] for the string tension in an $SU(2)$ lattice gauge theory in four dimensions up to 12th order in β was in good agreement with early Monte Carlo calculations at strong and intermediate gauge couplings [148].

The second application of polymer expansions refers to gauge theories with Higgs fields or quark fields. Polymers in this case are sets of links and sets of plaquettes which satisfy certain conditions. Again these conditions are determined by the factorization properties of integrals over the Boltzmann factors into integrals of the same form for each "connected" component. These connected components define the polymers. Here it is convenient to associate a graphical representation not only with the clusters but also with the polymers. The very definition of a polymer is then formulated as a set of conditions on the associated polymer graphs. The support of polymers consists of sites and links of the lattice which are vertices in the polymer graph. For the precise definition of polymers we refer to the original reference [149].

The activity functions generalize to products of integrals on sites and links on matter and gauge fields, respectively. In addition to one-plaquette functions $f_p(U)$ we have "one-link" functions $g_b(U, \psi)$, defined as

$$g_b(U, \psi) \equiv e^{-\mathcal{L}_b(U, \psi)} , \qquad (4.1.48)$$

where $\mathcal{L}_b(U, \psi)$ is the matter part of the action, restricted to an interaction of gauge fields $U(b)$ and matter fields $\psi(x), \psi(y)$ along one link $b \equiv (x, y)$.

Finally we have two rather than one expansion parameters, $\beta \equiv g_0^{-2}$ with g_0 the strong bare gauge coupling and K, the coupling of the interaction between the gauge and matter fields. In the $SU(2)$ gauge theory with quark or Higgs fields the interest was in the region of the phase diagram with small β (strong gauge coupling) and small K, in the $Z(2)$ gauge theory with Higgs fields with large β (or small gauge coupling) and small K. Therefore the polymer representations of Z are combined with strong coupling (small β) and large-β expansions, respectively. (We avoid to call the large-β expansions weak coupling expansions or low temperature expansions, because the expansions are formally quite similar to low temperature expansions in discrete spin models. Only the physical meaning of β in this work is that of a weak gauge coupling g_0, and not that of a low temperature.) Once suitable notions of polymers and attributes of being disjoint or connected are found, the polymer representation of the partition function is converted to a linked cluster expansion of the free energy of vortex configurations. This way the behavior of the vortex-free energy can be calculated in the various parameter regions. The original interest in this quantity was based on the conjecture that the vortex-free energy is a suitable disorder parameter to distinguish the phases of QCD even when matter fields are included. The Wilson-loop expectation value (that is the expectation value of the product of gauge fields along a rectangular contour in Euclidean spacetime) is not a suitable candidate, and, as it finally turned out, neither is the vortex-free energy. Cluster expansions of the vortex-free energy in the $Z(2)$ Higgs model showed that its large-volume behavior is the same in regions of the phase diagram that are separated by a line of phase transitions, but distinguishes between the confinement ($g_0 \gg 1, K \ll 1$) and screening ($K \gg 1$, g_0 arbitrary) region that are known to be analytically connected [150].

So far the results refer to the phase structure at zero physical temperature. Another interesting result for the $Z(2)$ Higgs model concerns the effect of a small, but finite physical temperature $T > 0$. For zero temperature the free energy $\propto \ln Z$ is analytic in several regions of phase space, in particular for $g_0 \ll 1, K \ll 1$, the so-called deconfinement phase. The analyticity in the deconfinement phase follows from the result of [151] that the linked cluster expansion of $\ln Z$ is absolutely convergent for sufficiently small g_0^2 and sufficiently small K. The convergence is uniform in the lattice volume. However, their proof does not generalize to systems at finite temperature $T > 0$. The absolute convergence of the cluster expansion is destroyed by a certain class of "finite temperature" graphs which contribute to $\ln Z$ only because of the periodic boundary conditions in the Euclidean time direction on the matter fields [152].

4.1.4 *Strong coupling expansions*

In statistical physics polymer representations of the partition function are combined with high- or low-temperature expansions. In particle physics the physical meaning of the expansion parameter is different, but the expansion scheme is quite similar. At the end of the last section we mentioned polymer representations of the $Z(2)$-Higgs model combined with strong coupling or "small" coupling expansions (to avoid confusion with weak coupling perturbation theory). In this section we focus on some details of strong coupling expansions, typically for an expectation value of an observable O. A standard example is the calculation of the Wilson-loop expectation value

$$< W > = \frac{\int \mathcal{D}U \ W \ e^{1/g_0^2 \sum_{p \in \Lambda} \left[\text{tr } U(\partial p) + \text{tr } U^\dagger(\partial p) \right]}}{\int \mathcal{D}U \ e^{1/g_0^2 \sum_{p \in \Lambda} \left[\text{tr } U(\partial p) + \text{tr } U^\dagger(\partial p) \right]}} \ , \qquad (4.1.49)$$

where

$$W = \prod_{b \in \mathcal{C}} U(b) \qquad (4.1.50)$$

is the product of U along the boundary of the loop $\mathcal{C}$ with spatial extension L and "time-like" extension T. (In contrast, the Wilson *line* or Polyakov loop is the product of U along a string in timelike direction that is finite at finite temperature.) The action in (4.1.49) has the generic form of an $SU(N_c)$ gauge theory with N_c denoting the number of colors. For $N_c = 2$ it is equivalent to (2.3.172) of section 2.3 up to a normalization constant. Naively, the first step could be an expansion of the Boltzmann factor in powers of $1/g_0^2$

$$e^{1/g_0^2 \sum_{p \in \Lambda}[\mathrm{tr}\, U(\partial p) + \mathrm{tr}\, U^\dagger(\partial p)]}$$

$$= \prod_{p \in \Lambda} \sum_{n=0}^{\infty} \frac{1}{n!} \left(\frac{1}{g_0^2}\right)^n \cdot \left[\mathrm{tr}\, U(\partial p) + \mathrm{tr}\, U^\dagger(\partial p)\right]^n$$

$$= \sum_{Q \subset \Lambda} \prod_{p \in Q} \sum_{n=1}^{\infty} \frac{1}{n!} \left(\frac{1}{g_0^2}\right)^n \left[\mathrm{tr}\, U(\partial p) + \mathrm{tr}\, U^\dagger(\partial p)\right]^n \,, \quad (4.1.51)$$

where Q are all possible subsets of plaquettes belonging to Λ, and $|\Lambda_p|$ is the number of plaquettes of the lattice. In principle (4.1.51) should now be inserted in (4.1.49), (4.1.50), and the group integrals should be performed.

In practice, however, it is much more convenient to represent the Boltzmann factor of a single plaquette as a generalized Fourier expansion on the gauge group. The Fourier expansion amounts to a character expansion, since the trace in the one-plaquette action is invariant under inner automorphisms, i.e. $tr(hgh^{-1}) = tr(g)$ for all $g, h \in G$, where G is the gauge group. It is based on the following results of group theory.

Any function $f \in L_2(G, dg)$, the set of square integrable functions on G with group measure dg (here the normalized Haar measure), allows for an expansion with respect to an orthogonal set of matrix elements $\tau_{ij}(g)$ that belong to irreducible representations τ of a group element $g \in G$. (For a compact group all irreducible representations are equivalent to unitary representations). More precisely the expansion reads

$$f(g) = \sum_{\tau} \sum_{i,j=1}^{d_\tau} c_{ij}^\tau \, \tau_{ij}(g) \,, \quad (4.1.52)$$

where d_τ denotes the dimension of the representation and the coefficients c_{ij}^τ are given by

$$c_{ij}^\tau = d_\tau \int_G dg \, \bar{\tau}_{ij}(g) \, f(g) \,. \quad (4.1.53)$$

The functions $\tau_{ij}(g)$ satisfy the orthogonality relation

$$\int_G dg \, \bar{\tau}_{ij}(g) \, \tau'_{kl}(g) = \delta_{\tau \tau'} \delta_{ik} \delta_{jl} \, (d_\tau)^{-1} \,. \quad (4.1.54)$$

If f is invariant under inner automorphisms, that is

$$f(hgh^{-1}) = f(g) \quad (4.1.55)$$

for all $g, h \in G$, it follows from Schur's lemma that

$$c_{ij}^\tau = \delta_{ij} \, c^\tau \tag{4.1.56}$$

so that

$$f(g) = \sum_\tau c^\tau \, \chi_\tau(g) \tag{4.1.57}$$

with characters

$$\chi_\tau(g) = \sum_{i=1}^{d_\tau} \tau_{ii}(g) \tag{4.1.58}$$

and

$$c^\tau = \int_G dg \, \overline{\chi}_\tau(g) \, f(g) \, . \tag{4.1.59}$$

Equation(4.1.54) then implies the orthogonality relation for characters

$$\int_G dg \, \chi_\tau(hg^{-1}) \, \chi_{\tau'}(g) = \delta_{\tau\tau'} \cdot (d_\tau)^{-1} \cdot \chi_\tau(h) \, , \tag{4.1.60}$$

which is quite useful for reducing the number of contributions to the power-series expansion. Since the Boltzmann factor for a single plaquette is invariant under inner group automorphisms, or, stated differently, is defined over equivalence classes, i.e.

$$e^{1/g_0^2 \left[\operatorname{tr} (U(\partial p)) + \operatorname{tr} (U^\dagger(\partial p)) \right]} = e^{1/g_0^2 \left[\operatorname{tr} (VU(\partial p)V^{-1}) + \operatorname{tr} (V^{-1}U^\dagger(\partial p)V) \right]} \, , \tag{4.1.61}$$

we can write it as a character expansion

$$e^{-S_p(U(\partial p))} \equiv e^{2/g_0^2 \operatorname{Re} \operatorname{tr} U(\partial p)}$$

$$= \sum_\nu c_\nu(\frac{1}{g_0^2}) \, \chi_\nu(U(\partial p)) \, . \tag{4.1.62}$$

According to (4.1.57) the sum ν extends over the set of nonequivalent irreducible unitary representations of the compact group G. Using the orthogonality relations (4.1.60) of group characters, we find for the coefficients

$$c_\nu(\frac{1}{g_0^2}) = \int dU \, \chi_\nu(U^{-1}) \, e^{-S_p(U)} \, , \tag{4.1.63}$$

where dU stands for the Haar measure on G. Here $U = U(\partial p)$, the group element associated with the boundary of the plaquette p. As two examples we consider the gauge groups $Z(2)$ and $SU(2)$.

- For the one-plaquette action of the $Z(2)$-gauge theory we write

$$-S_p = \frac{1}{g_0^2} \sigma(\partial p) , \qquad (4.1.64)$$

where $\sigma(\partial p)$ denotes the product of gauge fields along the boundary of the plaquette p, $\sigma(\partial p) = \pm 1$. The group integral $\int dU$ reduces to the normalized sum

$$\int dU = \frac{1}{2} \sum_{\{\sigma(\partial p) = \pm 1\}} . \qquad (4.1.65)$$

It is then easily checked that the character expansion reads

$$e^{-S_p} = \cosh(\frac{1}{g_0^2}) + \sinh(\frac{1}{g_0^2}) \cdot \sigma(\partial p) . \qquad (4.1.66)$$

- For an $SU(2)$ Yang Mills theory the one-plaquette action reads

$$-S_p = \frac{1}{2} \cdot \frac{1}{g_0^2} \, tr \, U(\partial p) ,$$

$$U(\partial p) = \prod_{b \in \partial p} U(b) , \qquad (4.1.67)$$

$U(b) \in SU(2)$ in the fundamental representation, i.e.

$$tr \, U(\partial p) = \chi_{1/2}(U(\partial p)) . \qquad (4.1.68)$$

In the fundamental representation any $U \in G$ can be parameterized as

$$U = U_0 + i \vec{U} \cdot \vec{\sigma} \qquad (4.1.69)$$

with

$$|U_0|^2 + |\vec{U}|^2 = 1 , \qquad (4.1.70)$$

or

$$U = \cos(\frac{\phi}{2}) + i \sin(\frac{\phi}{2}) \, \hat{n} \cdot \vec{\sigma} , \qquad (4.1.71)$$

where $\hat{n}$ is a unit vector in $\mathbf{R}^3$ and $\vec{\sigma}$ are the three Pauli matrices. In terms of ϕ

$$-S_p(U) = \frac{1}{g_0^2} \cos(\frac{\phi}{2}) . \qquad (4.1.72)$$

The character in the jth representation of $U \in SU(2)$ is given as

$$\chi_j(U) = \sum_{m=-j}^{j} e^{im\phi} = \frac{\sin((2j+1)\phi/2)}{\sin(\phi/2)} , \qquad (4.1.73)$$

and the normalized Haar measure on $SU(2)$ for one group integration

$$\int dU = \frac{1}{\pi} \int_0^{2\pi} d\phi \, (\sin(\phi/2))^2 . \qquad (4.1.74)$$

For the coefficients of the character expansion we obtain

$$\begin{aligned}
c_j(\frac{1}{g_0^2}) &= \frac{1}{\pi} \int_0^{2\pi} d\phi \, (\sin(\frac{\phi}{2}))^2 \frac{\sin((2j+1)\phi/2)}{\sin(\phi/2)} \cdot e^{1/g_0^2 \cos(\phi/2)} \\
&= 2 \cdot (2j+1) \, g_0^2 \, I_{2j+1}(\frac{1}{g_0^2}) , \qquad (4.1.75)
\end{aligned}$$

where I_{2j+1} are modified Bessel functions.

If we are interested in expectation values of observables, common prefactors factorize out. It is then more convenient to use

$$e^{-S_p} \propto 1 + \tanh(\frac{1}{g_0^2}) \cdot \sigma(\partial p) \qquad (4.1.76)$$

for $Z(2)$, and

$$e^{-S_p} \propto 1 + \sum_{j \neq 0} (2j+1) \frac{I_{2j+1}(\frac{1}{g_0^2})}{I_1(\frac{1}{g_0^2})} \chi_j(U(\partial p)) \qquad (4.1.77)$$

for $SU(2)$.

Notice that strong coupling expansions based on (4.1.76) and (4.1.77) are expansions in $\tanh(g_0^{-2})$ and ratios of modified Bessel functions, respectively, unless we re-expand $\tanh(g_0^{-2})$ and $I_{2j+1}(g_0^{-2})/I_1(g_0^{-2})$ in powers of g_0^{-2}. For small g_0^{-2} we have $\tanh(g_0^{-2}) \approx g_0^{-2}$ and $I_{2j+1}(g_0^{-2})/I_1(g_0^{-2}) \approx (g_0^{-2})^{2j}$. This way we re-obtain the naive strong coupling expansion (4.1.51) that formally corresponds to the high temperature expansion of statistical physics. This concludes our summary of the first step in the strong coupling expansion that was the

- character expansion of the Boltzmann factor for one plaquette. The next steps are

- inserting the representation (4.1.76), (4.1.77) of $e^{-S(U(\partial p))} \equiv e^{-S_p}$ into the product over all plaquettes of the lattice Λ, $\prod_{p \in \Lambda} e^{S(U(\partial p))}$, and rearranging the product into a sum over disjoint unions of polymers, which have to be suitably defined for this representation, cf. the last section,

- graphical representation of the polymer expansion,

- conversion of the polymer representation of Z into a polymer expansion of $\ln Z$ (also called linked cluster expansion), that is a sum over all subsets of connected polymers, if we are interested in $\ln Z$ or derived quantities.

Explicit calculations of strong coupling expansions can be found in the original references, see e.g. [153], where they are performed to some higher orders, e.g. to 12th order in g_0^{-2}, or in reviews on lattice gauge theories such as [154] or [155].

Lattice QCD (without matter fields) is exactly solvable for $g_0 \to \infty$. The strong coupling expansion is an expansion about this limit. In this limit and for sufficiently large couplings g_0 confinement is proved according to the Wilson criterion: the Wilson-loop expectation value decays exponentially with the area of the enclosed loop.

In physical units, however, the lattice is rather coarse grained for strong couplings. The physical interest is in the opposite limit $g_0 \to 0$, the continuum limit. One would like to prove confinement as a long-distance property of QCD (long distance in *physical* units) on a rather fine grained lattice $g_0 \ll 1$. Therefore the Wilson-loop expectation value was calculated up to 18th order in g_0^{-2} in order to get a reliable extrapolation towards the continuum limit. The area law is easily seen in the leading term. This behavior is not spoiled by higher order terms, if the strong coupling series is convergent and $g_0^{-2} < R$, where R is the radius of convergence. If this radius were infinite, the extrapolation to the continuum limit would be safe. The very fact that the radius of convergence is at least larger than zero is based on a proof of [156] that we reproduce in the next section.

4.1.5   *A theorem by Osterwalder and Seiler*

We consider a pure $SU(N)$-Yang Mills theory (or, more generally, a compact gauge group) on a lattice and prove for sufficiently large gauge coupling g_0 a theorem which has the following important consequences

- The infinite volume limit of expectation values of local observables exists.
- All n-point correlation functions are analytic functions of the complex variable g_0 if $1/|g_0|$ is sufficiently small.

We give an outline of the proof of the theorem, because by analogous methods one can show further important properties of Yang-Mils theories:

- the existence of a mass gap
- Osterwalder-Schrader positivity and invariance under lattice translations
- for nonabelian gauge theories and for sufficiently large g_0, the Wilson-loop expectation value has a convergent cluster (polymer) expansion. Furthermore, confinement holds according to the Wilson criterion if the coupling lies inside the domain of convergence.

The statement of the theorem is an inequality that tells us that for complex g_0 and $|g_0|$ large enough, the polymer expansion of the expectation value of a local observable $< \mathcal{F} >_\Lambda$ is absolutely convergent both for finite and for infinite Λ. For $|g_0|$ large enough there exist constants α, γ such that the inequality

$$\sum_{Q \subset \Lambda, Q \text{ con } Q_0, |Q| \geq K} \left| \int d\mu_\Lambda \ \mathcal{F} \cdot \left(\prod_{p \in Q} \rho_p \right) \frac{Z_{\Lambda \setminus \overline{Q \cup Q_0}}}{Z_\Lambda} \right| \leq \alpha \cdot \left(\frac{\gamma}{g_0^2} \right)^K \quad (4.1.78)$$

holds. In the following we explain the notations included in (4.1.78). This goes along with a partial derivation of this bound. The constant α only depends on the observable $\mathcal{F}$, the dimension D, the gauge group G and its representation χ, similarly $\gamma = \gamma(D, G, \chi)$, as we shall see below. The left hand side results from a polymer representation of the expectation value $< \mathcal{F} >_\Lambda$ of the observable $\mathcal{F}$. In general $< \cdots >$ depends on Λ as indicated by the subscript Λ. We write

$$< \mathcal{F} >_\Lambda = \frac{1}{Z_\Lambda} \int d\mu_\Lambda \ \mathcal{F} \ e^{-S_\Lambda} \quad (4.1.79)$$

with

$$Z_\Lambda = \int d\mu_\Lambda \ e^{-S_\Lambda}$$

$$-S_\Lambda = \frac{2}{g_0^2} \sum_{p \in \Lambda} \text{Re} \ [\chi(U(\partial p)) + \chi(\mathbf{1})] \ . \quad (4.1.80)$$

Here χ is the trace in a unitary representation of the gauge group $SU(N)$, otherwise the notation is the same as in (4.1.62). Note that the action differs by a constant from our choice (4.1.47). The constant is chosen such that the inequality (4.1.83) below is satisfied. Further ρ_p is the activity function of the plaquette p, defined as

$$\rho_p \equiv e^{2/g_0^2 \, \mathrm{Re} \, [\chi(U(\partial p)) + \chi(1)]} - 1 . \tag{4.1.81}$$

As we have shown in the previous section, in the polymer representation the Boltzmann factor is first rewritten according to

$$e^{-S_\Lambda} = \prod_{p \in \Lambda} (1 + \rho_p) = \sum_{Q \subset \Lambda} \prod_{p \in Q} \rho_p , \tag{4.1.82}$$

where the sum runs over all subsets Q of plaquettes p in Λ. Back to the notation in (4.1.78), $|Q|$ denotes the number of plaquettes of Q. Q_0 is the support of the observable $\mathcal{F}$, it consists of the set of links whose link variables belong to the domain of definition of $\mathcal{F}$. Further $d\mu_\Lambda$ is the Haar measure on the gauge group $SU(N)$. For sufficiently large g_0 it follows from (4.1.81) that

$$0 < \rho_p \le \frac{const}{g_0^2} . \tag{4.1.83}$$

From (4.1.79,4.1.80) and (4.1.82) we obtain

$$< \mathcal{F} >_\Lambda = \frac{1}{Z_\Lambda} \sum_{Q \subset \Lambda} \int d\mu_\Lambda \, \mathcal{F} \prod_{p \in Q} \rho_p . \tag{4.1.84}$$

Now we split Q into a subset Q_I which is the disjoint union of N connected components Q_i^c, $i = 1, ...N$, (the polymers) that have at least one bond in common with Q_0, and the remainder, called Q_{II}, so that $Q_{II} = Q \setminus Q_I$. Equation(4.1.84) then implies

$$< \mathcal{F} >_\Lambda = \sum_{Q \subset \Lambda} \left(\int d\mu_\Lambda \, \mathcal{F} \prod_{p \in Q_I} \rho_p \right) \frac{\int d\mu_\Lambda \prod_{p \in Q_{II}} \rho_p}{Z_\Lambda} . \tag{4.1.85}$$

Here we have used that $\int d\mu_\Lambda \cdot 1 = 1$, since the Haar measure can be normalized to one on a compact gauge group. Rather than summing over Q and Q_I under the constraint on $Q_{II} = Q \setminus Q_I$ we can sum over all Q_I and Q_{II} under the constraint that Q_{II} has no links in common with Q_0.

This way we obtain

$$< \mathcal{F} >_\Lambda = \sum_{Q_I \, con \, Q_0, \, Q_I \subset \Lambda} \int d\mu_\Lambda \, \mathcal{F} \prod_{p \in Q_I} \rho_p \, \frac{Z_{\Lambda \backslash \overline{Q_I \cup Q_0}}}{Z_\Lambda} \, , \qquad (4.1.86)$$

since

$$\sum_{Q_{II}} \int d\mu_\Lambda \prod_{p \in Q_{II}} \rho_p \, = \, Z_{\Lambda \backslash \overline{Q_I \cup Q_0}} \, , \qquad (4.1.87)$$

where $\overline{Q_I \cup Q_0}$ is the set of plaquettes having links in common with Q_I and Q_0. (4.1.86) and (4.1.87) constitute the polymer representation of $< \mathcal{F} >_\Lambda$. Thus the sum over all disjoint unions of polymers here amounts to a sum over all subsets of plaquettes belonging to the lattice Λ which are connected with the support Q_0 of the observable $\mathcal{F}$. (Contributions of subsets which are not connected with Q_0 factorize out.)

To show the absolute convergence of the series in (4.1.86) for sufficiently large $|g_0|$, we label the clusters Q with K plaquettes according to their size $|Q| \equiv K$ and estimate

$$\sum_{Q \subset \Lambda, Q \, con \, Q_0, |Q| \geq K} \left| \int d\mu_\Lambda \, \mathcal{F} \cdot \left(\prod_{p \in Q} \rho_p \right) \frac{Z_{\Lambda \backslash \overline{Q \cup Q_0}}}{Z_\Lambda} \right|$$

$$\leq \sum_{k \geq K} N(k) \cdot \sup_{Q, |Q| = k} \left| \int d\mu_\Lambda \, \mathcal{F} \prod_{p \in Q} \rho_p \right| \cdot \sup_{Q, |Q| = k} \left| \frac{Z_{\Lambda \backslash \overline{Q \cup Q_0}}}{Z_\Lambda} \right| \, , \qquad (4.1.88)$$

where $N(k)$ is the number of possible sets Q of plaquettes in Λ such that $|Q| = k$ and such that each connected component of Q is connected to Q_0. In the next steps we derive a bound on each of the three factors in the sum of the right hand side of (4.1.88).

- For $N(k)$ defined as above there exist constants c_2 and c_3 depending on Q_0 and D such that

$$N(k) \, \leq \, c_2 \cdot c_3^k \, . \qquad (4.1.89)$$

As it is claimed in [156], the proof is analogous to the one in [157].

- The second factor in (4.1.88) is bounded according to

$$\left| \int d\mu_\Lambda \, \mathcal{F} \prod_{p \in Q} \rho_p \right| \leq \| \mathcal{F} \|_\infty \cdot \left| \frac{c_1}{g_0^2} \right|^k \, , \qquad (4.1.90)$$

where $\| \cdot \|_\infty$ denotes the supremum norm and $|Q| = k$. Further we have used (4.1.83) and the fact that the Haar measure is normalized to one. The constant c_1 depends on the gauge group and its representation.

- The third factor satisfies the inequality

$$\left| \frac{Z_{\Lambda \backslash \overline{Q \cup Q_0}}}{Z_\Lambda} \right| \leq 2^{|\overline{Q \cup Q_0}|} \equiv c_4 \cdot 2^{|Q|} . \tag{4.1.91}$$

To prove (4.1.91) we will show that

$$2^{-|Q|} \leq \frac{Z_{\Lambda \backslash Q}}{Z_\Lambda} \leq 2^{|Q|} \tag{4.1.92}$$

or

$$\frac{1}{2} \leq \frac{Z_{\Lambda \backslash \{p_0\}}}{Z_\Lambda} \leq 2 , \tag{4.1.93}$$

because (4.1.93) implies (4.1.92), since for $Q = \cup_{i=1}^{k} p_i$ with pairwise distinct plaquettes p_i

$$\frac{Z_{\Lambda \backslash Q}}{Z_\Lambda} = \frac{Z_{\Lambda \backslash \{p_1 \cup p_2 \cup \cdots \cup p_k\}}}{Z_\Lambda} = \frac{Z_{(\Lambda \backslash p_1 \backslash p_2 \cdots) \backslash p_k}}{Z_\Lambda}$$

$$\leq 2 \cdot \frac{Z_{\Lambda \backslash p_1 \backslash p_2 \cdots \backslash p_{k-1}}}{Z_\Lambda} \leq 2 \cdot 2 \cdots \frac{Z_{\Lambda \backslash p_1}}{Z_\Lambda} \leq 2^{|Q|}, \tag{4.1.94}$$

and accordingly for the left hand side of the inequality. We will prove Eq. (4.1.93) by induction in $|\Lambda| = N$, the total number of plaquettes of the lattice Λ. For $N = 2$ (4.1.93) is satisfied. Let us assume that (4.1.93) holds for $|\Lambda| = N$. We write Λ_N if $|\Lambda| = N$. Using the induction assumption we prove next that

$$\left| \frac{Z_{\Lambda_{N+1}} - Z_{\Lambda_{N+1} \backslash \{p_0\}}}{Z_{\Lambda_{N+1} \backslash \{p_0\}}} \right| < \frac{1}{2} , \tag{4.1.95}$$

which immediately gives

$$\frac{1}{2} \leq \frac{Z_{\Lambda_{N+1} \backslash \{p_0\}}}{Z_{\Lambda_{N+1}}} \leq 2 , \tag{4.1.96}$$

or (4.1.93) for arbitrary N. (4.1.95) follows from

$$
Z_\Lambda - Z_{\Lambda\setminus\{p_0\}} = \int d\mu \left(\prod_{p\in\Lambda\setminus\{p_0\}} e^{-S_p} \right) \cdot \left(e^{-S_{p_0}} - 1 \right)
$$

(4.1.97)

$$
= \sum_{Q\subset\Lambda\setminus\{p_0\}} \left[\int d\mu_\Lambda \left(e^{-S_{p_0}} - 1 \right) \prod_{p\in Q_1} \rho_p \right] \cdot \left[\int d\mu_\Lambda \prod_{p\in Q_2} \rho_p \right]
$$

with Q_1 and Q_2 defined in analogy to above and Q_0 equals $\{p_0\} \equiv p_0$ here. Therefore

$$
\left| Z_\Lambda - Z_{\Lambda\setminus\{p_0\}} \right| \le \sum_{Q_1\subset\Lambda\setminus\{p_0\},\, Q_1\, con\{p_0\}} \int d\mu_\Lambda \left(\prod_{p\in Q_1} \rho_p \right) \left| \frac{c_1}{g_0^2} \right|
$$

$$
\cdot\, Z_{\Lambda\setminus\{p_0\}\setminus\overline{Q_1\cup\{p_0\}}}
$$

(4.1.98)

for arbitrary $|\Lambda|$, in particular $|\Lambda| = N+1$. Hence we find (similarly to (4.1.86))

$$
\left| \frac{Z_{\Lambda_{N+1}} - Z_{\Lambda_{N+1}\setminus\{p_0\}}}{Z_{\Lambda_{N+1}\setminus\{p_0\}}} \right| \le \sum_{Q_1\subset\Lambda\setminus\{p_0\},\, Q_1\, con\{p_0\}} \int d\mu_\Lambda \left(\prod_{p\in Q_1} \rho_p \right) \left| \frac{c_1}{g_0^2} \right|
$$

$$
\cdot \left| \frac{Z_{\Lambda_{N+1}\setminus\{p_0\}\setminus\overline{Q_1\cup\{p_0\}}}}{Z_{\Lambda_{N+1}\setminus\{p_0\}}} \right| ,
$$

(4.1.99)

or, with $\Lambda_{N+1}\setminus\{p_0\} \equiv \Lambda_N$, using the induction assumption and (4.1.83)

$$
\left| \frac{Z_{\Lambda_{N+1}} - Z_{\Lambda_{N+1}\setminus\{p_0\}}}{Z_{\Lambda_{N+1}\setminus\{p_0\}}} \right| \le \left| \frac{c_1}{g_0^2} \right| \cdot \sum_{Q\subset\Lambda\setminus\{p_0\},\, Q\, con\, \{p_0\}} \left| \frac{c_1}{g_0^2} \right|^{|Q|} \cdot 2^{|\overline{Q\cup\{p_0\}}|}
$$

$$
\le \frac{1}{|g_0^2|} \sum_{K=0}^{\infty} \left| \frac{c_1}{g_0^2} \right|^{K} N(K)\, 2^{|\overline{Q\cup\{p_0\}}|}
$$

$$
= \frac{1}{|g_0^2|} \sum_{K=0}^{\infty} \left| \frac{c_1}{g_0^2} \right|^{K} \widetilde{c_2}\, \widetilde{c_3}^{K}
$$

$$
= \frac{1}{|g_0^2|} \widetilde{c_2} \left(\frac{1}{1 - \left| \frac{c_1 \widetilde{c_3}}{g_0^2} \right|} \right)
$$

(4.1.100)

that becomes arbitrarily small for g_0 sufficiently large, in particular smaller than $\frac{1}{2}$ which proves (4.1.95).

Putting the bounds on the three factors in (4.1.88) together, we finally obtain

$$
\begin{aligned}
\sum_K \sum_{k \geq K} c_2 \, c_3^k \cdot \| \mathcal{F} \|_\infty & \left| \frac{c_1}{g_0^2} \right|^k c_4 \, 2^k \\
\equiv \sum_K \sum_{k \geq K} \widehat{c}_2 \, \| \mathcal{F} \|_\infty & \left| \frac{c_1 \widehat{c}_3}{g_0^2} \right|^k \\
\leq \sum_K \widehat{c}_2 \, \| \mathcal{F} \|_\infty & \left| \frac{c_1 \widehat{c}_3}{g_0^2} \right|^K \cdot \sum_{k=0}^{\infty} \left| \frac{c_1 \widehat{c}_3}{g_0^2} \right|^k \\
= \widehat{c}_2 \, \| \mathcal{F} \|_\infty & \left(\frac{1}{1 - \left| \frac{c_1 \widehat{c}_3}{g_0^2} \right|} \right)^2 .
\end{aligned} \tag{4.1.101}
$$

This concludes the proof of the absolute convergence of the series in (4.1.86), and for fixed K leads to a bound of the form (4.1.78). Since the bound in (4.1.101) is independent of $|\Lambda|$, the absolute convergence is uniform in $|\Lambda|$ and the thermodynamic limit of $< \mathcal{F} >_\Lambda$ exists.

For a proof of the important implications of this theorem (existence of a mass gap, Osterwalder-Schrader positivity) we refer to the original references [156, 157].

What we have considered so far are convergent series which are generated by a polymer representation of the partition function or expectation values of observables. The polymer expansions are combined with strong or small coupling expansions. Linked cluster expansions in this context are induced by a resummation of the polymer representation of Z. The resulting series are series in activities, but not primarily in the strong or small couplings themselves. In the next section we turn to convergent series expansions which result from a Taylor expansion of $\ln Z$ about a completely decoupled theory. This way we obtain power series in the inverse strong or small coupling parameters that allow an extrapolation to the critical region including a determination of the critical coupling.

4.1.6 *Linked cluster expansions as convergent expansions*

First we introduce some notations and specify the variety of systems to which linked cluster expansions apply. Let X denote the support of the degrees of freedom. X may consist of sites, links or plaquettes of a lattice Λ where the lattice should be understood as a cell complex (consisting of 0-cells (sites), 1-cells (links), 2-cells (plaquettes)). For the physical degrees of freedom we use the notation of scalar fields $\Phi(x)$, $x \in X$, but Φ can stand for scalar, fermion and gauge fields unless it is explicitly specified. Finally let $J(x)$ denote external currents. The partition function is then written as

$$
\begin{aligned}
Z(J,v) &\equiv e^{W(J,v)} \\
&= \int \prod_{x \in X} d\mu(\Phi(x)) \, e^{-S(\Phi,v) + \sum_{x \in \Lambda} J(x)\Phi(x)} \, .
\end{aligned} \quad (4.1.102)
$$

The parameter v denotes a generic coupling between n degrees of freedom at pairwise distinct so-called sites $x_1, x_2, ..., x_n \in X$. For $v = 0$ the system decouples completely. Therefore the action takes the form

$$
S(\Phi, v = 0) = \sum_{x \in X} \mathring{S}(\Phi(x)) \, , \quad (4.1.103)
$$

where the circle on S shall indicate that $\mathring{S}$ is an ultra-local quantity depending on Φ at one site x. Accordingly the partition function factorizes as

$$
Z(J, v = 0) = \prod_{x \in X} \mathring{S}(J(x)) \, , \quad (4.1.104)
$$

where the ultra-local partition function

$$
\mathring{Z} = \int d\mu(\Phi) \, e^{-\mathring{S}(\Phi) + J \cdot \Phi} \quad (4.1.105)
$$

can be solved "exactly", exactly in the sense of either analytically (in exceptional cases) or within the numerical accuracy. It follows for the logarithm W of $Z(J, v = 0)$

$$
W(J, v = 0) = \sum_{x \in X} \mathring{W}(J(x)) \, ,
$$
$$
\mathring{W} \equiv \ln \mathring{Z} \, , \quad (4.1.106)
$$

and for the connected n-point functions

$$\frac{\partial^n W(J,v)}{\partial J(x_1)...\partial J(x_n)}\bigg|_{v=0} = \frac{\partial^n \mathring{W}}{\partial J(x_1)^n} \, \delta_{x_1=...=x_n} \, . \qquad (4.1.107)$$

At $v = 0$ the connected n-point functions become ultra-local quantities. Usually one is interested in evaluating (4.1.107) at $J = 0$ after taking the derivative.

The linked cluster expansion (LCE) is then generated from a Taylor expansion of W in v about this decoupled system

$$W(J,v) = \left[\exp v \frac{\partial}{\partial \tilde{v}}\right] W(J,\tilde{v})\bigg|_{\tilde{v}=0} \equiv \sum_{L \geq 0} c_L \, v^L \, . \qquad (4.1.108)$$

At this place we do not worry about convergence properties, but understand (4.1.108) as a formal series. Hence LCEs are *power* series in v.

We give four examples.

- **Two-point interaction between scalars**
  $X \equiv \Lambda = \mathbf{Z}^D$ is a hyper-cubic lattice in D dimensions. The action is that of a Φ^4-theory

$$S(\Phi,v) = \sum_{x \in \Lambda} \left[\Phi(x)^2 + \lambda \left(\Phi(x)^2 - 1\right)^2\right] - \frac{1}{2} \sum_{x,y \, nn} (2\kappa) \, \Phi(x)\Phi(y) \, , \qquad (4.1.109)$$

so that v must be identified as

$$v(x,y) = 2\kappa \sum_{\mu=0}^{D-1} (\delta_{x,y+\hat{\mu}} + \delta_{x,y-\hat{\mu}}) \, , \qquad (4.1.110)$$

it couples two nearest neighbor (nn) sites x,y. In view of generalizations we call the interaction term proportional to $v(x,y)$ a two-point interaction. $\hat{\mu}$ is a unit vector in the positive μ-direction. Depending on the dimension and the renormalizability of the theory higher order monomials in $\Phi(x)$ may be included in the ultra-local part of the action. Spin models like Ising or Potts models fall into the same class of two-point interactions between scalars. It is this class of models for which LCEs are developed to the highest order in the expansion parameter. The accuracy is achieved by a computer aided, algorithmic generation of graphs in the associated graphical expansion. This way, in scalar models even the critical region becomes available by LCEs. This is the reason why we spend a separate section on LCEs in scalar models.

The following examples are much less elaborated to date, but with nowadays computer facilities a higher order accuracy seems to be within the range of feasibility.

- **Two-point interaction in Yukawa models**

  We consider the $Z(2)$ Yukawa model with Wilson fermions on the lattice and specify the action directly in a parametrization which is adapted to the application of linked cluster expansions. It consists of a bosonic part S_b and a fermionic part S_f

$$S = S_b + S_f \,, \qquad (4.1.111)$$

where S_b is given by (4.1.109), i.e. a scalar field with Φ^4-self-interaction, but the scalars here are coupled to two-flavor Wilson fermions with Yukawa couplings y included in the fermionic part S_f

$$S_f = \sum_{x \in \Lambda} \left(\sum_{\pm} \left[\overline{\psi}_\pm(x)\psi_\pm(x) - \chi P_\pm(x) \right] \right.$$

$$\left. + y \left[\overline{\psi}_+(x)\Phi(x)\psi_-(x) + \overline{\psi}_-(x)\Phi(x)\psi_+(x) \right] \right)$$

$$P_\pm(x) = \sum_{\mu=0}^{D-1} \overline{\psi}_\pm(x) \left[(r_0 - \gamma_\mu)\psi_\pm(x + \widehat{\mu}) + (r_0 + \gamma_\mu)\psi_\pm(x - \widehat{\mu}) \right].$$

$$(4.1.112)$$

The subscript $\pm$ refers to two flavors, γ_μ are Dirac matrices in Euclidean space, r_0 is the Wilson parameter (conveniently chosen as 1). Finally χ is the hopping parameter for the Wilson fermions. Now it is possible to perform combined linked cluster expansions in small κ and χ by Taylor expanding $\ln Z$ about $\kappa = \chi = 0$. The essential difference to pure scalar models does not arise because of the doubling of the expansion parameters, but comes from the graphical rules for fermions. While the hopping "propagator" for a scalar field

$$v(x, y) \equiv 2\kappa \sum_{\mu=0}^{D-1} (\delta_{x, y+\widehat{\mu}} + \delta_{x, y-\widehat{\mu}}) \qquad (4.1.113)$$

is not directed in the sense that the contribution to the partition function is independent on whether $y = x - \widehat{\mu}$ or $y = x + \widehat{\mu}$, the hopping propagator for the fermions

$$w(x,y) \equiv \chi \cdot \sum_{\mu=0}^{D-1} [\delta_{y,x+\widehat{\mu}}(r_0 - \gamma_\mu) + \delta_{y,x-\widehat{\mu}}(r_0 + \gamma_\mu)] \quad (4.1.114)$$

is directed, because $y = x + \widehat{\mu}$ is associated with $-\gamma_\mu$ and $y = x - \widehat{\mu}$ with $+\gamma_\mu$. Accordingly, in the graphical expansion the scalar hopping propagator is represented as an undirected line, the fermionic hopping propagator as a directed line. Directed lines make the calculation of the weights of the graphs more involved, but not unfeasible. (The reason is that the embedding factors do no longer decouple from the internal symmetry factors.) Therefore the development of efficient algorithms for such Yukawa-like systems is a challenge for future work.

Back in the nineties there was some interest in the phase structure of Yukawa models on the lattice. The interest in the phase structure of Yukawa models was based on the search for a nontrivial fixed point in the electroweak standard model in order to avoid a trivial continuum limit. The interested reader will find some references in [161].

- **Point-link-point interaction between two sorts of fields**
  The support $X = \Lambda \cup \Lambda_1$ is the union of the sets of sites of a hyper-cubic lattice with the set of links, $\Lambda_1 = \overline{\Lambda \times \Lambda} = \{\overline{(x,y)}/x, y \in \Lambda\}$. The bar denotes unoriented links. The generic form of the action is

$$S(\Phi, U, v) = \sum_{x \in \Lambda} \mathring{S}(\Phi(x)) + \sum_{(x,y) \in \Lambda_1} \mathring{\overset{1}{S}}(U(x,y))$$

$$- \frac{1}{2} \sum_{x,y} v(x,y)\Phi(x)U(x,y)\Phi(y), \quad (4.1.115)$$

where $v(x,y)$ can be chosen as above in Eq. (4.1.113), but is not necessarily restricted to next-neighbor interactions. Notice that in contrast to (4.1.109) not only the fields Φ have a self-interaction described by $\mathring{S}(\Phi(x))$, but also the fields U. The local (here in contrast to "ultralocal") interaction between the Φ-fields is no longer described by a constant parameter κ, but gets its own dynamics via $U(x,y)$. This is the reason why we called the associated graphical expansions *dynamical linked cluster expansions* (DLCEs) [158].

The action (4.1.115) includes two important classes of models, *gauged Higgs models* and *spin glasses*. We have encountered gauged Higgs models in connection with improved variational estimates for the $SU(2)$

Higgs model, cf. section 2.3. If the usual gauge invariant Wilson term of the $SU(2)$ Higgs model is treated within a variational approximation, one ends up with a model of the form of (4.1.115) with Φ the Higgs fields and U the gauge fields. In spin glasses the Φ have the meaning of spins and the U the meaning of dynamical couplings, usually called $J(x, y)$. An essential new ingredient in the graphical expansion is an additional notion of connectivity. Lines can be connected via some new kind of vertices not related to lattice sites. The full set of graphs which contribute to $\ln Z$ and n-point correlation functions was classified, their algorithmic generation worked out. The computer implementation of these algorithms is under construction. For further details we refer to the original paper [158].

- **Four-link interaction between gauge fields**

This time $X = \Lambda_1$, the set of links of a hyper-cubic lattice. The action is that of an $SU(N)$ lattice gauge theory

$$-S(U) = \frac{1}{g_0^2} \sum_{p \in \Lambda_2} \left[tr(U(\partial p)) + tr(U^\dagger(\partial p)) \right], \qquad (4.1.116)$$

where

$$U(\partial p) = \prod_{b \in \partial p} U(b) \equiv U(b_1) \, U(b_2) \, U(b_3) \, U(b_4) \qquad (4.1.117)$$

is the ordered product of U along the four links of the boundary ∂p of a plaquette p, as in (4.1.47). Here we have no counterpart to the ultra-local actions in (4.1.115). Notice that a linked cluster expansion in small g_0^{-2} is again some kind of strong coupling expansion, but very different from those of the previous section for observables like the Wilson-loop expectation value. It is an expansion directly for the free energy $\propto \ln Z$ and a power series in g_0^{-2} by construction. To our knowledge, this kind of strong coupling-linked cluster expansion for four-link interactions has not been promoted to higher orders so far. As a first step, appropriate graphical rules must be formulated, again with a new notion of connectivity. The expansion goes along with a considerable proliferation of graphs at a given order in g_0^{-2} that is a handicap for "hand-made" series. But with a computer-aided generation of graphs the proliferation is no longer an obstacle to the actual performance. The algorithmic organization of this generation of graphs is another challenge for future work.

This concludes our review of the variety of applications of generic linked cluster expansions. We turn next to an analysis of the coefficients of high-order linked cluster series and show what information about critical behavior is contained in these coefficients if the parameters are in the vicinity of a phase transition.

4.1.7 Convergent power series and the critical region: radius of convergence and physical singularity

Let us consider a function $f(z)$, $z \in \mathbf{C}$, which has a representation as a power series

$$f(z) = \sum_{L=0}^{\infty} a_L \, z^L, \qquad z \in \mathbf{C} \tag{4.1.118}$$

for $|z| < R$, where R is the radius of convergence. Usually the radius of convergence is either determined from the root criterion according to

$$R^{-1} = \limsup_{L \to \infty} |a_L|^{1/L} , \tag{4.1.119}$$

or from the ratio criterion as

$$R = \lim_{L \to \infty} \left| \frac{a_L}{a_{L+1}} \right| \tag{4.1.120}$$

if the limit exists. The function f is then analytic for $|z| < R$, but has at least one singularity on the circle $|z| = R$. Under certain conditions f can be analytically continued to a subregion of $|z| > R$. The analytic continuation is obtained by various methods, for example by a sequence of series representations of f with intersecting domains of convergence, by a closed representation of f like $(1-z)^{-1}$ $(= \sum_{L \geq 0} z^L$ for $|z| < 1)$ that has a single isolated pole at $z = 1$, or by an appropriate integral representation of f. If for $L \geq L_0$, $L_0 \in \mathbf{N}$, all coefficients satisfy $a_L \geq 0$ (or all coefficients satisfy $a_L \leq 0$), the strongest singularity for $|z| = R$ lies on the positive *real* axis. Therefore, in this case, the singularity next to $z = 0$ on the positive real axis can be obtained from the radius of convergence of the power series representation of f. Now, the question of interest is when the radius of convergence is related to a *physical* singularity. Let us assume that we have a series representation of the free energy $\propto \ln Z$ in powers of κ, $\kappa \in \mathbf{R}$, in the infinite volume with radius of convergence $R > 0$ and that all expansion coefficients a_L are positive. In this case the strongest singularity on the positive real axis, $\kappa_c \in \mathbf{R}$, is a singularity of the free energy and, as such, marks a phase transition point. The phase transition can be of first, second or higher order.

In a *finite* volume there is no true singularity of the free energy. Since f is then analytic for real κ, the singularity must lie on the circle $|z| = R$ somewhere in the complex plane, but the singularity moves towards the real axis as the volume goes to infinity and $\kappa \to \kappa_c = R$.

Now let us assume that we expect a *second*-order phase transition and calculate the two-point susceptibility with linked cluster expansions as a power series in 2κ to some finite order L_h

$$\chi_2 \approx \sum_{L=0}^{L_h} a_L \, (2\kappa)^L \, . \tag{4.1.121}$$

On the other hand we know from the renormalization-group analysis that the leading singular behavior of χ_2 takes the form

$$\chi_2 \propto \mathcal{A}_{\chi_2} \left(1 - \frac{\kappa}{\kappa_c} \right)^{-\gamma} \qquad \text{as } \kappa \to \kappa_c^- \, , \tag{4.1.122}$$

where the critical exponent γ characterizes the universality class. Next we apply the binomial theorem to (4.1.122) with binomial coefficients

$$\binom{x}{\nu} = \left(\prod_{\mu=0}^{\nu-1} (x - \mu) \right) \frac{1}{\nu!} \, , \qquad x \in \mathbf{R}, \ \nu \in \mathbf{N} \tag{4.1.123}$$

and obtain as leading terms for the ratio of two succeeding coefficients

$$\frac{a_L}{a_{L-1}} = \frac{1}{2\kappa_c} \left(1 + \frac{\gamma - 1}{L} + o(L^{-1}) \right) \qquad \text{as } L \to \infty \, . \tag{4.1.124}$$

In principle κ_c and γ can be determined from a fit of a number of ratios a_L/a_{L-1} for large enough L according to (4.1.124). A rough estimate of how large the highest order L_h in the linked cluster expansion should be is $L_h \propto |\kappa - \kappa_c|^{-1}$. Practically $L_h = 18$ for LCEs in scalar $O(N)$ models if the critical coupling shall have an accuracy of four digits [160]. There is also some kind of lower bound L_l on L such that coefficients a_L with $L < L_l$ should be discarded for the fit. A concrete example illustrates why. Let us assume that χ_2 has the form

$$\chi_2 = (1 - 2\kappa)^{-1} + e^{-4\kappa} \tag{4.1.125}$$

with a leading singularity at $\kappa_c = 1/2$, $\gamma = 1$ and an analytic part $e^{-4\kappa}$. The analytic part shall represent the regular contributions to χ_2 in a closed form. Naively one would expect that the analytic part does not affect the leading singular terms which are extracted from a power-series expansion of (4.1.125). The coefficients of the series of (4.1.125) are

$$a_L = 1 + \frac{(-2)^L}{L!} \qquad (4.1.126)$$

if we write (4.1.125) as a power series in 2κ and

$$\frac{a_L}{a_{L-1}} = 1 + o(L^{-1}), \qquad (4.1.127)$$

so that $2\kappa_c = 1$. What we actually find for a_L/a_{L-1} in the first few orders is shown in Table 4.1.1.

Table 4.1.1   Example for ratios a_L/a_{L-1} up to order $L = 8$

L	1	2	3	4	5	6	7	8
$\dfrac{a_L}{a_{L-1}}$	$-\dfrac{1}{2}$	-3	$-\dfrac{1}{9}$	-5	$\dfrac{11}{25}$	$\dfrac{49}{33}$	$\dfrac{307}{343}$	$\dfrac{317}{307}$

There is still an error of $\sim 3\%$ in κ_c if we extract the radius of convergence from a_8/a_7. In scalar $O(N)$ models it turns out that L should be larger than 8 or 10 when a_L shall be used for the fit. Even an estimate of κ_c from the highest order coefficients, e.g. a_{20}/a_{19}, is not as reliable as the extrapolated value from a fit for $L \to \infty$. The neglected terms in (4.1.124) are summarized by $o(L^{-1})$. Since there is no prediction of the analytic form of $o(L^{-1})$, e.g. of $o(L^{-1}) = 1/(L \ln L)$ or $o(L^{-1}) = 1/(L^2)$, several ansätze are made, and the ansatz which yields the best fit is used in the final extrapolation.

Two further critical exponents, ν and η, can be directly obtained from LCE series, in an analogous way to (4.1.124). The exponent ν follows from a series for $1/m_R^2$, where m_R is the renormalized mass, defined via the small-momentum behavior of the two-point vertex correlation

$$\widetilde{\Gamma}_{ab}^{(2)}(p, -p) = -\frac{1}{Z_R}\left(m_R^2 + p^2 + O(p^4)\right)\delta_{ab} \qquad \text{as } p = (p_0 = 0, \vec{p} \to \vec{0})$$

$$(4.1.128)$$

for zero or finite temperature. The series for m_R^{-2} is constructed from the

LCE series for the two-point susceptibility χ_2

$$\delta_{ab}\,\chi_2 \;=\; \sum_{x_2} \; <\Phi_a(0)\Phi_b(x_2)>^c \qquad (4.1.129)$$

and the second moment μ_2, for zero temperature given by

$$\mu_2 \;=\; \sum_{x} \left(\sum_{i=0}^{D-1} x_i^2\right) \; <\Phi_1(0)\Phi_1(x)>^c , \qquad (4.1.130)$$

for finite temperature given by (4.1.130) with $i = 1, \ldots, D - 1$; it is then constructed according to

$$\frac{1}{m_R^2} \;=\; \frac{1}{2D}\,\frac{\mu_2}{\chi_2} . \qquad (4.1.131)$$

Equation (4.1.131) follows from (4.1.128) in coordinate space together with (4.1.129) and (4.1.130).

Since the correlation length ξ defined as m_R^{-1} equals the correlation length defined as the inverse pole mass m_P^{-1} up to higher orders in m_P, and ξ diverges with the exponent ν at a second-order phase transition, we obtain

$$\xi^2 \;\equiv\; \frac{1}{m_R^2} \;\propto\; \left(1 - \frac{\kappa}{\kappa_c}\right)^{-2\nu} , \qquad (4.1.132)$$

where the series for the left hand side is taken from (4.1.131). Once the exponent ν is known, the exponent η can be read off from a fit of the series for the wave-function renormalization constant Z_R via

$$Z_R \;\propto\; \left(1 - \frac{\kappa}{\kappa_c}\right)^{\nu\eta} . \qquad (4.1.133)$$

In principle, the series for Z_R can be obtained from the series for m_R^{-2} and χ_2, since

$$Z_R \;=\; \chi_2 \cdot m_R^2 , \qquad (4.1.134)$$

as it follows from (4.1.128) and (4.1.129). In practice, however, η is determined from ν, γ and the scaling relations. Therefore, one first has to determine accurate values for ν and γ that follow from fits of the form (4.1.124) if κ_c was determined with high accuracy before. Here it is advantageous to use the series for the 1PI-susceptibilities. In particular, the connected two-point susceptibility χ_2 is related to the 1PI-susceptibility χ_2^{1PI} according to

$$\chi_2 = \frac{\chi_2^{1PI}}{1 - 2D2\kappa\chi_2^{1PI}} . \qquad (4.1.135)$$

Rather than determining κ_c from a fit (4.1.124) of χ_2, κ_c also follows from a fit of the zero of the denominator in (4.1.135) if χ_2 diverges at a second-order phase transition. For further details we refer to [159].

From the renormalization-group analysis we know that two exponents are sufficient to determine all others if there are two relevant scaling fields like the temperature and an external field. Therefore, if actually three exponents γ, ν, η are determined in the symmetric phase, it provides a check of the scaling relations or, vice versa, a check of the truncation errors in the LCE series.

Linked cluster expansions are available in the symmetric phase and only there. The expansion is about the high-temperature limit (the decoupled system). It can be maximally extended down to the critical region that is approached for $T \to T_c$ from above, or $\kappa \to \kappa_c$ from below. Therefore all critical exponents in the broken (low temperature/large κ) phase, in particular the exponent δ cannot be calculated directly from LCE series, but only via the scaling relations and the fact that the exponents in the symmetric phase agree with those in broken phase when the exponents are defined in both phases.

Further alternatives on the extrapolation of the infinite-order result for κ_c and the critical exponents are provided by

- shifting the singularity at $(-\kappa_c)$ to $(-\infty)$ by means of a suitable transformation of variables,
- using Padé approximants.

If the LCE series are determined in the infinite volume and the coefficients satisfy $a_L \geq 0$ for $L > L_0$ for some L_0, and if κ_c is the transition point of a first-order phase transition, κ_c can still be determined as the radius of convergence of the series for the free energy or for any derived quantity like an order parameter susceptibility which is singular at κ_c.

Even the question of the very order of the phase transition (first or second) can be decided within linked-cluster series expansions. Here it is quite convenient to go to a finite volume V and study the monotonicity behavior of the coefficients a_L as a function of V. As precursors of a

δ-function or power-law singularity in the infinite volume, a decreasing or increasing of the coefficients a_L with V signals a first- or second-order phase transition, respectively. This behavior is observed in a certain neighborhood of $\kappa_c(V = \infty)$ that was specified in section 2.6.

The tiny finite-temperature shift of the critical line $\kappa_c(\lambda)$ in a Φ^4-theory in four dimensions from $T = 0$ to $T > 0$ is hidden in the embedding factors which contribute to the coefficients a_L. In contrast to this tiny effect the universality class of this transition changes from $T = 0$ to $T > 0$ or from four to three dimensions. It is therefore important to calculate the critical exponents with an accuracy which is sufficient to distinguish between Gaussian and Ising exponents for $N = 1$, or $O(N)$ exponents for $N > 1$. In order to obtain such an accuracy in the critical exponents it turned out that the LCE series should be expanded including the 18th order in κ to get a sufficient accurate value of κ_c if the extension in temperature direction was chosen as $L_0 = 4$ [160].

For a variety of scalar models (N-component Φ^4-theories with $N = 1$ or $N = 4$ in three or four dimensions in the infinite volume, also for $(\Phi^4 + \Phi^6)$ theories with $N = 1$ or $N = 4$ in three dimensions in the infinite volume) we observed a uniform (positive) sign of the coefficients a_L of the LCE series, including the highest order in the expansion (18th and 20th, respectively). Therefore the identification of the radius of convergence with a singularity on the positive real axis and therefore with a physical singularity is justified modulo the uncertainty of the sign of higher order terms. In this sense the identification of a universality class by means of an LCE analysis is not a rigorous result (in contrast to the convergence proof of strong coupling expansions of local observables in the previous section). It gives, however, a strong hint that may be considered as a safe result if other indications, e.g. from Monte Carlo simulations or ϵ-expansions, point into the same direction.

In the next section we explain in detail how the LCE series for scalar models are actually generated and how it becomes possible to handle of the order of billions of graphs which contribute to the coefficients at higher orders in κ such as κ^{20}.

4.2   Linked cluster expansions in more detail

4.2.1   *Historical remarks*

Linked cluster expansions (LCE) are convergent series expansions of the free energy and connected correlation functions about completely decoupled lattice systems. The expansion parameter is the hopping parameter that couples fields at different lattice sites. In the literature it is often called κ. Only connected graphs, so-called *linked clusters*, contribute to the graphical expansion which is associated with the analytic terms of the series.

In many applications of statistical physics, the expansion parameter has the physical meaning of an inverse temperature. More generally, for instance in field theories at finite temperature T, κ is related to T in such a way that $\kappa = 0$ corresponds to $T = \infty$. Therefore linked cluster expansions are high temperature expansions. If no confusion can arise, we use *linked cluster-*, *high temperature-* and *hopping parameter expansion* as synonymous names.

Usually the analytic expansions are obtained as graphical expansions. Because of the progress in computer facilities and the development of efficient algorithms for generating graphs, it is nowadays possible to handle of the order of billions of graphs. The wide range from high temperatures towards the critical region becomes available, and thermodynamic quantities like critical temperatures and critical indices are determined with high precision. (The precision is comparable to high-quality Monte Carlo results if results are available for the same set of parameters.)

LCE have a long tradition in statistical physics [162]. Back in the thirties of the last century they were developed for classical fluids, back in the fifties for magnetic systems. Their generalization and application in the context of particle physics was pioneered by Lüscher and Weisz [163]. Lüscher and Weisz studied the issue of a trivial continuum limit of a lattice $\lambda_0 \Phi^4$-theory in four dimensions, at zero temperature and in the infinite volume. In this application, for sufficiently small lattice spacing a, the hopping parameter κ is related to a according to

$$\kappa_c(\lambda_0) - \kappa \sim a^2 \, ,$$

where $\kappa_c(\lambda_0)$ is the critical hopping parameter. If the continuum limit of the Φ^4-theory turns out to be trivial in the sense that the renormalized couplings λ_R vanish on all scales as the UV-cutoff a^{-1} is sent to infinity (and not only on an asymptotic scale), an interacting Φ^4-theory must be understood as an *effective* theory with an intrinsic large but finite cutoff. As an effective theory it keeps its predictive power for physics on energy scales much smaller than the cutoff scale. (Related to a trivial continuum limit of a Φ^4-theory is the supposed trivial continuum limit of the electroweak theory. Triviality of the continuum limit there implies an upper bound on the physical Higgs mass, and again, the postulate of an intrinsic UV-cutoff at some high-energy scale.)

Using LCE Lüscher and Weisz computed the renormalized quartic coupling λ_R on the lattice, with λ_R defined by the four-point vertex function at zero momentum, and the renormalized mass in lattice units $m_R \simeq a$, defined by the two-point vertex function at zero momentum. They computed both quantities in a region close to but still sufficiently apart from the critical line $\kappa_c(\lambda_0)$ where $m_R \simeq 0.5$. The values of λ_R turned out to be below the tree-level unitarity bound. This is a strong indication that renormalized perturbation theory, that is, perturbation theory in λ_R, can be used to evaluate physical quantities in the vicinity of the critical line. The perturbative β-function is then used to evolve $\lambda_R(\lambda_0, \kappa)$ as $\kappa \to \kappa_c(\lambda_0)$, or equivalently as $m_R \to 0$ for fixed λ_0. Triviality of the Φ^4-theory then holds if $\lambda_R(\lambda_0, \kappa_c(\lambda_0))$ vanishes for all values of the bare λ_0, that is for $0 \leq \lambda_0 \leq \infty$. More precisely, Lüscher and Weisz stated the triviality bound on λ_R saying that for given (large) cutoff a^{-1} or m_R^{-1}, λ_R must be chosen so small that the inequality

$$\ln \frac{1}{a} \propto \ln \frac{1}{m_R} \leq \frac{c_1}{\lambda_R} - c_2 \ln \lambda_R + c_3$$

holds. The constants c_1 and c_2 are the coefficients of the perturbative two-loop β-function. The constant c_3 is a nonperturbative number obtained by LCE.

In order to reach the range where $m_R \simeq 0.5$ and to verify that this range belongs already to the validity range of perturbation theory, one had to use LCE to order κ^{14}. Therefore Lüscher and Weisz developed a computer-aided algorithmic generation of graphs, which is the only practicable procedure when LCE are needed to such high orders.

A further step in the development of LCE is the extension to systems with torus-topology in one direction by Reisz [164]. The physical reason for this generalization are field theories at *finite* temperature, for which the inverse temperature is determined by the length of the torus. Two important questions concerning the phase transition of the four-dimensional lattice $\lambda_0 \Phi^4$-theory were answered in the framework of LCE. The first question concerns the order of the transition from the broken to the symmetric phase. Does it stay of second order at finite temperature? From the discussion of the constraint effective potential in perturbation theory in chapter 4.5 (with the prediction that the order of the transition depends on N, the number of components of the scalar field), the answer is open. Application of LCE showed that it actually stays of second order for all N. The second question concerns the universality class of the transition. The effect of the torus length L_0, that is of the number of lattice links in temperature direction, on the critical coupling $\kappa_c(\lambda_0, L_0)$ is rather weak, even for $L_0 = 4$. It deviates from the zero-temperature value $\kappa_c(\lambda_0, \infty)$ only by a few percent. But it is this tiny shift in κ_c that must be resolved in order to discover different critical exponents and therefore a change of the universality class. For the required accuracy of κ_c, LCE must be computed to order κ^{18}. In order to handle the graphs that contribute up to this order, it was necessary to optimize the computer-aided generation of graphs. As a result, the universality class of the second-order transition changes from Gaussian at zero temperature to the universality class of the three-dimensional $O(N)$ models at finite temperature. This result confirms the naive expectation based on classical dimensional reduction that the high temperature behavior (here even down to T_c) of an $O(N)$-scalar field theory in four dimensions is well described by the corresponding theory in three dimensions at zero temperature.

Finally we mention two generalizations of LCE by Meyer-Ortmanns and Reisz. The first one [165] refers to LCE for a system with finite extensions in *all* directions. LCE in a finite volume allow to identify first- and second-order transition regions in phase space, and to locate tricritical lines with high precision. They further allow to perform a finite-size scaling analysis and to directly compare the results with Monte Carlo simulations. We will discuss LCE in a finite volume later in this chapter.

The second generalization [166] endows the former constant hopping parameter with its own dynamics, leading to what we call *dynamical linked cluster expansions* (DLCE). In connection with variational estimates they have applications in QCD and in the $SU(2)$ Higgs model.

To date high-order linked cluster expansions are only available for scalar models such as Ising models, Potts models or $O(N)$-symmetric Φ^4-theories, $N \geq 0$, as indicated in section 4.1. In principle, LCE are applicable also to fermionic systems, although an algorithmic generation of graphs to comparable orders as for scalar models remains to be done for this case. In combination with variational estimates and Monte Carlo simulations, LCE can be utilized in *effective* scalar models for $SU(N)$ gauge theories with matter fields. After these introductory remarks we now go into detail.

4.2.2 Introduction to linked cluster expansions

In this section we introduce the very definitions of LCE and explain their graphical realizations. We discuss some central issues of the computer-aided generation of graphs. For a more detailed discussion we refer the interested reader to the original literature.

We consider a D-dimensional hypercubic lattice Λ, which is either unbounded in all directions or has a finite number L_0 of lattice links in temperature direction,

$$x = (x_0, \mathbf{x}) \; , \quad x_0 = 0, 1, \dots, L_0 - 1 \; , \; \mathbf{x} \in \mathbb{Z}^{D-1}. \qquad (4.2.1)$$

The lattice spacing is set to unity for simplicity. Periodic boundary conditions are imposed along the torus. Also, we assume that L_0 is an even number, which leads to a number of simplifications.

The class of models we discuss are described by the partition function

$$Z(J, v) = \int \prod_{x \in \Lambda} d^N \Phi(x) \, \exp\left(-S(\Phi, v) + \sum_{x \in \Lambda} J(x) \cdot \Phi(x)\right), \qquad (4.2.2)$$

where Φ denotes a real, N-component scalar field, J is an external source, and

$$J(x) \cdot \Phi(x) = \sum_{a=1}^{N} J_a(x) \Phi_a(x).$$

The action $S(\Phi, v)$ is of the form

$$S(\Phi, v) = \sum_{x \in \Lambda} \mathring{S}(\Phi(x)) - \frac{1}{2} \sum_{x \neq y \in \Lambda} \sum_{a,b=1}^{N} \Phi_a(x) v_{ab}(x, y) \Phi_b(y). \qquad (4.2.3)$$

The "ultra-local", single-site action $\mathring{S}$ is assumed to be $O(N)$ invariant and guarantees stability of the partition function (4.2.2) for sufficiently small v_{ab}. For example, for a $(\Phi^4 + \Phi^6)$-interacting theory, the canonical form is given by

$$\mathring{S}(\Phi) = \Phi^2 + \lambda(\Phi^2 - 1)^2 + \sigma(\Phi^2 - 1)^3, \qquad (4.2.4)$$

with $\sigma > 0$ or $\sigma = 0$, $\lambda \geq 0$. We emphasize, however, that the general techniques we present below are not restricted to this case. The factor $\exp(-\mathring{S}(\Phi))$ may have a very general form, in particular it happens that it is only known as a table or an integral when the effective models are obtained upon integration of partial degrees of freedom, analytically, or by means of Monte Carlo methods.

Fields at different lattice sites are coupled by the hopping term $v_{ab}(x,y)$. For the case of nearest-neighbor interactions,

$$v_{ab}(x,y) = \begin{cases} 2\kappa\,\delta_{a,b}\,, & x,y \text{ nearest neighbor,} \\ 0\,, & \text{otherwise.} \end{cases} \qquad (4.2.5)$$

The generating functional of connected correlation functions (the free energy functional) is given by

$$W(J,v) = \ln Z(J,v), \qquad (4.2.6)$$

$$W^{(2n)}_{a_1...a_{2n}}(x_1,\ldots,x_{2n}) = <\Phi_{a_1}(x_1)\cdots\Phi_{a_{2n}}(x_{2n})>^c$$

$$= \left.\frac{\partial^{2n}}{\partial J_{a_1}(x_1)\cdots\partial J_{a_{2n}}(x_{2n})}W(J,v)\right|_{J=0}.$$

Odd correlation functions vanish in the symmetric phase because of the $O(N)$ symmetry. In the following we are mainly concerned with the connected two-point function at zero momentum $\widetilde{W}^{(2)}_{ab}(0,0)$, that is the susceptibility χ_2 and the moment μ_2, defined according to

$$\delta_{a,b}\,\chi_2 = \widetilde{W}^{(2)}_{ab}(0,0) = \sum_x <\Phi_a(x)\Phi_b(0)>^c,$$

$$\delta_{a,b}\,\mu_2 = \sum_x g(x)\ <\Phi_a(x)\Phi_b(0)>^c = -D_r\left.\frac{\partial^2}{\partial p_1^2}\ \widetilde{W}^{(2)}_{ab}(p,-p)\right|_{p=0}, \qquad (4.2.7)$$

with $g(x) = \sum_{i=0}^{D} x_i^2$ and $D_r = D$ on the lattice that is unbounded in all directions, or with $g(x) = \sum_{i=1}^{D} x_i^2$ and $D_r = D-1$ on the finite-temperature lattice. The renormalized mass m_R (in lattice units) and the wave-function renormalization constant Z_R are obtained from χ_2 and μ_2 by

$$m_R^2 = 2D_r\frac{\chi_2}{\mu_2}\,, \quad Z_R = 2D_r\frac{\chi_2^2}{\mu_2}. \qquad (4.2.8)$$

LCE are series expansions of W and of $W^{(2n)}$ in the hopping parameter κ, that is of W and of $W^{(2n)}$ about completely decoupled lattice systems. Completely decoupled lattice systems are obtained for $v = 0$. In this case,

$S(\Phi, v = 0) = \sum_x \mathring{S}(\Phi(x))$, the partition function factorizes, and in turn $W(J, v = 0) = \sum_x \mathring{W}(J(x))$. In particular,

$$W^{(2n)}_{a_1 \ldots a_{2n}}(x_1, \ldots, x_{2n}) = \begin{cases} \frac{\mathring{v}^c_{2n}}{(2n-1)!!} C_{2n}(a_1, \ldots, a_{2n}), & \text{for } x_1 = \cdots = x_{2n}, \\ 0 & , \quad \text{otherwise} \end{cases}$$

(4.2.9)

with

$$\mathring{v}^c_{2n} = \left. \frac{\partial^{2n}}{\partial J_1^{2n}} \mathring{W}(J) \right|_{J=0}.$$

(4.2.10)

C_{2n} are the totally symmetric O(N)-invariant tensors, given recursively by

$$C_2(a, b) = \delta_{a,b},$$

(4.2.11)

$$C_{2n}(a_1, \ldots, a_{2n}) = \sum_{i=2}^{2n} \delta_{a_1, a_i} C_{2n-2}(a_2, \ldots, \widehat{a}_i, \ldots, a_{2n}), \quad n \geq 2,$$

where ($\widehat{}$) denotes omission of the corresponding entry. For all indices equal, we have

$$C_{2n}(1, \ldots, 1) = (2n - 1)!!,$$

(4.2.12)

where $(2n - 1)!! = 1 \cdot 3 \cdot 5 \cdots (2n - 1)$. The linked cluster expansion of the free-energy functional is the Taylor expansion with respect to $v(x, y)$ about this decoupled case,

$$W(J, v) = \left. \left(\exp \sum_{x,y} \sum_{a,b} v_{ab}(x, y) \frac{\partial}{\partial \widehat{v}_{ab}(x, y)} \right) W(J, \widehat{v}) \right|_{\widehat{v}=0},$$

(4.2.13)

with obvious generalizations to connected correlation functions according to (4.2.6). Multiple derivatives of W with respect to $v(x, y)$ are computed by the generating equation

$$\frac{\partial W}{\partial v_{ab}(x, y)} = \frac{1}{2} \left(\frac{\partial^2 W}{\partial J_a(x) \partial J_b(y)} + \frac{\partial W}{\partial J_a(x)} \frac{\partial W}{\partial J_b(y)} \right).$$

(4.2.14)

Let us compute the free energy density $w = W/V$ to fourth order in κ, by applying (4.2.13) to fourth order in v and (4.2.14) four times. Using (4.2.5), (4.2.6), (4.2.9) and translation invariance, we obtain after some tedious algebra

$$
w = \overset{\circ}{v_0^c} + \frac{1}{4}\left(\overset{\circ}{v_2^c}\right)^2 \sum_x v_{11}(0,x)^2 \sum_{a,b=1}^{N} C_2(a,b)C_2(a,b)
$$

$$
+\frac{1}{8}\left(\overset{\circ}{v_2^c}\right)^2 \frac{\overset{\circ}{v_4^c}}{3} \sum_x v_{11}(0,x)^2 \sum_y v_{11}(0,y)^2
$$

$$
\cdot \sum_{a_1,\ldots,a_4=1}^{N} C_2(a_1,a_2)C_4(a_1,\ldots,a_4)C_2(a_3,a_4)
$$

$$
+\frac{1}{2\cdot 4!}\left(\frac{\overset{\circ}{v_4^c}}{3}\right)^2 \sum_x v_{11}(0,x)^4 \tag{4.2.15}
$$

$$
\cdot \sum_{a_1,\ldots,a_4=1}^{N} C_4(a_1,\ldots,a_4)C_4(a_1,\ldots,a_4)
$$

$$
+\frac{1}{8}\left(\overset{\circ}{v_2^c}\right)^4 \sum_{x,y,z} v_{11}(0,x)v_{11}(x,y)v_{11}(y,z)v_{11}(z,0)
$$

$$
\cdot \sum_{a_1,\ldots,a_4=1}^{N} C_2(a_1,a_2)C_2(a_2,a_3)C_2(a_3,a_4)C_2(a_4,a_1)
$$

$$
+O(v^5)\,.
$$

Working out the summations of x, y, z over the lattice sites and the summation of $a_1, a_2, \ldots$ over the symmetry labels, we obtain with (4.2.11)

$$
w = \overset{\circ}{v_0^c} + \left[\frac{1}{4}\left(\overset{\circ}{v_2^c}\right)^2 (2D)\,N\right](2\kappa)^2
$$

$$
+ \left[\frac{1}{8}\left(\overset{\circ}{v_2^c}\right)^2 \frac{\overset{\circ}{v_4^c}}{3}(2D)^2\,N\,(N+2) + \frac{1}{48}\left(\frac{\overset{\circ}{v_4^c}}{3}\right)^2 (2D)\,N\,(N+2)\right.
$$

$$
\left. + \frac{1}{8}\left(\overset{\circ}{v_2^c}\right)^4 \bar{I}N\right](2\kappa)^4 + O(\kappa^5)\,, \tag{4.2.16}
$$

where $\bar{I}$ is given by

$$
\bar{I} = \begin{cases} 6D(2D-1) & , \quad \text{for } L_0 > 4, \\ 6D(2D-1)+2, & \text{for } L_0 = 4. \end{cases} \tag{4.2.17}
$$

Already to this low order, a lengthy computation is required to obtain the result (4.2.15). It is obvious that we cannot proceed to much higher orders in this way. To manage the rapidly increasing complexity of the analytic expressions, a diagrammatic approach is very convenient. The free energy density and the susceptibilities then become sums over (equivalence classes of) connected graphs Γ, with every graph contributing with weight $w(\Gamma)$. The following choice of rules is convenient. Every Γ consists of internal lines

$$\underset{x\ a \qquad b\ y}{\bullet\!\!-\!\!\!-\!\!\!-\!\!\!-\!\!\bullet} \qquad = \qquad v_{ab}(x,y) \qquad\qquad (4.2.18)$$

and of vertices

$$\overset{a_1 \quad a_2 \qquad a_{2n}}{\underset{x}{\bigvee\!\!\cdots\!\!}} \qquad = \qquad \frac{\overset{\circ c}{v}_{2n}}{(2n-1)!!}C_{2n}(a_1,\ldots,a_{2n}), \qquad (4.2.19)$$

where the right hand side turns out to be x-independent. To obtain $w(\Gamma)$, all symmetry labels of the internal lines are summed over. Furthermore, the vertices are placed at lattice sites, and the sum is taken over all possible placements, except for one vertex which is kept fixed, to absorb the extensive volume factor in the denominator. There is no explicit exclusion principle that forbids two different vertices to be placed at the same lattice site. There is only an implicit constraint that depends on the particular form of $v_{ab}(x,y)$. For (4.2.5), two vertices which are connected by a line must be placed at nearest-neighbor lattice sites. Finally, with every graph Γ one associates a topological symmetry number S_Γ that accounts for the topological symmetry with respect to exchange of lines and vertices. As the name suggests, the topological symmetry refers to purely graphical properties (saying which vertices are connected and with how many lines). The weight of Γ has a factor $1/S_\Gamma$. With these rules, (4.2.15) becomes

$$w \quad = \quad \bullet \quad + \quad \bigcirc\!\!\!\!\bigcirc \quad + \quad \bigcirc\!\!\!\!\bigcirc\!\!\!\!\bigcirc$$

$$+ \quad \bigcirc\!\!\!\!\bigcirc \quad + \quad \bigcirc\!\!\!\!\bigcirc \quad + \quad O(\kappa^5), \qquad (4.2.20)$$

where as usual every graph stands for its full weight. Comparing with (4.2.15) we realize that the five graphs of (4.2.20) have the topological symmetry numbers 1, 4, 8, $2\cdot 4!$ and 8, respectively.

In a similar way we obtain the series expansions for connected correlation functions and susceptibilities, by applying (4.2.13) and (4.2.14) together with additional derivatives with respect to external currents J according to (4.2.6). Graphs that contribute to susceptibilities have external lines, in addition to internal lines and vertices. External lines have only one endpoint vertex. For instance, graphs contributing to χ_2 and μ_2, Eq. (4.2.7), have two external lines

$$\underset{\substack{\\ \text{x}}}{\underset{1}{\rule[0.5ex]{6cm}{0.4pt}}\!\bullet} \tag{4.2.21}$$

attached. Because of the internal $O(N)$ symmetry, it is sufficient to choose the symmetry labels of the external lines equal to 1. In general the series representation of χ_2 reads

$$\chi_2 = \sum_{\nu \geq 0} \sum_{\Gamma \in \mathcal{G}_2(\nu)} w(\Gamma). \tag{4.2.22}$$

$\mathcal{G}_2(\nu)$ denotes the set of (equivalence classes of) connected graphs with two external lines and ν internal lines. Every $\Gamma \in \mathcal{G}_2(\nu)$ contributes to the order κ^ν. Hence the internal sum over $\mathcal{G}_2(\nu)$ in (4.2.22) is proportional to $(2\kappa)^\nu$. More precisely, every $\Gamma \in \mathcal{G}_2(\nu)$ contributes to χ_2 the weight

$$w(\Gamma) = \frac{1}{S_\Gamma} \mathcal{E}_\Gamma \, I_\Gamma \, C_\Gamma \left(\prod_w \frac{\mathring{v}^c_{l(w)}}{(l(w)-1)!!} \right) \cdot (2\kappa)^\nu, \tag{4.2.23}$$

where the various factors have the following meaning:

- The product $\prod_w$ is over all vertices w of Γ, and $l(w)$ denotes the total number of lines entering w. This is the only place where the particular form of the ultra-local action $\mathring{S}(\Phi(x))$ of (4.2.3) enters.
- C_Γ is the internal, i.e. the $O(N)$-symmetry number of Γ. It is obtained by summation over all symmetry labels of the internal lines. The particular symmetry group only enters the coefficients $\mathring{v}^c_{l(w)}$ and this factor C_Γ.
- I_Γ denotes the lattice-embedding number of Γ, that is the number of possibilities to put Γ onto the lattice, in accordance with the topology of the graph and of the lattice. One vertex is kept fixed (otherwise the extensive volume factor would drop out, since the contribution would be normalized over the volume), and two vertices with a line in common

are placed onto nearest neighbor lattice sites, according to (4.2.5). The particular geometry and shape of the lattice enters only the factor I_Γ. Because we consider a hyper-cubic lattice and assume L_0 to be an even number, I_Γ can be different from zero only if every loop of Γ has an even number of lines.

For weighted susceptibilities such as the second moment μ_2 of Eq. (4.2.7), every particular embedding on the lattice has to be weighted appropriately. Every embedding of a graph Γ contributing to μ_2 is weighted by $g(x)$, where x is the lattice distance of the two vertices that have external lines attached. (We have $x = 0$ and in turn $g(x) = 0$ if both external lines are attached to the same vertex.)

- $\mathcal{E}_\Gamma$ is a permutation factor of the external lines. It counts the number of possibilities to enumerate the external lines, while the relative enumeration of lines which enter the same vertex is omitted, that is, for n external lines

$$\mathcal{E}_\Gamma = \frac{n!}{\prod_w n(w)!}, \qquad (4.2.24)$$

where $n(w)$ is the number of external lines at the vertex w. For the case of $n = 2$, we have $\mathcal{E} = 2$ if the two external lines are attached to different vertices, and $\mathcal{E} = 1$ if they are both attached to the same vertex.

- S_Γ is the topological symmetry factor of Γ.

Because of the $O(N)$ symmetry, a graph Γ has a nonvanishing weight only if all vertices of Γ have an even number of lines attached. Furthermore, on the hyper-cubic lattice and for even L_0, every graph with an odd number of lines in a loop has a vanishing embedding number. In the following we denote by $\mathcal{G}_n^{\mathrm{ev}}(\nu)$ the set of (equivalence classes of) connected graphs with n external and ν internal lines, which have only vertices with an even number of lines attached and with an even number of lines in every loop.

Next we come to the topological symmetry factor S_Γ whose inverse enters the weight $w(\Gamma)$. Let us assume that the vertices are labelled in a way that two distinct vertices get a different label. As operations on the graph we allow an exchange of unlabelled lines between the vertices and a labelling of vertices (independently of the external lines which may be associated with these vertices). The symmetry factor then counts the number of possibilities to exchange lines and relabel vertices such that the connectivity remains the same. This means that the same vertices as before are connected with each other and by the same number of lines. A more thorough discussion would require a more precise notion of a graph and of the equivalence of any two graphs which is beyond the scope of this introduction. Let us instead explain the appearance of the symmetry factor by a simple example. We consider the graph

$$(4.2.25)$$

that contributes to the order κ^3 to χ_2. For unlabelled external lines we label the left vertex say with 1 and the right vertex with 2. (Imagine that vertex 1 and vertex 2 have each three "open ends" that they can use to produce the graph of (4.2.25). The first open end of vertex 1 has three possibilities for joining an open end of vertex 2 to form a link, the second open end has two choices and the third just 1, leading to 3! possibilities of permutations.) So there are 3! possibilities of permuting the lines such that the vertices 1 and 2 remain connected by means of three lines and two possibilities of placing vertex 1 and 2 (left or right of the graph) such that the connectivity is the same. Thus $S_\Gamma = 3! \cdot 2$.

Analytically the graph is obtained from the third order term of (4.2.13) by differentiating it twice with respect to J. The term that leads to (4.2.25) is given by

$$\sum_z \frac{\partial}{\partial J_1(0)} \frac{\partial}{\partial J_1(z)} \left(\frac{1}{3!} \frac{1}{2^3} \left(\prod_{i=1}^{3} \sum_{x_i y_i} \sum_{a_i b_i} \right) 4 \frac{\partial^3 W(J,\widehat{v})}{\partial J_{a_1}(x_1) \partial J_{a_2}(x_2) \partial J_{a_3}(x_3)} \right.$$

$$\left. \cdot \prod_{i=1}^{3} v_{a_i b_i}(x_i, y_i) \cdot \frac{\partial^3 W(J,\widehat{v})}{\partial J_{b_1}(y_1) \partial J_{b_2}(y_2) \partial J_{b_3}(y_3)} \right)_{J=0,\widehat{v}=0} .$$

$$(4.2.26)$$

The factor $1/3!$ comes from the Taylor expansion to third order, and the $1/2^3$ from the prefactor of the hopping term of the action (4.2.3), or equivalently from (4.2.14). The factor 4 accounts for the combinatorics of the differentiations with respect to J according to the product rule. Applying the "external" differentiations with respect to $J_1(0)$ and $J_1(z)$, we obtain

$$
\frac{1}{3!}\frac{1}{2}\,2\left(\sum_z\sum_{\{a_ib_i\}}\frac{\partial^4 W(J,\widehat{v}=0)}{\partial J_1(0)\partial J_{a_1}(0)\partial J_{a_2}(0)\partial J_{a_3}(0)}\right.
$$

$$
\left.\cdot\prod_{i=1}^3 v_{a_ib_i}(0,z)\cdot\frac{\partial^4 W(J,\widehat{v}=0)}{\partial J_1(z)\partial J_{b_1}(z)\partial J_{b_2}(z)\partial J_{b_3}(z)}\right)_{J=0}
$$

$$
=\frac{1}{3!2}\,2\left(\frac{\overset{\circ}{v}{}^c_4}{3}\right)^2\left(\sum_{z\text{ NN of }0}1\right) \tag{4.2.27}
$$

$$
\cdot\left(\sum_{a_1a_2a_3}C_4(1,a_1,a_2,a_3)C_4(1,a_1,a_2,a_3)\right)\cdot(2\kappa)^3
$$

$$
=\frac{1}{3!2}\,2\left(\frac{\overset{\circ}{v}{}^c_4}{3}\right)^2\cdot(2D)\cdot 3(N+2)\cdot(2\kappa)^3,
$$

where NN stands for nearest neighbor lattice sites. Hence, for the graph (4.2.25) we have

$$
S_\Gamma\ =\ 3!\,2\ ,\ \mathcal{E}_\Gamma\ =\ 2\ ,\ I_\Gamma\ =\ 2D\ ,\ C_\Gamma\ =\ 3(N+2)\ . \tag{4.2.28}
$$

4.2.3 *Classification of graphs*

Now, since we have defined a convenient way to represent the lengthy analytic expressions of LCE series in terms of graphs and their weights, let us spell out the graphical representation of χ_2 to the first four orders. It is given by

$$\chi_2 \;=\; \text{(graphs)} \tag{4.2.29}$$

$$+\, O(\kappa^5).$$

Every bracket in (4.2.29) includes graphs of the same order in κ. The number of graphs rapidly increases already to low orders. We recall that for a reliable investigation of the phase structure of the models (4.2.3), orders between 10 and 20 in κ are required. The task is to find all connected graphs with as many internal lines as possible and to compute their weights, graph by graph. As it turns out rather soon, this straightforward approach is not practicable to such high orders.

In view of a systematic generation of the LCE series of χ_2 and related susceptibilities, the solution is to introduce more specific graph classes where the specification refers to more than just to the property of being connected. The number of graphs in the restricted classes is less high, and it is easier to compute their contributions.

Usually, at least two steps are involved in this process. First, 1PI-(one-particle irreducible) graphs are introduced together with associated susceptibilities. 1PI-susceptibilities and (connected) susceptibilities are related by an appropriate Legendre transformation. The second step is the so-called vertex renormalization (not to be confused with renormalization in field theory). Vertex renormalization represents 1PI-graphs in terms of 1VI-(one-vertex irreducible) graphs and so-called renormalized-moment graphs of which there are considerably less. In fact, only the last two graph classes are actually generated by appropriate algorithms. 1PI-susceptibilities and the χ_n are finally computed analytically from them. We now discuss these steps in some more detail.

The notion of 1PI is taken from Feynman diagrams in particle physics. A graph Γ is called 1PI if it has the following property: Remove an arbitrary, single internal line of Γ. The resulting graph has then only one connected component which has external lines attached. In our present case, every graph has only vertices with an even number of lines (internal and external) entering. This implies that after the removal of a single internal line of a 1PI-graph, the resulting graphs is still connected, i.e., it has only one connected component.

We denote the subset of graphs of $\mathcal{G}_n(\nu)$ which are 1PI by $\mathcal{G}_n^{1\mathrm{PI}}(\nu)$. 1PI-susceptibilities are defined in analogy to (4.2.22), by restricting the internal sum to 1PI-graphs, e.g.

$$\chi_2^{1\mathrm{PI}} = \sum_{\nu \geq 0} \sum_{\Gamma \in \mathcal{G}_2^{1\mathrm{PI}}(\nu)} w(\Gamma), \qquad (4.2.30)$$

and similarly for weighted susceptibilities.

Let us introduce the graphical notation

$$W_{ab}^{(2)}(x,y) = \quad \text{} \qquad (4.2.31)$$

for the connected two-point function and

$$R_{ab}^{(2)}(x,y) = \quad \text{} \qquad (4.2.32)$$

for the sum of all two-point LCE graphs that are 1PI. The site labels at the vertices imply that the vertices with attached external lines are placed at fixed lattice sites. The connected correlation function $W^{(2)}$ is then obtained from the 1PI-correlation function $R^{(2)}$ by the geometric series

$$(4.2.33)$$

or

$$
\begin{aligned}
W^{(2)}_{a_1 a_2}(0, x) &= R^{(2)}_{a_1 a_2}(0, x) \\
&+ \sum_{b_1 b_2} \sum_{y_1 y_2} R^{(2)}_{a_1 b_1}(0, y_1)\, v_{b_1 b_2}(y_1, y_2)\, R^{(2)}_{b_2 a_2}(y_2, x) \\
&+ \sum_{b_1 \dots b_4} \sum_{y_1 \dots y_4} R^{(2)}_{a_1 b_1}(0, y_1)\, v_{b_1 b_2}(y_1, y_2)\, R^{(2)}_{b_2 b_3}(y_2, y_3) \\
&\qquad \cdot v_{b_3 b_4}(y_3, y_4)\, R^{(2)}_{b_4 a_2}(y_4, x) + \cdots .
\end{aligned}
$$

$$(4.2.34)$$

For the two-point susceptibilities we obtain by summation over x

$$
\begin{aligned}
\chi_2 &= \chi_2^{\mathrm{1PI}} + \chi_2^{\mathrm{1PI}} \left(\sum_x v_{11}(0, x) \right) \chi_2^{\mathrm{1PI}} + \cdots \\
&= \chi_2^{\mathrm{1PI}} \left(1 + (2D)(2\kappa)\chi_2^{\mathrm{1PI}} + \cdots \right) \\
&= \frac{\chi_2^{\mathrm{1PI}}}{1 - 4D\kappa\chi_2^{\mathrm{1PI}}} .
\end{aligned}
$$

$$(4.2.35)$$

Similarly, we obtain for the moment μ_2

$$
\mu_2 = \frac{\chi_2^{\mathrm{1PI}} + 4D\kappa \left(\chi_2^{\mathrm{1PI}} \right)^2}{\left(1 - 4D\kappa\chi_2^{\mathrm{1PI}} \right)^2} .
$$

$$(4.2.36)$$

The relation between $W^{(2)}$ and $R^{(2)}$ generalizes easily to higher correlation functions. It can be shown that the general relation has the form of a Legendre transformation. This implies an analytic relation between the $\{\chi_m\}$ and the $\{\chi_n^{\mathrm{1PI}}\}$, so that the $\{\chi_m\}$ are computable from the $\{\chi_n^{\mathrm{1PI}}\}$. It is thus sufficient to find methods to generate the graphical representation of the 1PI-susceptibilities. Even the generation of all 1PI-graphs can be avoided, and this is related to the so-called vertex renormalization. Vertex renormalization amounts to some kind of composition of 1PI-graphs in terms of graphs with stronger topological constraints.

Consider the graph

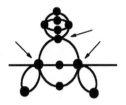

$$(4.2.37)$$

that represents a typical high-order 1PI-graph. The arrows point to the so-called articulation vertices of Γ. If one of the articulation vertices is removed, together with all the internal and external lines attached to it, Γ decomposes into various connected components, and there is at least one connected component that does not have any external line attached.

A graph without articulation vertices is called 1VI (1-vertex irreducible). We denote the subset of all $\Gamma \in \mathcal{G}_n^{1PI}(\nu)$ which are also 1VI by $\mathcal{S}_n(\nu)$. Every 1VI-graph with more than one vertex has at least two vertices with external lines attached. On the other hand, we denote the set of graphs $\Gamma \in \mathcal{G}_n^{1PI}(\nu)$ which have one and only one vertex with external lines attached, by $\mathcal{Q}_n(\nu)$ and call these graphs *renormalized moments*. The idea of vertex renormalization is to generate all 1PI-graphs from the 1VI-ones, by attaching renormalized-moment graphs to their vertices in all possible ways, as indicated in the graph (4.2.38).

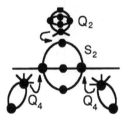

$$(4.2.38)$$

With these two notions, susceptibilities are represented as

$$\chi_2^{1PI} = \sum_{\nu \geq 0} \sum_{\Gamma \in \mathcal{S}_2^{1PI}(\nu)} w'(\Gamma), \qquad (4.2.39)$$

where $w'(\Gamma)$ is computed in the same way as $w(\Gamma)$ in (4.2.23) with the following exception. The vertex couplings $\overset{\circ}{v}{}^c_{2m}$ are replaced by a sum over renormalized moments,

$$\overset{\circ}{v}{}^c_{2m} \ \to \ v^c_{2m} \ = \ \sum_{\nu \geq 0} \ \sum_{\Gamma \in \mathcal{Q}_{2m}(\nu)} w(\Gamma). \qquad (4.2.40)$$

At this stage the only graphs that must be constructed are the 1VI-ones and the renormalized moments. It is still a challenge to develop efficient algorithms for generating all graphs of $\mathcal{S}_2$ and $\mathcal{Q}_2$ in higher orders. The interested reader is referred to the literature. In Table (4.2.1) we list the number of (pairwise inequivalent) graphs of the classes $\mathcal{S}_2(\nu)$ and $\mathcal{Q}_2(\nu)$, up to and including order $\nu = 20$, these are graphs with at most twenty internal lines. Obviously, $\mathcal{Q}_{2n}(\nu)$ for any n is obtained once $\mathcal{Q}_2(\nu)$ is known, just by attaching the additional external lines to the one vertex that has external lines already attached.

Table 4.2.1   The number of pairwise inequivalent graphs of the classes $\mathcal{S}_2(\nu)$ and $\mathcal{Q}_2(\nu)$, for $0 \leq \nu \leq 20$.

ν	$\mathcal{S}_2(\nu)$	$\mathcal{Q}_2(\nu)$	ν	$\mathcal{S}_2(\nu)$	$\mathcal{Q}_2(\nu)$
0	1	1	10	68	439
1	0	0	11	247	0
2	0	1	12	470	2877
3	1	0	13	1779	0
4	0	4	14	3937	20507
5	2	0	15	14801	0
6	3	15	16	35509	161459
7	8	0	17	135988	0
8	9	79	18	350614	1376794
9	40	0	19	1361878	0
			20	3705558	12694212

4.2.4   Towards a computer implementation of graphs

The next part of this section is devoted to computer implementation of graphs. In order to handle graphs by the aid of computers, to develop and implement fast algorithms for graph generation and computation of their weights, we need a convenient algebraic representation. Linked lists, incidence matrices and other possibilities are at our disposal. For our purposes incidence matrices turn out to be most convenient.

We recall that LCE graphs do not have self-lines, i.e they do not have internal lines with both end points entering the same vertex. Let Γ be an LCE graph and V_Γ the number of its vertices. An incidence matrix I_Γ of Γ is obtained as follows.

- Enumerate the vertices of Γ, so that the set of vertices is given by

$$\mathcal{B}_\Gamma = \{v_i \mid i = 1, \ldots, V_\Gamma\}. \tag{4.2.41}$$

  In particular, $v_i \neq v_j$ for $i \neq j$.
- For $i, j = 1, \ldots, V_\Gamma$, $i \neq j$, define

$$I_\Gamma(i, j) = m(v_i, v_j), \tag{4.2.42}$$

  where $m(v_i, v_j)$ is the number of lines between v_i and v_j.
- The diagonal elements are given by the number of external lines attached to the vertices,

$$I_\Gamma(i, i) = n(v_i), \quad i = 1, \ldots, V_\Gamma. \tag{4.2.43}$$

The incidence matrix is a symmetric $V_\Gamma \times V_\Gamma$ matrix. Below we give an example of a graph and its associated incidence matrix.

$$\simeq \begin{pmatrix} 0 & 1 & 0 & 1 \\ 1 & 1 & 2 & 0 \\ 0 & 2 & 0 & 2 \\ 1 & 0 & 2 & 1 \end{pmatrix}.$$

$$\tag{4.2.44}$$

All the required graph-theoretical operations such as "searching through a graph" (e.g. the so-called depth-first search) are easily implemented in terms of incidence matrices. Furthermore, operations like adding and removing vertices as well as internal and external lines of a graph are easily translated to operations on incidence matrices.

The above construction does not provide a unique definition of an incidence matrix I_Γ of a graph Γ. The vertices are enumerated in an arbitrary way. Any other enumeration corresponds to a nontrivial permutation $\pi \in \Pi_{V_\Gamma}$ and leads to the incidence matrix

$$I_\Gamma^\pi(i,j) = I_\Gamma(\pi(i), \pi(j)). \tag{4.2.45}$$

I_Γ is unique only modulo simultaneous permutations of its rows and columns. Permutations which leave I_Γ unchanged constitute a finite subgroup of Π_{V_Γ}. Let us write $S_{\mathrm{perm}}(\Gamma)$ for the number of elements of this group. The topological symmetry number of Γ is easily obtained as

$$S_\Gamma = S_{\mathrm{perm}}(\Gamma) \cdot \prod_{i<j} I_\Gamma(i,j)!. \tag{4.2.46}$$

Two graphs Γ_1 and Γ_2 are topologically equivalent if and only if they have the same number of vertices, say V, and their respective incidence matrices (with respect to an arbitrary initial labelling of the vertices) satisfy

$$I_{\Gamma_1}^\pi = I_{\Gamma_2}, \tag{4.2.47}$$

for some $\pi \in \Pi_V$. We consider topologically equivalent graphs as equal. This is because the sums in (4.2.22), (4.2.30), (4.2.39) and (4.2.40) are over equivalence classes of graphs, so that for each equivalence class one and only one member is required for its representation.

To identify two graphs by their incidence matrices, one must check all $V!$ permutations according to (4.2.47). This is very inefficient and must be avoided by further specification of incidence matrices. The representation of graphs via incidence matrices can be made unique if a total order relation on $M(V \times V)$ is chosen for every V, and in turn the maximum over all simultaneous permutations of rows and columns is taken. Then, for an arbitrary initial labelling, we get a canonical representation by

$$\widehat{I}_\Gamma := \max_{\pi \in \Pi_{V_\Gamma}} I_\Gamma^\pi, \tag{4.2.48}$$

where the maximum is taken with respect to the total order relation. A convenient choice is the following. Let $A, B \in M(V \times V)$. Then

$$A > B \overset{\text{DEF}}{\Longleftrightarrow} \begin{array}{l} \text{there is } i,j \in \{1, \ldots, V\} \text{ with } A(i,j) > B(i,j), \text{ and} \\ A(k,l) = B(k,l) \text{ for all } k = i, l < j \text{ and for all } k < i. \end{array}$$

This defines a unique (canonical) $\widehat{I}_\Gamma$ for every Γ. Still there are large factorials $V!$ involved in constructing the canonical form of incidence matrices.

The solution here is to introduce an order relation for the vertices of a graph first and then to enumerate the vertices according to this order. Here also topological properties of the graph are taken into account. For instance, let us consider the graph (4.2.44). The vertices with numbers 1 and 2 are easily distinguished because vertex 2 has one external line attached, whereas vertex 1 has not. Similarly, vertices 1 and 3 are distinguished because vertex 3 has 2 lines in common with its neighbors, whereas vertex 1 has only 1 line in common. On the other hand, vertices 2 and 4 are not distinguished. If we take these features into account, a possible order relation on the vertices leads to a change of enumeration according to

$$1 \to 1 , \, 3 \to 2 , \, 2 \to 3 , \, 4 \to 4 , \tag{4.2.49}$$

i.e.

$$\simeq \begin{pmatrix} 0 & 0 & 1 & 1 \\ 0 & 0 & 2 & 2 \\ 1 & 2 & 1 & 0 \\ 1 & 2 & 0 & 1 \end{pmatrix}$$

$$\tag{4.2.50}$$

Ordering cannot be complete, at least such vertices must stay unordered relative to each other whose exchange corresponds to a symmetry of the graph. The cost of pre-ordering the vertices increases only with some low power of V ($V^{1.5}$ or even $V \cdot \ln V$). Once the order relation is chosen, a canonical incidence matrix can be defined by subsequently allowing only those vertices for being exchanged in (4.2.48) which stay unordered relative to each other. In (4.2.50), only one permutation is left (3 $\leftrightarrow$ 4). As a by-product, the number of permutations which yields the (same) canonical incidence matrix is precisely the topological symmetry factor $S_{\text{perm}}(\Gamma)$. For the example (4.2.50), $S_{\text{perm}}(\Gamma) = 2$. Enhanced pre-ordering is essential to achieve high orders of the graphical expansion.

4.2.5 *Application to phase transitions and critical phenomena*

In contrast to the usual weak coupling expansions (expansion in powers of λ for the $\lambda\Phi^4$ theory, for example), the LCE series for the susceptibilities such as in (4.2.2) are **convergent** series for small hopping parameter κ. For the following considerations let us assume that the susceptibilities χ_2 and μ_2 have the series representations

$$\chi_2 = \sum_{\nu \geq 0} a_\nu \, (2\kappa)^\nu \,,$$

$$\mu_2 = \sum_{\nu \geq 0} b_\nu \, (2\kappa)^\nu \,. \tag{4.2.51}$$

The series for m_R^{-2} and Z_R^{-1} are then obtained from (4.2.51) by (4.2.8). All dependence on the local couplings like λ, on the lattice topology and on the symmetry group are included in the coefficients a_ν and b_ν. As power series, (4.2.51) are convergent in a circle of the complex κ-plane. The radius of convergence with respect to κ is obtained e.g. by the root or the ratio criterion according to

$$r = \liminf_{\nu \to \infty} |a_\nu|^{1/\nu} = \liminf_{\nu \to \infty} \left| \frac{a_\nu}{a_{\nu+1}} \right|, \tag{4.2.52}$$

respectively, that is, (4.2.51) are convergent for $2\kappa < r$.

On the boundary of the circle of convergence there is at least one singularity. If, for sufficiently large ν, all coefficients a_ν are of the same sign, the singularity which is closest to the origin $\kappa = 0$ lies on the positive real κ-axis. In the following we assume that this is the case. The model then has a phase transition at the critical point κ_c with

$$2\kappa_c = \lim_{\nu \to \infty} \left| \frac{a_\nu}{a_{\nu+1}} \right|, \tag{4.2.53}$$

because κ_c corresponds to a nonanalytic behavior of the free energy of the model. Let us assume that there is a second-order phase transition at κ_c and that the leading singular behavior of χ_2 at κ_c is of the form

$$\chi_2 \simeq \mathcal{A} \left(1 - \frac{\kappa}{\kappa_c} \right)^{-\gamma} \quad \text{as } \kappa \to \kappa_c - \tag{4.2.54}$$

with unknown critical amplitude $\mathcal{A}$, unknown critical point κ_c and unknown critical exponent γ. The values of $\mathcal{A}$, γ and κ_c are then determined by the coefficients of the series representation of χ_2. A straightforward calculation shows that

$$\frac{a_\nu}{a_{\nu-1}} \simeq \frac{1}{2\kappa_c} \left(1 + \frac{\gamma - 1}{\gamma + \nu - 1} \right) = \frac{1}{2\kappa_c} \left(1 + \frac{\gamma - 1}{\nu} + O(\nu^{-2}) \right) \tag{4.2.55}$$

as $\nu \to \infty$. Thus, κ_c and γ are determined by the high-order behavior of the series coefficients a_ν. In general, the error term $O(\nu^{-2})$ is influenced by the regular part of χ_2 and by further subleading singularities, which can increase the error to $o(\nu^{-1})$. To determine the values of γ and κ_c from this relation to a high accuracy, the coefficients a_ν must be known to sufficiently high order. For example, consider the function

$$\chi_2 = \frac{1}{1 - 2\kappa} + e^{8\kappa}. \tag{4.2.56}$$

This function has $2\kappa_c = 1$ and $\gamma = 1$. Furthermore, χ_2 has the series representation (4.2.51) with

$$a_\nu = 1 + \frac{4^\nu}{\nu!}. \tag{4.2.57}$$

The ratios of the coefficients are

$$\frac{a_\nu}{a_{\nu-1}} = \frac{143}{175}, \frac{301}{429}, \frac{1339}{2107}, \frac{827}{1339}, \tag{4.2.58}$$

for $\nu = 5, \ldots, 8$, respectively. To order $\nu = 8$, there is still a deviation of 39% from the limiting value 1. The convergence becomes worse if there are further but weaker singularities. For instance, the function

$$\chi_2 = \frac{1}{1 - 2\kappa} - 2 \cdot \frac{1}{(1 - 2\kappa)^{1/2}} + e^{8\kappa} \tag{4.2.59}$$

leads to

$$\frac{a_\nu}{a_{\nu-1}} = \frac{17359}{21350}, \frac{143717}{208308}, \frac{1236001}{2012038}, \frac{11522543}{19776016}, \tag{4.2.60}$$

for $\nu = 5, \ldots, 8$, respectively. To order $\nu = 8$, there is a deviation of 42% from 1.

These simple examples already illustrate how important sufficiently high orders in the series representation become in order to get precise data of the critical parameters. By experience orders between 14-20 turn out to be appropriate for continuous spin models, such as $O(N)$-symmetric lattice models.

The simple ratio criterion can be supplemented by various other techniques. For example, one can take into account weaker or subleading singularities, use Padé approximants or Legendre transforms as in the following example. The two-point susceptibility χ_2 has the representation

$$\chi_2 = \frac{\chi_2^{1PI}}{1 - 4D\kappa\chi_2^{1PI}}, \qquad (4.2.61)$$

in terms of the 1PI-two-point susceptibility. For a second-order transition, the critical point is obtained as the root of the denominator of (4.2.61). In many cases this gives a better convergence to κ_c than the ratio criterion.

Although usually high temperature series are applied to critical phenomena and to models on a lattice with infinite extension in all directions, they are not restricted to these cases. In particular, they apply also to field theories at finite temperature, in which one dimension is compactified to a torus, with torus length equal to the inverse temperature in lattice units. Furthermore, LCE series expansions apply to the investigation of first-order transitions. Such applications were only recently studied. Again, if the series coefficients a_ν are of the same sign for sufficiently large ν, the phase transition point κ_c is obtained by the (root or) ratio criterion

$$\liminf_{\nu \to \infty} \frac{a_\nu}{a_{\nu-1}} = \frac{1}{2\kappa_c}. \qquad (4.2.62)$$

LCE can be actually used to *determine* the order of a phase transition if they are combined with a finite-size scaling analysis. For this case, the series must be determined for a finite lattice (finite in all directions). The question then immediately arises of how the finite volume or, more generally, the lattice shape enters the graphical expansion. Recall that the graphical representation of χ_2 is given by (4.2.22). The weight $w(\Gamma)$ which is associated with each graph is given by (4.2.23). The included factor I_Γ denotes the lattice embedding number. Except for a trivial volume factor, it counts the number of ways Γ can be placed on the lattice. It is the only factor that depends on the topology of the lattice, in particular on the lattice size. In a finite volume with periodic boundary conditions, a graph can wind around the torus and therefore make up an additional closed loop, this way increasing its embedding number.

For instance, let us consider a lattice of size 4^D with periodic boundary conditions in all D directions. A simple, low-order graph like

$$\text{\raisebox{-0.5em}{\includegraphics}} \quad , \quad I_\Gamma = 2D , \qquad (4.2.63)$$

does not "feel" the finite volume, thus its embedding number is the same as on a lattice with infinite extension in all directions. On the other hand, the graph

$$(4.2.64)$$

has the embedding number

$$I_\Gamma = 6D(2D - 1) \qquad (4.2.65)$$

on the ∞^D lattice, but on the 4^D-lattice it can wind around the torus in every direction, leading to a total embedding number of

$$I_\Gamma = 6D(2D - 1) + 2D. \qquad (4.2.66)$$

Of course, higher order graphs with loops made of many internal lines are more sensitive to the lattice size and shape. Probing finite-size effects and finite-temperature behavior therefore lead to a further reason why it is worthwhile to perform LCE series to a high order in the expansion parameter.

Table 4.2.2 Critical exponents of the $\Phi^4 - O(N)$ lattice field models for $N = 1, \ldots, 4$, both in four dimensions at finite temperature (upper values) and in three dimensions at zero temperature (lower values).

N	γ	ν	$\nu\eta$
1	1.2400(87)	0.6300(49)	0.0193(132)
	1.2406(36)	0.6301(18)	0.0183(52)
2	1.3238(139)	0.6694(64)	0.0206(150)
	1.3250(52)	0.6734(28)	0.0199(76)
3	1.4032(156)	0.7167(76)	0.0246(169)
	1.4029(85)	0.7131(40)	0.0232(87)
4	1.4469(225)	0.7356(93)	0.0257(200)
	1.4504(113)	0.7361(68)	0.0213(123)

4.2.6   Some results

As an example for results we consider the four-dimensional $O(N)$ symmetric Φ^4-lattice field theories at finite temperature. These models have a second-order phase transition from the $O(N)$ symmetry-broken phase at low temperature (large κ) to the symmetric phase at high temperature. In Table 4.2.2 we list some critical exponents in these models in comparison to the critical exponents of the corresponding three-dimensional models. All data of Table 4.2.2 are obtained by LCE series, on lattices of size 4×18^3 and 18^3. It can be verified that the critical data agree for every N. The results confirm the conjecture based on dimensional reduction that the critical behavior of the four-dimensional models at finite temperature is the same as the critical behavior of the corresponding three-dimensional models at zero temperature.

4.3   Renormalization, perturbation theory and universality at zero temperature — the continuum limit

4.3.1   Generalities

For a lattice field theory, the inverse lattice spacing a^{-1} provides an UV-regularization. The lattice cutoff is nonperturbative in the sense that it does not refer to any perturbative expansion scheme in contrast to other cutoff schemes such as dimensional regularization. A lattice field theory provides a mathematically well defined framework from the very beginning. It is thus a good starting point to derive properties of field theories in a rigorous way.

For finite lattice spacing a, the correlation functions are well defined for momenta well below the UV-cutoff a^{-1}. An essential task of renormalization in quantum field theory is to prove the existence of correlation functions in the continuum limit $a \to 0$ with certain properties. In this limit a sequence of axioms must be satisfied in order to construct a Euclidean quantum field theory. These are the well-known Osterwalder-Schrader axioms. In many important cases convergence is achieved by appropriately adjusting a finite number of bare parameters of the action and by a rescaling of the fields. The bare parameters become functions of corresponding, renormalized coupling constants. The renormalized coupling constants are defined by normalization conditions imposed on renormalized correlation functions at fixed Euclidean or on-shell momenta. Renormalizability implies that all renormalized correlation functions, considered as functions of the renormalized parameters, stay well defined in the continuum limit, for all finite momenta p. The influence of the cutoff is suppressed by some power of pa and $m_R a$, with m_R the renormalized mass in physical units. In general it is the issue of renormalization theory to show that a given field theory can be re-parameterized in such a way that it stays finite if the UV-cutoff is removed and that the symmetry of the theory is preserved. This is a task common to all regularizations, such as simple momentum cutoff, Pauli-Villars, dimensional regularization and lattice cutoff. For a detailed discussion of renormalization theory we refer to the literature. Renormalization is intimately related to local quantum field theories, both at zero and finite temperature. Whenever we are interested in questions of how relevant results about the phase structures of particle physics are for continuum physics, although they were obtained for a finite lattice cutoff, we have to deal with renormalization. Therefore we shall recapitulate renormalization theory, but only to the extent which is needed for our discussions below.

At zero temperature strong theorems are available. In perturbation theory, correlation functions are represented as a sum of Feynman integrals. A central ingredient here is the existence of a power-counting theorem. It allows one to determine the convergence of Feynman integrals in the large cutoff limit "simply" by counting suitably defined UV-divergence degrees. The Feynman diagram associated with a Feynman integral, together with all its subdiagrams, are required to have negative UV-divergence degrees so that the Feynman integral is (absolutely) convergent. Generically, Feynman integrals are not convergent *a priori*. However, once we have a so-called power-counting theorem at our disposal, we can formulate a general renormalization prescription by "subtracting" the UV-divergencies, order by order in perturbation theory, preserving locality. An approved possibility is, for example, the BPHZ-subtraction scheme in continuum field theories (the Bogoliubov-Parasiuk-Hepp-Zimmermann finite part prescription). It applies to the integrand of momentum space Feynman integrals and can completely dispense with an UV-cutoff. On the other hand, for a cutoff theory the subtractions are arranged in such a way that they result from local counterterms to the action. These counterterms provide the map between bare and renormalized coupling constants and fields. They become uniquely determined by imposing normalization conditions on the Green functions. What is usually called "renormalizability by power counting" then means that only a finite number of renormalized parameters must be fixed, to all orders of perturbation theory. In addition, if a theory is rendered UV-finite already by a few low-order counterterms, it is called superrenormalizable.

A further issue is to show that a theory subject to a symmetry can be renormalized in such a way that the symmetry is preserved. This is highly nontrivial for theories which are subject to a symmetry under a nonlinear and/or local symmetry transformation. This concerns field theories like Yang-Mills theories, QCD and the electroweak standard model. The symmetries are manifest in Ward identities that are integro-differential equations imposed on the Green functions. Ward identities must be preserved under renormalization in the sense that the renormalized correlation functions should fulfill the Ward identities. It turns out that nonlinear symmetry transformations become renormalized themselves.

4.3.2   *Renormalization*

The basic aspects of perturbative renormalization theory show up already in the relatively simple example of a scalar Φ^4 theory in four dimensions. This theory is renormalizable. We consider a Φ^4-theory on the four-dimensional hypercubic lattice $a\mathbf{Z}^4$, described by the action

$$
S_0(\Phi_0|m_0,\lambda_0) = a^4 \sum_{x\in\Lambda_4} \left(\frac{1}{2}\sum_{\mu=0}^{3} \left(\frac{1}{a}\widehat{\partial}_\mu\Phi_0(x) \right)^2 + \frac{1}{2}m_0^2\Phi_0(x)^2 \right.
$$
$$
\left. + \frac{\lambda_0}{4!}\Phi_0(x)^4 \right),
\tag{4.3.1}
$$

with bare mass m_0 and bare quartic coupling constant λ_0. $\widehat{\partial}_\mu$ denotes the lattice-forward derivative,

$$
\widehat{\partial}_\mu\Phi_0(x) = \Phi_0(x+\widehat{\mu}a) - \Phi_0(x).
\tag{4.3.2}
$$

Throughout this section we use natural units, in which the bare and renormalized masses m_0 and m_R have engineering dimension one, for example, and the quartic coupling λ_0 is dimensionless. (Note that usually in the context of the lattice regularization the bare mass would be chosen dimensionless and related to m_0 by $m_0 \cdot a$, where $m_0 \cdot a$ is a number.) The form (4.3.1) of the lattice action immediately reveals the relation to its classical continuum limit. If m_0 and λ_0 are taken to be independent of the lattice spacing a, the continuum limit of (4.3.1) is given by

$$\lim_{a \to 0} S_0(\Phi_0) = \int d^4x \left[\frac{1}{2} \sum_{\mu=0}^{3} (\partial_\mu \Phi_0(x))^2 + \frac{1}{2} m_0^2 \Phi_0(x)^2 \right.$$
$$\left. + \frac{\lambda_0}{4!} \Phi_0(x)^4 \right], \tag{4.3.3}$$

which describes a classical scalar field theory with mass m_0 and quartic coupling constant λ_0. Beyond the level of classical field theory, the bare parameters must be tuned such that the continuum limit of the field theory exists.

The generating functional of the full and the connected correlation functions $Z_0(J, \chi)$ and $W_0(J, \chi)$ is given by

$$Z_0(J, \chi) = \exp(W_0(J, \chi)) = \int \prod_x d\Phi_0(x) \, \exp\left\{ -S_0(\Phi_0) \right.$$
$$\left. + a^4 \sum_x \left(J(x)\Phi_0(x) + \chi(x)\Phi_0(x)^2 \right) \right\}, \tag{4.3.4}$$

respectively. We have included a source term for the local, composite operator $\Phi_0(x)^2$. Composite operators require their own renormalization. The connected correlation functions are then obtained as

$$< \Phi_0(x_1) \ldots \Phi_0(x_n)\Phi_0(z_1)^2 \ldots \Phi_0(z_l)^2 >^c$$
$$= \frac{a^{-4(n+l)} \partial^{n+l}}{\partial J(x_1) \cdots \partial J(x_n) \partial \chi(z_1) \cdots \partial \chi(z_l)} W_0(J, \chi) \bigg|_{J=\chi=0}. \tag{4.3.5}$$

To obtain the corresponding vertex functions we introduce

$$\phi_0(x) = \frac{\partial W_0(J)}{\partial a^4 J(x)} \tag{4.3.6}$$

and the Legendre transform

$$W_0(J) = \Gamma_0(\phi_0) + a^4 \sum_x J(x)\Phi_0(x), \tag{4.3.7}$$

where J is to be expressed as a functional of ϕ_0 by solving (4.3.6). $\Gamma_0(\phi_0, \chi_0)$ is then the generating functional of the bare (i.e. unrenormalized) vertex functions $\Gamma_0^{(n,l)}(x_1, \ldots, x_n | z_1, \ldots, z_l; a)$,

$$
\begin{aligned}
\Gamma_0^{(n,l)}&(x_1, \ldots, x_n | z_1, \ldots, z_l; a) \\
&= \frac{a^{-4(n+l)} \partial^{n+l}}{\partial \phi_0(x_1) \cdots \partial \phi_0(x_n) \partial \chi(z_1) \cdots \partial \chi(z_l)} \, \Gamma_0(\phi_0, \chi) \bigg|_{\phi_0 = \chi = 0} \\
&= \int_{-\frac{\pi}{a}}^{\frac{\pi}{a}} \frac{d^4 p_1}{(2\pi)^4} \cdots \frac{d^4 p_{n-1}}{(2\pi)^4} \frac{d^4 q_1}{(2\pi)^4} \cdots \frac{d^4 q_l}{(2\pi)^4} \\
&\quad \cdot \exp\left(i \sum_{j=1}^{n-1} p_j \cdot (x_j - x_n) + i \sum_{j=1}^{l} q_j \cdot (z_j - x_n)\right) \\
&\quad \cdot \widetilde{\Gamma}_0^{(n,l)}(p_1, \ldots, p_{n-1} | q_1, \ldots q_l; a).
\end{aligned}
\tag{4.3.8}
$$

The vertex functions $\widetilde{\Gamma}_0^{(n,l)}$ vanish for odd n. $\widetilde{\Gamma}_0^{(2,0)}$ is the negative inverse full propagator, whereas for all other values of (n, l), $\widetilde{\Gamma}_0^{(n,l)}$ are the 1PI-parts of the connected n-point functions with amputated external Φ_0-field legs, and with l insertions of the composite operator Φ_0^2.

The continuum limit of the bare vertex functions does not exist. In what is called bare perturbation theory (i.e. perturbation theory in terms of bare parameters), the $\widetilde{\Gamma}_0^{(n,l)}$ are expanded in powers of the bare coupling constant λ_0,

$$
\begin{aligned}
\widetilde{\Gamma}_0^{(n,l)}&(p_1, \ldots, p_{n-1} | q_1, \ldots, q_l; a) \\
&= \sum_{\nu \geq n/2 - 1} \lambda_0^{\nu} \, \widetilde{\Gamma}_0^{(n,l)(\nu)}(p_1, \ldots, p_{n-1} | q_1, \ldots, q_l; a).
\end{aligned}
\tag{4.3.9}
$$

The coefficient functions are finite sums of lattice Feynman integrals. In a diagrammatic representation every Feynman integral is represented by a Feynman diagram. For instance, to subleading order in λ_0, the two- and four-point functions become

$$
\begin{aligned}
\widetilde{\Gamma}_0^{(2,0)}(p; a) &= -(\widehat{p}^2 + m_0^2) + \lambda_0 \, \widetilde{\Gamma}_0^{(2,0)(1)}(p; a) + O(\lambda_0^2) \\
&= -(\widehat{p}^2 + m_0^2) - \frac{\lambda_0}{2} \int_{-\pi/a}^{\pi/a} \frac{d^4 k}{(2\pi)^4} \frac{1}{\widehat{k}^2 + m_0^2} + O(\lambda_0^2)
\end{aligned}
\tag{4.3.10}
$$

$$
= -(\widehat{p}^2 + m_0^2) \quad + \quad \bigcirc\!\!\!- \quad + \, O(\lambda_0^2),
$$

where $\widehat{p}$ is defined according to

$$\widehat{p}_\mu = \frac{2}{a}\,\sin\frac{p_\mu a}{2}$$

$$\widehat{p}^2 = \sum_{\mu=0}^{3}\widehat{p}_\mu^2\,, \tag{4.3.11}$$

and

$$\widetilde{\Gamma}_0^{(4,0)}(p_1,p_2,p_3;a) = -\lambda_0 + \lambda_0^2\,\widetilde{\Gamma}_0^{(4,0)(2)}(p_1,p_2,p_3;a) + O(\lambda_0^3)$$

$$= -\lambda_0 + \frac{\lambda_0^2}{2}\int_{-\pi/a}^{\pi/a}\frac{d^4k}{(2\pi)^4}\,\frac{1}{\widehat{k}^2+m_0^2}\left[\frac{1}{\widehat{(k+p_1+p_2)}^2+m_0^2}\right.$$

$$\left.+\frac{1}{\widehat{(k+p_1+p_3)}^2+m_0^2}+\frac{1}{\widehat{(k+p_2+p_3)}^2+m_0^2}\right] \tag{4.3.12}$$

$$+\,O(\lambda_0^3)$$

$$= \quad \bigtimes \; + \; \left(\; \bigtimes\!\!\!\!\bigcirc\!\!\!\!\bigtimes \; + \; 2\ \text{perm.}\right) \; + \; O(\lambda_0^3).$$

The leading-order tree-level contributions have a finite continuum limit. On the other hand, the higher-order terms do not. For instance, at zero external momenta, the subleading, one-loop contribution to the four-point function has a small a-expansion of the form

$$\widetilde{\Gamma}_0^{(4,0)(2)}(0,0,0;a) = c_1\log(m_0 a)^2 + c_2 + O((m_0 a)^2\log(m_0 a)^2), \tag{4.3.13}$$

with constants c_1, c_2. As $a \to 0$ this expression diverges logarithmically. Similarly, the one-loop contribution to the two-point function $\widetilde{\Gamma}_0^{(2,0)(1)}(p;a)$ diverges quadratically as $a \to 0$.

What is required to have a finite limit are the so-called renormalized vertex functions. Towards their definition one first introduces a renormalized mass m_R, a renormalized coupling constant λ_R and a wave-function renormalization constant Z_R of the field Φ_0 by

$$\widetilde{\Gamma}_0^{(2,0)}(p;a) = -\frac{1}{Z_R}\left(m_R^2 + p^2 + O(p^4)\right) \qquad \text{as } p \to 0,$$

$$\widetilde{\Gamma}_0^{(4,0)}(0,0,0;a) = -\frac{1}{Z_R^2}\,\lambda_R, \tag{4.3.14}$$

$$\widetilde{\Gamma}_0^{(2,1)}(0,0,0;a) = \frac{1}{Z_R Z_{\Phi^2}}.$$

We have included also a wave-function renormalization constant for the composite operator $\Phi_0(x)^2$. So far, (4.3.14) is just a definition of m_R, λ_R and Z_R as a function of the bare parameters m_0 and λ_0 and of the cutoff a^{-1}, given by the zero-momentum behavior of the bare vertex functions $\widetilde{\Gamma}_0^{(2,0)}$ and $\widetilde{\Gamma}_0^{(4,0)}$. Actually, m_R and λ_R are related to the physical mass m_{phys} and physical coupling constant λ_{phys} that are defined by the pole of the renormalized propagator in the complex energy plane and the on-shell renormalized four-point function, respectively, by finite transformations.

The renormalized vertex functions $\Gamma_R^{(n,l)}$ are defined by

$$\widetilde{\Gamma}_R^{(0,1)} = 0,$$

$$\widetilde{\Gamma}_R^{(0,2)}(q,-q;a) = Z_{\Phi^2}^2 \left(\widetilde{\Gamma}_0^{(0,2)}(q,-q;a) - \widetilde{\Gamma}_0^{(0,2)}(0,0;a) \right), \quad (4.3.15)$$

$$\widetilde{\Gamma}_R^{(n,l)}(p_1,\ldots,p_{n-1}|q_1,\ldots,q_l;a)$$
$$= Z_R^{n/2} Z_{\Phi^2}^l \widetilde{\Gamma}_0^{(n)}(p_1,\ldots,p_{n-1}|q_1,\ldots,q_l;a) \quad \text{for all other } n,l.$$

We shall always consider the renormalized vertex functions $\widetilde{\Gamma}_R^{(n,l)}$ as functions of the momenta, the renormalized mass m_R and the coupling constant g_R, and of course of the UV-cutoff a^{-1}. This is possible if the map between the bare parameters m_0, λ_0 and the renormalized ones m_R, λ_R defined by (4.3.14) is invertible in the parameter range considered. Globally this is not necessarily the case. The following considerations of renormalizability do not crucially depend on the condition that the renormalized vertex functions are single-valued.

By construction, the renormalized vertex functions satisfy the normalization conditions

$$\widetilde{\Gamma}_R^{(2,0)}(p;a) = - \left(m_R^2 + p^2 + O(p^4) \right) \qquad \text{as } p \to 0$$
$$\widetilde{\Gamma}_R^{(4,0)}(0,0,0;a) = - \lambda_R$$
$$\widetilde{\Gamma}_R^{(2,1)}(0|0;a) = 1 \qquad\qquad\qquad (4.3.16)$$
$$\widetilde{\Gamma}_R^{(0,1)}(a) = \widetilde{\Gamma}_R^{(0,2)}(0;a) = 0.$$

The task of renormalization theory is to prove that the renormalized vertex functions $\widetilde{\Gamma}_R^{(n,l)}$ exist as the lattice spacing $a \to 0$, with m_R and λ_R kept finite and fixed. A nonperturbative proof is hard and has to face the problem of triviality of this particular model. Since we want to recapitulate the basic ideas of perturbative renormalization, we focus from now on to perturbation theory.

4.3.3   *Perturbative renormalization*

For small renormalized coupling constant λ_R, the renormalized vertex functions can be expanded as a power series in λ_R,

$$
\begin{aligned}
&\widetilde{\Gamma}_R^{(n,l)}(p_1,\ldots,p_{n-1}|q_1,\ldots,q_l;a)\\
&= \sum_{\nu \geq n/2-1} \lambda_R^\nu \, \widetilde{\Gamma}_R^{(n,l)(\nu)}(p_1,\ldots,p_{n-1}|q_1,\ldots,q_l;a).
\end{aligned} \tag{4.3.17}
$$

The coefficient functions $\widetilde{\Gamma}_R^{(n,l)(\nu)}$ are again finite sums of lattice Feynman integrals. A central theorem of renormalization theory on the lattice is the following. All coefficient functions $\widetilde{\Gamma}_R^{(n,l)(\nu)}$ have a finite continuum limit. For the normalization conditions (4.3.16) the limit is given by the BPHZ-finite part prescription that does not refer to any UV-cutoff at all,

$$
\begin{aligned}
&\lim_{a\to 0} \widetilde{\Gamma}_R^{(n,l)(\nu)}(p_1,\ldots,p_{n-1}|q_1,\ldots,q_l;a)\\
&= \widetilde{\Gamma}_{BPHZ}^{(n,l)(\nu)}(p_1,\ldots,p_{n-1}|q_1,\ldots,q_l;m_R,\lambda_R).
\end{aligned} \tag{4.3.18}
$$

In the continuum limit, the renormalized vertex functions only depend on the renormalized mass and the renormalized coupling constant λ_R. No reference to the bare parameters of the action is left. Finiteness of the vertex functions implies finiteness of the other Green functions as well. In particular, the connected correlation functions

$$
\lim_{a\to 0} \frac{1}{Z_R^{n/2}} Z_{\Phi^2}^l \; < \Phi_0(x_1)\cdots\Phi_0(x_n)\,\Phi_0(z_1)^2\cdots\Phi_0(z_l)^2 >^c \tag{4.3.19}
$$

exist for given m_R, to all order of perturbation theory in λ_R.

As an example let us consider the two- and four-point function to one-loop order. The bare vertex functions are given by (4.3.10) and (4.3.12). From (4.3.14) we first compute the map between the bare and the renormalized parameters,

$$
\begin{aligned}
Z_R &= 1 + O(\lambda_0^2)\\
m_R^2 &= m_0^2 - \lambda_0 \, \widetilde{\Gamma}_0^{(2,0)(1)}(0;a) + O(\lambda_0^2)\\
\lambda_R &= \lambda_0 - \lambda_0^2 \, \widetilde{\Gamma}_0^{(4,0)(2)}(0,0,0;a) + O(\lambda_0^3)
\end{aligned} \tag{4.3.20}
$$

with inversion

$$Z_R = 1 + O(\lambda_R^2)$$
$$m_0^2 = m_R^2 + \lambda_R \, \widetilde{\Gamma}_0^{(2,0)(1)}(0;a)_{m_0 \to m_R} + O(\lambda_R^2) \qquad (4.3.21)$$
$$\lambda_0 = \lambda_R + \lambda_R^2 \, \widetilde{\Gamma}_0^{(4,0)(2)}(0,0,0;a)_{m_0 \to m_R} + O(\lambda_R^3).$$

In the bare vertex functions on the right hand side of (4.3.21) the bare mass m_0 must be replaced by the renormalized mass m_R. We insert (4.3.21) into the right hand side of the definition (4.3.15) of the renormalized vertex functions. Using (4.3.10) and (4.3.12) and arranging terms of equal order in λ_R, we obtain

$$
\begin{aligned}
\widetilde{\Gamma}_R^{(2,0)}(p;a) &= Z_R \, \widetilde{\Gamma}_0^{(2,0)}(p;a) \\
&= -(p^2 + m_R^2) + \lambda_R \left(\widetilde{\Gamma}_0^{(2,0)(1)}(p;a) - \widetilde{\Gamma}_0^{(2,0)}(0;a) \right)_{m_0 \to m_R} \\
&\quad + O(\lambda_R^2) \qquad\qquad\qquad (4.3.22) \\
&= -(p^2 + m_R^2) + O(\lambda_R^2),
\end{aligned}
$$

and

$$
\begin{aligned}
\widetilde{\Gamma}_R^{(4,0)}(p_1,p_2,p_3;a) &= Z_R^2 \, \widetilde{\Gamma}_0^{(4,0)}(p_1,p_2,p_3;a) \\
&= -\lambda_R + \lambda_R^2 \left(\widetilde{\Gamma}_0^{(4,0)(2)}(p_1,p_2,p_3;a) - \widetilde{\Gamma}_0^{(4,0)(2)}(0,0,0;a) \right)_{m_0 \to m_R} \\
&\quad + O(\lambda_R^3) \qquad\qquad\qquad (4.3.23) \\
&= -\lambda_R + \frac{\lambda_R^2}{2} \int_{-\pi/a}^{\pi/a} \frac{d^4 k}{(2\pi)^4} \frac{1}{\widehat{k}^2 + m_R^2} \\
&\quad \cdot \left[\left(\frac{1}{\widehat{(k+p_1+p_2)}^2 + m_R^2} - \frac{1}{\widehat{k}^2 + m_R^2} \right) + 2 \text{ perm.} \right] \\
&\quad + O(\lambda_R^3).
\end{aligned}
$$

By construction, the renormalized vertex functions (4.3.22) and (4.3.23) satisfy the normalization conditions (4.3.16) (to the order to which they are computed). The "subtracted" momentum integrals remain finite as $a \to 0$, as can be seen by explicit computation or by referring to the lattice power-counting theorem of the next section. The continuum limit of $\widetilde{\Gamma}_R^{(4,0)}$

becomes

$$\lim_{a \to 0} \widetilde{\Gamma}_R^{(4,0)}(p_1, p_2, p_3; a) = -\lambda_R + \frac{\lambda_R^2}{2} \int_{-\infty}^{\infty} \frac{d^4 k}{(2\pi)^4} \frac{1}{k^2 + m_R^2} \quad (4.3.24)$$

$$\cdot \left[\left(\frac{1}{(k + p_1 + p_2)^2 + m_R^2} - \frac{1}{k^2 + m_R^2} \right) + 2 \text{ perm.} \right]$$

$$+ O(\lambda_R^3).$$

Renormalization theory predicts that UV-finiteness extends to all renormalized vertex functions, including those which have composite operator Φ^2 insertions. Furthermore, it extends to all orders of perturbation theory. The central ingredient is the availability of a power-counting theorem, which allows us to determine the convergence of Feynman integrals by counting of UV-divergence degrees. We will outline the idea behind power counting in the next section.

The very fact that the renormalized Green functions $\widetilde{\Gamma}_R^{(n,l)}$ as functions of the renormalized parameters, defined via (4.3.15), are actually finite to all orders of perturbation theory in λ_R may appear as an ad hoc-statement within this approach. To convince ourselves that it is justified we rederive $\widetilde{\Gamma}_R^{(n,l)}$ within the counterterm approach.

4.3.4 *The counterterm approach*

In the counterterm approach we start from a renormalized action that includes the most general counterterm from the point of view of power counting. At the moment we just state it as a result that the action (4.3.26) below has this form. We then derive the relation between bare and renormalized functionals. As a special case we obtain (4.3.15) and the relations between the wave-function renormalization constants in both approaches.

An equivalent description of the renormalized theory is to write it from the very beginning exclusively in terms of renormalized quantities, at the cost of introducing counterterms to the action. The above relations (4.3.4)-(4.3.7) of the bare theory between the full, connected, and 1PI-correlation functions remain valid for the renormalized correlation functions. It is instructive to see how this approach works. The generating functionals $\mathcal{Z}_R(J, \chi)$ and $W_R(J, \chi)$ of the renormalized full and connected correlation functions are obtained as

$$
\mathcal{Z}_R(J, \chi) = \exp(W_R(J, \chi)) = \int \prod_x d\Phi_R(x) \, \exp\left\{-S_R(\Phi_R, \chi | m_R, \lambda_R)\right.
$$
$$
\left. +a^4 \sum_x \left(J(x)\Phi_R(x) + \chi(x)\Phi_R(x)^2\right)\right\}. \tag{4.3.25}
$$

The renormalized action S_R includes all the required counterterms. On account of the lattice power-counting theorem and the lattice symmetries, S_R is given by

$$
S_R(\Phi_R, \chi | m_R, \lambda_R) = a^4 \sum_x \left\{\left[\frac{1}{2}\left(\frac{1}{a}\widehat{\partial}_\mu \Phi_R(x)\right)^2 + \frac{1}{2}m_R^2 \Phi_R(x)^2\right.\right.
$$
$$
\left. +\frac{\lambda_R}{4!}\Phi_R(x)^4\right] + \frac{1}{2}(Z_\Phi - 1)\left(\frac{1}{a}\widehat{\partial}_\mu \Phi_R(x)\right)^2
$$
$$
+\frac{1}{2}(Z_{m^2}Z_\Phi - 1)m_R^2 \Phi_R(x)^2 + (Z_\lambda Z_\Phi^2 - 1)\frac{\lambda_R}{4!}\Phi_R(x)^4
$$
$$
\left. -\chi(x)(Z_{\Phi^2}Z_\Phi - 1)\Phi_R(x)^2 - \chi(x)(Z_{01}) - \frac{\chi(x)^2}{2}(Z_{02})\right\}
$$
$$
= a^4 \sum_x \left\{\left[\frac{1}{2}Z_\Phi\left(\frac{1}{a}\widehat{\partial}_\mu \Phi_R(x)\right)^2 + \frac{1}{2}Z_{m^2}Z_\Phi m_R^2 \Phi_R(x)^2\right.\right. \tag{4.3.26}
$$
$$
\left. +Z_\Phi^2 Z_\lambda \frac{\lambda_R}{4!}\Phi_R(x)^4\right] - \chi(x)(Z_{\Phi^2}Z_\Phi - 1)\Phi_R(x)^2
$$
$$
\left. -\chi(x)(Z_{01}) - \frac{\chi(x)^2}{2}(Z_{02})\right\}
$$
$$
= S_0(Z_\Phi^{1/2}\Phi_R | Z_{m^2}^{1/2}m_R, Z_\lambda \lambda_R) - a^4 \sum_x \left\{\chi(x)(Z_{\Phi^2}Z_\Phi - 1)\Phi_R(x)^2\right.
$$
$$
\left. +\chi(x)(Z_{01}) + \frac{\chi(x)^2}{2}(Z_{02})\right\}.
$$

S_0 denotes the bare lattice action given by (4.3.1). The renormalization constants Z_Φ, Z_{m^2}, Z_λ, Z_{Φ^2}, Z_{01} and Z_{02} are power series in λ_R, with coefficient functions depending on m_R and the lattice spacing a. The theory is made finite by field-, mass- and coupling-constant renormalization. The last three counterterms are required to render correlation functions with composite operator $\Phi_R(x)^2$ insertions UV finite. There are six independent renormalization constants in the renormalized action (4.3.26). They are uniquely determined by the six independent normalization conditions

(4.3.16), order by order in perturbation theory. The notion of a "renormalized" action is a bit misleading. S_R does not stay finite as the cutoff is removed, but it is made divergent precisely in such a way that the renormalized correlation functions obtained from S_R (as outlined below) are finite and satisfy the normalization conditions.

Writing

$$m_0^2 = Z_{m^2} m_R^2, \ \lambda_0 = Z_\lambda \lambda_R, \ \Phi_0 = Z_\Phi^{1/2} \Phi_R, \qquad (4.3.27)$$

and

$$J_0 = Z_\Phi^{-1/2} J, \qquad \chi_0 = Z_{\Phi^2} \chi, \qquad (4.3.28)$$

we obtain for the generating functional of the renormalized full Green functions

$$\mathcal{Z}_R(J,\chi) = \exp\left\{ a^4 \sum_x \left(\chi(x) Z_{01} + \frac{1}{2}\chi(x)^2 Z_{02} \right) \right\} \int \prod_x d\Phi_R(x)$$

$$\cdot \exp\left\{ -S_0(\Phi_0|m_0,\lambda_0) + a^4 \sum_x \left(J_0(x)\Phi_0(x) + \chi_0(x)\Phi_0(x)^2 \right) \right\} \quad (4.3.29)$$

$$= \exp\left\{ a^4 \sum_x \left(\chi(x) Z_{01} + \frac{1}{2}\chi(x)^2 Z_{02} - \frac{1}{2a^4} \ln Z_\Phi \right) \right\} Z_0(J_0,\chi_0),$$

and

$$W_R(J,\chi) = a^4 \sum_x \left(\chi(x) Z_{01} + \frac{1}{2}\chi(x)^2 Z_{02} - \frac{1}{2a^4} \ln Z_\Phi \right) + W_0(J_0,\chi_0).$$
$$(4.3.30)$$

Finally, with

$$\phi_R(x) = \frac{\partial W_R(J)}{\partial J(x)} \qquad (4.3.31)$$

the generating functional of the renormalized vertex functions $\widetilde{\Gamma}_R^{(n,l)}$ is obtained by the Legendre transform

$$\Gamma_R(\phi_R,\chi) = W_R(J,\chi) - a^4 \sum_x J(x)\Phi_R(x), \qquad (4.3.32)$$

so that

$$\Gamma_R^{(n,l)}(x_1,\ldots,x_n|z_1,\ldots,z_l;a) \tag{4.3.33}$$

$$= \frac{a^{-4(n+l)}\partial^{n+l}}{\partial\phi_R(x_1)\cdots\partial\phi_R(x_n)\partial\chi(z_1)\cdots\partial\chi(z_l)}\,\Gamma_R(\phi_R,\chi)\bigg|_{\phi_R=\chi=0}.$$

Upon Fourier transform, the renormalized vertex functions become related to the bare ones by

$$\widetilde{\Gamma}_R^{(n,l)}(p_1,\ldots,p_{n-1}|q_1,\ldots,q_l;a)$$
$$= Z_\Phi^{n/2}Z_{\Phi^2}^l\,\widetilde{\Gamma}_0^{(n,l)}(p_1,\ldots,p_{n-1}|q_1,\ldots,q_l;a) \tag{4.3.34}$$
$$+\delta_{n0}\left(\delta_{l1}Z_{01}+\delta_{l2}Z_{02}\right).$$

The normalization conditions (4.3.16) imply that

$$Z_\Phi = Z_R$$
$$Z_{01} = -Z_{\Phi^2}\widetilde{\Gamma}^{(0,1)}(a) \tag{4.3.35}$$
$$Z_{02} = -Z_{\Phi^2}^2\widetilde{\Gamma}^{(0,2)}(0;a).$$

We conclude this section with a statement about universality. According to (4.3.24), the continuum limit of the renormalized vertex functions is given by the continuum vertex functions that are renormalized for example according to the BPHZ finite-part prescription. This implies universality of the continuum limit in the sense that the continuum limit to a large extent does not depend on the details of the lattice action. For instance, one might add to the bare lattice action a classically irrelevant interaction of the form

$$a^4\sum_{x\in\Lambda_4}a^2\,\frac{\tau_0}{6!}\,\Phi_0(x)^6 \tag{4.3.36}$$

with dimensionless coupling constant τ_0. This does not change the continuum limit of the $\widetilde{\Gamma}_R^{(n,l)(\nu)}$ as a function of m_R and g_R, to all orders of g_R, i.e. (4.3.18) remains unchanged.

However, what is actually modified by such an irrelevant interaction is the size of scaling violation

$$\left|a\frac{\partial}{\partial a}\widetilde{\Gamma}_R^{(n,l)}\right|, \tag{4.3.37}$$

so the irrelevant interaction has influence on the speed of convergence towards the continuum limit. This is utilized in so-called improvement programs in order to accelerate convergence to the continuum limit. Furthermore, irrelevant interactions change the relation between renormalized and bare parameters. They shift the bare parameters for which the continuum limit is constructed.

The fact that (4.3.36) is an irrelevant interaction crucially depends on lattice-power counting rules. The operator (4.3.36) does not change UV-divergence degrees. The counting of the UV-divergence degrees implies that the most general counterterm is still of the form given in (4.3.26). In particular, no tuning or renormalization of τ_0 is required, it is just an irrelevant coupling constant in the sense that it drops out in the continuum limit, to all orders of perturbation theory. Let us mention two further examples of terms which also vanish in the naive continuum limit, in common with (4.3.36), but where one may not jump to the conclusion that they can be either left out from the beginning or kept in view of an improvement of the convergence properties. Instead, a careful handling is necessary. The first example are Wilson fermions on the lattice. Although the difference to the naive fermions is proportional to the lattice constant a and thus corresponds to an irrelevant term in the above sense, this irrelevant term serves for lifting the species doubling, making power counting applicable. Even more, in case of gauge theories it ensures that the axial anomaly is correctly reproduced in the continuum limit. The reason is that the "irrelevant" Wilson term modifies the mapping between bare and renormalized parameters just in the appropriate way.

The second example refers to operators like

$$a^4 \sum_{x \in \Lambda_4} a \, \frac{\tau_1}{6!} \, \Phi_0(x)^6 \quad \text{and} \quad a^4 \sum_{x \in \Lambda_4} a^3 \, \frac{\tau_2}{6!} \, \Phi_0(x)^8 \qquad (4.3.38)$$

with dimensionful coupling constants τ_1 and τ_2. Such terms must be actually excluded from the beginning, as they render the theory unrenormalizable, although they vanish in the classical continuum limit. The reason is that their inclusion enhances UV-divergence degrees and in turn implies over-subtractions compared to the pure Φ^4-theory.

4.3.5 *Power counting in the continuum*

Perturbative calculations in local quantum field theories are affected by short distance singularities. The reason is that no fundamental physical length scale is assumed to exist which must be implemented as an ultimate cutoff. In coordinate-space representation, fields are tempered distributions, and the correlation functions are related to products of them. The product of distributions, however, is an ill-defined concept in general. (While a product of δ-distributions $\delta(x) \cdot \delta(y) \cdot \delta(z)$ (in a shorthand notation) may be well defined, $\delta(x)^2$ at the same spacetime point cannot be defined.) It is easier to tackle this problem by taking the Fourier transform to momentum space. In momentum-space representation, the short-distance singularities manifest themselves in divergencies of Feynman integrals for large internal momenta, i.e. for momenta which are integrated over. They are known as UV-divergencies.

UV-behavior of Feynman integrals in momentum space is controlled by so-called (UV) power-counting theorems. They relate the large-momentum behavior to divergence degrees of Feynman diagrams and subdiagrams. On the lattice, the formulation of such a theorem is rather peculiar because of the periodicity in momentum space with a period given by the Brillouin zone. Therefore we discuss power counting first in the continuum.

4.3.6 *Power counting for Feynman integrals in momentum space: UV-divergence for rational functions and integrals*

In the following we consider Feynman integrals in momentum space in D dimensions. In the continuum, a Feynman integral with m loops is of the form

$$\mathcal{I}(p) \;=\; \int_{-\infty}^{\infty} \prod_{i=1}^{m} d^D k_i \; I(k,p),$$

$$I(k,p) \;=\; \frac{P(k,p)}{\prod_{i=1}^{r} \left(l_i(k,p)^2 + \mu^2\right)} \qquad (4.3.39)$$

with $\mu^2 > 0$, p represents all external momenta, and the l_i are the line momenta

$$l_i(k,p) \;=\; K_i(k) \;+\; Q_i(p) \;=\; \sum_{j=1}^{m} C_{ij} k_j \;+\; Q_i(p). \qquad (4.3.40)$$

$P(k,p)$ is a polynomial of the loop momenta k and the external momenta p. In general, the integrand is a rational function of all momenta. This general form includes or easily generalizes to more complicated cases such as pure gauge theories and QCD, it includes subtracted Feynman integrals, cf. below. We have chosen the nonvanishing mass $\mu^2 > 0$ because we focus on the UV-behavior. Theories with massless modes will be considered in a subsequent section. The functions K_i are defined via 4.3.40, Q_i are linear combinations of external momenta p.

Divergencies of $\mathcal{I}(p)$ may arise if some or all loop momenta become large. The decay of the rational integrand should be sufficiently fast in all directions in k-space so that the Feynman integral is convergent. The directions are classified by so-called Zimmermann subspaces, and the decay rates by the associated UV-divergence degrees. Toward their definition we first introduce the notion of the UV-degree (or simply degree) of a multi-variable polynomial $Q(u, v)$. The UV-degree of $Q(u, v)$ with respect to u, say, is defined by $\overline{\mathrm{degr}}_u Q(u, v) = \omega$, with ω given by

$$Q(\lambda u, v) \simeq A(u, v) \, \lambda^\omega \quad \text{as } \lambda \to \infty, \tag{4.3.41}$$

$A(u, v) \not\equiv 0$. Let us call

$$\mathcal{L} = \{k_1, \ldots, k_m, l_1, \ldots, l_r\} \tag{4.3.42}$$

the set of all loop momenta and of all line momenta. Let $u_1, \ldots, u_d$, $v_1, \ldots, v_{m-d}$ be m linearly independent elements of $\mathcal{L}$ with respect to k, which means that their homogeneous parts in k are linearly independent. Keeping $v_1, \ldots, v_{m-d}$ fixed, this defines a class H of affine subspaces of the space of integration momenta, a Zimmermann subspace. With $k = k(u, v, p)$ the UV-divergence degree of $\mathcal{I}(p)$ with respect to H is defined by

$$\overline{\mathrm{degr}}_H \mathcal{I} = Dd + \overline{\mathrm{degr}}_u I(k(u, v, p), p), \tag{4.3.43}$$

with

$$\overline{\mathrm{degr}}_u I(k(u, v, p), p) = \overline{\mathrm{degr}}_u P(k(u, v, p), p) - \sum_{i=1}^r \overline{\mathrm{degr}}_u l_i (k(u, v, p), p)^2. \tag{4.3.44}$$

Notice that the UV-divergence degrees do not depend on particular values of the external momenta p.

The power-counting theorem now states the following. If for all Zimmermann subspaces H

$$\overline{\mathrm{degr}}_H \mathcal{I} < 0, \tag{4.3.45}$$

the Feynman integral $\mathcal{I}(p)$ is absolutely convergent.

For example, consider the four-dimensional one loop integral

$$\mathcal{J}_1(p_1, p_2) \;=\; \int_{-\infty}^{\infty} \frac{d^4k}{(2\pi)^4} \, \frac{1}{k^2 + \mu^2} \, \frac{1}{(k + p_1)^2 + \mu^2} \, \frac{1}{(k + p_1 + p_2)^2 + \mu^2}.$$
$$(4.3.46)$$

There is only one Zimmermann subspace H to one loop order, parameterized by the loop momentum k, and

$$\overline{\mathrm{degr}}_H \mathcal{J}_1 \;=\; 4 \,-\, \sum_{i=1}^{3} \overline{\mathrm{degr}}_k(k^2 + \mu^2) \;=\; -2. \qquad (4.3.47)$$

Hence, $\mathcal{J}_1$ is absolutely convergent. On the other hand,

$$\mathcal{J}_2(p) \;=\; \int_{-\infty}^{\infty} \frac{d^4k}{(2\pi)^4} \, \frac{1}{k^2 + \mu^2} \, \frac{1}{(k + p)^2 + \mu^2} \qquad (4.3.48)$$

has $\overline{\mathrm{degr}}_H \mathcal{J}_2 = 0$, so the power-counting theorem does not apply. Actually, $\mathcal{J}_2(p)$ does not exist. In higher loop order, divergencies do also arise if only parts of the loop momenta become large.

4.3.7   *Power counting for Feynman diagrams: UV-divergence degrees of propagators, vertices, fields and diagrams*

For a Feynman diagram γ it is useful to define divergence degrees of γ as well as for all of its subdiagrams. It is sufficient and convenient to consider 1PI graphs only. The Feynman integral $\mathcal{I}_\gamma$ associated with a 1PI-Feynman graph γ is obtained according to the Feynman rules. They are very specific for the theory which is under consideration. The generic form of $\mathcal{I}_\gamma$ is given by

$$\mathcal{I}_\gamma(p) \;=\; \int_{-\infty}^{\infty} \prod_{i=1}^{m(\gamma)} \frac{d^D k_i}{(2\pi)^D} \, \prod_{L \in \mathcal{L}_\gamma} \Delta_L(l_L(k, p)) \cdot \prod_{B \in \mathcal{B}_\gamma} P_B(k, p), \qquad (4.3.49)$$

where $m(\gamma)$ denotes the number of loops of γ, and $\mathcal{B}_\gamma$ and $\mathcal{L}_\gamma$ denotes the set of vertices and of internal lines of γ, respectively, p collects the external momenta of γ, more precisely the momenta of the external lines of γ; usually momentum conservation is implicitly assumed. Every vertex $B \in \mathcal{B}_\gamma$ carries a polynomial P_B of degree $d(B)$, say, of the momenta of the lines attached to B (modulo momentum conservation) and hence of k and

p. For instance, an interaction term or a composite operator of the form

$$\int d^D x \, \partial_x^{n_1} \Phi(x)^{n_2} \tag{4.3.50}$$

generates a vertex B with n_2 lines attached and with $d(B) = n_1$. For every internal line $L \in \mathcal{L}_\gamma$ with l_L denoting its momentum, there is one propagator $\Delta_L(l_L)$. For simplicity we assume only one type of propagator. The generalization is obvious. Suppose that the inverse free propagator has the large momentum behavior

$$\Delta(\lambda l)^{-1} \simeq A(l) \, \lambda^{(D - 2\omega_\Phi)} \quad \text{as } \lambda \to \infty \tag{4.3.51}$$

with $A(l) \not\equiv 0$. Furthermore there should exist a constant $c_\Delta > 0$ so that $\Delta(l)^{-1} \geq c_\Delta$ in order to avoid possible IR problems. For instance, if $\Delta(l)^{-1} = l^2 + \mu^2$ we get $\omega_\Phi = (D - 2)/2$. In general, the relation (4.3.51) defines the **canonical (UV) dimension** of the field Φ with free propagator $\Delta(l)$.

Furthermore it is convenient to introduce the quantities $e(\gamma)$, $L(\gamma)$ and $B(\gamma)$, which denote the number of external lines, internal lines and the number of vertices of γ, respectively.

Now let τ be any sub-diagram of γ, including the case $\tau = \gamma$. Notions like $m(\gamma)$, $\mathcal{L}_\gamma$ and $\mathcal{B}_\gamma$ generalize to subgraphs in the obvious way. The UV-divergence degree of τ is then defined by

$$\omega(\tau) = Dm(\tau) - (D - 2\,\omega_\Phi)L(\tau) + \sum_{B \in \mathcal{B}_\tau} d(B). \tag{4.3.52}$$

It is possible to show that for a Feynman integral $\mathcal{I}_\gamma$ of the form (4.3.49) the power-counting theorem can be rephrased as follows. If all 1PI-subdiagrams τ of γ have negative UV-divergence degree,

$$\omega(\tau) < 0, \tag{4.3.53}$$

the Feynman integral $\mathcal{I}_\gamma(p)$ is absolutely convergent.

The advantage of the definition (4.3.52) of the UV-divergence degree of graphs is that it is directly related to the canonical dimensions of the fields, the interaction part of the action and the composite operators involved. The reason is the following. Using the topological relations

$$2L(\tau) + e(\tau) = \sum_{B \in \mathcal{B}_\tau} n(B),$$

$$m(\tau) = L(\tau) - B(\tau) + 1 \,, \tag{4.3.54}$$

which hold for any connected Feynman (sub-) diagram τ, it is straightforward to see that

$$\omega(\tau) = D + \sum_{B \in \mathcal{B}_\tau} (\omega_B(B) - D) - \omega_\Phi \, e(\tau) \qquad (4.3.55)$$

with

$$\omega_B(B) = \omega_\Phi \, n(B) + d(B). \qquad (4.3.56)$$

The identity (4.3.55) together with (4.3.56) relates UV-divergence degrees of Feynman diagrams to the interactions and composite operators of the theory they are constructed of. As will be seen below, in the most common cases they allow us in a neat way to classify perturbatively renormalizable field theories, by counting powers of field and numbers of derivatives in coordinate space.

4.3.8   *Renormalization or: Removing UV-divergencies under the integral sign*

The power-counting criterion predicts convergence of momentum-space Feynman integrals if the divergence degrees with respect to all Zimmermann subspaces are less than zero. If the Feynman integral is associated with a Feynman graph in the sense of (4.3.49), this is fulfilled if all subdiagrams have a negative UV-divergence degree. In most of the interesting cases, however, Feynman diagrams have a nonnegative degree of divergence. A renormalization procedure must be invoked to render the corresponding Feynman integrals UV-finite. This requires to "subtract" the UV-divergencies in such a way that locality is preserved and such that the criteria of the power-counting theorem are fulfilled for the subtracted Feynman integral. It is achieved by applying Taylor subtractions with respect to external momenta to the integrand.

To outline the idea, let us consider the one-loop integral associated with a 1PI diagram γ,

$$\mathcal{I}_\gamma(p) = \int_{-\infty}^{\infty} d^D k \; I(k, p),$$

$$I(k, p) = \frac{P(k, p)}{\prod_{i=1}^{r} ((k + p_i)^2 + \mu^2)}, \qquad (4.3.57)$$

with $\mu^2 > 0$ and P as before. Suppose that γ has a nonnegative, overall-divergence degree

$$\omega(\gamma) \;=\; D \,+\, \overline{\mathrm{degr}}_{kp} I(k,p) \;=\; D \,+\, \overline{\mathrm{degr}}_{kp} P(k,p) \;-\; 2r \geq 0. \quad (4.3.58)$$

We denote by T_p^δ the Taylor expansion about zero momentum with respect to the external momenta p of order δ. Application of $1 - T_p^\delta$ to the integrand $I(k,p)$ reduces the divergence degree

$$\overline{\mathrm{degr}}_k \left(1 - T_p^\delta\right) I(k,p) \;\leq\; \overline{\mathrm{degr}}_{kp} I(k,p) - (\delta + 1). \quad (4.3.59)$$

Here we have used the rational structure of the integrand and the fact that every propagator which depends on some external momentum also depends on the loop momentum k, because the diagram is 1PI by assumption. Choosing $\delta \geq \omega(\gamma)$, we have

$$D \,+\, \overline{\mathrm{degr}}_k \left(1 - T_p^\delta\right) I(k,p) \;<\; 0. \quad (4.3.60)$$

This means that the subtracted Feynman integral

$$\int_{-\infty}^{\infty} d^D k \;\; \left(1 - T_p^\delta\right) I(k,p) \quad (4.3.61)$$

becomes convergent.

For example, consider the four-dimensional Feynman integral $\mathcal{J}_2$, given by (4.3.48), which turned out to be divergent with an overall UV-divergence degree $\overline{\mathrm{degr}}_H \mathcal{J}_2 = 0$. Applying $(1 - T_p^0)$ to the integrand of $\mathcal{J}_2$, we obtain

$$\int_{-\infty}^{\infty} \frac{d^4 k}{(2\pi)^4} \frac{1}{k^2 + \mu^2} \left(\frac{1}{(k+p)^2 + \mu^2} - \frac{1}{k^2 + \mu^2} \right)$$
$$= \int_{-\infty}^{\infty} \frac{d^4 k}{(2\pi)^4} \frac{k^2 - (k+p)^2}{(k^2 + \mu^2)^2((k+p)^2 + \mu^2)}, \quad (4.3.62)$$

which is easily checked to have a negative UV-divergence degree. Hence, (4.3.62) is finite.

This procedure generalizes to Feynman diagrams with an arbitrary number of loops. Beyond the overall divergence occurring when all loop momenta are made large simultaneously, this case requires to remove all divergencies which occur when only parts of the loop momenta become large. These are associated with divergencies of lower order diagrams. For every 1PI-subgraph τ with a nonnegative UV-divergence degree $\omega(\tau)$ given by (4.3.55), a Taylor operation is associated with respect to the external momenta of the sub-diagram, of order $\omega(\tau)$. The Taylor operations are organized according to the forest formula of Zimmermann and applied to the integrand. One can prove that the resulting integral meets the criteria of the power-counting theorem and thus is absolutely convergent.

A detailed discussion of the forest formula is beyond the scope of this book. The procedure just outlined is the so-called BPHZ-finite part prescription. If a Feynman graph γ has an UV-divergence degree $\omega(\gamma) \geq 0$, its BPHZ-renormalized Feynman integral vanishes at zero momentum together with its derivatives up to and including order $\omega(\gamma)$. This automatically imposes normalization conditions on vertex functions. Let $\widetilde{\Gamma}$ be any vertex function in momentum space and assume that $\omega_\Gamma = \max_\gamma \omega(\gamma)$ is a nonnegative finite number, where the maximum is taken over all Feynman graphs contributing to $\widetilde{\Gamma}$. The BPHZ-subtraction then amounts to a renormalized $\widetilde{\Gamma}$ that vanishes at zero momentum together with its derivatives up to and including order ω_Γ. Actually, except for simplicity, there is no deeper reason that the subtractions are applied at zero external momenta. Other normalizations work equally well, and different ones are related by finite renormalizations.

In this framework it is now easy to give a general criterion for power-counting renormalizability. First, every vertex function $\widetilde{\Gamma}$ should have a finite UV-divergence degree ω_Γ, as defined above. The emphasis here is on the existence of an upper bound on $\omega(\gamma)$ for all γ contributing to $\widetilde{\Gamma}$, to all orders of the loop expansion. Second, $\omega_\Gamma < 0$ if $\widetilde{\Gamma}$ is a vertex function with a sufficiently large number of fields.

4.3.9   *Renormalization or: Removing UV-divergencies in the counterterm approach*

It is quite useful to rephrase power counting renormalizability in terms of divergence degrees of Feynman diagrams as given by (4.3.55) and (4.3.56). It is sufficient to consider only graphs without composite operator vertices. A sufficient condition for power counting renormalizability then is that for all vertices B induced by the interactions of the theory satisfy

$$\omega_B(B) \equiv \omega_\Phi\, n(B) + d(B) \leq D, \qquad (4.3.63)$$

so that for any 1PI diagram γ we have $\omega(\gamma) \leq D - \omega_\Phi\, e(\gamma)$, independently of the order it contributes to. Furthermore, $\omega_\Phi > 0$ implies that $\omega(\gamma) \geq 0$ only if $e(\gamma) \leq D/\omega_\Phi$. For instance, any interaction of the from

$$\int d^D x\, \partial_x^{n_1} \Phi(x)^{n_2} \qquad (4.3.64)$$

generates a vertex B with $n(B) = n_2$ lines attached and with a polynomial of order $d(B) = n_1$ in momentum space. With ω_Φ the canonical UV dimension of the field Φ, the interaction (4.3.64) is renormalizable if $n_1 + n_2 \omega_\Phi \le D$.

Suppose now that the field theory is regularized by an UV-cutoff in the continuum, such as a momentum cutoff, Pauli-Villars or dimensional regularization. After regularization it then makes sense to talk about a bare theory, at least in perturbation theory, if the cutoff is intrinsically perturbative. The interaction part of the regularized action then generates the vertices of the corresponding Feynman diagrams. The Feynman integrals of the regularized theory are UV finite. The role of renormalization now is to render them finite even if the UV-cutoff is removed.

The BPHZ-subtraction scheme works in this framework of a regularized theory in the continuum too. However, the presence of a cutoff provides several advantages. In particular, one can prove that the subtractions generated by the Taylor operations and arranged according to the forest formula of Zimmermann can be understood as originating from local counterterms to the regularized action. For a power-counting renormalizable field theory, the most general counterterm part of the action which is required is of the form

$$\sum_{n=0}^{[D/\omega_\Phi]} \sum_{j=0}^{D-\omega_\Phi n} \delta Z_{nj} \, \partial_x^j \Phi(x)^n, \qquad (4.3.65)$$

where $[\,]$ denotes the integer part of D/ω_Φ. This is further restricted by the internal and external symmetries of the theory. The finitely many coefficients are obtained from the loop expansion and are uniquely determined by imposing a finite number of normalization conditions on the vertex functions with nonnegative UV-divergence degree, order by order.

4.3.10 *Power counting in lattice field theory*

From the point of view of perturbation theory, the lattice cutoff is peculiar, because the Feynman integrals have a very specific structure. The vertices and propagators are of a more complex form. Often one has to consider "irrelevant" Feynman diagrams that originate in so-called irrelevant vertices, these are Feynman diagrams which vanish in the classical continuum limit. The Feynman rules strongly depend on the specific choice of the lattice action. Last but not least, in momentum space the integration domain is compact and the integrand is periodic with respect to the loop momenta. Because of these reasons the continuum power-counting theorem does not apply, as it assumes that the integrand is a rational function as in (4.3.39). Furthermore, for continuum Feynman integrals, divergence degrees are defined via the behavior of a diagram for large loop momenta. On the lattice this limit does not make sense *a priori*, because of the periodicity of momentum-space correlation functions with the period of the Brillouin zone.

A modified notion of a divergence degree is required on the lattice, and a different, adapted power-counting theorem is needed to control the behavior of lattice Feynman integrals in the continuum limit. It should work under very general conditions in order to apply to the various lattice realizations of a field theory.

The first and most natural conjecture for a criterion is that a lattice Feynman integral is convergent in the continuum limit $a \to 0$ if the continuum limit of the integrand is absolutely integrable, since the typical lattice propagator behaves as

$$\frac{1}{\widehat{k}^2 + \mu^2} \simeq \frac{1}{k^2 + \mu^2}, \qquad (4.3.66)$$

with momentum k in a sufficiently small neighborhood of the origin, and otherwise

$$\frac{1}{\widehat{k}^2 + \mu^2} \leq \text{const } a^2. \qquad (4.3.67)$$

Unfortunately, this criterion turns out to fail. For instance, the lattice integral

$$\widehat{\mathcal{J}} = \int_{-\pi/a}^{\pi/a} \frac{d^4 k}{(2\pi)^4} \, \frac{\sum_{\nu=0}^{3} (1 - \cos k_\nu a)}{\widehat{k}^2 + \mu^2} \qquad (4.3.68)$$

has an integrand that vanishes in the classical continuum limit. However, by substituting $k \to k/a$, it is easily seen that $\mathcal{I}$ is actually quadratically divergent as $a \to 0$. Thus a sensible definition of a divergence degree must not only refer to the continuum limit of the integrand.

4.3.11 *Preliminaries of lattice-power counting and lattice UV-divergence degrees*

Before we can formulate such a definition of an UV-divergence degree on the lattice and the corresponding power-counting theorem, we have to introduce some definitions to make our notions more precise. Lattice power counting and renormalization (as described below) apply to lattice Feynman integrals of the following form

$$\widehat{\mathcal{I}}(p;\mu,a) = \int_{-\pi/a}^{\pi/a} d^D k_1 \cdots d^D k_m \ \widehat{I}(k,p;\mu,a) ,$$

$$\widehat{I}(k,p;\mu,a) = \frac{V(k,p;\mu,a)}{\prod_{i=1}^{r} C_i(l_i(k,p);\mu_i,a)} . \qquad (4.3.69)$$

The numerator V of the integrand is of the form

$$V(k,p;\mu,a) = a^{-\nu} F(ka,pa;\mu a), \qquad (4.3.70)$$

with ν a nonnegative integer number and F a smooth function that is 2π periodic in all momentum components of ka, pa and a polynomial in the mass parameters represented by μa. Furthermore it is assumed that

$$\lim_{a\to 0} V(k,p;\mu,a) = P(k,p,\mu) \qquad (4.3.71)$$

exists and is a (homogeneous) polynomial (of degree ν or identically zero). For further reference, the class of functions with the properties listed above is denoted by $\mathcal{C}_\nu$.

The denominators C_i belong to the function class $\mathcal{C}_2$ and are functions of the line momenta (4.3.40)

$$l_i(k,p) = K_i(k) + Q_i(p) = \sum_{j=1}^{m} C_{ij} k_j + Q_i(p), \qquad (4.3.72)$$

hence they are of the form

$$C_i(l_i;\mu_i,a) = a^{-2}\eta_i(l_i a;\mu_i a), \qquad (4.3.73)$$

with η_i smooth functions, 2π-periodic in the components of $l_i a$ and a polynomial of $\mu_i a$. We keep all $\mu_i > 0$ and consider the case with massless modes later. The functions C_i are subject to the following two conditions. First,

$$\lim_{a\to 0} C_i(l_i;\mu_i,a) = l_i^2 + \mu_i^2, \qquad (4.3.74)$$

second, for sufficiently small lattice spacing a there is a positive constant K such that

$$|C_i(l_i; \mu_i, a)| \geq K\left(\widehat{l}_i^{\,2} + \mu_i^2\right). \tag{4.3.75}$$

The properties (4.3.74) and (4.3.75) ensure that the denominators of the propagators are of the order of a^{-2} except for a neighborhood of zero line momentum (modulo the Brillouin zone), but otherwise behave as their continuum analogues. In particular, they imply the absence of species doubling.

We emphasize that the above assumptions on the structure of lattice Feynman integrals are quite general. They are matched by a large class of lattice field theories. Examples of propagators are given by

$$C_i(l; \mu, a) = \widehat{l}^2 + \mu^2, \tag{4.3.76}$$

as they occur in scalar and vector field theories, but also for Wilson fermions

$$C_i(l; \mu, a) = (1 + \mu a)\widehat{l}^2 + \mu^2 + \frac{1}{a^2}\left(J(l) - 1\right). \tag{4.3.77}$$

Even the more involved case of Ginsparg-Wilson fermions with propagator

$$C_i(l; \mu, a) = \widehat{l}^2 + \frac{1}{a^2}\left(J(l) - 1\right) - \frac{a^2}{4}\left(\widehat{l}^2\right)^2$$
$$+ J(l)\left[\mu + \frac{1}{a}\left(1 - \left(1 - \frac{a^2}{2}\widehat{l}^2\right)J(l)^{-1/2}\right)\right]^2, \tag{4.3.78}$$

where we set

$$J(l) = 1 + \frac{a^4}{2}\sum_{\nu < \rho = 0}^{D-1} \widehat{l}_\nu^2 \widehat{l}_\rho^2, \tag{4.3.79}$$

satisfies the above criteria. Notice that naive and staggered fermions are excluded because their propagators show multiple poles in the Brillouin zone in the sense that they violate (4.3.75).

Now the required notion of an UV-degree on the lattice is given as follows. Let $W(u, v; \mu, a)$ belong to some $\mathcal{C}_\nu$. The lattice UV-degree of W with respect to u is denoted by

$$\overline{\deg}_{\widehat{u}}\, W \tag{4.3.80}$$

and defined via the limit behavior of W for $u \to \infty$ and $a \to 0$ with limits taken simultaneously with the same rate. Let

$$W(\lambda u, v, q; \mu, \frac{a}{\lambda}) = A(u, v, p; \mu, a)\lambda^\delta + O(\lambda^{\delta-1}) \quad \text{as } \lambda \to \infty, \tag{4.3.81}$$

with $A(u, v, p; \mu, a) \not\equiv 0$. The lattice UV-degree of W with respect to u is then defined as

$$\overline{\mathrm{degr}}_{\widehat{u}} W = \delta. \tag{4.3.82}$$

Notice that even if W has a vanishing continuum limit, the lattice degree of W can be nonnegative. Also,

$$\overline{\mathrm{degr}}_u \lim_{a \to 0} W \leq \overline{\mathrm{degr}}_{\widehat{u}} W. \tag{4.3.83}$$

The particular form of the lattice propagators ensures that their lattice UV-degree coincides with the UV-degree of their continuum limits,

$$\overline{\mathrm{degr}}_{\widehat{l}} C_i(l; \mu, a) = \overline{\mathrm{degr}}_{\widehat{l}} \widehat{l}^2 = \overline{\mathrm{degr}}_l l^2 = 2. \tag{4.3.84}$$

Once we have defined the notion of the lattice UV-degree we can proceed and define lattice UV-divergence degrees for Feynman integrals $\widehat{\mathcal{I}}$, (4.3.69). For any Zimmermann subspace H parameterized by $u_1, \ldots, u_d$, say, the lattice UV-divergence degree of $\widehat{\mathcal{I}}$ is defined by

$$\overline{\mathrm{degr}}_{\widehat{H}} \widehat{\mathcal{I}} = Dd + \overline{\mathrm{degr}}_{\widehat{u}} \widehat{I}(k(u, v, p), p; \mu, a), \tag{4.3.85}$$

with

$$\overline{\mathrm{degr}}_{\widehat{u}} \widehat{I}(k(u, v, p), p; \mu, a) = \overline{\mathrm{degr}}_{\widehat{u}} V(k(u, v, p), p; \mu, a)$$
$$- \sum_{i=1}^{r} \overline{\mathrm{degr}}_{\widehat{u}} C_i(l_i(k(u, v, p), p; \mu_i, a). \tag{4.3.86}$$

Again, the various UV-divergence degrees do not depend on particular values of the external momenta p.

Next one may tentatively assume that the power-counting theorem on the lattice holds similarly as in the continuum but in terms of lattice divergence degrees. However, it turns out that one needs a further technical notion, the so-called naturalness of line momenta, which is related to the periodicity of momentum-dependent functions with the Brillouin zone. Let

$$\mathcal{M} = \{k_1, \ldots, k_m, K_1, \ldots, K_r\}, \tag{4.3.87}$$

where

$$K_i = \sum_{j=1}^{m} C_{ij} k_j \tag{4.3.88}$$

are the homogeneous parts of the line momenta, cf. (4.3.72). The first requirement of naturalness is that all C_{ij} are integer-valued. Furthermore, suppose that $w_1, \ldots, w_m$ are m linearly independent elements of $\mathcal{M}$ with respect to k. Then

$$k_i \; = \; \sum_{j=1}^{m} A_{ij} w_j \quad \text{with integer } A_{ij} \in \mathbf{Z}. \tag{4.3.89}$$

The emphasis here is on the fact that the coefficients A_{ij} are integer valued for every choice of basis of the loop momenta of $\mathcal{M}$. The condition implies that an arbitrary set of independent line momenta can be chosen as the integration variables without changing periodicity with the Brillouin zone. The condition becomes important whenever line momenta take values in higher Brillouin zones. To estimate the corresponding contributions for small lattice spacing, it is necessary to shift them back into the first Brillouin zone. Naturalness ensures that the shift can be performed by translation of the loop momenta by integer multiples of $2\pi/a$, without changing the integrand. Without this condition of naturalness, the definition of a divergence degree as given above would not be sufficient to describe the behavior in the continuum limit. Nevertheless, for integrals which correspond to lattice Feynman diagrams one can prove that naturalness always holds if the integration momenta are chosen from the set of line momenta. Now we are ready to state the lattice power-counting theorem.

4.3.12 *Power-counting theorem on the lattice*

Next let us formulate the power-counting theorem on the lattice. Let $\widehat{\mathcal{I}}$ be given by (4.3.69)ff. If for all Zimmermann subspaces H

$$\overline{\mathrm{degr}}_{\widehat{H}} \, \widehat{\mathcal{I}} \; < \; 0, \tag{4.3.90}$$

the continuum limit of $\widehat{\mathcal{I}}$ exists and is given by

$$\lim_{a \to 0} \widehat{\mathcal{I}}(p; \mu, a) \; = \; \int_{-\infty}^{\infty} d^D k_1 \cdots d^D k_m \; \frac{P(k, p, \mu)}{\prod_{i=1}^{r} \left(l_i(k, p)^2 + \mu_i^2 \right)} \, . \tag{4.3.91}$$

The integral on the right hand side of (4.3.91) is absolutely convergent on account of the continuum-power counting and the properties (4.3.83) and (4.3.84). Equation (4.3.91) states first of all that the continuum limit of the Feynman integral exists, and second that it is obtained by taking the continuum limit *under* the integral sign. As an important consequence the continuum limit is universal, i.e. it does not depend on the details of the lattice action as long as (4.3.90) holds.

Let us consider some examples such as the Feynman integral that we have encountered above

$$
\widehat{\mathcal{I}}(p; \mu, a) = \int_{-\pi/a}^{\pi/a} \frac{d^4 k}{(2\pi)^4} \frac{1}{\widehat{k}^2 + \mu^2} \left(\frac{1}{\widehat{(k+p)}^2 + \mu^2} - \frac{1}{\widehat{k}^2 + \mu^2} \right)
$$

$$
= \int_{-\pi/a}^{\pi/a} \frac{d^4 k}{(2\pi)^4} \frac{\widehat{k}^2 - \widehat{(k+p)}^2}{\left(\widehat{k}^2 + \mu^2\right)^2 \left(\widehat{(k+p)}^2 + \mu^2\right)},
$$

cf. (4.3.12) and (4.3.23), where we have claimed that its continuum limit exists. Indeed, by inspection,

$$
\overline{\mathrm{degr}}_{\widehat{k}} \left(\widehat{k}^2 - \widehat{(k+p)}^2 \right) = 1,
$$

$$
\overline{\mathrm{degr}}_{\widehat{k}} \left(\widehat{k}^2 + \mu^2 \right)^2 \left(\widehat{(k+p)}^2 + \mu^2 \right) = 6, \qquad (4.3.92)
$$

so that with respect to the one Zimmermann subspace H, which is parameterized by k,

$$
\overline{\mathrm{degr}}_{\widehat{H}} \, \widehat{\mathcal{I}}(p; \mu, a) = 4 + 1 - 6 < 0. \qquad (4.3.93)
$$

Hence the continuum limit of $\mathcal{I}$ exists and is obtained by taking the limit under the integral sign, which is given by (4.3.62). Another example is the integral $\widehat{\mathcal{J}}$ of (4.3.68). We have

$$
\overline{\mathrm{degr}}_{\widehat{k}} \sum_{\nu=0}^{3} (1 - \cos k_\nu a) = 0,
$$

$$
\overline{\mathrm{degr}}_{\widehat{k}} \left(\widehat{k}^2 + \mu^2 \right) = 2. \qquad (4.3.94)
$$

Hence $\overline{\mathrm{degr}}_{\widehat{H}} \, \widehat{\mathcal{J}} = 2$. Indeed, the integral is quadratically divergent as $a \to 0$. As a final example we consider

$$
\widehat{\mathcal{K}}(p; \mu, a) = \int_{-\pi/a}^{\pi/a} \frac{d^4 k}{(2\pi)^4} \frac{1}{\widehat{k}^2 + \mu^2} \frac{1}{\widehat{(k+p)}^2 + \mu^2} \tau_0 \, a^2, \qquad (4.3.95)
$$

where τ_0 denotes some coupling constant that does not depend on a. $\widehat{\mathcal{K}}$ satisfies $\overline{\mathrm{degr}}_{\widehat{H}} \, \widehat{\mathcal{K}} = -2$. Hence, upon taking the limit under the integral sign, $\lim_{a \to 0} \widehat{\mathcal{K}} = 0$.

4.3.13 *Renormalization on the lattice*

The power-counting theorem states the existence of lattice Feynman integrals in the continuum limit if the divergence degrees are less than zero. In general, Feynman integrals are singular so that the renormalization procedure must be invoked before the continuum limit can be taken. The combinatorial part of the renormalization program is actually very similar to the one in the continuum. The main difference is that Taylor subtraction operators T_p^δ must be replaced by lattice-subtraction operators $\widehat{T}_p^\delta$. The latter are required to preserve locality of the field theory on the lattice. In analogy to (4.3.59) a subtraction operator should satisfy

$$\overline{\mathrm{degr}_{\widehat{k}}} \left(1 - \widehat{T}_p^\delta\right) \widehat{I}(k, p; \mu, a) \leq \overline{\mathrm{degr}_{\widehat{kp}}} \, \widehat{I}(k, p; \mu, a) - (\delta + 1), \qquad (4.3.96)$$

whenever every propagator that depends on p also depends on k. This property holds if $\widehat{T}_p^\delta$ satisfies the relation

$$\left(1 - \widehat{T}_p^\delta\right) \widehat{I}(k, \lambda p; \mu, a) = O(\lambda^{\delta+1}) \quad \text{as } \lambda \to 0. \qquad (4.3.97)$$

Any lattice operator $\widehat{T}_p^\delta$ which is $2\pi/a$ periodic satisfies (4.3.97) and converges in the continuum limit to the Taylor operator $\widehat{T}_p^\delta$, is called a lattice-subtraction operator of order δ. Examples are Taylor operators with respect to "lattice momenta" $(\sin p_\mu a)/a$, but there are many more possibilities.

Apart from this modification of replacing the Taylor subtractions by lattice-subtraction operators, the renormalization procedure of continuum Feynman integrals is the same for lattice Feynman integrals. The subtractions are still organized according to the forest formula. Equally well they can be arranged in such away that they are equivalent to adding local counterterms to the lattice action. The order $\widehat{\omega}(\tau)$ of the subtraction operators $\widehat{T}$ applied to a (sub-)diagram τ is given by the lattice divergence degree $\widehat{\omega}(\tau)$ of τ. By the analogous topological rearrangements to the former ones, it is obtained as

$$\widehat{\omega}(\tau) = D + \sum_{B \in \mathcal{B}_\tau} (\widehat{\omega}_\mathcal{B}(B) - D) - \omega_\Phi \, e(\tau), \qquad (4.3.98)$$

where now

$$\widehat{\omega}_\mathcal{B}(B) = \omega_\Phi \, n(B) + d(B) - p(B). \qquad (4.3.99)$$

Here ω_Φ is the canonical UV-dimension of the field Φ defined by the lattice

UV-degree of the free lattice propagator, here called $\widehat{\Delta}$,

$$\overline{\operatorname{degr}_{\widetilde{l}}} \, \widehat{\Delta}(l; \mu, a)^{-1} \;=\; D - 2\omega_\Phi. \tag{4.3.100}$$

Further, $n(B)$ denotes the number of Φ-lines entering the vertex, $d(B)$ the number of lattice momenta at B and $p(B)$ the power of the lattice spacing at B.

Power-counting renormalizability requires that $\omega_\Phi > 0$ and that for all vertices B

$$\widehat{\omega}_B(B) \;\leq\; D. \tag{4.3.101}$$

In this case the most general counterterm which is required for the renormalization is local and has the form

$$\sum_{n=0}^{D/\omega_\Phi} \sum_{j=0}^{D-\omega_\Phi n} \delta Z_{nj} \, h_j \left(\frac{1}{a} \widehat{\partial}_x, \frac{1}{a} \widehat{\partial}_x^* \right) \Phi(x)^n, \tag{4.3.102}$$

where h_j is a homogeneous polynomial of order j in the forward- and backward-lattice difference operators. Further restrictions on the counterterm structure arise from the lattice symmetries. By construction, all the renormalized Feynman integrals satisfy the requirements of the lattice power-counting theorem. This implies universality of the continuum limit of the whole theory: The renormalized vertex functions converge to the continuum vertex functions which are renormalized according to a BPHZ finite part prescription with appropriate normalization conditions. In particular, all interactions of the lattice action which vanish in the classical continuum limit and satisfy (4.3.101) vanish also in the quantum theory after renormalization. One is free to add additional local counterterms as long as they are classically irrelevant and satisfy (4.3.101). Note that (4.3.101) was violated by the interactions (4.3.38). This freedom is very useful if symmetries on the lattice such as the very gauge symmetries must be exactly preserved, symmetries which include irrelevant lattice operators. A prominent example are lattice gauge theories with a finite number of irrelevant terms at every *finite* order in perturbation theory, cf. below.

Let us illustrate these features by a lattice Φ^4-theory in $D = 4$ dimensions, supplemented by some classically irrelevant interactions,

$$S(\Phi) = a^4 \sum_{x \in \Lambda_4} \left(\frac{1}{2} \sum_{\nu=0}^{3} \left(\frac{1}{a} \widehat{\partial}_\nu \Phi(x) \right)^2 + \frac{1}{2} \mu^2 \Phi(x)^2 + \frac{\lambda}{4!} \Phi(x)^4 \right.$$
$$\left. + \frac{\rho}{8} a^2 \sum_{\nu=0}^{3} \left(\frac{1}{a} \widehat{\partial}_\nu \Phi(x)^2 \right)^2 + \frac{\tau}{6!} a^2 \, \Phi(x)^6 \right) \tag{4.3.103}$$

with dimensionless coupling constants λ, ρ and τ. The propagator is given by the inverse of (4.3.76), so that $\omega_\Phi = (D-2)/2 = 1$. There are three types of vertices,

$$B_0 = \quad \times \quad = -\lambda \,,$$

$$B_1 = \quad \times\!\!\times \quad = -\rho \, a^2 \left[\widehat{(p_1 + p_2)}^2 + \widehat{(p_1 + p_3)}^2 + \widehat{(p_1 + p_4)}^2 \right],$$

$$\tag{4.3.104}$$

$$B_2 = \quad \times\!\!\!\times \quad = -\tau \, a^2 \,.$$

Each vertex has UV-degree four, $\widehat{\omega}_{\mathcal{B}}(B_0) = \widehat{\omega}_{\mathcal{B}}(B_1) = \widehat{\omega}_{\mathcal{B}}(B_2) = 4$. Hence, for all 1PI graphs γ, $\widehat{\omega}(\gamma) = 4 - e(\gamma)$. The theory is (strictly) power-counting renormalizable. Taking into account the internal and the Euclidean lattice symmetries, the required counterterm becomes

$$a^4 \sum_{x \in \Lambda_4} \left[\delta Z_\Phi \frac{1}{2} \sum_{\nu=0}^{3} \left(\frac{1}{a} \widehat{\partial}_\nu \Phi(x) \right)^2 + \frac{1}{2} \delta \mu^2 \Phi(x)^2 + \frac{\delta \lambda}{4!} \Phi(x)^4 \right]. \tag{4.3.105}$$

The three renormalization constants δZ_Φ, $\delta \mu^2$ and $\delta \lambda$ are uniquely fixed by imposing three independent normalization conditions, e.g. by the first two equalities of (4.3.16). The irrelevant coupling constants τ and ρ do not get renormalized. The continuum limit does not depend on them. For instance, the six-point function receives a contribution from the Feynman diagram

$$\tag{4.3.106}$$

with one irrelevant vertex. It has UV-divergence degree less than 0. Hence it vanishes as $a \to 0$, cf. (4.3.95). Also, the three-loop Feynman diagram

$$p \to \qquad\qquad\qquad\qquad\qquad\qquad\qquad\qquad\qquad (4.3.107)$$

with two irrelevant vertices has only an overall divergence with UV-divergence degree 0. The renormalized Feynman integral vanishes as $a \to 0$,

$$
\lim_{a \to 0} \int_{-\pi/a}^{\pi/a} \frac{d^4 k_1}{(2\pi)^4} \frac{d^4 k_2}{(2\pi)^4} \frac{d^4 k_3}{(2\pi)^4} \left(1 - \widehat{T}_p^0 \right) \left(a^2 \tau \right)^2
$$

$$
\cdot \, \frac{1}{\widehat{k_1}^2 + \mu^2} \frac{1}{\widehat{k_2}^2 + \mu^2} \frac{1}{\widehat{k_3}^2 + \mu^2} \frac{1}{\left(k_1 + \widehat{k_2 + k_3} + p \right)^2 + \mu^2}
$$

$$
= \int_{-\infty}^{\infty} \frac{d^4 k_1}{(2\pi)^4} \frac{d^4 k_2}{(2\pi)^4} \frac{d^4 k_3}{(2\pi)^4} \left(1 - T_p^0 \right) \lim_{a \to 0} \left(a^2 \tau \right)^2 \qquad (4.3.108)
$$

$$
\cdot \, \frac{1}{\widehat{k_1}^2 + \mu^2} \frac{1}{\widehat{k_2}^2 + \mu^2} \frac{1}{\widehat{k_3}^2 + \mu^2} \frac{1}{\left(k_1 + \widehat{k_2 + k_3} + p \right)^2 + \mu^2}
$$

$$
= 0 \, .
$$

This example explicitly shows that the six-point vertices do influence the renormalization of the four-point coupling, hence they enter the map between the bare and the renormalized parameters, or, equivalently, change the counterterm coefficients. Otherwise they yield a vanishing contribution to the renormalized four-point function in the continuum limit (as to any other renormalized Green function).

4.3.14 *Massless fields and IR-power counting*

Let us start with an overview of IR-problems, which are encountered in the context of this book, and indicate proposals for their solutions.

IR-problems may occur whenever a propagator has a vanishing mass, or whenever a critical theory shall be described. A critical theory is a theory with vanishing renormalized mass m_R, where the renormalized mass m_R is defined by the renormalized two-point function at zero momentum, for example. In particular, the vanishing of a mass may be a dynamical phenomenon in a second-order phase transition with a divergent correlation length. In a temperature-driven transition this mass becomes a function of temperature. In four dimensions at zero temperature, both for scalar field theories and for gauge theories it is possible to satisfy the normalization condition

$$\widetilde{\Gamma}_R^{(2)}(p = 0) \; = \; m_R \; = \; 0 \qquad\qquad (4.3.109)$$

with $\Gamma_R^{(2)}$ the renormalized two-point vertex function, to all orders of perturbation theory by introducing suitable mass counterterms. The same
recipe is no longer sufficient if we consider a scalar or gauge field theory in
four dimensions at finite temperature or in three dimensions. On the other
hand, one may utilize the mechanism of a dynamical mass generation. Let
us call the dynamically generated renormalized mass $m(T)$. From now on
we drop the index R of the mass. In section 4.4 we will show that a suitable
resummation cures the IR problem in scalar field theories, but not in gauge
theories. The dynamically generated, resummed mass $m(T)$ turns out to
be proportional to $\lambda(T)^{1/2}$, if $\lambda(T)$ denotes the quartic renormalized scalar
self-coupling. The ratio $\lambda(T)/m(T) \sim \lambda(T)^{1/2}$ becomes an effective expansion parameter for a loop expansion. The resummation recipe is sufficient
as long as $m(T) \neq 0$, but $m(T)$ goes to zero at a second-order transition.
Therefore one might think of a coupling resummation in addition with the
aim that $\lambda(T) \to 0$ when $m(T) \to 0$ so that $\lambda(T)/m(T)$ stays finite. We
are not aware of a solution along these lines. An alternative solution is provided by the renormalization group (cf. section 4.7 and 4.6). Again for ϕ^4
theories, but not for gauge theories, it can be achieved that $\lambda_\Lambda(T) \to 0$ with
$m_\Lambda(T) \to 0$. Now $\lambda_\Lambda(T)$ denotes an effective coupling constant depending
on the scale Λ, flowing according to the evolution of the renormalization-
group equations. The failure of the same procedure in gauge theories is due
to the magnetic screening mass that vanishes to one-loop order. Therefore
the ratio of renormalized gauge coupling $g(T)$ and $m(T)$ fails as a suitable
expansion parameter, since $g(T)/m(T) \sim O(1)$.

In this section we discuss IR-problems for field theories at zero temperature. We point out how they are cured by appropriate renormalization prescriptions. In a later chapter we will see that the well established
methods of this section are no longer sufficient for field theories at finite
temperature. In that case, infinite resummations must be applied.

In the first parts of this section, we have considered Feynman integrals
with massive propagators. They are only affected by short-distance or,
equivalently, large-momentum divergencies. Field theories which satisfy
the requirements of power counting renormalizability are rendered free of
singularities by considering their correlation functions as functions of a finite
number of renormalized coupling constants. For instance, in the BPHZ-

approach, these coupling constants are defined by normalization conditions on the vertex functions at zero external momentum.

In the case of a massless field theory, IR-singularities can appear already for a finite UV-cutoff. In momentum space, the propagator of a massless field becomes singular at vanishing line momentum. The singularity should be integrable, otherwise the Feynman integral does not exist. It is well known that IR-convergence depends on the particular values of the external momenta. For instance, consider the following Feynman integral in a massless Φ^4 theory,

$$p \to \bowtie = \int_{-\pi/a}^{\pi/a} \frac{d^D k}{(2\pi)^D} \frac{1}{\widehat{k}^2} \frac{1}{\widehat{(k+p)}^2} . \qquad (4.3.110)$$

For $D \geq 3$, as long as the external momentum $p \neq 0$, the integral is convergent. However, as $p \to 0$, it becomes logarithmically divergent even for finite lattice spacing a. The divergence is an IR singularity related to the small-momentum behavior.

This simple example already shows an important point for massless theories. They require a careful *disentanglement of UV and IR singularities*. In a massive Φ^4-theory the renormalized quartic coupling constant can be defined by the renormalized four-point vertex function at zero momentum, cf. (4.3.16). This amounts to remove the UV-divergencies as $a \to 0$ by a zero momentum subtraction applied to the one loop diagrams, (4.3.23), and similarly for higher order graphs. For the massless theory this is not possible because the four-point function is IR-divergent at zero momentum. Instead, the coupling constant should be defined at some IR-secure normalization point, a so-called nonexceptional momentum.

In generalization of this example, nonexceptional momenta are defined as follows. Let $\widetilde{G}^{(n)}(p_1, \ldots, p_{n-1})$ denote a connected n-point function, where the external momenta are all considered as incoming. The momentum configuration $p_1, \ldots, p_{n-1}$ is called nonexceptional if every partial sum of the p_i is nonvanishing,

$$\sum_{j \in I} p_j \neq 0 \quad \text{for any } I \subseteq \{1, \ldots, n-1\}. \qquad (4.3.111)$$

In the following we outline some peculiarities of IR-power counting and renormalization of massless field theories, but only to the extent that is required for a later understanding of problems arising in finite temperature perturbation theory with massless fields. A finite temperature effectively reduces the IR-dimension by one, this implies an enhancement of IR-singularities compared to zero temperature. Field theories which are IR-finite (renormalizable) at $T = 0$ do not need to stay renormalizable at $T > 0$.

We would like to provide some understanding of the zero-temperature case first. Proving renormalizability of massless field theories proceeds mainly in two steps. The first step is to formulate an IR-power counting theorem. It ensures IR-convergence if a certain set of IR-power counting criteria are satisfied. Secondly, the renormalization procedure must be adjusted in such a way that both the IR- and UV-power counting criteria are satisfied. As we shall see, this imposes strong constraints on the interactions of a massless field theory. If these constraints are matched, one can be sure that the renormalized correlation functions are both IR- and UV-convergent for nonexceptional external momenta. We furthermore keep the lattice as an UV-cutoff, although the formulation of an IR power-counting theorem is not specific for a lattice cutoff - apart from some technicalities. Consider again the momentum space integral $\widehat{\mathcal{I}}$, (4.3.69)ff, where we now allow some of the masses μ_i to vanish. An IR-degree is defined as follows. Let W denote the numerator V or any of the denominators C_i of (4.3.69). Let $\mathcal{L}$ as in (4.3.42) denote the set of loop and line momenta and $u_1, \ldots, u_d$, $v_1, \ldots, v_{m-d}$ be m linearly independent elements of $\mathcal{L}$ with respect to k, so that $k = k(u, v, p)$. As before, variable u and constant v define a Zimmermann subspace H. For a given set of external momenta p, the IR-degree of W with respect to u is written as

$$\underline{\mathrm{degr}}_{\widehat{u}|\widehat{v}} \, W(k(u, v, p), p; \mu, a) \, . \tag{4.3.112}$$

It is defined as the number s_u occurring in the small-u expansion of W according to

$$W(k(\lambda u, v, p), p; \mu, a) \; = \; B(u, v, p; \mu, a) \, \lambda^{s_u} + O(\lambda^{s_u+1}) \quad \text{as } \lambda \to 0, \tag{4.3.113}$$

with $B \not\equiv 0$ as a function of u, v, μ, a, and with p kept fixed.

The IR-divergence degree of $\widehat{\mathcal{I}}$ with respect to H is defined by

$$\underline{\mathrm{degr}}_{\widehat{H}} \, \widehat{\mathcal{I}} \; = \; Dd \; + \; \underline{\mathrm{degr}}_{\widehat{u}|\widehat{v}} \, \widehat{I}(k(u, v, p), p; \mu, a) \tag{4.3.114}$$

with

$$
\underline{\text{degr}}_{\widehat{u}|\widehat{v}}\, \widehat{I}(k(u,v,p),p;\mu,a) = \underline{\text{degr}}_{\widehat{u}|\widehat{v}}\, V(k(u,v,p),p;\mu,a)
$$
$$
- \sum_{i=1}^{r} \underline{\text{degr}}_{\widehat{u}|\widehat{v}}\, C_i(l_i(k(u,v,p),p;\mu_i,a). \tag{4.3.115}
$$

For momentum space integrals $\widehat{\mathcal{I}}$ of the form (4.3.69), with possibly massless propagators, the power counting criteria are as follows. If for all Zimmermann subspaces H

$$
\underline{\text{degr}}_{\widehat{H}}\, \widehat{\mathcal{I}} > 0, \tag{4.3.116}
$$

$\widehat{\mathcal{I}}$ is absolutely convergent for any $a > 0$. If in addition for all H

$$
\overline{\text{degr}}_{\widehat{H}}\, \widehat{\mathcal{I}} < 0, \tag{4.3.117}
$$

the continuum limit of $\widehat{\mathcal{I}}$ exists and can be determined by taking the limit under the integral sign in (4.3.91).

What do theories with massless fields imply for the renormalization procedure? The IR-divergence degrees (4.3.116) depend on the particular values of the external momenta p. A generic renormalization implies IR-convergence only for nonexceptional momenta. One might guess that the only change consists in imposing normalization conditions at nonexceptional momenta instead of zero momentum, for example. Unfortunately the situation is more involved. Actually, for particular vertex functions one is even *forced* to impose normalization conditions at zero momentum.

For example, consider the two-point vertex function $\widetilde{\Gamma}_R^{(2,0)}$ in a four-dimensional Φ^4-theory, represented according to

$$
\equiv \quad \Pi_R^{(2,0)}(p) = \widetilde{\Gamma}_R^{(2,0)} + \widehat{p}^2\,. \tag{4.3.118}
$$

The one- and two-loop contributions are given by

$$
= \quad \text{CT}, \tag{4.3.119}
$$

where CT represents all counterterm contributions. These graphs occur as subdiagrams in higher orders, for instance in the four-point function,

$$(4.3.120)$$

However, because of the two massless propagators attached to $\Pi_R^{(2,0)}(p)$, IR finiteness of this graph requires to choose the counterterm such that $\Pi_R^{(2,0)}(p=0) = \widetilde{\Gamma}_R^{(2,0)}(p=0) = 0$. This amounts to a zero-momentum subtraction. It is equivalent to adjust the bare mass in such a way that the renormalized two-point function vanishes at zero momentum. This corresponds to a requirement on the two-point vertex function at exceptional momenta. In principle, this requirement may be in conflict with the IR-singularities. In particular, the two-point vertex function must be IR-finite at zero momentum.

Let us give an example for such a conflict that finally leads to discard the very theory, namely the Φ^3-theory in four dimensions. The one-loop contribution to the two-point function

$$
\text{---}\bigcirc\text{---} \;=\; \int_{-\pi/a}^{\pi/a} \frac{d^4k}{(2\pi)^4}\, \frac{1}{\widehat{k}^2}\, \frac{1}{\widehat{(k+p)}^2} \tag{4.3.121}
$$

diverges at zero momentum $p = 0$. There is no way to renormalize it such that the two-loop diagram

$$(4.3.122)$$

becomes IR-finite. Although the obstacle shows up only on a diagrammatic level, it makes the full massless Φ^3 theory ill defined, because IR- and UV-singularities cannot be properly disentangled.

The requirement of an IR-finite perturbation theory imposes constraints on the interactions. A sufficient set of conditions is given as follows. The canonical IR-dimension r_Φ of the field Φ is defined by the small-momentum behavior of its free propagator $\widehat{\Delta}(l; \mu, a)$,

$$
\underline{\text{degr}}_{\widehat{l}}\,\widehat{\Delta}(l; \mu, a)^{-1} \;=\; D - 2r_\Phi. \tag{4.3.123}
$$

For instance, for $\widehat{\Delta}(l; \mu, a) = \left(\widehat{l}^2 + \mu^2\right)^{-1}$, we have $r_\Phi = D/2$ if $\mu \neq 0$ and $r_\Phi = (D-2)/2$ for $\mu = 0$. To every vertex B one assigns an IR-degree

$$\widehat{r}_B(B) = r_\Phi \, n(B) + d(B) \, . \tag{4.3.124}$$

As before, $n(B)$ denotes the number of Φ-lines entering the vertex and $d(B)$ the number of lattice momenta at B, cf. (4.3.99). In contrast to the corresponding definition of an UV-degree, factors of the lattice spacing a do not influence the IR-degree of a vertex.

The supplementary requirements of renormalizability of a massless theory are now the following. If $r_\Phi > 0$ and

$$\widehat{r}_B(B) \geq D \tag{4.3.125}$$

for all vertices B, the field theory is IR-finite renormalizable, to all orders of the loop expansion. Universality arguments and the "counterterm philosophy" as described for the massive case can be applied to the massless case without severe problems.

As an example let us consider the four-dimensional lattice Φ^4-theory, with action (4.3.103), and with $\mu = 0$. The propagator is given by $\widehat{\Delta}(l; a) = \left(\widehat{l}^2\right)^{-1}$, so that $r_\Phi = 1$. There are three types of vertices, cf. (4.3.104). According to (4.3.124),

$$\widehat{r}_B(B_0) = 4 \, , \, \widehat{r}_B(B_1) = \widehat{r}_B(B_2) = 6 \, . \tag{4.3.126}$$

Hence the above IR criteria are satisfied. Since the UV conditions on power-counting renormalizability are matched as well, the massless theory is renormalizable. The most general required counterterm is still of the form (4.3.105). The bare mass squared is equal to $\delta\mu^2$ and must be adjusted in such a way that the renormalized two-point vertex function vanishes at zero momentum,

$$\widetilde{\Gamma}_R^{(2,0)}(0; a) = 0. \tag{4.3.127}$$

The other two renormalization constants δZ_Φ and $\delta\lambda$ are defined by imposing normalization conditions at nonexceptional momenta, e.g.

$$\frac{\partial}{\partial p_0} \left. \widetilde{\Gamma}_R^{(2,0)}(p; a) \right|_{p=(\overline{\mu},0,0,0)} = -2\overline{\mu} \, ,$$

$$\widetilde{\Gamma}_R^{(4,0)}(\overline{p}_1, \overline{p}_2, \overline{p}_3; a) = -\lambda \, , \tag{4.3.128}$$

$$\overline{p}_i \cdot \overline{p}_j = \frac{\overline{\mu}^2}{2} \, (4\delta_{ij} - 1)$$

with $\overline{\mu} \neq 0$.

The situation is drastically different in three dimensions, the case that is important for field theories at finite temperature. From the point of view of the IR-behavior, a four-dimensional field theory at finite temperature is expected to share the IR-peculiarities of the corresponding theory in three dimensions. Whereas the massive Φ^4 theory is superrenormalizable in $D = 3$, the massless model suffers from enhanced IR-singularities. In this case, $r_\Phi = 1/2$, and

$$\hat{r}_B \left(B_0 = \times \right) = 2, \qquad (4.3.129)$$

so that the condition (4.3.125) is violated now. The massless theory does not have an IR-finite loop expansion, neither is it made finite by tuning the bare mass to a critical value which would imply that the renormalized two-point vertex function at zero momentum vanishes order by order in perturbation theory, cf. (4.3.127).

Nonrenormalizable IR-divgencies imply that the massless theory is no longer defined in the loop expansion. There may be different reasons for that. One possibility is a *dynamical* mass generation. As we have seen above, the loop expansion only applies if a theory which is massless on the tree level stays massless to all orders. This is achieved by imposing the appropriate normalization conditions on the two-point function. Therefore the dynamical generation of a mass cannot be described by this expansion and, moreover, a massless theory with dynamical mass generation cannot be defined in the loop expansion. In this sense the standard perturbative procedure fails to make such a theory IR-finite. In three-dimensional non-abelian gauge theories, the strong IR singularities go along with a dynamical generation of a mass gap and are often taken as a signal of confinement, although IR-problems need not be correlated to the issue of confinement.

An appealing idea to describe the dynamical generation of a mass gap within the framework of perturbation theory is the application of appropriate resummation techniques. This is justified as long as the renormalized coupling constant defined on the scale of the mass is sufficiently small. The resulting perturbative series are no longer power series in the coupling constant(s). We shall come back to this issue in the next section (4.4) on perturbation theory at finite temperature.

4.3.15   Gauge theories

In the last section we have outlined principles of perturbation theory and renormalization procedures that also apply to pure gauge theories, QCD and other quantum field theories for particle physics. Nonabelian gauge theories with gauge group $SU(N)$ are asymptotically free. The weak gauge-coupling expansion is a powerful method for describing their short-distance behavior. Theoretical predictions of experiments can be extrapolated to the large-cutoff limit so that the renormalization procedure must be invoked before the cutoff is sent to infinity. Thus one must use renormalized perturbation theory.

Gauge theories exhibit all complications which are encountered when massless field theories are renormalized that we have discussed in the last sections. Even more, the subtleties are enhanced by the gauge symmetry. The weak-gauge coupling expansion requires gauge fixing. We outline the essential steps towards the derivation of the gauge-fixed effective action which is used for perturbative computations. Control of the arising complications are best explained on the lattice. In this context, perturbation theory amounts to a saddle-point expansion about the minimum of the action. The following considerations are independent of the inclusion of fermions. For simplicity we consider pure Yang-Mills theories, in $D = 4$ dimensions.

In contrast to the situation in a Φ^4-theory, the minimum of the action becomes degenerated. In the infinite volume, the classical vacuum configurations are given by the pure gauge orbit

$$U(x;\mu) \; = \; \Lambda(x)\,\Lambda(x + a\widehat{\mu})^{-1} \qquad (4.3.130)$$

with $\Lambda(x) \in SU(N)$. In the finite volume, additional degeneracies are possible, depending on the boundary conditions which are imposed. Field configurations which minimize the action enlarge to the so-called toron configurations which contain the pure gauge orbit as a submanifold, cf. section 4.4. In this section we restrict attention to the infinite-volume case. The vacuum manifold (4.3.130) is isomorphic to G/G_0, with

$$G \; = \; \bigotimes_{x \in \Lambda_4} SU(N) \qquad (4.3.131)$$

the group manifold of the gauge transformations and G_0 the submanifold of G which is defined by constant gauge transformations $\Lambda(x) \equiv \Lambda_0 \in SU(N)$. In the following we identify G/G_0 with the subset of G obtained by fixing the global gauge $\Lambda(x_0) = 1$ at some $x_0 \in \Lambda_4$,

$$G/G_0 \;=\; \{\Lambda \in G \mid \Lambda(x_0) = 1\}. \tag{4.3.132}$$

Let $I(U)$ be any gauge invariant observable. The perturbative evaluation of $< I(U) >$ amounts to an asymptotic power series expansion in the gauge coupling g. It is sufficient to integrate only over a small neighborhood Ω of the pure gauge orbit. All other contributions are exponentially suppressed in g.

$$< I(U) > \;\simeq\; \frac{\int_\Omega \mathcal{D}U \; I(U) \; \exp\left(-S_W(U)\right)}{\int_\Omega \mathcal{D}U \; \exp\left(-S_W(U)\right)} \;=\; \frac{[I(U)]}{[1]} \tag{4.3.133}$$

with

$$\mathcal{D}U \;\equiv\; \prod_{x,\mu} dU(x;\mu) \tag{4.3.134}$$

the invariant product-Haar measure. $S_W(U)$ denotes the Wilson action, for example.

All field configurations which are related by a gauge transformation

$$U^\Lambda(x;\mu) \;=\; \Lambda(x)U(x;\mu)\Lambda(x + a\widehat{\mu})^{-1} \tag{4.3.135}$$

are equivalent in the sense that $I(U^\Lambda) \equiv I(U)$. In order to lift the degeneracy we have to employ an appropriate parametrization of Ω. A gauge fixing procedure must be invoked such that the gauge degrees of freedom "factorize out". The basic idea of gauge fixing is to identify physical and gauge degrees of freedom first in a small neighborhood of $U(x;\mu) \equiv 1$. This neighborhood is then extended to include gauge transformations out of a finite neighborhood Ω of the gauge orbit. Here we give only a rough outline of the gauge-fixing procedure. The easiest way to understand it is by means of the Faddeev-Popov trick.

Let $\mathcal{G}$ denote the set of Lie-algebra $(su(N)$-)valued fields ω defined on the sites of the lattice Λ_4,

$$\omega : x \in \Lambda_4 \;\rightarrow\; \omega(x) \in su(N). \tag{4.3.136}$$

A gauge-fixing function F is a differentiable map

$$F : \Omega_1 \;\rightarrow\; \mathcal{G}, \tag{4.3.137}$$

where Ω_1 is a small neighborhood of all $U(x;\mu) \equiv 1$. Furthermore, let $L(U) : \mathcal{G} \to \mathcal{G}$ be the differential of F under infinitesimal gauge transformations of G/G_0. The gauge-fixing function F is called admissible if the following conditions hold.

- F is invariant under global gauge transformations $\Lambda(x) \equiv \Lambda_0$, that is

$$F(U^{\Lambda_0}) = F(U). \tag{4.3.138}$$

- For almost all sufficiently small $X \in \mathcal{G}$ and almost all $U \in \Omega$ there is a unique $\Lambda \in G/G_0$ such that

$$U^\Lambda \in \Omega_1 \quad \text{and} \quad F(U^\Lambda) = X. \tag{4.3.139}$$

- For almost all $U \in \Omega_1$, i.e. except for a set of U of relative measure zero, $\det L(U) \neq 0$.

In a sufficiently small neighborhood Ω_1 of $U(x;\mu) \equiv 1$ we can parameterize

$$U(x;\mu) = \exp a A_\mu(x) \tag{4.3.140}$$

in terms of Lie algebra-valued vector fields $A_\mu(x)$.

An example of an admissible gauge-fixing function is given by the Lorentz gauge

$$F(\exp aA) = \sum_{\mu=0}^{3} \frac{1}{a} \, \widehat{\partial}_\mu^* A_\mu(x). \tag{4.3.141}$$

Under an infinitesimal gauge transformation

$$\Lambda(x) = \exp \omega(x) \simeq 1 + \omega(x) , \tag{4.3.142}$$

the gauge field $A_\mu(x)$ transforms according to

$$e^{A_\mu(x)+\delta_\omega A_\mu(x)} = e^{\omega(x)} e^{A_\mu(x)} e^{\omega(x+a\widehat{\mu})} , \tag{4.3.143}$$

or

$$\delta_\omega A_\mu(x) = -\left(\operatorname{ad} A_\mu(x) + \frac{\operatorname{ad} A_\mu(x)}{1 - e^{-\operatorname{ad} A_\mu(x)}} \frac{1}{a} \, \widehat{\partial}_\mu \right) \omega(x). \tag{4.3.144}$$

For the Lorentz gauge (4.3.141), $L(U)$ is then given by

$$L(e^{aA}) = -\sum_{\mu=0}^{3} \frac{1}{a} \, \widehat{\partial}_\mu^* \left(\operatorname{ad} A_\mu(x) + \frac{\operatorname{ad} A_\mu(x)}{1 - e^{-\operatorname{ad} A_\mu(x)}} \frac{1}{a} \, \widehat{\partial}_\mu \right). \tag{4.3.145}$$

Another important gauge which we shall later encounter in finite-temperature perturbation theory is the static, time-averaged Landau gauge (STALG).

Let $\chi_{\Omega_1}(U)$ be the characteristic function of Ω_1, i.e.

$$\chi_{\Omega_1}(U) = \begin{cases} 1, & \text{if } U \in \Omega_1, \\ 0, & \text{otherwise.} \end{cases} \tag{4.3.146}$$

The Faddeev-Popov trick amounts to inserting the identity

$$E_{\Omega_1, X}(U) \equiv \int \mathcal{D}'\Lambda\, \chi_{\Omega_1}(U^\Lambda)\, \delta(F(U^\Lambda) - X)\, \det \mathrm{L}(U^\Lambda) \,,$$

$$\mathcal{D}'\Lambda = \prod_{x \neq x_0} d\Lambda(x) \tag{4.3.147}$$

under the integral signs of (4.3.133). $E_{\Omega_1, X}(U)$ is equal to 1 apart from a set of gauge field configurations U that are of relative measure zero with respect to $\mathcal{D}U$. Hence the insertion of (4.3.147) does not change the integrals $[I(U)]$ and [1] defined in (4.3.133). We obtain

$$[I(U)] = \int_\Omega \mathcal{D}U\, I(U)\, \exp\left(-S_W(U)\right)$$

$$= \int_\Omega \mathcal{D}U \int \mathcal{D}'\Lambda\, \chi_{\Omega_1}(U^\Lambda)\, \delta(F(U^\Lambda) - X)\, \det \mathrm{L}(U^\Lambda)\, \exp\left(-S_W(U)\right)$$

$$= \int_\Omega \mathcal{D}U\, \chi_{\Omega_1}(U)\, \delta(F(U) - X)\, \det \mathrm{L}(U)\, \exp\left(-S_W(U)\right). \tag{4.3.148}$$

For the last equality we have applied a gauge transformation $U \to U^{\Lambda^{-1}}$, used gauge invariance of $I(U)$ and $S_W(U)$ and the fact that

$$\int d\Lambda(x) = 1. \tag{4.3.149}$$

In Ω_1 we parameterize U by (4.3.140). This yields

$$\mathcal{D}U = \prod_{x, \mu} dA_\mu(x)\, \exp\left(-S_m(A)\right) \tag{4.3.150}$$

with S_m the measure part of the effective action,

$$S_m(A) = -\sum_{x, \mu} \mathrm{tr} \log \frac{1 - e^{-\operatorname{ad} aA_\mu(x)}}{\operatorname{ad} aA_\mu(x)} \,. \tag{4.3.151}$$

The integration range of the Lie-algebra valued gauge fields $A_\mu(x)$ can be extended to $\pm\infty$, the corrections are exponentially suppressed in the gauge coupling g. The Jacobian $\det L(\exp aA)$ is the Faddeev-Popov determinant. Introducing the Grassmann- and Lie-algebra valued Faddeev-Popov ghost fields c and $\bar{c}$, we write

$$\det L(e^{aA}) = \text{const} \int \prod_x (dc(x)d\bar{c}(x)) \exp\left(-S_{FP}(A,c,\bar{c})\right) \quad (4.3.152)$$

with the Faddeev-Popov action

$$S_{FP}(A,c,\bar{c}) = a^4 \sum_x \text{tr}\,\bar{c}(x)\left(L(e^{aA})c\right)(x). \quad (4.3.153)$$

Finally, the δ-function constraint is raised to the action in the usual way. $I[U]$ does not depend on the particular X. Averaging over $X \in \mathcal{G}$ with a Gaussian weight,

$$[I(U)] = \text{const} \int \prod_x dX(x)\,\exp\left(\frac{\lambda}{g^2}a^4 \sum_x \text{tr}\,X(x)^2\right)[I(U)], \quad (4.3.154)$$

where $\lambda > 0$ is some gauge fixing parameter, we obtain upon integration over X

$$[I(U)] = \text{const} \int \prod_x \left(dc(x)d\bar{c}(x)\prod_\mu dA_\mu(x)\right) I(e^{aA})\, e^{-S_0(A,c,\bar{c})} \quad (4.3.155)$$

with action S_0 given by

$$S_0(A,c,\bar{c}) = S_W(e^{aA}) + S_m(A) + S_{FP}(A,c,\bar{c}) + S_{gf}(A), \quad (4.3.156)$$

where the gauge fixing part S_{gf} of the action reads

$$S_{gf}(A) = -\frac{\lambda}{g^2}\,a^4 \sum_x \text{tr}\,F(e^{aA})(x)^2. \quad (4.3.157)$$

The same arguments which we have used for the derivation of $[I(U)]$ apply to the denominator [1] of $< I(U) >$ in (4.3.133), with the same proportionality constant in front of the integral.

Let us give an explicit example for the gauge-fixed effective action in the Lorentz gauge (4.3.141). Up to exponentially suppressed contributions in g we have for any gauge invariant observable $I(U)$

$$< I(U) > \simeq < I(U) >_{S_0}, \quad (4.3.158)$$

with

$$< I(U) >_{S_0} = \frac{\int \prod_x \left(dc(x) d\bar{c}(x) \prod_\mu dA_\mu(x) \right) I(e^{aA}) e^{-S_0(A,c,\bar{c})}}{\int \prod_x \left(dc(x) d\bar{c}(x) \prod_\mu dA_\mu(x) \right) e^{-S_0(A,c,\bar{c})}}.$$

$$(4.3.159)$$

The action is given by (4.3.156) with S_m and S_W as before, the gauge-fixing term given by

$$S_{gf}(A) = -\frac{\lambda}{g^2} a^4 \sum_x \text{tr} \left(\sum_\mu \frac{1}{a} \widehat{\partial}_\mu^* A_\mu(x) \right)^2,$$

$$(4.3.160)$$

and the Faddeev-Popov action given by

$$S_{FP}(A, c, \bar{c}) = -a^4 \sum_x \text{tr} \Bigg\{ \bar{c}(x) \sum_{\mu=0}^{3} \frac{1}{a} \widehat{\partial}_\mu^* \Bigg[\text{ad} \, A_\mu(x)$$

$$+ \frac{\text{ad} \, A_\mu(x)}{1 - e^{-\text{ad} \, A_\mu(x)}} \frac{1}{a} \widehat{\partial}_\mu \Bigg] c(x) \Bigg\}.$$

$$(4.3.161)$$

The relation (4.3.159) for the gauge invariant observable $I(U)$ can be taken as the starting point for perturbative computations. Because of gauge fixing, non-gauge-invariant quantities acquire nonvanishing expectation values. In particular, the two-point functions

$$< A_\mu(x_1) A_\nu(x_2) >_{S_0} \quad \text{and} \quad < c(x_1) \bar{c}(x_2) >_{S_0} \qquad (4.3.162)$$

have power series expansions in the gauge coupling g. Their leading terms are given by the free-field propagators. Similar expansions hold for the other basic field-correlation functions of A_μ, c and $\bar{c}$. The coefficients are finite sums of Feynman diagrams.

Before the continuum limit can be taken the renormalization procedure must be invoked. The power-counting theorem provides strong constraints on the general structure of the counterterms needed to render a quantum field theory UV finite. However, it does not tell us whether the theory can be renormalized in a gauge invariant (BRS-invariant) way. To this end, the symmetry of the regularized field theory itself must be taken into account. This concerns external symmetries such as invariance under Euclidean transformations, as well as internal symmetries. Invariance of the regularized theory under a continuous internal symmetry transformation manifests itself in terms of Ward identities. Ward identities are integro-differential equations that are satisfied by the correlation functions. In many cases it is convenient to write them in terms of vertex functions and their generating functional.

Ward identities imply further stringent constraints on the counterterms of the theory. The central question then is as to whether the symmetry is preserved under renormalization in the sense that the vertex functions of the renormalized theory satisfy the corresponding Ward identities. This is a nontrivial problem in particular for gauge theories which are subject to nonlinear and/or local symmetries.

Because of gauge fixing, gauge invariance is lost in perturbation theory. However, the gauge-fixed effective action and the measure exhibit invariance under BRS (Becchi-Rouet-Stora) transformations. These are nonlinear, Grassmann-like and nilpotent transformations that look like infinitesimal gauge transformations for the gauge fields. The ghost fields c and $\bar{c}$ transform in such a way that the sum of the gauge fixing part S_{gf} and the Faddeev-Popov part S_{FP} of the action stay invariant. Such a BRS-symmetry exists for every gauge imposed by an admissible gauge-fixing function.

Power counting and BRS-symmetry together impose strong constraints on the form of the renormalized theory. For the action S_0 of (4.3.156) one can prove that renormalization is multiplicative. This means that the renormalized action S_R is obtained from the bare one by renormalizing the fields and the coupling constant according to

$$S_R(A, c, \bar{c} \mid g, \lambda) = S_0(Z_A^{1/2} A, Z_c^{1/2} c, Z_c^{1/2} \bar{c} \mid Z_g g, Z_g^2 Z_A^{-1} \lambda). \quad (4.3.163)$$

In (4.3.163) we have indicated explicitly the dependence of the action on the gauge coupling g and the gauge fixing parameter λ. All basic field-correlation functions computed with S_R become UV-finite, to all orders of the weak coupling expansion in g_0.

The renormalized theory is still invariant under a (multiplicatively renormalized) BRS-transformation. The renormalization constants are uniquely fixed by imposing three independent normalization conditions. Because of these normalizations, the continuum limit of the vertex functions is universal. In particular, it does not depend on the particular choice of the Wilson action.

We remark that this resolves the problem of a possible tuning of a mass term for the gauge field mentioned in the last section. On the one hand, power counting alone does not exclude the occurrence of a mass term, but such a mass term would violate BRS-symmetry. The Ward identities associated with the BRS-symmetry forbid such a mass term. On the other hand, it is then the BRS-symmetry of the lattice theory which ensures that such a term is actually not required for renormalization. Therefore gauge theories (with massless gauge bosons) in four dimensions can be made UV-finite without introducing mass counterterms as in Φ^4 and without introducing intractable IR-problems.

4.4   Weak coupling expansion at finite temperature

4.4.1   *Motivation and problems*

As outlined in the introduction, a characteristic difference between phase transitions in particle physics and usual systems of condensed matter systems are several typical length scales that characterize the high-temperature phase. An essential task is to identify these length scales. Usually the various scales are taken from the classification that is suggested by perturbation theory, more precisely by scales at which perturbation theory breaks down to a certain degree. There is first the scale of the order of the temperature T which is available to the weak coupling expansion. Furthermore there are the regions of the electric and magnetic screening masses of order gT and g^2T, respectively. Because of the UV-asymptotic freedom of nonabelian gauge theories naive perturbation theory in g applies to high temperatures *and small* distances. Screening properties, however, refer to long-distance properties. Therefore even at high temperatures we must distinguish perturbative and nonperturbative regions. This makes the nonabelian plasmas of QCD and the electroweak theory so challenging to understand.

After all, perturbation theory (combined with resummation techniques) is one of the major analytical tools to investigate the high-temperature phases. It was extensively used over the last decades.

In principle we have to distinguish two tasks. The first one refers to "standard" perturbation theory at high temperatures and short distances. It turns out to be "nonstandard" because of the fact that the IR-problems are those of effectively three-dimensional theories. What is needed are infinite resummations. The second one refers to an extension of the range of applicability of perturbation theory. Also this extension goes along with infinite resummations. In the following we discuss both topics in more detail.

In asymptotically free field theories like QCD, the weak coupling expansion is one of the most powerful tools for investigating the short distance properties. Let us assume that this feature remains valid at finite temperature. At finite temperature we have to deal in principle with two independent scales which enter a renormalized running coupling, the *scales of T and the spatial distance R*. Perturbation theory then applies for high T and momenta large compared to the temperature or, equivalently, scales R small compared to the inverse temperature.

It is challenging to use perturbation theory also for larger distances; this amounts to an extension of the applicability range of the short-distance expansion. However, if one leaves the validity range of the loop expansion, infinite resummations become obligatory. But even without this extension, perturbation theory in finite-temperature QCD is delicate and rather nontrivial. The reason is that short- and long-range scales are not properly disentangled in the very loop expansion, a phenomenon that is manifest in IR-divergencies of Feynman integrals even for high momenta.

As we have seen in the previous section, already at zero temperature perturbation theory for massless fields requires a special procedure in order to properly disentangle UV and IR properties. In the loop expansion, Feynman integrals with massless propagators require special IR renormalizations to become IR finite. For instance, the bare mass must be tuned in such a way that the two-point vertex function of the massless field vanishes at zero momentum, to all orders of the loop expansion. This condition may interfere with strong IR-divergencies because zero momenta are exceptional. It implies strong restrictions on the possible interactions of a massless field theory. They are phrased in terms of IR-power counting conditions. If these criteria are satisfied, the large and small momentum scales can be properly disentangled. If in addition certain UV power-counting criteria hold, the field theory becomes IR-finite and renormalizable.

The constraints which must be imposed on the interactions depend, however, on the very dimension of space. Whereas the massless Φ^4-theory in four dimensions is renormalizable in perturbation theory, it is nonrenormalizable in three dimensions. IR-singularities are considerably enhanced in lower dimensions. For zero-temperature QCD, gauge invariance (or BRS-invariance of the gauge fixed theory, respectively) ensures that the gluons stay massless to all orders, and the theory is renormalized in an IR finite way, without adding a mass counterterm to the action. At finite temperature this behavior drastically changes. From the point of view of IR power counting, the IR-singularities are related to those of three-dimensional QCD, but QCD in three dimensions does not satisfy the criteria for IR-finite renormalizability. Actually, the loop expansion in the infinite volume breaks down, on all scales.

Promising ideas to circumvent these problems are separation of scales and their associated degrees of freedom in a way which does not rely on perturbative approaches. The most interesting ones are dimensional reduction and the renormalization group. These techniques are described in very detail in later sections. Once the separation is achieved, different methods can be applied to the various scales. For instance, perturbation theory can be applied to the short-distance region, while large distances can be treated by Monte Carlo methods.

In this section we focus on purely perturbative approaches. There are various attempts to cure the IR-problems in perturbation theory. A systematic approach should proceed along the following lines. The first step is to introduce an IR-cutoff. This can be achieved by introducing masses for all fields or introducing a finite volume in such a way that the symmetries of the theory are preserved. Together with the UV-regularization, the field theoretical framework becomes well defined. In particular, the standard perturbative expansion applies, and the renormalization prescription is invoked along the usual lines. The resulting perturbative series turns out to be nonuniform in the IR-cutoff in the sense that we cannot remove the IR-cutoff order by order in the expansion without re-encountering IR-divergencies. This is then the place where resummation should be applied. In the Feynman diagrammatic approach resummation usually implies a summation over particular subsets of Feynman diagrams. The resulting resummed perturbation series should become uniform in the IR-cutoff, so that the cutoff can be finally removed without introducing new IR-divergencies. In contrast to the original expansion, we would expect for the resummed expansion that an infinite number of Feynman diagrams contributes to a given order in the coupling constant as it usually does for resummed series. A systematic approach along the above lines is still missing. It requires the construction of a perturbation series which works to all orders of the gauge coupling and becomes free of IR-divergencies in the infinite volume. There are, however, promising steps in this direction.

- **Finite-volume perturbation theory.** In finite-volume perturbation theory an IR-cutoff is provided by the finite volume so that pure Yang-Mills theories and QCD are considered in the finite volume as a first step. The volume cutoff is gauge invariant and not restricted to perturbative methods. On the lattice, the weak coupling expansion is recovered as the saddle-point expansion about the minimum of the action. Without any resummation, one obtains a short-distance perturbation theory which is free of IR-singularities. Correlation functions decay with powers of the distance in this region. Exponential decay and screening masses belong to distances above the perturbative horizon in this scheme. They are conveniently determined by means of Monte Carlo methods once it is known where the power-like perturbative regime is left.

The price one has to pay in this approach are supplementary zero modes apart from the pure gauge orbit. They behave in a rather complicated way. In the so-called coupled large-volume/small-gauge coupling expansion [175] they are systematically included in the loop expansion. The volume must satisfy an upper bound as a function of the gauge coupling. This bound is still sufficiently large so that both the perturbative and the screening regimes fit into it.

- **Resummation in the infinite volume.** The starting point here is the formal perturbation series in the infinite volume. IR-divergencies which are encountered at some order are removed by summation of appropriate Feynman diagrams. Typically this procedure generates masses dynamically for the *a priori* massless modes. If all particles acquire a mass the IR-divergencies disappear.

In this approach, it is a nontrivial task to verify that the new, resummed expansion still generates an expansion for small gauge coupling. The reason is that the resummed, thermal masses are small, they are suppressed by some power of the gauge coupling, because they are generated by quantum corrections. If they turn out to be too small the perturbative expansion breaks down again, at least at some higher order. This is actually what happens in QCD and what is known under the name of the magnetic-mass problem. To avoid this problem, more sophisticated resummations are required than those which are available to date. For instance, one should introduce classes of momentum-space Feynman integrals which are integrated only over some subsets in momentum space. In some sense this would realize the separation of scales that we have mentioned above.

In practice the available techniques are well suited to remove the IR-divergencies which are encountered to some low orders of the expansion. For explicit computations they are quite useful. It is an open problem which of these methods corresponds to a systematic weak coupling expansion in the sense that the divergencies are removed to all orders. The question of the extension to *all* orders is less academic than one may think at a first view. It is required in order not to invalidate the results obtained at lower orders. (If the corrections which were neglected so far turn out to be nonrenormalizable and divergent in the end, the finite leading terms may be meaningless numbers in the worst case.)

In this section we start with some preliminaries for perturbative computations at finite temperature. We then outline resummation techniques in some more detail. A thorough description of all perturbative approaches which are currently used would fill a book on its own. We therefore restrict the discussion to the basic ideas of resummation and to the problems one has to face. Finally we give a brief introduction to hard-thermal loop expansions [180]. Hard-thermal loop expansions do not attempt to solve the magnetic-mass problem, they were developed to improve the convergence of perturbation theory at low orders in nonabelian gauge theories and to derive an effective action for the interaction of quasiparticles in a nonabelian plasma.

4.4.2 Finite-temperature Feynman rules and renormalization

In momentum space, Feynman rules at finite temperature differ from their zero temperature counterparts in the fact that the momenta of lines or propagators are properly constrained. The components of the momenta in temperature direction are discrete. Their values depend on the boundary conditions which are imposed on the fields in this direction in configuration space. We identify this direction with the 0th one. Fields which are subject to periodic boundary conditions like scalar and gauge fields have the Fourier representation

$$\Phi(x) = T \sum_{k_0} \int \frac{d^{D-1}\mathbf{k}}{(2\pi)^{D-1}} e^{ik\cdot x} \widetilde{\Phi}(k), \qquad k_0 = 2n\pi T, n \in \mathbf{Z}. \quad (4.4.1)$$

k_0 is an even multiple of πT, including 0. On the other hand, fermions are subject to anti-periodic boundary conditions. This implies for their Fourier representation

$$\Psi(x) = T \sum_{k_0} \int \frac{d^{D-1}\mathbf{k}}{(2\pi)^{D-1}} e^{ik\cdot x} \widetilde{\Psi}(k), \qquad k_0 = (2n+1)\pi T, n \in \mathbf{Z}, \quad (4.4.2)$$

k_0 is an odd multiple of πT. In particular, k_0 is always different from 0.

It is convenient to introduce the notions of static and nonstatic modes. Modes with vanishing k_0, i.e. with $k_0 = 0$, are called static. They describe configurations that are constant in T-direction. All other modes, with $k_0 \neq 0$, are called nonstatic. These modes have an intrinsic IR-cutoff proportional to T. For massless fields the IR-problematic modes are the static ones. They are purely bosonic, since anti-periodicity implies that fermions are purely nonstatic with an IR-cutoff πT.

Except for this modification of discretizing the thermal momenta, the other Feynman rules remain the same at finite temperature as those at $T = 0$. In particular, the topological properties of Feynman diagrams such as loop counting and topological symmetry numbers remain unchanged. Although finite temperature amplitudes become products of momentum sums and integrals, we follow the convention and keep the notion of Feynman integrals.

For example, the four-dimensional scalar Φ^4 theory at finite T is described by the action

$$S = \int_0^{1/T} dx_0 \int d^3x \left[\frac{1}{2} m^2 \Phi(x)^2 + \frac{1}{2} \sum_{\mu=0}^{3} (\partial_\mu \Phi)(x)^2 + \frac{1}{4!} \lambda \Phi(x)^4 \right] \quad (4.4.3)$$

with Φ subject to periodic boundary conditions

$$\Phi(x + nT^{-1}(1, \mathbf{0})) = \Phi(x), \qquad n \in \mathbf{Z}. \quad (4.4.4)$$

For simplicity we regularize the model by a simple momentum cutoff Λ_0. In momentum space, the Feynman rules are

$$\bullet\!\!-\!\!\!-\!\!\!-\!\!\!-\!\!\!-\!\!\bullet \quad = \quad \frac{1}{p^2 + m^2}, \qquad p^2 = p_0^2 + \mathbf{p}^2, \quad (4.4.5)$$

$p^2 \leq \Lambda_0^2$, for the propagator and

$$\times \quad = \quad -\lambda \quad (4.4.6)$$

for the four-point vertex.

For gauge theories, some care is required if we impose a gauge-fixing procedure that is T-dependent itself. In this case, the propagators in momentum space become explicitly T-dependent. An example of such a gauge is the static, time averaged Landau gauge (STALG). It turns out to be very convenient for finite-T calculations in gauge theories. We shall discuss this gauge below in the framework of Yang-Mills theories in a finite volume.

Needless to say, regularization and renormalization stay important issues at finite temperature as at zero temperature. The renormalization procedure must be invoked before the large UV-cutoff limit can be taken.

Recall that at zero temperature, UV-divergencies are controlled by UV-power counting theorems. These theorems characterize UV-divergencies of momentum space Feynman integrals in terms of divergence degrees. In turn, the renormalization procedure amounts to adding local counterterms to the regularized action such that the correlation functions stay finite in the large-cutoff limit, order by order in perturbation theory. The regularized action with all counterterms included then composes the bare action, expressed in terms of renormalized coupling constants and fields. The counterterms are uniquely determined by imposing normalization conditions on renormalized correlation functions. Renormalizability then implies that only a finite number of normalization conditions must be imposed.

A similar procedure should work at finite temperature, at least in the case where all fields have a mass gap. At zero temperature strong theorems on power counting and the renormalization procedure are available. Actually, a generalization of these statements to field theories at finite temperature along the same lines is still missing. It would require, for example, a generalization of the power-counting theorems for momentum-space Feynman integrals to discrete momentum sums. By different methods (flow equations based on the perturbative renormalization group), but on the same level of rigor, a proof of renormalizability of massive scalar field theories at *finite* temperature was only recently given [176].

On the other hand, the experience from explicit computations to leading order in field theories tells us that the local counterterms which render a field theory UV-finite at zero T make it finite at any finite temperature $T > 0$ as well. This means that the UV-divergencies of the bare theory are actually T-independent.

Let us consider the above example of the Φ^4-theory. As we have seen in the previous section, the model requires mass-, wave-function and coupling-constant renormalization. The most general required counterterms to the action have the form

$$\delta S = \int_0^{1/T} dx_0 \int d^3x \left[\frac{1}{2}\delta m^2 \Phi(x)^2 + \frac{1}{2}\delta Z \sum_{\mu=0}^{3}(\partial_\mu \Phi)(x)^2 + \frac{1}{4!}\delta\lambda\Phi(x)^4 \right].$$

(4.4.7)

The action $S + \delta S$ with S given by (4.4.3) then induce the renormalized vertex functions $\widetilde{\Gamma}_R^{(n)}$. We fix the counterterm δS uniquely by imposing the zero temperature normalization conditions

$$\widetilde{\Gamma}_R^{(2)}(p|T=0) = -(m^2 + p^2 + O(p^4)) \quad \text{as } p \to 0,$$

$$\widetilde{\Gamma}_R^{(4)}(p=0|T=0) = -\lambda$$

(4.4.8)

with renormalized, zero temperature mass m and coupling constant λ. We know that at zero temperature and to all orders in λ, all vertex functions $\widetilde{\Gamma}_R^{(n)}$ are UV-finite. Let us check to one loop explicitly that the above vertex functions are UV-convergent at finite T as well. We have

$$\widetilde{\Gamma}_R^{(2)}(p|T) = -(p^2 + m^2) + \;\bigcirc\; - \delta m^2 + O(\lambda^2)$$

(4.4.9)

$$= -(p^2 + m^2) - \frac{\lambda}{2}T\sum_{k_0}\int^{\Lambda_0} \frac{d^3k}{(2\pi)^3}\frac{1}{k^2 + m^2} - \delta m^2 + O(\lambda^2).$$

The normalization condition (4.4.8) requires that

$$\delta m^2 = \;\bigcirc\Big|_{T=0} = -\frac{\lambda}{2}\int^{\Lambda_0}\frac{d^4k}{(2\pi)^4}\frac{1}{k^2 + m^2},$$

(4.4.10)

so that

$$\tilde{\Gamma}_R^{(2)}(p|T) = -(p^2 + m^2) \ + \ \bigcirc \ - \ \bigcirc\Big|_{T=0} + O(\lambda^2)$$

$$= -(p^2 + m^2) - \frac{\lambda}{2}\Bigg[T\sum_{k_0}\int^{\Lambda_0}\frac{d^3k}{(2\pi)^3}\frac{1}{k^2+m^2}$$

$$-\int^{\Lambda_0}\frac{d^4k}{(2\pi)^4}\frac{1}{k^2+m^2}\Bigg] + O(\lambda^2)$$

$$= -(p^2 + m^2) - \frac{\lambda}{2}\Bigg[\frac{T^2}{12} - \frac{mT}{4\pi} - \frac{m^2}{8\pi^2}\left(\ln\frac{m}{4\pi T} - \frac{1}{2} - \psi(1)\right)$$

$$+ O(\Lambda_0^{-1}, \frac{m^4}{T^2})\Bigg] + O(\lambda^2).$$

For the third equality sign we have used the expansion in small ratios m/T. Similarly, for the four-point function at zero momentum we obtain

$$\delta\lambda \ = 3 \ \bowtie\Big|_{p=T=0} = \frac{3\lambda^2}{2}\int^{\Lambda_0}\frac{d^4k}{(2\pi)^4}\frac{1}{(k^2+m^2)^2}, \tag{4.4.11}$$

hence

$$\tilde{\Gamma}_R^{(4)}(0,0,0|T) = \ \times \ + \ \left(\ \bowtie \ + 2\text{ perm}\right)_{p=0}$$

$$-3 \ \bowtie\Big|_{p=T=0} + O(\lambda^3) \tag{4.4.12}$$

$$= -\lambda + \frac{3\lambda^2}{2}\Bigg[T\sum_{k_0}\int^{\Lambda_0}\frac{d^3k}{(2\pi)^3}\frac{1}{(k^2+m^2)^2}$$

$$-\int^{\Lambda_0}\frac{d^4k}{(2\pi)^4}\frac{1}{(k^2+m^2)^2}\Bigg] + O(\lambda^3)$$

$$= -\lambda + \frac{3\lambda^2}{2}\Bigg[\frac{T}{8\pi m} + \frac{1}{8\pi^2}\left(\ln\frac{m}{4\pi T} - \psi(1) - \frac{\zeta(3)m^2}{8\pi^2 T^2}\right)$$

$$+ O(\Lambda_0^{-1}, \frac{m^4}{T^4})\Bigg] + O(\lambda^3).$$

The computations above verify that at least to one loop the renormalized vertex functions are made UV-finite by the zero temperature counterterms.

We can use the above relations to obtain the temperature behavior of renormalized mass and coupling constant, as predicted by perturbation theory. With the definitions

$$\widetilde{\Gamma}_R^{(2)}(0, \mathbf{p}|T) = -(m(T)^2 + \mathbf{p}^2 + O(\mathbf{p}^4)) \quad \text{as } \mathbf{p} \to 0,$$

$$\widetilde{\Gamma}_R^{(4)}(p = 0|T) = -\lambda(T) , \qquad (4.4.13)$$

we get to one loop

$$m(T)^2 = \frac{\lambda(T)}{2} \left(\frac{T^2 - T_0^2}{12} - \frac{m(T)}{4\pi}(T - T_0) - \frac{m(T_0)^2}{8\pi^2} \ln \frac{T}{T_0} \right.$$
$$\left. + O(\frac{m(T_0)^4}{T^2}) \right) \qquad (4.4.14)$$

and

$$\lambda(T) = -\frac{3\lambda(T_0)}{2} \left(\frac{T - T_0}{8\pi m(T_0)} \right.$$
$$+ \frac{1}{8\pi^2} \left[\ln \frac{T}{T_0} - \frac{\zeta(3)m(T_0)^2}{8\pi^2} \left(\frac{1}{T^2} - \frac{1}{T_0^2} \right) \right]$$
$$\left. + O(\frac{m(T_0)^4}{T^4}) \right), \qquad (4.4.15)$$

where $m(T_0)$ and $\lambda(T_0)$ are renormalized mass and renormalized coupling constant at temperature T_0.

The T-independence of the divergent parts of the counterterms allows one to calculate the T-dependence of correlation functions within renormalized perturbation theory. This feature is analogous to zero temperature, where the divergent part of the counterterms is independent of the choice of the normalization point. Therefore correlation functions in terms of renormalized parameters are finite on all momentum scales once they are finite at one scale, and the flow of couplings as a function of scale can be calculated within perturbation theory. Back to finite temperature: once the correlation functions in terms of renormalized parameters are finite at one temperature they are finite for all temperatures. As to whether perturbation theory is actually reliable over the whole temperature range from low to high temperatures is a different question. Indeed there is no reason to expect that at all. But it implies that the renormalized coupling constants can be defined at some fixed temperature T_0 (where perturbation theory is applicable) as an external input. In terms of T_0 the theory becomes UV-finite not just at T_0, but at any T.

The T-independence of the UV-divergencies should extend to all higher orders. We expect this extension whenever the IR-divergencies can be properly removed such as for massive field theories and for gauge theories in a finite volume.

In the following we assume that in the IR-safe case, power-counting criteria and renormalizability generalize to finite T in the sense that they hold under the same conditions at finite T as for zero T whenever there is an IR-cutoff.

4.4.3   *IR-divergencies and resummation techniques*

In this section we explain the idea behind partial resummations of perturbation series to circumvent their IR-problems for massless field theories.

As we have seen already at zero temperature, perturbation theory for massless fields requires special care. The free propagator becomes singular at zero momentum. Particular normalization conditions must be imposed in order to avoid IR-problems. In the previous section on perturbative renormalization at zero temperature we have outlined general conditions which interactions must satisfy so that momentum-space Feynman integrals become IR-finite. Convergence holds in general only for nonexceptional external momenta. This fact must be taken into account in the renormalization procedure. In order to properly disentangle UV- and IR-singularities, renormalized coupling constants must be defined at nonexceptional momenta. In addition, supplementary IR-renormalizations must be imposed. For instance, the four-dimensional Φ^4-theory requires the particular normalization condition that the renormalized two-point vertex function vanishes at zero momentum, i.e.

$$\widetilde{\Gamma}_R^{(2)}(p = 0 | T = 0) \; = \; 0. \qquad (4.4.16)$$

This means that the mass counterterm must be chosen in such a way that the renormalized mass vanishes, to every order of perturbation theory. In QCD, the very gauge invariance ensures that the supplementary constraints are fulfilled without tuning the bare parameters. In particular, no mass counterterm is required for the gluons.

At finite temperature, the analogous procedure to zero temperature does no longer lead to a cancellation of IR-divergencies in general. Again it is instructive to see how this happens at a simple example like the massless

Φ^4-theory. The action is given by $S + \delta S$, with S from (4.4.3), where we set $m = 0$. The counterterm part of the action is given by (4.4.7). We assume that the counterterms were chosen in such a way that the following (conventional) zero-temperature normalization conditions are satisfied

$$\widetilde{\Gamma}_R^{(2)}(p = 0 | T = 0) = 0,$$

$$\frac{\partial}{\partial p_0} \widetilde{\Gamma}_R^{(2)}(p | T = 0)\Big|_{p = (\bar{\mu}, 0, 0, 0)} = -2\,\bar{\mu}, \qquad (4.4.17)$$

$$\widetilde{\Gamma}_R^{(4)}(\bar{p}_1, \bar{p}_2, \bar{p}_3 | T = 0) = -\lambda,$$

where $\bar{\mu} \neq 0$ and for $i = 1, 2, 3$

$$\bar{p}_i \cdot \bar{p}_j = \frac{\bar{\mu}^2}{2}(4\delta_{ij} - 1). \qquad (4.4.18)$$

These normalization conditions at nonexceptional momenta uniquely determine the wave function and the quartic coupling-renormalization constants δZ and $\delta\lambda$. The first condition of (4.4.17) implies that the renormalized mass vanishes at $T = 0$, to all orders in λ. It uniquely determines δm^2.

The vertex function $\widetilde{\Gamma}_R^{(2)}(p | T)$ varies with temperature. Even at zero momentum it becomes nonvanishing once $T > 0$. On the other hand, the free propagator stays massless. To one loop order we have

$$\widetilde{\Gamma}_R^{(2)}(p|T) \;=\; -p^2 \;+\; \bigcirc \;-\; \delta m^2$$

$$=\; -p^2 \;+\; \bigcirc \;-\; \bigcirc\Big|_{T=0}$$

$$=\; -p^2 - \frac{\lambda}{2}\left[T\sum_{k_0} \int^{\Lambda_0} \frac{d^3 k}{(2\pi)^3}\,\frac{1}{k^2} - \int^{\Lambda_0} \frac{d^4 k}{(2\pi)^4}\,\frac{1}{k^2} \right]$$

$$=\; -p^2 - \frac{\lambda T^2}{24}. \qquad (4.4.19)$$

In higher orders, the one-loop diagrams of (4.4.19) appear as subgraphs. As part of a loop, with massless propagators attached, they lead to IR-divergencies. For instance, let us write for the renormalized one-loop

contribution

$$\underset{R}{\bigcirc} \;=\; \bigcirc \;-\; \bigcirc\bigg|_{T=0} \;=\; -\frac{\lambda T^2}{24}. \qquad (4.4.20)$$

The Feynman integral which corresponds to the Feynman graph

$$(4.4.21)$$

with n insertions of (4.4.20) is then given by

$$T\sum_{k_0}\int^{\Lambda_0}\frac{d^3k}{(2\pi)^3}\,T\sum_{l_0}\int^{\Lambda_0}\frac{d^3l}{(2\pi)^3}\,\frac{1}{l^2}\,\frac{1}{(k+l+p)^2}\,\frac{1}{k^2}$$
$$\cdot\left[\frac{1}{k^2}\left(-\frac{\lambda T^2}{24}\right)\right]^n. \qquad (4.4.22)$$

It has a nonintegrable singularity at $k = 0$ whenever $n \geq 1$. The IR-divergence becomes the stronger the larger n.

The idea of resummation is to sum particular classes of Feynman diagrams or even appropriate parts of the associated Feynman integrals. The aim is to do it in such a way that the IR-divergencies are removed not necessarily term by term, but such the IR-divergencies of the various diagrams cancel each other in the sum over the diagrams.

In the example above, the IR-singularities arise because $\widetilde{\Gamma}_R^{(2)}(p = 0|T)$ does not vanish, whereas the attached free propagator stays massless. This problem is circumvented when the applied resummations generate "thermal" masses for the *a priori* massless field Φ. Such a procedure is well known from massless Φ^4-theory in three dimensions. Before chains of the one-loop diagram (4.4.20) of the two-point function are inserted as subgraphs into higher order diagrams, they are first summed according to a geometric series, yielding the propagator

$$(4.4.23)$$

$$= \frac{1}{p^2} \sum_{\nu \geq 0} \left(-\frac{\lambda T^2}{24} \frac{1}{p^2} \right)^\nu = \left(p^2 + m_T^2 \right)^{-1}.$$

Here, $m_T^2 = (\lambda T^2)/24$ is the square of the thermal mass of the Φ-field (to leading order), generated by the interaction with the heat bath.

In terms of the action, the resummation we have just applied amounts to writing the action as

$$
\begin{aligned}
S + \delta S = \int_0^{1/T} dx_0 \int d^3x & \left[\left\{ \frac{1}{2} m_T^2 \Phi(x)^2 + \frac{1}{2} \sum_{\mu=0}^{3} (\partial_\mu \Phi)(x)^2 \right\} \right. \\
& + \left(\frac{1}{4!} \lambda \Phi(x)^4 - \frac{1}{2} m_T^2 \Phi(x)^2 + \frac{1}{2} \delta m^2 \Phi(x)^2 \right. \\
& \left. \left. + \frac{1}{2} \delta Z \sum_{\mu=0}^{3} (\partial_\mu \Phi)(x)^2 + \frac{1}{4!} \delta \lambda \Phi(x)^4 \right) \right].
\end{aligned}
\tag{4.4.24}
$$

The first part, in curly braces, is taken as the free action,

$$
S_2 = \int_0^{1/T} dx_0 \int d^3x \left\{ \frac{1}{2} m_T^2 \Phi(x)^2 + \frac{1}{2} \sum_{\mu=0}^{3} (\partial_\mu \Phi)(x)^2 \right\}.
\tag{4.4.25}
$$

It is quadratic in the field Φ and leads to the free propagator (4.4.23). The remaining part is considered as interaction

$$
\begin{aligned}
S_{int} = \int_0^{1/T} dx_0 \int d^3x & \left(\frac{1}{4!} \lambda \Phi(x)^4 - \frac{1}{2} m_T^2 \Phi(x)^2 + \frac{1}{2} \delta m^2 \Phi(x)^2 \right. \\
& \left. + \frac{1}{2} \delta Z \sum_{\mu=0}^{3} (\partial_\mu \Phi)(x)^2 + \frac{1}{4!} \delta \lambda \Phi(x)^4 \right).
\end{aligned}
\tag{4.4.26}
$$

For any correlation function, the resummed expansion now is equivalent to an expansion of the Boltzmann factor $\exp(-S_{int})$ in powers of S_{int}.

The Feynman rules for (4.4.25) and (4.4.26) become

$$\bullet\!-\!-\!-\!\bullet \qquad = \left(p^2 + m_T^2\right)^{-1}, \qquad\qquad (4.4.27)$$

for the free propagator, which is (4.4.23), while

$$\times\!\!\!\!\bullet\!\!\!\!\times \qquad = \quad -\lambda \qquad\qquad\qquad (4.4.28)$$

for the four-point vertex remains unchanged. There is an additional two-point vertex

$$-\!-\bullet\!-\!- \qquad = m_T^2 = \frac{\lambda T^2}{24}, \qquad\qquad (4.4.29)$$

which is the negative of (4.4.20).

To make things more clear let us compute the two- and four-point function in the resummed scheme to subleading order in λ. The renormalized two-point vertex function becomes

$$\widetilde{\Gamma}_R^{(2)}(p|T) = -\left(p^2 + m_T^2\right) + \widetilde{\Pi}_R(p|T), \qquad\qquad (4.4.30)$$

with

$$\widetilde{\Pi}_R(p|T) \equiv \quad \text{(diagram)} \; \Pi$$

$$= \quad \text{(loop diagram)} \quad + \quad \text{(line diagram)} \quad - \delta m^2 + O(\lambda^2)$$

$$= -\frac{\lambda}{2}T\sum_{k_0}\int^{\Lambda_0}\frac{d^3k}{(2\pi)^3}\frac{1}{k^2+m_T^2} + m_T^2 - \delta m^2 + O(\lambda^2)$$

$$= -\frac{\lambda}{2}T\sum_{k_0}\int^{\Lambda_0}\frac{d^3k}{(2\pi)^3}\left[\frac{1}{k^2+m_T^2}-\frac{1}{k^2}\right] + O(\lambda^2) \qquad (4.4.31)$$

$$= -\frac{\lambda}{2}\left(-\frac{m_T T}{4\pi}\right) + O(\lambda^2\ln\lambda)$$

$$= \frac{\lambda^{3/2}T^2}{8\pi24^{1/2}} + O(\lambda^2\ln\lambda).$$

The subleading contribution is proportional to λm_T or $\lambda^{3/2}$. For the four-point vertex function we obtain at zero momentum

$$\widetilde{\Gamma}_R^{(4)}(0,0,0|T) = \quad \text{(diagram)} \quad + 3\left(\text{(diagram)}\right) - \delta\lambda + O(\lambda^3)$$

$$= -\lambda + \frac{3\lambda^2}{2}T\sum_{k_0}\int^{\Lambda_0}\frac{d^3k}{(2\pi)^3}\frac{1}{(k^2+m_T^2)^2} - \delta\lambda + O(\lambda^3) \qquad (4.4.32)$$

$$= -\lambda + \frac{3\lambda^2}{2}\left[\frac{1}{8\pi^2}\left(\ln\frac{\Lambda_0}{m_T}-\frac{1}{2}+O(\Lambda_0^{-2})\right)\right.$$

$$\left.+\frac{T}{8\pi m_T}+\frac{1}{8\pi^2}\left(\ln\frac{m_T}{4\pi T}-\psi(1)-\frac{\zeta(3)m_T^2}{8\pi^2 T^2}\right)\right.$$

$$\left.+O(\frac{m_T^4}{T^4},\Lambda_0^{-1})\right] - \delta\lambda + O(\lambda^3)$$

$$= -\lambda + \frac{3}{16\pi}24^{1/2}\lambda^{3/2} + O(\lambda^2\ln\lambda).$$

As a result of the thermal-mass resummation, this is a finite expression. Without mass resummation, $\widetilde{\Gamma}_R^{(4)}(0,0,0|T)$ would be IR-divergent because 0 is an exceptional external momentum. For nonexceptional momenta p, the one-loop contribution to $\widetilde{\Gamma}_R^{(4)}(p|T)$ will be of the order of $O(\lambda^2\ln\lambda)$. We notice that the subleading term of (4.4.32) is again of order $\lambda^{3/2}$, although the contributing diagrams have two vertices, every vertex contributing a factor λ. It is the momentum integration which generates a term proportional to $m_T^{-1} \simeq \lambda^{-1/2}$.

The role of the additional vertex (4.4.29) is twofold. First of all, it ensures that diagrams which are already summed into the propagator (4.4.27) are not over-counted. Secondly, as it always occurs in combination with the massive one-loop diagram according to

$$- \delta m^2 \tag{4.4.33}$$

$$= \lambda \frac{m_T T}{8\pi} + O(\lambda^2 \ln \lambda),$$

the subleading contribution to $\widetilde{\Gamma}_R^{(2)}(p|T)$ in the resummed expansion becomes proportional to m_T, as we have seen in (4.4.31). The same is true whenever this combination occurs as subgraphs of a higher-loop diagram. Together with $m_T = c_T \lambda^{1/2}$ and $c_T \neq 0$ this is an essential precondition for the expansion in S_{int} to be still a weak-λ expansion.

For example, let us consider the analogous diagram to (4.4.21), but now within the resummed expansion scheme

$$\tag{4.4.34}$$

With n one-loop insertions of $\widetilde{\Pi}_R(p|T)$, (4.4.31), in a loop of massive propagators (4.4.27), this graph has $n + 2$ vertices of S_{int} and contributes to order $n/2 + 2$ in λ

$$\lambda^2 T \sum_{k_0} \int^{\Lambda_0} \frac{d^3k}{(2\pi)^3} \, T \sum_{l_0} \int^{\Lambda_0} \frac{d^3l}{(2\pi)^3} \frac{1}{l^2 + m_T^2} \frac{1}{k^2 + m_T^2}$$

$$\cdot \frac{1}{(k+l+p)^2 + m_T^2} \left[\frac{1}{k^2 + m_T^2} (\lambda m_T) \right]^n \tag{4.4.35}$$

$$= O(\lambda^{n/2+2}).$$

Thus the order of λ to which the Feynman integral contributes increases with the number of vertices as it should.

The above examples already show an important consequence of the resummation. The resummed expansion is no longer "analytic" with respect to the coupling λ in the sense that it does not generate series in integral powers of λ. Fractional powers of λ occur as well as logarithmic factors. This is due to the fact that the thermal mass m_T in the propagator depends on λ according to $m_T \simeq \lambda^{1/2}$.

4.4.4 *Resummation leading to a reasonable weak coupling expansion*

In general it becomes an important issue to show that the resummed expansion still generates a reasonable weak coupling expansion, reasonable in the sense that terms are suppressed the more the higher the order in the loop expansion. In particular, it should be still asymptotic (or even convergent). It should be IR-finite to all orders. The counterterms of the original theory should be resummable in a similar way and cancel the UV-divergencies order by order also in the new expansion. A thorough investigation of these questions would proceed along the lines summarized in the introduction to this chapter.

For the above example of a Φ^4-theory, IR-finiteness is obvious. The propagator of the Φ-field becomes massive by the one loop-mass resummation. It is no longer singular at zero momentum.

It is less easy to see that the expansion with respect to S_{int} is still an expansion in small λ. So far we have seen this feature only for the two-point function in (4.4.35), where the diagram with $n + 2$ vertices contributes to the order $\lambda^{n/2+2}$.

In general we must determine the order in the coupling to which a Feynman diagram contributes and relate it to the number of its vertices and loops. This requires an estimate of a Feynman integral I_Γ associated with a Feynman graph Γ for the case in which the masses of the propagators become small. It is based on IR-power counting criteria and relates divergent behavior for small masses to IR-divergence degrees of Γ and its subdiagrams, or equivalently, to all Zimmermann subspaces. For a precise definition of this notion we refer to the previous chapter on perturbation theory at zero temperature. In the present context, in which we are interested in the small-mass behavior in m_T, the IR-divergence degrees are counted as if all propagators were massless.

It is sufficient to consider 1PI-diagrams with only four-point vertices. Let Γ be a 1PI-diagram with $e_\Gamma \geq 0$ external lines and $l_\Gamma \geq 0$ loops. By an elementary topological relation, the number of vertices of Γ is given by $b_\Gamma = l_\Gamma - 1 + e_\Gamma/2$. Every vertex contributes a factor λ, so that I_Γ gets a total factor of

$$\lambda^{l_\Gamma - 1 + e_\Gamma/2} \tag{4.4.36}$$

from its vertices. The small-m_T behavior of I_Γ is estimated as follows. Suppose that

$$\underline{\mathrm{degr}}_H I_\Gamma > -\delta_\Gamma, \tag{4.4.37}$$

for all Zimmermann subspaces H of Γ. If $\delta_\Gamma \leq 0$, I_Γ will be convergent in the massless limit $m_T \to 0$. If $\delta_\Gamma > 0$, I_Γ will diverge for small m_T at worst as

$$I_\Gamma \simeq m_T^{-\delta_\Gamma + 1} \left(\log m_T\right)^{l_\Gamma}. \tag{4.4.38}$$

We know that m_T becomes small with λ according to $m_T \simeq \lambda^{1/2}$. Therefore the small-λ behavior of I_Γ is obtained by multiplying (4.4.36) and (4.4.38).

Next we need an upper bound on δ_Γ to make use of this estimate. Let us assume that all the external momenta of Γ are nonexceptional. The subleading part of the two-point vertex function always appears in the combination (4.4.33). For later consistency we count this as a one-loop contribution. At zero momentum it becomes proportional to λm_T. If we furthermore use that $m_T = c\lambda^{1/2}$ with $c \neq 0$, we can strictly prove by means of techniques known from renormalization theory of massless field theories (at zero temperature) that

$$\delta_\Gamma \leq l_\Gamma - 1. \tag{4.4.39}$$

Inserting (4.4.39) into (4.4.38) and including the factor (4.4.36) from the vertices of Γ, I_Γ behaves for small m_T according to

$$I_\Gamma \simeq \lambda^{e_\Gamma/2 + 1} \left(\frac{\lambda}{m_T}\right)^{l_\Gamma - 2} \left(\log \lambda\right)^{l_\Gamma}, \tag{4.4.40}$$

whenever Γ has at least two loops, i.e. $l_\Gamma \geq 2$. Finally, making use of $m_T \simeq \lambda^{1/2}$, we obtain the order in λ to which I_Γ contributes, namely

$$\begin{aligned} I_\Gamma &\simeq \lambda^{(e_\Gamma + l_\Gamma)/2} \left(\log \lambda\right)^{l_\Gamma} \\ &\simeq \lambda^{(1+b_\Gamma)/2 + e_\Gamma/4} \left(\log \lambda\right)^{b_\Gamma + 1 - e_\Gamma/2}. \end{aligned} \tag{4.4.41}$$

This shows that the expansion of correlation functions and of the free energy with respect to S_{int} is still a perturbative expansion for small λ. Furthermore, the loop ordering is preserved. Of two graphs which contribute to the same correlation function, the graph with more loops contributes to higher order in λ than the graph with less loops. We have seen an explicit example of this behavior in (4.4.35), with $l_\Gamma = b_\Gamma = n + 2$.

To summarize, the one-loop mass resummation in the massless Φ^4-theory is sufficient to remove all IR-divergencies at finite temperature. For all correlation functions as well as for the free energy, the resummed series stay small λ expansions. They are, however, no longer power series in λ, but series in $\lambda^{1/2}$, with logarithmic corrections.

We have not discussed the issue of renormalizability of the resummed perturbation theory. Of course, the structure of S_2 and S_{int} ensure renormalizability by adding appropriate local counterterms to the resummed action. The question, however, remains whether the UV-divergencies still cancel each other without adding new counterterms by hand, or, equivalently, by enforcing that the old normalization conditions still hold. If they do so, this means that the $T = 0$-counterterms of the action (4.4.24) or δS, (4.4.7), that ensure the normalization conditions (4.4.17), are still sufficient for UV-finiteness. Eventually, UV-finiteness will require appropriate resummations of the counterterm action as well. To our knowledge, a systematic proof of this important issue is still missing.

4.4.5 Gauge theories and the magnetic-mass problem

The infrared problems of perturbation theory which we have just encountered for the Φ^4-theory at finite temperature are common to field theories without mass gap at the tree level, including pure gauge theories and QCD. We could try to remove the IR-divergencies in the general case in a similar way as we removed them for the Φ^4-theory. Unfortunately this procedure fails for gauge theories. Below we shall point out why. A main reason is related to the so-called *magnetic-mass problem*.

In the Φ^4-theory two conditions are essential to obtain an IR-finite perturbation theory by mass resummation. First, one defines a thermal mass

m_T by the zero momentum value of the two-point function,

$$m_T^2 = - \quad \underline{\textcircled{R}} \quad + o(\lambda)$$

$$= c\lambda + o(\lambda),$$

(4.4.42)

with the one-loop coefficient $c > 0$, cf. (4.4.20). Second, with the thermal mass resummed into the propagator (4.4.27), the resummed perturbation theory should satisfy

$$\frac{\widetilde{\Gamma}_R^{(2)}(p = 0|T) + m_T^2}{m_T^2} = O(\lambda^\alpha), \quad \alpha > 0. \qquad (4.4.43)$$

Actually we verified that $\alpha = 1/2$.

Let us start with generalizing this procedure to gauge theories. For simplicity we consider the pure $SU(3)$ gauge theory. The inclusion of dynamical fermions does not invalidate the considerations below. The action is given by

$$S = S_{gauge} + S_{gf} + S_{gh}, \qquad (4.4.44)$$

where S_{gauge} is the gauge field action

$$S_{gauge} = \int_x \sum_{\mu,\nu=0}^{3} \sum_{d=1}^{3} \frac{1}{4} \mathcal{F}_{\mu\nu}^d(x) \mathcal{F}_{\mu\nu}^d(x), \qquad (4.4.45)$$

$$\mathcal{F}_{\mu\nu}^d(x) = \partial_\mu W_\nu^d(x) - \partial_\nu W_\mu^d(x) + g \sum_{e,f=1}^{8} f_{def} W_\mu^e(x) W_\nu^f(x),$$

$d = 1, \ldots, 8$, and the f_{def} are the structure constants of the Lie algebra $su(3)$. For the general Lorentz gauge, the gauge fixing action becomes

$$S_{gf} = \int_x \frac{1}{2} \xi \sum_{d=1}^{8} \mathcal{G}_d(x) \mathcal{G}_d(x),$$

$$\mathcal{G}_d(x) = \sum_{\mu} \partial_\mu W_\mu^d(x) \qquad (4.4.46)$$

with $\xi > 0$ the gauge fixing parameter. The ghost-field part of the action reads

$$S_{gh} = \int_x \sum_{d,e=1}^{8} \sum_{\mu=0}^{3} \bar{c}_d(x) \partial_\mu \left(\delta_{de} \partial_\mu + g f_{ade} W_\mu^a(x) \right) c_e(x). \qquad (4.4.47)$$

We skip the explicit form of the counterterm action for simplicity.

To define thermal gluon masses we need the general structure of the full gluon propagator $\widetilde{G}$. With $\widetilde{D}$ as the free gluon propagator, $\widetilde{G}$ is related to the vacuum polarization tensor $\widetilde{\Pi}$ by

$$\widetilde{G}(k) = \widetilde{D}(k) \sum_{n=0}^{\infty} \left(\widetilde{\Pi}\widetilde{D} \right)^n (k). \qquad (4.4.48)$$

It is the tensor structure of $\widetilde{D}$, $\widetilde{\Pi}$ and $\widetilde{G}$ that makes things more complicated here than in the scalar case. The tensors are diagonal in color space, but not with respect to the Lorentz indices. Let us choose the Landau gauge $\xi = \infty$. As a result, $\widetilde{G}(k)$ acquires a particularly simple form. In Landau gauge, the free propagator (derived from the quadratic part of the action) becomes

$$\widetilde{D}_{\mu\nu}(k) = \frac{1}{k^2} \left(\delta_{\mu\nu} - \frac{k_\mu k_\nu}{k^2} \right). \qquad (4.4.49)$$

The vacuum polarization tensor $\widetilde{\Pi}(k)_{\mu\nu}$ depends on two independent vectors, on the momentum k and on $u = (1, \mathbf{0})$, the unit vector in temperature direction. Hence $\widetilde{\Pi}(k)_{\mu\nu}$ is written as the sum of four linearly independent tensors. A convenient choice of these tensors is given by

$$(P_T)_{\mu\nu} = \sum_{i,j=1}^{3} \delta_{\mu i} \left(\delta_{ij} - \frac{k_i k_j}{\mathbf{k}^2} \right) \delta_{j\nu} \qquad (4.4.50)$$

(which is transversal with respect to $\mathbf{k}$ and only nonvanishing for spatial components),

$$(P_L)_{\mu\nu} = \left(\delta_{\mu\nu} - \frac{k_\mu k_\nu}{k^2} \right) - (P_T)_{\mu\nu} = \frac{k^2}{\mathbf{k}^2} u_\mu^T u_\nu^T,$$

$$u_\mu^T = u_\mu - k_\mu \frac{u \cdot k}{k^2}, \; (u_\mu^T)^2 = \frac{\mathbf{k}^2}{k^2} \qquad (4.4.51)$$

(which is transversal with respect to the four-vector k and called temporal). Moreover,

$$(P_G)_{\mu\nu} = \frac{k_\mu k_\nu}{k^2} \qquad (4.4.52)$$

is longitudinal with respect to k. Finally we have

$$S_{\mu\nu} = \frac{1}{\sqrt{2\mathbf{k}^2}} \left(k_\mu u_\nu^T + k_\nu u_\mu^T \right).$$ (4.4.53)

P_T, P_L and P_G are orthogonal projection operators satisfying

$$P_T^2 = P_T, \ P_L^2 = P_L, P_G^2 = P_G,$$
$$P_T P_L = P_T P_G = P_L P_G = 0.$$ (4.4.54)

Furthermore, S satisfies the relations

$$(SP_L)_{\mu\nu} = \frac{1}{\sqrt{2\mathbf{k}^2}} k_\mu u_\nu^T, \ P_L S P_L = 0$$
$$SP_T = 0$$
$$(SP_G)_{\mu\nu} = \frac{1}{\sqrt{2\mathbf{k}^2}} u_\mu^T k_\nu, \ P_G S P_G = 0$$ (4.4.55)
$$S^2 = \frac{1}{2} (P_G + P_L).$$

In terms of these tensors the general form of $\widetilde{\Pi}(k)$ is written as

$$\widetilde{\Pi} = \widetilde{\Pi}_L \, P_L + \widetilde{\Pi}_T \, P_T + \widetilde{\Pi}_G \, P_G + \widetilde{\Pi}_S \, S.$$ (4.4.56)

Using the relations (4.4.54) and (4.4.55) it is straightforward to compute for every integer $n \geq 1$

$$\left(\widetilde{\Pi}\widetilde{D} \right)^n = \frac{1}{(k^2)^n} \left(\widetilde{\Pi}_L^n \, P_L + \widetilde{\Pi}_T^n \, P_T + \widetilde{\Pi}_L^{n-1} \, \widetilde{\Pi}_S \, S \, P_L \right)$$ (4.4.57)

and thus

$$\widetilde{G}(k) = \frac{1}{k^2} \left(\left[1 - \frac{\widetilde{\Pi}_L}{k^2} \right]^{-1} P_L + \left[1 - \frac{\widetilde{\Pi}_T}{k^2} \right]^{-1} P_T \right)$$
$$= \frac{1}{k^2 - \widetilde{\Pi}_L(k)} P_L(k) + \frac{1}{k^2 - \widetilde{\Pi}_T(k)} P_T(k).$$ (4.4.58)

From the form (4.4.58) it is easy to extract the perturbative representation of what is called the plasma masses, "plasma masses", because it is these masses which are currently used in perturbation theory to determine the long-range screening behavior in a gluon plasma at high temperatures.

Electric (longitudinal) and magnetic (transversal) plasma masses are obtained as

$$m_L^2 = -\widetilde{\Pi}_L(0)$$
$$m_T^2 = -\widetilde{\Pi}_T(0).$$

(4.4.59)

An explicit computation shows that

$$\widetilde{\Pi}(0)_{00} = \widetilde{\Pi}_L(0),$$
$$\sum_{i=1}^{3} \widetilde{\Pi}(0)_{ii} = 2\,\widetilde{\Pi}_T(0) + \widetilde{\Pi}_G(0).$$

(4.4.60)

Inserting (4.4.60) into (4.4.59) we obtain

$$m_L^2 = -\widetilde{\Pi}(0)_{00},$$
$$m_T^2 = -\frac{1}{2}\left(\sum_{i=1}^{3} \widetilde{\Pi}(0)_{ii} - \Pi_G(0)\right).$$

(4.4.61)

Eqs. (4.4.61) are the perturbative representations of the electric (m_L) and magnetic (m_T) plasma masses obtained in Landau gauge. By experience, the explicit one-loop expressions for m_L^2 and m_T^2 are pretty independent of the choice of the gauge-fixing function, but this independence does not extend to higher orders.

Let us now see whether the mass resummation which rendered the $\lambda\Phi^4$-theory IR-finite to all orders of perturbation theory generalizes to gauge theories. The one-loop condition (4.4.42) for the scalar theory translates into the two conditions

$$m_L^2 = c_L g^2 + o(g^2)\,, \quad c_L > 0,$$
$$m_T^2 = c_T g^2 + o(g^2)\,, \quad c_T > 0.$$

(4.4.62)

An explicit computation shows that $c_L = N_c/3T^2$ with $N_c = 3$ for $SU(3)$. This is fine. On the other hand, the analogous computation for m_T reveals that $c_T = 0$. The magnetic mass vanishes to one loop. Actually, without explicit computation one can show that the vanishing of m_T to one loop order is a consequence of the Ward identities that express the gauge invariance of the theory (or, more precisely, the BRS-symmetry after gauge fixing.) This is the formal expression of the magnetic-mass problem. It prevents a generalization of the established methods for Φ^4-theories to gauge theories.

Apart from the fact that the gap equations (4.4.61) and (4.4.62) are violated by the magnetic gauge fields in the sense that $c_T = 0$, there are other obstructions against a pure mass resummation. A mass resummation would lead to a finite magnetic mass, giving a chance of curing the IR-problems, but a pure resummation of the two-point correlation function would violate the Ward identities. Ward identities are hardly dispensable in proofs of renormalizability or gauge independence of the scattering matrix. The Ward identities are nonlinear relations between Green functions. In order to preserve them, the mass resummation should therefore be supplemented by corresponding resummations of higher correlation functions as well. This remark may indicate the complexity of the magnetic-mass problem.

4.4.6 *Hard-thermal loop resummation*

So far we were looking for procedures for finite-temperature gauge theories that lead to an IR-finite perturbation expansion in the infinite spatial volume. No procedure is known to date that works to *all* orders of the gauge coupling. Procedures which have been proposed so far fail due to the magnetic-mass problem. Typically, observables stay IR-finite to low orders of perturbation theory. For instance, let us look at the free energy density f, a basic quantity for thermodynamic considerations. For this f, the first IR divergencies arise at order g^6. Up to and including order g^5, f stays IR-finite. More explicitly, after daisy-diagram resummations the IR-finite part of f becomes

$$
f \simeq -\frac{8\pi^2}{45} T^4 [1 - \frac{15}{16\pi^2} g^2 + \frac{15}{4\pi * 3} g^3
$$
$$
+ \frac{135}{32\pi^4} \left(\ln \frac{g^2}{4\pi^2} - \frac{11}{36} \ln \frac{\mu}{2\pi T} + 3.5 \right) g^4
$$
$$
+ \frac{495}{64\pi^5} \left(\ln \frac{\mu}{2\pi T} - 3.23 \right) g^5] . \tag{4.4.63}
$$

To one loop, the $\overline{MS}$ coupling constant $g = g(\mu)$ is given by

$$
g^2(\mu) = \frac{16\pi^2}{11} \frac{1}{\ln \left(\mu^2 / \Lambda_{\overline{MS}}^2 \right)} . \tag{4.4.64}
$$

The remainder of the expression (4.4.63), which is formally of the order $O(g^6)$, actually does not exist in the sense that it is IR-divergent. Its convergence is bad in the sense that the numerical values of subsequent terms decrease quite slowly. Also they are alternating in sign. In particular, the order-g^5 term is not a small correction to f up to and including order g^4. The very meaning of the "finite part" of such an expression is questionable as we have argued above. The correction term is infinite for every nonzero value of g. Of course, this sheds some doubts on the very weak coupling expansion that has been used.

The poor convergence of the free-energy density is not specific for QCD, but has a similar reason as in the case of massless scalar field theory with Φ^4 interactions (where large corrections come from the gT-scale as well). Only the tools of resummed or "screened" perturbation theory by which the poor convergence properties of the free energy in massless Φ^4 theories are cured do not apply to gauge theories like QCD. In contrast to conventional perturbation theory about an ideal gas of massless particles, resummed perturbation theory amounts to an expansion about an ideal gas of quasiparticles with temperature dependent masses. A local mass term proportional to Φ^2 is added and subtracted to the Lagrangian, the added term treated nonperturbatively and the subtracted term included in the perturbative expansion. There were early indications that such a reorganization of perturbation theory makes also sense for QCD, because models in terms of quasiparticles, i.e. quarks and gluons with temperature dependent masses, were able to fit results from lattice QCD for thermodynamic quantities (see for example [181] for a short review on such attempts). In terms of quasiparticles, the effective Lagrangian is expected to simplify as compared to a description in terms of the original degrees of freedom: quarks with current quark masses and massless gluons. Nevertheless, a mere copy of screened perturbation theory in scalar field theories will fail for QCD, because a local mass term for the gluons would violate gauge invariance.

Now it is the very hard-thermal loop (HTL) expansion that provides an expansion about an ideal gas of quarks-and gluon-quasiparticles. Rather than adding and subtracting a local mass term for gluons, one adds and subtracts HTL-correction terms to the action while *maintaining local gauge invariance.* "Expansion" means a systematic inclusion of interactions between these quasiparticles up to and including next-to-next-to-leading order in HTL-perturbation theory. (Such an inclusion of interactions between quasiparticles is a novel feature of this approach as compared to quasiparticle descriptions on a mere phenomenological level.) Beyond this order, the magnetic mass problem shows up again, here in the form of nonlocal correction terms whose renormalization is not solved to date and whose solution is even conceptually not clear. From a pragmatic point of view, however, one can take the HTL-approach to the extent that it is well defined, i.e. below a certain order in the coupling, and compare its predictions for quasiparticle masses, screening of the gauge interaction and Landau damping with experimental results of real-time processes in relativistic heavy-ion collisions. A final agreement would a posteriori justify the expansion even if there is still a lack in understanding HTL-expansion from a systematic point of view.

Later in section 4.6, we will study a very different approach of circumventing the severe IR-problems of QCD at finite temperature. We then first derive a dimensionally reduced action for QCD and next treat the modes causing the IR-problems fully nonperturbatively via Monte Carlo simulations. This approach proceeds in Euclidean quantum field theory and therefore relies on the analytic continuation to imaginary time, while the HTL-expansion has the clear advantage of being directly applicable to real-time processes in order to predict signatures of the quark-gluon plasma.

Our following discussion will focus on some basic definitions, supplemented by a concrete example of a hard-thermal loop and the result for the effective HTL action. For further details and the derivation of the HTL-action or the free energy density of a hot gluon plasma, to leading order in HTL-perturbation theory, we refer to the original literature [180, 181], respectively.

What are hard-thermal loops? Let us consider a Feynman diagram like (4.4.21) or (4.4.34), but now in gauge theories rather than in Φ^4. Typically, this graph stands for a Feynman integral of the form

$$g^4 T \sum_{p_0} \int \frac{d^3p}{(2\pi)^3} \, T \sum_{l_0} \int \frac{d^3l}{(2\pi)^3} \frac{1}{l^2} \frac{1}{(p+l+q)^2} \frac{1}{p^2}$$
$$\cdot \left[\frac{1}{p^2} \Pi(p) \right]^n. \qquad (4.4.65)$$

$\Pi(p)$ represents the one-loop vacuum polarization, which is proportional to g^2. Since color and vector indices are not essential for the following considerations, they are suppressed.

Let us consider the convergence speed of (4.4.65) that we have mentioned above. For a reliable weak coupling expansion with increasing n, the expression (4.4.65) should rapidly decrease by some power of g. This is the case if

$$\frac{1}{p^2} \, \Pi(p) \; = \; O(g^\alpha) \tag{4.4.66}$$

for some $\alpha > 0$. On the other hand, for momenta p for which (4.4.66) is violated, the convergence is bad, in particular for momenta p for which

$$\Pi(p) \; \geq \; p^2. \tag{4.4.67}$$

The idea of hard-thermal loop resummation is to identify the momenta p which are problematic, i.e. those which satisfy (4.4.67), next to identify the loop momenta of Π which are responsible for (4.4.67), and finally to resum the corresponding contributions into the free part of the action.

It is easy to identify those p, the external momenta of $\Pi(p)$, which are problematic. Because $\Pi(p)$ is a sum of one-loop Feynman integrals, it satisfies

$$\Pi(p) \; \simeq \; g^2 T^2. \tag{4.4.68}$$

If p is "hard", i.e. $p \simeq T$ or larger,

$$\frac{\Pi(p)}{p^2} \; \leq \; \frac{g^2 T^2}{p^2} \; \leq \; g^2. \tag{4.4.69}$$

If p is "soft", i.e. $p \simeq gT$ or smaller,

$$\frac{\Pi(p)}{p^2} \; \geq \; \frac{g^2 T^2}{p^2} \; \geq \; 1. \tag{4.4.70}$$

From this estimate of the order of magnitude we infer that it is the soft external momenta $p \simeq gT$ that are the troublemakers.

4.4.6.1 *Example of a hard-thermal loop*

How can we identify the loop momenta which cause the trouble? To indicate the answer to this question, we give just one example of a Feynman integral that contributes to $\Pi(p)$, in order to trace back from which part of the loop integration the inequality (4.4.67) originates. As it turns out, the problematic loop momenta, say k, are hard ones, i.e. $k \geq T$.

By definition, hard-thermal loops are the part of one-loop Feynman integrals in momentum space which contribute to the same order in the gauge coupling g as the corresponding tree-level amplitude. All external momenta p are considered as soft momenta, of the order of gT, $p \simeq gT$.

To see the origin of hard-thermal loops we consider a Feynman integral that has the form of a typical contribution to the two-point function in gauge theories. The term

$$
I = \ \text{—}\bigcirc\hspace{-0.3em}\bullet\text{—}
$$

$$
= g^2 T \sum_{\substack{k_0=2\pi Tn \\ n\in\mathbf{Z}}} \int \frac{d^3\mathbf{k}}{(2\pi)^3}\ \mathbf{k}^2\Delta(k)\Delta(k+p) \qquad (4.4.71)
$$

contributes to the gluonic vacuum polarization $\sum_{i=1}^{3}\Pi_{ii}(p)$, (color and vector indices suppressed,) where

$$
\Delta(k) = \frac{1}{k^2} = \frac{1}{k_0^2+k_s^2}\,,
$$
$$
k = (k_0,\mathbf{k})\,,\ k_s = |\mathbf{k}|\,, \qquad (4.4.72)
$$

k_0 and p_0 are integer multiples of $2\pi T$.

In the following manipulations we closely follow Braaten and Pisarski [180], for example in continuing p_0 to more general values, in particular to soft ones.

We shall write the integral (4.4.71) in a form that allows to identify the hard-thermal loop contributions. The first step is a Fourier transform with respect to the 0-component of the free propagator $\Delta(k)$,

$$
\Delta(k) = \int_0^{1/T} d\tau\ e^{ik_0\tau}\ \Delta_0(\tau,\mathbf{k}), \qquad (4.4.73)
$$

with inversion

$$
\Delta_0(\tau,\mathbf{k}) = T\sum_{k_0} e^{-ik_0\tau}\Delta(k). \qquad (4.4.74)
$$

The summation can be converted to a contour integral. With $\Delta(-k_0,\mathbf{k})=\Delta(k_0,\mathbf{k})$, $\Delta_0(\tau,\mathbf{k})$ is written as

$$
\Delta_0(\tau,\mathbf{k}) = \int_{-\infty-i\epsilon}^{\infty+i\epsilon} \frac{dk_0}{2\pi}\ \Delta(k_0,\mathbf{k})\left\{ e^{-ik_0\tau} + \frac{e^{-ik_0\tau}+e^{ik_0\tau}}{e^{ik_0/T}-1}\right\}. \qquad (4.4.75)
$$

$\Delta(k_0,\cdot)$ has simple poles at $k_0=\pm ik_s$ with residue $\pm 1/(2ik_s)$. Together with the decay properties of $\Delta(k_0,\mathbf{k})$ for large k_0, this allows us to compute the integral (4.4.75) by deformation of the integration path, using Cauchy's theorem, cf. Fig. 4.4.1. We obtain after some algebra

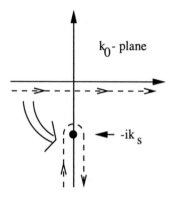

Fig. 4.4.1   Deformation of the integration path

$$\Delta_0(\tau, \mathbf{k}) = \frac{1}{2k_s} \left\{ (1 + n(k_s)) \, e^{-k_s \tau} + n(k_s) e^{k_s \tau} \right\}, \qquad (4.4.76)$$

where we defined

$$n(k_s) = \frac{1}{e^{k_s/T} - 1}. \qquad (4.4.77)$$

Inserting (4.4.73) into (4.4.71) we obtain

$$I = g^2 T \sum_{k_0} \int \frac{d^3 \mathbf{k}}{(2\pi)^3} \, \mathbf{k}^2 \int_0^{1/T} d\tau_1 \, e^{ik_0 \tau_1} \Delta_0(\tau_1, \mathbf{k})$$
$$\cdot \int_0^{1/T} d\tau_2 \, e^{i(k_0 + p_0)\tau_2} \Delta_0(\tau_2, \mathbf{k} + \mathbf{p}) \, . \qquad (4.4.78)$$

Using (4.4.76) and the fact that

$$T \sum_{k_0} e^{ik_0 \tau} = \sum_{n \in \mathbf{Z}} \delta(\tau - \frac{n}{T}) \qquad (4.4.79)$$

as the periodic δ-distribution, we obtain

$$I = g^2 \int \frac{d^3 \mathbf{k}}{(2\pi)^3} \, \mathbf{k}^2 \int_0^{1/T} d\tau \, e^{ip_0 \tau} \frac{1}{2|\mathbf{k}| \, 2|\mathbf{k} + \mathbf{p}|}$$
$$\cdot \left\{ (1 + n(|\mathbf{k}|)) \, e^{-|\mathbf{k}|\tau} + n(|\mathbf{k}|) \, e^{|\mathbf{k}|\tau} \right\} \qquad (4.4.80)$$
$$\cdot \left\{ (1 + n(|\mathbf{k} + \mathbf{p}|)) \, e^{-|\mathbf{k} + \mathbf{p}|\tau} + n(|\mathbf{k} + \mathbf{p}|) \, e^{|\mathbf{k} + \mathbf{p}|\tau} \right\} .$$

The τ-integration is straightforward and yields

$$
I = g^2 \int \frac{d^3\mathbf{k}}{(2\pi)^3} \frac{k_s^2}{2k_s 2(k+p)_s}
$$
$$
\left\{ (1+n(k_s)) \, (1+n((k+p)_s)) \, \frac{e^{(ip_0 - k_s - (k+p)_s)/T} - 1}{ip_0 - k_s - (k+p)_s} \right.
$$
$$
+ n(k_s) \, n((k+p)_s) \, \frac{e^{(ip_0 + k_s + (k+p)_s)/T} - 1}{ip_0 + k_s + (k+p)_s}
$$
$$
+ (1+n(k_s)) \, n((k+p)_s) \, \frac{e^{(ip_0 - k_s + (k+p)_s)/T} - 1}{ip_0 - k_s + (k+p)_s}
$$
$$
\left. + n(k_s) \, (1+n((k+p)_s)) \, \frac{e^{(ip_0 + k_s - (k+p)_s)/T} - 1}{ip_0 + k_s - (k+p)_s} \right\}. \qquad (4.4.81)
$$

In order to simplify this expression, we use the following relations

$$
1 + n(k_s) + n((k+p)_s) = \left(e^{(k_s + (k+p)_s)/T} - 1 \right) n(k_s) n((k+p)_s),
$$
$$
(1+n(k_s)) \, (1+n((k+p)_s)) = e^{(k_s + (k+p)_s)/T} \, n(k_s) n((k+p)_s),
$$

so that

$$
(1+n(k_s)) \, (1+n((k+p)_s)) \, \left(e^{-(k_s + (k+p)_s)/T} - 1 \right)
$$
$$
= - (1+n(k_s) + n((k+p)_s)),
$$
$$
(1+n(k_s)) \, n((k+p)_s) = e^{k_s/T} \, n(k_s) n((k+p)_s),
$$
$$
n(k_s) - n((k+p)_s) = \left(e^{(k+p)_s/T} - e^{k_s/T} \right) n(k_s) n((k+p)_s),
$$
$$
(1+n(k_s)) \, n((k+p)_s) \, \left(e^{-(k_s + (k+p)_s)/T} - 1 \right) = n(k_s) - n((k+p)_s).
$$

Using these relations in (4.4.81) yields

$$
I = g^2 \int \frac{d^3\mathbf{k}}{(2\pi)^3} \frac{k_s^2}{2k_s 2(k+p)_s} \left\{ (1+n(k_s) + n((k+p)_s)) \right.
$$
$$
\cdot \left[\frac{1}{ip_0 + k_s + (k+p)_s} - \frac{1}{ip_0 - k_s - (k+p)_s} \right]
$$
$$
+ (n(k_s) - n((k+p)_s))
$$
$$
\left. \cdot \left[\frac{1}{ip_0 - k_s + (k+p)_s} - \frac{1}{ip_0 + k_s - (k+p)_s} \right] \right\}. \qquad (4.4.82)
$$

We have used that

$$
e^{ip_0/T} = 1. \qquad (4.4.83)
$$

Following Braaten and Pisarski, the integral expression I, as it is now given in (4.4.82), is analytically continued to complex p, including soft momenta $p \simeq kT$. Below we shall identify the hard-thermal loop contributions to (4.4.82), this is the range of loop momenta $\mathbf{k}$ which yields a contribution to (4.4.82) proportional to $p^2 \simeq g^2 T^2$.

We distinguish three contributions: from zero temperature, soft loop momenta and hard thermal-loop momenta.

- The zero-temperature contribution to I is given by

$$
I_{T=0} = g^2 \int \frac{d^3\mathbf{k}}{(2\pi)^3} \frac{k_s^2}{2k_s 2(k+p)_s}
$$
$$
\cdot \left[\frac{1}{ip_0 + k_s + (k+p)_s} - \frac{1}{ip_0 - k_s - (k+p)_s} \right].
$$
(4.4.84)

After zero-temperature renormalization, the integral becomes proportional to p^2. This is a consequence of the Ward identities that follow from gauge invariance. Hence, this contribution is always soft, that is,

$$
I_{T=0} \simeq g^2 p^2,
$$
(4.4.85)

so it is suppressed by a factor of g^2 compared to the tree-level contribution p^2.

- Soft loop momenta $k \simeq gT$ contribute to I according to

$$
I_{soft} \simeq g^2 \int_{k \simeq gT} \frac{d^3\mathbf{k}}{(2\pi)^3} \frac{k_s^2}{2k_s 2(k+p)_s} \left\{ (n(k_s) + n((k+p)_s)) \right.
$$
$$
\cdot \left[\frac{1}{ip_0 + k_s + (k+p)_s} - \frac{1}{ip_0 - k_s - (k+p)_s} \right]
$$
$$
+ (n(k_s) - n((k+p)_s))
$$
$$
\left. \cdot \left[\frac{1}{ip_0 - k_s + (k+p)_s} - \frac{1}{ip_0 + k_s - (k+p)_s} \right] \right\}.
$$
(4.4.86)

Both k and p are soft, hence

$$
n(k_s) = \frac{1}{e^{k_s/T} - 1} \simeq \frac{T}{k_s} \simeq \frac{1}{g}.
$$
(4.4.87)

The integral measure contributes a factor $(gT)^3$, and (4.4.86) becomes

$$
I_{soft} \simeq g^2 (gT)^3 \frac{1}{g^2 T} \simeq g (gT)^2 \simeq g p^2.
$$
(4.4.88)

The contribution I_{soft} of soft internal momenta is suppressed by a factor of g compared to the tree level contribution p^2.

- For the hard-thermal loop momentum $k \geq T$,

$$I_{hard} \simeq g^2 \int_{k \geq T} \frac{d^3\mathbf{k}}{(2\pi)^3} \frac{k_s^2}{2k_s 2(k+p)_s} \left\{ (n(k_s) + n((k+p)_s)) \right.$$

$$\cdot \left[\frac{1}{ip_0 + k_s + (k+p)_s} - \frac{1}{ip_0 - k_s - (k+p)_s} \right]$$

$$+ (n(k_s) - n((k+p)_s))$$

$$\left. \cdot \left[\frac{1}{ip_0 - k_s + (k+p)_s} - \frac{1}{ip_0 + k_s - (k+p)_s} \right] \right\}. \qquad (4.4.89)$$

In this case we approximate

$$ip_0 \pm [k_s + (k+p)_s] \simeq \pm 2k_s \simeq T,$$

$$ip_0 \pm [k_s - (k+p)_s] \simeq ip_0 \pm p_s \cos\theta \simeq gT,$$

$$n(k_s) - n((k+p)_s) \simeq n(k_s)(1 + n(k_s)) \frac{p_s}{T} \cos\theta,$$

$$n(k_s) + n((k+p)_s) \simeq 2n(k_s),$$

where θ is the angle between $\mathbf{p}$ and $\mathbf{k}$. This rough estimate in the integrand yields

$$I_{hard} \simeq g^2 \int_{k \geq T} \frac{d^3\mathbf{k}}{(2\pi)^3} \left[\frac{2n(k_s)}{k_s} - \frac{n(k_s)(1 + n(k_s))}{T} \cdot \frac{(p_s \cos\theta)^2}{p_0^2 + p_s^2} \right]$$

$$\simeq g^2 T^2 \simeq p^2, \qquad (4.4.90)$$

that is of the same order as the tree level amplitude.

Thus we have seen that the region of loop momenta k which is responsible for I of (4.4.82) to be of the order $g^2 T^2$ for external momenta $p \simeq gT$, consists of the hard momenta $k \geq T$.

4.4.6.2 *Effective HTL-action*

We have just isolated the HTL part of the one-loop diagram for the vacuum polarization, i.e. for the gluonic two-point function. More generally, HTL contributions to gluonic n-point functions evolve as well, they are of order $g^2 T^2/p^2$ times the corresponding connected tree-level amplitude, where p stands for a generic external momentum.

Now it is possible to sum all HTL-contributions into an effective action S_{HTL}. The procedure is analogous to the resummation of the scalar Φ^4

theory, cf. the Eqs. (4.4.24)-(4.4.26). The gauge-field action S_g (including gauge fixing and ghost action) is written as

$$S_g = S_g + S_{HTL} - S_{HTL}$$
$$= \left\{ S_g^{(2)} + S_{HTL} \right\} + [S_{g,int} - S_{HTL}] , \qquad (4.4.91)$$

where $S_g^{(2)}$ is the part of S_g which is quadratic in the fields (i.e. quadratic in the gauge fields and ghosts), and $S_{g.int} = S_g - S_g^{(2)}$ is the interaction part of S_g. In HTL-perturbation theory, the term in braces of 4.4.91 is the leading order part, whereas the term in brackets counts as the HTL-interaction part. In higher orders of HTL-perturbation theory, the Boltzmann factor $\exp\left(-[S_{g,int} - S_{HTL}]\right)$ is expanded in powers of $[S_{g,int} - S_{HTL}]$.

Anderson et al. [181] formulate the HTL-Lagrangian in Minkowski space and in terms of hermitian gauge fields. If we translate this form to Euclidean space and to anti-hermitian gauge fields (according to our usual convention), we end up with a HTL-action for the pure SU(3) gauge theory of the form

$$S_{HTL} = \frac{m_e^2}{2} \int d^4x d^4y \sum_{\mu,\alpha,\beta} \text{tr} \, F_{\mu\alpha}(x) \mathcal{H}_{\alpha\beta}(x-y) F_{\mu\beta}(y) . \qquad (4.4.92)$$

This is the action for interacting quasiparticles. Equation (4.4.92) is consistent with a Boltzmann factor of the form $\exp\left(-S_{HTL}\right)$. Furthermore we have used $x^0 \to -ix^0$, $x^i \to x_i$ for x, analogously for all four-vectors, and in addition for the gauge fields $-iA_0 \to -A_0$ and $A_i \to -iA_i$, because of anti-hermitian gauge fields. F denotes the field strength tensor

$$F_{\mu\nu}(x) = \partial_\mu A_\nu(x) - \partial_\nu A_\mu(x) + [A_\mu(x), A_\nu(x)] , \qquad (4.4.93)$$

and

$$\mathcal{H}_{\alpha\beta}; = < \frac{n_\alpha n_\beta}{(n \cdot D)^2} >_n , \qquad (4.4.94)$$

where $D = \partial + gA$ denotes the gauge-covariant derivative, $n = (1, \vec{n})$ with three-dimensional unit vector $\vec{n}$, $< \cdot >_n$ denotes the equal-weight average over the spatial directions of $\vec{n}$, that is for some function $F(n)$

$$< F(n) >_n = \frac{1}{4\pi} \int d\Omega_2(\vec{n}) \, F(n) . \qquad (4.4.95)$$

In Eq. (4.4.92) m_e^2 is the perturbative expression for the electric Debye-screening mass squared $m_e^2 = g^2 T^2$. S_{HTL} involves arbitrary powers of the gauge field A. In this sense it describes *interacting* quasiparticles.

With (4.4.92) as starting point, the free energy density of the gluon plasma was calculated in [181] using the following approximations

- $S_{g,int} - S_{HTL}$ was set to zero, that is keeping only the leading HTL contribution.
- In S_{HTL} only the quadratic part in the gauge fields A was kept.
- A particular choice of new temperature-dependent counterterms was chosen to remove the UV-divergencies of the free energy density.

The result for the free energy agrees well with Monte Carlo data of [182]. From a phenomenological point of view it appears rather appealing to derive thermodynamic quantities from an effective action like S_{HTL} in terms of interacting quasiparticles.

From the systematic point of view, several question remain open, for example

- Are the HTL-corrections, arising from an expansion of the exponential $\exp -(S_{g,int} - S_{HTL})$, really small for temperatures out of an intermediate range?
- What are the corrections from higher-order terms in A of the leading HTL-contribution $S_g^{(2)} + S_{HTL}$ that were dropped to arrive at the above result?
- Do the former counterterms (those without resummation) eventually cancel the HTL-UV-divergencies, after appropriate HTL-resummations? If not, how do the values and scale dependencies of the renormalized coupling constants change because of the new renormalization scheme?

4.5 Constraint effective potential and gap equations

4.5.1 *Generalities*

In chapter II we have introduced the constraint effective potential $\mathcal{V}_{\mathrm{CEP}}$ as a useful alternative concept to the effective potential $\mathcal{V}_{\mathrm{stat}}$ in statistical physics and field theory (for the definitions cf. section 2.2). $\mathcal{V}_{\mathrm{CEP}}$ determines the probability density for the system to be in a state which is characterized by a certain value of the order parameter, and therefore has a direct physical interpretation for the phase structure of the system.

Although $\mathcal{V}_{\text{stat}}$ remains well defined when the volume is finite and the classical potential is nonconvex, it is sometimes difficult to calculate it both perturbatively or nonperturbatively. The loop expansion runs into problems, and numerical simulations via Monte Carlo are hampered by the external current J, the Legendre transformation and the supremum to be taken in the end, i.e. by the very definition of $\mathcal{V}_{\text{CEP}}$.

In the infinite-volume limit, $\mathcal{V}_{\text{CEP}}$ agrees with $\mathcal{V}_{\text{stat}}$ (cf. section 2.2), but does not make use of a Legendre transformation in its very definition, and is therefore more convenient for a nonperturbative evaluation in Monte Carlo simulations. If the numerical computation does not involve further truncations, the nonconvex shape of $\mathcal{V}_{\text{CEP}}$ in the symmetric phase and in a finite volume is just an indication of a first-order phase transition in the infinite-volume limit. On the other hand, if $\mathcal{V}_{\text{CEP}}$ stays nonconvex in the infinite-volume limit, the shape is an artifact of the involved approximations and may or may not indicate the first-order nature of the phase transition.

In this chapter we discuss a *perturbative* evaluation of $\mathcal{V}_{\text{CEP}}$ in view of the fact that the electroweak model is in the realm of perturbation theory for small Higgs masses. But as we will soon see, it takes a long way to derive a constraint effective potential that is actually conclusive for the order of the electroweak phase transition.

In the next section we start with a one-component scalar field theory and calculate $\mathcal{V}_{\text{CEP}}$ for small scalar self-coupling λ. The presentation is detailed. As we shall see, the small-λ expansion is not a loop expansion. To one-loop order this manifests itself by the appearance of an imaginary renormalized mass and complex $\mathcal{V}_{\text{CEP}}$ for small order parameter. There is actually an infinite number of Feynman graphs that contributes to the same order of λ. On the way of identifying all graphs which contribute to the leading nontrivial order we encounter a gap equation for the renormalized mass. Solving the gap equation amounts to a resummation of "thermal" masses. From section 4.4 we know that such resummations cure IR problems of massless scalar fields. Here the resummation is required to pick up all contributions of order λ and $\lambda^{3/2}$ in $\mathcal{V}_{\text{CEP}}$. We also state the result for $\mathcal{V}_{\text{CEP}}$ for an N-component scalar field theory. As a perturbative artifact, the potential appears to be nonconvex (even in the infinite volume).

Next we use the constraint effective potential, computed to order $\lambda^{3/2}$, to study the high-temperature phase transition. The surprising result is the indication of a first-order phase transition for $N < 4$ and a second-order transition for $N \geq 4$. However, we know e.g. from reliable lattice calculations that the transition is of second-order for all N. Therefore we must interpret the indicated first-order as a perturbative artifact, going along with its nonconvex shape, or the truncation at order $\lambda^{3/2}$. We give arguments that a computation of $\mathcal{V}_{\text{CEP}}$ to order λ^2 is ultimately required.

In the third part of this section we turn to $\mathcal{V}_{\text{CEP}}$ for the $SU(2)$ Higgs model to discuss the order of the electroweak phase transition. In principle, to order $\lambda^{3/2}$ and g^3, with g the gauge coupling constant, the calculation of $\mathcal{V}_{\text{CEP}}$ proceeds along the same lines as for the scalar case. As a result, there is no radiative correction to the transverse plasmon mass. This implies that the conditions for a first-order transition are always satisfied, even for arbitrarily large Higgs masses. This may be cured by higher-order corrections, but this *a priori* requires the solution of the magnetic mass problem. The problem is unsolved to date. Instead, to estimate the effect of such a mass, we follow Buchmüller et al. [183] by adding a proposed transverse plasmon mass of the order of g^2 by hand. The first-order transition then disappears for sufficiently large Higgs masses.

4.5.2 Scalar field theories

In this section we describe in detail how to compute the constraint effective potential in perturbation theory for a scalar field theory with quartic self-interaction λ. The two leading orders will be computed that turn out to be of order λ and $\lambda^{3/2}$. We discuss the issue of renormalization to obtain a finite result. We will see the natural occurrence of gap equations and their solution as an appropriate way to collect all Feynman diagrams which contribute to the same order in the coupling constant. For simplicity, we choose a simple momentum cutoff $k^2 \leq \Lambda_0^2$.

The action is given by

$$S(\phi) = S_{cl}(\phi) + S_{CT}(\phi). \qquad (4.5.1)$$

S_{cl} is the classical action

$$S_{cl}(\phi) = \int_x \left(\frac{1}{2} \sum_{\nu=0}^{3} (\partial_\nu \phi)(x)^2 + \frac{\lambda}{4!} \phi(x)^4 - \mu^2 \phi(x)^2 \right), \qquad (4.5.2)$$

with $\mu^2 > 0$ so that $S_{cl}(\phi)$ has a nontrivial minimum at $\phi(x)^2 \equiv v^2$ with

$$v^2 = \frac{12\mu^2}{\lambda} > 0. \qquad (4.5.3)$$

In the following we consider the value of v as given. For the electroweak standard model, $v = 246[GeV]$. S_{CT} represents the counterterms,

$$S_{CT}(\phi) = \int_x \left(\frac{1}{2}(Z_2 - 1) \sum_{\nu=0}^{3} (\partial_\nu \phi)(x)^2 + (Z_\lambda - 1)\frac{\lambda}{4!}\phi(x)^4 \right.$$
$$\left. -(Z_{\mu^2} - 1)\mu^2\phi(x)^2 + Z_V \right). \qquad (4.5.4)$$

We have written

$$\int_x \equiv \int_0^{T^{-1}} dx_0 \int d^3x. \qquad (4.5.5)$$

The renormalization constants Z_2, Z_λ, Z_{μ^2} and Z_V are determined by imposing four independent normalization conditions at temperature $T = 0$. Only Z_V and Z_{μ^2} enter our calculation below, whereas $Z_2 - 1$, $(Z_\lambda - 1)\lambda = O(\lambda^2)$.

We want to compute the constraint effective potential $\mathcal{V}_{\text{CEP}}$, obtained from

$$\exp\left(-V_4 \mathcal{V}_{\text{CEP}}(\overline{\phi})\right) = \int \mathcal{D}\phi \, \delta\left(\frac{1}{V_4}\int_x \phi(x)\right) \exp\left(-S(\phi + \overline{\phi})\right), \quad (4.5.6)$$

where $V_4 = T^{-1}V$, in the infinite spatial volume limit $V \to \infty$. $\overline{\phi}$ is a constant background field and ϕ a fluctuation field. According to the δ-constraint the fluctuation field has no zero-momentum contribution, $\widetilde{\phi}(k = 0) = 0$. We have

$$S(\phi + \overline{\phi}) = V_4 \left(\frac{\lambda}{4!}\overline{\phi}^4 - \frac{\lambda}{12}v^2\overline{\phi}^2 \right)$$
$$+ \int_x \left[\frac{1}{2}\sum_\nu (\partial_\nu \phi)(x)^2 + \frac{1}{2}m^2\phi(x)^2 + \frac{\lambda\overline{\phi}}{3!}\phi(x)^3 + \frac{\lambda}{4!}\phi(x)^4 \right]$$
$$+ V_4 \left((Z_\lambda - 1)\frac{\lambda}{4!}\overline{\phi}^4 - (Z_{\mu^2} - 1)\frac{\lambda}{12}v^2\overline{\phi}^2 + Z_V \right) \qquad (4.5.7)$$
$$+ \int_x \left[\frac{1}{2}(Z_2 - 1)\sum_\nu (\partial_\nu \phi)(x)^2 + \frac{1}{2}\left\{ (Z_\lambda - 1)\frac{\lambda}{2}\overline{\phi}^2 \right. \right.$$
$$\left. -(Z_{\mu^2} - 1)\frac{\lambda}{6}v^2 \right\}\phi(x)^2$$
$$\left. + (Z_\lambda - 1)\frac{\lambda\overline{\phi}}{3!}\phi(x)^3 + (Z_\lambda - 1)\frac{\lambda}{4!}\phi(x)^4 \right].$$

Here, the tree-level mass m of the fluctuation field ϕ is given by

$$m^2 = \frac{\lambda}{6}\left(3\overline{\phi}^2 - v^2\right). \tag{4.5.8}$$

It depends on the order parameter field $\overline{\phi}$ and is of $O(\lambda)$.

Using the Gaussian integration formula

$$\int \mathcal{D}\phi\, e^{-\frac{1}{2}\int_{x,y}\phi(x)A(x,y)\phi(y)} = e^{-\frac{1}{2}\int_{x}(\log\frac{A}{2\pi})(x,x)} \tag{4.5.9}$$

for positive (semi-)definite A, we obtain up to a constant

$$\mathcal{V}_{\mathrm{CEP}}(\overline{\phi}) = \mathcal{V}_{\mathrm{CEP}}^{(0)}(\overline{\phi}) + \mathcal{V}_{\mathrm{CEP}}^{(1)}(\overline{\phi}) + \mathcal{V}_{\mathrm{CEP}}^{(r)}(\overline{\phi}) \tag{4.5.10}$$

with classical part

$$\mathcal{V}_{\mathrm{CEP}}^{(0)}(\overline{\phi}) = \left(\frac{\lambda}{4!}\overline{\phi}^4 - \frac{\lambda}{12}v^2\overline{\phi}^2\right)$$
$$+ \left[(Z_\lambda - 1)\frac{\lambda}{4!}\overline{\phi}^4 - (Z_{\mu^2} - 1)\frac{\lambda}{12}v^2\overline{\phi}^2 + Z_V\right], \tag{4.5.11}$$

containing the (overall-) counterterms with one-loop contribution

$$\mathcal{V}_{\mathrm{CEP}}^{(1)}(\overline{\phi}) = \frac{1}{2}\int_k^{\sim} \ln(k^2 + m^2), \tag{4.5.12}$$

where

$$\int_k^{\sim} = T \sum_{\substack{k_0 = 2\pi T n \\ n \in \mathbf{Z}}} \int_{k^2 \le \Lambda_0^2} \frac{d^3k}{(2\pi)^3}, \tag{4.5.13}$$

and $\mathcal{V}_{\mathrm{CEP}}^{(r)}(\overline{\phi})$ represents all 1PI-vacuum graphs with at least two loops, together with their counterterm graphs.

For a more explicit representation of the one-loop part of $\mathcal{V}_{\mathrm{CEP}}$ we use the high-temperature expansion of

$$B^{(4)}(T^{-1}, m, \Lambda_0) \equiv \int_k^{\sim} \ln(k^2 + m^2), \tag{4.5.14}$$

leading to

$$B^{(4)}(T^{-1}, m, \Lambda_0) = B^{(4)}(\infty, m, \Lambda_0) + B_{conv}^{(4)}(T^{-1}, m, \Lambda_0), \tag{4.5.15}$$

where the first term on the right hand side, the zero-temperature part, involves all UV-divergencies of $B^{(4)}(T^{-1}, m, \Lambda_0)$. For large UV-cutoff Λ_0

$$B^{(4)}(\infty, m, \Lambda_0) = \frac{1}{16\pi^2}\left(\frac{\Lambda_0^4}{2}\ln\Lambda_0^2 - \frac{\Lambda_0^4}{4} + \Lambda_0^2 m^2 - \frac{m^4}{4} - \frac{m^4}{2}\ln\frac{\Lambda_0^2}{m^2} + O(\Lambda_0^{-1})\right).$$
(4.5.16)

The temperature-dependent part $B_{conv}^{(4)}(T^{-1}, m, \Lambda_0)$ has the small m/T-expansion

$$\overline{B}^{(4)}(T^{-1}, m) \equiv \lim_{\Lambda_0 \to \infty} B_{conv}^{(4)}(T^{-1}, m, \Lambda_0)$$

$$= -\frac{\pi^4}{45}T^4 + \frac{1}{12}m^2 T^2 - \frac{1}{6\pi}m^3 T$$

$$- \frac{m^4}{32\pi^2}\left(\ln\frac{m^2}{(4\pi T)^2} - \frac{3}{2} - 2\psi(1)\right) + O(\frac{m^6}{T^2}).$$
(4.5.17)

Similar expansions are derived for

$$B_n^{(4)}(T^{-1}, m, \Lambda_0) \equiv \int_k^{\sim} \frac{1}{(k^2 + m^2)^n}$$
(4.5.18)

by using

$$B_n^{(4)}(T^{-1}, m, \Lambda_0) = \frac{(-1)^{n-1}}{(n-1)!}\frac{\partial^n}{\partial(m^2)^n}B^{(4)}(T^{-1}, m, \Lambda_0)$$
(4.5.19)

together with (4.5.16) and (4.5.17). For later use we collect some results. For $m = O(\lambda^{1/2})$, cf. (4.5.8), we have

$$B_1^{(4)}(\infty, m, \Lambda_0) = \frac{\Lambda_0^2}{16\pi^2} + O(\lambda),$$

$$B_1^{(4)}(T^{-1}, m, \Lambda_0) = \frac{\Lambda_0^2}{16\pi^2} + \frac{T^2}{12} - \frac{1}{4\pi}mT + O(\lambda),$$
(4.5.20)

$$B_2^{(4)}(T^{-1}, m, \Lambda_0) = \frac{T}{8\pi}\frac{1}{m} + O(\lambda^0).$$

With (4.5.17), we obtain for the one-loop part $\mathcal{V}_{\text{CEP}}{}^{(1)}(\overline{\phi})$ of the constraint effective potential

$$\mathcal{V}_{\text{CEP}}{}^{(1)}(\overline{\phi}) = \frac{1}{16\pi^2}\left(\frac{\Lambda_0^4}{4}\ln\Lambda_0^2 - \frac{\Lambda_0^4}{8} + \frac{1}{2}\Lambda_0^2 m^2\right)$$

$$- \frac{\pi^4}{90}T^4 + \frac{1}{24}m^2 T^2 - \frac{1}{12\pi}m^3 T + O(\lambda^2).$$
(4.5.21)

In order to obtain a finite one-loop result of order λ, the counterterm constants Z_V and Z_{μ^2} must be chosen appropriately. We impose on the two-point vertex function the zero-temperature normalization condition

$$\widetilde{\Gamma}_2^{(2)}(p = 0|T = 0) = -m^2. \tag{4.5.22}$$

Writing

$$\text{—▫—} = (Z_{\mu^2} - 1)\frac{\lambda}{6}v^2 \tag{4.5.23}$$

for the mass-counterterm vertex, the renormalization condition (4.5.22) means that

$$\text{—▫—} + \left(\frac{1}{2}\,\bigcirc\!\!\!\!\!\!- + \frac{1}{2}\,-\!\!\bigcirc\!\!\bigcirc\!\!-\right)_{p=T=0}$$

$$= \text{—▫—} + \frac{1}{2}\,\bigcirc\!\!\!\!\!\!-\,\Big|_{T=0} + O(\lambda^2) \tag{4.5.24}$$

$$= O(\lambda^2);$$

this implies that to order λ,

$$\text{—▫—} \simeq -\frac{1}{2}\,\bigcirc\!\!\!\!\!\!-\,\Big|_{T=0} = -\frac{1}{2}(-\lambda)\int_k^{\sim} \frac{1}{k^2 + m^2}\Big|_{T=0}$$

$$= \frac{\lambda}{2}B_1^{(4)}(\infty, m, \Lambda_0) \simeq \frac{1}{16\pi^2}\frac{\lambda}{2}\Lambda_0^2. \tag{4.5.25}$$

Here and in the remainder of this chapter, we write the topological symmetry factors of the various graphs explicitly. This is particularly useful for the various vacuum graphs we will encounter below and simplifies the notation. Together with the (zero-temperature) vacuum counterterm constant

$$Z_V = Z_V^{(0)} = \frac{1}{16\pi^2}\left(-\frac{\Lambda_0^4}{4}\ln\Lambda_0^2 + \frac{\Lambda_0^4}{8} + \frac{\Lambda_0^2}{12}\lambda v^2\right) \tag{4.5.26}$$

we finally obtain

$$\mathcal{V}_{\text{CEP}}(\overline{\phi}) = \mathcal{V}_{\text{CEP}}^{(0)+(1)}(\overline{\phi}) + \mathcal{V}_{\text{CEP}}^{(r)}(\overline{\phi}) \tag{4.5.27}$$

with

$$\mathcal{V}_{\text{CEP}}{}^{(0)+(1)}(\overline{\phi}) = \left(\frac{\lambda}{4!}\overline{\phi}^4 - \frac{\lambda}{12}v^2\overline{\phi}^2\right)$$
$$-\frac{\pi^4}{90}T^4 + \frac{1}{24}m^2T^2 - \frac{1}{12\pi}m^3T + O(\lambda^2\ln\lambda). \tag{4.5.28}$$

Formula (4.5.28) represents $\mathcal{V}_{\text{CEP}}$ in one-loop order. There are some apparent problems with this expression. First, for small $\overline{\phi}$, the mass m, (4.5.8), becomes imaginary, and $\mathcal{V}_{\text{CEP}}$ becomes complex. This has no physical meaning. The reason for this strange behavior is that the integral (4.5.9) actually does not exist because A is no longer positive. Second, even for $m^2 > 0$, the potential $\mathcal{V}_{\text{CEP}}(\overline{\phi})$ is not convex in $\overline{\phi}$, although we know that the correct constraint effective potential has to be convex in the infinite volume (and agrees with the effective potential $\mathcal{V}_{\text{stat}}$). This indicates that the small-λ expansion of $\mathcal{V}_{\text{CEP}}$ is not uniform in the volume. That is, the weak coupling expansion does not give the correct $\mathcal{V}_{\text{CEP}}$ for large volumes.

These conceptual problems shed some doubt on the perturbative computation of $\mathcal{V}_{\text{CEP}}$. We cannot solve the second problem. As a matter of fact, the nonconvexity of $\mathcal{V}_{\text{CEP}}$ computed for small λ is a perturbative artifact. One can at least hope that the local minima of $\mathcal{V}_{\text{CEP}}$ which also belong to the convex hull of $\mathcal{V}_{\text{CEP}}$ are physically significant in the sense that they are related to the ground states.

On the other hand, the first problem, namely that $m^2 < 0$ for small $\overline{\phi}$, can be ascribed to a one-loop artifact and is traced back to the fact that the loop expansion is not an expansion for small λ. This is because $m^2 = O(\lambda)$ and not $O(\lambda^0)$. The mass enters the denominators of the propagators in the Feynman diagrams. The standard simple counting of vertices or loops to obtain the order of a Feynman graph does not apply. For $\mathcal{V}_{\text{CEP}}(\overline{\phi})$ this implies that there are Feynman graphs with two or more loops which are of the same order λ and $\lambda^{3/2}$ as the one-loop part. And there are infinitely many of them! For instance, every diagram of the form

$$(4.5.29)$$

contributes to the order λ to $\mathcal{V}_{\text{CEP}}$. To obtain the correct small-λ expansion we have to find and sum all graphs of the same order. As we shall see, this cures the problem of the negative mass squared.

Instead of the straightforward way of identifying and summing all diagrams which contribute to order λ and $\lambda^{3/2}$ to $\mathcal{V}_{\mathrm{CEP}}(\overline{\phi})$, we can do better. We proceed along the lines which remove the IR-problem in high-temperature massless ϕ^4-theory, as described in section 4.4. Let us briefly recapitulate. The IR divergencies are removed by summing appropriate diagrams which provide a mass for the *a priori* massless field. The resummed propagator then becomes

$$(p^2 + m_T^2)^{-1}, \qquad (4.5.30)$$

with $m_T^2 = c_T \lambda$ and $c_T = T^2/24 > 0$. One then must verify that the new Feynman-diagrammatic expansion with free propagators (4.5.30) becomes a systematic expansion for small λ. That is, to every finite order in λ there are only finitely many graphs that contribute. This is shown on behalf of IR-power counting criteria. The following sufficient condition is essential: the one-loop correction to the two-point vertex function at zero momentum is suppressed by some power of λ, i.e.

$$\frac{\widetilde{\Gamma}^{(2)}(p=0|T) + m_T^2}{m_T^2} = O(\lambda^\alpha) \qquad (4.5.31)$$

with some $\alpha > 0$. Equation (4.5.31) is a so-called gap equation.

Our present case is only slightly more complicated than in section 4.4, since we have also three-point vertices, but the arguments otherwise apply here, too. Before we proceed, we convince ourselves that the analogue of (4.5.31),

$$\frac{\widetilde{\Gamma}^{(2)}(p=0|T) + m^2}{m^2} = O(\lambda^\alpha), \quad \alpha > 0, \qquad (4.5.32)$$

with m^2 given by (4.5.8), is actually violated, so that a further mass resummation is inevitable. For, we have to one-loop order

$$\widetilde{\Gamma}^{(2)}(p = 0|T) + m^2$$

$$= \frac{1}{16\pi^2} \frac{\lambda}{2} \Lambda_0^2 - \frac{\lambda}{2} B_1^{(4)}(T^{-1}, m, \Lambda_0) + \frac{1}{2}(\lambda\overline{\phi})^2 B_2^{(4)}(T^{-1}, m, \Lambda_0)$$

$$= \frac{\lambda}{2} \frac{\Lambda_0^2}{16\pi^2} - \frac{\lambda}{2}\left(\frac{\Lambda_0^2}{16\pi^2} + \frac{T^2}{12}\right) + O(\lambda^{3/2}) \qquad (4.5.33)$$

$$= -\frac{\lambda}{24}T^2 + O(\lambda^{3/2}),$$

so that

$$\frac{\widetilde{\Gamma}^{(2)}(p = 0|T) + m^2}{m^2} = -\frac{T^2}{4(3\overline{\phi}^2 - v^2)} \qquad (4.5.34)$$

is not small for small λ.

4.5.2.1 *Mass resummation and gap equations*

A relation of the form (4.5.32) is achieved if all contributions to $\widetilde{\Gamma}^{(2)}(p = 0|T)$ which are of order λ are summed into the propagator mass. Mass resummation is equivalent to replace the propagator

$$(p^2 + m^2)^{-1} \qquad (4.5.35)$$

by

$$(p^2 + m^2 + \delta m^2)^{-1} \qquad (4.5.36)$$

and simultaneously add to the counterterm action a contribution

$$-\int_x \frac{1}{2}\delta m^2 \phi(x)^2. \qquad (4.5.37)$$

The new free propagator that enters the Feynman diagrams is now given by

$= \left(p^2 + m^2 + \delta m^2\right)^{-1}, \qquad (4.5.38)$

and we have an additional mass-counterterm vertex

$$\underline{\quad\times\quad} \;\; = \;\; \delta m^2 \, . \tag{4.5.39}$$

In order to achieve that after resummation the Feynman-diagram expansion becomes a small-λ expansion with only a finite number of diagrams contributing to a given order in λ, δm^2 must be found such that the gap equation

$$\widetilde{\Gamma}^{(2)}(p = 0|T) + (m^2 + \delta m^2) \;\stackrel{!}{=}\; O(\lambda^{1+\alpha} \ln^{\#} \lambda) \tag{4.5.40}$$

holds for some $\alpha > 0$. Here and below the notation $\ln^{\#} \lambda$ implies some generic power of $\ln \lambda$ (typically $(\ln \lambda)^{l_\Gamma}$, with l_Γ the number of loops of a graph Γ which contributes to $\widetilde{\Gamma}^{(2)}$). In graphical terms, we search for a solution of

$$\underline{\quad(R)\quad} \;\equiv\; \underline{\quad\times\quad} + \left(\frac{1}{2} \; \bigcirc \; + \frac{1}{2} \; \bigcirc \right)_{p=0}$$

$$+ \quad \underline{\quad\square\quad} \;=\; O(\lambda^{1+\alpha} \ln^{\#} \lambda) \, . \tag{4.5.41}$$

Below we will solve the gap equation for $\alpha = 1$. Let us assume for a moment that we have done it already. $\mathcal{V}_{\mathrm{CEP}}$ then becomes

$$\mathcal{V}_{\mathrm{CEP}}(\overline{\phi}) \;=\; \left(\frac{\lambda}{4!}\overline{\phi}^4 - \frac{\lambda}{12}v^2\overline{\phi}^2 \right) + Z_V - (Z_{\mu^2} - 1)\frac{\lambda}{12}v^2\overline{\phi}^2 \tag{4.5.42}$$

$$+ \int_k \ln(k^2 + (m^2 + \delta m^2)) + \overline{U}_{3/2}(\overline{\phi}) + O(\lambda^2 \ln^{\#} \lambda).$$

$\overline{U}_{3/2}(\overline{\phi})$ consists of 1PI-vacuum graphs with at least two loops, together with their counterterm vertices (including (4.5.39)), that contribute to order $\lambda^{3/2}$. We know that there are only very few graphs contributing to $\overline{U}_{3/2}(\overline{\phi})$.

To find the graphs of $\overline{U}_{3/2}(\overline{\phi})$ we adapt the power counting rules of section 4.4 to the present case. The general rule to find the order of a Feynman diagram is as follows. Let Γ be any 1PI-vacuum diagram, with $V_{\Gamma,3}$ and $V_{\Gamma,4}$ the number of its three- and four-point vertices, respectively, and l_Γ the number of loops of Γ. Then the Feynman integral I_Γ associated with Γ contributes to the following order in λ.

- If for all Zimmermann subspaces H,

$$\underline{\mathrm{degr}}_H I_\Gamma > 0 \tag{4.5.43}$$

with the IR-degrees counted as if all propagators were massless, we have

$$I_\Gamma \lesssim \lambda^{V_{\Gamma,3}+V_{\Gamma,4}}. \tag{4.5.44}$$

- In general, for $l_\Gamma \geq 2$ we have the worse bound

$$I_\Gamma \lesssim \lambda^{\frac{3}{4}V_{\Gamma,3}+\frac{1}{2}V_{\Gamma,4}+\frac{1}{2}}(\ln \lambda)^{l_\Gamma}. \tag{4.5.45}$$

Applying these rules, we find that only the following two graphs contribute to $\overline{U}_{3/2}(\overline{\phi})$,

$$\tag{4.5.46}$$

together with their counterterm graphs, cf. below, but not, for instance, the graphs

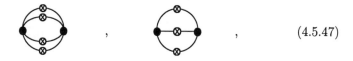

$$\tag{4.5.47}$$

which are of order λ^2 and $\lambda^2 \ln^2 \lambda$, respectively. Actually, a careful consideration shows that the second graph of (4.5.46), together will all its counterterm graphs, is of order $\lambda^2 \ln^\# \lambda$ as a result of the gap equation (4.5.40) with $\alpha > 1$.

In summary, we get

$$
V_{\text{CEP}}(\overline{\phi}) = \left(\frac{\lambda}{4!}\overline{\phi}^4 - \frac{\lambda}{12}v^2\overline{\phi}^2 \right) + Z_V - (Z_{\mu^2} - 1)\frac{\lambda}{12}v^2\overline{\phi}^2
$$

$$
+ \int_k^{\sim} \ln(k^2 + (m^2 + \delta m^2))
$$

$$
- \frac{1}{8} \; \text{[diagram]} \; - \frac{1}{2} \; \text{[diagram]} \; - \frac{1}{2} \; \text{[diagram]}
$$

$$
+ O(\lambda^2 \ln^{\#} \lambda)
$$

$$
= \left(\frac{\lambda}{4!}\overline{\phi}^4 - \frac{\lambda}{12}v^2\overline{\phi}^2 \right) + Z_V - \frac{1}{16\pi^2}\frac{\lambda}{2}\Lambda_0^2\frac{\overline{\phi}^2}{2}
$$

$$
+ \int_k^{\sim} \ln(k^2 + (m^2 + \delta m^2))
$$

$$
+ \frac{1}{8} \; \text{[diagram]} \; + \frac{1}{4} \; \text{[diagram]}
$$

$$
- \frac{1}{2}\left(\; \text{[diagram]} \; + \frac{1}{2} \; \text{[diagram]} \; + \frac{1}{2} \; \text{[diagram]} \right.
$$

$$
\left. + \; \text{[diagram]} \; \right) \; + O(\lambda^2 \ln^{\#} \lambda).
$$

The box around the subdiagram indicates that it is taken at zero external momentum. By the gap equation (4.5.41) the last bracket is

$$
O(\lambda^{1+\alpha} \ln^{\#} \lambda) \cdot O(\lambda^0) = O(\lambda^{1+\alpha} \ln^{\#} \lambda). \tag{4.5.48}
$$

Hence, because we solve the gap equation with $\alpha = 1$, we obtain

$$
V_{\text{CEP}}(\overline{\phi}) = \left(\frac{\lambda}{4!}\overline{\phi}^4 - \frac{\lambda}{12}v^2\overline{\phi}^2 \right) + Z_V - \frac{1}{16\pi^2}\frac{\lambda\overline{\phi}^2\Lambda_0}{4}
$$

$$
+ \int_k^{\sim} \ln(k^2 + (m^2 + \delta m^2))
$$

$$
+ \frac{1}{8} \; \text{[diagram]} \; + \frac{1}{4} \; \text{[diagram]}
$$

$$
+ O(\lambda^2 \ln^{\#} \lambda). \tag{4.5.49}
$$

It remains to solve (4.5.41) with $\alpha = 1$ and to compute the two diagrams of (4.5.49).

To solve the gap equation (4.5.41) with $\alpha = 1$, we compute

$$
\begin{aligned}
\text{—(R)—} \;=\; & \delta m^2 + \frac{1}{2}(-\lambda)B_1^{(4)}(T^{-1}, \sqrt{m^2 + \delta m^2}, \Lambda_0) \\
& + \frac{1}{2}(\lambda\bar\phi)^2 B_2^{(4)}(T^{-1}, \sqrt{m^2 + \delta m^2}, \Lambda_0) + \frac{1}{2}\lambda B_1^{(4)}(\infty, m, \Lambda_0) \\
=\; & \delta m^2 - \frac{\lambda}{2}\left[\frac{\Lambda_0^2}{16\pi^2} + \frac{T^2}{12} - \frac{T}{4\pi}\sqrt{m^2 + \delta m^2} + O(\lambda\ln\lambda)\right] \\
& + \frac{1}{2}(\lambda\bar\phi)^2\left[\frac{T}{8\pi}\frac{1}{\sqrt{m^2 + \delta m^2}} + o(1)\right] + \frac{\lambda}{2}\frac{\Lambda_0^2}{16\pi^2} \\
=\; & \delta m^2 + (-\lambda)\left[\frac{T^2}{24} - \frac{T}{8\pi}\sqrt{m^2 + \delta m^2} - \frac{T}{16\pi}\frac{\lambda\bar\phi^2}{\sqrt{m^2 + \delta m^2}}\right] \\
& + O(\lambda^2\ln\lambda) \\
\overset{!}{=}\; & O(\lambda^2\ln^\#\lambda) \, .
\end{aligned}
\tag{4.5.50}
$$

The solution is

$$
\delta m^2 = \lambda\frac{T^2}{24} - \lambda\frac{T}{8\pi}\left[\left(m^2 + \lambda\frac{T^2}{24}\right)^{1/2} + \frac{\lambda\bar\phi^2}{2}\left(m^2 + \lambda\frac{T^2}{24}\right)^{-1/2}\right]
$$
$$
+ O(\lambda^2\ln^\#\lambda) \, .
\tag{4.5.51}
$$

Furthermore,

$$
\begin{aligned}
\frac{1}{4}\;\text{⊗(⊗▢⊗)}\;=\; & \frac{\lambda\bar\phi^2}{4}B_1^{(4)}(T^{-1}, \sqrt{m^2 + \delta m^2}, \Lambda_0) \\
& \cdot B_2^{(4)}(T^{-1}, \sqrt{m^2 + \delta m^2}, \Lambda_0) \\
=\; & \frac{(\lambda\bar\phi)^2}{4}\left[\frac{\Lambda_0^2}{16\pi^2} + \frac{T^2}{12} - \frac{T}{4\pi}\sqrt{m^2 + \delta m^2} + O(\lambda\ln^\#\lambda)\right] \\
& \cdot\left[\frac{1}{16\pi^2}\left(-1 - 2\psi(1) + \ln\frac{\Lambda_0^2}{(4\pi T)^2}\right) + \frac{T}{8\pi}\frac{1}{\sqrt{m^2 + \delta m^2}} + O(\lambda)\right] \\
=\; & \frac{(\lambda\bar\phi)^2}{4}\frac{T}{8\pi}\frac{1}{\sqrt{m^2 + \delta m^2}}\left(\frac{\Lambda_0^2}{16\pi^2} + \frac{T^2}{12}\right) + O(\lambda^2\ln^\#\lambda) \, .
\end{aligned}
\tag{4.5.52}
$$

Finally,

$$\frac{1}{8}\;\bigotimes\!\bigotimes\; = \;-\frac{\lambda}{8}\left[B_1^{(4)}(T^{-1},\sqrt{m^2+\delta m^2},\Lambda_0)\right]^2$$

$$= -\frac{\lambda}{8}\left[\frac{\Lambda_0^2}{16\pi^2}+\frac{T^2}{12}-\frac{T}{4\pi}\sqrt{m^2+\delta m^2}+O(\lambda\ln\lambda)\right]^2$$

$$= -\frac{\lambda}{8}\left[\left(\frac{\Lambda_0^2}{16\pi^2}+\frac{T^2}{12}\right)^2-\left(\frac{\Lambda_0^2}{16\pi^2}+\frac{T^2}{12}\right)\frac{T}{2\pi}\sqrt{m^2+\delta m^2}\right]$$

$$+\,O(\lambda^2\ln^{\#}\lambda).\tag{4.5.53}$$

Inserting (4.5.51), (4.5.52) and (4.5.53) into (4.5.49) and writing $Z_V = Z_V^{(0)}+\lambda Z_V^{(0)}$, with $Z_V^{(0)}$ given by (4.5.26), we obtain up to terms which vanish as $\Lambda_0\to\infty$

$$\mathcal{V}_{\text{CEP}}(\overline{\phi}) = \left(\frac{\lambda}{4!}\overline{\phi}^4-\frac{\lambda}{12}v^2\overline{\phi}^2\right)+Z_V-\frac{1}{16\pi^2}\frac{\lambda\overline{\phi}^2}{4}\Lambda_0^2$$

$$+\left[\frac{1}{16\pi^2}\left(\frac{\Lambda_0^4}{4}\ln\Lambda_0^2-\frac{\Lambda_0^4}{8}+\frac{1}{2}\Lambda_0^2(m^2+\delta m^2)\right)\right.$$

$$\left.-\frac{\pi^4}{90}T^4+\frac{1}{24}(m^2+\delta m^2)T^2-\frac{T}{12\pi}(m^2+\delta m^2)^{3/2}\right]$$

$$-\frac{\lambda}{8}\left[\left(\frac{\Lambda_0^2}{16\pi^2}+\frac{T^2}{12}\right)^2-\left(\frac{\Lambda_0^2}{16\pi^2}+\frac{T^2}{12}\right)\frac{T}{2\pi}\sqrt{m^2+\delta m^2}\right]$$

$$+\frac{(\lambda\overline{\phi})^2}{4}\frac{T}{8\pi}\frac{1}{\sqrt{m^2+\delta m^2}}\left(\frac{\Lambda_0^2}{16\pi^2}+\frac{T^2}{12}\right)+O(\lambda^2\ln^{\#}\lambda)$$

$$= \left(\frac{\lambda}{4!}\overline{\phi}^4-\frac{\lambda}{12}v^2\overline{\phi}^2\right)+\lambda Z_V^{(1)}$$

$$-\frac{\pi^4}{90}T^4+\frac{T^2}{24}(m^2+\frac{\lambda T^2}{24})-\frac{T}{12\pi}(m^2+\frac{\lambda T^2}{24})^{3/2}$$

$$-\frac{\lambda}{8}\left(\frac{\Lambda_0^2}{16\pi^2}\right)^2-\frac{\lambda}{1152}T^4+O(\lambda^2\ln^{\#}\lambda).\tag{4.5.54}$$

Choosing the vacuum counterterm $Z_V^{(1)}$ as

$$\lambda Z_V^{(1)} = \frac{\lambda}{8}\left(\frac{\Lambda_0^2}{16\pi^2}\right)^2\tag{4.5.55}$$

(independent of T and $\overline{\phi}$), we finally obtain the (renormalized) constraint effective potential $\mathcal{V}_{\mathrm{CEP}}(\overline{\phi})$ as

$$
\mathcal{V}_{\mathrm{CEP}}(\overline{\phi}) = \left(\frac{\lambda}{4!} \overline{\phi}^4 + \frac{1}{2} \overline{m}_T^2 \overline{\phi}^2 \right) - \frac{T}{12\pi} \left(\frac{\lambda}{2} \overline{\phi}^2 + \overline{m}_T^2 \right)^{3/2}
$$
$$
- \left(\frac{\pi^4}{90} + \frac{\lambda}{1152} \right) T^4 + \frac{\overline{m}_T^2 T^2}{24} + O(\lambda^2 \ln^\# \lambda) , \tag{4.5.56}
$$

where

$$
\overline{m}_T^2 = \frac{\lambda}{6} \left(-v^2 + \frac{T^2}{4} \right) . \tag{4.5.57}
$$

Equation (4.5.56) is the constraint effective potential for the one-component ϕ^4-theory, (4.5.1) f, to order $\lambda^{3/2}$.

For the following considerations we give also the result of $\mathcal{V}_{\mathrm{CEP}}(\overline{\phi})$ for the N-component, $O(N)$-symmetric scalar theory. For $N = 4$ this is the scalar part of the electroweak standard model.

$$
\mathcal{V}_{\mathrm{CEP}}(\overline{\phi}) = \left(\frac{\lambda}{4!} \overline{\phi}^4 + \frac{1}{2} \overline{m}_T^2 \overline{\phi}^2 \right)
$$
$$
- \frac{T}{12\pi} \left(\left[\frac{\lambda}{2} \overline{\phi}^2 + \overline{m}_T^2 \right]^{3/2} + (N-1) \left[\frac{\lambda}{6} \overline{\phi}^2 + \overline{m}_T^2 \right]^{3/2} \right)
$$
$$
+ N \left\{ \frac{\overline{m}_T^2 T^2}{24} - \left(\frac{\pi^4}{90} + \frac{\lambda}{1152} \frac{N+2}{3} \right) T^4 \right\}
$$
$$
+ O(\lambda^2 \ln^\# \lambda) , \tag{4.5.58}
$$

where now

$$
\overline{m}_T^2 = \frac{\lambda}{6} \left(-v^2 + \frac{T^2}{4} \frac{N+2}{3} \right) . \tag{4.5.59}
$$

For $N = 1$, this agrees with (4.5.56) and (4.5.57). The derivation of (4.5.58) is not considerably more involved than that of (4.5.56) for the one-component model. The main complication is that instead of a single gap equation we now have to solve a system of two coupled gap equations. Their solution provides $\overline{\phi}$-dependent different mass-shifts for the Higgs field and the $N - 1$ Goldstone fields. They generate the two "cubic" terms in (4.5.58).

In the next section we investigate what the constraint effective potential (4.5.58) predicts for the order of the high-temperature phase transition.

4.5.3   *The order of the phase transition*

Let us consider a hypothetical constraint effective potential of the form

$$
\mathcal{V}_{\mathrm{CEP}}(\overline{\phi}) = \left(\frac{\lambda}{4!}\,\overline{\phi}^4 + \frac{a}{2}\,T^2\,\overline{\phi}^2 \right) - \sum_{i=1}^{n} \frac{b_i T}{3} \left(\overline{\phi}^2 + c_i^2\,T^2 \right)^{3/2} \tag{4.5.60}
$$

with a, b_i and c_i dimensionless but T-dependent parameters. All constraint effective potentials we consider in this section fit into this form.

Which order does a $\mathcal{V}_{\mathrm{CEP}}(\overline{\phi})$ of the above form predict for the high-temperature phase transition? In view of the electroweak phase transition that we will discuss in the next section, we formulate and use the following criterion for the phase transition to be of first order.

Barrier condition. There is a temperature for which

$$
\left. \frac{\partial^2 \mathcal{V}_{\mathrm{CEP}}(\overline{\phi})}{\partial \overline{\phi}^2} \right|_{\overline{\phi}=0} = 0. \tag{4.5.61}
$$

Let us denote by T_b the largest temperature for which (4.5.61) holds and call it the barrier temperature.

First- order condition. Let the temperature be $T = T_b$. The transition is of first order if and only if there is a real $\overline{\phi}_0 \neq 0$ where $\mathcal{V}_{\mathrm{CEP}}(\overline{\phi} = \overline{\phi}_0)$ has an absolute minimum.

This behavior is illustrated in Fig. 4.5.1.

The potential (4.5.60) is symmetric with respect to $\overline{\phi} \to -\overline{\phi}$, $\mathcal{V}_{\mathrm{CEP}}(-\overline{\phi}) = \mathcal{V}_{\mathrm{CEP}}(\overline{\phi})$. With

$$
\frac{\partial \mathcal{V}_{\mathrm{CEP}}(\overline{\phi})}{\partial \overline{\phi}^2} = \frac{\lambda}{12}\,\overline{\phi}^2 + \frac{a}{2}\,T^2 - \sum_{i=1}^{n} \frac{b_i T}{2} \left(\overline{\phi}^2 + c_i^2\,T^2 \right)^{1/2} \tag{4.5.62}
$$

and

$$
\frac{\partial^2 \mathcal{V}_{\mathrm{CEP}}(\overline{\phi})}{\partial \left(\overline{\phi}^2 \right)^2} = \frac{\lambda}{12} - \sum_{i=1}^{n} \frac{b_i T}{4} \left(\overline{\phi}^2 + c_i^2\,T^2 \right)^{-1/2}, \tag{4.5.63}
$$

the barrier condition (4.5.61) becomes

$$
a - \sum_{i=1}^{n} b_i c_i = 0 \tag{4.5.64}
$$

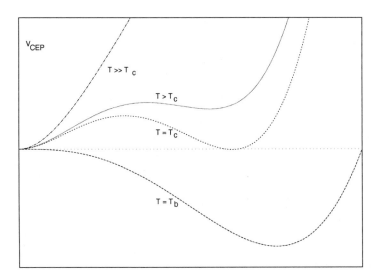

Fig. 4.5.1 A qualitative plot of the constraint effective potential $\mathcal{V}_{\mathrm{CEP}}(\overline{\phi})$ for various temperatures and for a first-order phase transition. At the barrier temperature $T = T_b$, $\mathcal{V}_{\mathrm{CEP}}(\overline{\phi})$ has a nontrivial minimum. Above the phase transition temperature T_c, the absolute minimum resides at $\overline{\phi} = 0$.

to be satisfied for the barrier temperature $T = T_b$. Without loss of generality we have assumed that $c_i \geq 0$. The finite-temperature phase transition is then of first order if we have at $T = T_b$ (where (4.5.64) holds)

$$\lambda < \sum_{i=1}^{n} 3\,\frac{b_i}{c_i}. \tag{4.5.65}$$

We apply this criterion to the $O(N)$-symmetric ϕ^4-theory. The constraint effective potential is given by (4.5.58). For general N we have $n = 2$,

$$a = \frac{\overline{m}_T^2}{T^2} = \frac{\lambda}{6}\left(\frac{1}{4}\frac{N+2}{3} - \frac{v^2}{T^2}\right), \tag{4.5.66}$$

and

$$b_1 = \frac{1}{4\pi} \left(\frac{\lambda}{2} \right)^{3/2}, \, c_1 = \left(\frac{2}{\lambda} \frac{\overline{m}_T^2}{T^2} \right)^{1/2},$$

$$b_2 = \frac{N-1}{4\pi} \left(\frac{\lambda}{6} \right)^{3/2}, \, c_2 = \left(\frac{6}{\lambda} \frac{\overline{m}_T^2}{T^2} \right)^{1/2}. \quad (4.5.67)$$

The barrier-temperature condition (4.5.64) yields

$$\overline{m}_T = \frac{1}{4\pi} \frac{\lambda}{2} \frac{N+2}{3} T, \quad (4.5.68)$$

and the first-order criterion becomes

$$\lambda < \lambda \frac{N+8}{2(N+2)} \quad (4.5.69)$$

or, whenever $\lambda \neq 0$,

$$N < 4. \quad (4.5.70)$$

This is a surprising result. The constraint effective potential (4.5.58) of the $O(N)$ models predicts that the high-temperature transition is of second order for $N = 4, 5, \ldots$, but is of first order for $N = 1, 2, 3$. On the other hand it is strongly supported by various other methods that the transition is actually of second order for all N. These methods include the renormalization group and high order ϵ-expansion, flow equations, dimensional reduction and high-order linked cluster expansions. We will discuss some of these approaches in detail later in this book. The question remains why does $\mathcal{V}_{\text{CEP}}(\overline{\phi})$, Eq. (4.5.58), give such a strange result?

Let us summarize what we have gained so far. We computed the constraint effective potential to subleading order in λ. We saw that the straight-forward loop expansion leads to an expression (4.5.27) (4.5.28) that does not collect all contributions to order $\lambda^{3/2}$, not even to order λ. This manifests itself in imaginary masses and a complex valued potential for small order parameter $\overline{\phi}$. A possible solution of the problem is to identify and sum the infinitely many Feynman graphs which contribute to $\mathcal{V}_{\text{CEP}}(\overline{\phi})$ to order λ and $\lambda^{3/2}$. A more systematic way that is also easier to manage is to sum all diagrams which contribute to the two-point vertex function $\widetilde{\Gamma}^{(2)}(p = 0|T)$ in the same order as the tree level mass m^2, so that the new free propagator becomes

$$\left(p^2 + m^2 + \delta m^2 \right)^{-1}. \quad (4.5.71)$$

This amounts to finding δm^2 as a solution of the so-called gap equation

$$\frac{\widetilde{\Gamma}^{(2)}(p = 0|T) + (m^2 + \delta m^2)}{m^2 + \delta m^2} \overset{!}{=} O(\lambda^\alpha \ln^\# \lambda) \qquad (4.5.72)$$

for some $\alpha > 0$. Here we have to consider only a few graphs. If in addition δm^2 satisfies

$$m^2 + \delta m^2 = c\lambda, \qquad (4.5.73)$$

with $c \neq 0$ some λ-independent constant, the new diagrammatic expansion is a small λ-expansion, with only a finite number of Feynman diagrams contributing to every finite order of λ. This is satisfied for the ϕ^4-theory, cf. (4.5.51). In turn, we identify the few diagrams that contribute to $\mathcal{V}_{\mathrm{CEP}}(\overline{\phi})$ in order $\lambda^{3/2}$, which yields (4.5.58).

The perturbative computation yields a constraint effective potential that is not convex. We know that the true $\mathcal{V}_{\mathrm{CEP}}(\overline{\phi})$ is convex in the infinite volume, so the nonconvexity is a perturbative artifact. Nevertheless, the absolute minimum is usually related to the stable ground state of the system because it belongs to the convex hull.

If we consider $\mathcal{V}_{\mathrm{CEP}}(\overline{\phi})$ to order λ only, i.e. ignore in (4.5.58) all terms of order $\lambda^{3/2}$, there are no "cubic" terms. The high temperature phase transition inferred from it is of second order. If the $\lambda^{3/2}$-contributions are included, which provide the first nontrivial fluctuations, it stays of second order for $N \geq 4$, but becomes of first order for $N < 4$. We conclude that the next contributions to $\mathcal{V}_{\mathrm{CEP}}(\overline{\phi})$ of order $\lambda^2 \ln^\# \lambda$ will become important. One should see if their inclusion cures the strange behavior for $N < 4$ and leads to a second-order prediction for all N.

There is a second, independent reason to include at least the terms of order $\lambda^2 \ln^\# \lambda$. The result of $\mathcal{V}_{\mathrm{CEP}}(\overline{\phi})$ to order $\lambda^{3/2}$ is completely independent of the precise definition of the renormalized coupling constant λ. We have already seen, however, how important it becomes for perturbative computations in particular close to phase transitions to use a convenient definition of λ. The first order in which the renormalization of the coupling constant λ enters is the order $\lambda^2 \ln^\# \lambda$.

So far $\mathcal{V}_{\mathrm{CEP}}(\overline{\phi})$ was not (yet) calculated to order λ^2. Ultimately this computation should be done, because it becomes important for the whole concept of a perturbative computation of the constraint effective potential close to the phase transition.

### 4.5.4	*The weak-electroweak phase transition*

In this section we investigate the electroweak phase transition by means of the constraint effective potential. The pure scalar part of the standard model is an $O(4)$ symmetric self-interacting ϕ^4 theory. From the above considerations we know that the constraint effective potential predicts a second-order high temperature transition for this model. Below we will see how this is changed by the interaction with the gauge fields. We closely follow the work of Buchmüller, Fodor, Helbig and Walliser [183].

We consider the SU(2) Higgs model described by the following action

$$S = S_{gauge} + S_{Higgs} + S_{gf} + S_{gh}. \qquad (4.5.74)$$

S_{gauge} is the gauge field action

$$S_{gauge} = \int_x \sum_{\mu,\nu=0}^{3} \sum_{d=1}^{3} \frac{1}{4} \mathcal{F}_{\mu\nu}^d(x) \mathcal{F}_{\mu\nu}^d(x), \qquad (4.5.75)$$

$$\mathcal{F}_{\mu\nu}^d(x) = \partial_\mu W_\nu^d(x) - \partial_\nu W_\mu^d(x) + g \sum_{e,f=1}^{3} \epsilon_{def} W_\mu^e(x) W_\nu^f(x),$$

$d = 1, 2, 3$. The Higgs-field action is given by

$$S_{Higgs} = \int_x \left\{ \frac{1}{2} \sum_{\mu=0}^{3} (D_\mu \Phi)(x)^\dagger (D_\mu \Phi)(x) \right.$$
$$\left. - \frac{\lambda}{12} v^2 \Phi(x)^\dagger \Phi(x) + \frac{\lambda}{4!} \left(\Phi(x)^\dagger \Phi(x) \right)^2 \right\}, \qquad (4.5.76)$$

with Φ an SU(2) doublet, i.e. a complex, two-component scalar field

$$\Phi(x) = \begin{pmatrix} \chi_1 + i\chi_2 \\ \bar{\phi} + \phi + i\chi_3 \end{pmatrix}(x), \qquad (4.5.77)$$

$$D_\mu \Phi = \left(\partial_\mu - i\frac{g}{2} \sum_{d=1}^{3} \sigma_d W_\mu^d \right) \Phi,$$

and σ_d, $d = 1, 2, 3$ the Pauli matrices

$$\sigma_1 = \begin{pmatrix} 0 & 1 \\ 1 & 0 \end{pmatrix}, \ \sigma_2 = \begin{pmatrix} 0 & -i \\ i & 0 \end{pmatrix}, \ \sigma_3 = \begin{pmatrix} 1 & 0 \\ 0 & -1 \end{pmatrix}. \qquad (4.5.78)$$

For the Lorentz gauge, the gauge-fixing action is

$$S_{gf} = \int_x \frac{1}{2} \xi \sum_{d=1}^{3} \mathcal{G}_d(x) \mathcal{G}_d(x),$$

$$\mathcal{G}_d(x) = \sum_{\mu} \partial_\mu W_\mu^d(x) - \frac{1}{2\xi} g\overline{\phi} \, \chi'^d(x),$$

$$(\chi_1', \chi_2', \chi_3') = (\chi_2, \chi_1, -\chi_3), \tag{4.5.79}$$

where $\xi > 0$ is the gauge fixing parameter. The term proportional to ξ^{-1} of $\mathcal{G}_d$ is chosen such that it cancels the mixed quadratic term proportional to χW of the Higgs field action (4.5.76). In the following we choose the Landau gauge $\xi = \infty$, which implies a sequence of simplifications. Finally, for $\xi = \infty$ the ghost field action becomes

$$S_{gh} = \int_x \sum_{d,e,\mu} \overline{c}_d(x) \partial_\mu \left(\delta_{de} \partial_\mu + g\epsilon_{ade} W_\mu^a(x) \right) c_e(x). \tag{4.5.80}$$

Here and in the following we do not further specify the counterterm action for simplicity.

The $\overline{\phi}$-dependent masses of the gauge field W, the Higgs field ϕ and the Goldstone field χ are obtained from the zero-momentum part of the action which is quadratic in these fields. It is given by

$$\int_x \left\{ -\frac{\lambda}{12} v^2 \left(\chi^2 + \phi^2 \right) + \frac{\lambda}{24} \left(2\overline{\phi}^2 (\chi^2 + \phi^2) + 4\overline{\phi}^2 \phi^2 \right) \right.$$

$$\left. + \frac{g^2 \overline{\phi}^2}{8} \sum_{\mu,d} (W_\mu^d)^2 \right\} (x). \tag{4.5.81}$$

Hence, the tree-level masses which enter the propagators are given as

$$m^2 = \frac{\lambda}{6} \left(3\overline{\phi}^2 - v^2 \right), \quad \phi \text{ Higgs boson},$$

$$M^2 = \frac{\lambda}{6} \left(\overline{\phi}^2 - v^2 \right), \quad \chi \text{ Goldstone bosons}, \tag{4.5.82}$$

$$\mathcal{M}^2 = \frac{1}{4} g^2 \overline{\phi}^2, \quad \text{gauge boson}.$$

The connection between the coupling constants λ, g and the physical, zero-temperature masses m_H of the Higgs field and m_W of the W-bosons is obtained from (4.5.82) for $\overline{\phi} = v$,

$$m_H^2 = \frac{\lambda}{3} v^2 , \quad \text{Higgs-boson mass}$$

$$m_W^2 = \frac{g^2}{4} v^2 , \quad \text{gauge-boson mass} , \qquad (4.5.83)$$

with $v = 246[GeV]$.

The free gauge-field propagator becomes in Landau gauge and in momentum space

$$\widetilde{D}_{\mu\nu}(k) = \frac{1}{k^2 + \mathcal{M}^2} \left(\delta_{\mu\nu} - \frac{k_\mu k_\nu}{k^2} \right) . \qquad (4.5.84)$$

For the definition of longitudinal and transverse W-boson masses we need the full gauge field propagator. With $\widetilde{\Pi}(k)$ the vacuum polarization(-tensor), the full gauge field propagator is given by

$$\widetilde{G}(k) = \widetilde{D}(k) \sum_{n=0}^{\infty} \left(\widetilde{\Pi}\widetilde{D} \right)^n (k). \qquad (4.5.85)$$

$\widetilde{\Pi}(k)$ depends on two independent vectors, the momentum k and $u = (1, \mathbf{0})$, the unit vector in temperature direction. Hence, in general $\widetilde{\Pi}(k)$ is a sum of four linearly independent tensors. As we have seen in section 4.4, in Landau gauge $\widetilde{G}(k)$ takes a particularly simple form,

$$\widetilde{G}(k) = \frac{1}{k^2 + \mathcal{M}^2 - \Pi_L(k)} P_L(k) + \frac{1}{k^2 + \mathcal{M}^2 - \Pi_T(k)} P_T(k), \quad (4.5.86)$$

where the orthogonal projection operators P_L and P_T are defined by

$$(P_T)_{\mu\nu} = \sum_{i,j=1}^{3} \delta_{\mu i} \left(\delta_{ij} - \frac{k_i k_j}{\mathbf{k}^2} \right) \delta_{j\nu}, \qquad (4.5.87)$$

which is orthogonal to $\mathbf{k}$, and by

$$(P_L)_{\mu\nu} = \left(\delta_{\mu\nu} - \frac{k_\mu k_\nu}{k^2} \right) - (P_T)_{\mu\nu} = \frac{k^2}{\mathbf{k}^2} u_\mu^T u_\nu^T , \qquad (4.5.88)$$

with

$$u_\mu^T = u_\mu - k_\mu \frac{u \cdot k}{k^2} \qquad (4.5.89)$$

orthogonal to k. $\Pi_L(k)$ and $\Pi_T(k)$ are defined by

$$\Pi_L(k) = \sum_{\mu=0}^{3} \left(\tilde{\Pi}(k) P_L(k) \right)_{\mu\mu} \quad , \quad \Pi_T(k) = \frac{1}{2} \sum_{\mu=0}^{3} \left(\tilde{\Pi}(k) P_T(k) \right)_{\mu\mu} .$$
$$(4.5.90)$$

In perturbation theory, electric (longitudinal) and magnetic (transversal) W-boson masses are defined by

$$m_L^2 = \mathcal{M}^2 - \Pi_L(0) \quad , \quad m_T^2 = \mathcal{M}^2 - \Pi_T(0). \qquad (4.5.91)$$

In terms of P_L and P_T, the free gauge field propagator becomes

$$\tilde{D}(k) = \frac{1}{k^2 + \mathcal{M}^2} \left(P_L + P_T \right). \qquad (4.5.92)$$

To compute the constraint effective potential to subleading order in the scalar self coupling λ and the gauge coupling g, one proceeds in quite analogy to the pure scalar case. In this way one finally obtains $\mathcal{V}_{\text{CEP}}(\overline{\phi})$ to order $\lambda^{3/2}$ and g^3.

First, working out the one-loop contribution, the constraint effective potential becomes

$$\mathcal{V}_{\text{CEP}}(\overline{\phi}) = \mathcal{V}_{\text{CEP}}^{(0)+(1)}(\overline{\phi}) + \mathcal{V}_{\text{CEP}}^{(r)}(\overline{\phi}), \qquad (4.5.93)$$

with

$$\mathcal{V}_{\text{CEP}}^{(0)+(1)}(\overline{\phi}) = \frac{\lambda}{4!} \overline{\phi}^4 + \left[\left(\frac{3g^2}{16} + \frac{\lambda}{12} \right) T^2 - \frac{\lambda}{6} v^2 \right] \frac{\overline{\phi}^2}{2}$$
$$- \frac{T}{12\pi} \left(m^3 + 3M^3 + 9\mathcal{M}^3 \right), \qquad (4.5.94)$$

where $\overline{\phi}$-independent, additive contributions have been omitted. $\mathcal{V}_{\text{CEP}}^{(r)}(\overline{\phi})$ includes all 1PI-vacuum graphs with at least two loops.

The one-loop result (4.5.94) suffers from the same problems for small values of $\overline{\phi}$ which we encountered already for the pure scalar case. Furthermore, it is not the complete contribution up to and including order $\lambda^{3/2}$ and g^3. To obtain the latter we define resummed masses by

$$m_\phi^2 = m^2 + \delta m^2 \, , \, m_\chi^2 = M^2 + \delta M^2,$$
$$m_L^2 = \mathcal{M}^2 + \delta m_L^2 \, , \, m_T^2 = \mathcal{M}^2 + \delta m_T^2. \qquad (4.5.95)$$

They must be inserted into the free propagators, in addition they introduce new mass counterterms. In particular, the gauge field propagator now becomes

$$\tilde{D}(k) = \frac{1}{k^2 + m_L^2} P_L + \frac{1}{k^2 + m_T^2} P_T, \qquad (4.5.96)$$

and we have a new mass counterterm for the gauge fields, which in momentum space is given by

$$P_L \delta m_L^2 + P_T \delta m_T^2 = P_L(m_L^2 - \mathcal{M}^2) + P_T(m_T^2 - \mathcal{M}^2). \qquad (4.5.97)$$

According to [183], the four gap equations which must be fulfilled to order $\lambda^{3/2}$ and g^3 are then given by

$$
\begin{aligned}
m_\phi^2 - m^2 &= \left(\frac{3g^2}{16} + \frac{\lambda}{12}\right) T^2 \\
&\quad - \frac{3g^2}{16\pi} T\left(m_L + 2m_T + \mathcal{M}^2\left(\frac{1}{m_L} + \frac{2}{m_T}\right)\right) \\
&\quad - \frac{\lambda}{8\pi} T\left(m_\phi + m_\chi + \frac{\lambda \bar{\phi}^2}{6}\left(\frac{3}{m_\phi} + \frac{1}{m_\chi}\right)\right), \\
m_\chi^2 - \mathcal{M}^2 &= \left(\frac{3g^2}{16} + \frac{\lambda}{12}\right) T^2 - \frac{3g^2}{16\pi} T(m_L + 2m_T) \\
&\quad - \frac{\lambda}{24\pi} T\left(m_\phi + 5m_\chi + \frac{2\lambda\bar{\phi}^2}{3}\frac{1}{m_\phi + m_\chi}\right), \qquad (4.5.98) \\
m_L^2 - \mathcal{M}^2 &= \frac{11g^2}{6} T^2 \\
&\quad - \frac{g^2}{16\pi} T\left(\frac{4\mathcal{M}^2}{m_L + m_\phi} + m_\phi + 3m_\chi + 16m_T\right), \\
m_T^2 - \mathcal{M}^2 &= \frac{g^2}{3\pi} m_T T \\
&\quad - \frac{g^2}{6\pi} T\left(\frac{\mathcal{M}^2}{m_T + m_\phi} - \frac{1}{8}\frac{(m_\phi - m_\chi)^2}{m_\phi + m_\chi}\right).
\end{aligned}
$$

This system of gap equations must be solved for m_ϕ, m_χ, m_L and m_T. After some tedious algebra, the constraint effective potential becomes to the order $\lambda^{3/2}$ and g^3

$$
\begin{aligned}
\mathcal{V}_{\text{CEP}}(\bar{\phi}) &= \frac{\lambda}{4!}\bar{\phi}^4 + \left[\left(\frac{3g^2}{16} + \frac{\lambda}{12}\right) T^2 - \frac{\lambda}{6}v^2\right]\frac{\bar{\phi}^2}{2} \\
&\quad - \frac{T}{12\pi}\left(m_\phi^3 + 3m_\chi^3 + 3m_L^3 + 6m_T^3\right), \qquad (4.5.99)
\end{aligned}
$$

with masses

$$m_\phi^2 = \left[\left(\frac{3g^2}{16} + \frac{\lambda}{12}\right) T^2 - \frac{\lambda}{6} v^2\right] + \frac{\lambda}{2}\overline{\phi}^2,$$

$$m_\chi^2 = \left[\left(\frac{3g^2}{16} + \frac{\lambda}{12}\right) T^2 - \frac{\lambda}{6} v^2\right] + \frac{\lambda}{6}\overline{\phi}^2,$$

$$m_L^2 = \frac{g^2}{4}\overline{\phi}^2 + \frac{11g^2}{6}T^2,$$

$$m_T^2 = \frac{g^2}{4}\overline{\phi}^2.$$

(4.5.100)

Let us now consider what the potential (4.5.99) predicts for the high-temperature phase transition of the $SU(2)$ Higgs model. $\mathcal{V}_{\text{CEP}}(\overline{\phi})$ fits into the form (4.5.60), so that we test for the barrier condition (4.5.64) and the first-order constraint (4.5.65). In the notation of (4.5.60), we have $n = 4$,

$$a = \left(\frac{3g^2}{16} + \frac{\lambda}{12}\right) - \frac{\lambda}{6}\frac{v^2}{T^2},$$

(4.5.101)

and (cf. (4.5.67) with $N = 4$ for the pure scalar part)

$$b_1 = \frac{1}{4\pi}\left(\frac{\lambda}{2}\right)^{3/2}, \ c_1 = \left(\frac{2}{\lambda}a\right)^{1/2},$$

$$b_2 = \frac{3}{4\pi}\left(\frac{\lambda}{6}\right)^{3/2}, \ c_2 = \left(\frac{6}{\lambda}a\right)^{1/2},$$

$$b_3 = \frac{3}{4\pi}\left(\frac{g}{2}\right)^3, \ c_3 = \sqrt{\frac{22}{3}},$$

$$b_4 = \frac{6}{4\pi}\left(\frac{g}{2}\right)^3, \ c_4 = 0.$$

(4.5.102)

With these identifications the barrier condition (4.5.64) becomes

$$a = \frac{1}{4\pi}\left(\lambda\sqrt{a} + \frac{3}{4}\sqrt{\frac{11}{6}}g^3\right),$$

(4.5.103)

or

$$\sqrt{a} = \frac{\lambda}{8\pi} + \sqrt{\left(\frac{\lambda}{8\pi}\right)^2 + \frac{3}{16\pi}\sqrt{\frac{11}{6}}g^3},$$

(4.5.104)

to be realized at some barrier temperature $T = T_b$. Because of $c_4 = 0$, the condition (4.5.65) becomes

$$\lambda < \infty,$$

(4.5.105)

that is, for any finite Higgs self-coupling λ the phase transition is of first order.

Whereas for the pure scalar model the high-temperature transition is of second order, we now have the result that $\mathcal{V}_{\text{CEP}}$ predicts the electroweak phase transition to be of first order. The conversion is induced by the interactions with the gauge fields. The prediction agrees with results obtained by other methods. Some of them will be presented later in this book.

The above results are obtained from the perturbative $\mathcal{V}_{\text{CEP}}(\overline{\phi})$ computed to order $\lambda^{3/2}$ and g^3. They are reliable for sufficiently small coupling constants only. Indeed, we have already argued in section 3.1.2.6 that the phase transition does not persist for arbitrary large λ. According to (4.5.83), small λ implies small (zero temperature) Higgs mass m_H. In the section on dimensional reduction (4.6) we will see that the first-order transition actually ends in a critical endpoint for some critical Higgs mass.

Why does $\mathcal{V}_{\text{CEP}}(\overline{\phi})$ predict a first-order transition? The reason is that $\overline{\phi} = 0$ implies a vanishing transverse plasmon mass $m_T = 0$ to order $g^{3/2}$. We should expect that the weak constraint (4.5.105) becomes replaced by a stronger and more reliable one if higher orders corrections are taken into account. They should generate a transverse mass $m_T \simeq g^2$ by solving the appropriate gap equations. A consistent computation requires to take into account all contributions to $\mathcal{V}_{\text{CEP}}(\overline{\phi})$ of order g^4. But then, unfortunately, we are running into the magnetic mass problem of high-temperature gauge theories. We have discussed this problem in section 4.4. Because of $m_T^2 = O(g^4)$, the IR-singularities of perturbation theory in the infinite volume are not removable in higher orders by mass resummation as in the case of scalar fields. For instance, the free energy is not computable to order g^4 by the resummation methods which are available so far.

For the $SU(2)$ Higgs model, as long as $\overline{\phi}$ is sufficiently large, the constraint effective potential $\mathcal{V}_{\text{CEP}}(\overline{\phi})$ does not suffer from the IR-problems. This is the case deeply in the broken phase with $\overline{\phi} \simeq v$. But in order to investigate the phase transition we need to consider $\mathcal{V}_{\text{CEP}}(\overline{\phi})$ close to $\overline{\phi} \simeq 0$.

The absence of a nontrivial solution for m_T of the gap equations (4.5.98) at $\overline{\phi} = 0$, i.e. of

$$m_T^2 = \frac{g^2}{3\pi} T m_T \qquad (4.5.106)$$

to order $g^{3/2}$ is another manifestation of the magnetic mass problem.

In order to estimate the effect of a nonvanishing transverse mass m_T for

$\overline{\phi} = 0$ we follow the reference [183]. We add to the $\overline{\phi}$-dependent solution of m_T^2, (4.5.100), the nontrivial solution of (4.5.106), weighted by some parameter ρ. For completeness we also introduce the top-quark mass m_t. $\mathcal{V}_{\text{CEP}}(\overline{\phi})$ then becomes

$$
\mathcal{V}_{\text{CEP}}(\overline{\phi}) = \frac{\lambda}{4!}\overline{\phi}^4 + \left[\left(\frac{3g^2}{16} + \frac{\lambda}{12} + \frac{m_t^2}{2v^2}\right)T^2 - \frac{\lambda}{6}v^2\right]\frac{\overline{\phi}^2}{2}
$$
$$
- \frac{T}{12\pi}\left(m_\phi^3 + 3m_\chi^3 + 3m_L^3 + 6m_T^3\right), \tag{4.5.107}
$$

where now the masses are given by

$$
m_\phi^2 = \left[\left(\frac{3g^2}{16} + \frac{\lambda}{12} + \frac{m_t^2}{2v^2}\right)T^2 - \frac{\lambda}{6}v^2\right] + \frac{\lambda}{2}\overline{\phi}^2,
$$
$$
m_\chi^2 = \left[\left(\frac{3g^2}{16} + \frac{\lambda}{12} + \frac{m_t^2}{2v^2}\right)T^2 - \frac{\lambda}{6}v^2\right] + \frac{\lambda}{6}\overline{\phi}^2,
$$
$$
m_L^2 = \frac{g^2}{4}\overline{\phi}^2 + \frac{11g^2}{6}T^2, \tag{4.5.108}
$$
$$
m_T^2 = \frac{g^2}{4}\overline{\phi}^2 + \rho\left(\frac{g^2 T}{3\pi}\right)^2.
$$

The barrier condition then becomes

$$
\sqrt{a} = \frac{\lambda}{8\pi} + \sqrt{\left(\frac{\lambda}{8\pi}\right)^2 + \frac{1}{4\pi}c_\rho(g)} \tag{4.5.109}
$$

with

$$
c_\rho(g) = \frac{3}{4}\sqrt{\frac{11}{6}}g^3\left(1 + \rho g\frac{2}{3\pi}\sqrt{\frac{6}{11}}\right). \tag{4.5.110}
$$

The condition for a first-order transition reads

$$
\lambda < \frac{\lambda^2}{4\pi}\frac{1}{\sqrt{a}} + d_\rho(g), \tag{4.5.111}
$$

where a is given by (4.5.109) and

$$
d_\rho(g) = \frac{27}{32}\frac{g^2}{\rho}\left(1 + \rho g\frac{1}{6\pi}\sqrt{\frac{6}{11}}\right). \tag{4.5.112}
$$

Let us plug in some realistic numbers. Using (4.5.83), we write

$$
\lambda = \frac{3}{4}g^2 r, \qquad r = \left(\frac{m_H}{m_W}\right)^2. \tag{4.5.113}
$$

For $m_W \simeq 80[GeV]$ and with the parameter ρ and the ratio $\sqrt{r}$ of Higgs-
and W-boson mass of the order of 1, we obtain as the barrier condition

$$\frac{3g^2}{16} + \frac{g^2}{16}\, r + \frac{m_t^2}{2v^2} - \frac{g^2}{8}\, r\, \frac{v^2}{T_b^2} \simeq \frac{3}{16\pi}\, \sqrt{\frac{11}{6}}\, g^3\,, \qquad (4.5.114)$$

or, for the barrier temperature T_b

$$\frac{T_b^2}{v^2} \simeq \frac{g^2}{8}\, r\, \left(\frac{m_t^2}{2v^2} + \frac{3}{16}\left(1 + \frac{r}{3}\right) g^2 - \frac{3}{16\pi}\, \sqrt{\frac{11}{6}}\, g^3 \right). \qquad (4.5.115)$$

Because of

$$\frac{\lambda^2}{4\pi}\, \frac{1}{\sqrt{a}} \lesssim d_\rho(g) \simeq \frac{27}{32}\, \frac{g^2}{\rho}, \qquad (4.5.116)$$

the condition for a first-order transition becomes

$$\frac{3}{4}\, g^2\, r < \frac{27}{32}\, \frac{g^2}{\rho} \qquad (4.5.117)$$

or

$$r < \frac{9}{8}\, \frac{1}{\rho}. \qquad (4.5.118)$$

There is an upper bound on the Higgs mass beyond which the first-order
transition disappears. The bound depends on the input parameter ρ.

Let us summarize. The constraint effective potential computed in the
small coupling expansion predicts the following scenario for the electroweak
phase transition. The pure scalar part of the SU(2) Higgs model is equivalent to an $O(4)$ symmetric ϕ^4-theory which has a second-order phase transition. For the gauged Higgs model and the electroweak standard model, the
transition becomes of first order. The first-order is induced by the interaction of the scalar fields with the $SU(2)$ gauge field. The reason is that the
transverse plasmon mass is small or even vanishes. This picture is reliable
for sufficiently small Higgs masses.

In order to describe the high-temperature transition for large Higgs
masses, higher orders of perturbation theory become important. Their
inclusion requires to tackle the problem of the magnetic mass that, however,
is unsolved so far. An estimate of the effect of a magnetic mass leads to the
conclusion that the first-order transition disappears for sufficiently large
Higgs masses.

4.6 Dimensional reduction at high temperature

4.6.1 *Generalities*

We are interested in the temperature dependence of QCD and the electroweak part of the standard model over a wide range of scales, starting from zero temperature (recall that the temperature of the universe nowadays is slightly above zero) up to temperatures well above the strong and electroweak interaction scales. Therefore it is natural to ask what happens in the extreme case of infinite temperatures and then to carefully extrapolate to lower temperatures, if possible down to the critical region. As usual one would expect a considerable simplification in such an extreme limit. If we recall that a finite temperature in the framework of a Euclidean quantum field theory is described by a finite extension T^{-1} in the Euclidean time direction, a simple geometric argument suggests the dimensional reduction of the field theory in D dimensions to $D - 1$ dimensions, since the temperature direction shrinks to zero as T goes to infinity. For example, let us consider the action of a pure $SU(3)$ Yang-Mills theory

$$S = -\frac{1}{2g^2} \int_0^{T^{-1}} dx_0 \int d^3x \sum_{\mu,\nu=0}^{3} \operatorname{tr} F_{\mu\nu}(x) F_{\mu\nu}(x),$$

$$F_{\mu\nu}(x) = \partial_\mu A_\nu(x) - \partial_\nu A_\mu(x) + [A_\mu(x), A_\nu(x)]. \tag{4.6.1}$$

For high temperatures T it seems legitimate to completely neglect the Euclidean time-dependence of the fields $A_\mu(x)$ and to write

$$S \simeq -\frac{1}{2G^2} \int d^3x \sum_{\mu,\nu=0}^{3} \operatorname{tr} F_{\mu\nu}(\mathbf{x}) F_{\mu\nu}(\mathbf{x})$$

$$\simeq -\frac{1}{2G^2} \int d^3x \left(\sum_{i,j=1}^{3} \operatorname{tr} F_{ij}(\mathbf{x}) F_{ij}(\mathbf{x}) + 2 \sum_{k=1}^{3} \left(D_k^{\mathrm{adj}} A_0(\mathbf{x}) \right)^2 \right),$$

$$\tag{4.6.2}$$

in which $G^2 = g^2 T$ is a dimensionful coupling constant and

$$D_k^{\mathrm{adj}} A_0(\mathbf{x}) = \partial_k A_0(\mathbf{x}) + [A_k(\mathbf{x}), A_0(\mathbf{x})]. \tag{4.6.3}$$

This is the action of a three-dimensional gauge theory with the same gauge group, coupled to a scalar field $A_0(\mathbf{x})$ that transforms under the adjoint representation. The scalar field has its origin in the T-component A_0 of the original four-dimensional gauge field A_μ, $\mu = 0, \ldots, 3$. The fields in (4.6.2) are called the static modes because they do not depend on the T-component anymore. The only temperature dependence that is left is hidden in the coupling constant G.

The action of (4.6.2) cannot resolve fluctuations on length scales $R \lesssim T^{-1}$, but heavy modes are assumed to decouple on length scales $R \gtrsim T^{-1}$. We call the steps from (4.6.1) to (4.6.2) and (4.6.3) *classical dimensional reduction*, because it proceeds on the level of actions, that is in classical field theory.

If the dimensional reduction survives the path integration over the non-static modes in the form of (4.6.2), i.e. without any remnant of the underlying nonstatic degrees of freedom, apart from an overall factor T^{-1} as in (4.6.2), we call dimensional reduction *complete*. Complete dimensional reduction also implies that no additional interactions of the static modes are induced by the nonstatic modes, or if so, they would be cancelled in the UV-renormalization of the theory.

Complete dimensional reduction has similar features as the complete decoupling of heavy mass modes in a zero-temperature field theory, in which the decoupling is predicted by the Appelquist-Carazzone theorem if certain conditions are satisfied (see below). Since the nonstatic modes have an intrinsic mass gap proportional to the temperature, the Appelquist-Carazzone theorem seems to suggest the complete decoupling of the nonstatic modes at high temperatures. Originally, on account of this theorem, it was widely believed that the classical picture above does generalize to quantum field theories. However, as we will see later in this section, the application of the theorem is not allowed. In a careful renormalization analysis it turns out that one may not even expect dimensional reduction in the complete form. In general the integration upon the nonstatic modes will induce additional, even nonlocal terms in the $D-1$ dimensional action and renormalized, temperature-dependent couplings. Nonstatic modes do influence the $D-1$ dimensional theory even at high temperatures and in the far-infrared region. Nevertheless we will show that dimensional reduction works (although not complete) for renormalizable quantum field theories. This includes such important cases as QCD and the $SU(2)$ Higgs model, also the Gross-Neveu model and scalar field theories. By "works" we mean that at high temperatures and under well controlled expansions, the theories reduce to **local, renormalizable** quantum field theories in $D-1$ dimensions with T-dependent couplings and a finite number of additional local interaction terms between the static modes. In the path-integral approach the effective interaction is obtained by integrating out the nonstatic fluctuations. Since the nonstatic modes have an intrinsic IR-cutoff of the order of T, the couplings of the effective $D - 1$ dimensional action allow an analytic expansion about zero-field momentum. IR-power counting then is used to select the interactions which contribute at high temperature and small momenta. This way one is led to a local quantum field theory in one less dimension. The reduced theories may serve as starting point for independent investigations in $D - 1$ dimensions.

In previous sections we have mentioned the IR-problems at finite temperature, in particular in the vicinity of second-order phase transitions. An integration upon the nonstatic modes turns out to be IR-safe, because these modes have an intrinsic mass gap proportional to the temperature. The resulting $(D - 1)$-dimensional theory is then further studied as a zero-temperature field theory with the same IR-behavior.

Several advantages are at hand if we proceed in the indicated order.

- From a physical point of view one may hope to filter the truly important degrees of freedom and the relevant interaction terms at high temperatures, if the approximations are well under control.

- From a practical or numerical point of view one saves computer time and disc space, since the total number of degrees of freedom is reduced from D to $D - 1$ dimensions. Monte Carlo simulations in four dimensions at finite temperature are CPU-time consuming because it must be ensured that $1 << L_0 << L_s$ (if L_s denotes the extension of the lattice in spacelike directions and L_0 in T-direction). In particular the second inequality should be satisfied for disentangling finite-size from finite-temperature effects.

- A welcome feature of dimensionally reduced theories with dynamical fermions is the absence of fermions in the $D - 1$ dimensional theories. Fermionic modes are intrinsically nonstatic because of the anti-periodic boundary conditions. Dimensionally reduced QCD with dynamical fermions is, for example, an adjoint $SU(3)$ Higgs model in three dimensions. Also it is possible to start with the electroweak standard model with chiral fermions in four dimensions at $T > 0$ in a continuum formulation and to treat the dimensionally reduced model without fermions on the lattice. This way one avoids the conceptual problem of putting chiral fermions on the lattice.

So far we have not quantified what we actually mean by *high* temperatures. Certainly we are not interested in temperatures so high that even complete dimensional reduction may work, temperatures far above all mass scales of a specific model. More concretely the question is as to whether T_c of QCD is high enough that dimensional reduction works down to or in the vicinity of T_c. (By T_c we mean a temperature of the order of the QCD-scale, say of 150 [MeV]. This gives the right order for the temperature range, at which at least a rapid crossover phenomenon is expected to happen if there is no true phase transition with a well defined T_c.) So far the answer for QCD is positive for $T \geq 2T_c$, but not known below $2T_c$. For the Gross-Neveu model dimensional reduction works down to T_c. This may indicate that it does depend on specific features of the model whether an extrapolation from high to critical temperatures by means of dimensional reduction works, or the approximation scheme breaks down somewhere above T_c. Furthermore, dimensional reduction is not restricted to weakly coupled models.

The section is organized as follows. First we recall the Appelquist-Carazzone theorem to show why it is not applicable to the decoupling of nonstatic modes. We further argue why complete dimensional reduction can neither be achieved by absorbing all D-dimensional remnants of nonstatic modes into counterterms. In section 4.6.3 we then describe a generic way of effectively reducing the dimension of the theory in a well controlled way. The reduction proceeds in a certain combination of a high-temperature expansion (yielding a criterion for truncating the reduced action) and a small-momentum expansion (making the effective action local). The reduction in the dimension which is obtained this way is not complete, but works in the sense that we have described above. For the computation of the effective action various perturbative expansions are applied. We first illustrate the procedure in a scalar Φ^4-theory from four to three dimensions in which the loop expansion amounts to an expansion in the quartic coupling constant. Next we outline the essential steps in dimensional reduction of a pure $SU(2)$ gauge theory and of lattice QCD with dynamical fermions, again from four to three dimensions. The reduced action is computed in perturbation theory with respect to the renormalized gauge coupling in four dimensions. We further use the dimensionally reduced theory, i.e. the adjoint $SU(3)$ Higgs model in $D = 3$ to calculate the electric screening mass, one of the important signatures for the deconfinement phase.

In the third example we consider the Gross-Neveu model. In three dimensions it serves as an effective model for the chiral phase transition of QCD. Here the reduction occurs between dimensions three and two. The perturbatively small parameter is $1/N$ rather than a coupling constant, N

denotes the number of fermionic species that are called "flavor" in analogy to QCD. As a particular feature of the Gross-Neveu model we indicate why dimensional reduction works down to T_c. Furthermore we solve a pretended contradiction concerning the universality class of the analogue of the chiral transition in the Gross-Neveu model. The contradicting predictions arise if two hand-waving lines of arguments -both of which appear as plausible- are taken for granted. In a combination of the large-N expansion, dimensional reduction, and linked cluster expansions for the resulting effective theory we will find out which of the plausibility arguments leads to the correct result in the end.

Finally we sketch dimensional reduction in the electroweak standard model. The dimensionally reduced $SU(2)$ Higgs model in three dimensions was frequently investigated by means of Monte Carlo simulations to study the strength of the first-order transition in this model as a function of the Higgs mass.

Fig. 4.6.1   The world of a finite-temperature field theory in the Euclidean (imaginary time) formalism, in which the horizontal direction stands for $D-1$ spacelike directions and the timelike direction has extension T^{-1}.

4.6.2   *Matsubara decomposition and the Appelquist-Carazzone decoupling theorem*

We consider a local, renormalizable quantum field theory at finite temperature in $D \geq 3$ dimensions. The theory shall describe a system in thermal equilibrium. In the functional-integral formalism such a field theory is formulated in a D-dimensional volume with one dimension compactified to a torus of the length of the inverse temperature (Fig. 4.6.1). Boundary conditions must be imposed on the physical fields in this direction because of the trace operation of the partition function. Bosonic degrees of freedom are subject to periodic boundary conditions, fermionic degrees of freedom to anti-periodic boundary conditions.

The basic idea behind dimensional reduction is based on the geometrical picture that at high temperatures T the length of the temperature torus T^{-1} becomes small. As the temperature increases, the D-dimensional system becomes more and more $D - 1$-dimensional on spatial length scales R large compared to T^{-1}. In the following we discuss the implications for a renormalizable quantum field theory.

Consider, for simplicity, a scalar model with an action of the form

$$S(\Phi) = \int_0^{T^{-1}} dz_0 \int d^{D-1}\mathbf{z} \left(\frac{1}{2} \sum_{\mu=0}^{D-1} (\partial_\mu \Phi(z))^2 + V(\Phi(z)) \right) \qquad (4.6.4)$$

with periodic boundary conditions in z_0 imposed on $\Phi(z)$, and V the self-interaction of the field. Periodicity implies that the corresponding momentum components in this direction are discrete. A Matsubara decomposition is a Fourier transform with respect to this zeroth momentum component,

$$\Phi(z) = T^{1/2} \sum_{k_0 = 2\pi T n, n \in \mathbf{Z}} e^{ik_0 z_0} \phi_n(\mathbf{z}). \qquad (4.6.5)$$

The kinetic part of the action becomes

$$\int_0^{T^{-1}} dz_0 \int d^{D-1}\mathbf{z} \frac{1}{2} \sum_{\mu=0}^{D-1} (\partial_\mu \Phi(z))^2 = \int d^{D-1}\mathbf{z} \frac{1}{2} \sum_{i=1}^{D-1} (\partial_i \phi_0(\mathbf{z}))^2$$

$$+ \sum_{n \neq 0} \int d^{D-1}\mathbf{z} \frac{1}{2} \left(\sum_{i=1}^{D-1} (\partial_i \phi_n(\mathbf{z}))^2 + (2\pi n T)^2 \phi_n(\mathbf{z})^2 \right). \qquad (4.6.6)$$

Apparently the field components with $n \neq 0$, the so-called nonstatic modes, acquire a thermal mass proportional to T. This provides an infrared cutoff to the nonstatic modes mediated by the interaction with the heat bath. The only modes which do not acquire a T-dependent mass term are the ϕ_0. They are called the static modes because they are precisely the field components which do not fluctuate in T-direction, their momentum k_0 is zero. Classically, i.e. on the tree-level action, they are the only degrees of freedom which survive dynamically on scales $R \gtrsim T^{-1}$. All other modes decouple. If this decoupling survives the path integration over the nonstatic modes, we call it *complete dimensional reduction* of the quantum field theory as indicated in the introduction.

How does the classical picture change under quantization? We have to investigate the full partition function given by

$$Z = \int \mathcal{D}\Phi \, \exp(-S(\Phi)) \quad ; \quad \mathcal{D}\phi = \prod_z d\Phi(z) \qquad (4.6.7)$$

with an action $S(\Phi)$ as in (4.6.4).

To make the path integral (4.6.7) well defined from the very beginning we introduce an UV-cutoff Λ. To a large extent the following considerations are independent of a particular cutoff scheme. For simplicity we choose a simple momentum cutoff. For more involved cases such as pure gauge theories (Yang-Mills theories) and QCD, other convenient choices are a Pauli-Villars cutoff, dimensional regularization, or a lattice cutoff.

From the Matsubara-decomposition (4.6.5) we see that the finite-temperature field theory of the $\Phi(z)$ in D dimensions can be viewed as a $D-1$-dimensional field theory of the $\phi_n(\mathbf{z})$ at zero temperature. There are multiple fields ϕ_n now, almost all of which become massive. The only effect of the temperature is a renormalization of the interaction. We write (4.6.7) as

$$Z = \int \left(\prod_{n \in \mathbf{Z}} \mathcal{D}\phi_n(\mathbf{z}) \right) \exp\left(-S_{MD}(\{\phi_n\})\right). \qquad (4.6.8)$$

The Matsubara-decomposed action S_{MD} is of the form

$$S_{MD}(\{\phi_n\}) = S_2(\{\phi_n\}) + S_I(\{\phi_n\}) \qquad (4.6.9)$$

with a quadratic part S_2 given by the right hand side of (4.6.6), and S_I determined by the interaction $V(\Phi)$ upon insertion of (4.6.5). For instance, for $D = 4$ and $V(\Phi(z)) = (m^2/2)\Phi(z)^2 + (\lambda/4!)\Phi(z)^4$, we have

$$S_{MD}(\{\phi_n\}) = \int d^3\mathbf{z} \left[\frac{1}{2} \sum_{i=1}^{D-1} (\partial_i \phi_0(\mathbf{z}))^2 + \frac{m^2}{2}\phi_0(\mathbf{z})^2 + \frac{\lambda T}{4!}\phi_0(\mathbf{z})^4 \right.$$

$$+ \sum_{n \neq 0} \frac{1}{2} \left(\sum_{i=1}^{D-1} (\partial_i \phi_n(\mathbf{z}))^2 + \left[m^2 + (2\pi n T)^2 \right] \phi_n(\mathbf{z})^2 \right) \qquad (4.6.10)$$

$$\left. + \frac{\lambda T}{4!} \sum_{(n_1,n_2,n_3) \neq 0} \phi_{n_1}(\mathbf{z})\phi_{n_2}(\mathbf{z})\phi_{n_3}(\mathbf{z})\phi_{-n_1-n_2-n_3}(\mathbf{z}) \right].$$

One might guess that the heavy modes decouple on the infrared scale as in the classical case, so that the only degrees of freedom which are left are the static fields $\phi_0(\mathbf{z})$, but the nonstatic modes will leave some remnants in the interactions between the static modes.

There are well known statements in zero-temperature quantum field theory on the large-mass behavior of a theory. The Appelquist-Carazzone theorem predicts the decoupling of massive fields from the low-energy scale of the theory under quite general conditions. An important ingredient is the renormalizability of the theory and the order of limits. First the cutoff must be sent to infinity, next the masses. We discuss this theorem to some extent.

Let us consider a model which describes (one or more) light particles of mass $m > 0$ interacting with a heavy particle of mass $M \gg m$. Decoupling of the heavy particle takes place if the low energy observables are reliably described by the action of light particles only.

To understand the large-mass behavior we consider Feynman graphs with heavy-mass propagators. The Feynman amplitude of a connected Feynman graph Γ with n loops and N propagators typically is of the form

$$I_\Gamma(p, m, M) = \int \frac{d^D k_1}{(2\pi)^D} \cdots \frac{d^D k_n}{(2\pi)^D} \, P(k, p)$$

$$\cdot \prod_{i=1}^{N_1} \frac{1}{l_i(k, p)^2 + m^2} \prod_{i=N_1+1}^{N} \frac{1}{l_i(k, p)^2 + M^2}, \qquad (4.6.11)$$

where p stands for the external momenta. P is a generic polynomial of both the internal or loop momenta $k_1, \ldots, k_n$ and of the external momenta p. The l_i are the line momenta, so that momentum conservation holds at each vertex. Let $N_1 < N$ so that there is at least one heavy propagator with mass M in Γ.

If the Feynman diagram Γ and all its subdiagrams have negative UV divergence degrees, the amplitude I_Γ is convergent without UV-cutoff. (This is the generic statement of a power counting theorem.) In this case, as $M \to \infty$, $I_\Gamma(p, m, M) \to 0$, as is easily seen by applying the dominated convergence theorem. For large mass $M \geq m$ the integrand is uniformly estimated by

$$|P(k, p)| \prod_{i=1}^{N} \frac{1}{l_i(k, p)^2 + m^2}, \qquad (4.6.12)$$

that is absolutely integrable by assumption. Hence the limit $M \to \infty$ of I_Γ may be shifted under the integral sign, leading to a vanishing limit.

Convergent diagrams are bounded by some inverse power of M, thus they vanish as $M \to \infty$. Divergent diagrams, on the other hand, behave differently. Let us regularize the theory by some UV-cutoff Λ. If only the overall divergence degree of Γ is nonnegative, $\omega(\Gamma) \geq 0$, the UV-finite part of I_Γ will increase with increasing M according to

$$I_\Gamma(p, m, M) \simeq P_{\omega(\Gamma)}(p, M) \, (\log M)^n, \qquad (4.6.13)$$

in which $P_{\omega(\Gamma)}$ is a polynomial of order $\omega(\Gamma)$.

In general, Γ or some of its subdiagrams have nonnegative divergence degrees. In this case the above procedure does not work, but requires renormalization before it makes sense to study the large-mass behavior. The bare parameters must be rewritten in terms of renormalized ones, defined by appropriate normalization conditions on Green functions, for example on vertex functions at zero momentum. Conditions have to be imposed on all nontrivial vertex functions with a nonnegative overall degree of divergence (in which all loop momenta become large simultaneously). Renormalizable field theories require only a finite number of normalization conditions. To all orders of the loop expansion this map between bare and renormalized parameters renders the renormalized correlation functions UV-finite. Equivalently, renormalization amounts to adding local counterterms to the cutoff action (the action in terms of bare parameters) if the bare parameters are simultaneously replaced by renormalized parameters. This way, m and M become the renormalized masses of the particles. A study of the large mass behavior is then understood in the sense that M satisfies $m << M << \Lambda$. For $M >> \Lambda$ the decoupling is trivial, according to the discussion above. For a lattice cutoff $\Lambda = a^{-1}$, masses with $M >> \Lambda$ just fall through the meshes of the lattice.

Normalization conditions on Green functions in momentum space naturally translate to every Feynman integral which contributes to the Green functions. They are equivalent to Taylor subtractions applied to the integrand according to the forest formula of Zimmermann. If all subtractions are made at zero momentum, we obtain the BPHZ-subtraction scheme. It is the BPHZ-scheme that enters the Appelquist-Carazzone theorem.

The essence of the decoupling theorem is that the subtractions applied to the Feynman integrals, or, equivalently, the local counterterms added to the action, are chosen in such a way that the large-mass divergence (4.6.13) is removed. Obviously it strongly depends on the renormalization scheme whether decoupling takes place or not. This is not surprising. Normalization conditions are scale dependent and the Appelquist Carazzone theorem is a statement about the decoupling of heavy masses on the low-energy scale. Minimal subtraction is for example a scheme without decoupling of heavy masses. On the other hand, the BPHZ-subtraction is optimal in the sense that normalization conditions are imposed at zero momentum, so that all contributions of the form (4.6.13) are absorbed into the counterterms.

The precise statement of the Appelquist-Carazzone decoupling theorem is then the following. Let Γ be a free-propagator amputated, connected

Feynman diagram with at least one heavy particle propagator. The BPHZ subtracted Feynman amplitude I_Γ of Γ satisfies the bound

$$|I_{\Gamma,\text{BPHZ}}(p, m, M)| \leq C_\Gamma(p, m) M^{-q} (\log M)^n.\qquad(4.6.14)$$

Here n is the number of loops of Γ, $C_\Gamma(p, m)$ is a function that is uniformly bounded for p in a compact domain. Furthermore,

$$q = \max(1, \min_{\gamma \in \Gamma_M} [-\omega(\gamma)])\qquad(4.6.15)$$

if Γ_M denotes the set of subgraphs of Γ (including Γ itself) which contain *all* massive lines. We remark that $q \geq 1$, so the BPHZ-renormalized Feynman amplitudes vanish in the large-mass limit $M \to \infty$ if they have at least one heavy particle propagating.

The statement remains true if normalization conditions are imposed at nonzero momenta $\bar{p}$ which are kept in the small-momentum regime $|\bar{p}|/M << 1$. Such a scheme is related to the BPHZ approach by a small renormalization-group transformation. For instance, the on-shell renormalization scheme at $\bar{p}^2 = -m^2$ also yields decoupling. In this case m is the pole mass of the lightest particle.

Let $\widetilde{G}_R^{(n\ light)}(p, m, M)$ denote the renormalized n-point Green function of the light particles, and let $\widetilde{G}_{R\ \text{light}}^{(n\ light)}(p, m, M)$ be the same n-point function but for which, in addition, only interactions between the light particles are taken into account. To every order of a loop expansion, the Green functions are composed of Feynman amplitudes. Diagrams contributing to $\widetilde{G}_R^{(n\ light)}(p, m, M)$ have only light external lines. In addition the graphs of $\widetilde{G}_{R\ \text{light}}^{(n\ light)}(p, m, M)$ have only light internal propagators. The bound (4.6.14) on the heavy-mass diagrams easily translates into the decoupling theorem of renormalized Green functions. Let us assume that the purely light Green functions $\widetilde{G}_{R\ \text{light}}^{(n\ light)}(p, m, M)$ do not vanish. For large M we have

$$\widetilde{G}_R^{(n\ light)}(p, m, M) = \widetilde{G}_{R\ \text{light}}^{(n\ light)}(p, m, M) \left(1 + O(\frac{|p|}{M}, \frac{m}{M})\right).\qquad(4.6.16)$$

Eq. (4.6.16) makes precise what it means that the Green functions on the low-energy scale are described by the interaction of merely light particles.

To summarize, the Appelquist-Carazzone theorem works in the framework of zero-temperature quantum field theory. It states general sufficient conditions for the decoupling of heavy masses from the low-energy scale of the theory. The pure light-particle sector should interact nontrivially, which means that the purely light Green functions do not vanish identically. All (renormalized) coupling constants of the model have to be defined on the low momentum scale. This includes on-shell renormalization conditions at $p^2 = -m^2$ and BPHZ-subtractions at zero momentum. They are related by renormalization-group transformations over a change of scale much smaller than M. Thus complete decoupling in the sense of (4.6.16) works because it is possible to absorb the large-M dependencies completely in the bare parameters.

The decoupling theorem requires that all Green functions with nonnegative divergence degree are subtracted independently of each other. Problems arise for the case of local or nonlinear symmetries if the symmetries must be preserved under renormalization. The renormalized Green functions are subjected to Ward identities that restrict the set of free normalization conditions. Typical examples in which the Appelquist-Carazzone theorem fails are theories with a gauge symmetry or an anomaly.

Let us now try to utilize the decoupling theorem for the study of the high-temperature behavior of the renormalizable field theory (4.6.4), (4.6.7). With the Matsubara decomposition (4.6.5) we obtained (4.6.8)-(4.6.10). This is a field theory of the ϕ_n-degrees of freedom in one less dimension, in which all the ϕ_ns with $n \neq 0$ acquire a mass proportional to the temperature T. In $D - 1$ dimensions the theory is superrenormalizable, thus it requires counterterms only to some low orders in the loop expansion. Applying the BPHZ-prescription in this $(D - 1)$-dimensional model one might expect on account of the Appelquist-Carazzone theorem that at high temperatures the (purely static) correlations of the ϕ_0-fields are described by the purely static action

$$S_0(\phi_0) = \int d^{D-1}\mathbf{z} \left[\frac{1}{2} \sum_{i=1}^{D-1} (\partial_i \phi_0(\mathbf{z}))^2 + \frac{1}{T} V(T^{1/2} \phi_0(\mathbf{z})) \right] \qquad (4.6.17)$$

whenever $|\mathbf{p}|/T$ and m/T are small.

At the first sight this looks like an appealing result and suggests complete dimensional reduction of the quantum field theory so that we were done. Unfortunately it is wrong. It must be wrong, because otherwise the UV-problem of renormalizing the theory in D dimensions would considerably simplify at high temperatures, since only a few low order counterterms must be added to the $(D-1)$-dimensional superrenormalizable theory. Such a simplification cannot be true, because the UV-and IR behavior are expected to decouple at least in massive theories. The apparent contradiction is easily solved if we recall the right order of limits. First we must remove the UV-cutoff and send $\Lambda \to \infty$, next $T \to \infty$ and not vice versa. In this order of limits the Matsubara formalism has to deal with a countable but infinite number of heavy modes just as in (4.6.10), whereas the Appelquist-Carazzone decoupling theorem applies to models with one or at most a finite number of heavy masses. It makes no prediction in case of an unbounded number of heavy particles. This would require additional large-mass bounds that are uniform in the number of particles.

Thus we must decide by other means whether quantum fluctuations preserve or destroy complete dimensional reduction. The problem was investigated by Landsman in a careful perturbative renormalization-group analysis. The generic answer is negative. To understand why, we have to deal with renormalization aspects again. Let $\widetilde{G}(\mathbf{p}, T)$ denote a generic connected correlation function, $\mathbf{p}$ a generic spatial momentum. $\widetilde{G}$ gets contributions both from the static modes ϕ_0 and from all nonstatic modes ϕ_n, $n \neq 0$. Similarly as in the heavy-mass case, we expect from the behavior of single diagrams which contribute to $\widetilde{G}$ that at large T

$$\widetilde{G} \simeq T^\delta \, (\log T)^n \tag{4.6.18}$$

with appropriate exponents δ and n. (This will be made more precise in the next section. As we shall see, the behavior will be even more divergent than in the heavy-mass case.) Let us assume that we succeed in renormalizing the theory in such a way that the terms (4.6.18) which increase with T are removed. In this case we get

$$\widetilde{G}(\mathbf{p}, T) = \widetilde{G}_0(\mathbf{p}, T) \left(1 + O(\frac{\mathbf{p}}{T}, \frac{m}{T})\right), \tag{4.6.19}$$

where $\widetilde{G}_0$ denotes the purely static contribution to $\widetilde{G}$. The point now to be made is that in order to remove the large-T behavior (4.6.18), it is not sufficient to impose normalization conditions at some fixed temperature T_0. A T-dependent renormalization scheme is required, which in turn implies that the coupling constants become dependent on T. In particular, the renormalized mass m that enters the correction term in (4.6.19) usually becomes proportional to T in this scheme, so the "correction" term becomes of the order of 1. It is at best suppressed by some power of the coupling constant. The conclusion is that it is not possible to absorb all effects from the nonstatic modes into appropriate counterterms, or if one tries to do so, the correction term $O(\mathbf{p}/T, m/T)$ of (4.6.19) actually is not small. In particular we must expect remnants of the nonstatic modes in an effective action which is derived from the original one by integrating out the nonstatic modes.

4.6.3 General outline of dimensional reduction

As we have learned from the last section we may not expect that dimensional reduction in a quantum field theory at finite temperature is complete. The nonstatic modes do influence the high-temperature and IR-behavior beyond a purely multiplicative, T-dependent renormalization of the D-dimensional input parameters. The question then is as to whether the IR-behavior still can be described by a local field theory, obtained by a reliable high-temperature expansion. The answer is yes! Let us see how. In the following section we describe the generic procedure that applies to any renormalizable quantum field theory, just for simplicity we use the notation of a scalar field theory.

We write the decomposition (4.6.5) of the fields into static and nonstatic components as

$$\Phi(z) = T^{1/2}\phi_{st}(\mathbf{z}) + \phi_{ns}(z), \tag{4.6.20}$$

with

$$\int_0^{1/T} dz_0\, \phi_{ns}(z) = 0. \tag{4.6.21}$$

The rescaling of the field ϕ_{st} with $T^{1/2}$ is chosen so that the kinetic term of the effective action for the ϕ_{st}-modes has a prefactor of 1, cf. Eq. (4.6.29) below. With this decomposition the partition function becomes

$$Z = \int \mathcal{D}\phi_{st}\, \exp\left(-S_{eff}(\phi_{st})\right), \tag{4.6.22}$$

with effective Boltzmann factor

$$\exp\left(-S_{eff}(\phi_{st})\right) = \int \mathcal{D}\phi_{ns}\, \exp\left(-S(T^{1/2}\phi_{st} + \phi_{ns})\right), \tag{4.6.23}$$

obtained by integrating out the nonstatic modes ϕ_{ns}. We write the effective action as

$$S_{eff}(\phi_{st}) = S_0(\phi_{st}) + (\delta_{ns}S)(\phi_{st}), \tag{4.6.24}$$

in which S_0 is the classically reduced action (4.6.17) of the temporal zero modes ϕ_{st}. The action S_0 is superrenormalizable. The quantum fluctuations of the nonstatic modes are described by $\delta_{ns}S$. Their induced interaction is nonlocal in the sense that it couples fields $\phi_{st}(\mathbf{x})$ and $\phi_{st}(\mathbf{y})$ at arbitrary distances $|\mathbf{x} - \mathbf{y}|$. Moreover, we know from the last section that $\delta_{ns}S$ cannot be neglected even at very high temperatures and for momenta $|\mathbf{p}| << T$. Hence, at the first glance the effective model of the static modes is a theory in $(D-1)$-dimensions, but rather complex and nonlocal. If we show a reliable way how to truncate the nonlocal terms we are done.

We know that the nonstatic fields ϕ_{ns} have an intrinsic IR-cutoff of the order of the temperature T. Thus the integration of (4.6.23) actually amounts to solve a noncritical system with a mass gap of the order of T. This implies that in momentum space $\delta_{ns}S$ is analytic about zero momentum. With increasing T, at fixed momentum, the effective static action $S_{eff}(\phi_{st})$ becomes more and more local.

Therefore the solution to obtain a dimensionally reduced field theory which describes the IR-behavior of the full theory at high temperature, and that is at the same time a local field theory, is to generate the high-temperature reduction by the zero-momentum expansion of $\delta_{ns}S$. This provides an expansion in powers of $\mathbf{p}/T$ and $\mathbf{p}/\Lambda$ with UV-cutoff $\Lambda >> T$ (similarly to $\Lambda >> M$ in the context of the Appelquist-Carazzone theorem). We emphasize that no terms of the form $\mathbf{p}/m$ arise, if m denotes the renormalized mass, because $\delta_{ns}S$ stays analytic as $m \to 0$.

For the following considerations it is convenient to work in the momentum-space representation. Fourier transformed fields are defined by

$$\widetilde{\Phi}(k) = \int d^D x \, e^{-ik \cdot x} \, \Phi(x),$$

$$\widetilde{\phi}_{st}(\mathbf{k}) = \int d^{D-1}\mathbf{x} \, e^{-i\mathbf{k} \cdot \mathbf{x}} \, \phi_{st}(\mathbf{x}), \qquad (4.6.25)$$

$$\widetilde{\phi}_{ns}(k) = \int d^D x \, e^{-ik \cdot x} \, \phi_{ns}(x).$$

The decomposition (4.6.20) into static and nonstatic modes then reads

$$\widetilde{\phi}(k) = T^{-1/2}\widetilde{\phi}_{st}(\mathbf{k})\delta_{k_0,0} + \widetilde{\phi}_{ns}(k)\,(1 - \delta_{k_0,0}). \qquad (4.6.26)$$

The effective action $S_{eff}(\phi_{st})$ (4.6.24) acquires the form

$$S_{eff}(\phi_{st}) = \sum_{n \geq 2} \frac{1}{n!} \int \frac{d^{D-1}\mathbf{p}_1}{(2\pi)^{D-1}} \cdots \frac{d^{D-1}\mathbf{p}_{n-1}}{(2\pi)^{D-1}}$$
$$\cdot \widetilde{\phi}_{st}(\mathbf{p}_1) \cdots \widetilde{\phi}_{st}(\mathbf{p}_{n-1})\widetilde{\phi}_{st}(-\mathbf{p}_1 - \cdots - \mathbf{p}_{n-1}) \qquad (4.6.27)$$
$$\cdot \widetilde{S}_{eff}^{(n)}(\mathbf{p}_1, \ldots, \mathbf{p}_{n-1}),$$

where we have used translation invariance. From the very definition (4.6.23) we see that $S_{eff}(\phi_{st})$ can be considered as the negative generating functional of the free-propagator amputated (connected) correlation functions of the purely nonstatic system, with purely static external fields. In a perturbative approach, except for tree level contributions, $(-\widetilde{S}_{eff}^{(n)}(\mathbf{p_1}, \ldots, \mathbf{p_{n-1}}))$ consists of connected momentum-space Feynman diagrams with n external lines, all of them are static, and all internal lines carry nonstatic momenta $(k_0, \mathbf{k})$ with $k_0 \neq 0$. Because of momentum conservation, the purely nonstatic Feynman graphs are 1PI, thus they are denoted as $\widetilde{\Gamma}_{ns}^{(n)}(\mathbf{p_1}, \ldots, \mathbf{p_{n-1}})$. Making the purely static part $\widetilde{S}_0^{(n)}$ of $\widetilde{S}_{eff}^{(n)}$ explicit, we have

$$\widetilde{S}_{eff}^{(n)}(\mathbf{p}_1, \ldots, \mathbf{p}_{n-1}) = \widetilde{S}_0^{(n)}(\mathbf{p}_1, \ldots, \mathbf{p}_{n-1}) - \widetilde{\Gamma}_{ns}^{(n)}(\mathbf{p}_1, \ldots, \mathbf{p}_{n-1}). \quad (4.6.28)$$

Note that because of this splitting, apart from the counterterm vertices, the $\widetilde{\Gamma}_{ns}^{(n)}(\mathbf{p}_1, \ldots, \mathbf{p}_{n-1})$ have no tree-level contribution in contrast to the full n-point vertex functions. They are renormalized vertex functions assuming that the effective action S_{eff} is written in terms of renormalized parameters and counterterms. We remark that in general $\widetilde{\Gamma}_{ns}^{(n)}$ are not completely UV-finite (as their attribute "renormalized" may suggest), but contain some additive local counterterms. This is because of definition (4.6.28). $\widetilde{\Gamma}_{ns}^{(n)}$ are computed from the full action that includes all counterterms (static and nonstatic) which are required to render the correlation functions of the full theory UV-finite. Part of these counterterms remove the UV-divergencies of some low-order contributions which are purely static. Because the static UV-divergencies do not show up in $\widetilde{\Gamma}_{ns}^{(n)}$, these parts survive upon integration of the nonstatic modes. The role of the divergent terms in $\widetilde{\Gamma}_{ns}^{(n)}$ is then to cancel the UV divergencies left over from D dimensions in the dimensionally reduced model upon integration of the static modes.

For the action (4.6.4), $\widetilde{S}_0^{(n)}$ is given by

$$\widetilde{S}_0^{(2)}(\mathbf{p}) = \mathbf{p}^2 + m^2,$$
$$\widetilde{S}_0^{(n)}(\mathbf{p}_1, \ldots, \mathbf{p}_{n-1}) = \lambda_n T^{n/2-1}, \qquad n \geq 3, \qquad (4.6.29)$$

where

$$m^2 = \left. \frac{\partial^2 V(\phi)}{\partial \phi^2} \right|_{\phi=0},$$
$$\lambda_n = \left. \frac{\partial^n V(\phi)}{\partial \phi^n} \right|_{\phi=0}, \qquad n \geq 3. \qquad (4.6.30)$$

Because of their analytic behavior, we expand the $\widetilde{\Gamma}_{ns}^{(n)}(\mathbf{p_1}, \ldots, \mathbf{p_{n-1}})$ about zero momenta, and keep the high-temperature contributions. As we have seen in section 4.4, in a renormalizable field theory the high-temperature behavior of the renormalized n-point functions $\widetilde{\Gamma}_{ns}^{(n)}$ is expected to be given by

$$\widetilde{\Gamma}_{ns}^{(n)}(\mathbf{p_1}, \ldots, \mathbf{p_{n-1}}|T) = T^{\rho_n} \cdot P_{\rho_n}(\frac{\mathbf{p_1}}{T}, \ldots, \frac{\mathbf{p_{n-1}}}{T}) + o(1) \qquad (4.6.31)$$

as $T \to \infty$. P_ν denotes a (multidimensional) polynomial in $\mathbf{p}/T$ of degree ν for $\nu \geq 0$, with coefficients depending at most logarithmically on T, and $P_\nu = 0$ for $\nu < 0$. The degrees ρ_n only depend on n and the dimension D, but, as a consequence of renormalizability, they are otherwise independent of the order to which $\widetilde{\Gamma}_{ns}^{(n)}$ is computed in a perturbative (e.g. weak coupling) expansion. They are determined by UV-power counting rules. We have

$$\rho_n = (D - 1) - n\frac{D - 3}{2}. \qquad (4.6.32)$$

In the asymptotic formula (4.6.31) we have absorbed the scaling violations $O(\mathbf{p}/\Lambda)$ in the remainder of $o(1)$. Equations (4.6.31) and (4.6.32) show that the contributions to $\widetilde{\Gamma}_{ns}^{(n)}$ which are not suppressed at high T are local in the sense that they are polynomials in momentum space. This leads to the following criterion for the truncation of the effective action (4.6.27).

Let us write $\mathcal{T}_{\mathbf{p}}^{\delta} f(\mathbf{p})$ for the (multidimensional) Taylor expansion about $\mathbf{p} = 0$ to the order δ of the function $f(\mathbf{p})$,

$$\mathcal{T}_{\mathbf{p}}^{\delta} f(\mathbf{p_1}, \ldots, \mathbf{p_s}) = \sum_{l=0}^{\delta} \frac{1}{l!} \left(\sum_{i=1}^{s} \sum_{j=1}^{D-1} (p_i)_j \frac{\partial}{\partial(\widetilde{p}_i)_j} \right)^l f(\widetilde{\mathbf{p}}_1, \ldots, \widetilde{\mathbf{p}}_s) \Bigg|_{\widetilde{\mathbf{p}}=0}.$$
$$(4.6.33)$$

In order to truncate the effective action $S_{eff}(\widetilde{\Phi}_{st})$ to local terms we have to Taylor expand the contributions $\widetilde{\Gamma}_{ns}^{(n)}$ of (4.6.31) about $\mathbf{p} = 0$. Furthermore, to keep the finite-T contributions to $\delta_{ns}S$ on the infrared scale, we have to include all n for which $\rho_n \geq 0$ in (4.6.32). Therefore the maximal degree of the Taylor polynomial of $\widetilde{\Gamma}_{ns}^{(n)}$ in p is then given by ρ_n. Thus we replace

$$\widetilde{\Gamma}_{ns}^{(n)}(\mathbf{p_1}, \ldots, \mathbf{p_{n-1}}|T) \to \mathcal{T}_{\mathbf{p}}^{\rho_n - q_n} \widetilde{\Gamma}_{ns}^{(n)}(\mathbf{p_1}, \ldots, \mathbf{p_{n-1}}|T) \qquad (4.6.34)$$

whenever $\rho_n \geq 0$, with $0 \leq q_n \leq \rho_n$, q_n an integer, and set all other $\widetilde{\Gamma}_{ns}^{(n)}$ to 0. The value of q_n determines down to which length scales the high-temperature behavior of $\widetilde{\Gamma}_{ns}^{(n)}$ is resolved. If $q_n = \rho_n$ in (4.6.34), only the leading high-T behavior of $\widetilde{\Gamma}_{ns}^{(n)}(\mathbf{p_1}, \ldots, \mathbf{p_{n-1}}|T)$ is kept with spatially constant vertex functions, so that possible corrections of the order of $\mathbf{p}/T$ are ignored. This makes sense in the far IR. On the other hand, for $q_n = 0$ the complete high-T part of (4.6.31) is kept. With decreasing q_n the effective action describes more and more smaller length scales of the high-temperature behavior.

Note an important difference between $D \geq 4$ and $D = 3$. For $D = 3$ $\rho_n = 2$ is independent of n, hence the leading contribution in T is independent of the number n of external legs of $\widetilde{\Gamma}_{ns}^{(n)}$. For $D \geq 4$, the high T behavior of $\widetilde{\Gamma}_{ns}^{(n)}$ becomes weaker with increasing n. Only nonstatic correlations $\widetilde{\Gamma}_{ns}^{(n)}$ with $\rho_n \geq 0$ contribute to $\delta_{ns}S$ at high temperature. For $n > 2(D-1)/(D-3)$, $\widetilde{\Gamma}_{ns}^{(n)}$ vanish as $T \to \infty$. Clearly one may truncate the effective action at smaller values of n than this limiting case if the temperature is sufficiently high. For example, for $D = 4$, in the reduction down to three dimensions we may omit ϕ_{st}^6-interactions induced by the nonstatic modes if the temperature is large. ϕ_{st}^6-terms are suppressed by a factor of T^{-1} compared to the quartic ϕ_{st}^4-contributions.

In any case the zero momentum expansion (4.6.34) generates a polynomial $\widetilde{S}_{eff}^{(n)}(\mathbf{p_1}, \ldots, \mathbf{p_{n-1}})$ in momentum space. Hence, we end up with a locally interacting model of the fields $\phi_{st}(\mathbf{z})$ in one less dimension.

If we compare different choices of q_n for fixed n, the least local contributions are obtained for $q_n = 0$ with the highest powers in $\mathbf{p}/T$ taken into account, but the higher the power in $\mathbf{p}/T$, the more suppressed is this contribution for high temperatures.

If we set $q_n = 0$ in (4.6.34) and keep the full high-temperature part of $\widetilde{\Gamma}_{ns}^{(n)}$, the effective model becomes a strictly (power counting-) renormalizable field theory in $D - 1$ dimensions. If we are interested in keeping only the leading term for every correlation function with $\rho_n \geq 0$, we set $q_n = \rho_n$ in (4.6.34) and obtain

$$\widetilde{\Gamma}_{ns}^{(n)}(\mathbf{p_1}, \ldots, \mathbf{p_{n-1}}|T) \; \to \; T_{\mathbf{p}}^0 \widetilde{\Gamma}_{ns}^{(n)}(\mathbf{p_1}, \ldots, \mathbf{p_{n-1}}|T) \equiv \widetilde{\Gamma}_{ns}^{(n)}(\mathbf{0}, \ldots, \mathbf{0}|T) \tag{4.6.35}$$

for all amplitudes $\widetilde{\Gamma}_n$ with $\rho_n \geq 0$. This restriction makes sense at spatial separations which are sufficiently large compared to the inverse temperature T^{-1}, for instance for the investigation of universal critical behavior at a second-order phase transition. Corrections are suppressed by the order of $\mathbf{p}/T$.

We should emphasize at this point that (4.6.35) is not equivalent to keeping only those effective interactions that stay superrenormalizable in

$D-1$ dimensions (except for the reduction from $D = 3$ to $D - 1 = 2$). We still obtain a renormalizable theory. In general, in order to get an even superrenormalizable reduced model, contributions from nonstatic correlation functions $\widetilde{\Gamma}_{ns}^{(n)}$ with $\rho_n = 0$ must be omitted in addition. This is only allowed if the temperature is sufficiently large. Without a further justification this truncation may be not appropriate for the investigation of finite-temperature phase transitions.

Now let us assume that we do keep only the effective interactions (mediated by the nonstatic modes) which preserve superrenormalizability of the (D-1)-dimensional effective model. The reduced model is then made ultraviolet finite by the D-dimensional counterterms of the original theory, projected in the same way as in (4.6.34). In the more general case, supplementary normalization conditions must be imposed. They are obtained by matching conditions to the D-dimensional theory.

Let us summarize the state of the art for a local renormalizable quantum field theory at finite temperature. After dimensional reduction along the lines described above its infrared behavior on spatial length scales $R \gtrsim T^{-1}$ is quantitatively described by a local, renormalizable or even superrenormalizable field theory of the purely static modes $\phi_{st}(\mathbf{z})$. This is a field theory in one less dimension and at zero temperature. The nonstatic degrees of freedom $\phi_{ns}(z)$, which account for the oscillations along the temperature torus, have an intrinsic mass gap of the order of T. Hence they do not survive as dynamical degrees of freedom in this region. However, they induce local interactions of the static fields that inevitably have to be taken into account in order to describe the infrared properties in a correct way. This concerns both the characteristics of the high-temperature phase as well as finite-temperature phase transitions.

A remarkable property of dimensional reduction is the fact that fermions do not survive the reduction process as dynamical degrees of freedom. The anti-periodic boundary conditions imposed on the fermion fields in the T-direction provide a mass gap and identify fermions as purely nonstatic modes. Their "only" effect (by means of δS_{eff}) on the infrared scale are the induced interactions between the static fields that are otherwise purely bosonic.

The effective action of the static fields ϕ_{st} can be calculated in various ways. Renormalized perturbation theory in a weak coupling was extensively used for scalar field theories, four-dimensional Yang-Mills theories, QCD and the electroweak standard model. For the Gross-Neveu model we will perform dimensional reduction in a large-N expansion.

4.6.4 Dimensional reduction of a Φ^4-theory from four to three dimensions

Now it is time to illustrate dimensional reduction with several examples. Let us start with a one-component massive scalar Φ^4-theory in four dimensions at finite temperature. As UV-regularization we choose a sharp-momentum cutoff for simplicity. We want to derive the effective action (4.6.27) in three dimensions in which all nonvanishing contributions at high temperature are kept so that we avoid any *a priori* hand-waving truncations. As criterion for a truncation in the order n of the renormalized n-point functions $\widetilde{\Gamma}_{ns}^{(n)}$ that contribute to (4.6.27) we use Eq. (4.6.31) with (4.6.32). In $D = 4$ dimensions we have $\rho_n \geq 0$ for $n \leq 6$, thus we have to determine $\widetilde{\Gamma}_{ns}^{(n)}$ for $n = 2, 4, 6$. Let us derive $S_{eff}(\Phi_{st})$, starting from the Φ^4- theory in four dimensions with bare action given as

$$
\begin{aligned}
S_0(\Phi_0) = \int_0^{1/T} dx_0 \int d^3x \left(\frac{1}{2} \sum_{\mu=0}^{3} (\partial_\mu \Phi_0(x))^2 \right. \\
\left. + \frac{M_0^2}{2} \Phi_0(x)^2 + \frac{\lambda_0}{4!} \Phi_0(x)^4 \right) .
\end{aligned} \tag{4.6.36}
$$

Here M_0 is the bare mass and λ_0 the bare coupling. To reexpress $S(\Phi_0)$ in terms of renormalized quantities, we rescale

$$
\begin{aligned}
M_0^2 &= Z_{M^2} M^2 \\
\lambda_0 &= Z_\lambda \lambda \\
\Phi_0 &= Z_\Phi^{1/2} \Phi ,
\end{aligned} \tag{4.6.37}
$$

so that the bare n-point vertex functions $\Gamma_0^{(n)}$ are related to the renormalized vertex functions $\Gamma_R^{(n)}$ according to

$$
\Gamma_0^{(n)}(p_1, \ldots, p_{n-1}, \lambda_0, M_0 | T) = Z_\Phi^{-n/2} \Gamma_R^{(n)}(p_1, \ldots, p_{n-1}, \lambda, M | T) . \tag{4.6.38}
$$

The renormalization constants are uniquely determined via normalization conditions. A standard choice for a Φ^4-theory is

$$
\begin{aligned}
\Gamma_R^{(2)}(0 | T = 0) &= -M^2 \\
\left. \frac{\partial \Gamma_R^{(2)}}{\partial p^2}(p | T = 0) \right|_{p=0} &= -1 \\
\Gamma_R^{(4)}(0, 0, 0 | T = 0) &= -\lambda .
\end{aligned} \tag{4.6.39}
$$

In terms of the renormalized zero-temperature quantities λ and M, the action (4.6.36) becomes the renormalized one and takes the following form

$$
S_0(\Phi_0) \;=\; S_R(\Phi) \;=\; \int_0^{1/T} dx_0 \int d^3x \;\left(\frac{1}{2} \sum_{\mu=0}^3 (\partial_\mu \Phi(x))^2 \;+\; \frac{M^2}{2}\Phi(x)^2 \right.
$$
$$
\left. +\; \lambda \frac{\Phi(x)^4}{4!} + \frac{1}{2}\delta Z_\Phi \sum_{\mu=0}^3 (\partial_\mu \Phi(x))^2 + \frac{1}{2}\delta M^2 \,\Phi(x)^2 + \delta\lambda \frac{\Phi(x)^4}{4!} \right)
$$

$$(4.6.40)$$

with the following definitions

$$
\delta M^2 = (Z_{M^2} Z_\Phi - 1)\, M^2
$$
$$
\delta\lambda = (Z_\lambda Z_\Phi^2 - 1)\, \lambda
$$
$$
\delta Z_\Phi = Z_\Phi - 1.
$$

$$(4.6.41)$$

For the Φ^4 theory the loop expansion amounts to the weak coupling expansion in λ. Evaluating the normalization conditions to one-loop order, we obtain

$$
\delta M^2 = -\frac{\lambda}{2} B_1^{(4)}(\beta = \infty, M, \Lambda_0)
$$
$$
\delta\lambda = 3\frac{\lambda^2}{2} B_2^{(4)}(\beta = \infty, M, \Lambda_0)
$$
$$
\delta Z_\Phi = 0 \,,
$$

$$(4.6.42)$$

where Λ_0 denotes the sharp-momentum cutoff. In (4.6.42) we introduced the notation

$$
B_n^{(4)}(\beta, M, \Lambda_0) \;=\; \frac{1}{\beta} \sum_{k_0 = 2\pi n T, n \in \mathbf{Z}} \int^{\Lambda_0} \frac{d^3 k}{(2\pi)^3} \frac{1}{(k^2 + M^2)^n}
$$

$$(4.6.43)$$

with $\beta = 1/T$, $M > 0$ for simplicity, and

$$
k^2 \;=\; k_0^2 + \mathbf{k}^2 \;\leq\; \Lambda_0^2 \,.
$$

$$(4.6.44)$$

For $\beta = \infty$ this becomes

$$
B_n^{(4)}(\beta = \infty, M, \Lambda_0) \;=\; \int^{\Lambda_0} \frac{d^4 k}{(2\pi)^4} \frac{1}{(k^2 + M^2)^n}.
$$

$$(4.6.45)$$

For further use we also introduce the notation

$$B_{n,ns}^{(4)}(\beta, M, \Lambda_0) = \frac{1}{\beta} \sum_{k_0=2\pi nT, n\in\mathbf{Z}, n\neq 0} \int^{\Lambda_0} \frac{d^3\mathbf{k}}{(2\pi)^3} \frac{1}{(k^2+M^2)^n}. \quad (4.6.46)$$

From now on we distinguish between static and nonstatic modes according to (4.6.20). The effective action S_{eff} as a function of the three-dimensional static modes $\phi_{st}(\mathbf{x})$ is then defined via

$$\begin{aligned}
\exp\left(-S_{eff}(\phi_{st})\right) &= \int \mathcal{D}\phi_{ns} \exp\left(-S(T^{\frac{1}{2}}\phi_{st}+\phi_{ns})\right) \\
&\equiv \exp\left(-S_{cl}(T^{\frac{1}{2}}\phi_{st})\right) \cdot \int \mathcal{D}\phi_{ns} \\
&\quad \cdot \exp\left[-\left(S(T^{1/2}\phi_{st}+\phi_{ns}) - S_{cl}(T^{1/2}\phi_{st})\right)\right] \\
&\equiv \exp\left(-S_{cl}(T^{\frac{1}{2}}\phi_{st})\right) \cdot \exp\left(\Gamma_{ns}(\phi_{st})\right). \quad (4.6.47)
\end{aligned}$$

The classically reduced action that is split off in (4.6.47) is given by

$$\begin{aligned}
S_0(\phi_{st}) = S_{cl}(T^{1/2}\phi_{st}) &= \frac{1}{T}\int d^3\mathbf{x} \left(\frac{1}{2}\sum_{i=1}^{3}\left(T^{1/2}\partial_i\phi_{st}(\mathbf{x})\right)^2 \right. \quad (4.6.48) \\
&\quad \left. + \frac{M^2}{2}\left(T^{1/2}\phi_{st}(\mathbf{x})\right)^2 + \frac{\lambda}{4!}\left(T^{1/2}\phi_{st}(\mathbf{x})\right)^4\right).
\end{aligned}$$

Note that ϕ_{st} in (4.6.20) was scaled in a way that the prefactor of the kinetic term in (4.6.48) is temperature independent. $\Gamma_{ns}(\phi_{st})$ generates the renormalized, nonstatic connected n-point correlation functions with static external legs in four dimensions. Note that in contrast to the vertex functions in (4.6.39), the classical contributions to the correlation functions which are generated by Γ_{ns} are subtracted.

Thus in momentum space we write for the effective action

$$\begin{aligned}
S_{eff}(\phi_{st}) &= \sum_{n\geq 2}\frac{1}{n!}\prod_{i=1}^{n-1}\left(T\sum_{p_{i0}=2\pi n_0 T, n_0\in\mathbf{Z}}\int \frac{d^3\mathbf{p}_i}{(2\pi)^3} T^{-1/2}\widetilde{\phi}_{st}(\mathbf{p}_i)\delta_{p_{i0},0}\right) \\
&\quad \cdot \left(T^{-1/2}\widetilde{\phi}_{st}(-\sum_{i=1}^{n-1}\mathbf{p}_i)\right) \quad (4.6.49) \\
&\quad \cdot \left(\widetilde{S}_{cl}^{(n)}(\mathbf{p}_1,\cdots,\mathbf{p}_{n-1}) - \widetilde{\Gamma}_{ns,R}^{(n)(4D)}(\mathbf{p}_1,\cdots,\mathbf{p}_{n-1})\right)
\end{aligned}$$

$$= \sum_{n \geq 2} \frac{1}{n!} \int \frac{d^3 \mathbf{p}_1}{(2\pi)^3} \cdots \int \frac{d^3 \mathbf{p}_{n-1}}{(2\pi)^3} \, \widetilde{\phi}_{st}(\mathbf{p}_1) \cdots \widetilde{\phi}_{st}\left(-\sum_{i=1}^{n-1} \mathbf{p}_i\right)$$

$$\cdot \, T^{n/2-1} \left(\widetilde{S}_{cl}^{(n)}(\mathbf{p}_1, \cdots, \mathbf{p}_{n-1}) - \widetilde{\Gamma}_{ns,R}^{(n)(4D)}(\mathbf{p}_1, \cdots, \mathbf{p}_{n-1}) \right).$$

$$(4.6.50)$$

Here,

$$\widetilde{S}_{cl}^{(n)}(\mathbf{p}_1, \cdots, \mathbf{p}_{n-1}) = \begin{cases} \mathbf{p}_1^2 + M^2, & n = 2, \\ \lambda, & n = 4, \\ 0, & \text{otherwise}. \end{cases} \qquad (4.6.51)$$

The superscript $(4D)$ of the vertex functions $\widetilde{\Gamma}_{ns,R}^{(n)(4D)}$ in (4.6.49), (4.6.50) indicates that it is computed with the Feynman rules in four dimensions without rescaling by a T-dependent factor.

(4.6.50) is exactly of the form of (4.6.27) with (4.6.28) if we identify

$$\widetilde{S}_0^{(n)}(\mathbf{p}_1, \cdots, \mathbf{p}_{n-1}) = T^{n/2-1} \cdot \widetilde{S}_{cl}^{(n)}(\mathbf{p}_1, \cdots, \mathbf{p}_{n-1}) \qquad (4.6.52)$$

and

$$\widetilde{\Gamma}_{ns,R}^{(n)}(\mathbf{p}_1, \cdots, \mathbf{p}_{n-1}) = T^{n/2-1} \cdot \widetilde{\Gamma}_{ns,R}^{(n)(4D)}(\mathbf{p}_1, \cdots, \mathbf{p}_{n-1}). \qquad (4.6.53)$$

Next we derive the Feynman rules for the computation of S_{eff} in the form of (4.6.50), but later we have to keep an eye on the right rescaling according to (4.6.52), (4.6.53), since we are finally interested in the contributions to S_{eff} in the form of (4.6.27) that is expressed merely in 3-dimensional terms.

Let us evaluate $S_{eff}(\phi_{st})$ to one-loop order in perturbation theory. As argued above, only 1PI graphs will contribute to $\Gamma_{ns,R}^{(n)(4D)}(\phi_{st})$, with Feynman rules determined by

$$
-\left(S(T^{1/2}\phi_{st} + \phi_{ns}) - S_{cl}(T^{1/2}\phi_{st})\right)
$$

$$
= -\frac{1}{T} \int d^3\mathbf{x} \left(\frac{1}{2}\delta M^2 \left(T^{1/2}\phi_{st}(\mathbf{x})\right)^2\right.
$$

$$
+ \frac{1}{2}\delta Z_\Phi \sum_{i=1}^{3} \left(\partial_i(T^{1/2}\phi_{st})(\mathbf{x})\right)^2 + \frac{\delta\lambda}{4!} \left.\left(T^{1/2}\phi_{st}(\mathbf{x})\right)^4\right)
$$

$$
- \int_0^{1/T} dx_0 \int d^3\mathbf{x} \left(\frac{1}{2}(M^2 + \delta M^2)\phi_{ns}(x)^2\right.
$$

$$
+ \frac{1}{2}(1 + \delta Z_\phi) \sum_{\mu=0}^{3} (\partial_\mu\phi_{ns})(x))^2
$$

$$
+ \frac{\lambda + \delta\lambda}{4!} \left[\phi_{ns}(x)^4 + 6\left(T^{1/2}\phi_{st}(\mathbf{x})\right)^2 \phi_{ns}(x)^2\right.
$$

$$
+ 4\left(T^{1/2}\phi_{st}(\mathbf{x})\right)\left.\left.\phi_{ns}(x)^3\right]\right), \tag{4.6.54}
$$

where we have used (4.6.20). Note that (4.6.54) contains counterterms for both the static and nonstatic modes. Thus the renormalized n-point functions $\Gamma_{ns,R}^{(n)(4D)}(\phi_{st})$ which are generated by the action (4.6.54) do still include divergent terms (in contrast to what the attribute "renormalized" may suggest, cf. our comment on this point in section 4.3. These terms will precisely cancel the divergencies of the effective theory when $\int \mathcal{D}\phi_{st}$ is performed in perturbation theory and only superrenormalizable terms are kept as the effective action is derived.

In momentum space the associated Feynman rules are given by

$$
\bullet\!\!-\!\!-\!\!-\!\!\bullet \quad = \quad (k^2 + M^2)^{-1} (1 - \delta_{k_0,0}) \tag{4.6.55}
$$

for the nonstatic propagator and

$$\multimap\hspace{-0.3em}\circ\hspace{-0.3em}\multimap \quad = \quad -(\delta M^2 + \delta Z_\Phi k^2) \tag{4.6.56}$$

$$\text{---}\circ\text{---} \quad = \quad -(\delta M^2 + \delta Z_\Phi \mathbf{k}^2) \tag{4.6.57}$$

$$\times\hspace{-0.8em}\boxtimes\hspace{-0.8em}\times \quad = \quad -\delta\lambda \tag{4.6.58}$$

$$\times\hspace{-0.8em}\bullet\hspace{-0.8em}\times \quad = \quad -\lambda \tag{4.6.59}$$

for the counterterm and interaction vertices. A solid line represents the nonstatic field $\widetilde{\phi}_{ns}(p)$, a dashed line the static field $\widetilde{\phi}_{st}(\mathbf{p})$ according to (4.6.25) and (4.6.26). A dashed-dotted line is either static or nonstatic with the rule that there is never a single nonstatic line at a vertex because of momentum conservation.

Now we are ready to explicitly evaluate the contributions to $S_{eff}^{(n)}(\phi_{st})$ for $n = 2, 4, 6$. Since we want to keep the full high-temperature behavior of $\widetilde{\Gamma}_{ns,R}^{(n)}(\mathbf{p}_1, \cdots, \mathbf{p}_{n-1}|T)$, we choose $q_n = 0$ in (4.6.34). Thus we expand

$$\widetilde{\Gamma}_{ns,R}^{(2)}(\mathbf{p}|T) = T_p^2 \widetilde{\Gamma}_{ns,R}^{(2)}(\mathbf{p}|T) + O(p^3/T)$$

$$\widetilde{\Gamma}_{ns,R}^{(4)}(\mathbf{p}_1, \mathbf{p}_2, \mathbf{p}_3|T) = T_p^1 \widetilde{\Gamma}_{ns,R}^{(4)}(\mathbf{p}_1, \mathbf{p}_2, \mathbf{p}_3|T) + O(p^2/T)$$

$$\widetilde{\Gamma}_{ns,R}^{(6)}(\mathbf{p}_1, \cdots, \mathbf{p}_5|T) = T_p^0 \widetilde{\Gamma}_{ns,R}^{(6)}(\mathbf{p}_1, \cdots, \mathbf{p}_5|T) + O(p/T) . \tag{4.6.60}$$

For $n = 2$ we find

$$\widetilde{S}_0^{(2)}(\mathbf{p}) - T_p^2 \widetilde{\Gamma}_{ns,R}^{(2)}(\mathbf{p})$$

$$= (\mathbf{p}^2 + M^2) - \quad \text{---}\circ\text{---} \quad - \quad \text{loop}$$

$$= (\mathbf{p}^2 + M^2) + (\delta M^2 + \delta Z_\Phi \mathbf{p}^2) \tag{4.6.61}$$

$$+ \frac{1}{2}\lambda T \sum_{\substack{k_0 = 2\pi nT \\ \neq n \in \mathbf{Z}}} \int^{\Lambda_0} \frac{d^3\mathbf{k}}{(2\pi)^3} \frac{1}{\mathbf{k}^2 + M^2}.$$

Note that the one-loop graph is independent of p and $\delta Z_\Phi = 0$ to one-loop, so that $\mathcal{T}_P^2 \widetilde{\Gamma}^{(2)}_{ns,R}(\mathbf{p}) = \widetilde{\Gamma}^{(2)}_{ns,R}(\mathbf{p} = 0)$. It follows that

$$\widetilde{S}_0^2(\mathbf{p}) - \widetilde{\Gamma}^{(2)}_{ns,R}(\mathbf{p})$$
$$= (\mathbf{p}^2 + M^2) - \frac{\lambda}{2} \left(B_1^{(4)}(\beta = \infty, M, \Lambda_0) - B_{1,ns}^{(4)}(\beta, M, \Lambda_0) \right) . \tag{4.6.62}$$

For $n = 4$ we need $\mathcal{T}_p^1 \widetilde{\Gamma}^{(4)}_{ns,R}(\mathbf{p})$. Because of the invariance of $\Gamma^{(n)}_{ns,R}$ under $\mathbf{p} \to -\mathbf{p}$ it follows that $\mathcal{T}_p^1 \widetilde{\Gamma}^{(4)}_{ns,R}(\mathbf{p}) = \widetilde{\Gamma}^{(4)}_{ns,R}(\mathbf{0})$. Thus

$$\widetilde{S}_0^{(4)}(\mathbf{p}_1, \mathbf{p}_2, \mathbf{p}_3) - \widetilde{\Gamma}^{(4)}_{ns,R}(\mathbf{0}, \mathbf{0}, \mathbf{0})$$
$$= -T \left[\times\!\!\!\times + \times\!\!\!\times + \left\{ \times\!\!\!\bigcirc\!\!\!\times + 2 \text{ perm} \right\} \right]_{p=0}$$
$$= T \left(\lambda + \delta\lambda - \frac{3}{2}\lambda^2 T \sum_{0 \neq k_0 = 2\pi nT, n \in \mathbf{Z}} \int^{\Lambda_0} \frac{d^3\mathbf{k}}{(2\pi)^3} \frac{1}{(\mathbf{k}^2 + M^2)^2} \right)$$
$$= \lambda T - \frac{3}{2}\lambda^2 T \left(B_{2,ns}^{(4)}(\beta, M, \Lambda_0) - B_2^{(4)}(\beta = \infty, M, \Lambda_0) \right) . \tag{4.6.63}$$

Finally, for $n = 6$, there is only one contribution to one-loop, given by

$$\widetilde{S}_0^{(6)}(\mathbf{p}_1, \cdots, \mathbf{p}_5) - \widetilde{\Gamma}^{(6)}_{ns,R}(\mathbf{0}, \cdots, \mathbf{0})$$
$$= -T^2 \left(\bigcirc + 14 \text{ perm} \right)_{\mathbf{p}=0}$$
$$= -T^2(-\lambda)^3 \, 15 \left(T \sum_{0 \neq k_0 = 2\pi nT, n \in \mathbf{Z}} \int^{\Lambda_0} \frac{d^3\mathbf{k}}{(2\pi)^3} \frac{1}{(\mathbf{k}^2 + M^2)^3} \right)$$
$$= 15 T^2 \lambda^3 B_{3,ns}^{(4)}(\beta, M, \Lambda_0) . \tag{4.6.64}$$

Putting our results (4.6.62), (4.6.63), (4.6.64) together, we obtain for the effective action $S_{eff}(\phi_{st})$ up to terms of o(1) in p/T (i.e. which vanishes as

$\mathbf{p}/T \to 0$)

$$S_{eff}(\phi_{st}) = \frac{1}{2!} \int \frac{d^3\mathbf{p}}{(2\pi)^3} \, \widetilde{\phi}_{st}(\mathbf{p})\widetilde{\phi}_{st}(-\mathbf{p})$$

$$\cdot \left[\mathbf{p}^2 + M^2 - \frac{\lambda}{2} \left(B_1^{(4)}(\beta = \infty, M, \Lambda_0) - B_{1,ns}^{(4)}(\beta, M, \Lambda_0) \right) \right]$$

$$+ \frac{1}{4!} \int \frac{d^3\mathbf{p}_1}{(2\pi)^3} \cdots \int \frac{d^3\mathbf{p}_3}{(2\pi)^3} \, \widetilde{\phi}_{st}(\mathbf{p}_1)\widetilde{\phi}_{st}(\mathbf{p}_2)\widetilde{\phi}_{st}(\mathbf{p}_3)\widetilde{\phi}_{st}(-\sum_{i=1}^{3}\mathbf{p}_i)$$

$$\cdot \left[\lambda T + \frac{3}{2}\lambda^2 T \left(B_2^{(4)}(\beta = \infty, M, \Lambda_0) - B_{2,ns}^{(4)}(\beta, M, \Lambda_0) \right) \right]$$

$$+ \frac{1}{6!} \int \frac{d^3\mathbf{p}_1}{(2\pi)^3} \cdots \int \frac{d^3\mathbf{p}_5}{(2\pi)^3} \, \widetilde{\phi}_{st}(\mathbf{p}_1) \cdots \widetilde{\phi}_{st}(\mathbf{p}_5)\widetilde{\phi}_{st}(-\sum_{i=1}^{5}\mathbf{p}_i)$$

$$\cdot \left(15\lambda^3 T^2 B_{3,ns}^{(4)}(\beta, M, \Lambda_0) \right) . \tag{4.6.65}$$

Clearly we would like to see the temperature-, mass- and cutoff-dependence of the coefficient functions more explicitly. In particular we would like to verify the generic result of formula (4.6.31) that was derived quite independently of any explicit high-temperature expansion and based on dimensional analysis and power counting in a renormalizable field theory.

4.6.4.1 *Large-cutoff and high-temperature expansions of one-loop integrals*

Ultimately we are interested in the large-cutoff limit of the high-temperature behavior of the correlation functions described by the effective action (4.6.65). It is determined by the coefficient functions $B_n^{(4)}(\beta = \infty, M, \Lambda_0) - B_{n,ns}^{(4)}(\beta, M, \Lambda_0)$ for $n = 1, 2, 3$.

We calculate the nonstatic contributions $B_{n,ns}^{(4)}(\beta, M, \Lambda_0)$ according to

$$B_{n,ns}^{(4)}(\beta, M, \Lambda_0) = B_n^{(4)}(\beta, M, \Lambda_0) - B_{n,st}^{(4)}(\beta, M, \Lambda_0) \tag{4.6.66}$$

with $B_n^{(4)}(\beta, M, \Lambda_0)$ given by (4.6.43) and

$$B_{n,st}^{(4)}(\beta, M, \Lambda_0) = \beta^{-1} \int^{\Lambda_0} \frac{d^3\mathbf{k}}{(2\pi)^3} \frac{1}{(\mathbf{k}^2 + M^2)^n} . \tag{4.6.67}$$

The "static" integrals $B_{n,st}^{(4)}$ are proportional to the temperature. Their large-cutoff behavior is easily derived with the results

$$B_{1,st}^{(4)}(\beta, M, \Lambda_0) = \frac{1}{2\pi^2\beta}\left(\Lambda_0 - \frac{\pi}{2}M + O(\frac{M^2}{\Lambda_0})\right),$$

$$B_{2,st}^{(4)}(\beta, M, \Lambda_0) = \frac{1}{8\pi M\beta}\left(1 + O(\frac{M}{\Lambda_0})\right),$$ (4.6.68)

$$B_{3,st}^{(4)}(\beta, M, \Lambda_0) = \frac{1}{32\pi M^3\beta}\left(1 + O(\frac{M}{\Lambda_0})\right).$$

The nontrivial temperature dependence of S_{eff} comes from the first term in (4.6.66), $B_n^{(4)}(\beta, M, \Lambda)$. We split $B_n^{(4)}(\beta, M, \Lambda)$ into a zero-temperature part $B_n^{(4)}(\beta = \infty, M, \Lambda_0)$ which contains the divergent term for $\Lambda_0 \to \infty$, and a temperature-dependent finite remainder $B_{n,conv}^{(4)}(\beta, M, \Lambda_0)$

$$B_n^{(4)}(\beta, M, \Lambda_0) = B_n^{(4)}(\beta = \infty, M, \Lambda_0) + B_{n,conv}^{(4)}(\beta, M, \Lambda_0).$$ (4.6.69)

For $n = 1, 2$, the $B_n^{(4)}(\beta = \infty, M, \Lambda_0)$ drop out of the r.h.s. of S_{eff} in (4.6.65), and

$$B_3^{(4)}(\beta = \infty, M, \Lambda_0) = \frac{1}{32\pi^2 M^2} + O(\Lambda_0^{-2}).$$ (4.6.70)

The high-T and large-Λ_0 behavior of $B_{n,conv}^{(4)}(\beta, M, \Lambda)$ is traced back to the corresponding limits of $B_{conv}^{(4)}(\beta, M, \Lambda_0)$ with

$$B_{conv}^{(4)}(\beta, M, \Lambda_0) = B^{(4)}(\beta, M, \Lambda_0) - B^{(4)}(\beta = \infty, M, \Lambda_0),$$ (4.6.71)

where

$$B^{(4)}(\beta, M, \Lambda_0) = \beta^{-1} \sum_{\substack{k_0=2\pi nT \\ n\in\mathbf{Z}}} \int^{\Lambda_0} \frac{d^3\mathbf{k}}{(2\pi)^3} \ln(k^2 + M^2).$$ (4.6.72)

We have

$$B_{conv}^{(4)}(\beta, M, \Lambda_0) = \overline{B}^{(4)}(\beta, M) + O(\frac{\ln \Lambda_0}{\Lambda_0}),$$

$$\overline{B}^{(4)}(\beta, M) = \lim_{\Lambda_0 \to \infty} B_{conv}^{(4)}(\beta, M, \Lambda_0).$$ (4.6.73)

The high-T expansion of $\overline{B}^{(4)}(\beta, M)$ is given by

$$
\begin{aligned}
\overline{B}^{(4)}(\beta, M) = -\frac{8}{\pi^2 \beta^4} & \left[\frac{\pi^4}{360} - \frac{\pi^2}{96}(\beta M)^2 + \frac{\pi}{48}(\beta M)^3 \right. \\
& + \frac{(\beta M)^4}{128}\left(\ln \frac{\beta M}{4\pi} - \frac{3}{4} - \psi(1) \right) \\
& \left. - \frac{(\beta M)^6}{3072\pi^2}\,\zeta(3) + O((\beta M)^8) \right].
\end{aligned}
\tag{4.6.74}
$$

The high-temperature behavior of $B^{(4)}_{n,conv}(\beta, M)$ then follows from (4.6.74) and

$$
B^{(4)}_{n,conv}(\beta, M, \Lambda_0) = \frac{(-1)^{n-1}}{(n-1)!}\frac{d^n}{d(M^2)^n}B^{(4)}_{conv}(\beta, M, \Lambda_0),
\tag{4.6.75}
$$

leading to

$$
\begin{aligned}
B^{(4)}_{1,conv}(\beta, M, \Lambda_0) = {} & \frac{1}{12\beta^2} - \frac{M}{4\pi\beta} - \frac{M^2}{8\pi^2}\left[\ln \frac{\beta M}{4\pi} - \frac{1}{2} - \psi(1) \right] \\
& + \frac{\zeta(3)\beta^2 M^4}{128\pi^4} + O(\beta^4 M^6, \Lambda_0^{-1} \ln \Lambda_0) \\
B^{(4)}_{2,conv}(\beta, M, \Lambda_0) = {} & \frac{1}{8\pi\beta M} + \frac{1}{8\pi^2}\left[\ln \frac{\beta M}{4\pi} - \psi(1) - \frac{\zeta(3)\beta^2 M^2}{8\pi^2} \right] \\
& + O(\beta^4 M^4, \Lambda_0^{-1} \ln \Lambda_0)
\end{aligned}
\tag{4.6.76}
$$

and

$$
\begin{aligned}
B^{(4)}_{3,conv}(\beta, M, \Lambda_0) = {} & \frac{1}{32\pi M^3 \beta}\left[1 - \frac{M\beta}{\pi}\left(1 - \frac{\zeta(3)\beta^2 M^2}{4\pi^2} \right) \right] \\
& + O(\beta^4 M^2, \Lambda_0^{-1} \ln \Lambda_0).
\end{aligned}
\tag{4.6.77}
$$

Finally we obtain for the effective action (4.6.65)

$$
\begin{aligned}
S_{eff}(\phi_{st}) = {} & \frac{1}{2!}\int \frac{d^3\mathbf{p}}{(2\pi)^3}\,\widetilde{\phi}_{st}(\mathbf{p})\widetilde{\phi}_{st}(-\mathbf{p})\,(\mathbf{p}^2 + M^2 + \delta M^2_{\Lambda_0}) \\
& + \frac{1}{4!}\int \frac{d^3\mathbf{p}_1}{(2\pi)^3}\cdots\frac{d^3\mathbf{p}_3}{(2\pi)^3}\,\widetilde{\phi}_{st}(\mathbf{p}_1)\widetilde{\phi}_{st}(\mathbf{p}_2)\widetilde{\phi}_{st}(\mathbf{p}_3)\widetilde{\phi}_{st}(-\sum_{i=1}^{3}\mathbf{p}_i)\,\lambda_{eff}T \\
& + \frac{1}{6!}\int \frac{d^3\mathbf{p}_1}{(2\pi)^3}\cdots\frac{d^3\mathbf{p}_5}{(2\pi)^3}\,\widetilde{\phi}_{st}(\mathbf{p}_1)\cdots\widetilde{\phi}_{st}(\mathbf{p}_5)\widetilde{\phi}_{st}(-\sum_{i=1}^{5}\mathbf{p}_i)\,\sigma_{eff},
\end{aligned}
\tag{4.6.78}
$$

where the coupling constants $\delta M_{\Lambda_0}^2$, λ_{eff} and σ_{eff} for $M \lesssim T \lesssim \Lambda_0$ are given by

$$\delta M_{\Lambda_0}^2 \simeq -\frac{\lambda}{2}\left(\frac{T\Lambda_0}{2\pi^2} - \frac{T^2}{12} + \frac{M^2}{8\pi^2}\left[\ln\frac{M}{4\pi T} - \frac{1}{2} - \psi(1)\right]\right),$$

$$\lambda_{eff} \simeq \lambda - \frac{3}{2}\lambda^2\left(\frac{1}{8\pi^2}\left[\ln\frac{M}{4\pi T} - \psi(1) - \frac{\zeta(3)M^2}{8\pi^2 T^2}\right]\right), \qquad (4.6.79)$$

$$\sigma_{eff} \simeq \frac{15\lambda^3\zeta(3)}{128\pi^4}.$$

Note that for the two-point function the leading term in T is $O(T^2)$ as predicted by Eq. (4.6.31). Furthermore $\delta M_{\Lambda_0}^2$ contains an additive term that is linearly divergent as $\Lambda_0 \to \infty$. This counterterm will cancel the divergence in the one-loop static two-point function when the integration $\int \mathcal{D}\phi_{st}$ over the static modes is performed. To every order of the loop expansion, as long as we keep only the superrenormalizable ϕ^2- and ϕ^4-terms in (4.6.78), the mass counterterm inherited from four dimensions provides a contribution to S_{eff} that precisely cancels the divergencies from "static" loops in three dimensions. If we keep, in addition, the strictly renormalizable ϕ^6-term in S_{eff}, the induced renormalized n-point functions stay finite at one-loop, but at higher order in perturbation theory additional counterterms must be included that are determined by additional normalization conditions in three dimensions and cannot be traced back to normalization conditions in four dimensions. This means that supplementary matching conditions have to be imposed on the effective theory.

The reason why one may keep the ϕ^6-term in three dimensions is that it is $O(T^0)$. The quartic term is proportional to T, so that at sufficiently high temperature the six-point interaction is negligible. However, at smaller T this may not be justified. Certainly such a ϕ^6-interaction has some impact on the generic phase structure of the effective theory in three dimensions. It allows for regions of first order phase transitions in addition to the familiar second-order regions of a pure ϕ^4-theory.

4.6.4.2 *The steps after dimensional reduction*

In (4.6.78) we have derived the effective action to one-loop, in which the divergent terms were regularized with a sharp momentum cutoff Λ_0. Let us assume that we want to further investigate the phase structure of the three-dimensional model in a Monte Carlo simulation for the same set of renormalized parameters M, λ and T. What is then the right choice of the bare parameters for the Monte Carlo simulation? This is a typical problem that poses itself when analytical calculations shall be combined with Monte Carlo simulations, or, more generally, when the cutoff scheme is changed. We have to find the lattice version of (4.6.78). Let us restrict the action to superrenormalizable terms to get a closed expression for the mapping between bare and renormalized parameters. In terms of renormalized parameters the lattice action finally will be of the form

$$S_{eff,lat}(\Phi, M, \lambda; a) = a^3 \sum_{x \in \Lambda_3} \Big(\sum_{i=1}^{3} \frac{1}{2} \overline{Z} \, (\frac{1}{a}\widehat{\partial}_i \Phi(x))^2$$
$$+ \frac{1}{2}(M^2 + \delta M_{lat}^2)\Phi(x)^2 + \overline{\lambda}T \, \frac{\Phi(x)^4}{4!} \Big) \qquad (4.6.80)$$

with lattice constant a, lattice derivative $\widehat{\partial}_i \Phi(x) = \Phi(x + a\hat{i}) - \Phi(x)$, $i = 1, 2, 3$, sites x on the three-dimensional cubic lattice Λ_3. As a consequence of superrenormalizability, $\overline{Z}$ and $\overline{\lambda}$ are UV-finite to any order, whereas δM_{lat}^2 gets divergent contributions up to two-loop order. In the following we evaluate δM_{lat}^2 to one loop. To this order, $\overline{Z} = 1$ and $\overline{\lambda} = \lambda_{eff}$. What is the appropriate normalization condition? We impose the condition that the renormalized T-dependent mass $M(T)$ in the full model and in the effective model coincide. This mass $M(T)$ is defined via the renormalized two-point vertex function in four dimensions at zero momentum

$$\Gamma_R^{(2)(4D)}(p = 0|T) = -M^2(T) . \qquad (4.6.81)$$

For $T = 0$, $M^2(T) = M^2$ from (4.6.39). For finite temperature, $M^2(T)$ is given by $(\lambda T^2/24 - (\lambda MT)/(8\pi) - (\lambda M^2)/(16\pi^2)(\ln(M/(4\pi T)) - 1/2 - \psi(1)))$, as we have just calculated with a sharp momentum cutoff Λ_0. The corresponding normalization condition in three dimensions then reads

$$\Gamma_{eff,R}^{(2)(3D)}(p = 0|T) = -M^2(T) . \qquad (4.6.82)$$

Here $\Gamma_{eff,R}^{(2)(3D)}$ denotes the renormalized two-point vertex function for the effective action (4.6.80) in three dimensions. For the sharp momentum cutoff Λ_0, $M^2(T)$ was given by

$$M(T)^2 = M^2 + \delta M_{\Lambda_0}^2 + \frac{1}{2}\lambda T \int^{\Lambda_0} \frac{d^3\mathbf{k}}{(2\pi)^3} \frac{1}{\mathbf{k}^2 + M^2} . \qquad (4.6.83)$$

Apart from scaling-violating terms (that vanish as $\Lambda_0 \to \infty$) the l.h.s. of (4.6.83) is independent of the choice for the cutoff, thus we also have

$$M(T)^2 = M^2 + \delta M_{lat}^2 + \frac{1}{2}\lambda T \, I_{lat} \qquad (4.6.84)$$

with the lattice one-loop integral

$$I_{lat} = \int_{-\pi/a}^{\pi/a} \frac{d^3\mathbf{k}}{(2\pi)^3} \frac{1}{\widehat{\mathbf{k}}^2 + M^2} \qquad (4.6.85)$$

and lattice momentum

$$\widehat{\mathbf{k}}^2 = \sum_{i=1}^{3} \widehat{k}_i^2, \quad \widehat{k}_i = \frac{2}{a} \sin\left(\frac{k_i a}{2}\right). \qquad (4.6.86)$$

From (4.6.83)-(4.6.86) it follows that to order λ

$$\delta M_{lat}^2(T) = \frac{\lambda}{2}\left(\frac{T^2}{12} - \frac{TM}{4\pi} - \frac{M^2}{8\pi^2}\ln\left(\frac{M}{4\pi T} - \frac{1}{2} - \psi(1)\right) - T\frac{1}{a}\widehat{I}_{lat}\right) \qquad (4.6.87)$$

with $\widehat{I}_{lat} = aI_{lat}$ a finite dimensionless integral. Hence the linearly divergent term is now proportional to a^{-1} as $a \to 0$ in the continuum limit. This concludes the derivation of the effective lattice action (4.6.80) in terms of renormalized parameters M, λ, and T.

We remark that the normalization conditions (4.6.81) and (4.6.82) work well in the massive case $M > 0$. On the other hand, for the massless theory with $M = 0$, IR-divergencies of the static sector prevent this matching to work in higher orders. This is best circumvented if the lattice regularization is chosen from the very beginning. In this case no matching is required, and only the nonstatic modes are involved in the computation of the effective action S_{eff}.

Next we write the renormalized action in terms of bare dimensionless parameters M_0, λ_0 and $L_0 = (aT)^{-1}$. Rescaling to dimensionless fields ϕ_0 according to

$$\phi(x) = a^{-\frac{1}{2}} \phi_0(x), \qquad (4.6.88)$$

the action becomes

$$S_{eff,lat}(\phi, M, \lambda; a) = S_{eff,lat}^{bare}(\phi_0, M_0, \lambda_0)$$

$$= \sum_{x \in \Lambda_3} \left(\frac{1}{2}\sum_{i=1}^{3}\left(\widehat{\partial}_i\phi_0(x)\right)^2 + \frac{1}{2}M_0^2\phi_0(x)^2 + \frac{\lambda_0}{4!}\phi_0(x)^4\right) \qquad (4.6.89)$$

with dimensionless bare parameters

$$M_0^2 = \left(M^2 + \delta M_{lat}^2\right) a^2$$

$$= M^2 a^2 + \frac{\lambda}{2} \left(\frac{1}{12}(Ta)^2 - \frac{1}{4\pi}(Ma)\,(Ta)\right.$$

$$\left. - \frac{M^2 a^2}{8\pi^2}\left[\ln \frac{Ma}{4\pi Ta} - \frac{1}{2} - \psi(1)\right] - (Ta)\,\widehat{I}_{lat}\right) \qquad (4.6.90)$$

and

$$\lambda_0 = \lambda_{eff}\,Ta$$

$$= (Ta)\left[\lambda - \frac{3}{2}\lambda^2 \frac{1}{8\pi^2}\left(\ln \frac{Ma}{4\pi Ta} - \psi(1)\right)\right]. \qquad (4.6.91)$$

Lattice results from (4.6.89)-(4.6.91) may be interpreted as continuum results for fixed physical parameters M, λ and T if M_0^2 and λ_0 are sent to zero as $a \to 0$ precisely in the form which is given in (4.6.90) and (4.6.91). A naive translation of the counterterm $\delta M_{\Lambda_0}^2$ (that was derived with a sharp momentum cutoff) onto the lattice would lead to $(-\lambda Ta/(4\pi^2))$ rather than $-\lambda Ta\hat{I}_{lat}/2$, thus it would be wrong by a factor $(\hat{I}_{lat})/(2\pi^2)$, so that a lattice simulation with a counterterm that was not appropriately adjusted effectively amounts to a different choice of renormalized parameters M, λ, and T. From (4.6.90) it is also obvious what happens if the lattice counterterm would be dropped completely. It gives the leading contribution to the continuum limit as $a \to 0$, since the other terms of the action are suppressed by one order in a. Also it has the opposite sign. Hence a check of scaling in two simulations with lattice constants a_1 and a_2 may lead to completely misleading results unless the bare parameters are tuned according to (4.6.90) and (4.6.91), with a replaced by a_1 and a_2, respectively. A similar matching problem between different cutoff schemes occurs as the SU(2)-Higgs model for the electroweak transition is dimensionally reduced from four to three dimensions, cf. section 4.6.7.

4.6.5 *Pure gauge theories and QCD*

In view of applications in particle physics nonabelian Yang-Mills theories (i.e. pure $SU(N_c)$ gauge theories) and QCD were the first realistic theories to which dimensional reduction was successfully applied. In the framework of dimensional reduction it was possible to show that screening is a characteristic property of the high-temperature phase, in case of QCD of the quark-gluon plasma phase, and to quantitatively calculate the values of the screening masses. Similarly to confinement in the low-temperature phase, screening is one of the challenging properties of QCD in the high-temperature phase, because it is a long-range property where strong interactions are actually strong. (In contrast to that, strong interactions at low energies are weak and chiral perturbation theory makes use of that fact, cf. section 6.1.) In particular screening of quarks implies deconfinement in the sense that the gluonic string breaks at large distances and the color charges of the quarks are shielded.

Usually the (color-electric) screening mass is calculated from the vacuum polarization tensor of the gauge fields in analogy to QED. In QED the infrared properties are available within perturbation theory, thus the polarization tensor can be calculated perturbatively. The polarization tensor must be evaluated in a fixed gauge. In the end it must be checked, that the result is independent of the choice of gauge.

Here we choose a definition of the screening mass via the large-distance behavior of the potential between two static quarks. The definition is gauge invariant and independent of any perturbative scheme.

4.6.5.1 *Pure gauge theories*

Let us consider the case of pure gauge theories first. In the following we consider the gauge group $SU(N_c)$. The large-distance behavior of the quark potential is obtained from the correlation function of Polyakov loops $L(\mathbf{x})$ that are defined by

$$L(\mathbf{x}) = \frac{1}{N_c} \operatorname{tr} P \exp\left(\int_0^{1/T} dx_0 A_0(x_0, \mathbf{x})\right), \qquad (4.6.92)$$

where P denotes path ordering. In section 3.3.8 we have shown that the correlation function $< L(\mathbf{x}_1) \cdots L(\mathbf{x}_n) >$, up to a normalization factor, is the partition function of n static quarks, located at $\mathbf{x}_1, \ldots, \mathbf{x}_n$, in a system of dynamical gauge fields. For pure Yang-Mills theories, the expectation value $< L(\mathbf{x}) >$ provides an order parameter. For temperatures below the phase-transition temperature T_c, $< L(\mathbf{x}) >= 0$, whereas $< L(\mathbf{x}) > \neq 0$ above T_c. The free energy $F(\mathbf{x})$ of a static quark-antiquark pair is given by

$$\exp\left(-\frac{F(\mathbf{x})}{T}\right) = < L(\mathbf{x})L^*(\mathbf{0}) > . \qquad (4.6.93)$$

Now we know that the truncated Polyakov-loop correlation functions satisfy the cluster property

$$P(\mathbf{x}) \equiv < L(\mathbf{x})L^*(0) >_c \equiv < L(\mathbf{x})L^*(0) > - | < L(0) > |^2$$
$$\rightarrow 0 \quad \text{as } |\mathbf{x}| \rightarrow \infty. \tag{4.6.94}$$

The issue of confinement versus deconfinement is then related to the value of $< L(0) >$ being zero or not. If $< L(0) >= 0$, (4.6.94) implies that $F(\mathbf{x}) \rightarrow \infty$ as $|\mathbf{x}| \rightarrow \infty$. This implies confinement, since $F(\mathbf{x})$ converges to the free energy of two isolated quarks as $|\mathbf{x}| \rightarrow \infty$ and the free energy of two isolated quarks diverges. On the other hand, if $< L(0) >\neq 0$, $F(\mathbf{x})$ has the same large-distance behavior as $P(\mathbf{x})$,

$$-\frac{F(\mathbf{x})}{T} = \ln | < L(0) > |^2 + \frac{P(\mathbf{x})}{| < L(0) > |^2} + O(P(\mathbf{x})^2). \tag{4.6.95}$$

Hence $F(\mathbf{x})$ saturates at large $\mathbf{x}$. For $< L(\mathbf{x}) >\neq 0$, the static quark-antiquark potential $V(\mathbf{x})$ is defined by normalizing the free energy according to

$$\exp\left(-\frac{V(\mathbf{x})}{T}\right) = \frac{< L(\mathbf{x})L^*(0) >}{| < L(0) > |^2} = 1 + \frac{P(\mathbf{x})}{| < L(0) > |^2}. \tag{4.6.96}$$

Screening then means that the potential $V(\mathbf{x})$ and hence $P(\mathbf{x})$ decay exponentially fast at large distances,

$$\frac{V(\mathbf{x})}{T} \simeq |\mathbf{x}|^{-\alpha} \exp\left(-\mu|\mathbf{x}|\right) \tag{4.6.97}$$

as $|\mathbf{x}| \rightarrow \infty$. Here the decay parameter μ is called the electric screening mass, "electric" or more precisely color-electric screening, because $\mu \neq 0$ implies screening of stationary colored charges.

4.6.5.2   *QCD with fermions*

In QCD, $< L(\mathbf{x}) >$ is no longer a true order parameter. Nevertheless, in the high-temperature phase, $< L(\mathbf{x}) >\neq 0$ together with (4.6.97) still implies electric screening of quarks in the quark-gluon plasma. In the confined phase, a nonvanishing but considerably smaller value of $< L(\mathbf{x}) >$ is taken as an indication of gluon-string breaking.

Let us first see how we could calculate screening by standard methods. The short-distance region of QCD is perturbative because of asymptotic freedom. Weak gauge-coupling expansion gives a quantitatively reliable description whenever g^2RT is small if R denotes the spatial length scale and g the renormalized coupling constant. In this region the Polyakov-loop correlation decays power like in $R = |\mathbf{x}| \lesssim T^{-1}$.

On the other hand, screening is a large-scale property. For calculating the screening masses the quark potential must be determined outside the range of perturbation theory. This is quite cumbersome in asymptotically free gauge theories like QCD. Let us recall why.

Large-scale properties of QCD are usually studied by Monte Carlo simulations. On a lattice of size $L_0 \times L_s^3$ the temperature and the spatial size are given by $T = 1/(L_0 a)$ and $\mathcal{L}_s = L_s a$, respectively. To simulate finite-temperature QCD close to its continuum limit, two conditions must be satisfied.

- L_0 should be large to keep lattice artifacts small. These effects can be rather strong if there are only a few lattice links in temperature direction. In case of QCD with dynamical fermions such lattice artifacts are even more pronounced than for pure gauge theories.
- $L_s \gg L_0$ should be satisfied in order to describe a system at finite temperature and to distinguish it from a zero-temperature system in a small volume. The spatial finite-size effects should be small.

We remark that one should carefully distinguish between finite-size effects from the finite extension in Euclidean time and finite-size effects from the finite extent in space $L_s < \infty$. $L_s < \infty$ induces the familiar IR-cutoff effects which can be controlled by means of a finite-size scaling analysis if L_s is sufficiently large.

The finite extension $1/T$ of the temperature torus is a physical constraint in order to have a finite-T quantum system. On the lattice, a finite number L_0 of lattice links in this direction should describe a field theory at temperature $T = 1/(L_0 a)$. Ideally an infinite number of L_0 should cover the finite distance in time direction. Thus a finite number of time slices L_0 is an UV-cutoff effect as long as the lattice spacing is finite. The continuum limit is ultimately obtained as $a \to 0$ with $T = 1/(L_0 a)$ fixed so that L_0 must be sent to infinity too, and apart from that, to obtain the infinite-volume limit of the continuum limit, L_s should be send to infinity as well.

Investigation of screening requires measurements of the Polyakov-loop correlation $P(\mathbf{x})$ at spatial distances $R \geq L_0$ in lattice units. For smaller distances, $P(\mathbf{x})$ decays power-like and is well described by perturbation theory.

Table 4.6.1   The fermionic part of the vacuum polarization at zero momentum, $\Pi_{00}(0)_f$, to one-loop order, obtained on $L_0 \times L_s^3$ lattices with $L_s/L_0 = 4$ for various values of L_0, the extent of the torus in the temperature direction, and normalized against the scaling form $\Pi_{00}(0)_{f,sl} = -N_f/(6L_0^2)$ for N_f flavors.

L_0	$\Pi_{00}(0)_f/\Pi_{00}(0)_{f,sl}$ Wilson, $r = 1$	$\Pi_{00}(0)_f/\Pi_{00}(0)_{f,sl}$ staggered
4	4.17	2.23
8	1.52	1.47
12	1.16	1.16
16	1.08	1.08
20	1.05	1.05
24	1.03	1.03

In the section on perturbation theory we have actually seen large cutoff effects produced by a small value for L_0 in the example of the $SU(2)$ Yang-Mills theory. The effects are even more drastic for fermions, in particular for the fermionic part of the vacuum polarization at zero momentum. This is a quantity we will later need for calculating the effective action in the framework of dimensional reduction. In Table 4.6.1 we list some values, obtained from perturbation theory on the lattice, for small and moderate lattice sizes, both for Wilson and staggered fermions. The value of L_0 is in the scaling region, that is sufficiently large for Wilson or staggered fermions, if $\Pi_{00}(0)_f$ is close to $\Pi_{00}(0)_{f,sl}$, where $\Pi_{00}(0)_{f,sl} = -N_f/(6L_0^2)$ denotes the scaling value of the fermionic part of the vacuum polarization. Even for moderate number of lattice links in T-direction such as 8 or 12, the effects are nonnegligible.

Estimates for four-flavor QCD require a minimal lattice size of 12×72^3 in order to simultaneously satisfy the conditions of

- simulating high-T QCD close to its continuum limit,
- having small finite-(spatial-)volume effects.

Screening then should be observed at spatial distances $R \geq 12$ in lattice units. In spite of all progress during the last decades, such large lattices are still outside the range of nowadays computer capabilities. (Actually, these large lattice volumes may no longer be needed when so-called improved actions are used that are constructed to reduce the UV-artifacts, cf. section 4.7.5.)

This is the place where dimensional reduction enters the game. By means of dimensional reduction we will circumvent simulations on large lattices in four dimensions. What we will need are Monte Carlo simulations of *short*-range properties in four dimensions and long-range behavior in three dimensions. As we shall see, the IR-properties of high-temperature QCD can be computed in an effective three-dimensional lattice field theory. We shall demonstrate Debye screening and measure the screening masses. A further advantage of the effective model is the absence of dynamical fermions. As it turns out, a 24^3-lattice is sufficient to determine the screening masses in full QCD.

All temperature effects are absorbed in the coupling constants of the effective three-dimensional model. We must first compute the coupling constants of the effective action and identify which observables in the three-dimensional model correspond to the observables in full QCD. Because the first part of the computation proceeds analytically (via renormalized perturbation theory on the lattice), lattice artifacts can be removed from the finite-T torus by dropping terms of order $O(a)$ or smaller.

The renormalized perturbation theory on the lattice is not the standard one. Let us recall why. In section 4.4 we have seen that conventional perturbation theory in finite-T gauge theories is plagued by severe IR-problems in the infinite spatial volume. IR- and UV-singularities are not properly disentangled unless appropriate resummation techniques are applied. Such approaches in the infinite volume often intend a conversion of UV-perturbation theory into IR- perturbation theory when screening masses shall be determined.

The approach which we have in mind works in the finite volume. In an asymptotically-free theory like QCD, the primary realm of perturbation theory is the UV-region. At least one would like to make this region IR-safe, since in massless theories IR-singularities do not automatically decouple from the UV-behavior. This is the reason why we directly work in a finite volume. Fortunately, for all practical purposes, the volume can be made sufficiently large for small gauge coupling according to

$$\mathcal{L}_s \leq c(gT)^{-1}, \qquad (4.6.98)$$

where it turns out that typically $c \simeq 8$. Equation (4.6.98) reflects the intrinsic entanglement of UV-and IR-properties in massless gauge theories.

The resulting perturbation theory is the so-called coupled large volume-/small gauge coupling expansion. It does not require any resummation, but is still an expansion in the number of loops. The price one has to pay for keeping the volume finite is to control additional zero momentum modes that arise because of further degeneracies of the field configurations which minimize the action in the finite volume. The expansion works quite well whenever $g^2 RT$ is small as we have explicitly seen for the quark potential. Here g denotes the temperature dependent renormalized gauge coupling constant defined in this perturbative scheme on the scale $R \simeq T^{-1}$, for example by matching the quark-antiquark potential $V(\mathbf{x})$ calculated in perturbation theory for fixed $|\mathbf{x}_0| = R_0 = 1/T$ with the value measured in a Monte Carlo simulation in four dimensions.

The computations mentioned so far are performed in four dimensions. They are independent of any separation between static and nonstatic modes. The perturbative scheme remains the same in a computation of the dimensionally reduced effective theory. The nonstatic modes themselves have an intrinsic IR-cutoff proportional to T^{-1}. IR-problems arising when the volume is made large are well separated from the nonstatic modes. Thus they are IR-safe. The IR-singularities only concern the static degrees of freedom including the zero-momentum modes. The only requirement to apply perturbation theory is that the renormalized gauge coupling g is small. The coupling constants of the effective model become *power* series in g without fractional or logarithmic terms, reflecting the fact that the nonstatic modes are free of IR-singularities.

4.6.5.3 *Dimensional reduction for pure $SU(N_c)$ gauge theories*

We first consider the case of pure $SU(N_c)$ gauge theories. Fermionic degrees of freedom will be added later. In perturbation theory the expectation value of a gauge invariant observable $I(U)$ is obtained as

$$< I(U) > = \frac{1}{Z} \int_{\mathcal{O}} \mathcal{D}U I(U) \exp\left(-S_W(U)\right) \qquad (4.6.99)$$

with $< 1 >= 1$. $\mathcal{O}$ denotes a neighborhood in field space of the toron configurations, the set of configurations which minimize the gauge field action $S_W(U)$ in the finite volume. Furthermore,

$$\mathcal{D}U = \prod_{x,\mu} dU(x;\mu) \qquad (4.6.100)$$

denotes the product Haar measure, and $S_W(U)$ is the gauge field action. In the following we consider the Wilson action

$$S_W(U) = \beta \sum_{x \in \Lambda_4} \sum_{\mu < \nu = 0}^{3} \left(1 - \frac{1}{N_c} \operatorname{Re} \operatorname{tr} U(x; \mu) U(x + a\widehat{\mu}; \nu) \right.$$

$$\left. U(x + a\widehat{\nu}; \mu)^{-1} U(x; \nu)^{-1} \right) \tag{4.6.101}$$

with $\beta = (2N_c)/g_0^2$, g_0 denotes the bare gauge coupling constant.

Because of gauge fixing $< I(U) >$ may be computed just from a neighborhood of $U(x; \mu) \equiv 1$ if we parameterize

$$U(x; \mu) = \exp(a A_\mu(x)), \tag{4.6.102}$$

with the Lie algebra valued gauge field $A_\mu(x)$ satisfying a set of admissible gauge fixing constraints. In static time-averaged Landau gauge (STALG) these constraints are

$$A_0(x_0, \mathbf{x}) \equiv A_0(\mathbf{x}),$$

$$\frac{1}{L_0 a} a \sum_{x_0} \sum_{i=1}^{3} \frac{1}{a} \widehat{\partial}_i^* A_i(x_0, \mathbf{x}) = 0. \tag{4.6.103}$$

The expectation value $< I(U) >$ then becomes

$$< I(U) > = \frac{1}{Z} \int \prod_x \left(d'b(x) d'\overline{b}(x) \prod_\mu dA_\mu(x) \right) \prod_{\mathbf{x}} (d'c(\mathbf{x}) d'\overline{c}(\mathbf{x}))$$

$$\cdot I(\exp aA) \exp(-S(A, b, \overline{b}, c, \overline{c})). \tag{4.6.104}$$

Here $b, \overline{b}, c, \overline{c}$ are the ghost fields. The action S is given by

$$S(A, b, \overline{b}, c, \overline{c}) = S_W(\exp aA) + S_m(A) + S_{gf}(A)$$

$$+ S_{FP}^{(1)}(A_0, b, \overline{b}) + S_{FP}^{(2)}(A_0, c, \overline{c}). \tag{4.6.105}$$

Here S_m denotes the part originating from the Haar measure (4.6.100) upon the parametrization (4.6.102) and (4.6.103). S_{gf} is the gauge fixing action, $S_{FP}^{(1)}$ and $S_{FP}^{(2)}$ are the Faddeev-Popov actions.

The prime at the measures indicates that the zero modes are removed in the integration upon the Fadeev-Popov ghost fields. Finally the constant field modes in the gauge field A must be separated in the finite volume. Here we do not spell out the details, but refer to section 4.4.

The decomposition into static and nonstatic modes proceeds for the gauge field according to

$$A_\mu(x) = T^{1/2} A_\mu^{st}(\mathbf{x}) + A_\mu^{ns}(x), \qquad \sum_{x_0=0}^{L_0} A_\mu^{ns} = 0, \qquad (4.6.106)$$

and similarly for the ghost fields. For the Fourier-transformed fields

$$\widetilde{A}_\mu(k) = a^4 \sum_{x \in \Lambda_4} \exp\left(-ik \cdot x\right) \exp\left(-ik_\mu a/2\right) A_\mu(x) \qquad (4.6.107)$$

this implies

$$\widetilde{A}_\mu(k) = T^{-1/2} \widetilde{A}_\mu^{st}(\mathbf{k}) \delta_{k_0,0} + \widetilde{A}_\mu^{ns}(k) \left(1 - \delta_{k_0,0}\right) \qquad (4.6.108)$$

with

$$\widetilde{A}_\mu^{st}(\mathbf{k}) = a^3 \sum_{\mathbf{x} \in \Lambda_3} \exp\left(-i\mathbf{k} \cdot \mathbf{x}\right) \exp\left(-ik_\mu a/2\right) A_\mu^{st}(\mathbf{x}),$$

$$\widetilde{A}_\mu^{ns}(k) = a^4 \sum_{x \in \Lambda_4} \exp\left(-ik \cdot x\right) \exp\left(-ik_\mu a/2\right) A_\mu^{ns}(x). \qquad (4.6.109)$$

We remark that ghost fields, unlike fermions, are not subject to anti-periodic boundary conditions, although they are Grassmann variables. In general they involve both static and nonstatic modes.

As we have already seen in section 4.4, STALG provides several advantages for finite-temperature calculations. In particular, the T-component A_0 of the gauge field is purely static, $A_0^{ns}(x) \equiv 0$. Therefore the Polyakov loop acquires a particularly simple form,

$$L(\mathbf{x}) = \frac{1}{N_c} \operatorname{tr} \prod_{x_0=0}^{L_0} U\left((x_0, \mathbf{x}); 0\right) = \frac{1}{N_c} \operatorname{tr} \exp\left(aL_0 A_0(\mathbf{x})\right), \qquad (4.6.110)$$

it is purely static itself. Furthermore, the ghost-degrees of freedom cleanly separate into static and nonstatic ghosts, since the Faddeev-Popov determinant factorizes accordingly. The ghosts c, $\bar{c}$, which only couple to the spatial components of the gauge field, are purely static. The ghosts b, $\bar{b}$, which couple to the static $A_0(\mathbf{x})$ fields, are purely nonstatic. Thus we finally arrive at the following representation of the expectation value $< I(U) >$

$$< I(U) > = \frac{1}{Z_{eff}} \int \prod_{\mathbf{x}} \left(dc(\mathbf{x}) d\bar{c}(\mathbf{x}) \prod_\mu dA_\mu^{st}(\mathbf{x}) \right)$$

$$\cdot I_{eff}(A^{st}, c, \bar{c}) \exp\left(-S_{eff}(A^{st}, c, \bar{c})\right) \qquad (4.6.111)$$

with

$$Z_{eff} = Z = \int \prod_{\mathbf{x}} \left(dc(\mathbf{x}) d\bar{c}(\mathbf{x}) \prod_{\mu} dA_{\mu}^{st}(\mathbf{x}) \right) \exp\left(-S_{eff}(A^{st}, c, \bar{c})\right).$$
(4.6.112)

Here the effective action is determined by

$$\exp\left(-S_{eff}(A^{st}, c, \bar{c})\right) = Z_{ns}(A^{st}, c, \bar{c}) = \int \prod_{x} \left(db(x) d\bar{b}(x) \prod_{\mu} dA_{\mu}^{st}(\mathbf{x}) \right)$$
$$\cdot \exp\left(-S(T^{1/2}A^{st} + A^{ns}, b, \bar{b}, T^{1/2}c, T^{1/2}\bar{c})\right),$$
(4.6.113)

and the effective observable I_{eff} is obtained as

$$I_{eff}(A^{st}, c, \bar{c}) = \frac{1}{Z_{ns}(A^{st}, c, \bar{c})} \int \prod_{x} \left(db(x) d\bar{b}(x) \prod_{\mu} dA_{\mu}^{st}(\mathbf{x}) \right)$$
$$\cdot I\left(\exp\left(T^{1/2}A^{st} + A^{ns}\right) \right)$$
$$\cdot \exp\left(-S(T^{1/2}A^{st} + A^{ns}, b, \bar{b}, T^{1/2}c, T^{1/2}\bar{c})\right).$$
(4.6.114)

In general, the relation between the observable I_{eff} in the effective model and in the original model I need not be simple. An example for a complicated relation is given by spatial Wilson loops (that is products of gauge fields along spatial loops). For Polyakov loops (that is products of gauge fields along a line in timelike direction) and their correlations it is, however, quite simple. Since Polyakov loops are purely static observables, they factorize out of the integral in (4.6.114), so that

$$L_{eff}(A^{st})(\mathbf{x}) = L(T^{1/2}A^{st})(\mathbf{x}) = \frac{1}{N_c} \operatorname{tr} \exp\left((aL_0)^{1/2}A_0^{st}(\mathbf{x})\right), \quad (4.6.115)$$

and similarly do their correlation functions.

4.6.5.4 *Renormalization and zero-momentum projection*

So far we have merely written $< I(U) >$ as expectation value in terms of static and nonstatic fields. In the next steps we renormalize the four-dimensional theory and apply the zero-momentum projections to the effective action $S_{eff}(A^{st}, c, \bar{c})$, as discussed in the last section.

Renormalization amounts to express the bare parameters and fields in terms of renormalized ones in such a way that the correlation functions have a well defined continuum limit in terms of renormalized coupling constants, and in a way that the required symmetries are preserved. For multiplicatively renormalizable gauge theories this is achieved by renormalizing the fields according to

$$A \to Z_A^{1/2} A, \quad b \to Z_c^{1/2} b, \quad \bar{b} \to Z_b^{1/2} \bar{b}, \quad \text{etc.,} \tag{4.6.116}$$

and the gauge coupling according to

$$g_0(g) = Z_g g, \quad Z_g = 1 + O(g). \tag{4.6.117}$$

The renormalization constants are uniquely defined by imposing independent normalization conditions on correlation functions. In perturbation theory they are (not necessarily analytic) series in the renormalized gauge coupling g. The maps (4.6.116) and (4.6.117) can be realized by adding local counterterms to the lattice action.

We know that at zero temperature the vacuum polarization $\Pi_{\mu\nu}(k)$ vanishes at zero momentum,

$$\Pi_{\mu\nu}(k = 0) = 0, \quad \text{for } T = 0, \tag{4.6.118}$$

as a consequence of the BRS-symmetry of the gauge-fixed action. According to the conventional lore, the $T = 0$-counterterms render the theory finite also at any finite T, therefore we do not need additional T-dependent mass subtractions. The Polyakov loop gets an additional T-dependent renormalization

$$L(A)(\mathbf{x}) \to L(Z_A^{1/2} A)(\mathbf{x}) \cdot \exp\left(b_0 + T^{-1} b_1\right) \tag{4.6.119}$$

that drops out of the quark potential defined by (4.6.96), in which we are interested here, so we do not discuss the Polyakov-loop renormalization in more detail.

The next steps are to compute the effective action S_{eff} in the (large volume/small coupling-) perturbation theory on the lattice and to apply the zero momentum expansion to S_{eff} according to the rules of dimensional reduction. We present the results of a perturbative calculation to one-loop.

The leading tree-level part of the effective action is given by the classical level of dimensional reduction. It corresponds to the part denoted by S_0 in the last section. We have

$$\begin{aligned}
S_{eff}^{(0)}(A^{st}, c, \bar{c}) = {} & S_W^{3d}(U) + S_{adj}(U, A_0^{st}) + S_{FP}(A_i^{st}, c, \bar{c}) \\
& + S_{gf}(A_i^{st}) + S_m(A_i^{st}).
\end{aligned} \tag{4.6.120}$$

$A_0^{st}(\mathbf{x})$ still is a Lie-algebra valued (here $su(N_c)$-valued) field. Let us see what the various terms are. In the following we write

$$U(\mathbf{x}; i) = \exp\left(aT^{1/2}A_i^{st}(\mathbf{x})\right),$$
$$D_i^{adj}(U)A_0^{st}(\mathbf{x}) = U(\mathbf{x}; i)A_0^{st}(\mathbf{x} + a\widehat{i})U(\mathbf{x}; i)^{-1} - A_0^{st}(\mathbf{x}) \quad (4.6.121)$$

for $i = 1, 2, 3$ and for any Lie-algebra valued field X

$$J(X)(\mathbf{x}) = \frac{1 - \exp\left(-\operatorname{ad} X\right)}{\operatorname{ad} X}(\mathbf{x})$$
$$= 1 - \frac{1}{2}\operatorname{ad} X(\mathbf{x}) + \frac{1}{6}\left(\operatorname{ad} X(\mathbf{x})\right)^2 + O((\operatorname{ad} X(\mathbf{x}))^3). \quad (4.6.122)$$

The function J occurs in the Jacobian of the gauge-field measure, cf. section 4.3. Then

$$S_W^{3d}(U) = \frac{2N_c}{g^2 T}\, a \sum_{\mathbf{x} \in \Lambda_3} \sum_{i<j=1}^{3}\left(1 - \frac{1}{N_c}\operatorname{Re} \operatorname{tr} U(\mathbf{x}; i)U(\mathbf{x} + a\widehat{i}; j)\right.$$
$$\left. \cdot U(\mathbf{x} + a\widehat{j}; i)^{-1}U(\mathbf{x}; j)^{-1}\right), \quad (4.6.123)$$

$$S_{adj}(U, A_0^{st}) = -\frac{1}{g^2}\, a^3 \sum_{\mathbf{x}}\sum_{i=1}^{3} \operatorname{tr}\left(\frac{1}{a}D_i^{adj}(U)A_0^{st}(\mathbf{x})\right)^2. \quad (4.6.124)$$

Faddeev-Popov and gauge-fixing actions become

$$S_{FP}(A_i^{st}, c, \overline{c}) = -a^3 \sum_{\mathbf{x}} \operatorname{tr}\left(\overline{c}(\mathbf{x})\right.$$
$$\left. \cdot \sum_{i=1}^{3}\frac{1}{a}\partial_i^*\left(\operatorname{ad} A_i^{st}(\mathbf{x}) + J^{-1}(aT^{1/2}A_i^{st})(\mathbf{x})\frac{1}{a}\partial_i\right) c(\mathbf{x})\right), \quad (4.6.125)$$

$$S_{gf}(A_i^{st}) = -\frac{\lambda}{g^2}\, a^3 \sum_{\mathbf{x}} \operatorname{tr}\left(\sum_{i=1}^{3}\frac{1}{a}\partial_i A_i^{st}(\mathbf{x})\right)^2. \quad (4.6.126)$$

The measure part is given by the standard expression

$$S_m(A_i^{st}) = -\sum_{\mathbf{x}}\sum_{i=1}^{3} \operatorname{tr} \ln J(aT^{1/2}A_i^{st})(\mathbf{x}). \quad (4.6.127)$$

For small lattice spacing a we obtain the expressions for the gauge-field action that are familiar from continuum formulations

$$S_W^{3d}(U) \simeq -\frac{1}{2g^2} a^3 \sum_{\mathbf{x}} \left(\sum_{i,j=1}^{3} (\operatorname{tr} F_{ij} F_{ij})(\mathbf{x}) + O(a) \right),$$

$$F_{ij}(\mathbf{x}) = \left(\frac{1}{a} \left(\partial_i A_j^{st} - \partial_j A_i^{st} \right) + T^{1/2}[A_i^{st}, A_j^{st}] \right)(\mathbf{x}), \quad (4.6.128)$$

the adjoint Higgs-field action

$$S_{adj}(U, A_0^{st}) = -\frac{1}{g^2} a^3 \sum_{\mathbf{x}} \left(\sum_{i=1}^{3} \operatorname{tr} \left(\frac{1}{a} \partial_i A_0^{st} + T^{1/2}[A_i^{st}, A_0^{st}] \right)^2 (\mathbf{x}) \right.$$

$$\left. + O(a) \right), \qquad (4.6.129)$$

and the Faddeev-Popov action

$$S_{FP}(A_i^{st}, c, \bar{c}) = -a^3 \sum_{\mathbf{x}} \sum_{i=1}^{3} \left(\operatorname{tr} \bar{c}(\mathbf{x}) \left(\frac{1}{a} \partial_i [A_i^{st}, c] + \frac{1}{a^2} \partial_i^2 c \right) (\mathbf{x}) + O(a) \right).$$

$$(4.6.130)$$

Here g always denotes the renormalized gauge-coupling constant in four dimensions.

Next we include the nonstatic one-loop corrections $S_{eff}^{(1)}$ to $S_{eff}^{(0)}$. We will include only those effective interactions which do not spoil the superrenormalizability of the effective three-dimensional model. This is justified if the temperature is sufficiently large. Since the BRS-invariance is partially preserved by the procedure of dimensional reduction, the most general reduced effective action should be BRS-invariant again. It must be an *adjoint* Higgs model because $A_0^{st}(\mathbf{x})$ transforms under the adjoint representation of the gauge group. This expectation is confirmed by the one-loop computation. BRS-invariance implies that all nonstatic corrections $\Delta_1 S_{eff}^{(1)}$, which are of the form of the classically reduced action, are given by

$$\Delta_1 S_{eff}^{(1)} = \left(\frac{r_{As}}{2} \frac{\partial}{\partial \tau_{As}} + \frac{r_{A0}}{2} \frac{\partial}{\partial \tau_{A0}} + \frac{r_c}{2} \frac{\partial}{\partial \tau_c} + r_g \frac{\partial}{\partial \tau_g} \right)$$

$$\cdot S_{eff}^{(0)}(\tau_{As} A_i^{st}, \tau_{A0} A_0^{st}, \tau_c c, \tau_c \bar{c}, \tau_g g) \Big|_{\tau \equiv 1} \qquad (4.6.131)$$

with $r_{As}, r_{A0}, r_c, r_g = O(g^2)$. The form of the various contributions to

$\Delta_1 S_{eff}^{(1)}$ imply finite multiplicative renormalizations

$$A_i^{st} \to Z_{As,fin}^{1/2} A_i^{st},$$
$$A_0^{st} \to Z_{A0,fin}^{1/2} A_0^{st},$$
$$c(\bar{c}) \to Z_{c,fin}^{1/2} c(\bar{c}), \qquad (4.6.132)$$
$$g \to Z_{g,fin} g$$

with $Z_{As} = 1 + r_{As}$, $Z_{A0} = 1 + r_{A0}$, $Z_c = 1 + r_c$, $Z_g = 1 + r_g$. Terms of $S_{eff}^{(1)}$ which are not of this form are pure self-interactions of the adjoint Higgs field A_0^{st},

$$\Delta_2 S_{eff}^{(1)} = a^3 \sum_{\mathbf{x}} \left(\Pi_{00}^{ns}(0) \, \mathrm{tr} \left(A_0^{st}(\mathbf{x}) \right)^2 \right. \qquad (4.6.133)$$
$$\left. + \frac{T}{12\pi^2} \left[N_c \, \mathrm{tr} \left(A_0^{st}(\mathbf{x}) \right)^4 + 3 \left(\mathrm{tr} \left(A_0^{st}(\mathbf{x}) \right)^2 \right)^2 \right] \right).$$

$\Pi_{00}^{ns}(\mathbf{k})$ denotes the nonstatic part of the vacuum polarization of the A_0-field to one-loop order. (We write $\Pi_{00}^{ns}(\mathbf{k})$ as a number because it is diagonal in color space.) Note that the effective couplings in (4.6.133) were determined by the nonstatic vertex functions $\widetilde{\Gamma}_{ns}$ at zero momentum. The small a expansion of the vacuum polarization is given by

$$\Pi_{00}^{ns}(0) = \frac{1}{a} 2 N_c T \frac{1}{L_s^3} \sum_{\mathbf{k} \neq 0} \frac{1}{4 \sum_{i=1}^{3} \sin^2 (k_i/2)} - \frac{N_c}{3} T^2 + O(a). \quad (4.6.134)$$

The quartic self-interaction is UV-finite, we have inserted the scaling form to remove cutoff effects of a small number of lattice links in T-directions, as discussed above. Therefore, after absorbing (4.6.131) into $S_{eff}^{(0)}$, the effective action in three dimensions takes the following form to one-loop order

$$S_{eff} = S_{eff}^{(0)} + a^3 \sum_{\mathbf{x}} \left(\Pi_{00}^{ns}(0) \, \mathrm{tr} \left(A_0^{st}(\mathbf{x}) \right)^2 \right.$$
$$\left. + \frac{T}{12\pi^2} \left[N_c \, \mathrm{tr} \left(A_0^{st}(\mathbf{x}) \right)^4 + 3 \left(\mathrm{tr} \left(A_0^{st}(\mathbf{x}) \right)^2 \right)^2 \right] \right). \quad (4.6.135)$$

The nonstatic part $S_{eff}^{(1)}$ of the effective action inherits the one-loop counterterms of the four-dimensional theory. Most of them just absorb the UV singularities of the effective coupling constants induced by the nonstatic modes and ensure that the normalization conditions imposed on the full theory are satisfied by the effective theory as well. However, although the three-dimensional effective model is superrenormalizable, according to the truncation we have imposed on S_{eff}, still some low-order counterterms are required to render the model UV-finite. These counterterms are actually provided as those parts of the four-dimensional counterterms which are not needed for absorbing the nonstatic divergencies. An example is the vacuum polarization $\Pi_{00}^{ns}(0)$ above. It is the bare nonstatic vacuum polarization plus the mass counterterm that vanish according to the renormalization conditions we have imposed. The UV-divergence left in $\Pi_{00}^{ns}(0)$ is precisely the counterterm required to make $\Pi_{00}^{st}(\mathbf{p})$ UV-finite in three dimensions, because of $\Pi_{00}^{st}(0) + \Pi_{00}^{ns}(0) = \Pi_{00}(0)$, that is finite. Note that in contrast to the Φ^4-example, here no matching condition is required to determine the counterterm, because we do not change the UV-cutoff.

So far we have stated results for the effective action which were based on several criteria for truncating further terms. Perturbation theory was performed including graphs to one-loop order. The zero-momentum expansion was truncated after the leading zero momentum contribution to the n-point functions, and only those terms were kept that do not spoil superrenormalizability. Of course we must check, whether these truncations are justified for the actual observable (here the quark potential) that we want to calculate. The answer depends on the employed perturbation scheme (here the large volume/small coupling expansion), on the renormalization scheme (specified by the normalization conditions), and last but not least on the observables themselves. For example, the one-loop order is only sufficient if the renormalized coupling constant g is small, so that there is a chance for good asymptotic convergence. In this sense the coupled large volume/small gauge coupling expansion provides a good perturbation scheme, cf. section 4.4. With zero-temperature mass renormalization it works down to at least two times the phase-transition temperature T_c (and even below if temperature-dependent mass subtractions are involved). This means that observables are available in this expansion whenever $g^2 RT \lesssim 1$, if R represents the spatial length scale. Examples are Wilson loops and Polyakov-loop correlations. Generically, g depends on two scales. One scale is the temperature, the other one is a spatial scale R, say, $g = g(R, T)$. We define a renormalized, temperature-dependent, running coupling $g(T)$ in four dimensions by means of the quark potential $V(\mathbf{x}_0)$ at distance $|\mathbf{x}_0| = R_0 \simeq T^{-1}$, more precisely at $R_0 = 1/(4T)$. This means that $V(\mathbf{x}_0)$ is computed perturbatively and matched to its "experimental" (in the sense of computer-experimental) value. In contrast to appropriate observables at zero temperature, $V(\mathbf{x}_0)$ is not measurable in a real experiment, but in a Monte Carlo simulation in four dimensions. This determines the running $g(T)$ for the particular perturbation scheme used, in our case the coupled large volume/small g expansion.

Perturbation theory is reliable for $R \lesssim T^{-1}$. On the other hand, the truncation of the zero-momentum expansion of the effective action is only justified if one is interested in large-distance properties. In principle it depends on the observable whether higher orders in the p/T expansion must be included to ensure an overlap with the region to which perturbation theory applies. It turns out that in fact there is a region, in which both perturbation theory and dimensional reduction work

$$c_1 \leq RT \leq \frac{c_2}{g(T)^2}, \tag{4.6.136}$$

in which the values of both c_1, c_2 depend on the observable. We illustrate these bounds with results for the Polyakov-loop correlation, calculated on the one hand side in perturbation theory to order g^6 in the effective model, and on the other hand -as a check of whether dimensional reduction works- also in four dimensions in the original model. The terms of S_{eff} which contribute to this order in g are given by

$$S_{eff} = S_{eff}^{(0)} + a^3 \sum_{\mathbf{x}} \left(\Pi_{00}^{ns}(0) \ \mathrm{tr} \left(A_0^{st}(\mathbf{x}) \right)^2 + \Sigma_{00}^{ns}(\tau) \sum_{i=1}^{3} \mathrm{tr} \left(\frac{1}{a} \partial_i A_0^{st}(\mathbf{x}) \right)^2 \right), \tag{4.6.137}$$

where $S_{eff}^{(0)}$ is the tree-level part (4.6.120)f, and Σ_{00}^{ns} is the renormalized nonstatic part of the wave function renormalization constant

$$\Sigma_{00}^{ns}(\tau) = \frac{1}{2} \left(\frac{\partial^2}{\partial \widehat{k}_1^2} \Pi_{00}^{ns}(\mathbf{k}) \bigg|_{\mathbf{k}=0} - \frac{\partial^2}{\partial \widehat{k}_1^2} \Pi_{00}(\mathbf{k}) \bigg|_{\mathbf{k}=\tau} \right) \tag{4.6.138}$$

with $\tau = \pi T \cdot (1,1,1)$. Σ_{00}^{ns} is UV finite. Note that (4.6.137) includes a subleading term in the zero momentum expansion proportional to the wave function renormalization constant. For details on the computation of Π_{00} and on the employed renormalization scheme we refer to section 4.4.

With the effective Polyakov loop $L_{eff}(\mathbf{x})$ given by (4.6.115), the correlation between Polyakov loops in the effective model is obtained as

$$P_{eff}(\mathbf{x}) = \ < L_{eff}(A^{st})(\mathbf{x}) L_{eff}(A^{st})^*(\mathbf{0}) >_c^{eff} \tag{4.6.139}$$

$$= \ < L_{eff}(A^{st})(\mathbf{x}) L_{eff}(A^{st})^*(\mathbf{0}) >^{eff} - | < L_{eff}(A^{st})(\mathbf{0}) >^{eff} |^2,$$

where

$$< L_{eff}(A^{st})(\mathbf{x}) L_{eff}(A^{st})^*(\mathbf{0}) >^{eff}$$

$$= \frac{1}{Z_{eff}} \int \prod_{\mathbf{x}} \left(dA^{st}(\mathbf{x}) dc(\mathbf{x}) d\bar{c}(\mathbf{x}) \right) \qquad (4.6.140)$$

$$\cdot L_{eff}(A^{st})(\mathbf{x}) L_{eff}(A^{st})^*(\mathbf{0}) \, \exp\left(-S_{eff}(A^{st}, c, \bar{c}) \right).$$

Now we compare this correlation in the effective model with $P(\mathbf{x})$, the Polyakov-loop correlation of the full four-dimensional theory. In Table 4.6.2 we summarize results in an SU(2) gauge theory. Compared are perturbative results on the on-axis correlations $P(R)$ and $P_{eff}(R)$ for two temperatures above T_c, computed on lattices of size 4×24^3 and 24^3, respectively. The results agree even down to distances $R = 1/(4T)$, so that $c_1 \leq 0.25$ for Polyakov loop correlations. As we have seen in section 4.4, for our temperatures the perturbative horizon is given by $R = T^{-1}$. This means that there is a large overlap region of distances where both perturbation theory and dimensional reduction simultaneously work. (Similarly it can be shown that $c_1 \leq 0.75$ for spatial Wilson loops.) Without the wave function renormalization the quantitative agreement would be less good.

Table 4.6.2 On-axis correlation functions $P(R)$ and $P_{eff}(R)$ for two temperatures above the phase transition T_c of the SU(2) Yang-Mills theory, obtained by renormalized perturbation theory.

R	$T \simeq 4T_c$ $g^2(T) = 1.28$ $P(R)$	$P_{eff}(R)$	$T \simeq 6T_c$ $g^2(T) = 1.19$ $P(R)$	$P_{eff}(R)$
1	0.01628	0.01712	0.01405	0.01472
2	0.002960	0.003127	0.002578	0.002714
3	0.000806	0.000855	0.000714	0.000754
4	0.000280	0.000298	0.000253	0.000268
5	0.000114	0.000121	0.000105	0.000111
6	0.000053	0.000056	0.00050	0.000052

The influence of the wave-function term of $S_{eff}^{(1)}$ is a measure of the quality of the perturbation scheme which is used both for the computation of S_{eff} and for the definition of the renormalized coupling constant $g(T)$. This part of $S_{eff}^{(1)}$ contributes to the quark potential (defined via the Polyakov-loop correlations) to lower order than the chromomagnetic corrections in S_{eff}, but provides already small contributions. It acts as a coupling-constant renormalization of the effective model

$$g^2 \rightarrow g^2 \left(1 + \alpha g^2\right), \qquad \text{with } \alpha = -\Sigma_{00}^{ns}(\tau). \qquad (4.6.141)$$

4.6.5.5 *Nonperturbative verification of dimensional reduction and determination of the screening masses*

Screening masses are obtained from the long-range behavior of the static quark potential in the effective model. We know that for $|\mathbf{x}| \leq T^{-1}$ the quark potential $V(\mathbf{x})$ obtained from $P(\mathbf{x})$ by (4.6.96) is well described by perturbation theory in this region. It has as a power-like decay with the distance. At the perturbative horizon screening sets in, that is, $V(\mathbf{x})$ decays exponentially fast outside the range of perturbation theory. Here the screening mass must be determined (nonperturbatively) by Monte Carlo simulations. Towards this end we "revert" the Faddeev-Popov trick after dimensional reduction, i.e. in the effective model, because it is still BRS-invariant. This step amounts to the following. As shown previously, all parts of the partition function which depend on the ghosts and on the gauge fixing are recombined to a constant factor, i.e. they drop out of the effective action which is used in the Monte Carlo simulations. This form effectively amounts to apply perturbation theory only to the nonstatic modes of the theory.

The precise form of this effective action is given as follows. For the Monte Carlo simulations the effective model is written in terms of bare parameters. For an $SU(N_c)$ gauge theory, the partition function of the effective model is given by

$$Z_{eff} = \int \prod_{\mathbf{x},i} dU(\mathbf{x}; i) \prod_{\mathbf{x}} dA_0(\mathbf{x}) \, \exp\left(-S_{eff}(U, A_0)\right) \qquad (4.6.142)$$

with the following effective action S_{eff}. In lattice units with $a = 1$, up to higher order corrections in g^2, we have

$$S_{eff}(U, A_0) = S_W^{3d}(U) + S_{adj}(U, A_0) + S_{eff}^{(1)}(U, A_0), \qquad (4.6.143)$$

where

$$S_W^{3d}(U) = \beta_3 \sum_{\mathbf{x} \in \Lambda_3} \sum_{i<j=1}^{3} \left(1 - \frac{1}{N_c} \text{Re tr} \, U(\mathbf{x}; i) U(\mathbf{x} + \widehat{i}; j) \right.$$

$$\left. \cdot U(\mathbf{x} + \widehat{j}; i)^{-1} U(\mathbf{x}; j)^{-1} \right), \tag{4.6.144}$$

$$S_{adj}(U, A_0) = -\frac{1}{g^2} \sum_{\mathbf{x}} \sum_{i=1}^{3} \text{tr} \left(D_i^{adj}(U) A_0(\mathbf{x}) \right)^2, \tag{4.6.145}$$

and

$$S_{eff}^{(1)}(A_0) = \frac{1}{g^2} \sum_{\mathbf{x}} \left[-h \, \text{tr} \, (A_0(\mathbf{x}))^2 \right.$$

$$\left. + \kappa_1 \, \text{tr} \, (A_0(\mathbf{x}))^4 + \kappa_2 \, \left(\text{tr} \, A_0(\mathbf{x})^2 \right)^2 \right]. \tag{4.6.146}$$

We have written A_0 for A_0^{st}. The bare coupling constants are given by

$$\beta_3 = \frac{2 N_c L_0}{g^2 (1 + \alpha g^2)}, \qquad \text{with } \alpha = -\Sigma_{00}^{ns}(\tau_0),$$

$$h = -g^2 \Pi_{00}^{ns}(0), \tag{4.6.147}$$

$$\kappa_1 = \frac{N_c g^2}{12 \pi^2 L_0} , \quad \kappa_2 = \frac{g^2}{4 \pi^2 L_0} .$$

$\Pi_{00}^{ns}(0)$ and $\Sigma_{00}^{ns}(\tau_0)$ are computed on the $L_0 \times L_s^3$ lattice with $\tau_0 = (1, 1, 1) \pi / L_0$, cf. (4.6.138). According to (4.6.134), we have for sufficiently large L_0

$$\Pi_{00}^{ns}(0) = \frac{2 N_c}{L_0} \frac{1}{L_s^3} \sum_{\mathbf{k} \neq 0} \frac{1}{4 \sum_{i=1}^{3} \sin^2 (k_i/2)} - \frac{N_c}{3 L_0^2}. \tag{4.6.148}$$

The effective Polyakov loop is given by

$$L_{eff}(A_0)(\mathbf{x}) = \exp (L_0^{1/2} A_0(\mathbf{x})). \tag{4.6.149}$$

Recall that the value of the renormalized coupling constant $g = g(T)$ was obtained by the quark potential $V(\mathbf{x})$ in four dimensions at distance $|\mathbf{x}| = 1/(4T)$, that is at $R = L_0/4$ in lattice units.

We remark that L_0 serves as a bare parameter. Like the bare coupling g_0 it sets the scale of the lattice spacing a in the three-dimensional model, but has no "intrinsic" meaning as in four dimensions. The lattice constant a in four dimensions was related to the temperature T by $L_0 = 1/(aT)$. Lattice results for four and three dimensions can be compared also for different values of L_0 if the corresponding change of scale is taken into account. The continuum limit of the effective model at temperature T is achieved as $L_0 \to \infty$ with $g = g(T)$ kept fixed.

Above we have explicitly verified that dimensional reduction works quite well for the quark potential if the quark potential, both in the full and in the effective model, is computed perturbatively. As the next step we verify that this agreement is not a perturbative artifact. We check that the agreement extends beyond the perturbative horizon into the region $R \geq T^{-1}$, for which screening is observed. For the case of a pure SU(2) Yang-Mills theory, the screening range $R \geq T^{-1}$ is measurable in four dimensions even on lattices with small extent $L_0 = 4$ of lattice links in temperature direction. The cutoff effects due to this small extension are minimized for $L_s = 24$. The quark potential is then measured both in four dimensions and in the effective model. Later we will use this experience to measure screening in full QCD via the corresponding dimensionally reduced model, although in full QCD a measurement of screening in the original four-dimensional model by means of the static quark potential lies beyond the nowadays computer facilities, as we have indicated above.

Table 4.6.3   Effective coupling constants at two temperatures for the SU(2) lattice gauge theory with T-extension $L_0 = 4$. The renormalized coupling constant $g = g(T)$ is defined by the quark potential at distance $R_0 = 1/(4T)$.

	$\simeq T/T_c$	g^2	β_3	h	$\kappa_1 + 2\kappa_2$
4×24^3	4.0	1.28	13.54	-0.26	0.0217
$(\alpha = -0.059)$	6.0	1.19	14.48	-0.24	0.0200

Table 4.6.3 shows one set of values of the effective couplings for two temperatures and for the gauge group $SU(2)$. Note that in this case

$$\text{tr}\, A_0(\mathbf{x})^4 = \frac{1}{2} \left(\text{tr}\, A_0(\mathbf{x})^2 \right)^2, \qquad (4.6.150)$$

so that the effective quartic couplings κ_1 and κ_2 only occur in the combination $\kappa_1 + 2\kappa_2$, i.e. (4.6.146) becomes

$$S_{eff}^{(1)}(A_0) = \frac{1}{g^2} \sum_{\mathbf{x}} \left[-h\, \text{tr}\, (A_0(\mathbf{x}))^2 + (\kappa_1 + 2\kappa_2)\, \text{tr}\, (A_0(\mathbf{x}))^4 \right]. \qquad (4.6.151)$$

Numerical results of the quark potential $V_{eff}(\mathbf{x})$ are shown in Table 4.6.4, measured by a mixed heat-bath/Metropolis algorithm. The table also includes the four-dimensional potential $V(\mathbf{x})$ determined with Monte Carlo and with perturbation theory. According to (4.6.96), the definition of $V_{eff}(\mathbf{x})$ includes the expectation value $< L(\mathbf{0}) >$ of the Polyakov loop. As an "ultra-local" quantity in space direction, $< L(\mathbf{0}) >$ cannot be trusted when it is computed by dimensional reduction. Therefore we normalize the ratio of $V_{eff}(\mathbf{x_0})$ to $V(\mathbf{x_0})$ at $|\mathbf{x_0}| = 1/(2T)$. However, the spatial decay of the Polyakov-loop correlation and the potential can be computed as long-range properties by dimensional reduction.

Table 4.6.4 Comparison of the on-axis effective static quark potential $-V_{eff}(R)/T$ for $T \simeq 4T_c$ with the quark potential $-V(R)/T$ determined either by Monte Carlo simulations in four dimensions (measured at $\beta = 2.8$ on the 4×24^3 lattice) (second column) or within perturbation theory (third column). $V_{eff}(R)$ is normalized at distance $RT = 0.5$ to the potential of the original model. The Polyakov-loop expectation value is given by $< L(\mathbf{0}) >=0.42649(9)$.

$4RT$	$-V_{eff}(R)/T$ $24^3,\ \beta = 13.54$	$-V(R)/T$ $4 \times 24^3,\ \beta_4 = 2.8$	$-V(R)/T$ pert.
1	0.09128(36)	0.08953(18)	0.08995
2	0.01683(14)	0.016832(46)	0.017045
3	0.004661(79)	0.004508(21)	0.004690
4	0.001564(64)	0.001476(17)	0.001640
5	0.000579(50)	0.000556(14)	0.000665
6	0.000243(54)	0.000223(14)	0.000308

Obviously the good agreement of potentials measured in the full and in the effective model extends from the perturbative region to the range of Debye screening. We therefore conclude that dimensional reduction is realized in the whole distance region which is either accessible via perturbation theory or via Monte Carlo simulations.

We should emphasize how important the nonstatic contributions to S_{eff} are in achieving this agreement. In Fig. 4.6.2 we compare the four-dimensional potential with an effective potential obtained by an effective model in which all nonstatic interactions are ignored, i.e. with $h = \kappa = 0$, hence $S_{eff}^{(1)}$ set to zero. This model would be obtained by complete dimensional reduction. The discrepancy in the data points is pronounced, so the classical picture, in which dimensional reduction is complete, can be ruled out. Dimensional reduction, however, holds on the quantum level. The quantum corrections induced by the nonstatic modes lead to renormalizations of the static interactions. For a quantitative description these renormalizations are important, as the data told us. Above the phase transition these effects can be calculated in perturbation theory.

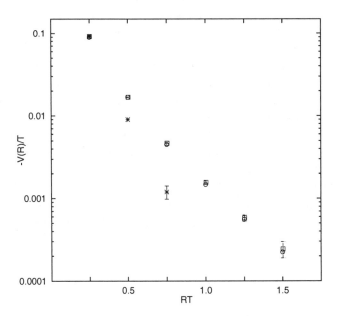

Fig. 4.6.2 Comparison of the quark potentials for the SU(2) gauge theory at $T \simeq 4T_c$. The potential for the four-dimensional theory (o) is well reproduced by the effective model (□). The interactions generated by the nonstatic modes are important. Without them, i.e. setting $h = \kappa = 0$, one would obtain the data (∗).

4.6.5.6 *Dimensional reduction in QCD with dynamical fermions*

The data we presented above were obtained for the $SU(2)$ Yang-Mills theory. The gauge group $SU(2)$ was chosen for simplicity. Dimensional reduction works as well for other gauge groups such as $SU(3)$. The above results have further encouraged us to study the IR-behavior of QCD with dimensional reduction, in particular to determine Debye screening masses in the quark-gluon plasma. As a remarkable property of dimensional reduction the fermionic sector is only represented in the form of local interactions of an otherwise purely bosonic effective theory. This is due to the anti-periodic boundary conditions in temperature direction that must be imposed on the quark-degrees of freedom. In addition, in our framework the only place where the fermions of the original model (say QCD) enter the game is the normalization condition for fixing the renormalized gauge coupling $g(T)$. Below we discuss the case of four massless quarks. The determination of $g(T)$ and of the effective coupling constants become more complicated in the presence of nonvanishing current quark masses, but the main steps in deriving the screening masses along the lines we indicate below remain the same.

A further advantage of the above procedure is that cutoff effects of a small number L_0 of time slices in temperature direction can be removed analytically by computing the scaling values of the effective coupling constants in lattice perturbation theory.

The effective model for QCD looks quite similar to the effective model for the pure gauge theory with the same gauge group $SU(N_c)$, but with supplementary interactions induced by the fermionic degrees of freedom. It is again of the form (4.6.142) and (4.6.143), with $S_W^{3d}(U)$, $S_{adj}(U, A_0)$ and $S_{eff}^{(1)}(A_0)$ given by (4.6.144), (4.6.145) and (4.6.146), respectively. If N_f denotes the number of fermionic flavors, the coupling constants are now given by

$$
\beta_3 = \frac{2N_c L_0}{g^2(1 + \alpha g^2)}, \qquad \alpha = -\Sigma_{00}^{ns}(\tau_0),
$$

$$
h = -g^2 \Pi_{00}^{ns}(0), \tag{4.6.152}
$$

$$
\kappa_1 = \frac{N_c g^2}{12\pi^2 L_0} \left(1 - \frac{N_f}{N_c}\right), \qquad \kappa_2 = \frac{g^2}{4\pi^2 L_0}.
$$

Both $\Pi_{00}^{ns}(0)$ and $\Sigma_{00}^{ns}(\tau_0)$ involve fermionic contributions. The effective coupling constants are independent of the current quark masses, because the perturbative calculations were only performed for four staggered fermions in the chiral limit. In particular,

$$
\Pi_{00}^{ns}(0) = \Pi_{00}^{ns}(0)_g + \Pi_{00}^{ns}(0)_f \tag{4.6.153}
$$

with the pure gauge part $\Pi_{00}^{ns}(0)_g$ as before, (4.6.148). The fermionic part $\Pi_{00}^{ns}(0)_f$ has a small-volume dependence, is UV-finite and vanishes at zero temperature. Its scaling form is given by

$$\Pi_{00}^{ns}(0)_f = -\frac{N_f}{6L_0^2}. \tag{4.6.154}$$

Using this scaling form removes strong cutoff effects for small L_0.

The way of proceeding for extracting the screening masses is then the following.

(1) The first step is a simulation of four-dimensional QCD (including dynamical fermions) to measure *short*-range properties of correlation functions, in particular the quark potential $V(\mathbf{x})$ in four dimensions. These simulations may be performed on relatively small lattices. This is also the place, where the number of fermionic flavors and the values of the bare quark masses enter. The final value for the screening mass certainly will depend on these parameters, but for high temperatures it is likely less sensitive than the nature of the phase transition itself. In our results below we refer to simulations for four staggered fermions and bare quark masses of $m_0 = 0.025$.

(2) The second step is a perturbative calculation of the effective action and of the quark potential $V(\mathbf{x})$ at short distances. The only unknown parameter, the value of $g(T)$, is obtained by matching the perturbative $V(\mathbf{x}_0)$ for some $\mathbf{x}_0$ to its value which was determined in the first step in a Monte Carlo simulation. In lattice units, a possible choice is $|\mathbf{x}_0| = L_0/4$ which is well within the perturbative region. All effective coupling constants are then determined in perturbation theory, and the cutoff artifacts are removed by taking their scaling forms (4.6.152).

(3) Finally the effective theory in three dimensions, which was obtained in the first two steps, is simulated with Monte Carlo to measure the IR-behavior of the full theory via the IR-behavior of the effective theory. By construction dimensional reduction should leave the IR-behavior of the original model unchanged.

Let us show some results on the Polyakov-loop correlation function $P(\mathbf{x})$ for four-flavor QCD ($N_f = 4$) at temperature $T \simeq 2.5 T_c$ and bare quark masses given by $m_0 = 0.025$. The data of Figure 4.6.3 are obtained in the effective theory on a 24^3-lattice. The bare parameter values were computed according to the above relations and are summarized in Table (4.6.5).

The value of the renormalized gauge coupling $g(T)$ is determined in lattice QCD with one staggered fermion (corresponding to four flavors in the continuum limit) with bare mass $m_0 = 0.025 = 0.1T$ on the 4×16^3 lattice.

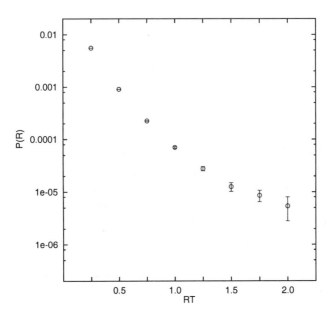

Fig. 4.6.3   The connected correlation function of Polyakov loops for QCD measured in the effective model on a 24^3-lattice.

Table 4.6.5   Effective coupling constants for four-flavor QCD. The renormalized coupling constant $g = g(T)$ is defined by the quark potential at distance $R_0 = 2$ in lattice units on a 4×16^3 lattice at $\beta = 5.40$, yielding $g(T)^2 = 0.790$.

L_0	α	β_3	h	κ_1	κ_2
4	-0.104	33.10	-0.205	-0.00168	0.00500

For $\beta = 5.4$, the system is at a temperature $T \simeq 2.5T_c$. By T_c we mean the scale set by the critical temperature of the deconfinement transition in the pure $SU(3)$ gauge theory, since it is not clear whether the phase transition persists for our simulation with four flavors in the continuum and a bare mass of $m_0 = 0.025$. The Polyakov-loop correlation at short distances was measured with a standard hybrid Monte Carlo algorithm, for $R = 1, 2$ in lattice units. Matching to the perturbative expansion of the Polyakov loop yields $g(T)^2 = 0.790$. The effective interactions are then computed perturbatively, the cutoff artifacts due to the temperature torus are removed. Finally, the large-distance behavior of $P(\mathbf{x})$ is measured similarly as before for the case of a pure gauge theory. In our actual simulations we chose a mixed heat-bath/Metropolis algorithm.

Screening masses are obtained from distances beyond the perturbative horizon. For the current case of four-flavor QCD at about 2.5 T_c, the perturbative horizon is slightly larger than for the pure gauge theory. Perturbation theory works quite well for $RT \leq 1.25$. Beyond this range, the screening mass μ is obtained from a fit of the Monte Carlo data of $P(\mathbf{x})$ against an exponentially decaying function such as

$$f(R) = c + b \left(\frac{\exp(-\mu R)}{R^\gamma} - \frac{\exp(-\mu(L_s - R))}{(L_s - R)^\gamma} \right) \qquad (4.6.155)$$

for the on-axis correlations, or, more appropriately, a fit which accounts naturally for the violation of $O(3)$ rotation symmetry and periodicity,

$$f(\mathbf{x}) = \left(\sum_k \frac{\exp(i\mathbf{k} \cdot \mathbf{x})}{\widehat{\mathbf{k}}^2 + (\mu/\gamma)^2} \right)^\gamma . \qquad (4.6.156)$$

Some results are summarized in Table (4.6.6).

Table 4.6.6 Screening masses calculated in the effective theory, for two gauge groups and QCD, choosing $\gamma = 2$.

	$\simeq T/T_c$	μ/T
SU(2)	4	2.05(10)
	6	2.02(7)
SU(3)	3	1.68(8)
QCD	2.5	2.22(24)

The value for the screening mass in full QCD certainly should not be considered as a final result. It still remains to correctly account for the *physical* quark masses both in the Monte Carlo simulations in four dimensions and in the perturbative part of the calculations. Since the main steps of the derivation will be the same, we have outlined the derivation of the preliminary result in some detail. We expect that away from the phase-transition region an implementation of realistic values for the quark masses will slightly change the renormalized coupling in four dimensions and the couplings of the effective model in three dimensions. In the phase transition region, however, the effect will be pronounced.

At present the most important result is a qualitative one that the electric screening mass is different from zero. The quantitative value of the screening mass is important for finally predicting the experimentally observed cross sections of heavy-quark bound states produced in the quark-gluon plasma phase. In this case it will still take some time to match theory with experiment.

In the literature results for electric screening masses are usually compared to the screening masses obtained by means of *resummed* perturbation theory in QCD. The aim of the resummation is twofold. At a first place resummation provides a perturbative scheme which becomes IR-finite in the infinite spatial volume. Second, resummation should convert a short-distance expansion of an UV-asymptotically free theory to an expansion scheme which applies to the IR-region in which screening should be observed. The advantage then is that screening masses can be calculated in (resummed) perturbation theory.

Within such a resummation scheme, Rebhahn [192] calculated the electric screening mass to subleading order in the gauge coupling. The Debye mass m_D is obtained as the root of the vacuum polarization $\Pi_{00}(\bar{p}) + \bar{p}^2 = 0$ with $\bar{p} = \overline{(p_0, \mathbf{p})} = (0, 0, 0, i m_D)$. This definition is gauge independent (but, of course, not gauge invariant). As a result,

$$\mu = 2m_D = 2\widehat{N}g_\Sigma T \left(1 - \frac{N_c}{4\pi} \frac{g_\Sigma}{\widehat{N}} \ln \frac{g_\Sigma}{\widehat{N}} + O(g_\Sigma) \right). \qquad (4.6.157)$$

Here g_Σ is the renormalized gauge coupling of the resummation scheme, which was used in computing (4.6.157), and

$$\widehat{N} = \left(\frac{N_c}{3} + \frac{N_f}{6} \right)^{1/2}. \qquad (4.6.158)$$

The factor of 2 arises because the first nontrivial order of the quark potential is given by a two-gluon exchange diagram.

In order to compare the nonperturbative results for the screening mass to (4.6.157), the coupling constant g_Σ of the scheme used in (4.6.157) must be either determined itself or related to other known schemes. To our knowledge this relation is not yet established. The coupling g_Σ is based on an infinite resummation and refers to a different length scale. Therefore we cannot expect that g_Σ equals either the coupling constant of the $\overline{\text{MS}}$ scheme, or of the coupled large volume/small g scheme which we used above. What we can do instead is to turn around the argument and match (4.6.157) against the Monte Carlo data. This defines g_Σ and gives a hint on the validity of the techniques that were used in deriving (4.6.157). As a result we find that the subleading term provides a correction of at most 20% for $N_c = 2$ and becomes even smaller than 10% for $SU(3)$ and QCD. This is a promising result and encourages further investigations of other observables by means of resummed perturbation theory.

We conclude with some remarks. First, we emphasize again that for a correct determination of screening masses, the quark potential $V(\mathbf{x})$ has to be measured at distances $|\mathbf{x}|$ larger than those of the perturbative short-distance region. For temperatures slightly above $2T_c$ the perturbative region extends over distances $|\mathbf{x}| \leq T^{-1}$. In this range $V(\mathbf{x})$ is the sum of power-like functions with possible logarithmic corrections. If the perturbative range is naively included in an exponential fit like (4.6.155), screening masses are typically overestimated by a factor of about two.

Let us recall why it was legitimate to calculate the effective interactions perturbatively. The nonstatic modes have an intrinsic mass gap $m \gtrsim T$, thus they are accessible in perturbation theory. Nevertheless, the choice of a convenient perturbation scheme for the computation of $S_{eff}^{(1)}$ is essential. Different schemes show considerably different (asymptotic) convergence behavior. For instance, the $\overline{\text{MS}}$ scheme appears to be slowly convergent for the temperature region considered here, so that higher-order computations are unavoidable. This is indicated by a rather large coupling $g_{\overline{\text{MS}}}(T)^2 \simeq 3$. Because of the rather weak T-dependence of g^2, temperatures which are orders of magnitude larger than T_c are required to make the coupling sufficiently small. On the other hand, the coupled large volume/small gauge coupling expansion rapidly converges on scales $|\mathbf{x}| \leq T^{-1}$ already for temperatures above $2T_c$. The value of the coupling constant $g(T)^2$ in this scheme is defined by the quark potential at distance scale $|\mathbf{x}| = 1/(4T)$. The T-dependence of $g(T)^2$ closely follows the one-loop β-function. Furthermore, it is about three times smaller than $g_{\overline{\text{MS}}}(T)^2$. Therefore the one-loop computation is already very accurate. From a systematic point of view the finite volume in the large volume/small gauge coupling expansion may appear as a disadvantage, because one is finally interested in the infinite volume limit and the IR-cutoff should be removed in the very end. For practical purposes, however, the finite volume in the coupled expansion can be made large enough to match and compare the perturbative results with the Monte Carlo results (which are unavoidably obtained in a finite volume).

It may be instructive to recall how we circumvented a conversion from perturbation theory in the UV to the IR that is otherwise attempted to be achieved by means of appropriate resummations. The main input from the four-dimensional theory was the renormalized coupling $g(T)$ measured in the UV. By means of perturbation theory in the UV we derived an effective action in one less dimension according to the rules of dimensional reduction. The IR-properties are then investigated by Monte Carlo simulations that become feasible from a practical point of view (because of one less dimension) and essentially correspond to a nonperturbative method.

Two advantages compared to standard Monte Carlo simulations of full QCD in four dimensions were the absence of fermions in the effective theory and the absence of cutoff effects of a finite L_0 because of the analytic derivation.

Furthermore we have seen that the nonstatic modes are important to make dimensional reduction work. The interactions induced by these quantum corrections in the static sector are quite important to obtain the correct large-scale properties in three dimensions.

It is tempting to extrapolate the validity range of the reduced effective model down to the phase transition region. What kind of difficulties would be encountered? So far dimensional reduction was realized by means of renormalized perturbation theory. The effective interactions of the static sector generated by the nonstatic modes are computed as expansions with respect to the renormalized gauge coupling $g(R, T)$, defined at temperature T and length scale $R \simeq T^{-1}$. Furthermore only effective interactions were kept which preserve superrenormalizability. This is legitimate if the temperature is sufficiently large. For nonabelian pure gauge theories and QCD these assumptions turned out to be justified for temperatures at least twice as large as the deconfinement-transition temperature. For lower temperatures two problems may arise.

- Strictly renormalizable interactions become important. For instance, the adjoint scalar self-interaction should be supplemented by a term proportional to $A_0(\mathbf{x})^6$.
- The renormalized gauge coupling increases so that perturbation theory finally breaks down.

The two points are not necessarily related. In particular, one expects that perturbation theory in QCD breaks down on distance scales $R \simeq T^{-1}$ close to the chiral phase transition. However, dimensional reduction may still hold at the transition in the sense that there exists an effective three-dimensional theory for the phase transition region. The only problem then is its explicit construction, i.e. its derivation by nonperturbative methods. To date this is an open problem.

In the next section we study the strongly interacting Gross-Neveu model. As we shall see, dimensional reduction works down to the finite-temperature phase transition of this model if the number of fermionic species is large.

4.6.6 *The Gross-Neveu model in three dimensions*

The reason why we consider the Gross-Neveu model in the context of dimensional reduction is twofold. The first reason is its similarity to QCD with respect to the chiral phase transition for two massless flavors. The second is the closed analytic control on the Gross-Neveu model so that we can illustrate the ideal case of how dimensional reduction works down to T_c. It works down to T_c in spite of a strong four-fermion interaction, because dimensional reduction can be realized independently of any perturbative expansion in a small coupling. We emphasize this point, since the failure in a perturbative approach does not necessarily imply the failure of dimensional reduction if other expansion techniques are at hand to calculate the reduced effective action. In the Gross-Neveu model it is the large-N expansion that allows an explicit calculation of the effective two-dimensional theory.

Let us turn to the first point now. What is the relation to QCD? We recall the Columbia plot of section 3.1 and focus on the chiral corner of two massless flavors $m_u = m_d = 0$. The chiral transition is supposed to be of first order below some (hypothetical) value m_s^* of the bare strange quark mass and of second order for $m_s > m_s^*$. The conjecture was based on an analysis of Pisarski and Wilczek in the framework of effective models for QCD. Let us recall how the effective model was constructed in terms of a local order-parameter field. For the chiral transition the order parameter field is the chiral condensate $\overline{\psi}(x)\psi(x)$ of QCD, in analogy to the free-energy functional à la Landau and Ginzburg. We write for two flavors ψ_i,

$i = 1, 2$

$$\int_0^{T^{-1}} dx_0 \overline{\psi}_{Ri}(x)\psi_{Lj}(x) \sim i\sum_{\alpha=1}^{3} \phi_\alpha(\mathbf{x}) (\sigma_\alpha)_{ij} + \phi_0(\mathbf{x})\delta_{ij}, \qquad (4.6.159)$$

where R and L refer to right- and left-handed fermions. The σ_α, $\alpha = 1, 2, 3$ are the Pauli matrices acting in flavor space.

$$\Phi(\mathbf{x}) = (\phi_0(\mathbf{x}), \ldots, \phi_3(\mathbf{x})) \qquad (4.6.160)$$

is a four-component real-valued scalar field. As we have indicated in section 3.1, the chiral symmetries of QCD translate into an $O(4)$ symmetry of the action in terms of $\Phi(\mathbf{x})$, constructed à la Landau-Ginzburg in three dimensions. We shall further comment on that action in section 6.1. Accordingly, the chiral phase transition is supposed to be described by $O(4)$ critical exponents.

The assumption then is that the composite fermionic substructure of $\Phi(\mathbf{x})$ may be completely ignored in the vicinity of the chiral phase transition and only the effective scalar degrees of freedom determine the universality class of the chiral transition. What is missing is an actual *derivation* of the $O(4)$ model from QCD in the limit of two massless flavors for small momenta, since only long-wavelength modes are relevant in the vicinity of a second-order transition.

Kocic and Kogut [196] criticized the composite-field approach. In contrast to Pisarski and Wilczek [195] they claimed that the fermionic substructure may not be ignored and even may change the type of universality class to mean-field behavior. Their conjecture is based on a supposed analogy to the $\mathbf{Z}_2$-symmetric Gross-Neveu model in three dimensions for a large but finite number N of fermionic species. Their results for the Gross-Neveu model, obtained by numerical simulations, seem to confirm the universality class with mean-field behavior that is observed at $N = \infty$. It is then natural to explain this behavior by arguing that the symmetry determines the universality class and the symmetry is the same at finite and infinite N. However, there is a flaw in the symmetry argument. By the same argument one would expect Gaussian behavior for a three-dimensional Φ^4-theory for all values of the coupling λ, although one finds Gaussian exponents for $\lambda = 0$, but Ising exponents otherwise. As we shall derive below, similarly to the change from $\lambda = 0$ to $\lambda \neq 0$ in a Φ^4-theory, the transition from $N = \infty$ to finite N is not smooth in the sense of leaving the critical exponents unchanged. In the ideal case, from our analysis below we would expect to observe a sharp gap in the exponents as N becomes finite. Practically the gap will be smoothed: in a numerical simulation because of the finite volume, in a series expansion because of the finite truncation. To anticipate our results: after all, we will uniquely identify Ising exponents for large and finite N in the framework of an effective two-dimensional scalar model which is derived from the original Gross-Neveu model by means of dimensional reduction. Depending on the parameter range, this effective model also allows both a Gaussian and an Ising universality class, but the parameter range which is selected by the underlying Gross-Neveu model corresponds to the Ising case. It may well be that the discrepancy of the results of Kocic and Kogut for finite N are due to rounding effects in the numerical data because of the influence of the Gaussian universality class that (from a practical point of view) is fully suppressed only in the immediate vicinity of the critical point. Meanwhile it was confirmed that the width of the critical region with non-Gaussian behavior shrinks to zero proportional to N^{-x}, $x > 0$, for $N \to \infty$ [194].

So far the universality class of the QCD chiral transition in the limit of two massless flavors is an open question, although the question is more about $O(2)$- or $O(4)$-critical exponents that are actually measured in Monte Carlo simulations rather than mean-field exponents.

In the background of this controversy in QCD let us focus on the finite-temperature transition of the three-dimensional Gross-Neveu model with a large number N of fermionic species (in analogy to QCD called flavors). We consider the model in the range of strong four-fermion coupling. What is the chiral symmetry for QCD is the global $Z(2)$ parity symmetry in the Gross-Neveu model which we consider. As we shall see below, the action is invariant under $Z(2)$, a mass term of the form $m\overline{\psi}\psi$ would explicitly break $Z(2)$, but in the massless case the symmetry is spontaneously broken at zero temperature and restored at high temperature by a second-order transition. Immediately we can run into contradicting claims about the universality class of this transition if we take the arguments by Kocic and Kogut (presented above) and adapt the arguments of Pisarski and Wilczek to the Gross-Neveu model. According to Pisarski and Wilczek the Gross-Neveu model would be described by an effective $Z(2)$-symmetric scalar model in two dimensions in the transition region and share the universality class of the Ising model in two dimensions. In the remainder of this section we will solve the apparent contradiction by showing the following steps. Actually one can derive an effective two-dimensional scalar model by means of dimensional reduction in a large-N expansion combined with a small-momentum expansion. In principle this model still allows both universality classes under discussion, but the parameter range which corresponds to the original Gross-Neveu model selects the Ising universality class.

We consider the Gross-Neveu model as a field theory of N two-component Dirac spinor fields, that is of N fermion species, in three dimensions at finite temperature T

$$\psi(z) = (\psi_{\alpha i}(z))|_{\alpha=1,2;i=1,\ldots,N}$$
$$\overline{\psi}(z) = (\overline{\psi}_{\alpha i}(z))|_{\alpha=1,2;i=1,\ldots,N} \qquad (4.6.161)$$

with

$$z = (z_0, z_1, z_2) \equiv (z_0, \mathbf{z}) \in \mathbf{R}/T^{-1} \times \mathbf{R}^2. \qquad (4.6.162)$$

The Grassmann-valued fields ψ, $\overline{\psi}$ are subject to anti-periodic boundary conditions in the temperature direction

$$\psi(z_0 \pm T^{-1}, \mathbf{z}) = -\psi(z_0, \mathbf{z}), \quad \overline{\psi}(z_0 \pm T^{-1}, \mathbf{z}) = -\overline{\psi}(z_0, \mathbf{z}). \qquad (4.6.163)$$

Thus the fermionic fields have the Fourier representation

$$\psi(z) = \sum_{k_0=\pi T(2n+1), n \in \mathbf{Z}} e^{ik_0 z_0} \, \widetilde{\psi}_n(\mathbf{z}), \qquad (4.6.164)$$

so that $|k_0| \geq \pi T$. The action of the model is given by

$$S_{GN}(\psi, \overline{\psi}) = \int_z (\overline{\psi}(z) D \psi(z)) - \frac{\lambda^2}{N} \int_z \left(\overline{\psi}(z) \psi(z) \right)^2 \qquad (4.6.165)$$

with the shorthand notation

$$\int_z \equiv \int_{\mathbf{R}^2} d^2\mathbf{z} \int_{-\frac{1}{2}T^{-1}}^{\frac{1}{2}T^{-1}} dz_0. \qquad (4.6.166)$$

The Dirac operator D is given by

$$D = \sum_{i=0}^{2} \gamma_i \frac{\partial}{\partial z_i}, \qquad (4.6.167)$$

$\gamma_0, \gamma_1, \gamma_2$ denote the three basis elements of a two-dimensional Dirac representation in three dimensions. We normalize the γ-matrices according to

$$\gamma_i \gamma_j + \gamma_j \gamma_i = 2\delta_{ij} \text{ with } i, j = 0, 1, 2. \qquad (4.6.168)$$

Here and in the following we suppress spinor and flavor indices of $\overline{\psi}(z)$, $\eta(z)$ in expressions like

$$\overline{\psi}(z)\eta(z) \equiv \sum_{\alpha=1,2} \sum_{i=1,\ldots,N} \overline{\psi}_{\alpha i}(z)\eta_{\alpha i}(z). \tag{4.6.169}$$

The partition function (generating functional of full correlation functions) is then given by

$$Z(\eta,\overline{\eta}) = \int \mathcal{D}\psi\mathcal{D}\overline{\psi} \ \exp\left(-S_{gn}(\overline{\psi},\psi) + \int_z (\overline{\psi}(z)\eta(z) + \overline{\eta}(z)\psi(z))\right), \tag{4.6.170}$$

where

$$\int \mathcal{D}\psi\mathcal{D}\overline{\psi} = \prod_{z,\alpha,i} d\psi_{\alpha i}(z)d\overline{\psi}_{\alpha i}(z). \tag{4.6.171}$$

The fields η and $\overline{\eta}$ are external sources introduced in order to derive correlation functions from $Z(\eta,\overline{\eta})$ by differentiation.

For simplicity we regularize the Gross-Neveu model with a simple momentum cutoff Λ_0. According to power-counting rules of ultraviolet divergencies, the three-dimensional model turns out to be nonrenormalizable in an expansion in powers of the coupling constant λ. We need more and more counterterms with increasing order of λ to render the ultraviolet limit finite. Thus one condition for the applicability of dimensional reduction, renormalizability, seems to be violated, but the situation is different for a large number N of fermionic flavors. Within the large-N expansion the model becomes renormalizable to all orders in $1/N$. The large-N expansion will be explained in detail below. Furthermore, we consider the model for large λ to ensure that parity symmetry is spontaneously broken at small T.

As indicated above, the model (4.6.165) serves as a toy model for studying a chiral phase transition of QCD because it reveals essential similarities to massless QCD. It has a symmetry which is explicitly broken by a fermionic mass term of the form

$$m\overline{\psi}(z)\psi(z). \tag{4.6.172}$$

Actually, in three dimensions this symmetry is not a true chiral symmetry. A chiral symmetry would require the existence of a γ-matrix, the γ_5, which anti-commutes with all generating elements $(\gamma_0, \gamma_1, \gamma_2)$ of a faithful representation of the associated Clifford algebra. Such a γ_5 does not exist in three dimensions. Instead, the model is invariant under the global parity transformation

$$\psi(z) \to \psi(-z), \qquad \overline{\psi}(z) \to -\overline{\psi}(-z). \qquad (4.6.173)$$

Thus a mass term like (4.6.172) is not invariant under (4.6.173), but both the action (4.6.165) and the measure (4.6.171) are invariant under this transformation. As we shall see below, for sufficiently strong coupling constant λ, the symmetry under transformations (4.6.173) is spontaneously broken at $T = 0$, and it becomes restored at high temperature by a second-order phase transition. The parity condensate

$$\frac{1}{N} < \overline{\psi}(z)\psi(z) >$$

$$= \frac{1}{Z(0,0)} \int \mathcal{D}\psi \mathcal{D}\overline{\psi} \, \frac{1}{N} \left(\overline{\psi}(z)\psi(z) \right) \exp\left(-S_{gn}(\overline{\psi}, \psi)\right) \quad (4.6.174)$$

serves as an order parameter for the transition.

We remark that it is a discrete symmetry which is spontaneously broken in the two-dimensional effective model. Therefore there is no contradiction to the Mermin-Wagner theorem that excludes the spontaneous breaking of a continuous symmetry in $d \le 2$ dimensions.

Note that the model has also a continuous global symmetry in flavor space, defined by

$$\psi_{\alpha i}(z) \to \sum_{j=1}^{N} U_{ij} \psi_{\alpha j}(z) \,, \quad \overline{\psi}_{\alpha i}(z) \to \sum_{j=1}^{N} U_{ji}^{\dagger} \overline{\psi}_{\alpha j}(z) \qquad (4.6.175)$$

with $U \in U(N)$. This symmetry is neither explicitly broken by a mass term (4.6.172) nor does it get broken spontaneously. In particular this symmetry implies fermion-number conservation.

4.6.6.1 *Large N*

Next we represent the Gross-Neveu model in a form which is particularly suited when the number of fermionic species is sufficiently large. This form avoids the flaws related to the small λ-expansion as mentioned above. The idea is to sum the graphs which contribute to the same order in $1/N$. Let us see how this is achieved. We first consider a subset of graphs which contributes to the four-point vertex function, cf. Figure 4.6.4. This is the set of the so-called planar graphs. It turns out that they give the leading contribution to the $1/N$ expansion we are after. If we associate with each fermion loop a factor N, each four-fermion vertex contributes a factor of

$2\lambda^2/N$. If we write c for the combinatoric factor and the trace in spinor and momentum space, every graph of the left hand side of Figure 4.6.4 contributes to the same order in $1/N$. Their sum is proportional to

$$\frac{\lambda^2}{N} \left(1 - cN\frac{\lambda^2}{N} + \left(cN\frac{\lambda^2}{N} \right)^2 + \cdots \right) = \frac{1}{N} \left(\frac{1}{\lambda^2} + c \right)^{-1}. \qquad (4.6.176)$$

It is natural to represent the sum by the right hand side of Figure 4.6.4, i.e. the chain of the fermionic interactions is represented as an effective interaction mediated by an auxiliary scalar field σ.

Fortunately, the resummation which we have just illustrated at the example of the four-point vertex function, can be implemented on the level of the generating functional $Z(\eta, \overline{\eta})$ in a closed form. There is a standard trick to quadratize a quartic interaction that is also commonly used in statistical mechanics and runs under the name of Hubbard-Stratonovich transformation . For a four-fermion interaction the transformation reads

$$\exp \left(\frac{\lambda^2}{N} \left(\overline{\psi}(z)\psi(z) \right)^2 \right) = \left(\frac{N}{4\pi\lambda^2} \right)^{\frac{1}{2}}$$
$$\cdot \int_{-\infty}^{\infty} d\sigma(z) \exp \left(-\frac{1}{2} \left(\frac{N}{2\lambda^2} \right) \sigma(z)^2 \pm \sigma(z) \, \overline{\psi}(z)\psi(z) \right). \qquad (4.6.177)$$

What do we gain from this transformation? Inserting (4.6.177) into (4.6.170) yields

$$Z(\eta, \overline{\eta}) = \int \mathcal{D}\psi \mathcal{D}\overline{\psi} \mathcal{D}\sigma \, \exp \left(-S_0(\overline{\psi}, \psi, \sigma) + \int_z (\overline{\psi}(z)\eta(z) + \overline{\eta}(z)\psi(z)) \right) \qquad (4.6.178)$$

with

$$S_0(\overline{\psi}, \psi, \sigma) = \int_z \left(\frac{N}{4\lambda^2}\sigma(z)^2 + \overline{\psi}(z)(D + m + \sigma)\psi(z) \right). \qquad (4.6.179)$$

The action S_0 is of the Yukawa-type and is at most bilinear in the fermionic fields. The price is that we have now to deal with an additional auxiliary field σ. Thus the identity is only useful if the path integral (4.6.178) over $\psi, \overline{\psi}$ and the auxiliary field σ can be evaluated in a reliable way. For every finite λ, λ^2/N is small for sufficiently large N. Upon integrating out the fermion fields, the inverse of this factor appears as a large overall prefactor in front of the effective action for the auxiliary field σ (cf. (4.6.182) below). Thus the saddle-point expansion amounts to an expansion in $1/N$ in a quite natural way, the so-called large-N expansion.

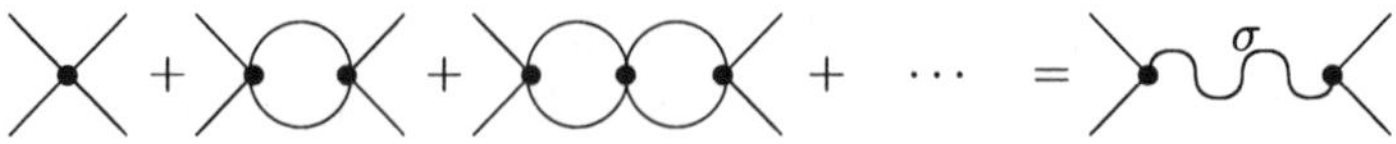

Fig. 4.6.4   Resummation of planar graphs.

The representation (4.6.179) allows to prove renormalizability of the model for large N. We will come back to this representation (4.6.178), (4.6.179) below when we apply dimensional reduction to the Gross-Neveu model.

To discuss the phase structure in the large-N limit it is more convenient to first integrate out the fermions by applying the identity

$$\int \mathcal{D}\psi \mathcal{D}\overline{\psi} \, \exp\left(-\int_z \int_{z'} \overline{\psi}(z) \, Q(z,z')\psi(z') + \int_z (\overline{\psi}(z)\eta(z) + \overline{\eta}(z)\psi(z))\right)$$

$$= \det(-Q) \, \exp\left(\int_z \int_{z'} \overline{\eta}(z)Q^{-1}(z,z')\eta(z')\right). \qquad (4.6.180)$$

In our case, Q is flavor-diagonal, i.e. $Q = \mathbf{1} \otimes K$ with

$$K = D + \sigma \cdot 1 \,, \qquad (4.6.181)$$

acting on the product of spinor and configuration space. We set $m = 0$. D is defined by (4.6.167). As the remnant of the anti-periodic boundary conditions (4.6.163) imposed on the fermionic fields, the momentum-space representation of D has nonvanishing components only for the discrete thermal momenta $k_0 = \pi T(2m+1)$, $m \in \mathbf{Z}$. Up to a normalization constant we then obtain the following representation for the partition function (4.6.170)

$$Z(\eta, \overline{\eta}) = \text{const} \int \mathcal{D}\sigma \, \exp\left(-N \, S_{eff}(\sigma)\right)$$

$$\cdot \exp\left(\int_z \int_{z'} \overline{\eta}(z)(1 \otimes K^{-1})(z,z')\eta(z')\right) \qquad (4.6.182)$$

with an effective action $S_{eff}(\sigma)$ given by

$$S_{eff}(\sigma) = \int_z \left(\frac{\sigma(z)^2}{4\lambda^2} - \text{tr}_s \, (\log K)(z,z)\right). \qquad (4.6.183)$$

The trace tr_s runs over spinor indices. For convenience we impose periodic boundary conditions on the auxiliary field σ in T-direction. (The choice is a matter of convenience. Anti-periodic boundary conditions would no longer lead to a translation-invariant minimum of the action.)

So far no approximation was involved. The partition function was just rewritten in a form that is most convenient if the number N of fermions becomes large. The parity condensate (4.6.174) becomes the expectation value of $- \operatorname{tr}_s K^{-1}(z, z)$

$$
\begin{aligned}
\frac{1}{N} < \overline{\psi}(z)\psi(z) > &= \frac{1}{N} \frac{1}{Z(0,0)} \frac{\partial}{\partial(-\eta(z))} \frac{\partial}{\partial\overline{\eta}(z)} Z(\eta,\overline{\eta})\bigg|_{\eta=\overline{\eta}=0} \\
&= \frac{\int \mathcal{D}\sigma \ \left(- \operatorname{tr}_s K^{-1}(z,z)\right) \exp\left(-NS_{eff}(\sigma)\right)}{\int \mathcal{D}\sigma \ \exp\left(-NS_{eff}(\sigma)\right)} \\
&= - < \operatorname{tr}_s K^{-1}(z,z) >_{eff} .
\end{aligned}
\tag{4.6.184}
$$

Similar relations hold for condensate correlations. It is straightforward to show to all orders of $1/N$ that a vanishing condensate (4.6.174) is equivalent to a vanishing expectation value of the auxiliary field σ. Moreover, we can read off the singular behavior of the condensate susceptibility

$$
\widehat{\chi}_2 = \int_x < \frac{1}{N} \operatorname{tr}_{s,c} \overline{\psi}(x)\psi(x); \frac{1}{N} \operatorname{tr}_{s,c} \overline{\psi}(0)\psi(0) >,
\tag{4.6.185}
$$

where ; stands for the connected part of the correlation in the two-point susceptibility of the scalar field σ. We come back to this point below. The auxiliary field $\sigma(z)$ itself is parity odd. The parity symmetry of the original model implies invariance under

$$
\sigma(z) \rightarrow -\sigma(-z)
\tag{4.6.186}
$$

in the effective model. The only place where the number N enters in (4.6.182) is as the prefactor of the action. The explicit form (4.6.182) reveals that the standard saddle point expansion is convenient to study the Gross-Neveu model for large N. The large-N expansion is performed in two steps.

- Localization of the translation invariant minima of the effective action S_{eff}. This yields the so-called gap equation

$$
\frac{\delta S_{eff}(\sigma)}{\delta\sigma(z)}\bigg|_{\sigma(z)\equiv\mu} = 0
\tag{4.6.187}
$$

and the positivity condition

$$\frac{\delta^2 S_{eff}}{\delta\sigma(z)\delta\sigma(z')}\bigg|_{\sigma(z)\equiv\sigma(z')\equiv\mu} > 0. \tag{4.6.188}$$

- Expansion about the minimum solutions according to

$$\sigma(z) = \mu + \frac{\phi(z)}{N^{\frac{1}{2}}}. \tag{4.6.189}$$

The rescaling of Φ with $N^{1/2}$ is chosen such that the part of the action which is quadratic in ϕ becomes of the order of 1. Existence of stable solutions of the gap equation and their properties will be discussed below. We first insert the ansatz (4.6.189) into the effective action (4.6.183) and expand for large N. To this end we first introduce some notations we will use in the following. Fermionic and bosonic thermal momenta are defined as the sets

$$\mathcal{F} = \{\frac{\pi}{\beta}(2m+1) \mid m \in \mathbf{Z}\}, \qquad \mathcal{B} = \{\frac{\pi}{\beta}2m \mid m \in \mathbf{Z}\}. \tag{4.6.190}$$

For the inverse temperature we write $\beta = T^{-1}$. For $x, \beta \geq 0$ and $n = 0, 1, 2, \ldots$ let

$$J_n(x,\beta) = \frac{1}{\beta} \sum_{q_0 \in \mathcal{F}} \int' \frac{d^2\mathbf{q}}{(2\pi)^2} \frac{1}{(q^2+x)^n},$$

$$Q_n(x,\beta) = \frac{1}{\beta} \sum_{q_0 \in \mathcal{F}} \int' \frac{d^2\mathbf{q}}{(2\pi)^2} \frac{\mathbf{q}^2}{(q^2+x)^n}. \tag{4.6.191}$$

The prime at the integral sign shall indicate that the sum and integral are confined to the region with

$$q^2 \equiv q_0^2 + \mathbf{q}^2 \leq \Lambda_0^2. \tag{4.6.192}$$

For $\beta < \infty$ the $J_n(x,\beta)$ are always infrared-finite for every $x \geq 0$, whereas

$$J_n(x,\infty) = \int_{q^2 \leq \Lambda_0^2} \frac{d^3q}{(2\pi)^3} (q^2+x)^{-n} \tag{4.6.193}$$

is infrared-singular as $x \to 0+$ for $n \geq 2$. Furthermore, the ultraviolet singularities of the J_n are temperature independent, which means that

$$\lim_{\Lambda_0 \to \infty} (J_n(x,\beta) - J_n(x,\infty)) < \infty, \tag{4.6.194}$$

for each $x \geq 0$. This is a property that is shared by a large class of one-loop Feynman integrals, i.e. they can be rendered ultraviolet-finite by imposing the appropriate zero-temperature normalization conditions. What we need of these integrals are large-cutoff, high-temperature, and small x-expansions.

Let us now come back and insert (4.6.189) into the effective action (4.6.183). We then obtain the following large-N representation. Let

$$K_0 = D + \mu. \tag{4.6.195}$$

The partition function then becomes

$$Z(\eta, \overline{\eta}) = \mathcal{N} \int \mathcal{D}\phi \, \exp\left(-S(\phi) + \int_z \int_{z'} \overline{\eta}(z)(1 \otimes \widehat{K}^{-1})(z, z')\eta(z')\right) \tag{4.6.196}$$

with normalization

$$\mathcal{N} = \text{const } \exp\left(-N \int_z (4\lambda^2\mu^2(J_1(\mu^2, \beta))^2 - tr_s \log(K_0)(z, z))\right) \tag{4.6.197}$$

and

$$\widehat{K} = K_0 + \frac{\phi}{N^{\frac{1}{2}}}, \tag{4.6.198}$$

and with action

$$S(\phi) = \int_z \left(\frac{\phi(z)^2}{4\lambda^2} - N \, tr_s \left(\log(1 + \frac{1}{N^{\frac{1}{2}}}\phi K_0^{-1}) - \frac{1}{N^{\frac{1}{2}}}\phi K_0^{-1}\right)(z, z)\right)$$

$$= \sum_{n \geq 2} \frac{1}{n!} S_n(\phi). \tag{4.6.199}$$

Here, we introduced for $n \geq 2$

$$S_n(\phi) = \int_z \left(\frac{\phi(z)^2}{2\lambda^2} \delta_{n,2} + \frac{(-1)^n}{N^{n/2-1}} (n-1)! \, tr_s \left(\phi K_0^{-1}\right)^n (z, z)\right). \tag{4.6.200}$$

$\widehat{K}^{-1}(x, y)$ has the large-N-expansion

$$\widehat{K}^{-1}(x, y) = \left[K_0^{-1} \sum_{n \geq 0} \left(-\frac{1}{N^{1/2}}\phi K_0^{-1}\right)^n\right](x, y)$$

$$= K_0^{-1}(x, y) + O(1/N^{1/2}). \tag{4.6.201}$$

The n-point interaction $S_n(\phi)$ generates n-point vertices which are of the order of $O(N^{1-n/2})$. In the first term of (4.6.197) we have replaced $\mu(\beta)^2/4\lambda^2$

by $4\lambda^2\mu(\beta)^2 J_1(\mu^2,\beta)^2$, where $\mu(\beta)$ is the saddle-point solution, because this form is valid also for $m \neq 0$ with m as in (4.6.179). For later convenience we write the quadratic part of the action in its momentum-space representation that is obtained as the Fourier transform. With

$$\phi(x) = \frac{1}{\beta}\sum_{k_0\in\mathcal{B}}\int'\frac{d^2\mathbf{k}}{(2\pi)^2}\,e^{ik\cdot x}\,\widetilde{\phi}(k)\qquad(4.6.202)$$

and the free-fermion propagator in momentum space written as

$$\widetilde{K}_0^{-1}(q) = (\mu+i\gamma\cdot q)^{-1} = \frac{\mu-i\gamma\cdot q}{\mu^2+q^2},\qquad(4.6.203)$$

we obtain

$$\begin{aligned}
S_2(\phi) &= \int_z\left[\frac{\phi(z)^2}{2\lambda^2}+\mathrm{tr}_s\left(\phi K_0^{-1}\right)^2(z,z)\right]\\
&= \frac{1}{\beta}\sum_{k_0\in\mathcal{B}}\int'\frac{d^2\mathbf{k}}{(2\pi)^2}\,|\widetilde{\phi}(k)|^2\left[\frac{1}{2\lambda^2}\right.\\
&\quad+\frac{1}{\beta}\sum_{q_0\in\mathcal{F}}\int'\frac{d^2\mathbf{q}}{(2\pi)^2}\,\mathrm{tr}_s\,\frac{\mu-i\gamma\cdot q}{\mu^2+q^2}\,\frac{\mu-i\gamma\cdot(q+k)}{\mu^2+(q+k)^2}\Big]\\
&= \frac{1}{\beta}\sum_{k_0\in\mathcal{B}}\int'\frac{d^2\mathbf{k}}{(2\pi)^2}\,|\widetilde{\phi}(k)|^2\left[\frac{1}{2\lambda^2}\right.\\
&\quad+2\frac{1}{\beta}\sum_{q_0\in\mathcal{F}}\int'\frac{d^2\mathbf{q}}{(2\pi)^2}\,\frac{-q(q+k)+\mu^2}{(q^2+\mu^2)((q+k)^2+\mu^2)}\Big].
\end{aligned}$$

Hence

$$S_2(\phi) = \frac{1}{\beta}\sum_{k_0\in\mathcal{B}}\int'\frac{d^2\mathbf{k}}{(2\pi)^2}\,|\widetilde{\phi}(k)|^2\widetilde{\Delta}(k,\mu,\beta)^{-1}\qquad(4.6.205)$$

with the inverse free propagator of the fluctuation field ϕ

$$\begin{aligned}
\widetilde{\Delta}(k,\mu,\beta)^{-1} &= 2\left(\frac{1}{4\lambda^2}-J_1(\mu^2,\beta)\right)\\
&\quad+\frac{1}{\beta}\sum_{q_0\in\mathcal{F}}\int'\frac{d^2\mathbf{q}}{(2\pi)^2}\left[\frac{k^2+4\mu^2}{(q^2+\mu^2)((q+k)^2+\mu^2)}\right.\\
&\quad-\left(\frac{1}{(q+k)^2+\mu^2}-\frac{1}{q^2+\mu^2}\right)\Big].
\end{aligned}\qquad(4.6.206)$$

In (4.6.206) we split the inverse of the free propagator in a k-independent mass term and a k-dependent part that is written in a Taylor-subtracted form. Then it becomes manifest that the one-loop integral is at most logarithmically divergent in the UV-limit $\Lambda_0 \to \infty$. Further note that (4.6.206) is (up to a prefactor) just the explicit form of the inverse of the right hand side of Eq. (4.6.176). We remark that $\widetilde{\Delta}(k, \mu, \beta)^{-1}$ increases at most linearly for large momenta k. Thus the Gross-Neveu model is only strictly renormalizable and not superrenormalizable.

The full propagator of the fluctuation field ϕ is given by the connected two-point function of ϕ

$$
\begin{aligned}
\widetilde{W}^{(2)}(k) &= < \widetilde{\phi}(k)\widetilde{\phi}(-k) > \\
&= \frac{\int \mathcal{D}\phi \; \widetilde{\phi}(k)\widetilde{\phi}(-k) \exp\left(-S(\phi)\right)}{\int \mathcal{D}\phi \; \exp\left(-S(\phi)\right)}.
\end{aligned} \tag{4.6.207}
$$

The wave-function renormalization constant Z_R and the renormalized mass m_R (the inverse correlation length) are then defined by the small-momentum behavior of $\widetilde{W}^{(2)}(k)^{-1}$ according to

$$
\widetilde{W}^{(2)}(k_0 = 0, \mathbf{k})^{-1} = \frac{1}{Z_R}\left(m_R^2 + \mathbf{k}^2 + O(\mathbf{k}^4)\right) \quad \text{as } \mathbf{k} \to 0. \tag{4.6.208}
$$

Before we discuss the phase transition for infinite and finite N, let us argue why it is sufficient to consider correlations in terms of σ or ϕ rather than $\overline{\psi}\psi$. It is straightforward to show to all orders in $1/N$ that a vanishing condensate (4.6.174) is equivalent to a vanishing expectation value of the auxiliary field σ. In the symmetric phase we have $\mu = 0$ and $K_0 = D$ in (4.6.201). From (4.6.201) it is evident that an even power in K_0^{-1} is accompanied by an odd power of ϕ, the corresponding ϕ correlations being zero. On the other hand, an odd power in K_0^{-1} is accompanied by an even power of ϕ, the corresponding ϕ correlations are symmetric in momentum space, the trace over the γ-matrices is antisymmetric, thus $< \text{tr}_s K^{-1}(z, z) >_{eff}$ and hence $< \overline{\psi}(z)\psi(z) >$ vanishes term by term.

It is also instructive to check that it is actually sufficient to read off the leading singular behavior of the condensate susceptibilities at the second-order phase transition from the two-point susceptibility $\chi_2 \equiv \widetilde{W}^{(2)}(k = 0)$ in terms of the scalar field, with $\widetilde{W}^{(2)}$ defined as in (4.6.207). In terms of the original fermionic variables the susceptibility of the order parameter is

$$
\begin{aligned}
\widehat{\chi}_2 &= \int_x < \frac{1}{N} \text{tr}_{s,c} \overline{\psi}(x)\psi(x); \frac{1}{N} \text{tr}_{s,c} \overline{\psi}(0)\psi(0) > \tag{4.6.209} \\
&= \int_x < \text{tr}_s \widehat{K}^{-1}(x, x); \text{tr}_s \widehat{K}^{-1}(0, 0) >_{eff} - \frac{1}{N} < \text{tr}_s \widehat{K}^{-2}(0, 0) >_{eff}.
\end{aligned}
$$

Inserting the large-N expansion (4.6.201) of $\widehat{K}^{-1}(x,x)$ we obtain

$$
\begin{aligned}
\widehat{\chi}_2 = {} & \frac{1}{N} \int_z <\mathrm{tr}_s\ \phi(z)P^2(z,z)\ ;\ \mathrm{tr}_s\ \phi(0)P^2(0,0)>_{eff} \\
& + 2\frac{1}{N^2} \int_{z,w_1,w_2} <\mathrm{tr}_s\ \phi(z)P^2(z,z)\ ;\ \mathrm{tr}_s\ P(0,w_1)\phi(w_1) \\
& \qquad\qquad\qquad P(w_1,w_2)\phi(w_2)P^2(w_2,0)>_{eff} \qquad (4.6.210) \\
& + \frac{1}{N^2} \int_{z,z_1,w} <\mathrm{tr}_s\ \phi(z)P(z,z_1)\phi(z_1)P^2(z_1,z)\ ; \\
& \qquad\qquad\qquad \mathrm{tr}_s\ \phi(0)P(0,w)\phi(w)P^2(w,0)>_{eff} \\
& + O(\frac{1}{N^3}) - \frac{1}{N} <\mathrm{tr}_s\ \widehat{K}^{-2}(0,0)>_{eff},
\end{aligned}
$$

where we have written $P = K_0^{-1}$. Up to a constant factor, the first term is the two-point susceptibility χ_2 of the auxiliary field ϕ. If we anticipate that at the phase transition the connected two-point function behaves as

$$
W^{(2)}(z) = <\phi(z);\phi(0)>_{eff} \sim \frac{1}{|z|^\alpha} \qquad (4.6.211)
$$

for large z, the singularity of χ_2 comes from the nonintegrability of $W^{(2)}(z)$. The question then arises why the singular behavior of the further terms of (4.6.210) is not worse than the singularity of the first term. The reason are the thermal fermion masses in the internal fermion loops. For instance, the third term in (4.6.210) can be estimated according to

$$
\begin{aligned}
& \int_{z,z_1,w} W^{(2)}(z)\ W^{(2)}(z-w)\ \exp -m_T|z-z_1|\ \exp -m_T|w| \\
& \qquad \leq \mathrm{const} \int_{z,x} W^{(2)}(z)\ W^{(2)}(x)\ \exp -m_T|z-x|\ , \qquad (4.6.212)
\end{aligned}
$$

where m_T is a thermal fermion mass proportional to T. This is not worse divergent than the leading term in (4.6.210).

Here we have demonstrated the argument for a specific contribution to $\widehat{\chi}_2$ as the susceptibility in $\overline{\psi}\psi$, but the argument generalizes to all contributions of n-point functions in ϕ that contribute to $\widehat{\chi}_2$. The result is that the leading singularity is determined by χ_2, now as the susceptibility in ϕ, as one would have naively expected.

4.6.6.2 *Phase structure at infinite N*

Now we are prepared to consider the phase structure of the Gross-Neveu model in the limit of an infinite number of fermionic flavors. (The case of $N < \infty$ will be discussed separately below.) This includes a discussion of the stability of the saddle-point solutions, cf. (4.6.188).

The gap equation (4.6.187) reads

$$\mu \left(\frac{1}{4\lambda^2} - J_1(\mu^2, \beta) \right) = 0, \qquad (4.6.213)$$

its solution $\mu(\beta)$ yields the condensate as a function of β. We are mainly interested in the phase structure at finite temperature, but it is instructive to include the case $\beta = \infty$ of zero temperature. Obviously $\mu = 0$ is always a solution of (4.6.213). For fixed β, $J_1(\mu^2, \beta)$ is monotonically decreasing with μ^2. Hence, for $1/(4\lambda^2) > J_1(0, \beta)$, $\mu = 0$ is the only solution, and it is a minimum of the action. This is most easily seen from (4.6.205), (4.6.206). On the other hand, for $1/(4\lambda^2) < J_1(0, \beta)$ there are two nontrivial stable solutions $\sigma(z) = \pm\mu(\beta)$ of (4.6.187) with $\mu(\beta) > 0$ satisfying (4.6.213).

Now suppose that the bare coupling constant λ is so large that the parity-broken phase is realized at zero temperature. We define a zero-temperature mass scale M by

$$J_1(M^2, \infty) \equiv \frac{1}{4\lambda^2} < J_1(0, \infty). \qquad (4.6.214)$$

For fixed μ^2, as a function of β, $J_1(\mu^2, \beta)$ is monotonically increasing in β. Decreasing β from $\beta = \infty$, there is a unique critical value β_c at which the broken phase becomes unstable, cf. Figure 4.6.5. This β_c is given by

$$J_1(M^2, \infty) = J_1(0, \beta_c) \quad \text{or} \quad \beta_c = \frac{2 \ln 2}{M}. \qquad (4.6.215)$$

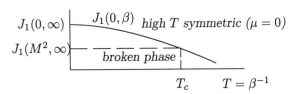

Fig. 4.6.5 Finite-temperature phase structure of the three-dimensional Gross-Neveu model at $N = \infty$. For given mass scale M, the symmetric high-temperature phase becomes unstable at $T_c = M/(2 \ln 2)$, the model undergoes a second-order phase transition to the broken phase.

In order to show that the symmetry actually does get broken spontaneously in the low temperature region $\beta > \beta_c$, we introduce an explicit parity-breaking term (4.6.172) into the original fermionic action (4.6.165). We choose $m > 0$. Then we first remove the IR-cutoff (send the volume to infinity) and next let $m \to 0+$ to see whether a non trivial minimum remains. The only modification which the fermionic mass term implies for the derivation of the large-N representation of the partition function (as compared to the last section) is the replacement of (4.6.181) by

$$K = D + (m + \sigma) \cdot 1. \tag{4.6.216}$$

The gap equation (4.6.213) is replaced by

$$(\mu - m) \frac{1}{4\lambda^2} - \mu J_1(\mu^2, \beta) = 0. \tag{4.6.217}$$

For small m and $\beta > \beta_c$ there are two stable solutions of (4.6.217) given by $\mu(m, \beta) = \pm\mu(\beta) + \delta$, where $\mu(\beta) > 0$ is the "unperturbed" solution of (4.6.213) and

$$\delta = \frac{1}{8\lambda^2 \mu(\beta)^2 J_2(\mu(\beta)^2, \beta)} m + o(m). \tag{4.6.218}$$

Inserting $\mu(m, \beta)$ into (4.6.197), we see that for small δ and m, the negative solution $-\mu(\beta)$ of the gap equation gets exponentially suppressed in the thermodynamic limit for positive m. We are left with a unique absolute minimum of the action, breaking the parity symmetry.

The nontrivial minimum of the broken phase goes continuously to zero as $T \to T_c$ from below, and the correlation length m_R^{-1} diverges at β_c. Based on the above arguments we then define the critical exponents in the standard way, namely via the two-point susceptibility χ_2 in the order-parameter field ϕ of the effective model

$$\begin{aligned} \chi_2 &\equiv \widetilde{W}^{(2)}(k = 0) \simeq (\beta_c - \beta)^{-\gamma}, \\ m_R &\simeq (\beta_c - \beta)^{\nu}, \qquad \text{as } \beta \to \beta_c - \\ Z_R &\simeq (\beta_c - \beta)^{\nu\eta}, \end{aligned} \tag{4.6.219}$$

and similarly the primed exponents ν', γ' by approaching the phase transition from the broken phase. To leading order in N^{-1} the two-point vertex function behaves as

$$\widetilde{W}^{(2)}(k_0 = 0, \mathbf{k})^{-1} = \widetilde{\Delta}(k_0 = 0, \mathbf{k}, \mu = 0, \beta)^{-1}$$
$$= 2\left(J_1(M^2, \infty) - J_1(0, \beta)\right) + \mathbf{k}^2 \, 2 \, Q_3(0, \beta) + O(\mathbf{k}^4) \qquad (4.6.220)$$
$$= \frac{\ln 2}{\pi \beta \beta_c}(\beta_c - \beta) + \mathbf{k}^2 \frac{\beta}{16\pi} + O(\mathbf{k}^4) \qquad (\text{as } \mathbf{k} \to 0)$$

in the symmetric phase, and

$$\widetilde{W}^{(2)}(k_0 = 0, \mathbf{k})^{-1} = \widetilde{\Delta}(k_0 = 0, \mathbf{k}, \mu, \beta)^{-1}$$
$$= 4\mu^2 J_2(\mu^2, \beta) + \mathbf{k}^2 \, 2\left(J_2(\mu^2, \beta) - 2\mu^2 J_3(\mu^2, \beta)\right.$$
$$\left. + 4\mu^2 Q_4(\mu^2, \beta) - Q_3(\mu^2, \beta)\right) + O(\mathbf{k}^4) \, (\text{as } \mathbf{k} \to 0) \qquad (4.6.221)$$
$$\simeq \frac{2\ln 2}{\pi \beta_c^2}(\beta - \beta_c) + \mathbf{k}^2 \frac{\beta_c}{16\pi} + O(\mathbf{k}^4) \qquad (\text{as } \beta \to \beta_c+)$$

in the broken phase. For the last equality in (4.6.220) and the last approximation in (4.6.221) we used the expansions of the integrals defined in (4.6.191). We thus obtain $\gamma = \gamma' = 1$, $\nu = \nu' = 1/2$ and $\eta = 0$. Other critical exponents are determined in the standard way with the result that $\alpha = \alpha' = 0$, $\delta = 3$ and $\beta = 1/2$.

4.6.6.3 *Dimensional reduction for finite N*

So far we know the phase structure of the finite-temperature Gross-Neveu model in the limit of an infinite number of flavors. Now we are interested in the phase structure of the model if there is a large but finite number N of fermionic species. The parity symmetry remains broken at small temperature and is restored at sufficiently high temperature. The universality class of the proposed second-order transition is controversial if we confront two different lines of plausibility arguments.

On the one hand, one may follow the reasoning proposed by Pisarski and Wilczek [195]. The length of the temperature torus is given by the inverse temperature β. In the high-temperature regime this length is rather small, hence the geometry is essentially two-dimensional on length scales large compared to β. Furthermore, due to the anti-periodic boundary conditions that must be imposed on the fermionic degrees of freedom, fermions are subject to an infrared cutoff, cf. (4.6.164). Hence, in the far infrared, only the scalar modes survive. It is natural to assume that the effective scalar model belongs to the universality class of the Ising model in two dimensions.

In this case the critical exponents are predicted as $\nu = 1$, $\gamma = 1.75$, $\eta = 0.25$ in contrast to the case of $N = \infty$.

On the other hand, the reasoning proposed by Kocic and Kogut [196] was roughly the following. The model has a $Z(2)$ symmetry. This symmetry does not depend on the number of fermionic flavors. For $N = \infty$ the model is Gaussian ($\nu = 1/2$, $\gamma = 1$), hence it may stay so for finite N. Arguing in this way uses the basic assumption that it is the symmetry pattern alone which determines the universality class of a phase transition. But the symmetry argument may be misleading. An example is a classical spin model in three dimensions as described by an action of the type $\phi^2 + \lambda\phi^4$. It has a global symmetry $\phi \to -\phi$ that does not depend on a particular value of the quartic coupling λ. For $\lambda = 0$ the model is purely Gaussian in contrast to the critical behavior at finite λ that is Ising-like. Similarly, for the Gross-Neveu model, the large-N behavior is not predictive for finite N, at least not without closer inspection.

Which scenario is the right one? In general, it is neither the symmetry pattern alone which determines the universality class of a transition nor is the classical reduction argument sufficient to exclude an influence of the fermionic substructure on the universality class.

In this section we will give an answer in the framework of dimensional reduction in a renormalizable quantum field theory. The Gross-Neveu model is such a renormalizable field theory for a large number N of flavors so that dimensional reduction applies. It is instructive to start with the representation (4.6.178), (4.6.179) in which the fermions are not yet integrated out. We insert (4.6.189) with $\mu = 0$ in the high-temperature phase and introduce renormalization constants Z_ψ, Z_N and Z_{M^2}. The partition function in terms of renormalized parameters and counterterms then becomes

$$Z(\eta, \overline{\eta}) = \int \mathcal{D}\psi \mathcal{D}\overline{\psi}\mathcal{D}\phi \, \exp\left(-S_0(\overline{\psi}, \psi, \phi) + \int_z (\overline{\psi}(z)\eta(z) + \overline{\eta}(z)\psi(z))\right) \tag{4.6.222}$$

with renormalized action

$$S_0(\overline{\psi}, \psi, \phi) = \int_z \left[Z_\psi \overline{\psi}(z) \left(D + Z_N \frac{1}{N^{1/2}}\phi(z)\right) \psi(z) \right.$$
$$\left. + Z_{M^2} J_1(M^2, \infty)\phi(z)^2 \right]. \tag{4.6.223}$$

For later convenience we separate the counterterms from the tree-level action,

$$
\begin{aligned}
S_0(\overline{\psi}, \psi, \phi) &= \int_z \left[\overline{\psi}(z) \left(D + \frac{1}{N^{1/2}} \phi(z) \right) \psi(z) + J_1(M^2, \infty) \phi(z)^2 \right] \\
&+ \int_z \left[\overline{\psi}(z) \left((Z_\psi - 1)D + (Z_\psi Z_N - 1) \frac{1}{N^{1/2}} \phi(z) \right) \psi(z) \right. \\
&\left. + (Z_{M^2} - 1) J_1(M^2, \infty) \phi(z)^2 \right],
\end{aligned}
\tag{4.6.224}
$$

where $Z_\psi, Z_N, Z_{M^2} = 1 + O(N^{-1})$. Static and nonstatic modes are decomposed according to

$$
\phi(z) = \frac{(16\pi)^{1/2}}{\beta} \phi_{st}(\mathbf{z}) + \phi_{ns}(z), \quad \int dz_0 \, \phi_{ns}(z) = 0, \tag{4.6.225}
$$

or, equivalently in momentum space

$$
\widetilde{\phi}(k) = (16\pi)^{1/2} \widetilde{\phi}_{st}(\mathbf{k}) \delta_{k_0,0} + \widetilde{\phi}_{ns}(k)(1 - \delta_{k_0,0}), \tag{4.6.226}
$$

whereas the fermion field is purely nonstatic. The reason for the particular rescaling of the static field $\phi_{st}(\mathbf{z})$ is to obtain a prefactor of 1 for the full kinetic term in the effective action (4.6.241) below. Note that the engineering dimension of the auxiliary scalar field $\phi(z)$ is 1 in $D = 3$ and not $1/2$ according to $(D-2)/2$ as we used in Eq. (4.6.20). (The dimension of the auxiliary field follows for example from the Yukawa term in (4.6.223), since the dimension of the fermionic fields $\psi, \overline{\psi}$ is one in $D = 3$). The dimension of $\phi_{st}(\mathbf{z})$ is again standard, namely $(D-2)/2 = 0$ in $D = 2$.

The dimensionally reduced action is then obtained from

$$
S_{red}(\phi_{st}) = -\ln \int \mathcal{D}\psi \mathcal{D}\overline{\psi} \mathcal{D}\phi_{ns} \, \exp\left[-S_0 \left(\psi, \overline{\psi}, \frac{(16\pi)^{1/2}}{\beta} \phi_{st} + \phi_{ns} \right) \right]. \tag{4.6.227}
$$

In analogy to (4.6.27) we decompose the effective action S_{red} into components $S_{red,n}$ which are homogeneous in the fluctuation field ϕ_{st},

$$
S_{red}(\phi_{st}) = \sum_{n \geq 2} \frac{1}{n!} S_{red,n}(\phi_{st}), \tag{4.6.228}
$$

so that in momentum space

$$
\begin{aligned}
S_{red,n}(\widetilde{\phi}_{st}) = \int{}' \frac{d^2 \mathbf{p}_1}{(2\pi)^2} &\cdots \frac{d^2 \mathbf{p}_{n-1}}{(2\pi)^2} \, \widetilde{\phi}_{st}(\mathbf{p}_1) \cdots \widetilde{\phi}_{st}(\mathbf{p}_{n-1}) \widetilde{\phi}_{st}\left(-\sum_{i=1}^{n-1} \mathbf{p}_i \right) \\
&\cdot \widetilde{S}_{red}^{(n)}(\mathbf{p}_1, \ldots, \mathbf{p}_{n-1})
\end{aligned}
\tag{4.6.229}
$$

and

$$\widetilde{S}^{(n)}_{red}(\mathbf{p}_1,\ldots,\mathbf{p}_{n-1}) = (16\pi)^{n/2}\beta^{1-n}2J_1(M^2,\infty)\delta_{n,2}$$
$$-\widetilde{\Gamma}^{(n)}_{ns}(\mathbf{p}_1,\ldots,\mathbf{p}_{n-1}). \tag{4.6.230}$$

The factor β^{1-n} comes from the trivial integrations over the zeroth momentum component along with one overall momentum conservation. Actually, because of parity symmetry, only even powers $n = 2m$ of ϕ_{st} occur in (4.6.228). The explicit computation of the $S_{red,n}$ is performed by means of the large-N expansion. Therefore we write

$$\widetilde{\Gamma}^{(2m)}_{ns}(\mathbf{p}_1,\ldots,\mathbf{p}_{2m-1}) = \sum_{\nu \geq m-1} \widetilde{\Gamma}^{(2m)(\nu)}_{ns}(\mathbf{p}_1,\ldots,\mathbf{p}_{2m-1})\, N^{-\nu}. \tag{4.6.231}$$

Recall that we are finally interested in identifying the universality class of the second-order parity-restoring phase transition. It is then sufficient to study the range of long wavelengths or small momenta. Therefore we apply the high-temperature expansion combined with the small-momentum expansion in the form (4.6.35) to all nonstatic correlation functions $\widetilde{\Gamma}^{(n)}_{ns}(\mathbf{p}_1,\ldots,\mathbf{p}_{n-1})$.

$$\widetilde{\Gamma}^{(2m)}_{ns}(\mathbf{p}_1,\ldots,\mathbf{p}_{2m-1}) \;\to\; \frac{1}{N^{m-1}}\,\widetilde{\Gamma}^{(2m)(m-1)}_{ns}(\mathbf{0},\ldots,\mathbf{0}), \quad \text{for } m \geq 2, \tag{4.6.232}$$

but for $m = 2$

$$\widetilde{\Gamma}^{(2)}_{ns}(\mathbf{p}) \;\to\; T^2_{\mathbf{p}}\widetilde{\Gamma}^{(2)(0)}_{ns}(\mathbf{p}) + \frac{1}{N}\,\widetilde{\Gamma}^{(2)(1)}_{ns}(\mathbf{0}). \tag{4.6.233}$$

So we keep the leading nontrivial term of both the small-momentum and the large-N expansion for every $\widetilde{\Gamma}^{(2m)}_{ns}$ with $m \geq 2$, whereas we keep the leading as well as the subleading terms for the two-point function $\widetilde{\Gamma}^{(2)}_{ns}$. The reason for this is that the two-point function gets a one-loop contribution already on the "classical" level of the saddle-point equation (as a consequence of the resummation of planar graphs), thus the two-loop contributions in this case are the first nontrivial quantum fluctuations.

In terms of Feynman diagrams we have

$$\widetilde{\Gamma}^{(2m)(m-1)}_{ns} = (16\pi)^m\beta^{1-2m}\left(\;\begin{array}{c}{}^{2m}\\[-2pt]1-\!\!\!\!\!\!\bigcirc\!\!\!\!\vdots\\[-2pt]2\end{array}\; + \text{perm.}\right) \tag{4.6.234}$$

for $m \geq 2$ (there are $2m$ external lines and $(2m-1)!$ terms in total) and

$$\widetilde{\Gamma}^{(2)}_{ns} = \frac{16\pi}{\beta} \left(\text{---} \bigcirc \text{---} + \frac{1}{N} \left[\text{---} \ominus \text{---} \right. \right.$$

$$\left. \left. + \text{---} \ominus \text{---} + CT \right] + O(N^{-2}) \right),$$ (4.6.235)

where

$$\bullet\!\!-\!\!-\!\!-\!\!-\!\!\bullet \qquad = \quad \frac{-i\gamma \cdot p}{p^2} \qquad (4.6.236)$$

denotes the free fermion propagator (with momentum p) and

$$\bullet\text{---}\!\circ\!\text{---}\bullet \qquad = \quad (1 - \delta_{k_0,0})\, \widetilde{\Delta}(k, \mu = 0, \beta) \qquad (4.6.237)$$

denotes the nonstatic scalar propagator, for which the little circle in the middle of the scalar line stands for "nonstatic". CT collects all contributions which involve counterterm parts of the action. Fortunately we do not need to compute the $1/N$ contribution to the two-point function by the above two-loop diagrams. They are actually fixed by a normalization condition on the effective theory, cf. Eq. (4.6.245) below. With $p_i = (0, \mathbf{p}_i)$ one easily computes

$$\mathcal{T}^2_{\mathbf{p}} \widetilde{\Gamma}^{(2)(0)}_{ns}(\mathbf{p}) = -\frac{16\pi}{\beta} \mathcal{T}^2_{\mathbf{p}} \frac{1}{\beta} \sum_{q_0 \in \mathcal{F}} \int' \frac{d^2\mathbf{q}}{(2\pi)^2} \, \mathrm{tr}_s \, \frac{i\gamma \cdot q}{q^2} \frac{i\gamma \cdot (q+p)}{(q+p)^2}$$

$$= \frac{32\pi}{\beta} \left[J_1(0, \beta) - \mathbf{p}^2 \left(J_2(0, \beta) - Q_3(0, \beta) \right) \right] \qquad (4.6.238)$$

$$= \frac{32\pi}{\beta} \left[\left(\frac{\Lambda_0}{2\pi^2} - \frac{\ln 2}{2\pi\beta} \right) - \frac{\beta}{32\pi} \mathbf{p}^2 \right],$$

and for $m \geq 2$

$$\widetilde{\Gamma}_{ns}^{(2m)(m-1)}(\mathbf{p}_1 = 0, \ldots, \mathbf{p}_{2m-1} = 0) = -(16\pi)^m \beta^{1-2m}(2m-1)!$$

$$\cdot \frac{1}{\beta} \sum_{q_0 \in \mathcal{F}} \int' \frac{d^2 \mathbf{q}}{(2\pi)^2} \, \mathrm{tr}_s \left(\frac{i\gamma \cdot q}{q^2} \right)^{2m} \tag{4.6.239}$$

$$= (-1)^{m-1}(16\pi)^m \beta^{1-2m} 2 \, (2m-1)! \, J_m(0, \beta).$$

Writing $J_m(0, \beta) = \alpha_m \beta^{2m-3}$ with β-independent positive constants α_m, we finally obtain from (4.6.232)

$$\widetilde{S}_{red}^{(2)}(\mathbf{p}) = \mathbf{p}^2 + m_0^2, \tag{4.6.240}$$

$$\widetilde{S}_{red}^{(2m)}(\mathbf{p}_1, \ldots, \mathbf{p}_{2m-1}) = \frac{1}{\beta^2} \frac{(-1)^m}{N^{m-1}} (16\pi)^m \, 2 \, (2m-1)! \, \alpha_m, \quad m \geq 2,$$

and hence for the dimensionally reduced action

$$S_{red}(\phi_{st}) = \int d^2 \mathbf{z} \left[\frac{1}{2} \sum_{i=1}^{2} \left(\frac{\partial}{\partial z_i} \phi_{st}(\mathbf{z}) \right)^2 + \frac{1}{2} m_0^2 \phi_{st}(\mathbf{z})^2 \right.$$

$$\left. + \frac{1}{\beta^2} \sum_{m \geq 2} (-1)^m \frac{c_m}{N^{m-1}} \phi_{st}(\mathbf{z})^{2m} \right]. \tag{4.6.241}$$

The interactions are alternating in sign. The effective coupling constants c_m for the first few contributions with $m \geq 2$ and the bare mass squared m_0^2 are given by

$$c_2 = 8\,\pi, \quad c_3 = \frac{32}{9}\pi^2, \quad c_4 = \frac{128}{45}\pi^3,$$

$$m_0^2 = m^2 + \delta m^2 \tag{4.6.242}$$

with

$$m^2 = \frac{32\pi}{\beta} \left(\frac{\ln 2}{2\pi\beta} - \frac{M}{4\pi} \right). \tag{4.6.243}$$

Here $\delta m^2 = O(N^{-1})$ can be obtained from the above two-loop and counterterm diagrams. Actually we compute δm^2 more easily by the effective model itself. The $\mathbf{Z}_2$-symmetric model described by the action (4.6.241) is superrenormalizable. The UV-divergence degree ω_Γ of any Feynman diagram Γ is given by $\omega_\Gamma = 2 - 2V_\Gamma$, where V_Γ denotes the number of vertices of Γ. Hence ω_Γ becomes smaller than zero for $V_\Gamma \geq 2$. The $1/N$ expansion is an expansion in the number of loops. The only ultraviolet divergence occurs for the two-point function to one-loop order, that is the order of N^{-1}.

There is only one corresponding logarithmically divergent mass counterterm that must be included in the effective model. The role of the counterterm is played by δm^2. It is uniquely determined by imposing the normalization condition that the two-point functions $\widetilde{\Gamma}^{(2)}$ at zero momentum of the full and of the effective theory coincide up to a factor $16\pi/\beta$,

$$\widetilde{\Gamma}^{(2)}_{eff}(p = 0) = \frac{16\pi}{\beta} \, \widetilde{\Gamma}^{(2)}(\mathbf{p} = 0), \qquad (4.6.244)$$

where the prefactor has its origin in the scaling factor of the static fields of Eq. (4.6.225), cf. this scaling with Eq. (4.6.20). Imposing the condition that m^2 is the renormalized mass squared, that is $\widetilde{\Gamma}^{(2)}(\mathbf{p} = 0) = -m^2$, we have

$$\qquad\qquad - \delta m^2 = 0 \qquad\qquad (4.6.245)$$

or

$$\begin{aligned}
\delta m^2 &= - \frac{c_2}{N\beta^2} \frac{1}{2} \int' \frac{d^2\mathbf{k}}{(2\pi)^2} \frac{1}{\mathbf{k}^2 + m^2} \\
&= - \frac{4\pi}{N\beta^2} \int' \frac{d^2\mathbf{k}}{(2\pi)^2} \frac{1}{\mathbf{k}^2 + m^2} \, . \qquad (4.6.246)
\end{aligned}$$

In contrast to the reduced actions in four dimensions the complete interaction part of S_{red} in (4.6.241) has a prefactor of β^{-2}. Higher interactions with larger powers of the field ϕ_{st} are not suppressed with the temperature β^{-1}. This is a consequence of the power-counting criterion (4.6.32). For $D = 3$, we have $\rho_n = 2$, so that all interactions $\widetilde{\Gamma}^{(n)}_{ns}(0)$ contribute to the same order in the temperature. In the present case, however, the $2m$-point coupling constant is of the order of N^{1-m}. For large N, interactions $\phi_{st}(\mathbf{z})^{2m}$ with increasing m become more and more suppressed.

For $N = \infty$, only the quadratic part of the action survives and the model is purely Gaussian as expected. On the other hand, as long as N is large but finite, we obtain an interacting model.

Moreover, let us comment on one of the welcome features of the effective action (4.6.241). The action (4.6.241) was derived from the Gross-Neveu model in $D = 3$. The Gross-Neveu model depends on three parameters: the $T = 0$-mass scale M, the inverse temperature β and the number of flavors N. Therefore also (4.6.241), independently of how many terms in ϕ^m are kept before the expansion is truncated, does depend on the very same number of independent parameters. The coupling coefficients c_m take values which are fixed by the underlying theory. This is a big advantage as compared to a Landau-Ginzburg type of ansatz for the same effective theory as a scalar model in two dimensions. In such an ansatz the couplings c_m would be free, unknown parameters.

4.6.6.4 *Phase structure at finite N*

We are prepared now to investigate the parity-restoring phase transition of the Gross-Neveu model at finite N. This will be done by analyzing the phase structure of the associated two-dimensional effective model described by the action (4.6.241).

Two-dimensional field theories typically evolve a rich and complicated phase structure. Renormalization-group studies, for example, reveal that $Z(2)$ symmetric scalar models have complicated fixed point manifolds in field space. We have to expect more than just one (the Ising) universality class for two-dimensional scalar models, in general. The type of universality class depends to a large extent on the interactions and the values of the coupling constants.

For the following *nonperturbative* investigation we introduce a lattice cutoff and study the lattice version of (4.6.241). First of all, we truncate the action beyond the interaction ϕ_{st}^8. Terms of the order of N^{-4} are omitted. As explained above, this is legitimate for finite, but large-N we are interested in. On the lattice, the action becomes

$$S_{lat}(\phi) \;=\; a^2 \sum_{\mathbf{x}\in\Lambda_2} \left\{ \frac{1}{2} \sum_{i=1}^{2} \left(\frac{1}{a}\widehat{\partial_i}\phi(\mathbf{x}) \right)^2 + \frac{1}{2} m_{lat}^2 \phi(\mathbf{x})^2 \right.$$
$$\left. + \frac{1}{\beta^2} \left(\frac{c_2}{N}\phi(\mathbf{x})^4 - \frac{c_3}{N^2}\phi(\mathbf{x})^6 + \frac{c_4}{N^3}\phi(\mathbf{x})^8 \right) \right\}. \quad (4.6.247)$$

Here Λ_2 denotes the hyper-cubic lattice in two dimensions with lattice spacing a, and $\widehat{\partial_i}$ is the forward-lattice derivative

$$(\widehat{\partial_i}\phi)(\mathbf{x}) \;=\; \left(\phi(\mathbf{x}+a\widehat{i}) - \phi(\mathbf{x}) \right) \quad (4.6.248)$$

with $\widehat{i}$ as unit vector in the positive ith direction.

In passing we remark that we are actually allowed to switch from the simple momentum cutoff to a lattice cutoff. This is a consequence of the power-counting theorems on the lattice. In particular the statements about the renormalizability of the Gross-Neveu model remain valid, and the lattice model stays superrenormalizable. The continuum limit in which the lattice spacing is sent to zero exists to all orders of the $1/N$ expansion. Therefore we may use the same arguments as above for the simple momentum cutoff why we need nothing else but a mass counterterm: The only bare parameter which is related to the renormalized parameters by a function which does not become the identity as $a \to 0$ is the bare mass m_{lat}. The relation is given by a one-loop expression similar to (4.6.246). We have

$$
m_{lat}^2 = m^2 - \frac{4\pi}{N\beta^2} \int_{-\pi/a}^{\pi/a} \frac{d^2\mathbf{k}}{(2\pi)^2} \frac{1}{\widehat{\mathbf{k}}^2 + m^2}
\tag{4.6.249}
$$

with renormalized mass squared m^2 given by (4.6.243) and

$$
\widehat{\mathbf{k}}^2 = \sum_{i=1}^{2} \frac{4}{a^2} \sin^2 \frac{k_i a}{2}.
\tag{4.6.250}
$$

Equation (4.6.249) is obtained in the same way as (4.6.246). Next we rewrite the action in a form which is familiar from lattice-spin models. The lattice spacing a is set to unity in the following. We rescale the fields by introducing the so-called hopping parameter κ and define further ultra-local coupling constants τ, σ and ω by

$$
\phi(\mathbf{x}) = (2\kappa)^{1/2} \phi_0(\mathbf{x}),
$$
$$
\kappa(4 + m_{lat}^2 a^2) = 1 - 2\tau + 3\sigma - 4\omega,
$$
$$
4c_2\kappa^2 \frac{1}{\beta^2 N} a^2 = \tau - 3\sigma + 6\omega,
\tag{4.6.251}
$$
$$
-8c_3\kappa^3 \frac{1}{\beta^2 N^2} a^2 = \sigma - 4\omega,
$$
$$
16c_4\kappa^4 \frac{1}{\beta^2 N^3} a^2 = \omega.
$$

Up to an irrelevant constant normalization factor, we then obtain the partition function in the form

$$
Z = \int \prod_{\mathbf{x} \in \Lambda_2} d\phi_0(\mathbf{x}) \, \exp\left(-S_0(\phi_0)\right)
\tag{4.6.252}
$$

with action

$$S_0(\phi_0(\mathbf{x})) = \sum_{\mathbf{x} \in \Lambda_2} \left(S^{(0)}(\phi_0(\mathbf{x})) - (2\kappa) \sum_{\mu=1,2} \phi_0(\mathbf{x})\phi_0(\mathbf{x} + \widehat{\mu}) \right),$$

$$S^{(0)}(\phi_0(\mathbf{x})) = \phi_0(\mathbf{x})^2 + \tau(\phi_0(\mathbf{x})^2 - 1)^2 \qquad (4.6.253)$$
$$+ \sigma(\phi_0(\mathbf{x})^2 - 1)^3 + \omega(\phi_0(\mathbf{x})^2 - 1)^4.$$

The coupling constants τ, σ and ω are of the order of N^{-1}, N^{-2} and N^{-3}, respectively, and are given by

$$\tau = 36 \left(1 - \frac{4}{\mathcal{R}} + \frac{96}{5\mathcal{R}^2} \right) \bar{l},$$

$$\sigma = -\frac{48}{\mathcal{R}} \left(1 - \frac{48}{5\mathcal{R}} \right) \bar{l}, \qquad (4.6.254)$$

$$\omega = \frac{576}{5} \frac{1}{\mathcal{R}^2} \bar{l},$$

with

$$\bar{l} = \frac{8\pi\kappa^2}{9N\beta^2} \quad \text{and} \quad \mathcal{R} = \frac{3N}{2\pi\kappa}. \qquad (4.6.255)$$

Note that the three independent parameters in this formulation are β, N and κ. Later the parameter space will be scanned via ω and $\mathcal{R}$ rather than β and N (cf. the discussion below Eq. (4.6.264) in connection with the linked cluster expansion). The particular values of κ, β and N, for which $\bar{l} = 0$ and $\mathcal{R} = \infty$ simultaneously, correspond to the case of an infinite number of flavors, $N = \infty$. For $N = \infty$ the model is a free Gaussian model. For all other values it is convenient to parametrize the phase structure via the ratios $\widehat{\alpha}$, $\widehat{\beta}$ defined as

$$\widehat{\alpha} = \frac{\tau}{\omega} = \frac{15}{48}\mathcal{R}^2 \left(1 - \frac{4}{\mathcal{R}} + \frac{96}{5\mathcal{R}^2} \right),$$

$$\widehat{\beta} = \frac{\sigma}{\omega} = -\frac{5}{12}\mathcal{R} \left(1 - \frac{48}{5\mathcal{R}} \right). \qquad (4.6.256)$$

The finite-temperature phase transition and its properties are determined nonperturbatively by means of the linked cluster expansion (LCE) that we introduced in section 4.2. This technique provides convergent hopping-parameter series in κ for the free-energy density and the connected correlation functions of the model about completely disordered lattice systems. The critical behavior concerning both the location of the transition as well as the critical exponents and amplitudes is encoded in the high-order coefficients of the series.

4.6.6.5   *The strong-coupling limit*

Before we set the coupling constants to their prescribed values given by (4.6.254), it is instructive to discuss a particular strong-coupling limit first. As we discussed in section 4.2, in limiting cases of certain couplings, universality classes can be read off from the k-dependence of the vertex couplings $\overset{\circ}{v}_{2k}$ defined by

$$\overset{\circ}{v}_{2k} = \frac{\int d\phi_0\, \phi_0^{2k} \exp\left(-S^0(\phi_0)\right)}{\int d\phi_0 \exp\left(-S^0(\phi_0)\right)}. \tag{4.6.257}$$

In these cases it is not necessary to go through the complete analysis of the LCE series. For $\mathbf{Z}_2$-symmetric models the two cases we are interested in are the following. If there exists a positive real number z so that $\overset{\circ}{v}_{2k}$ behave like

$$\overset{\circ}{v}_{2k} = \frac{(2k-1)!!}{2^k}\, z^{2k} \tag{4.6.258}$$

for all $k = 0, 1, 2, \ldots$, the model belongs to the universality class of the Gaussian model. The critical hopping parameter κ_c is given by $\kappa_c z = 1/(2D)$ in D dimensions, and the critical exponents are $\gamma = 1$, $\nu = 1/2$ and $\eta = 0$. On the other hand, if a positive real number z exists so that the vertex couplings are given by

$$\overset{\circ}{v}_{2k} = \frac{(2k-1)!!}{2^k}\, z^{2k}\, \frac{\Gamma(\frac{1}{2})}{\Gamma(\frac{1}{2}+k)}, \tag{4.6.259}$$

the universality class is that of the Ising model. In two dimensions it can be shown that $\kappa_c z = (1/4)\ln\left(2^{1/2}+1\right)$, and $\gamma = 1.75$, $\nu = 1$ and $\eta = 0.25$.

The strong-coupling limit of (4.6.252), (4.6.253) is defined by sending $\omega \to \infty$ while the ratios $\widehat{\alpha} = \tau/\omega$ and $\widehat{\beta} = \sigma/\omega$ are kept fixed. The behavior of the vertices $\overset{\circ}{v}_{2k}$ is obtained by a saddle-point expansion. As a function of $\widehat{\alpha}$ and $\widehat{\beta}$ the model evolves a rather complicated phase structure. We do not need to discuss it in detail, but the following regions are of particular interest.

There are regions with Gaussian behavior, and regions with Ising behavior. For $\widehat{\alpha}/\widehat{\beta}^2 > 1/4$ we obtain (4.6.259) with $z = 1$. In this case, $\kappa_c = (1/4)\ln\left(2^{1/2}+1\right)$ independent of $\widehat{\alpha}$ and $\widehat{\beta}$. For $0 < \widehat{\alpha}/\widehat{\beta}^2 < 1/4$ and

$\widehat{\beta} < 0$ we again get (4.6.259) with

$$z = 1 - (3\widehat{\beta}/8)\left(1 + \left(1 - \frac{32}{9}\frac{\widehat{\alpha}}{\widehat{\beta}^2}\right)^{1/2}\right) > 0. \tag{4.6.260}$$

In both cases the transition is Ising like. On the other hand, in the region $0 < \widehat{\alpha}/\widehat{\beta}^2 < 1/4$ and $\widehat{\beta} > 0$, the stability condition from the saddle-point equation reads $\widehat{\alpha} < \min\left(\frac{3}{2}\widehat{\beta} - 2, \widehat{\beta} - 1\right)$. It implies (4.6.258) with $z \sim \omega^{-1}$. The model shows Gaussian behavior or, alternatively, it completely decouples over the lattice because of $z \to 0$. In Figure 4.6.6 we show a qualitative plot of the universality domains.

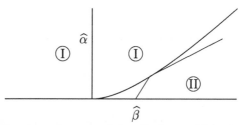

Fig. 4.6.6   Part of the phase structure of the effective model in the strong-coupling limit. In the regions I the critical exponents are those of the two-dimensional Ising model. Region II is the Gaussian domain, it corresponds to a decoupled system across the lattice.

In our case the ratios $\widehat{\alpha}$ and $\widehat{\beta}$ are fixed to a large extent. They are given by (4.6.256), and

$$\frac{\widehat{\alpha}}{\widehat{\beta}^2} = \frac{9}{5}\frac{1 - \frac{4}{\mathcal{R}} + \frac{96}{5\mathcal{R}^2}}{\left(1 - \frac{48}{5\mathcal{R}}\right)^2}. \tag{4.6.261}$$

Because $\mathcal{R}$ is proportional to N, for sufficiently large N we have both $\widehat{\beta} < 0$ and $\widehat{\alpha}/\widehat{\beta}^2 > 1/4$. This inevitably selects the Ising-universality class.

4.6.6.6   *Finite couplings and LCE-expansions*

Next we consider the case of finite coupling constants τ, σ and ω. We make use of LCE to construct the high-order series in κ of the susceptibilities χ_2

and μ_2, defined by

$$\chi_2 = \sum_{\mathbf{x} \in \Lambda_2} < \phi_0(\mathbf{x})\phi_0(\mathbf{0}) >^c,$$

$$\mu_2 = \sum_{\mathbf{x} \in \Lambda_2} (x_1^2 + x_2^2) < \phi_0(\mathbf{x})\phi_0(\mathbf{0}) >^c \qquad (4.6.262)$$

and of the correlation length m_R^{-1}. From the series coefficients we then extract the critical coupling κ_c and the critical exponents ν and γ. For example, from

$$\chi_2 = \sum_{L \geq 0} a_{L,2} \, (2\kappa)^L \qquad (4.6.263)$$

with real coefficients $a_{L,2} \geq 0$, we find the leading singular behavior of χ_2 close to κ_c according to

$$\frac{a_{L,2}}{a_{L-1,2}} = \frac{1}{2\kappa_c} \left(1 + \frac{\gamma - 1}{L} + o(L^{-1}) \right), \qquad (4.6.264)$$

extrapolated to large orders in L. The ansatz (4.6.264) is based on the assumption that χ_2 close to κ_c behaves as

$$\chi_2 \simeq \left(1 - \frac{\kappa}{\kappa_c} \right)^{-\gamma}. \qquad (4.6.265)$$

Now we choose $\mathcal{R}$ and ω as independent parameters with $\mathcal{R} = O(N)$ sufficiently large. The ratios $\widehat{\alpha}$ and $\widehat{\beta}$ are then fixed. For every $\mathcal{R}$ we vary the 8-point coupling ω from $\omega = \infty$ down to small values, the other couplings running according to $\tau = \widehat{\alpha}\omega$ and $\sigma = \widehat{\beta}\omega$. From the LCE series of χ_2 we obtain the critical hopping parameter κ_c as a function of τ, σ, ω. This in turn defines a value of N corresponding to the transition point

$$N = \frac{2\pi\kappa_c}{3} \mathcal{R}, \qquad (4.6.266)$$

since altogether only three parameters are independent. Actually the relation (4.6.266) imposes a consistency condition on the compatibility of the large-N expansion with the extrapolation in the LCE. The series coefficients are determined for given ω and $\mathcal{R} = \mathcal{R}(N, \kappa)$. For fixed κ a fixed large value of N corresponds to large $\mathcal{R}$, but for extrapolated κ_c and fixed $\mathcal{R}$ the "critical" value of N should still come out sufficiently large to be compatible with the very derivation of the model in a large-N expansion. The "critical" values of N associated with κ_c are listed in Table 4.6.7 below.

As a next step the critical exponents γ and ν are measured. Their values turn out to be independent of ω, and they are equal to the Ising exponents. The exception is a small region about the Gaussian model with $\tau = \sigma = \omega = 0$, in which we observe a smooth crossover to the Gaussian numbers. This is likely a truncation effect of the infinite series (in analogy to finite-size effects in numerical simulations, cf. our remarks in the introduction to this section). The above procedure of measuring κ_c and the critical exponents is then repeated for various values of $\mathcal{R}$ so that N varies over a large region.

In Table 4.6.7 we have collected some results for the critical exponents γ and ν obtained in this way. For all N the critical exponents agree with exponents of the two-dimensional Ising model. For small "critical" N one need not trust the results because we obtained the effective model within the large-N expansion. On the other hand, for large N the results are predictive.

N	$\gamma(N)$	$2\nu(N)$
46.1	1.751(2)	2.001(2)
24.3	1.752(3)	2.001(4)
4.68	1.749(2)	1.999(2)
1.41	1.750(2)	2.001(2)

Table 4.6.7   The critical exponents γ and ν of the three-dimensional finite-temperature Gross-Neveu model for various "critical" values N of flavors. (The critical values of N are determined via (4.6.266). The critical exponents agree with those of the two-dimensional Ising model.

4.6.6.7   *Summary*

Let us summarize this section. As the main result of the analysis we conclude that the three-dimensional Gross-Neveu model with a large but finite number N of fermionic flavors and at strong four-fermion coupling λ has a finite-temperature phase transition that belongs to the universality class of the two-dimensional Ising model with critical exponents $\gamma = 1.75$, $\nu = 1$ and $\eta = 0.25$. Based on plausibility arguments we had originally two candidates for the universality classes with mean field and Ising exponents. Mean-field behavior could be only ruled out by explicit calculations.

The results were obtained by a combination of three computational techniques: the large-N expansion, dimensional reduction and high-order linked cluster expansions.

As an advantage of the large-N representation the Gross-Neveu model becomes (strictly) renormalizable for large N, as it is manifest in the expansion in powers of N^{-1}.

Via dimensional reduction the Gross-Neveu model was mapped onto a two-dimensional effective field theory. Due to the anti-periodic boundary conditions on the fermionic fields, the resulting model was purely bosonic. The interaction terms were calculated at a first place in a large-N expansion, second in a p/T-expansion. This means, each term of the large-N expansion was evaluated in a p/T-expansion in which only the (sub)leading contributions were kept, because we were interested in the leading singular behavior of the model in the vicinity of a second-order phase transition. The resulting effective action was then local (because of the truncation in p/T) and superrenormalizable.

For a toy model like the Gross-Neveu model the number N of the fermionic flavors can be made sufficiently large, so large that the validity range of the effective scalar model extends down to T_c. This is in contrast to QCD in which there is no handle on keeping the renormalized gauge coupling g sufficiently small down to T_c (differently from $1/N$ here).

The effective model has a complex phase structure. In particular there are both regions in coupling-constant space with mean field and with Ising behavior. However, the coupling constants in our case were no longer free but computed by dimensional reduction. The relevant phase structure of the dimensionally reduced effective model was explored by means of the linked cluster expansion. This identified the Ising domain as the part which describes the finite-temperature Gross-Neveu model at large N.

What do we learn from these considerations for the chiral phase transition in QCD? The situation is similar to the Gross-Neveu model. In QCD, the effective model is a three-dimensional, purely bosonic theory. Fermions do not act as dynamical degrees of freedom on the infrared scale $R \gtrsim T^{-1}$, but they determine the bosonic interactions. As we have seen in the previous section, dimensional reduction is a powerful method to study the infrared properties of the QCD plasma phase. The reduction step from four to three dimensions is done by means of renormalized perturbation theory, i.e. by means of an expansion with respect to the renormalized gauge coupling constant $g(R, T)$ at temperature T and length scale $R \sim T^{-1}$. In this way there is analytic control of the cutoff dependence and the volume dependence.

The scheme works fine for temperatures at least twice above the phase transition temperature T_c. Compared to the Gross-Neveu model we have one more spatial dimension. From dimensional power counting this implies that interactions of higher operators are suppressed by inverse powers of the temperature. We need no longer rely on a large number of flavors in order to truncate higher operators. As mentioned above, however, approaching the phase transition, the renormalized gauge-coupling constant g becomes large, and perturbation theory breaks down. As we have seen in this section this does not imply the breakdown of dimensional reduction. It only implies that the relevant coupling constants of the effective model have to be computed by nonperturbative means.

4.6.7   *Dimensional reduction in the $SU(2)$ Higgs model*

As we have seen in earlier sections, the electroweak phase transition in the standard model is often studied in a reduced model, the SU(2) Higgs model in four dimensions. For a nonperturbative treatment one would think of Monte Carlo simulations. However, these simulations are CPU-time consuming because of the light mass modes close to T_c. If the extension in Euclidean time is of the order of T_c and the Higgs mass at T_c is less or equal $0.1T_c$, ratios of spatial (L_s) to temporal (L_0) lattice extensions L_s/L_0 of more than 10 are needed to obtain reliable results. Perturbative calculations in the continuum, on the other hand, break down for large lattice masses and small Higgs fields Φ. The Higgs-field expectation value can be perturbatively evaluated for large Φ by means of an effective potential. The calculation of the critical temperature, however, and the potential for small Φ are not available within the perturbative scope. Since equilibrium-quantum field theories at finite temperature are equivalent to zero-temperature field theories with finite fourth dimension (the higher the temperature, the smaller this extension), it is natural to think about dimensional reduction. Here dimensional reduction should particularly work, because the gauge coupling stays small in the phase transition region, in contrast to QCD, where it is of the order one.

However, the program of dimensional reduction as it was applied to the SU(2) Higgs model by Kajantie et al. [199], differs from the approaches of the previous sections in two aspects. The guiding principles for constructing the effective actions in three dimensions are different, and the "effective" integration over nonstatic modes is realized in a different way.

4.6.7.1 Guidelines for an alternative form of dimensional reduction

Let us first summarize the guidelines as they are used in the work of [199]. As we have seen in the earlier sections, the severe IR-problems are caused by the light modes with masses proportional to g^2T. The idea then is to push the integration upon all other modes as far as possible to derive an action merely in terms of the light modes. What are these light modes in contrast to the remaining modes? Here we recall the perturbative classification of energy scales according to T, gT, and g^2T. Consider a generic renormalizable field theory at high temperature with gauge fields A_μ, scalar fields Φ, and fermionic fields ψ, couplings g (gauge), λ (scalar) and g_Y (Yukawa), where it is assumed that $\lambda \sim g^2$ and $g_Y \propto g$ for the purpose of power counting. An expansion in terms of the Matsubara frequencies of the scalar and fermionic fields contains the three-dimensional tree-level masses for the bosonic and fermionic three-dimensional fields, being proportional to T, i.e. $\omega_n^{boson} = 2n\pi T$, $\omega_n^{fermion} = (2n + 1)\pi T$. The masses of the static modes Φ_0 of the Matsubara expansion get corrections from the nonstatic modes $\Phi_{n \neq 0}$ and ψ_n. The one-loop corrected, temperature dependent masses of the static modes then take the form

$$m_i^2(T) = \gamma_i T^2 + m_i^2, \quad \gamma_i \propto g^2 \,, \tag{4.6.267}$$

where i labels the various modes, m_i is the zero temperature mass, and γ_i depend on the modes as follows. It is zero for the spatial components of the gauge fields as well as m_i, for the temporal components of the gauge fields $m_i^2 = 0$, but $\gamma_i \neq 0$, for the scalar fields $\gamma_i \neq 0$ and $m_i^2 = 0$. Now the masses are classified according to their values at high temperature. First of all the three-dimensional masses of all fermionic modes and all static ($n \neq 0$) bosonic modes are proportional to T and called *superheavy*. The masses of the temporal modes of the gauge fields A_0 as well as a subset of scalar masses are proportional to gT, they are called *heavy*. The other subset of scalar masses has $m_i^2(T) \sim (\gamma_i T)^2 \propto g^4T^2$, these modes are called *light* as well as the spatial components of the gauge fields with $\gamma_i = 0$.

Dimensional reduction proceeds then in two steps. In the first step the superheavy modes are integrated out to obtain an action in terms of heavy and light modes in three dimensions. Actually only in this step the dimension is reduced from four to three. Moreover, the expansion is no longer a p/T expansion, as we used it for the Φ^4 theory or for QCD and the Gross Neveu model, but it is a combined small-momentum and small-coupling expansion in the sense that the momentum $p \sim gT$. The three-dimensional theory for light and heavy modes is valid up to momenta $p << T$, but as large as gT.

In the second step the heavy modes are integrated out to obtain a three-dimensional theory exclusively in terms of the light modes, valid up to p as large as g^2T. Finally this continuum version of the three-dimensional theory of light modes is translated into a lattice version to evaluate its predictions in a nonperturbative framework. Both three-dimensional actions describe the physics of the underlying four-dimensional theory only within a certain accuracy that depends on the truncation criteria of the three-dimensional actions, cf. below.

4.6.7.2 *Performing the integration step*

The second aspect in which the dimensional reduction of Kajantie et al. differs from the previous approaches concerns the so-called integration upon the superheavy and heavy modes. Actually, this step is not performed as an integration, but via matching conditions between the appropriate Green functions of the theories, for example the two-, three-, and four-point Green functions of the four-dimensional theory and the three-dimensional theory of heavy and light modes. Here the three-dimensional theory is postulated to be superrenormalizable. This amounts to a truncation of higher order terms in the fields, e.g. Φ^6 terms, and facilitates the evaluation; on the other hand it imposes an upper bound on the accuracy by which the physics of the four-dimensional theory is represented in the three-dimensional theory. Similarly, the coefficients of the three-dimensional light theory are determined via matching conditions on the appropriate Green functions of the three-dimensional heavy-light and light theories. Again the truncation in the light theory leads to a bound on the accuracy by which the three-dimensional heavy-light theory is represented by the light one.

A detailed derivation of the effective actions is beyond the scope of this book, it can be found in the original literature [199]. In the following we sketch only three examples for the matching between the Green functions of the four- and three-dimensional theories and one example for the matching between the heavy-light and light theories. The overall derivation in the original papers is technically quite demanding, but physically very well motivated by the separation of mass scales.

Apart from appropriate modifications for the description of the broken phase, we encountered the SU(2) Higgs model already in the context of

Buchmüller's gap equations. It is determined by the action

$$S = S_{gauge} + S_{Higgs} + \delta S. \qquad (4.6.268)$$

S_{gauge} is the gauge field action

$$S_{gauge} = \int_0^{1/T} dx_0 \int d^3\mathbf{x} \sum_{\mu,\nu=0}^{3} \sum_{d=1}^{3} \frac{1}{4} \mathcal{F}_{\mu\nu}^d(x) \mathcal{F}_{\mu\nu}^d(x) \qquad (4.6.269)$$

with

$$\mathcal{F}_{\mu\nu}^d(x) = \partial_\mu W_\nu^d(x) - \partial_\nu W_\mu^d(x) + g \sum_{e,f=1}^{3} \epsilon_{def} W_\mu^e(x) W_\nu^f(x) \ .$$

The Higgs-field action is given by

$$S_{Higgs} = \int_0^{1/T} dx_0 \int d^3\mathbf{x} \left\{ \frac{1}{2} \sum_{\mu=0}^{3} (D_\mu \Phi)(x)^\dagger (D_\mu \Phi)(x) \right.$$

$$\left. + \frac{m_s^2}{2} \Phi(x)^\dagger \Phi(x) + \frac{\lambda}{4!} \left(\Phi(x)^\dagger \Phi(x) \right)^2 \right\}, \qquad (4.6.270)$$

with Φ an SU(2) doublet, i.e. a complex, two-component scalar field

$$\Phi(x) = \begin{pmatrix} \chi_1 + i\chi_2 \\ \phi + i\chi_3 \end{pmatrix}(x), \qquad (4.6.271)$$

$$D_\mu \Phi = \left(\partial_\mu - i\frac{g}{2} \sum_{d=1}^{3} \sigma_d W_\mu^d \right) \Phi,$$

and σ_d, $d = 1, 2, 3$ the Pauli matrices. In (4.6.268) we have absorbed the gauge fixing, the ghost and the counter-term parts of the action in δS.

4.6.7.3 *Integrating upon the superheavy modes*

Let us consider the three-dimensional action of heavy and light modes arising from the four-dimensional Higgs model by "integration" upon the superheavy modes. As shown in earlier chapters, the most general effective, dimensionally reduced action in three dimensions, whose couplings do not vanish as $T \to \infty$, is strictly renormalizable. In momentum-space representation it takes the form

$$
S_{eff}(\varphi) = \sum_{n=2}^{6} \frac{1}{n!} \int \frac{d^3 \mathbf{p}_1}{(2\pi)^3} \cdots \frac{d^3 \mathbf{p}_{n-1}}{(2\pi)^3}
$$
$$
\cdot \ \widetilde{\varphi}(\mathbf{p}_1) \cdots \widetilde{\varphi}(\mathbf{p}_{n-1}) \widetilde{\varphi}(-\mathbf{p}_1 - \cdots - \mathbf{p}_{n-1}) \quad (4.6.272)
$$
$$
\cdot \ \widetilde{S}_{eff}^{(n)}(\mathbf{p}_1, \ldots, \mathbf{p}_{n-1}).
$$

Here $\widetilde{S}_{eff}^{(n)}$ is a polynomial in the momenta of order $3 - n/2$. In particular, for $n = 2$, the coefficient

$$
\frac{1}{2} \widetilde{S}_{eff}^{(2)}(\mathbf{p}) = m_3^2 + z_3 \, \mathbf{p}^2 \qquad (4.6.273)
$$

represents contributions from mass and wave function of the effective action

$$
\int d^3 \mathbf{x} \, \frac{1}{2} \left(m_3^2 \varphi(\mathbf{x})^2 + \sum_{i=1}^{3} z_3 \, (\partial_i \varphi) \, (\mathbf{x})^2 \right) .
$$

In (4.6.272) $\widetilde{\varphi}$ stands for the momentum-space representation of a general field in three dimensions, that is, A_0, A_i and Φ.

In the previous section the actual expansion parameter that was used to obtain the n-th coefficient $\widetilde{S}_{eff}^{(n)}$ was $\mathbf{p}/T$. In the framework of [199] $\mathbf{p}/T$ is replaced by g because of $\mathbf{p} \sim gT$. Furthermore, the scalar self coupling is counted as $\lambda \sim g^2$ for convenience. In this scheme, contributions of strictly renormalizable terms are suppressed compared to the contributions of superrenormalizable terms by some power of g. Therefore, to a certain accuracy in g it is sufficient to keep only superrenormalizable interactions in S_{eff}. In order to achieve an accuracy of relative order g^4, i.e.

$$
\frac{\Delta G}{G} = O(g^4) \qquad (4.6.274)
$$

for all Green functions G, it is required to keep in addition to the super-renormalizable terms also the wave function correction terms. They amount to a rescaling of the fields. This implies that a computation of the contributions $\widetilde{S}_{eff}^{(n)}$ as power series in g must be truncated according to (4.6.274). Combining these steps and taking into account the symmetries, one ends up with the following generic form of the effective action for the heavy and light modes in three dimensions

$$S_{eff} = \int d^3\mathbf{x} \bigg\{ z_A \sum_{i,j=1}^{3} \sum_{d=1}^{3} \frac{1}{4} \mathcal{F}_{ij}^d(\mathbf{x}) \mathcal{F}_{ij}^d(\mathbf{x})$$

$$+ z_\Phi \frac{1}{2} \sum_{i=1}^{3} (D_i\Phi)(\mathbf{x})^\dagger (D_i\Phi)(\mathbf{x}) + z_{A_0} \frac{1}{2} \sum_{i=1}^{3} \left(D_i^{adj} A_0 \right)(\mathbf{x})^2$$

$$+ V_3(\Phi, A_0)(\mathbf{x}) \bigg\} , \qquad (4.6.275)$$

where

$$V_3(\Phi, A_0)(\mathbf{x}) = m_3^2 \, \Phi(\mathbf{x})^\dagger \Phi(\mathbf{x}) + \lambda_3 \left(\Phi(\mathbf{x})^\dagger \Phi(\mathbf{x}) \right)^2$$

$$+ h_3 \, \Phi(\mathbf{x})^\dagger \Phi(\mathbf{x}) A_0(\mathbf{x})^2 + \frac{1}{2} \, m_D^2 A_0(\mathbf{x})^2 + \lambda_{A_0} A_0(\mathbf{x})^4 , \quad (4.6.276)$$

if we suppress symmetry labels. Now, the precise form of the coupling constants as a function of the parameters of the action in four dimensions is determined by the matching conditions as indicated above. To a specified accuracy in g, matching is postulated to hold between two-, three- and four-point functions in the theories in four and three dimensions at certain momenta. The generic form (4.6.275) together with the coefficients determined in this way ensures that all n-point Green functions of the effective theory at momenta of order gT (including the normalization points) describe the same physics as the corresponding n-point functions of the underlying theory, to a relative accuracy of $O(g^4)$. For example, matching the two-point functions in four dimensions

$$\Gamma_{4d}^{(2)}(\mathbf{k}^2) = -\mathbf{k}^2 - m_s^2 + \Pi_3(\mathbf{k}^2) + \overline{\Pi}(\mathbf{k}^2) \qquad (4.6.277)$$

and in three dimensions

$$\Gamma_{3d}^{(2)}(\mathbf{k}^2) = -\mathbf{k}^2 - m_3^2 + \Pi_3(\mathbf{k}^2) \qquad (4.6.278)$$

amounts to the following matching conditions

$$\Gamma_{4d}^{(2)}(\mathbf{k}^2 = 0) = \Gamma_{3d}^{(2)}(\mathbf{k}^2 = 0)$$
$$\Gamma_{4d}^{(2)\,\prime}(\mathbf{k}^2)|_{\mathbf{k}=0} = \Gamma_{3d}^{(2)\,\prime}(\mathbf{k}^2)|_{\mathbf{k}=0}, \qquad (4.6.279)$$

where the prime denotes differentiation with respect to $\mathbf{k}^2$. In (4.6.277) and (4.6.278), $\Pi_3(\mathbf{k}^2)$ is the contribution to the vacuum polarization of the heavy and light modes only, while $\overline{\overline{\Pi}}(\mathbf{k}^2)$ represents all other contributions. This leads to

$$z_\Phi = T\left(1 - \overline{\overline{\Pi}}'(0)\right),$$
$$m_3^2 = \left[m_s^2 + \overline{\overline{\Pi}}(0)\right]\left[1 - \overline{\overline{\Pi}}'(0)\right] \tag{4.6.280}$$

as coefficients in the effective action.

4.6.7.4 *Examples for the matching procedure to integrate upon the superheavy modes*

Now, let us consider in more detail how the guiding principles work. We illustrate with three examples the integration upon the superheavy modes. **1.** As the first example, the effective action S_{eff} contains a contribution of the form

$$S_1 \sim g^6 \int d^3\mathbf{p}_1 \cdots d^3\mathbf{p}_5 \cdot \tilde{\Phi}^\dagger(\mathbf{p}_1)\tilde{\Phi}(\mathbf{p}_2)$$
$$\cdot \tilde{A}_0(\mathbf{p}_3)\tilde{A}_0(\mathbf{p}_4)\tilde{A}_0(\mathbf{p}_5)\tilde{A}_0(-\mathbf{p}_1 - \mathbf{p}_2 - \mathbf{p}_3 - \mathbf{p}_4 - \mathbf{p}_5). \tag{4.6.281}$$

The form is contained in (4.6.272). In particular it is strictly renormalizable and of order g^6. This vertex contributes to the scalar two-point function in leading order the two-loop term

$$\sim g^6\left(\int \frac{d^3\mathbf{k}}{\mathbf{k}^2 + m_H^2}\right)^2 \sim g^6 m_H^2 \sim g^8 T^2, \tag{4.6.282}$$

where $m_H \sim gT$ represents a generic heavy mass. This order in g must be compared to the contribution of the superrenormalizable scalar mass term $(m_3^2/2)\Phi^\dagger\Phi$ of (4.6.275), where m_3^2 is a series in g. Except for T-independent parts, for the heavy scalars,

$$m_3^2 = \left(c_{H2}g^2 + c_{H4}g^4 + \cdots\right)T^2 \tag{4.6.283}$$

and for the light scalars

$$m_3^2 = \left(c_{L4}g^4 + c_{L6}g^6 + \cdots\right)T^2. \tag{4.6.284}$$

Therefore the contribution of the strictly renormalizable interaction (4.6.281) is actually suppressed by a relative factor $O(g^6)$ or $O(g^4)$ for the heavy and light scalar modes, respectively. Hence, the series for m_3^2 should be truncated at least at order g^6 to make the expansion self-consistent, cf. (4.6.274).

2. As a second example consider

$$S_2 \sim g^2 \int d^3\mathbf{p} \, \mathbf{p}^2 \, \widetilde{\Phi}^\dagger(\mathbf{p})\widetilde{\Phi}(-\mathbf{p}). \qquad (4.6.285)$$

This term is strictly renormalizable and contributes to the wave-function part

$$\frac{1}{2} \int d^3\mathbf{p} \, \mathbf{p}^2 \, \widetilde{\Phi}^\dagger(\mathbf{p})\widetilde{\Phi}(-\mathbf{p}), \qquad (4.6.286)$$

which originates in the classically reduced action and is of order g^0. The term S_2 must be kept in order to keep the accuracy $O(g^4)$ in accordance with (4.6.274).

3. As a third example,

$$S_3 \sim \frac{g^2}{T^2} \int d^3\mathbf{p} \, P_4(\mathbf{p}) \, \widetilde{\Phi}^\dagger(\mathbf{p})\widetilde{\Phi}(-\mathbf{p}), \qquad (4.6.287)$$

where P_4 denotes a polynomial of fourth order in $\mathbf{p}$. This is an example of a nonrenormalizable term as it is either seen by looking at (4.6.272) since such a term is not included there, or by directly counting the UV-dimension which is $(4 + 2 \cdot (1/2) = 5 > 3)$. Nevertheless, in the current scheme, on the scale $p \sim gT$, the term is not suppressed in T but becomes of order g^4 times the tree-level wave-function part (4.6.286). Therefore it is suppressed only by $O(g^4)$.

4.6.7.5 *Integrating upon the heavy modes*

Now we come to the second step, the integration upon the heavy modes. This step has no correspondence in the framework of dimensional reduction as it was outlined in the context of QCD. Again the aim is to derive an effective action that is superrenormalizable, but now it is a theory for the light modes only, i.e. it holds up to and including a momentum scale of the order of $p \sim g^2 T$. The former $\mathbf{p}/T$ expansion with $\mathbf{p} \sim gT$ in the step of dimensional reduction is now replaced by a heavy-mass expansion in the spirit of Appelquist and Carazzone [184]. It amounts to an expansion in terms of $\mathbf{p}/m_H$, with $m_H \sim gT$ and $\mathbf{p} \sim g^2 T$. Again, neglecting strictly renormalizable terms leads to constraints on the order of g to which this scheme is consistent, i.e. to which order in g the effective theory may be used consistently in perturbation theory. In particular, dropping these terms yields the order in g, up to which the coefficients of the superrenormalizable interactions must be expanded in g. The former accuracy (4.6.274) is now replaced by

$$\frac{\Delta G}{G} = O(g^3) \qquad (4.6.288)$$

for all Green functions G.

Let us illustrate the accuracy (4.6.288) with one example. We consider the contribution to the scalar two-point function within the light theory from the effective six-point scalar vertex. Upon the integration of the heavy modes we easily estimate that they provide a vertex of the form

$$\int d^3\mathbf{x}\, g^3 \left(\Phi(\mathbf{x})^\dagger \Phi(\mathbf{x})\right)^3 . \qquad (4.6.289)$$

An estimate reveals that

$$\sim g^6 T^3 \int \frac{d^3\mathbf{k}}{(\mathbf{k}^2 + m_H^2)^3} \sim g^6/m_H^3 \sim g^3, \qquad (4.6.290)$$

which shows that the factor g^6 from the vertices is partially compensated by the heavy masses in the propagators. The vertex (4.6.289) is strictly renormalizable. Within the light theory it provides a contribution to the scalar two-point function proportional to $g^3 m_L^2 \sim g^7 T^2$, as a similar computation as in (4.6.282) shows. Compared to the leading, temperature-dependent mass term of the light modes $g^4(\Phi(\mathbf{x})^\dagger \Phi(\mathbf{x}))^2$, it is therefore suppressed by $O(g^3)$.

Again, as in the first step, in which the dimension was reduced, the coefficients in the effective action of the light modes are not calculated by an actual integration upon the heavy modes, but via matching, now between the Green functions of the heavy-plus-light theory and the purely light theory, both in three dimensions. The resulting effective action takes the form of an $SU(2)$ Higgs model, here in three dimensions. Its form is given by

$$
S_{light} = \int d^3\mathbf{x} \left\{ \overline{z}_A \sum_{i,j=1}^{3} \sum_{d=1}^{3} \frac{1}{4} \mathcal{F}_{ij}^d(\mathbf{x}) \mathcal{F}_{ij}^d(\mathbf{x}) \right.
$$
$$
\left. + \overline{z}_\Phi \frac{1}{2} \sum_{i=1}^{3} (D_i\Phi)(\mathbf{x})^\dagger (D_i\Phi)(\mathbf{x}) + \overline{V}_3(\Phi)(\mathbf{x}) \right\}, \quad (4.6.291)
$$

where now the term $\overline{V}_3$ is of the form

$$
\overline{V}_3(\Phi)(\mathbf{x}) = \overline{m}_3^2 \, \Phi(\mathbf{x})^\dagger \Phi(\mathbf{x}) + \overline{\lambda}_3 \left(\Phi(\mathbf{x})^\dagger \Phi(\mathbf{x}) \right)^2. \quad (4.6.292)
$$

Equation (4.6.291) is the continuum form that must be translated to the lattice in order to study the phase structure by Monte Carlo simulations without (explicitly) running into IR-problems. Since (4.6.291) is superrenormalizable, it would be possible, as a next step, to calculate the mapping between bare and renormalized parameters by performing at most two-loop integrals, because the theory is UV-finite beyond the two-loop level. This finally leads to the lattice version of the three-dimensional $SU(2)$ Higgs model with bare parameters whose functional dependence in terms of the three-dimensional continuum parameters such as $\overline{m}_3^2$ and the like is explicitly known. We anticipated this lattice action already in the overview of the phase structure of the $SU(2)$ Higgs model in section 3.1.2.6.

4.7   Flow equations of Polchinski

4.7.1   *Generalities*

In a previous chapter we introduced the basics of the renormalization group in the framework of block spin transformations on a hyper-cubic lattice. As we have seen, it allows in principle even for a rigorous nonperturbative investigation of quantum field theories and critical phenomena, by approaching the involved ultraviolet and infrared limits in a controlled way.

Polchinski was the first to generalize these concepts to Euclidean quantum field theories in the continuum with a momentum cutoff [200]. In this framework the theories are described by an effective action S_Λ, depending on a scale Λ with $0 \leq \Lambda \leq \Lambda_0 < \infty$. Here Λ plays a similar role as an infrared cutoff, Λ_0 denotes the ultraviolet cutoff. S_Λ should satisfy the following conditions. At the ultraviolet cutoff $\Lambda = \Lambda_0$, S_Λ coincides with the bare action. Furthermore, for $\Lambda < \Lambda_0$, S_Λ is obtained upon integration of the field-degrees of freedom which propagate with momenta p between Λ and Λ_0. Finally, as Λ goes to 0, the action S_Λ approaches a theory without infrared cutoff. Formally $S_{\Lambda=0}$ is given by the (negative) generating functional of the connected, free propagator-amputated correlation functions of the full theory without infrared cutoff. Thus the final effective action is based on the full information of the original action that evolves under a change of scale.

Changing the infrared cutoff Λ leads to renormalization-group equations which describe the scale dependence of the effective theories on Λ in a compact way. Because Λ is varied continuously, the resulting flow equations are first-order differential equations in the infrared cutoff Λ. Their solution under appropriate boundary conditions (at $\Lambda = 0$ or $\Lambda = \Lambda_0$) amounts to compute effective low-energy interactions, or, more generally, to determine the infrared and ultraviolet properties of a field theory.

Polchinski and -somewhat later- Keller, Kopper and Salmhofer showed that these ideas also lead to a simplified proof of perturbative renormalizability of quantum field theories [200–203]. Usually, complete proofs of renormalizability are rather involved, because of the complex combinatorics of overlapping ultraviolet divergencies of a Feynman-diagrammatic approach. They require a power-counting theorem which ensures finiteness of multi-dimensional Feynman integrals by imposing the appropriate subtractions. In the framework of flow equations this complicated analysis is circumvented, at least for field theories without nonlinear symmetries. There was considerable progress also for gauge theories.

Increasing interest in flow equations was also raised from a point of view which is of interest in the context of this book. The aim is to find new approximation schemes, different from standard perturbation theory, in particular for situations where standard perturbation theory is plagued with severe infrared problems. An important case are finite-temperature field theories with zero-mass excitations. In the infinite volume, perturbation theory suffers from severe infrared divergencies in higher orders of the weak coupling expansion. As we have seen in section 4.4, these problems can be partially cured by appropriate resummations of the perturbative series, leading to a dynamical mass generation. Nevertheless, infrared problems remain at second-order phase transitions with vanishing mass gap. As an application below we consider $O(N)$ models which serve as prototypes of models with second-order phase transitions.

4.7.2 *Flow equations for effective interactions*

Let us first consider a scalar field theory in D Euclidean dimensions. At finite temperature T, the 0th dimension is the temperature torus of length T^{-1}, scalar fields are subject to periodic boundary conditions.

Let Λ and Λ_0 be two scale parameters with $0 \leq \Lambda \leq \Lambda_0 < \infty$. As before, Λ_0 serves as the ultraviolet cutoff, while Λ serves as flow parameter in momentum space. In the far infrared, Λ goes to zero. For Λ and Λ_0 constrained in this way, we next define the covariance with ultraviolet cutoff Λ_0 and infrared cutoff Λ, i.e. the free propagator or the free two-point correlation function by its Fourier transform as follows

$$\widetilde{D_\Lambda^{\Lambda_0}}(k) \;=\; \frac{1}{k^2 + m^2} \; (R(k, \Lambda_0) - R(k, \Lambda)) \,. \tag{4.7.1}$$

For all $\Lambda \leq \Lambda_0$, $R(k, \Lambda)$ is a smooth regularizing function that satisfies the following constraints. It rapidly vanishes for $k^2 > 2\Lambda^2$, whereas for $k^2 < \Lambda^2/2$, $R(k, \Lambda)$ approaches the identity sufficiently fast. In this way, $\widetilde{D_\Lambda^{\Lambda_0}}$ is essentially supported in the region of momenta given by

$$\frac{\Lambda^2}{2} \;\leq\; k^2 \;\leq 2\Lambda_0^2. \tag{4.7.2}$$

Furthermore, for $\Lambda = \Lambda_0$, the propagator vanishes,

$$\widetilde{D}_{\Lambda_0}^{\Lambda_0}(k) \;\equiv\; 0. \tag{4.7.3}$$

Apart from these conditions, there is a large amount of arbitrariness in the choice of R. For example, a convenient choice in the context of a proof of renormalizability is given by

$$R(k, \Lambda) \;=\; K\Big(\frac{k^2}{\Lambda^2}\Big), \tag{4.7.4}$$

where $K(x)$ is a smooth function with compact support in $[0, 2]$ and

$$K(x) \;=\; \begin{cases} 1, & x \leq \frac{1}{2}, \\ 0, & x \geq 2, \end{cases} \tag{4.7.5}$$

and $0 \leq K(x) \leq 1$. An alternative convenient choice is given by

$$R(k, \Lambda) \;=\; \exp\Big(-\frac{k^2}{\Lambda^2}\Big), \tag{4.7.6}$$

it has noncompact support but is rapidly decaying for large momenta.

Using the regularization (4.7.1), we define the (renormalized) generating functionals $Z_\Lambda^{\Lambda_0}(J)$ and $W_\Lambda^{\Lambda_0}(J)$ of the full and connected correlation functions by

$$
Z_\Lambda^{\Lambda_0}(J) = \exp W_\Lambda^{\Lambda_0}(J)
$$
$$
= \mathcal{N}_\Lambda^{\Lambda_0} \int \mathcal{D}\Phi \exp\left(-\frac{1}{2}\left(\Phi, (D_\Lambda^{\Lambda_0})^{-1}\Phi\right)\right) \exp\left(-S_I(\Phi) + (J, \Phi)\right).
$$

(4.7.7)

We have introduced the notation

$$
(J, \Phi) = \int_x J(x)\Phi(x) = \int_p \widetilde{J}(p)\widetilde{\Phi}(-p), \qquad (4.7.8)
$$

in particular

$$
\left(\Phi, (D_\Lambda^{\Lambda_0})^{-1}\Phi\right) = \int_{x,y} \Phi(x)\, (D_\Lambda^{\Lambda_0})^{-1}(x,y)\, \Phi(y)
$$
$$
= \int_k \widetilde{\Phi}(-k)\widetilde{D_\Lambda^{\Lambda_0}}(k)^{-1}\widetilde{\Phi}(k). \qquad (4.7.9)
$$

Integration is over all of space or all of momenta, according to

$$
\int_x \equiv \int d^D x \quad \text{and} \quad \int_p \equiv \int \frac{d^D p}{(2\pi)^D}, \qquad (4.7.10)
$$

for zero temperature, or for finite temperature T

$$
\int_x \equiv \int_0^{1/T} dx_0 \int d^{(D-1)}x \quad \text{and} \quad \int_p \equiv T\sum_{p_0} \int \frac{d^{D-1}\mathbf{p}}{(2\pi)^{D-1}}, \qquad (4.7.11)
$$

with $p_0 = 2\pi n T$, $n \in \mathbf{Z}$.

It should be mentioned that the path-integral representation of $Z_\Lambda^{\Lambda_0}$ in (4.7.7) remains *symbolic* without further specifications. First one should appropriately define the functional space in which $Z_\Lambda^{\Lambda_0}(J)$ is defined. The functional space should be identified as the appropriate subspace of tempered distributions depending on the bare action $S_I(\Phi)$. Otherwise one looses control over the validity of the formal manipulations.

$S_I(\Phi)$ defines the interaction of the model on the ultraviolet scale Λ_0. For later convenience we normalize S_I so that $S_I(0) = 0$. For instance, for the $\lambda\Phi^4$-theory, $S_I(\Phi)$ becomes

$$
S_I(\Phi) = \int_x \left(\frac{1}{2}\delta m^2\Phi(x)^2 + \frac{1}{2}\delta Z(\partial_\mu\Phi)(x)^2 + \frac{1}{4!}(\lambda + \delta\lambda)\Phi(x)^4\right). \qquad (4.7.12)
$$

Here λ is a renormalized coupling, the counterterms δm^2, δZ and $\delta\lambda$ are determined by imposing three independent normalization conditions on the correlation functions at vanishing infrared cutoff $\Lambda = 0$, for instance, as the first two conditions,

$$\widetilde{W}_{\Lambda=0}^{\Lambda_0\,(2)}(p) = \frac{1}{m^2 + p^2 + O(p^4)} \quad \text{as } p \to 0,$$

and, as the third condition,

$$\widetilde{W}_{\Lambda=0}^{\Lambda_0\,(4)}(p=0) = -\frac{\lambda}{(m^2)^4}. \tag{4.7.13}$$

In perturbation theory, δm^2 and δZ are both $O(\lambda)$ and $\delta\lambda = O(\lambda^2)$.

Note that S_I plays the role of the interaction part of the bare action with bare parameters that have been re-expressed in terms of the renormalized mass m, coupling λ and counterterms. We impose the normalization conditions at $\Lambda = 0$. This is in particular the limit in which we are interested for a second-order phase transition. $\Lambda \to 0$ corresponds to the thermodynamic limit in a lattice approach to the same model.

Finally, the normalization factor $\mathcal{N}_\Lambda^{\Lambda_0}$ in (4.7.7) is chosen such that $W_\Lambda^{\Lambda_0}(J=0) = 0$.

As a remark let us mention that some care is needed to make the generating functional $Z_\Lambda^{\Lambda_0}$ well defined in the infinite volume even at finite Λ. It may require an intermediate volume cutoff $V < \infty$ in addition to $\Lambda > 0$, and an approach of the thermodynamic limit $V \to \infty$ by imposing an appropriate constraint on the interaction $S_I(\Phi)$. This is related to the required positivity of $S_I(\Phi)$ or the existence of lower bounds on $S_I(\Phi)$ and the large-field problem, as we have described in the context of the block spin renormalization group.

Flow equations are usually formulated in terms of the interaction part of an action. There are two forms that are commonly used. The first one is the effective interaction, which is equal to the generating functional of the connected, free-propagator amputated correlation functions. The corresponding flow equations will be derived below. The second one is the so-called effective average potential, the associated flow equation is easily derived from the first one. It will be discussed later.

The effective interaction $S_{I,\Lambda}^{\Lambda_0}(\widehat{\Phi})$ on scale Λ is defined as a functional of $\widehat{\Phi} = D_\Lambda^{\Lambda_0} J$ by

$$\exp W_\Lambda^{\Lambda_0}(J) = \exp\left(\frac{1}{2}\left(J, D_\Lambda^{\Lambda_0} J\right)\right) \cdot \exp\left(-S_{I,\Lambda}^{\Lambda_0}(\widehat{\Phi})\right). \tag{4.7.14}$$

Because of $W_\Lambda^{\Lambda_0}(0) = 0$ we have $S_{I,\Lambda}^{\Lambda_0}(0) = 0$ as well. We recall that $W_\Lambda^{\Lambda_0}(J)$ is the generating functional of the connected correlation functions. The splitting into a free part and an interacting part according to (4.7.14) is natural. The reason why $S_{I,\Lambda}^{\Lambda_0}$ is considered as a functional of $\widehat{\Phi}$ is that the flow equations it obeys (and that we will derive below) become particularly simple. For $\widehat{\Phi} = D_\Lambda^{\Lambda_0} J$, for generic n apart from $n = 2$, $-S_{I,\Lambda}^{\Lambda_0}$ generates (by differentiation with respect to $\widehat{\Phi}$) connected n-point correlation functions with free propagators $D_\Lambda^{\Lambda_0}$ and vertices of S_I. External fields are free-propagator amputated in the sense that each field gets multiplied by the inverse propagator $(D_\Lambda^{\Lambda_0})^{-1}$. For $n = 2$, there is an additional subtraction of the inverse free propagator, i.e.

$$-S_{I,\Lambda}^{\Lambda_0\,(2)} = (D_\Lambda^{\Lambda_0})^{-1} W_\Lambda^{\Lambda_0\,(2)} (D_\Lambda^{\Lambda_0})^{-1} - (D_\Lambda^{\Lambda_0})^{-1}$$
$$-S_{I,\Lambda}^{\Lambda_0\,(n)} = (\otimes_{i=1}^n (D_\Lambda^{\Lambda_0})^{-1}) W_\Lambda^{\Lambda_0\,(n)} \qquad \text{for } n > 2, \qquad (4.7.15)$$

or, the second line more explicitly,

$$-S_{I,\Lambda}^{\Lambda_0\,(n)}(x_1, \cdots, x_n) = \prod_{i=1}^n \int_{y_i} \left((D_\Lambda^{\Lambda_0})^{-1}(x_i, y_i) \right) \cdot W_\Lambda^{\Lambda_0\,(n)}(y_1, \cdots, y_n).$$
$$(4.7.16)$$

Equation (4.7.15) implies that $S_{I,\Lambda}^{\Lambda_0} = 0$ if S_I vanishes, or, for the above example (4.7.12) of the Φ^4-theory that $S_{I,\Lambda}^{\Lambda_0} = O(\lambda)$. This explains why $S_{I,\Lambda}^{\Lambda_0}$ is called effective interaction.

The physical meaning of $\widehat{\Phi}$ becomes more evident in the representation of $S_{I,\Lambda}^{\Lambda_0}$ in terms of a functional integral. Starting with (4.7.7) and applying quadratic completion $\Phi \to \Phi + \widehat{\Phi}$ with $\widehat{\Phi} = D_\Lambda^{\Lambda_0} J$, we obtain from the definition (4.7.14)

$$\exp\left(-S_{I,\Lambda}^{\Lambda_0}(\widehat{\Phi})\right) = \mathcal{N}_\Lambda^{\Lambda_0} \int \mathcal{D}\Phi \exp\left(-\frac{1}{2}\left(\Phi, (D_\Lambda^{\Lambda_0})^{-1}\Phi\right)\right)$$
$$\cdot \exp\left(-S_I(\Phi + \widehat{\Phi})\right). \qquad (4.7.17)$$

The splitting of the field in a "background field" $\widehat{\Phi}$ and a fluctuation part Φ reminds us to a block-spin field and fluctuations about the block-spins in the framework of block-spin transformations. Upon integration over the fluctuations, the effective (inter)action does only depend on $\widehat{\Phi}$ or the block-spin variables, respectively.

We also see that $S_{I,\Lambda}^{\Lambda_0}$ provides a functional interpolation between the interaction part of the bare action on the ultraviolet scale $\Lambda = \Lambda_0$,

$$S_{I,\Lambda_0}^{\Lambda_0} = S_I, \tag{4.7.18}$$

and the generating functional of the full theory without infrared cutoff as $\Lambda \to 0$, $S_{I,0}^{\Lambda_0}$. In the ideal case, i.e. without further approximations, $S_{I,0}^{\Lambda_0}$ still contains the complete information of the quantum field theory with ultraviolet cutoff Λ_0.

Before we derive the flow equation of the effective interaction $S_{I,\Lambda}^{\Lambda_0}$, let us see how $S_{I,\Lambda}^{\Lambda_0}$ is obtained from the bare interaction S_I by applying the functional Laplace-like operator according to

$$\exp\left(-S_{I,\Lambda}^{\Lambda_0}(\widehat{\Phi})\right) = \frac{\mathcal{N}_\Lambda^{\Lambda_0}}{\mathcal{N}_{0,\Lambda}^{\Lambda_0}} \exp\left(\frac{1}{2}\left(\frac{\delta}{\delta\widehat{\Phi}}, D_\Lambda^{\Lambda_0}\frac{\delta}{\delta\widehat{\Phi}}\right)\right) \exp\left(-S_I(\widehat{\Phi})\right), \tag{4.7.19}$$

with the normalization factor $\mathcal{N}_{0,\Lambda}^{\Lambda_0}$ being defined by

$$\mathcal{N}_{0,\Lambda}^{\Lambda_0} \int \mathcal{D}\Phi \exp\left(-\frac{1}{2}\left(\Phi, (D_\Lambda^{\Lambda_0})^{-1}\Phi\right)\right) = 1. \tag{4.7.20}$$

In order to prove (4.7.19) we first rewrite the generating functional $Z_\Lambda^{\Lambda_0}(J)$, Eq. (4.7.7), in the standard form as

$$Z_\Lambda^{\Lambda_0}(J) = \frac{\mathcal{N}_\Lambda^{\Lambda_0}}{\mathcal{N}_{0,\Lambda}^{\Lambda_0}} \exp\left(-S_I(\frac{\delta}{\delta J})\right) \exp\left(\frac{1}{2}\left(J, D_\Lambda^{\Lambda_0}J\right)\right). \tag{4.7.21}$$

Assuming that $S_I(\Phi)$ allows for a power series representation in the field Φ (at least a formal one), the functional identity

$$\exp\left(-\frac{1}{2}\left(J, D_\Lambda^{\Lambda_0}J\right)\right) \frac{\delta^n}{\delta J(x_1)\cdots\delta J(x_n)} \exp\left(\frac{1}{2}\left(J, D_\Lambda^{\Lambda_0}J\right)\right)$$

$$= \exp\left(\frac{1}{2}\left(\frac{\delta}{\delta\widehat{\Phi}}, D_\Lambda^{\Lambda_0}\frac{\delta}{\delta\widehat{\Phi}}\right)\right) \widehat{\Phi}(x_1)\cdots\widehat{\Phi}(x_n)\Big|_{\widehat{\Phi}=D_\Lambda^{\Lambda_0}J}, \tag{4.7.22}$$

valid for all $n = 0, 1, 2, \ldots$, implies that

$$Z_\Lambda^{\Lambda_0}(J) = \frac{\mathcal{N}_\Lambda^{\Lambda_0}}{\mathcal{N}_{0,\Lambda}^{\Lambda_0}} \exp\left(\frac{1}{2}\left(J, D_\Lambda^{\Lambda_0}J\right)\right)$$

$$\cdot \exp\left(\frac{1}{2}\left(\frac{\delta}{\delta\widehat{\Phi}}, D_\Lambda^{\Lambda_0}\frac{\delta}{\delta\widehat{\Phi}}\right)\right) \exp\left(-S_I(\widehat{\Phi})\right)\Big|_{\widehat{\Phi}=D_\Lambda^{\Lambda_0}J}. \tag{4.7.23}$$

Equation (4.7.19) then follows from the definition (4.7.14).

So far the representation (4.7.19) is defined in the framework of formal power-series expansions in $\widehat{\Phi}$ or J of the functionals that are involved, whenever $S_I(\Phi)$ itself allows for such a representation in the fields, as, for instance, the action of a Φ^4-theory (4.7.12) both at zero and nonzero temperature.

As mentioned above, beyond the perturbative framework, the functional space in which (4.7.14) should be valid must be appropriately identified as the appropriate subspace of the space of tempered distributions. In this space (4.7.14) is then derived by functional Fourier transforms without referring to the expansion of $S_I(\Phi)$.

The flow equation for the effective interaction $S_{I,\Lambda}^{\Lambda_0}$ then follows by applying $\partial/\partial\Lambda$ on both sides of (4.7.19), keeping the ultraviolet cutoff Λ_0 fixed as well as the bare action S_I. We obtain

$$
\left(-\partial_\Lambda S_{I,\Lambda}^{\Lambda_0}(\widehat{\Phi})\right)\,\exp\left(-S_{I,\Lambda}^{\Lambda_0}(\widehat{\Phi})\right)
$$

$$
= \left[\partial_\Lambda\left(\frac{\mathcal{N}_\Lambda^{\Lambda_0}}{\mathcal{N}_{0,\Lambda}^{\Lambda_0}}\right) + \frac{\mathcal{N}_\Lambda^{\Lambda_0}}{\mathcal{N}_{0,\Lambda}^{\Lambda_0}}\frac{1}{2}\left(\frac{\delta}{\delta\widehat{\Phi}},\partial_\Lambda D_\Lambda^{\Lambda_0}\frac{\delta}{\delta\widehat{\Phi}}\right)\right]
$$

$$
\cdot\exp\left(\frac{1}{2}\left(\frac{\delta}{\delta\widehat{\Phi}},D_\Lambda^{\Lambda_0}\frac{\delta}{\delta\widehat{\Phi}}\right)\right)\,\exp\left(-S_I(\widehat{\Phi})\right)
$$

$$
= \left[\partial_\Lambda\ln\frac{\mathcal{N}_\Lambda^{\Lambda_0}}{\mathcal{N}_{0,\Lambda}^{\Lambda_0}} + \frac{1}{2}\left(\frac{\delta}{\delta\widehat{\Phi}},\partial_\Lambda D_\Lambda^{\Lambda_0}\frac{\delta}{\delta\widehat{\Phi}}\right)\right]\exp\left(-S_{I,\Lambda}^{\Lambda_0}(\widehat{\Phi})\right) \qquad (4.7.24)
$$

$$
= \exp\left(-S_{I,\Lambda}^{\Lambda_0}(\widehat{\Phi})\right)\left[\partial_\Lambda\ln\frac{\mathcal{N}_\Lambda^{\Lambda_0}}{\mathcal{N}_{0,\Lambda}^{\Lambda_0}} - \frac{1}{2}\left(\frac{\delta}{\delta\widehat{\Phi}},\partial_\Lambda D_\Lambda^{\Lambda_0}\frac{\delta}{\delta\widehat{\Phi}}\right)S_{I,\Lambda}^{\Lambda_0}\right.
$$

$$
\left.+ \frac{1}{2}\left(\frac{\delta S_{I,\Lambda}^{\Lambda_0}(\widehat{\Phi})}{\delta\widehat{\Phi}},\partial_\Lambda D_\Lambda^{\Lambda_0}\frac{\delta S_{I,\Lambda}^{\Lambda_0}(\widehat{\Phi})}{\delta\widehat{\Phi}}\right)\right],
$$

where $\partial_\Lambda \equiv \partial/\partial\Lambda$. Hence

$$
\partial_\Lambda S_{I,\Lambda}^{\Lambda_0}(\widehat{\Phi}) = \frac{1}{2}\left[\left(\frac{\delta}{\delta\widehat{\Phi}},\partial_\Lambda D_\Lambda^{\Lambda_0}\frac{\delta}{\delta\widehat{\Phi}}\right)S_{I,\Lambda}^{\Lambda_0}\right.
$$

$$
\left.- \left(\frac{\delta S_{I,\Lambda}^{\Lambda_0}(\widehat{\Phi})}{\delta\widehat{\Phi}},\partial_\Lambda D_\Lambda^{\Lambda_0}\frac{\delta S_{I,\Lambda}^{\Lambda_0}(\widehat{\Phi})}{\delta\widehat{\Phi}}\right)\right] - \partial_\Lambda\ln\frac{\mathcal{N}_\Lambda^{\Lambda_0}}{\mathcal{N}_{0,\Lambda}^{\Lambda_0}}. \qquad (4.7.25)
$$

More explicitly, the flow equation (4.7.25) of $S_{I,\Lambda}^{\Lambda_0}$ reads in configuration space

$$\partial_\Lambda S_{I,\Lambda}^{\Lambda_0}(\widehat{\Phi}) = \frac{1}{2} \int_{x,y} \partial_\Lambda D_\Lambda^{\Lambda_0}(x,y) \left[\frac{\delta^2 S_{I,\Lambda}^{\Lambda_0}(\widehat{\Phi})}{\delta\widehat{\Phi}(x)\delta\widehat{\Phi}(y)} - \frac{\delta S_{I,\Lambda}^{\Lambda_0}(\widehat{\Phi})}{\delta\widehat{\Phi}(x)} \frac{\delta S_{I,\Lambda}^{\Lambda_0}(\widehat{\Phi})}{\delta\widehat{\Phi}(y)} \right]$$
$$-\text{dto}(\widehat{\Phi}=0), \tag{4.7.26}$$

or, in momentum space, using translational invariance,

$$\partial_\Lambda S_{I,\Lambda}^{\Lambda_0}(\widehat{\Phi}) = \frac{1}{2} \int_k \partial_\Lambda \widetilde{D_\Lambda^{\Lambda_0}}(k) \left[\frac{\delta^2 S_{I,\Lambda}^{\Lambda_0}(\widehat{\Phi})}{\delta\widetilde{\widehat{\Phi}}(k)\delta\widetilde{\widehat{\Phi}}(-k)} - \frac{\delta S_{I,\Lambda}^{\Lambda_0}(\widehat{\Phi})}{\delta\widetilde{\widehat{\Phi}}(k)} \frac{\delta S_{I,\Lambda}^{\Lambda_0}(\widehat{\Phi})}{\delta\widetilde{\widehat{\Phi}}(-k)} \right]$$
$$-\text{dto}(\widehat{\Phi}=0). \tag{4.7.27}$$

The above differential equations describe the change of the effective interaction $S_{I,\Lambda}^{\Lambda_0}(\widehat{\Phi})$ under a change of the scale Λ. We recall that $-S_{I,\Lambda}^{\Lambda_0}(\widehat{\Phi})$ is the generating functional of the free-propagator amputated connected correlation functions, that is they are obtained from $-S_{I,\Lambda}^{\Lambda_0}(\widehat{\Phi})$ by differentiation with respect to $\widehat{\Phi}$ at $\widehat{\Phi}=0$. Let us write

$$S_{I,\Lambda}^{\Lambda_0} = \sum_{n\geq 1} \frac{1}{n!} \int_{p_1,\ldots,p_{n-1}} \widetilde{\widehat{\Phi}}(p_1)\cdots\widetilde{\widehat{\Phi}}(p_{n-1})\widetilde{\widehat{\Phi}}(-\sum_{i=1}^{n-1} p_i)\widetilde{S}_{I,\Lambda,n}^{\Lambda_0}(p_1,\ldots p_{n-1})$$
$$\tag{4.7.28}$$

with $\widetilde{S}_{I,\Lambda,n}^{\Lambda_0}$ totally symmetric in the momenta. Inserting (4.7.28) into the flow equations and comparing the coefficients of the monomials $\widetilde{\widehat{\Phi}}(p_1)\cdots\widetilde{\widehat{\Phi}}(p_n)$, we obtain

$$\partial_\Lambda \widetilde{S}_{I,\Lambda,n}^{\Lambda_0}(p_1,\ldots,p_{n-1})$$
$$= \frac{1}{2} \int_k \left(\partial_\Lambda \widetilde{D_\Lambda^{\Lambda_0}}(k) \right) \widetilde{S}_{I,\Lambda,n+2}^{\Lambda_0}(p_1,\ldots,p_{n-1},k,-k) \tag{4.7.29}$$
$$- \frac{1}{2} \sum_{\substack{l,m\geq 1 \\ (l+m=n+2)}} \frac{n!}{(l-1)!(m-1)!} \left[\widetilde{S}_{I,\Lambda,l}^{\Lambda_0}(p_1,\ldots,p_{l-1}) \right.$$
$$\left. (\partial_\Lambda \widetilde{D_\Lambda^{\Lambda_0}}(q)) \widetilde{S}_{I,\Lambda,m}^{\Lambda_0}(-q,p_l,\ldots,p_{n-1}) \right]_{\text{sym}}, \quad n=1,\ldots,\infty,$$

where $q = -\sum_{i=1}^{l-1} p_i$. The additional symmetrization bracket $[\ldots]_{sym}$ refers to the sum over all permutations of the momenta $p_1, \ldots, p_n$ among the two factors $\widetilde{S}$, normalized with the total number $n!$ of permutations of $p_1, \ldots, p_n$. (Permutations of the p_i's within the argument of each factor are trivial because of permutation symmetry, but nontrivial, if momenta of both factors are exchanged. This is taken into account by the prefactor.)

Note that (4.7.29) is an infinite set of integro-differential equations for the interaction functional of $\widehat{\Phi}$. The infinite number just reflects the fact that the flow equation (4.7.25) works in an infinite=dimensional function space. This may be bad news from a practical point of view. Good news, on the other hand, is the one-loop nature of Eq. (4.7.29). There is only one momentum integration in the first term of the right hand side of (4.7.29), and this is an exact result and not an artifact of a truncation. It is also because of this very aspect why the scheme of flow equations is sometimes called "exact" renormalization group. In practice, truncations enter the scheme, as in any feasible framework, but at different places. Clearly it is not surprising that good news and bad news go along with each other. The set of equations provides a realization of the renormalization group with infinitely high resolution of the scale dependence, described by an infinite set of equations, so, for an infinitesimally small change of scale, a one-loop calculation becomes sufficient, even more, it becomes exact.

4.7.3   *Effective average action*

Above we have derived the flow equations for the effective interaction $S_{I,\Lambda}^{\Lambda_0}$. $S_{I,\Lambda}^{\Lambda_0}$ was the generating functional of the free-propagator $D_\Lambda^{\Lambda_0}$ amputated, connected correlation functions. Alternatively, flow equations can be derived for connected correlation functions with generating functional $W_\Lambda^{\Lambda_0}$ or 1PI-correlation functions with functional $\Gamma_\Lambda^{\Lambda_0}$. A particular choice is a matter of convenience as long as no approximations are involved, because the functionals are uniquely mapped onto each other.

As soon as approximations enter the game, one type of flow equations may be superior to the other. Let us recall the experience we have made with linked cluster expansions. There it was very convenient to compute the series of quantities (1PI-susceptibilities) which were appropiately Legendre-transformed rather than the connected quantities. The technical and computational effort was considerably reduced this way, and values for the critical line and critical indices were even improved.

Similar considerations apply to approximation schemes of flow equations. Thus it turned out to be useful to study the scale dependencies in terms of the vertex functional $\Gamma_\Lambda^{\Lambda_0}$, the generating functional of the 1PI-correlation functions, or, closely related to $\Gamma_\Lambda^{\Lambda_0}$, in terms of the *effective average action* $\widehat{\Gamma}_\Lambda^{\Lambda_0}$ that will be defined below.

In this section we derive the flow equations of $\Gamma_\Lambda^{\Lambda_0}$ and $\widehat{\Gamma}_\Lambda^{\Lambda_0}$ from the flow equations for the effective interaction $S_{I,\Lambda}^{\Lambda_0}$, (4.7.25). The first step is to derive the analogous equations for the generating functional $W_\Lambda^{\Lambda_0}$ of connected correlation functions. From (4.7.14) and (4.7.25) we obtain with $\widehat{\Phi} = D_\Lambda^{\Lambda_0} J$ the relation

$$
\begin{aligned}
\partial_\Lambda W_\Lambda^{\Lambda_0}(J) = {} & \frac{1}{2}\left(J, \partial_\Lambda D_\Lambda^{\Lambda_0} J\right) - \partial_\Lambda S_{I,\Lambda}^{\Lambda_0}(\widehat{\Phi}) \\
& - \left(\partial_\Lambda D_\Lambda^{\Lambda_0} J, \frac{\delta S_{I,\Lambda}^{\Lambda_0}(\widehat{\Phi})}{\delta\widehat{\Phi}}\right) \\
= {} & \frac{1}{2}\left(\left[J - \frac{\delta S_{I,\Lambda}^{\Lambda_0}(\widehat{\Phi})}{\delta\widehat{\Phi}}\right], \partial_\Lambda D_\Lambda^{\Lambda_0}\left[J - \frac{\delta S_{I,\Lambda}^{\Lambda_0}(\widehat{\Phi})}{\delta\widehat{\Phi}}\right]\right) \quad (4.7.30) \\
& - \frac{1}{2}\left(\frac{\delta}{\delta\widehat{\Phi}}, \partial_\Lambda D_\Lambda^{\Lambda_0}\frac{\delta}{\delta\widehat{\Phi}}\right) S_{I,\Lambda}^{\Lambda_0} - \mathrm{dto}(J=0).
\end{aligned}
$$

Also from (4.7.14), we get

$$
\frac{\delta S_{I,\Lambda}^{\Lambda_0}(\widehat{\Phi})}{\delta\widehat{\Phi}} = (D_\Lambda^{\Lambda_0})^{-1}\widehat{\Phi} - (D_\Lambda^{\Lambda_0})^{-1}\frac{\delta W_\Lambda^{\Lambda_0}(J)}{\delta J}, \qquad (4.7.31)
$$

or

$$
J - \frac{\delta S_{I,\Lambda}^{\Lambda_0}(\widehat{\Phi})}{\delta\widehat{\Phi}} = (D_\Lambda^{\Lambda_0})^{-1}\frac{\delta W_\Lambda^{\Lambda_0}(J)}{\delta J}. \qquad (4.7.32)
$$

Furthermore, one more differentiation applied to (4.7.31) yields

$$
\begin{aligned}
& \left(\frac{\delta}{\delta\widehat{\Phi}}, \partial_\Lambda D_\Lambda^{\Lambda_0}\frac{\delta}{\delta\widehat{\Phi}}\right) S_{I,\Lambda}^{\Lambda_0}(\widehat{\Phi}) \\
& = \mathrm{tr}\left[\partial_\Lambda D_\Lambda^{\Lambda_0}\left((D_\Lambda^{\Lambda_0})^{-1} - (D_\Lambda^{\Lambda_0})^{-1}W_\Lambda^{\Lambda_0\,(2)}(J)(D_\Lambda^{\Lambda_0})^{-1}\right)\right] \quad (4.7.33) \\
& = \mathrm{tr}\left[\partial_\Lambda(D_\Lambda^{\Lambda_0})^{-1}W_\Lambda^{\Lambda_0\,(2)}(J) + \partial_\Lambda D_\Lambda^{\Lambda_0}(D_\Lambda^{\Lambda_0})^{-1}\right],
\end{aligned}
$$

where we have written $\mathrm{tr}[O] \equiv \int_x O(x,x)$ for the involved operators O and

$$
W_\Lambda^{\Lambda_0\,(2)}(J)(x,y) = \frac{\delta^2 W_\Lambda^{\Lambda_0}(J)}{\delta J(x)\delta J(y)}. \qquad (4.7.34)
$$

Inserting (4.7.33) into (4.7.30), we get the flow equation for $W_\Lambda^{\Lambda_0}(J)$. It reads

$$
\partial_\Lambda W_\Lambda^{\Lambda_0}(J) = \frac{1}{2}\left((D_\Lambda^{\Lambda_0})^{-1}\frac{\delta W_\Lambda^{\Lambda_0}}{\delta J}, \partial_\Lambda D_\Lambda^{\Lambda_0}(D_\Lambda^{\Lambda_0})^{-1}\frac{\delta W_\Lambda^{\Lambda_0}}{\delta J}\right)
$$

$$
- \frac{1}{2}\operatorname{tr}\left(\partial_\Lambda (D_\Lambda^{\Lambda_0})^{-1}W_\Lambda^{\Lambda_0\,(2)}(J) + \partial_\Lambda D_\Lambda^{\Lambda_0}(D_\Lambda^{\Lambda_0})^{-1}\right)
$$

$$
- \mathrm{dto}(J = 0)
$$

$$
= -\frac{1}{2}\left(\frac{\delta W_\Lambda^{\Lambda_0}}{\delta J}, \partial_\Lambda (D_\Lambda^{\Lambda_0})^{-1}\frac{\delta W_\Lambda^{\Lambda_0}}{\delta J}\right) \tag{4.7.35}
$$

$$
- \frac{1}{2}\operatorname{tr}\left(\partial_\Lambda (D_\Lambda^{\Lambda_0})^{-1}W_\Lambda^{\Lambda_0\,(2)}(J)\right)
$$

$$
- \mathrm{dto}(J = 0).
$$

Next we apply a Legendre transformation to the effective action in the conventional form, to obtain the generating functional of the 1PI-Green functions $\Gamma_\Lambda^{\Lambda_0}(\phi)$,

$$
\Gamma_\Lambda^{\Lambda_0}(\phi) = W_\Lambda^{\Lambda_0}(J) - (J, \phi)\;, \qquad \phi = \frac{\delta W_\Lambda^{\Lambda_0}(J)}{\delta J}, \tag{4.7.36}
$$

with inverse relation $J = -\delta\Gamma_\Lambda^{\Lambda_0}(\phi)/\delta\phi$. This yields the expected relation

$$
\partial_\Lambda \Gamma_\Lambda^{\Lambda_0}(\phi) = \partial_\Lambda W_\Lambda^{\Lambda_0} + \left(\partial_\Lambda J, \frac{\delta W_\Lambda^{\Lambda_0}(J)}{\delta J}\right) - (\partial_\Lambda J, \phi)
$$

$$
= \partial_\Lambda W_\Lambda^{\Lambda_0}(J). \tag{4.7.37}
$$

Inserting (4.7.35), and using that

$$
\Gamma_\Lambda^{\Lambda_0\,(2)}(\phi)(x,y) = \frac{\delta^2\Gamma_\Lambda^{\Lambda_0}(\phi)}{\delta\phi(x)\delta\phi(y)} = -\left(W_\Lambda^{\Lambda_0\,(2)}(J)\right)^{-1}(x,y), \tag{4.7.38}
$$

we end up with the flow equations for the vertex functional $\Gamma_\Lambda^{\Lambda_0}$

$$
\partial_\Lambda \Gamma_\Lambda^{\Lambda_0}(\phi) = -\frac{1}{2}\left(\phi, \partial_\Lambda (D_\Lambda^{\Lambda_0})^{-1}\phi\right)
$$

$$
+ \frac{1}{2}\operatorname{tr}\left[\partial_\Lambda (D_\Lambda^{\Lambda_0})^{-1}\Gamma_\Lambda^{\Lambda_0\,(2)\,-1}(\phi)\right] \tag{4.7.39}
$$

$$
- \mathrm{dto}(\phi = \overline{\phi}),
$$

$\overline{\phi}$ is defined by

$$
\frac{\delta\Gamma_\Lambda^{\Lambda_0}(\overline{\phi})}{\delta\overline{\phi}} = 0. \tag{4.7.40}
$$

It should be noticed that $\overline{\phi} = \overline{\phi}_\Lambda$, i.e. $\overline{\phi}$ does depend on the scale. Thus the last term in (4.7.39) is a scale dependent constant that affects the normalization of the vacuum energy, but is dropped in many applications so that the normalization condition $W_\Lambda^{\Lambda_0}(J = 0) = 0$ is violated. In what follows we will also ignore the correct normalization to reproduce some results of the literature, but not without adding a warning. From rigorous nonperturbative investigations by means of the block-spin renormalization group one knows that it is important to control the scale dependence of the vacuum energy as well, because it amounts to the most relevant term in the action. An appropriate renormalization of the vacuum energy under a change of scale may affect the very existence of the functional when the cutoffs are removed.

The effective average action $\widehat{\Gamma}_\Lambda^{\Lambda_0}$ is then defined by

$$\widehat{\Gamma}_\Lambda^{\Lambda_0}(\phi) = -\Gamma_\Lambda^{\Lambda_0}(\phi) - \frac{1}{2}\left(\phi, (D_\Lambda^{\Lambda_0})^{-1}\phi\right). \tag{4.7.41}$$

The definition is chosen such that it satisfies the initial condition

$$\widehat{\Gamma}_{\Lambda_0}^{\Lambda_0}(\phi) = S_I(\phi) \tag{4.7.42}$$

on the scale of the UV cutoff. Moreover the explicit ϕ-dependence of the flow equations drops out. Using (4.7.39) without the scale-dependent constant, $\widehat{\Gamma}_\Lambda^{\Lambda_0}$ obeys the flow equation

$$\partial_\Lambda \widehat{\Gamma}_\Lambda^{\Lambda_0}(\phi) = \frac{1}{2}\,\mathrm{tr}\left[\partial_\Lambda(D_\Lambda^{\Lambda_0})^{-1}\left(\widehat{\Gamma}_\Lambda^{\Lambda_0\,(2)}(\phi) + (D_\Lambda^{\Lambda_0})^{-1}\right)^{-1}\right]. \tag{4.7.43}$$

4.7.4 High-temperature phase transition of $O(N)$ models

In this section we illustrate how flow equations can be used to study the high-temperature transition in $O(N)$-symmetric scalar models in four dimensions. Of particular interest is the special case of $N = 4$. $O(4)$-scalar models constitute the scalar part of the electroweak standard model. $O(4)$ models are also supposed to share the universality class with QCD in the limit of two massless flavors in the vicinity of the chiral transition at finite temperature, cf. section 3.1.

As we have seen in section 4.4, higher orders in perturbation theory suffer from IR-divergencies when massless modes in the infinite volume are involved. Examples are gauge theories with massless gauge bosons (at zero temperature), and massless Φ^4-theory at finite temperatures. In the Φ^4-theory the divergencies are cured by applying infinite resummations of the perturbation series in such a way that a mass gap $m(T)$ is resummed into the free propagator. The resulting perturbative series are no longer power series in the coupling constant $\lambda(T)$. Fractional powers and even logarithms of the coupling constant occur. The expansion parameter becomes proportional to $\lambda(T)/m(T)$, with $m(T) \sim \lambda(T)^{1/2}$.

However, for the case of a second-order phase transition in which we are interested now, we are not done if we apply these procedures. Even if the resummation works, the IR-problem reappears if we approach the second-order phase transition at some T_c, where the renormalized mass vanishes, $m(T_c) = 0$. The expansion parameter diverges, and perturbation theory breaks down again. Therefore we should further reorganize the series expansion in terms of a renormalized coupling constant $\lambda(T)$ which itself vanishes rapidly fast as T approaches T_c. This means that $\lambda(T)/m(T)$ stays finite at the phase-transition temperature T_c. To our knowledge, no solution was attempted along these lines.

On the other hand, we mentioned already in sections 2.4 and 4.6 that the renormalization group provides a systematic approach to IR-problems, because a successive integration upon momenta in the order of starting from high down to low momenta provides a clear separation of UV- and IR-scales. From a practical point of view the renormalization group amounts to an improved resummation. For example, for a $\lambda\Phi^4$-theory in three dimensions one knows that the ratio of the scale-dependent, running couplings $\lambda_\Lambda/m_\Lambda$ of renormalized quartic coupling and mass has an IR-stable fixed point that is associated with a second-order phase transition. In each renormalization-group step one has to deal with a noncritical system free of UV- and IR singularities. If the renormalization group is realized in the framework of the flow equations, each step is infinitesimal. The flow of the generating functionals (such as $W_\Lambda^{\Lambda_0}$, $S_{I,\Lambda}^{\Lambda_0}$, or $\widehat{\Gamma}_\Lambda^{\Lambda_0}$) then induces a flow of couplings as a function of scale. Following this flow down to zero momenta yields a more accurate scale- and temperature-dependence of renormalized coupling constants than in standard perturbation theory.

For weakly-coupled models it is natural to consider a perturbative realization of the renormalization group. It should be noticed that the expansion parameter does depend on the scale Λ, in contrast to standard perturbation theory at one fixed scale. In standard perturbation theory the involved mapping between bare and renormalized couplings corresponds to a "one-step" renormalization-group transformation between the (ultraviolet) scale, at which the bare couplings are defined, and the scale of the renormalized couplings. If these scales are widely separated, the resolution of the scale dependence of running couplings is poor in a one-step relation, but becomes the higher the more renormalization-group steps are performed.

Flow equations are a particular realization of the renormalization group (here the perturbative one) with a continuous change of scale. In the ideal case they determine the flow of one of the generating functionals over a wide range of scales down to the far IR with an infinitely high resolution, but without encountering IR-problems.

Next we focus on the flow equations for the effective average action $\widehat{\Gamma}_\Lambda^{\Lambda_0}$, because the running couplings are conveniently defined in terms of the vertex functional $\Gamma_\Lambda^{\Lambda_0}$, and, within an appropriate truncation scheme, the infinite set of integro-differential equations for scale-dependent functionals $\widehat{\Gamma}_\Lambda^{\Lambda_0}$ reduces to a few evolution equations for a set of running couplings. Let us see how this scheme works for the example of scalar $O(N)$ models in four dimensions. We follow the reasoning of Tetradis and Wetterich [205].

First -for simplicity- the UV-cutoff Λ_0 is formally sent to infinity. We call this step formal, because it ultimately requires to show the convergence of the correlation functions on all length scales, or equivalently, convergence of their generating functional $\Gamma_\Lambda^{\Lambda_0}$ as $\Lambda_0 \to \infty$. The convergence is only guaranteed in perturbation theory. More seriously, one should keep in mind the issue of triviality. In the large-cutoff limit the scalar models are expected to converge to a free noninteracting continuum field theory. This means, for $\Lambda_0 \to \infty$, the resulting theory is a free field theory on *all* scales. Thus let us think of $O(N)$ models as effective models with a large but still finite cutoff Λ_0. In this case their IR-behavior is nontrivial, they serve as prototype for models with second-order phase transitions and symmetry restoration.

The flow equations are started at a scale $\overline{\Lambda}$, well below Λ_0, but still large enough compared to all relevant mass and temperature scales. The Λ_0-dependence is ignored in the following (but Λ_0 is kept finite and large). The fluctuation propagator (propagator of the ϕ-field) is chosen as

$$\widetilde{D}_\Lambda(k)_{ab} = \frac{1}{k^2} \frac{1 - R_\Lambda(k)}{R_\Lambda(k)} \delta_{a,b}, \qquad a, b = 1, \ldots, N, \qquad (4.7.44)$$

with

$$R_\Lambda(k) = \exp\left(-\alpha \left(\frac{k^2}{\Lambda^2}\right)^\beta\right). \tag{4.7.45}$$

The choice of positive constants α and β is a matter of convenience. Momenta that are small compared to Λ are strongly suppressed, $\widetilde{D}_\Lambda(k)_{ab}$ is essentially supported in $\Lambda^2/2 \le k^2$. In contrast to the fluctuation propagator of (4.7.1) we see that Λ_0 was sent to infinity and there is an additional $R_\Lambda(k)$ in the denominator. This has the advantage that momenta much larger than Λ are strongly suppressed in the flow equations, suppressed by $\widetilde{D_\Lambda^{-1}}$ and $\partial_\Lambda \widetilde{D_\Lambda^{-1}}$,

$$\partial_\Lambda \widehat{\Gamma}_\Lambda(\phi) = \frac{1}{2} \sum_{a,b=1}^N \int_k^{\sim} \left[\partial_\Lambda \widetilde{D_\Lambda^{-1}}(k)_{ab} \left(\widehat{\widetilde{\Gamma}}_\Lambda^{(2)}(\phi) + \widetilde{D_\Lambda^{-1}}\right)^{-1}(k)_{ba}\right]. \tag{4.7.46}$$

In order to investigate this functional equation, one has to make an ansatz for $\widehat{\Gamma}_\Lambda$. Thus, unavoidably at this point, a truncation scheme enters the game. A particular truncation defines the functional space or the space of coupling constants in which the flow will be considered. Clearly, the very ansatz decides about both properties of whether an application of the flow equations to this model is reliable and practicable. Usually it requires some intuition to select the right degrees of freedom which have prominent influence on the IR-behavior. An omission of these degrees of freedom leads to results that are an artifact of the truncation. On the other hand, the truncation should be drastic enough to strongly simplify the infinite set of integro-differential equations. We consider the following ansatz for $\widehat{\Gamma}_\Lambda$

$$\widehat{\Gamma}_\Lambda(\phi) = \int_x \left(U_\Lambda(\rho(x)) + \frac{Z_\Lambda(\rho(x))}{2} \sum_{\mu=0}^3 \sum_{a=1}^N (\partial_\mu \phi_a(x))^2\right.$$
$$\left. + \frac{Y_\Lambda(\rho(x))}{4} \sum_{\mu=0}^3 (\partial_\mu \rho(x))^2\right), \tag{4.7.47}$$

where

$$\rho(x) = \frac{1}{2} \sum_{a=1}^N \phi_a(x)^2. \tag{4.7.48}$$

U_Λ, Z_Λ and Y_Λ are three functions of Λ. The Λ-dependence will be determined by solving the flow equations. U_Λ is the effective average potential,

Z_Λ the effective wave-function renormalization constant of the field ϕ, and similarly Y_Λ the effective wave-function renormalization constant of the composite field ρ. The initial conditions at $\Lambda = \overline{\Lambda}$ are

$$Z_{\overline{\Lambda}}(\rho) \equiv 1, \quad Y_{\overline{\Lambda}}(\rho) \equiv 0 \quad \text{and} \quad U_{\overline{\Lambda}}(\rho) = \frac{\overline{\lambda}}{2}(\rho - \overline{\rho})^2 \qquad (4.7.49)$$

with "bare" parameters $\overline{\lambda}$ and $\overline{\rho}$ (i.e. parameters on the scale $\overline{\Lambda}$).

Why is the ansatz chosen in this way? First of all it is globally $O(N)$-symmetric as the original model. It is of a form that one obtains in an expansion of $\widehat{\Gamma}_\Lambda(\phi)$ to subleading order about constant fields or zero momenta. Constant field configurations are expected to constitute the ground state for spin and scalar models in $D \geq 3$ dimensions. Finally the structure of the ground state and its stability under fluctuations determine the phase structure which we are interested in.

In order to derive the flow equations of the coefficient functions U_Λ, Z_Λ and Y_Λ, we have to compute the second derivative of $\widehat{\Gamma}_\Lambda$ with respect to the field ϕ. To this end we parameterize ϕ about a constant field configuration according to

$$\phi_a(x) = \phi^c \delta_{a,1} + \tau_a(x), \quad \text{with} \quad \int_x \tau_a(x) = 0. \qquad (4.7.50)$$

Inserting (4.7.50) into $\widehat{\Gamma}_\Lambda$ and differentiating twice with respect to the field τ, we obtain for the constant field contribution to $\widetilde{\widehat{\Gamma}}_\Lambda^{(2)}(\phi)$

$$\widetilde{\widehat{\Gamma}}_\Lambda^{(2)\,c}(k)_{ab} = \left[\delta_{a,b}U_\Lambda'(\rho) + 2\rho\delta_{a,1}\delta_{b,1}U_\Lambda''(\rho)\right]$$
$$+ k^2\left[\delta_{a,b}Z_\Lambda(\rho) + \rho\delta_{a,1}\delta_{b,1}Y_\Lambda(\rho)\right], \qquad (4.7.51)$$

where $\rho = 1/2(\phi^c)^2$ for constant fields. The inverse of $(\widetilde{\widehat{\Gamma}}_\Lambda^{(2)\,c} + \widetilde{D}_\Lambda^{-1})$ that enters the flow equation of U_Λ is diagonal in internal symmetry space, thus we easily obtain

$$\partial_\Lambda U_\Lambda(\rho) = \frac{1}{2}\sum_{a,b=1}^N \int_k \left[\partial_\Lambda \widetilde{D}_\Lambda^{-1}(k)_{ab} \left(\widetilde{\widehat{\Gamma}}_\Lambda^{(2)\,c}(k) + \widetilde{D}_\Lambda^{-1}(k)\right)_{ba}^{-1}\right].$$
$$= \frac{1}{2}\int_k \partial_\Lambda \left(k^2 \frac{R_\Lambda(k)}{1 - R_\Lambda(k)}\right)\left(\frac{1}{P_r(k)} + \frac{N-1}{P_g(k)}\right). \qquad (4.7.52)$$

Here we have written

$$P_r(k) = (U'_\Lambda(\rho) + 2\rho U''_\Lambda(\rho)) + k^2 \left(Z_\Lambda(\rho) + \rho Y_\Lambda(\rho) + \frac{R_\Lambda(k)}{1 - R_\Lambda(k)} \right) \tag{4.7.53}$$

for the inverse radial-mode propagator, and

$$P_g(k) = U'_\Lambda(\rho) + k^2 \left(Z_\Lambda(\rho) + \frac{R_\Lambda(k)}{1 - R_\Lambda(k)} \right) \tag{4.7.54}$$

for the inverse longitudinal (or Goldstone-mode) propagator. Similar equations follow for the scale dependence of $Z_\Lambda(\rho)$ and $Y_\Lambda(\rho)$, again by taking into account fluctuations of $\widetilde{\Gamma}_\Lambda^{(2)}(\phi)$ about the constant field configurations.

We know that in $D \geq 3$ dimensions the anomalous dimension η has a rather small value, much smaller than 1. Typically, for $D = 3$, $\eta \simeq 0.05$ or smaller. η is closely related to the scale dependence of the wave function renormalization constant Z_Λ, at least for small Λ. Thus we may expect that a truncation scheme which keeps Z_Λ fixed will already give rather accurate results. Similar considerations apply to Y_Λ. In the following, we will keep Z_Λ and Y_Λ constant and equal to their initial values, $Z_\Lambda(\rho) \equiv 1$ and $Y_\Lambda(\rho) \equiv 0$ for all $\Lambda \leq \overline{\Lambda}$. The flow equations then degenerate to a single integro-differential equation for the effective average potential $U_\Lambda(\rho)$. Let us write

$$P_\Lambda(k) = k^2 + k^2 \frac{R_\Lambda(k)}{1 - R_\Lambda(k)} = \frac{1}{1 - R_\Lambda(k)} k^2. \tag{4.7.55}$$

In terms of $P_\Lambda(k)$ the flow equation (4.7.52) becomes

$$\partial_\Lambda U_\Lambda(\rho) = \frac{1}{2} \int_k \partial_\Lambda P_\Lambda(k) \left(\frac{1}{P_\Lambda(k) + M_{r,\Lambda}(\rho)} + \frac{N-1}{P_\Lambda(k) + M_{g,\Lambda}(\rho)} \right) \tag{4.7.56}$$

with

$$M_{r,\Lambda}(\rho) = U'_\Lambda(\rho) + 2\rho U''_\Lambda(\rho),$$
$$M_{g,\Lambda}(\rho) = U'_\Lambda(\rho). \tag{4.7.57}$$

We are interested in solving (4.7.56) in the vicinity of the scale-dependent minimum ρ_Λ of the effective average potential $U_\Lambda(\rho)$. Thus it is natural to consider an ansatz for $U_\Lambda(\rho)$ in powers of fluctuations about ρ_Λ.

Let us first consider the case with $\rho_\Lambda > 0$. This domain is called the broken regime ("regime" in contrast to phase, because $\rho_\Lambda > 0$ does not necessarily imply that we are in the phase of spontaneously broken symmetry; below we will further characterize the difference). ρ_Λ is then a local minimum of $U_\Lambda(\rho)$ with $U'_\Lambda(\rho_\Lambda) = 0$. We define a coupling λ_Λ as coefficient of the term quartic in the ϕ's by

$$\lambda_\Lambda = U''_\Lambda(\rho_\Lambda). \tag{4.7.58}$$

To obtain the flow equations for ρ_Λ and λ_Λ, we expand $U_\Lambda(\rho)$ about its minimum to fourth order in $\rho - \rho_\Lambda$,

$$U_\Lambda(\rho) = c_\Lambda + \frac{1}{2}\lambda_\Lambda(\rho - \rho_\Lambda)^2 + \frac{1}{3!}d_\Lambda(\rho - \rho_\Lambda)^3$$
$$+ \frac{1}{4!}e_\Lambda(\rho - \rho_\Lambda)^4 + O((\rho - \rho_\Lambda)^5). \tag{4.7.59}$$

We have explicitly spelled out all terms which will enter the equations for $\partial_\Lambda \rho_\Lambda$ and $\partial_\Lambda \lambda_\Lambda$. Thus we have

$$\partial_\Lambda U_\Lambda(\rho) = \partial_\Lambda c_\Lambda + \lambda_\Lambda(-\partial_\Lambda \rho_\Lambda)(\rho - \rho_\Lambda)$$
$$+ \frac{1}{2}(\partial_\Lambda \lambda_\Lambda - d_\Lambda \partial_\Lambda \rho_\Lambda)(\rho - \rho_\Lambda)^2 + O((\rho - \rho_\Lambda)^3),$$

$$U'_\Lambda(\rho) = \lambda_\Lambda(\rho - \rho_\Lambda) + \frac{1}{2}d_\Lambda(\rho - \rho_\Lambda)^2 + O((\rho - \rho_\Lambda)^3), \tag{4.7.60}$$

$$U''_\Lambda(\rho) = \lambda_\Lambda + d_\Lambda(\rho - \rho_\Lambda) + \frac{1}{2}e_\Lambda(\rho - \rho_\Lambda)^2 + O((\rho - \rho_\Lambda)^3).$$

Here, for the remainder estimate to be true, the expansion must be uniformly convergent, or at least uniform asymptotically convergent, with respect to Λ, otherwise a suppression of terms of $O((\rho - \rho_\Lambda)^3)$ on one scale would not guarantee their suppression under an evolution with Λ. Expanding the integrand on the right hand side of (4.7.56) to order $(\rho - \rho_\Lambda)^2$ and comparing coefficients, after some algebra we obtain **the flow equations in the broken regime**

$$\partial_\Lambda c_\Lambda = \frac{1}{2}\int_k^{\sim} (\partial_\Lambda P_\Lambda(k))\left[\frac{1}{P_\Lambda(k) + 2\rho_\Lambda \lambda_\Lambda} + \frac{N-1}{P_\Lambda(k)}\right],$$

$$\partial_\Lambda \rho_\Lambda = \frac{1}{2}\int_k^{\sim} (\partial_\Lambda P_\Lambda(k))\left[\frac{3 + 2\frac{\rho_\Lambda d_\Lambda}{\lambda_\Lambda}}{(P_\Lambda(k) + 2\rho_\Lambda \lambda_\Lambda)^2} + \frac{N-1}{P_\Lambda(k)^2}\right], \tag{4.7.61}$$

$$\partial_\Lambda \lambda_\Lambda = \int_k^{\sim} (\partial_\Lambda P_\Lambda(k))\left[\frac{\left(-1 + \frac{\rho_\Lambda d_\Lambda}{\lambda_\Lambda} - \frac{\rho_\Lambda e_\Lambda}{d_\Lambda}\right)d_\Lambda}{(P_\Lambda(k) + 2\rho_\Lambda \lambda_\Lambda)^2}\right.$$
$$\left. + \frac{(3\lambda_\Lambda + 2\rho_\Lambda d_\Lambda)^2}{(P_\Lambda(k) + 2\rho_\Lambda \lambda_\Lambda)^3} + \frac{N-1}{P_\Lambda(k)^3}\lambda_\Lambda^2\right].$$

The system of flow equations is not closed. We have three equations for five coefficients. It becomes closed if we ignore all terms of $U_\Lambda(\rho)$ of the order $O((\rho - \rho_\Lambda)^3)$ from the very beginning, in particular if we put $d_\Lambda \equiv e_\Lambda \equiv 0$. This amounts to a further projection of the renormalization-group flow onto a space of three couplings. Furthermore, we recall that the very derivation of the flow equation (4.7.43) for the effective average action $\widehat{\Gamma}_\Lambda$ ignored the normalization condition on the partition function Z_Λ. A scale-dependent constant will influence the absolute height of U_Λ or the constant c_Λ in the ansatz for $U_\Lambda(\rho)$. The height may be ignored as long as it does stay finite for $\Lambda \to 0$.

It is convenient to make the Λ-dependence of the right hand side of the flow equations more explicit. To this end let us rewrite (4.7.61) in terms of dimensionless integrals. We note that

$$P_\Lambda(k) = \Lambda^2 P_{\Lambda=1}(\frac{k}{\Lambda}), \qquad (4.7.62)$$

so that

$$(\partial_\Lambda P_\Lambda)(k) = \Lambda (\partial_\Lambda P_\Lambda)_{\Lambda=1}(\frac{k}{\Lambda}). \qquad (4.7.63)$$

For nonnegative x, y and $n \geq 1$ define

$$L_n^D(x,y) = \frac{n}{2v_D} \int_{k'}^{\widetilde{y}} \frac{(\partial_\Lambda P_\Lambda)_{\Lambda=1}(k')}{(P_{\Lambda=1}(k') + x)^{n+1}} \qquad (4.7.64)$$

with

$$v_D = \frac{1}{2} \frac{\Omega_D}{(2\pi)^D} = \left(2^D \pi^{D/2} \Gamma(\frac{D}{2})\right)^{-1}, \qquad (4.7.65)$$

where Ω_D denotes the volume of the unit sphere in D-dimensional Euclidean space, and

$$\int_{k'}^{\widetilde{y}} = y \sum_{k_0'} \int_{-\infty}^{\infty} \frac{d^{D-1}k'}{(2\pi)^{D-1}}, \qquad (4.7.66)$$

$k_0' = 2\pi y n$, $n \in \mathbf{Z}$. For y=0, this becomes

$$\int_{k'}^{\widetilde{y}=0} = \int_{-\infty}^{\infty} \frac{d^D k'}{(2\pi)^D}. \qquad (4.7.67)$$

In terms of the functions L_n^D we have for $w \geq 0$

$$\int_k^{\sim} \frac{(\partial_\Lambda P_\Lambda)(k)}{(P_\Lambda(k) + w)^{n+1}} = \frac{2v_D}{n} \Lambda^{D-2n-1} L_n^D(\frac{w}{\Lambda^2}, \frac{T}{\Lambda}). \qquad (4.7.68)$$

Because of $\partial_\Lambda P_\Lambda(k) \geq 0$, the functions L_n^D are nonnegative and dimensionless. In this notation the flow equations (4.7.61) (for $d_\Lambda \equiv e_\Lambda \equiv 0$) become

$$\partial_\Lambda \rho_\Lambda = v_D \Lambda^{D-3} \left[3L_1^D(\frac{2\rho_\Lambda \lambda_\Lambda}{\Lambda^2}, \frac{T}{\Lambda}) + (N-1)L_1^D(0, \frac{T}{\Lambda}) \right],$$

$$\partial_\Lambda \lambda_\Lambda = v_D \Lambda^{D-5} \lambda_\Lambda^2 \left[9L_2^D(\frac{2\rho_\Lambda \lambda_\Lambda}{\Lambda^2}, \frac{T}{\Lambda}) + (N-1)L_2^D(0, \frac{T}{\Lambda}) \right]. \qquad (4.7.69)$$

The right hand side of the equations (4.7.69) is always nonnegative. Both ρ_Λ and λ_Λ are driven to smaller values as Λ decreases.

Next we consider the case $\rho_\Lambda = 0$, in which the minimum of $U_\Lambda(\rho)$ is at the boundary of the allowed values for ρ_Λ. This domain is called the symmetric regime. If $\rho_\Lambda = 0$, it stays zero for all scales smaller than Λ. In this case, the running mass m_Λ and coupling λ_Λ are defined by

$$m_\Lambda = U_\Lambda'(0),$$
$$\lambda_\Lambda = U_\Lambda''(0). \qquad (4.7.70)$$

The flow equations for the couplings m_Λ and λ_Λ in the symmetric regime are quite similarly derived as for the couplings in the broken regime. Because of a nonvanishing first derivative of $U_\Lambda(\rho)$ at $\rho_\Lambda = 0$ we make an ansatz for the potential of the form

$$U_\Lambda(\rho) = c_\Lambda + m_\Lambda^2 \rho + \frac{1}{2}\lambda_\Lambda \rho^2, \qquad (4.7.71)$$

truncated after the term quadratic in ρ. Inserting (4.7.71) into (4.7.56) yields **the flow equations in the symmetric regime**

$$\partial_\Lambda m_\Lambda^2 = -(N+2)v_D \Lambda^{D-3} \lambda_\Lambda L_1^D(\frac{m_\Lambda^2}{\Lambda^2}, \frac{T}{\Lambda}),$$

$$\partial_\Lambda \lambda_\Lambda = (N+8)v_D \Lambda^{D-5} \lambda_\Lambda^2 L_2^D(\frac{m_\Lambda^2}{\Lambda^2}, \frac{T}{\Lambda}). \qquad (4.7.72)$$

As in the broken phase, λ_Λ is continuously driven to smaller values as Λ decreases, while m_Λ^2 increases towards the IR.

The notions of symmetric and broken "regimes" are convenient to distinguish the cases of $\rho_\Lambda = 0$ and $\rho_\Lambda \neq 0$ in contrast to the symmetric and broken "phases", characterized by the true physical expectation values $\rho_{\Lambda=0} = 0$ and $\rho_{\Lambda=0} \neq 0$ that are reached as $\Lambda \to 0$. A value $\rho_\Lambda \neq 0$ only anticipates the vacuum expectation value in the broken phase if the temperature is sufficiently low. Otherwise, say at a temperature T above T_c, a start deeply in the broken regime should end up in the symmetric phase. The flow then evolves according to (4.7.69) until it reaches the boundary between the two regimes at some finite cutoff $\Lambda_c > 0$ with $\rho_{\Lambda_c} = 0$ and $\lambda = \lambda_c$. Afterwards the flow is governed by (4.7.72), the evolution equations in the symmetric regime, down to $\Lambda = 0$ with initial conditions $m_{\Lambda_c}^2 = 0$ and λ continuously changing at Λ_c.

The flow equations are solved in D=4 dimensions for a given set of initial conditions in the broken regime $\rho_{\overline{\Lambda}} > 0$ and $\lambda_{\overline{\Lambda}}$ at the UV cutoff scale $\overline{\Lambda}$, and for a given temperature T. The ratio of $L_2^4(x, y)/L_2^4(x, 0)$ is plotted in Fig.4.7.1 as a function of y for various x, since L_2^4 enter Eq. (4.7.72).

The qualitative behavior is already obtained by the asymptotic properties of the functions L_n^D, defined in (4.7.64). First, for all $y \geq 0$, we have

$$L_n^D(x, y) \simeq x^{-(n+1)} \left(c_D(y) + O(x^{-1}) \right) \quad \text{as } x \to \infty, \qquad (4.7.73)$$

with n-independent coefficient function

$$c_D(y) = \int_{k'}^{\tilde{y}} (\partial_\Lambda P_\Lambda)_{\Lambda=1}(k'). \qquad (4.7.74)$$

Furthermore, for $x \neq 0$,

$$\left. \begin{array}{l} L_n^D(x, y) \simeq y \, L_n^{D-1}(x, 0), \\ c_D(y) \simeq y \, c_{D-1}(0) \end{array} \right\} \quad y \to \infty. \qquad (4.7.75)$$

Also, $L_n^D(0, 0)$ are D and n-dependent positive constants.

Following the flow down to $\Lambda = 0$, we obtain the physical state, parameterized by the pair $\rho_{\Lambda=0}$ and $\lambda_{\Lambda=0}$ in the broken phase, and by $m_{\Lambda=0}^2$ and $\lambda_{\Lambda=0}$ in the symmetric phase. The final state that is reached does depend on the temperature. In the symmetric phase, renormalized coupling constants are defined by

$$m_R^2(T) = m_{\Lambda=0}(T)^2, \quad \lambda_R(T) = \lambda_{\Lambda=0}(T). \qquad (4.7.76)$$

In the symmetric regime, the asymptotics of the flow equations is determined by the asymptotics of the functions L_n^D.

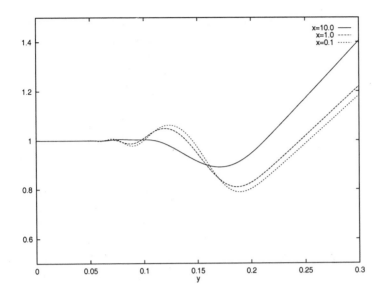

Fig. 4.7.1   Ratio of functions $L_2^4(x, y)/L_2^4(x, 0)$ plotted over y for various x.

In the broken regime, we define for all N nontrivial renormalized parameters by

$$\rho_R(T) = \rho_{\Lambda=0}(T), \quad \lambda_R(T) = \lambda_{\Lambda_R}(T), \qquad (4.7.77)$$

with $\Lambda_R = (2\lambda_R(T)\rho_R(T))^{1/2}$. This definition is also appropriate for the nonabelian case $N > 1$, where the situation is more complicated because of the presence of Goldstone modes. The asymptotic properties of the functions L_n^D stated above imply that the second part on the right hand side of the flow equations (4.7.69) becomes dominant. For $T > 0$,

$$\partial_\Lambda \lambda_\Lambda \simeq v_4 \Lambda^{-1} \lambda_\Lambda^2 \left((N-1)\frac{T}{\Lambda} L_2^3(0,0) \right)$$

$$= \left(v_4(N-1) L_2^3(0,0)T \right) \frac{\lambda_\Lambda^2}{\Lambda^2}, \qquad (4.7.78)$$

so that $\lambda_\Lambda \simeq \Lambda$ as $\Lambda \to 0$. (Similarly, $\lambda_\Lambda \simeq \log \Lambda$ at $T = 0$.) On the other hand, for $N = 1$ the second part on the right hand side of the flow equations (4.7.69) vanishes. The asymptotics of the first term implies that

λ_Λ is essentially constant and different from zero for scales Λ smaller than the mass of the massive $N = 1$-mode. On these scales the radial massive mode then decouples from the evolution process as it should.

Both in the symmetric and in the broken regime we can mainly distinguish three regions depending on the values for T and Λ.

(1) $T << \Lambda$. This region is always realized at zero temperature, but even up to high temperatures as long as Λ is of the order of $\overline{\Lambda}$, which we have chosen sufficiently large. The flow is then purely four-dimensional. The momentum sum over k_0 in (4.7.64) becomes an integral (4.7.67).

(2) $T \simeq \Lambda$. In the intermediate region the flow is complicated. It depends on details of the functions L_n^4, the flow must be determined numerically.

(3) $T >> \Lambda$. For moderate values of Λ this limit is only realized for high temperatures, but in the far IR even for small temperatures $T > 0$. The sum over k_0 is dominated by the $k_0 = 0$ term. The corrections are exponentially suppressed in T/Λ, because of the decay properties of P_Λ and $\partial_\Lambda P_\Lambda$ in momentum space,

$$L_n^4(x,y) \simeq y L_n^3(x,0). \qquad (4.7.79)$$

The flow both in the broken and in the symmetric regime is purely three-dimensional.

Finally, the symmetry-restoring phase transition of the $O(N)$ models is investigated in the following steps. First, following the flow at $T = 0$, one obtains the physical state described by $\rho_R(T = 0)$ and $\lambda_R(T = 0)$. One always stays in the region (1) of the above list. These zero-temperature numbers are used as reference values. With increasing temperature, the vacuum-expectation value $(\rho_R(T))^{1/2}$ of the field Φ decreases. At some critical temperature T_c, ρ_R will vanish. This signals the restoration of the $O(N)$-symmetry in a second-order phase transition. Above T_c, the renormalized mass $m_R(T) \neq 0$ increases from zero to positive values, while ρ_R stays zero.

The critical exponents of the transition are obtained from a fit of $\rho_R(T)$, $m_R(T)$ and $\lambda_R(T)$ close to T_c against the leading asymptotic behavior that is expected for a second-order transition

$$\rho_R(T) \simeq (T_c - T)^{2\beta} \quad \text{as } T \uparrow T_c$$

$$\left.\begin{array}{l} m_R(T) \simeq (T - T_c)^\nu \\ \lambda_R(T) \simeq (T - T_c)^\zeta \end{array}\right\} \quad \text{as } T \downarrow T_c. \qquad (4.7.80)$$

Some results for the exponents are listed in Table (4.7.1).

N	β	ν	ζ
1	0.25	0.50	0.50
3	0.37	0.75	0.75
4	0.40	0.81	0.81
10	0.46	0.92	0.92

Table 4.7.1 The critical exponents of the finite-temperature Φ^4-model as determined by the flow equations (4.7.69) and (4.7.72).

First of all we see that the critical exponents ν and ζ coincide, a result that is in agreement with the standard scaling relations. Perturbation theory in $\lambda_R(T)/m_R(T)$ thus is IR-finite even at the transition. Furthermore, within an error bar of about 20%, the critical exponents agree with the exponents of the corresponding three-dimensional models. Thus the universal behavior of the four dimensional models at finite temperature seems to be purely three-dimensional, an indication for what is called dimensional reduction.

If one is finally interested only in the far-IR region, it is tempting to start the flow equations directly in their asymptotically three-dimensional form. However, it then becomes even more important to have a sensible ansatz for the effective average action $\widehat{\Gamma}_\Lambda$ at the start, because it must have correctly absorbed renormalization effects along the evolution through the four-dimensional and the intermediate regions (1) and (2). This is an important point related to the issue of dimensional reduction. Dimensional reduction was taken for granted when Tetradis and Wetterich considered further improvements on the scheme directly in three dimensions. The improvement refers to the following points. Parameters which were kept fixed so far are now evaluated as functions of Λ. An example is given by the wave-function renormalization constant. Also higher powers in $(\rho - \rho_\Lambda)$ are included.

As an example for one of these improvements we list results for the critical exponents in Table 4.7.2 that were obtained in three dimensions by keeping powers of ρ up to and including ρ^6 in $U_\Lambda(\rho)$. To assess the quality of the results in comparison with others that were published in the literature, in Table 4.7.2 we also include results for the same critical exponents from high-order ϵ-expansions and high-order linked cluster expansions. Note that the results of these two methods agree quite well with each other, but somewhat less with the results of the flow equations. Thus the question arises how one could systematically improve on the results from flow equations? Here we want to comment on two points.

The first one concerns the inclusion of higher powers in $\rho - \rho_\Lambda$ in U_Λ. The improvement that is achieved this way is limited, because the ansatz for the effective average action $\widehat{\Gamma}_\Lambda$ as a power series in the fields is at best asymptotically convergent. This implies that for a finite coupling constant λ there is a limiting order beyond which the results become less accurate again. With increasing λ this limiting order will decrease, thus one cannot be sure that the additional work in the framework of flow equations actually leads to an improvement of the results. Alternatively one may combine the renormalization group with convergent expansion techniques to get a better control on the truncation and a reliable estimate for the neglected remainder.

The second point concerns the issue of dimensional reduction. Above we have distinguished three regions of the scale $0 \leq \Lambda \leq \overline{\Lambda}$ depending on the ratio of Λ/T. In the "small temperature and/or UV-region" $T/\Lambda << 1$ the flow was essentially T-independent (i.e. purely four-dimensional), while it was three-dimensional in the "high temperature or IR-region" $T/\Lambda >> 1$. In the intermediate range of $T \simeq \Lambda$ the scale dependence was rather complicated. It should be noticed that the quantitative extension of the intermediate range does sensitively depend on the analyticity and decay properties of the cutoff function $R(k, \Lambda)$ and the truncated right hand side of the flow equations. With the appropriate truncation, as long as there are finitely many terms kept, this picture does not change qualitatively. However, this does not imply the same behavior for an infinite system. In general it may be dangerous to ignore the influence of the flow in the four-dimensional and intermediate range of scales on $\widehat{\Gamma}_\Lambda$ and to directly start with the dimensionally reduced flow equations (in three dimensions) with an ansatz that is taken from the defining model say in the far-UV, but just in one less dimension. There are renormalization effects along the intermediate range between four to three dimensions, which are of particular importance for nonuniversal behavior, but also may modify the naive picture of universality classes. For $O(N)$ models the dimensional reduction, as it was used in the flow equations, seems to be safe, because also an investigation by means of linked cluster expansions showed that the critical IR-behavior of four-dimensional finite temperature $O(N)$ models *is* three-dimensional.

N	β	γ	ν	η
1	0.333	1.247	0.638	0.045
	0.327(2)	1.239	0.631	0.038
		1.241(4)	0.630(2)	0.029(8)
2	0.365	1.371	0.700	0.042
	0.348	1.315	0.670	0.039
		1.325(5)	0.673(3)	0.030(11)
3	0.390	1.474	0.752	0.038
	0.366	1.386	0.706	0.038
		1.403(8)	0.713(4)	0.033(12)
4	0.409	1.556	0.791	0.034
	0.382	1.449	0.738	0.036
		1.450(11)	0.736(7)	0.029(17)
10	0.461	1.795	0.906	0.019
	0.440	1.697	0.859	0.024

Table 4.7.2 The critical exponents of the three-dimensional Φ^4-model for various N. We compare the data obtained by the flow equations of the effective average action (upper value) with results obtained by means of the ϵ-expansion (second lines) to six-loop order, see S. A. Antoneko, A. I. Sokolov, Phys. Rev. E **51** (1995) 1894 and J. Zinn-Justin, *Quantum field theory and critical phenomena*, Clarendon Press (Oxford, 1996). The third lines, for $N = 1, \ldots, 4$, are results from high-order linked cluster expansions obtained by one of us, see T. Reisz, Nucl.Phys.B **450** (1995) 569; Phys. Lett. **360 B** (1995) 77.

Without this check by an independent method one could evolve the full flow equations in four dimensions, starting with $\widehat{\Gamma}_\Lambda$ in four dimensions, carrying along the renormalization effects down to the far IR and see whether the three-dimensional exponents are reproduced. But it is an open question and must be analyzed in each case separately, whether similar reductions hold in more realistic and complex theories, in particular for strongly interacting QCD. The problem is closely related to two other issues that will be discussed in section 4.6, the issues of whether dimensional reduction works down to the phase transition and whether it is complete or not.

A further step in the direction of systematic improvements of the flow equation approach first requires to show the very existence of the generating functional beyond the range of perturbation theory and independently of a lattice regularization. As long as further improvements appear inherently difficult both from a conceptual and practical point of view, it is the more important to study the same problems with alternative methods (convergent expansion techniques, Monte Carlo simulations and others) to get an independent control on the validity of the results.

4.7.5 *Appendix: Perturbative renormalization*

A promising application of the flow equations is the very proof of renormalizability of quantum field theories, without having to struggle with the complexities of an approach in terms of Feynman diagrams. The main reason for this simplification is the "1-loop" nature of (4.7.25), it considerably simplifies the inductive part of the proof.

An appealing new field of application and current research are field theories at *finite* temperature. As indicated in section 4.4, to our knowledge there are no strong proofs of renormalizability so far, i.e. proofs beyond some low order computations. From explicit examples one concludes that the counterterms at zero temperature render the field theory UV-finite at nonzero temperature as well. For a conceptual understanding of field theories at finite temperature it is important to know whether this conjecture extends to all orders of perturbation theory. One would expect such an extension if UV- and IR-divergencies decouple because of an intrinsic finite mass scale, by which UV- and IR scales are distinguished within the theory. Moreover, flow equations also appear as appropriate tool to circumvent the IR problems of the standard perturbative approach to gauge theories in the infinite volume.

For the reader who is interested in such conceptual questions concerning finite-temperature field theories, we summarize in this section the main ideas of a proof of perturbative renormalizability which is based on the flow equations. For simplicity we restrict to field theories at zero temperature. The generalization to finite temperatures appears as an appealing open problem that remains to be solved. Let us briefly recapitulate some basic knowledge.

Renormalizability of a quantum field theory implies that the unregularized Green functions

$$\lim_{\Lambda \to 0, \Lambda_0 \to \infty} \widetilde{S}_{I,\Lambda,n}^{\Lambda_0}(p_1, \ldots, p_{n-1})$$

exist in the sense that they are both IR and UV finite. Finite limits are achieved by imposing a finite set of normalization conditions on a physical scale that is independent of the UV cutoff Λ_0. In this way, renormalized (i.e. physical) coupling constants are defined, such as the renormalized mass and the strength of the interaction. The Green functions are functions of the momenta and of renormalized coupling constants. While the IR-and the UV-cutoffs are removed, the renormalized coupling constants are kept fixed.

For simplicity let us consider the case in which all fields are massive to avoid any IR problems for a moment. Proving renormalizability then becomes the issue of showing the existence of the large UV-cutoff limit $\Lambda_0 \to \infty$.

Conventional proofs of renormalizability are based on an approach in terms of Feynman diagrams. They are quite complicated and extended mainly because of two reasons.

First, we need a criterion to determine the convergence of a Feynman integral under quite general conditions. Such a criterion is called power-counting theorem as introduced in section 4.3. Second, a systematic procedure is required to achieve convergence to all orders in perturbation theory, that is to "subtract" the divergencies from all Feynman integrals of the theory. The subtraction is performed in such a way that all conditions of the power counting theorem are met. For a local quantum field theory, it amounts to adding counterterm polynomials in momentum space of the order that is given by the divergence degree. This renormalization can be described recursively or in a closed form, as described by the forest formula of Zimmermann.

It is the combinatorics of counting divergence degrees and of removing divergencies which make the proof rather complex, both for nested and overlapping divergencies. However, altogether the required subtractions should correspond to a finite number of normalization conditions imposed on the Green functions to all orders of perturbation theory.

Next we want to give some hints why a proof that is based on the flow equations, has a chance to circumvent the complicated combinatorics of the Feynman diagrammatic approach. As we have seen above, the flow equations are an infinite set of coupled differential equations. However, in every finite order of perturbation theory, the set becomes finite.

Let us consider the Φ^4-theory in four dimensions with quartic coupling constant λ. In perturbation theory the correlation functions $\widetilde{S}^{\Lambda_0}_{I,\Lambda,n}$ become (formal or asymptotic) power series in λ,

$$\widetilde{S}^{\Lambda_0}_{I,\Lambda,n} = \sum_{\nu \geq \nu_n} \lambda^\nu \, \widetilde{S}^{\Lambda_0(\nu)}_{I,\Lambda,n} \tag{4.7.81}$$

with $\nu_n = \max(1, (n-2)/2)$. The lower bound on the order in λ comes from the fact that every Feynman diagram which contributes to the connected n-point function has at least $(n-2)/2$ vertices (and that the effective interaction satisfies $S_{I,\Lambda}^{\Lambda_0} = O(\lambda)$). Upon inserting (4.7.81) into the flow equations (4.7.29), we get

$$
\partial_\Lambda \widetilde{S}_{I,\Lambda,n}^{\Lambda_0(\nu)}(p_1,\ldots,p_{n-1})
$$

$$
= \frac{1}{2} \int_k \left(\partial_\Lambda \widetilde{D_\Lambda^{\Lambda_0}}(k) \right) \widetilde{S}_{I,\Lambda,n+2}^{\Lambda_0(\nu)}(p_1,\ldots,p_{n-1},k,-k) \tag{4.7.82}
$$

$$
- \frac{1}{2} \sum_{\substack{l,m\geq 1 \\ (l+m=n+2)}} \sum_{\substack{\nu_1,\nu_2\geq 1 \\ (\nu_1+\nu_2=\nu)}} \frac{n!}{(l-1)!(m-1)!} \left[\widetilde{S}_{I,\Lambda,l}^{\Lambda_0(\nu_1)}(p_1,\ldots,p_{l-1}) \right.
$$

$$
\left. (\partial_\Lambda \widetilde{D_\Lambda^{\Lambda_0}}(q)) \, \widetilde{S}_{I,\Lambda,m}^{\Lambda_0(\nu_2)}(-q,p_l,\ldots,p_{n-1}) \right]_{\text{sym}} .
$$

To every order $\nu \geq 1$ of λ this is a finite set of flow equations, depending on a finite number of correlation functions

$$
\widetilde{S}_{I,\Lambda,n}^{\Lambda_0(\nu)} , \; n \leq 2\nu + 2, \tag{4.7.83}
$$

higher-order-n correlations vanish identically. A proof of renormalizability now amounts to show that

$$
\lim_{\Lambda\to 0,\Lambda_0\to\infty} \widetilde{S}_{I,\Lambda,n}^{\Lambda_0(\nu)}(p_1,\ldots,p_{n-1}) < \infty,
$$

for all $1 \leq \nu$, arbitrary ν and arbitrary n with $n \leq 2\nu + 2$. We have to impose normalization conditions to define a renormalized mass, wavefunction constant and coupling constant. A convenient choice is given by (4.7.13). In terms of the $\widetilde{S}$, (4.7.13) becomes

$$
\widetilde{S}_{I,\Lambda=0,2}^{\Lambda_0}(0) = \frac{\partial}{\partial p^2} \widetilde{S}_{I,\Lambda=0,2}^{\Lambda_0}(0) = 0,
$$

$$
\widetilde{S}_{I,\Lambda=0,4}^{\Lambda_0}(0) = \lambda. \tag{4.7.84}
$$

The existence proof is traced back to a proof of the following bounds on the correlation functions and their momentum derivatives, $\partial^w \widetilde{S}_{I,\Lambda,n}^{\Lambda_0(\nu)}$ $(p_1,\ldots, p_{n-1})$, as well as their cutoff dependence. Let $w = (w_{1,0},\ldots, w_{1,3},$ $w_{2,1},\ldots, w_{n-1,3})$ be an arbitrary multi-index, $|w| = \sum_{ij} w_{i,j}$ its length, and ∂^w be the corresponding product of multiple derivatives with respect to the momenta, $|w|$ in number,

$$\partial^w = \frac{\partial^{w_{1,0}}}{\partial p_{1,0}^{w_{1,0}}} \cdots \frac{\partial^{w_{n-1,3}}}{\partial p_{n-1,3}^{w_{n-1,3}}}.$$

Furthermore, let M be an arbitrary finite mass scale, $0 < M < \Lambda_0$. Then the following bounds are satisfied

$$|\partial^w \widetilde{S}_{I,\Lambda,n}^{\Lambda_0(\nu)}(p)| \leq \begin{cases} P_1(|p|) & , \quad 0 \leq \Lambda \leq M, \\ \Lambda^{4-n-|w|} P_2(\log \Lambda) P_3(\frac{|p|}{\Lambda}) & , \quad M \leq \Lambda \leq \Lambda_0. \end{cases}$$
$$(4.7.85)$$

$$|\partial_{\Lambda_0} \partial^w \widetilde{S}_{I,\Lambda,n}^{\Lambda_0(\nu)}(p)| \leq \begin{cases} \Lambda_0^{-2} P_4(|p|) P_5(\log \Lambda_0) , & 0 \leq \Lambda \leq M, \\ \Lambda_0^{-2} \Lambda^{5-n-|w|} P_6(\log \Lambda) P_7(\frac{|p|}{\Lambda}) , & M \leq \Lambda \leq \Lambda_0. \end{cases}$$
$$(4.7.86)$$

The P_i are polynomials with coefficients that depend on ν, n and w, but neither on the cutoffs Λ, Λ_0 nor on the momenta p.

Integration of the bound (4.7.86) over the UV cutoff Λ_0 immediately proves the convergence of the $\partial^w \widetilde{S}_{I,\Lambda,n}^{\Lambda_0(\nu)}(p_1,\ldots,p_{n-1})$ as $\Lambda_0 \to \infty$. Convergence is uniform for all the momenta within a compact region.

The bounds (4.7.85, 4.7.86) are proved by induction, using the flow equations (4.7.82) integrated over the cutoff Λ. A detailed description of the proof is beyond the scope of this book, but let us sketch the idea behind it.

Induction is with increasing $\nu = 1, 2, \ldots$, the order of the coupling λ, but for given ν with decreasing n, starting at $n = 2\nu + 3$ where $\widetilde{S}_{I,\Lambda,n}^{\Lambda_0(\nu)} = 0$. Finally, for given ν, n, one decreases $|w|$, starting at some $|w| \geq 3$. The induction works because on the right hand side of (4.7.82), there are only functions $\widetilde{S}_{I,\Lambda,n+2}^{\Lambda_0(\nu)}$ and $\widetilde{S}_{I,\Lambda,l}^{\Lambda_0(\nu')}$ with $1 \leq \nu' < \nu$. Hence they satisfy the bounds by the induction hypothesis.

Two cases of $(n, |w|)$ must be distinguished, corresponding to relevant and irrelevant coupling constants.

1. $n + |w| \geq 5$ (the irrelevant terms). In this case we realize that on the scale of the UV cutoff the effective interaction vanishes,

$$\partial^w \widetilde{S}_{I,\Lambda_0,n}^{\Lambda_0(\nu)}(p_1,\ldots,p_{n-1}) \equiv 0, \qquad (4.7.87)$$

i.e., there are no irrelevant terms in the bare action. It is thus appropriate to integrate up to Λ_0,

$$\partial^w \widetilde{S}_{I,\Lambda,n}^{\Lambda_0(\nu)}(p) = - \int_\Lambda^{\Lambda_0} d\Lambda' \, \partial^w (\partial_{\Lambda'} \widetilde{S}_{I,\Lambda',n}^{\Lambda_0(\nu)})(p).$$

On the right hand side the flow equation for $\partial_{\Lambda'} \widetilde{S}_{I,\Lambda',n}^{\Lambda_0(\nu)}$ is inserted. Using the induction hypothesis and appropriate bounds on the derivatives of the free propagator $\widehat{D_\Lambda^{\Lambda_0}}(k)$, some algebra reveals the desired bound on $\partial^w \widetilde{S}_{I,\Lambda,n}^{\Lambda_0(\nu)}(p)$.

2. $n + |w| \leq 4$ (the relevant terms). This is the place where the normalization conditions enter. Terms with $n + |w| \leq 4$ are subject to normalization conditions (4.7.84) at $\Lambda = 0$, i.e., on the "full" correlation functions (without IR cutoff). We thus integrate down to $\Lambda = 0$ to get

$$|\partial^w \widetilde{S}_{I,\Lambda,n}^{\Lambda_0(\nu)}(p)| \leq |\widetilde{S}_{I,0,n}^{\Lambda_0(\nu)}(p)| + \int_0^\Lambda d\Lambda' \, |\partial^w (\partial_{\Lambda'} \widetilde{S}_{I,\Lambda',n}^{\Lambda_0(\nu)})(p)|. \quad (4.7.88)$$

Again the flow equations should be inserted in the second term. An apparent complication arises by the first term, because the normalization conditions are imposed at some fixed momentum ($p = 0$ in our case), but we need an upper bound for all momenta. This is achieved by applying a Taylor expansion with remainder term about the normalization points. Terms which get accompanied by additional momentum derivatives are then estimated by the induction hypothesis or by applying the former arguments of 1.

The second pair of bounds (4.7.86) is proved in a similar way. To summarize: There are two central ingredients in the proof. First, normalization conditions must be imposed on all interactions $\partial^w \widetilde{S}_{I,\Lambda=0,n}^{\Lambda_0(\nu)}$ with $|w| + n \leq 4$ (except those which vanish or are related by a symmetry). These are precisely the relevant parts of the effective interaction. Second, irrelevant interactions, with $|w| + n \geq 5$, vanish on the UV scale Λ_0. Actually, (4.7.87) can be replaced by the weaker bound

$$|\partial^w \widetilde{S}_{I,\Lambda_0,n}^{\Lambda_0(\nu)}(p_1, \ldots, p_{n-1})| \leq Q_1(\log \Lambda_0) \, Q_2(\frac{|p|}{\Lambda_0}) \, \Lambda_0^{4-(n+|w|)}, \quad (4.7.89)$$

with polynomials Q_1 and Q_2 with w, n, ν dependent coefficients. This implies that the derivatives of $\widetilde{S}$ rapidly vanish as the UV cutoff is removed.

Chapter 5

Numerical Methods in Lattice Field Theories

5.1 Algorithms for numerical simulations in lattice field theories

Monte Carlo on the lattice. In a Monte Carlo procedure a set of field configurations $\{U(x;\mu)\}^{(1)}, \ldots, \{U(x;\mu)\}^{(N)}$ is generated such that the Boltzmann factor is absorbed in the selection of configurations. For a positive-definite and real action $S(U)$ the expectation value of an observable O defined as

$$\langle O \rangle = \frac{\int \prod_{x,\mu} dU(x;\mu) \, e^{-\beta S(U)} \, O(U)}{\int \prod_{x,\mu} dU(x;\mu) \, e^{-\beta S(U)}}, \tag{5.1.1}$$

cf. (3.3.45) (a lattice version of Eq. (3.3.5) for a pure gauge theory) equals

$$\langle O \rangle = \lim_{N \to \infty} \frac{1}{N} \sum_{i=1}^{N} O(U^{(i)}) \tag{5.1.2}$$

and is approximated by the arithmetic average over a finite number of configurations,

$$\langle O \rangle \approx \langle O \rangle_N = \frac{1}{N} \sum_{i=1}^{N} O(U^{(i)}). \tag{5.1.3}$$

The observables in Eq. (5.1.3) are evaluated on an ensemble of configurations which are representative of the coupling (temperature) β^{-1}. The configurations are selected according to the probability distribution

$$P(U)\mathcal{D}U \equiv \frac{e^{-\beta S(U)}}{Z_g} \mathcal{D}U \,, \tag{5.1.4}$$

where $\mathcal{D}U$ and Z_g are given by (3.3.40) and (3.3.42), respectively. The fundamental idea of the Monte Carlo method is stated by the law of large numbers, which tells us that under very general conditions on the probability measure $\lim_{N \to \infty} \langle O \rangle_N = \langle O \rangle$.

The sequence of configurations is generated with a Markov process . An easy way of constructing a Markov process, that has a particular distribution P as its fixed point, is to choose the transition probabilities $Q : (U^{(i)} \longrightarrow U^{(f)})$ from an initial configuration $U^{(i)}$ to a final configuration $U^{(f)}$ to satisfy the detailed balance condition

$$P(U^{(i)}) \, Q \left(U^{(i)} \longrightarrow U^{(f)} \right) \mathcal{D}U$$
$$= P(U^{(f)}) \, Q \left(U^{(f)} \longrightarrow U^{(i)} \right) \mathcal{D}U \,. \tag{5.1.5}$$

The fixed point distribution is in our application the Boltzmann equilibrium distribution. A particular way of realizing the detailed balance condition is the Metropolis algorithm [208].

The Metropolis algorithm
The Metropolis algorithm is an example of a local updating procedure. One starts from an initial configuration, which may be chosen ordered ($U(x; \mu) = 1$ for all links (cold start)) or disordered ($U(x; \mu) \in SU(N_c)$ random for all links (hot start)), or mixed. A single new link variable $U(x; \mu)'$ is then chosen randomly and always accepted as a replacement for the old variable if it lowers the action, i.e. $\Delta S = S(U') - S(U) < 0$. Otherwise the change is accepted with a conditional probability. For $\Delta S > 0$ a random number r with $0 \leq r \leq 1$ is selected. If $r < e^{-\Delta S}$, the new variable U' is still accepted for replacing the old one, otherwise not, one goes back to $\{U(x; \mu)\}$ and repeats the steps as indicated above. For readers who are interested in the actual source code we refer to [209].

In this way all links of the lattice are changed, either randomly or successively. Such a sweep through the entire lattice is counted as one Monte Carlo iteration. Usually hundreds or thousands of such iterations are necessary, before the equilibrium distribution is reached. The expectation value of the observable is then obtained according to Eq. (5.1.3), where the sum extends only over the last N equilibrated (or thermalized) configurations.

When fermions are included, one encounters a technical problem due to the nonlocality of the effective action Eq. (3.3.75). The nonlocality refers to the det Q-term or equivalently to the tr ln Q-term. For staggered fermions the matrix Q has $(3L_s^3 L_0)^2$ complex elements. In the local Metropolis algorithm this matrix should be calculated for every link in every iteration, which renders it impracticable.

The hybrid Monte Carlo algorithm

In the last decade much effort has been invested in improving algorithms for QCD with dynamical quarks (for a review see e.g. [210] or [211]. Integration over the fermions led to the factor $\det^l a^4 - (Q)$ in Eq. (3.3.73). While Q was a local operator, coupling only nearest and next to nearest neighbors on the lattice, ln Q is nonlocal. A calculation of det Q with $\exp\{-S_W\}$ as Boltzmann factor should be avoided. The idea is to bosonize det Q and to "lift" Q in the exponent. The bosonization is performed with the *pseudofermion method* [212].

The pseudofermion method

The pseudofermion method is based on the following formula for the determinant:

$$\det Q = \frac{1}{\det Q^{-1}} \int \mathcal{D}\lambda^* \, \mathcal{D}\lambda \, e^{-\sum_{x,y} \lambda_x^* Q_{xy}^{-1} \lambda_y} . \tag{5.1.6}$$

Here λ, λ^* are complex bosonic (pseudofermionic) variables. Note that Q has to be a positive-definite matrix for the Gaussian integral in Eq. (5.1.6) to converge. If we identify Q with the fermionic matrix Q of Eq. (3.3.70), Q has negative eigenvalues. This is the reason why Q is replaced by Q^+Q at the cost of doubling the number of fermionic flavors or (for compensating this doubling) by $\hat{Q}^+\hat{Q}$, where χ fields live on even and $\bar{\chi}$ fields on odd sites. (Another reason for the replacement is that it is convenient for easily generating the pseudofermion fields from a heat bath.) Via the doubling $(Q \to QQ^+)$ configurations leading to negative eigenvalues of Q are mapped on the same contribution to the effective action as those with positive eigenvalues, although ln det Q in the original effective action is not given as tr ln Q for negative eigenvalues. Thus the error entering this way is difficult to control.

Like ln Q, the action in Eq. (5.1.6) is nonlocal. Hence we see, when the bosonic part $(-\sum_{x,y} \lambda_x^* Q_{xy}^{-1} \lambda_y)$ is added to the gauge part of the QCD action to account for fermions, we need a nonlocal updating procedure. A practical method, which is free of uncontrolled errors and often used nowadays in QCD applications, is the *hybrid Monte Carlo algorithm* [213]. It has to be combined with the pseudofermion representation to yield a feasible simulation of QCD. The hybrid Monte Carlo algorithm is a stochastic hybrid of a Langevin and a microcanonical algorithm, for which an additional Monte Carlo acceptance test has been built in to remove the systematic errors of the stochastic hybrid. The stochastic hybrid algorithm combines the virtues of the Langevin and the microcanonical (or molecular dynamics) algorithms. Both algorithms use equations of motion to select new variables.

In the **Langevin algorithm** [214, 215] and [216], the equation of motion is a partial differential equation, which is of first-order in time. The finite step size in time entering the discretized version has to be extrapolated to zero by varying its size. This is time consuming and an unwanted feature.

The microcanonical algorithm

The microcanonical algorithm [217,218] is based on the observation that the Euclidean path integral of a quantum field theory in four dimensions can be written as a partition function for a system of classical statistical mechanics in four spatial dimensions with a canonical Hamiltonian that governs the dynamics in a fifth new 'time' variable. This time τ may be identified with the simulation time. The algorithm determines the new variables $\Phi(\tau, x)$ and their canonically conjugate momenta $\pi(\tau, x)$ in a fully deterministic way. Here one moves faster through phase space $(\phi, \pi) \rightarrow (\phi', \pi') \rightarrow \cdots$, but at the price of losing ergodicity.

The stochastic hybrid algorithm

The stochastic hybrid algorithm [219] interrupts the integration of Hamilton's equation of motion along a single trajectory with fixed energy. The interruption is made once in a while for a refreshment of momenta. The new parameter, which enters and should be optimized, is the frequency of momenta refreshment. If it is low, the algorithm is as slow as the Langevin algorithm, if it is too high, ergodicity will be violated. Yet there is a systematic error in the hybrid algorithm introduced by the finite time step in the integration procedure, which leads to a violation of energy conservation.

The idea of the *hybrid Monte Carlo algorithm* is to absorb this energy violation in a 'superimposed' Monte Carlo procedure. The phase-space configurations at the end of every molecular dynamics chain are subjected a Metropolis acceptance test. If the time step is too large, the energy violation too strong, the (Φ, π) configuration is likely to be rejected. When the energy is conserved, the configuration is always accepted. In this way the systematic error due to finite integration steps is eliminated.

Above we have sketched a number of intermediate steps in developing the hybrid Monte Carlo algorithm in order to give an idea of how complex the problem of finding efficient algorithms is. Although the hybrid Monte Carlo algorithm moves the system quickly through configuration space, it is slow due to a matrix inversion, which enters the equations of motion. The matrix inversion is usually performed with the conjugate gradient method. The number of conjugate gradient steps is proportional to (m_f^{-1}) and takes more than 90% of the CPU-time. The hybrid Monte Carlo acceptance rate is proportional to $(m_f^{-3/4})$ and the autocorrelation time $\propto m_f^{-1}$ [220]. For example, one lattice update on a $16^3 \times 8$ lattice with $m_f = 0.01$ took about 20 min on a 200 MFlop machine. Here we see why it is so difficult to work with small bare-quark masses m_f. Quark masses should be as small as possible to avoid an uncontrolled extrapolation to the chiral limit, but some extrapolation is unavoidable.

More generally we see why Monte Carlo simulations of lattice QCD are so time consuming when dynamical fermions are included. It is extremely hard to satisfy the following requirements simultaneously:

- The masses are small enough to guarantee a safe extrapolation to the chiral limit.
- The volume is large enough to avoid finite-size effects.
- The bare coupling is small enough to be in the asymptotic scaling regime.

The reader who is interested in further details, in particular in the source code written in C, will find the code, freely available, under [256].

Our selection is far from being complete. For the reader who is interested in getting involved in actual simulations of lattice gauge theory we mention a few other algorithms that belong to the standard repertoire nowadays. Working close to the continuum limit or in the vicinity of phase transitions implies large correlation lengths growing with the lattice size. Growing correlation in space goes along with growing correlation in time, but physical observables should be measured on independent field configurations. The number of iterations needed to obtain a new independent field configuration is called the autocorrelation time τ, τ is related to the correlation length ξ according to $\tau \propto \xi^z$. The autocorrelation time obviously depends on the algorithm. For local algorithms like the Metropolis or heat-bath algorithm $z = 2$, so for a correlation length of ten lattice units, only every hundred configurations one obtains an independent configuration that should enter the average of an observable. This phenomenon is called *critical slowing down*. Fight against it led to proposals of new algorithms with smaller exponent z. Such algorithms are cluster algorithms (applicable only to spin systems) [257], multigrid algorithms [258] with modest improvement for nonabelian gauge theories, and so-called overrelaxation algorithms , initiated by S.Adler in [259] and continued in [260, 261]. The overrelaxation is a generalization of the heat-bath algorithm, combined with microcanonical updates at a given mixing ratio. The algorithm contains a relaxation parameter ω, $0 < \omega < 2$, for $\omega = 1$ it reduces to the heat bath and for $\omega \to 0$ to the Langevin approach. Multicanonical algorithms have been developed in [221] . They are particularly useful for measuring the interface tension at first order phase transitions, for which the probability distribution of the order parameter typically shows a double peak structure with two peaks separated by a valley, corresponding to the coexisting phases at the transition. For large volumes it is notoriously difficult to generate configurations corresponding to the valley between the two peaks. These configurations are suppressed by the large amount of surface free energy that is just the quantity of interest. The reason is that usual local upgrading procedures, like the Metropolis or heat bath algorithms, are designed for single- peaked distributions $P(\Omega)$. Configurations are selected with an importance sampling according to their Boltzmann weight. A precise calculation of the maxima and minima of the double peak structure of $P(\Omega)$ requires frequent tunneling through the minimum, implying exponentially large autocorrelation times $\tau \propto \exp\{2\sigma L^{d-1}\}$ with L the linear size of the system and σ the surface tension. The multicanonical algorithm circumvents the problems of local algorithms. In [209] the interested reader finds detailed explanations with examples of the source code and further applications of the multicanonical approach.

5.2   Pitfalls on the lattice

- **Finite-size effects** Finite-size effects are specific for the Monte Carlo approach, which is frequently used in lattice calculations. They contaminate any numerical calculation which is performed in a finite volume, if the largest correlation length of the system is not small compared to the smallest linear size. In the case of first-order transitions we have presented a phenomenological approach (cf. section 2.6). The formulas of section 2.6 for the scaling of the specific heat or the susceptibility as function of the linear size L hold in the *zero external field limit* $h = 0$. As we have seen in section 3.3, Monte Carlo calculations of QCD with dynamical fermions are necessarily performed at finite bare quark masses. The updating time is estimated to increase proportional to $m_f^{-11/4}$. The chiral limit $m_f \to 0$ must be extrapolated. Thus a second pitfall is a contamination of data through finite mass effects.

- **Finite-mass effects.** In section 2.5 we have stressed the similarity between the finite volume L^3 and an external field h, and between an external field and finite quark masses. In the vicinity of T_c one may consider a *finite mass scaling analysis* at vanishing $1/L$ rather than a *finite size scaling analysis* at vanishing h. For a moment let us assume that the condition $1/L \to 0$ is satisfied to a sufficient accuracy. A mass-scaling analysis then provides a tool for controlling the rounding of algebraic singularities in correlation functions due to finite quark masses. The precondition is a second-order phase transition. (An analogous analysis may be performed for a first-order transition and rounding effects due to finite masses as well). The need of extrapolation to zero masses is specific for exploring the chiral limit, when dynamical fermions are included.

A third class of pitfalls which is known from the pure $SU(N)$ gauge theory and is common to any discrete version of a continuum model are artifacts due to the UV-cut-off.

- **Artifacts due to the UV-cut-off.** Here we distinguish three manifestations. The most obvious one occurs in the interpretation of any physical observable, when the lattice units are translated to physical units. In section 3.3 we have already argued about the ambiguity in predicting T_c [MeV]. Such an ambiguity occurs for the QCD transitions in the presence of dynamical quarks, when the quark mass in lattice units, e.g. $ma = 0.025$, shall be translated in units of [MeV], although the lattice is rather coarse grained and a splitting of a mass according to $m \cdot a$ is not unique.

A less obvious and more subtle UV-artifact is a change in the effective symmetry group of the lattice action as a function of the bare coupling. This artifact is lately a consequence of the 'No-Go'-theorems referring to lattice regularized actions for fermions. As we have mentioned in section 3.3, the lattice action with massless fermions has a global $U(n) \times U(n)$ symmetry for n species of staggered fermions. Only in the continuum limit the full $SU(N_f) \times SU(N_f)$ flavor symmetry will be restored with $N_f = 4 \cdot n$. Far outside the continuum region the lattice action has only the global $U(1) \times U(1)$ symmetry for $n = 1$.

A global $U(1) \times U(1)$ symmetry may trigger a second-order phase transition with $O(2)$- critical exponents, whereas the spontaneous breaking of the restored $SU(4) \times SU(4)$ symmetry in the corresponding continuum limit is supposed to induce a first-order transition [225]. Thus the order of the chiral transition may change when passing from strong to weak couplings. Such a possibility has been discussed by Boyd et al. (1992) [224] and will be the topic of a later section.

If one expects an $SU(2) \times SU(2)$ flavor symmetry for the continuum limit, the continuum phase transition would be also of second order. In that case only the critical exponents would change in passing from strong to weak couplings.

If one wants to describe an odd number of flavors in the continuum limit within the staggered fermion formulation, one usually "corrects" for the desired number of continuum flavors on the level of determinants. The correction is performed after integrating over the fermionic degrees of freedom, whereas the fermionic operator Q is that of the original lattice action. The flavor symmetry in the original action is broken and describes at least four flavors in the continuum limit. The representation of the prefactor of the determinant-term in Eq. (3.3.75) as $N_f/4$ cannot be derived. It should be considered as a *prescription* when N_f is not an integer multiple of 4. In this case it is even more difficult to infer the symmetry that drives the phase transition and leads to a vanishing condensate above T_c.

So far the UV-effects are supposed to modify the critical indices or the order of the chiral transition. A third manifestation of the UV-cut-off are *bulk transitions*. Here the very transition itself is an artifact of the lattice. Bulk transitions are not restricted to QCD with dynamical fermions. They occur in pure $SU(N)$ gauge theories as well. As an illustration we consider the example of $N_f = 8$-flavor-QCD in section 5.4.3. For eight flavors the bulk transition is either dominant or superimposed on the usual finite temperature transition of QCD.

5.3 Pure gauge theory: The order of the $SU(3)$-deconfinement transition

The deconfinement transition in a pure $SU(3)$ gauge theory is nowadays believed to be of first order (see [223] and references therein). Over several years there was a controversial discussion about the order of the transition in the literature. It was due to the subsequent use of different criteria to determine the order from a numerical analysis. We reported the history of this controversy in [262]. In the first part of this section we summarize two possible origins for such a debate.

We know the general criteria from finite size scaling analysis from the second chapter, sections 2.6, 2.5. In the second part of this section we also identify the specific observables (correlation functions of Polyakov loops) that were measured in the case of the pure $SU(3)$ gauge theory.

5.3.1 *The order of limits*

Let us recall criteria for first- and second-order phase transitions. Criteria for first-order transitions are signs of metastabilities, hysteresis effects, two-state-signals or jumps in thermodynamic quantities. In early simulations these criteria were incautiously used in the sense that the results for signatures were taken for granted from measurements for a single lattice size and on rather small lattices compared to present standards. A finite size scaling analysis may be applied to bulk and "other" quantities. Examples of bulk quantities are the internal-energy density ε, the specific heat c or the order parameter susceptibility χ. "Other" quantities refer to distribution functions of the order parameter $P(O)$, and derived moments or cumulants like the Binder cumulant. Finite size corrections to the average internal energy are *exponentially* suppressed. The shifting and rounding effects in the peaks of c and χ are of the order of $1/V$ in first-order transitions as we have derived in section 2.6. For identifying the order of the transition, quantities like the Binder cumulant have power-law finite-size corrections and are, therefore, applicable only for very large volumes, for which these power-law corrections are suppressed. Similarly, correlation lengths show power law corrections. Even their very definition in a finite volume is rather intricate compared to bulk quantities. Their practical utility turned out to be questionable as we show below.

5.3.2   Correlation lengths, mass gaps and tunneling events

We use a notation, which is independent of the application in a spin- or gauge model. We will distinguish *physical* and *tunneling correlation lengths* $\xi^{(p)}$ and $\xi^{(t)}$ with associated mass scales $m^{(p)}$ and $m^{(t)}$ both in the finite and the infinite volume. A physical or bulk correlation length in the infinite volume limit can be defined via the decay of the connected pair-correlation C_c as function of the distance z according to

$$C_c(z) = C(z) - B \xrightarrow[z \to \infty]{} A \cdot e^{-m^{(p)}z} \qquad (5.3.1a)$$

$$\xi^{(p)} \equiv 1/m^{(p)} . \qquad (5.3.1b)$$

Here C is the unsubtracted, disconnected pair correlation and B stands for the constant disconnected part, which is different from zero in the phase of spontaneously broken symmetry. The inverse correlation length $\xi^{(p)-1} = m^{(p)}$ is called the *physical mass gap* .

If we are now in a finite volume and have different degenerate vacua in the phase of broken symmetry, a second correlation length must be taken into account, the tunneling correlation length. Roughly speaking its physical meaning is the typical extension of regions of aligned spins (if we consider a spin model), that is the diameter of correlation volumes. More precisely, it arises in an analysis of the transfer matrix . One expects in general for the (disconnected pair) correlation function that it is a coherent sum of its eigenvalues. If we consider only the two smallest eigenvalues above the ground state, the correlation $C(z)$ may be written as

$$C(z) \sim a_1 \, e^{-z/\xi^{(p)}} + a_2 \, e^{-z/\xi^{(t)}} + \dots , \qquad (5.3.2)$$

which should be compared with Eq. (5.3.1a). Equation (5.3.2) defines the tunneling correlation length $\xi^{(t)}$, or more precisely, $\xi^{(t)}$ arises in the spectrum of the transfer matrix from the level splitting associated with tunneling between different degenerate vacua. As long as $\xi^{(t)}$ yields the dominating contribution in (5.3.2), a fit of $C(z)$ according to

$$C(z) \longrightarrow A \, e^{-mz} + B \qquad (5.3.3)$$

rather than according to (5.3.2) leads to a behavior of m which reflects the increase of $\xi^{(t)}$ instead of $\xi^{(p)}$, the physical correlation length with inverse mass gap $\xi^{(p)-1}$. This tunneling has a finite probability as long as the volume is finite, resulting in a finite $\xi_L^{(t)}$, but the potential barrier between the degenerate ground states becomes infinitely high in the infinite volume limit, leading to

$$\lim_{L\to\infty} \xi_L^{(t)} = \infty \,. \tag{5.3.4}$$

From Eq. (5.3.2) we recognize that as $L \to \infty$ the diverging $\xi^{(t)}$ gives rise to the constant disconnected term B in the correlation function Eq. (5.3.1a).

In a *finite volume* both length scales $\xi^{(p)}$ and $\xi^{(t)}$ are necessarily finite and (when considered at criticality, cf. Eq. (5.3.6)) both increase with increasing volume at a first-order transition. For the tunneling correlation length the divergence is clear, because the barrier between degenerate vacua goes to infinity in the infinite volume limit. Let us recapitulate when also the physical correlation length $\xi^{(p)}$ diverges for first-order transitions. The essential point is the order of limits $\lim_{\beta\to\beta_c}$ and $\lim_{L\to\infty}$.

In the conventional lore it is the infinite-volume correlation length which diverges as $\beta \to \beta_c$ for a second-order transition and stays finite in the first-order case.

$$\lim_{\beta\to\beta_c^-} \lim_{L\to\infty} \xi_L^{(p)}(\beta, L)$$

$$= \lim_{\beta\to\beta_c^-} \xi^{(p)}(\beta) \begin{cases} \to \infty & \text{2nd order} \\ \to \text{const} < \infty & \text{1st order.} \end{cases} \tag{5.3.5}$$

This is the order of limits to which the standard distinction between first- and second-order limits refers. Monte Carlo simulations are performed in a finite volume, so that $\lim_{L\to\infty}$ anyway can be realized only approximatively. In the opposite order of limits, however,

$$\lim_{L\to\infty} \lim_{\beta\to\beta_c^-} \xi_L^{(p)}(\beta, L)$$

$$= \lim_{L\to\infty} \xi_L^{(p)}(\beta_c, L) \begin{cases} \to \infty \text{ 2nd order} \\ \to \infty \text{ 1st order.} \end{cases} \tag{5.3.6}$$

The 'critical' correlation length $\xi_L^{(p)}(\beta_c, L)$ diverges at β_c in the infinite volume limit also in the case of a first-order transition [222]

This behavior is familiar, if we recall from the second chapter that singularities in second derivatives of thermodynamical potentials occur for first-order transitions as well. The only difference in the first-order case is the type of divergence: δ- function singularities may occur in the specific heat and the magnetic susceptibility, when the infinite volume limit is taken *at* criticality. Thus a diverging correlation length $\xi_L^{(p)}(\beta_c, L)$ for $L \to \infty$ is not conclusive for a second-order transition.

As we have seen, there is a twofold risk of misinterpreting an increasing correlation length: the order of limits has not been properly arranged, and $\xi^{(t)}$ is intertwined with $\xi^{(p)}$.

Adaptation to periodic boundary conditions

In the discussion so far we have not mentioned an adaptation of the ansatz for the correlation length (whatever its meaning is) to the periodic boundary conditions. A finite-volume ansatz for $C(z)$ which was inspired by the infinite volume decay and adapted to these boundary conditions on a cubic lattice is

$$C_i(z) = A_i \left[\exp(-z/\xi)/z^i + \exp(-[L-z]/\xi)/(L-z)^i\right], \qquad (5.3.7)$$

where ξ stands for a generic correlation length, and $i = 0, 1$. Equation (5.3.7) includes only the first image of a source in z-direction, the one which is a distance L apart. Multi-mirror images (accounting for images a distance $n \cdot L$ apart) should be also included, if the correlation can become large compared to the considered volume.

5.3.3 *Correlation functions in the pure $SU(3)$ gauge theory*

Next let us consider concrete realizations of the (disconnected) correlation function C in the pure $SU(3)$ gauge theory. In analogy to Potts models, where the spin expectation value plays the role of an order parameter and C refers to spin-spin correlations, we have now to deal with correlations of pairs of Polyakov loops. Let n denote the spacelike coordinates $n = (x_1, x_2, x_3)$, Ω_n be the Polyakov loop at n, i.e.

$$\Omega_n = \frac{1}{3} \operatorname{tr} \left(\prod_{\tau=1}^{L_0} U^{(0)}_{(n,\tau)}\right), \qquad (5.3.8)$$

that is the trace of the product over gauge field variables $U^{(0)}_{(n,\tau)} \in SU(3)$ associated with timelike links leaving the site n, τ in direction 0. To project on zero momentum, consider the average $\bar{\Omega}_i(x_i)$ of Polyakov loops Ω_n, $n = (x_i, x_j, x_k)$ over a spacelike plane, spanned in directions $j, k, j\!\!\!/, k\!\!\!/$, $i, j, k \in \{1, 2, 3\}$, located at spacelike coordinate x_i, where the sum in $\bar{\Omega}_i(x_i)$ runs over x_j, x_k, and correlate $\bar{\Omega}_i(x_i)$ with $\bar{\Omega}_i^\dagger(x_i + \hat{x})$, a distance $\hat{x}$ apart. The Polyakov-pair correlation then results from an average over the spacelike directions and the locations x_i, $i = 1, \ldots, N_s$ of the (j, k) planes according to

$$C_0(\widehat{x}) = \frac{1}{3N_s} \sum_{i=1}^{3} \sum_{x_i=1}^{N_s} < \bar{\Omega}_i(x_i) \, \bar{\Omega}_i^+(x_i + \widehat{x}) > \, , \qquad (5.3.9)$$

N_s denotes the extension in spacelike directions. The additive constant B of Eq. (5.3.1a), that can be split off in $C_0(\widehat{x})$, is given by $| < \Omega > |^2$. In the $SU(3)$ gauge theory, the susceptibility of the order parameter, i.e. the Polyakov loop, reads

$$\chi = V \left[< (Re\Omega)^2 > - < Re\Omega >^2 \right]. \qquad (5.3.10)$$

Here Ω is defined as in Eq. (5.3.8), where Re Ω is taken as the projection of $(\sum_n \Omega_n / V)$ onto the nearest $Z(3)$-axis on the complex Ω-plane. The results of [223] for the maximum in the susceptibility lead to a scaling of $\chi_{\max} \propto V^{\gamma/(d \cdot \nu)}$ with $\gamma/(d \cdot \nu) = 0.99(6)$ and a width $\Delta\beta \propto V^{-1/(d\nu)}$ with $1/(d\nu) = 0.95(5)$ in reasonable agreement with the expected values of 1 for a first-order transition. Further signals of a first-order transition were observed in [223] such as two-state signals . Likewise the double-peak structure in the order parameter distribution $P(|\Omega|)$ (with Ω the Polyakov loop of (5.3.8) was more pronounced for a larger lattice size. (Recall that in case of a second-order transition the double-peak structure (if there were any initially) would weaken and finally fade away for $V \to \infty$.) Therefore a variety of signatures in the Monte Carlo simulations of [223] supported the original conjecture of a first-order deconfinement transition in the $SU(3)$ gauge theory.

5.4 Including dynamical fermions

The ambiguity in studying the order of the deconfinement transition in the $SU(3)$ gauge theory was caused by the finite volume. A finite-size scaling analysis resolved the controversy. It is natural to attempt a finite-size scaling analysis in the presence of dynamical fermions as well. Other artifacts can superimpose on finite-size effects. First we report on the finite-size scaling analysis of the chiral transition of [263]. Possible effects of finite quark masses are disregarded, although finite masses amount to an ordering effect, which may well compete with the finite volume. Therefore we turn in the next subsection to a finite-mass scaling analysis. It serves, for example, to discriminate possible UV artifacts in the chiral transition. *Bulk transitions* are a further manifestation of UV-effects. We mention the example of $N_f = 8$-flavor QCD for bulk transitions. Results for the more realistic cases of two and three flavors are summarized afterwards.

5.4.1　*Finite-size scaling analysis*

For a study of the chiral transition in the presence of fermions one has to specify the number of flavors, the bare quark mass, entering the lattice action, the lattice extension in space (L_s) and in time (L_0) directions, the fermionic algorithm and its time step size, and last but not least, the order parameters. As order parameters are taken the chiral condensate $< \bar{\chi}\chi >$ and the Polyakov loop Ω, although they strictly have this meaning only in the limiting cases $m_f \to 0$ and $m_f \to \infty$, respectively. The chiral condensate can be expressed in terms of pseudofermionic fields h (these are complex three-vectors of Gaussian random numbers) that occur in the effective fermionic action (3.3.75). The Polyakov loop expectation value is calculated as

$$\langle Re\Omega \rangle = \frac{1}{L_s^3} \sum_n \langle Re\Omega(n) \rangle \tag{5.4.1}$$

with $n = x_i, x_j, x_k$, $i, j, k = 1, 2, 3$ the spacelike coordinates, $\Omega(n)$ a single Polyakov loop as in (5.3.8) and the expectation value $\langle \ldots \rangle$ calculated with the measure and the effective action of (3.3.75). In view of the finite-size scaling analysis the spatial size is varied between several values for otherwise fixed parameters. For a concrete example the reader may look at [264].

First one would look for signs of metastabilities and associated two-state signals in time histories of the Polyakov loop $Re\Omega$ for given choices of N_f, m_f, L_s and a given number of simulation steps. The time unit is set by one trajectory in the R-algorithm . The values of $Re\Omega$ will fluctuate as a function of time. Clear flip-flops indicate tunneling events between the "ordered" and "disordered" phases, metastabilities are visible as a two-state separation in the sense that the system remains in the ordered phase over thousands of iterations, starting from an ordered start and in the disordered phase starting from a random start. A crucial ambiguity is hidden in what is called "long time". A comparison of concrete data (see [264]) shows a significant difference in time scales. Fluctuations in the order parameter value look more irregular on a large scale. On a short time scale the same fluctuations may be (mis)interpreted as a two-state signal. A check of the dependence on the initial conditions is crucial as well. From a single run one can never exclude the possibility that a flip-flop will occur, if one waits for a long enough time. For large volumes the metastabilities can be so pronounced that the system is in one phase over the entire simulation. The phase is interpreted as a truly stable phase and a first-order transition is easily overlooked. Thus the choice of the volume has to be optimized. It should be neither too small to see distinct flip-flops nor too large to see them at all.

Further indications for the first-order transition are taken from the finite-size scaling behavior of various susceptibilities. Susceptibilities are 'magnetic' response functions that can be expressed in terms of an order parameter field $O(x)$ according to

$$\chi = \frac{1}{V} \sum_{x,y} ((O(x)O(y)) - (O(x))(O(y))) , \qquad (5.4.2)$$

where the notation has to be adapted to the choice of order parameter. In this context O may stay for $Re\Omega$ (Ω being the Polyakov loop), the chiral condensate, or the average plaquette $P \equiv trU(\partial P)$, where $U(\partial P)$ denotes the product of U-variables along the boundary of a plaquette. As we know from section 2.6, the peak in the susceptibility χ_{max} should scale with the volume according to

$$\chi_{max} = const + a \cdot V^p , \qquad (5.4.3)$$

where $p = 1$ for a first-order transition and $p < 1$ for a second order transition. The constant accounts for a contribution from the regular part of the free energy, which cannot be neglected as long as the volume is relatively small. Furthermore the Binder cumulant can be tested (see section 2.6, Eq. (2.6.62) with the energy density e replaced by the average plaquette P. Recall that the minimum should approach $2/3$ for a second-order transition as the volume increases.

To a certain extent the actual results of [264] for these observables were not yet conclusive. The authors of [264] suggested that the relatively large quark mass of $m_f = 0.025$ (in lattice units) might interfere with the finite size of the system. Such interference calls for a finite-mass scaling analysis, that is the topic of the next section.

5.4.2 *Finite-mass scaling analysis*

So far we were concerned about infrared artifacts induced by the finite volume. In this section we discuss a finite-mass scaling analysis that was used to study UV-artifacts of the chiral transition [224] dealing with possible changes in critical indices and a change in the order of the chiral transition as a function of the bare coupling. Mean-field calculations of [226, 227] and [228] for $SU(2)$ suggested that the chiral QCD transition is of second-order in the strong coupling limit. Second order transitions are immediately washed out in the presence of an external field, as we know from statistical physics. The singularities in thermodynamical functions will be rounded, when the infinite volume limit is taken at fixed, nonvanishing quark mass. An extrapolation to zero masses via a finite-mass scaling analysis allows to determine the universality class in the strong-coupling limit and to compare the results with those for weaker couplings. Such a comparison will answer the question as to whether the observed behavior at weaker couplings represents continuum physics or resembles strong-coupling artifacts. In particular, from the renormalization group analysis of Pisarski and Wilczek we expect a second-order chiral transition with $O(4)$-exponents for $SU(2) \times SU(2)$ continuum symmetry, while continuous behavior as a remnant of the strong coupling $U(1) \times U(1)$ symmetry restoration would lead to $O(2)$-exponents.

The critical index characterizing the "finite-mass scaling" at criticality is $1/\delta$ (see the dictionary at the end of section 3.1). It is defined as

$$\lim_{L_s \to \infty} \langle \bar{\chi}\chi \rangle (m, t, L_s)|_{t=0} \propto m^{1/\delta} , \qquad (5.4.4)$$

where t stands for the reduced temperature $t = (T - T_c)/T_c$ (in the actual simulations of [224] it was replaced by an analogous parameter called γ). The order parameter

$$\langle \bar{\chi}\chi \rangle = \frac{1}{L_s^3 L_0} \frac{\partial}{\partial m} \ln Z(m, \gamma) \qquad (5.4.5)$$

is the chiral condensate of strong coupling QCD, χ stands for the staggered fermion fields. It should not be confused with the susceptibility χ, but confusion can be avoided, since the fermion fields occur only in the combination $\bar{\chi}\chi$.

Further recall the critical index γ which specifies the singular behavior of the susceptibility χ, when T approaches T_c at zero field (here zero mass)

$$\chi(t) = \lim_{m \to 0} \lim_{L_s \to \infty} \chi(t, m, L_s) \propto t^{-\gamma}$$

$$\chi(t, m, N_\sigma) = \frac{\partial^2 \ln Z}{\partial m^2} . \qquad (5.4.6)$$

Note the order of limits. To measure γ, first the infinite volume limit has to be taken and next the zero mass limit. (In the opposite order the order parameter would vanish also in the broken phase due to tunneling events in the finite volume). Equation (5.4.6) is still a zero mass limit.

A finite-*mass* scaling analysis proceeds in complete analogy to a finite-*size* scaling analysis. The rounding and shifting effects on singularities are derived from a scaling ansatz for the free energy density. The nonanalytic part of the free-energy density of a generic statistical ensemble is the presence of an external field (the finite quark mass) is written as

$$f(t,m) = b^{-1} f \left(b^{y_t} \cdot t, b^{y_h} \cdot m \right) , \tag{5.4.7}$$

where b is an arbitrary scale factor as in section 2.4, y_t and y_h are the thermal and magnetic critical exponents ($y_t \equiv \lambda_1, y_h \equiv \lambda_2$ in our former notation. Choosing the scale factor

$$b = m^{-\frac{1}{y_h}} , \tag{5.4.8}$$

we find that the free-energy density $f = \frac{1}{L_0 L_s^3} \ln Z$ transforms to

$$f(t,m) = m^{\frac{1}{y_h}} f \left(tm^{-\frac{y_t}{y_h}}, 1 \right) . \tag{5.4.9}$$

Equation (5.4.9) implies the finite mass scaling behavior of the order parameter $\langle \bar{\chi}\chi \rangle(t,m)$ and the susceptibility $\chi(t,m)$ in the vicinity of T_c (the critical temperature)

$$\langle \bar{\chi}\chi \rangle(t,m) = m^{\frac{1}{\delta}} F \left(tm^{-\frac{y_t}{y_h}} \right) \tag{5.4.10}$$

and

$$\chi(t,m) \frac{1}{\delta} m^{\frac{1}{\delta}-1} \left[F \left(tm^{-\frac{y_t}{y_h}} \right) - \frac{y_t}{1-y_h} tm^{-\frac{y_t}{y_h}} F' \left(tm^{-\frac{y_t}{y_h}} \right) \right] , \tag{5.4.11}$$

where F and F' are scaling functions. It follows that the peak of the finite mass susceptibility occurs at

$$t_m = c \cdot m^{\frac{y_t}{y_h}} \tag{5.4.12}$$

and scales according to

$$\chi(t_0 = 0, m) \propto m^{\frac{1}{\delta}-1} . \tag{5.4.13}$$

Equation (5.4.13) should be compared with Eq. (5.4.3) for the finite-*size* scaling of the peak in the susceptibility. Equation (5.4.12) gives the shift in the critical temperature T_c or in the reduced critical temperature $t_0 \to t_m$ due to the finite quark mass.

With $\delta = y_h/(1 - y_h)$ it is seen that the exponent δ can be measured either from the shift in t_m according to Eq. (5.4.12) or from Eq. (5.4.11) for the susceptibility or the mass dependence of the order parameter Eq. (5.4.10) at $t = t_0 \equiv 0$.

In case of a second-order transition the exponent δ is characteristic for the universality class of the action. Thus a measurement of δ is of much interest for verifying that the restoration of the $U(1) \times U(1)$ symmetry drives the phase transition at T_c.

Boyd et al. (1992) have proposed a related quantity, the *chiral cumulant*, to measure δ and t_0 from finite mass calculations. It has analogous properties to the Binder cumulant and is defined as

$$\Delta(t, m) = \frac{m\chi}{\langle \bar{\psi}\psi \rangle} = \frac{1}{\delta} - \frac{y_t x F'(x)}{y_h F(x)} \tag{5.4.14}$$

with $x \equiv t \cdot m^{-y_t/y_h}$. It follows from Eqs. (5.4.10) and (5.4.11) that $\Delta(0, m)$ gives $1/\delta$. The slope of Δ increases with decreasing m

$$\left. \frac{\partial \Delta}{\partial t} \right|_{t=0} (m) \propto m^{-\frac{y_t}{y_h}} \tag{5.4.15}$$

in such a way that the ratio Δ itself has a Θ- function shape for vanishing m

$$\lim_{m \to 0} \Delta(t, m) = \begin{cases} 1 & t > 0 \\ 1/\delta & t = 0 \\ 0 & t < 0. \end{cases} \tag{5.4.16}$$

Equation (5.4.16) follows from the definition of Δ and Eqs. (5.4.10) and (5.4.11). When the ratios of $\Delta(t, m)$ are plotted as a function of t for various values of m, the curves cross at the $m = 0$ critical point t_0. The crossing comes from the fact that Δ increases with decreasing m for $T > T_c$ and decreases with decreasing m for $T < T_c$. This behavior is analogous to that of the Binder cumulant $(\langle O^4 \rangle / \langle O^2 \rangle^2)$, where O stands either for the order parameter or the internal energy e. In this way one may extrapolate the zero mass critical point t_0 from a series of finite mass measurements.

Such a determination of the zero mass critical coupling is in principle free of an uncontrolled extrapolation. However, corrections originate in irrelevant terms, which may not be sufficiently suppressed in the vicinity of T_c, and in contributions coming from the regular part of the free energy (cf. section 2.5). A further source which leads to interfering effects with the finite mass scaling behavior is the finite lattice volume used in the Monte

Carlo simulation. As we have seen above, the formulas (5.4.10), (5.4.11), (5.4.12), (5.4.16) hold in the infinite-volume limit. In general finite-mass- and finite-volume effects are competing ordering effects. It is a question of relative size of both effects whether one is allowed to neglect one with respect to the other.

Let us compare the mass scaling behavior of the peak in the suscepti- bility in the infinite volume (Eq. (5.4.11))

$$\chi_{peak} \propto m^{\frac{1}{\delta}-1} \quad \text{for } L \to \infty \tag{5.4.17}$$

with the finite-size scaling behavior in the zero mass limit

$$\chi_{peak} \propto L_s^{\frac{\gamma}{\nu}} \quad \text{for } m \to 0 . \tag{5.4.18}$$

The correction coming from $L_s < \infty$ as $m \to 0$ is small compared to the rounding due to $m > 0$ as $L_s \to \infty$, if

$$
\begin{aligned}
m &> const \cdot L_s^{-b} , \\
b &= \frac{\gamma \delta}{\nu(\delta-1)} ,
\end{aligned}
\tag{5.4.19}
$$

following from Eqs. (5.4.17) and (5.4.18). On the other hand the quark mass has to be chosen sufficiently small in order to keep the contributions from the regular part of the free energy density small. To keep the corrections small, the quark mass must be reduced proportional to $1/\sqrt{L_0}$, when L_0 is increased in the contributions from the regular part.

The caveat to see unwanted features of the strong-coupling symmetries also applies to the case of three flavors in the staggered fermion formulation. In the "worst" case, a crossover phenomenon might be a result of a second- order transition in the strong-coupling regime, which is immediately washed out when finite masses are included. The crossover phenomenon then would be an UV-artifact indicating that the bare coupling g was still too strong.

Once a second-order phase transition has been identified, a measurement of critical indices is not a minor detail for its further characterization. As we have seen, at finite temperatures and fixed L_0 the continuum limit coincides with the high temperature limit. Thus one has to increase L_0 to shift the critical coupling towards smaller values. The large-volume limit is necessary for the continuum limit at low temperatures. In Monte Carlo simulations, *extrapolations* to zero mass, zero lattice spacing and infinite volume are unavoidable. In the vicinity of a second-order phase transition the critical exponents enter the extrapolation formulas. Thus their correct identification and knowledge of their precise values are needed for taking the right limits.

If finite-mass- or finite-size scaling analysis are not practicable or not applicable, an alternative fermion formulation (Wilson fermions) should be explored to reestablish the type of phase transition which has been observed in the staggered formulation. In particular, when odd numbers of flavors are described in the staggered-fermion formulation, it is difficult to ascertain how the trick of doubling and reducing the flavor degrees of freedom affects the effective symmetry of the lattice action which triggers the phase transition.

5.4.3   *Bulk transitions*

Bulk transitions are phase transitions at zero temperature. Their very occurrence is a lattice artifact. Systems on the lattice are systems of statistical mechanics with their own dynamics. They do not care about a well defined continuum limit. Generic phase transitions can occur at some critical coupling, while the temperature is zero. Their physical meaning depends on the context. They have no physical relevance for continuum QCD, if they do not "survive" the continuum limit. Thus one would like to ignore them completely, but one is not allowed to do so.

The phenomenon of bulk transitions in lattice gauge theory is known from the pure gauge sector. According to the conventional lore on four-dimensional lattice gauge theories there should be no zero-temperature phase transition for nonabelian $SU(N)$ gauge groups separating the strong coupling ($g^2 \gg 1$) from the weak coupling ($g^2 \ll 1$) region. This is a desired feature, as it should guarantee that the continuum limit of lattice gauge theory includes both the confinement properties (proven on the lattice for strong couplings) and asymptotic freedom in the weak coupling regime. A bulk transition may in principle destroy the confinement properties of the strong coupling regime.

Monte Carlo calculations for the $SU(2)$- and $SU(3)$- gauge groups with fields in the fundamental representation have verified the conventional lore. For $N_c \geq 4$ first-order bulk transitions have been found for the $SU(4)$ lattice gauge theory by Moriarty [229] and for the $SU(5)$ theory by Creutz [230]. Creutz argues, however, that the confinement property is not lost at weak couplings in spite of the transition.

In $SU(2)$- and $SU(3)$ gauge theories one observes, instead, a rapid crossover phenomenon between the strong and weak coupling regime instead. This is explained by a nearby critical point in the (β, β_A) plane, where $\beta \equiv 2N_c/g_0^2$ as above and β_A denotes the coupling of a $\text{Tr}_A U(\partial P)$ term in the adjoint representation of $SU(N)$. For two colors we have a mixed $SU(2)$-$SO(3)$ lattice action. A small β_A leads to a bulk transition [231], which is absent for $\beta_A = 0$.

A bulk transition is signaled if the location of the phase boundary stays fixed in coupling parameter space, independently of the number of time slices. In contrast, continuum universality for the thermal transition requires a shift of the critical couplings towards smaller values, as L_0 is increased, cf. section 3.3. In the usual finite temperature transition with a correspondence in continuum field theory, the critical coupling (temperature)should scale with L_0 such that the physical transition temperature $T_c = L_0 a(g_0)^{-1}$ remains constant. Thus it is the absence of a shift in β_c when L_0 is increased which suggests a possible lattice artifact. The persistence of the transition on a symmetric ($T = 0$) lattice further supports an interpretation as a bulk transition.

For the $SU(3)$ mixed fundamental/adjoint action the bulk and thermal phase boundaries coalesce for $L_0 = 4$, but split into two lines for small enough couplings and larger values of L_0 ($L_0 = 6, 8$) [232]. The shift of the thermal transition line towards weaker couplings for increasing L_0 supports the hypothesis of continuum universality.

The peculiar behavior, which is sometimes found for the phase structure of QCD with dynamical fermions, has been attributed to bulk transitions in mixed fundamental/adjoint actions. These actions arise as effective actions from the integration over fermions.

Similarly the integration over the eight dynamical flavors in the staggered fermion scheme has been conjectured to induce an adjoint term of the $SU(2) - SO(3)$ mixed action in the effective action. This provides a possible explanation of the bulk transition, which is seen in $N_f = 8$-flavor QCD (for $L_0 \geq 8$-time slices) [233].

One may wonder why the special case of eight light flavors is of any interest at all, as only two (or three) quark flavors are approximately massless in nature. Nevertheless the reason for us is a physical one. Usually the strength of the chiral transition is argued to grow with an increasing number of flavors. A comparison of the transition for two, three, four and eight flavors reveals a strengthening of the transition as the number of flavors increases [234,265], where the range of time slices L_0 lies between four and eight.

Usually this tendency is interpreted as reflection of a physically plausible effect: The chiral transition gets more pronounced the higher the number of flavors is which drive the transition, whatever the "driving dynamics" in detail may be. Results about the $N_f = 8$-transition expose this tendency as a possible lattice artifact and give rise to further studies in effective models about the supposed flavor dependence of the chiral transition.

Obviously an accurate determination of the critical coupling is essential for an identification of the bulk transition. Here one has to face the usual dilemma, which is encountered for *strong first-order* transitions. If the volume is chosen too large, strong metastabilities make both phases stable within a large range of couplings. This requires long Monte Carlo runs on large lattices. Metastabilities are less pronounced when the volume is small, but then the data are contaminated with strong finite size effects.

This problem is solved in the simulations of [233] by starting with a mixed phase configuration. Small changes in the coupling cause the system to rapidly evolve into one of the phases, which by itself is a typical signal for a first-order transition. Starting from a mixed phase configuration, the evaluation of $\langle \bar{\chi}\chi \rangle$ is followed for several values of β to get upper and lower bounds on β_c. The first order of the bulk transition is concluded from the evolution of $\langle \bar{\chi}\chi \rangle$ starting from hot and cold starts. The first-order is signaled by the persistence of two phases over a time scale, which is considerably larger than the equilibration time. This criterion is applicable for strongly first-order transitions. A further evidence comes from the jump in the order parameter, which is interpreted as a tunneling event between coexisting phases in the transition region.

A more detailed discussion would demonstrate that the bulk transition does not merely replace the finite temperature transition and can be ignored as a lattice artifact. It seems to be superimposed on the structure of a finite temperature transition. For a smaller number of flavors ($N_f = 2, 3, 4$) no bulk transition has been observed. There it may be even more difficult to discriminate precursors of the bulk transition at $N_f = 8$ from continuum behavior.

Brown et al. [233] attempt an explanation for the bulk transition as an outgrowth of the rapid crossover region seen in the pure $SU(3)$ gauge theory for $\beta = 5.6$. When passing from strong to weak couplings in this region, a strong deviation from the scaling behavior predicted by the perturbative β-function is seen [235]. Adding light dynamical quarks to the pure gauge action, the crossover region narrows with increasing number of flavors. The sharper the crossover, the stronger the violation of perturbative scaling and the larger the increase in slope of L_0 vs $\beta = 6/g_0^2$. A plot of L_0 vs β_c for $N_f = 0$, 2, 4, and 8 flavors supports this view.

In concluding we summarize the phase structure, which is expected for various temporal extents. For small values of L_0 ($L_0 \leq 4$) one has to work with a coarse grained lattice to reach the transition region. The scale is controlled by the lattice spacing. A single finite-temperature transition is observed, but its relevance to a finite-temperature transition in the continuum is not obvious. At intermediate values of L_0 ($4 \leq L_0 \leq 16$) the corresponding critical couplings fall in the crossover region. A change occurs in what is called the relevant scale. At the end of the crossover region the scale becomes loosely related to the lattice spacing, but controlled by the continuum behavior. This rapid change of relevant scales is manifest in a bulk transition preventing any smooth change of L_0 vs β, both quantities are apparently unrelated. For larger values of L_0 (≥ 32) (and also a simultaneous extension of the spatial lattice size to mimic a finite temperature box) the transition region is expected to lie in a coupling regime, where the lattice is rather fine grained. The bulk transition has disappeared, a finite temperature transition recurs. This time it will be related to the chiral transition of continuum QCD (taking for granted that it does recur). Thus the $N_f = 8$-simulations – although far from modeling realistic QCD systems – allow an identification of lattice artifacts that may also influence more realistic lattice simulations ($N_f = 2$, 3) in a weakened form.

5.4.4 *Results for two and three flavors: The physical mass point*

After all the limiting cases of QCD that we have considered so far we should keep in mind that there is just one physical mass point in the Columbia plot of section 3.1, corresponding to QCD as it is realized in nature. In this section we therefore select computer experiments whose choice of parameters comes close to the physical mass point, or which allow at least a short extrapolation to this point.

The limiting cases we have discussed in the previous sections are not close to the realistic parameters of QCD. The pure gauge theory, the limit of four and eight massless flavors and the strong coupling approximation may be regarded as tools to gain some insight in the dynamical origin of the QCD transitions, when it is viewed from certain limiting cases. They can further give some hints about the stability of QCD results with respect to variations of input parameters. Of particular interest is the role of the quark masses. A crossover phenomenon for experimental quark masses could be understood as a result of mass values, which are too small to sustain the first-order deconfinement transition and too large to sustain the chiral transition. Such a conclusion can be drawn if unrealistic mass values have been studied before.

Let us turn now to the cases of two and three flavors, which come close to the experimental relation of two light ($m_u \sim m_d \sim 5-7$ [MeV]) and one less light ($m_s \sim 150-180$ [MeV]) flavor. An early attempt to address the nature of the physical mass point was the work of Brown et al. [266] in the framework of staggered fermions . Recent studies of the phase transition for nearly physical quark masses [267] are more involved (for example in the choice of action), they make more assumptions that need justification. For a final quantitative result they are not (yet) conclusive. In view of the scope of our book it is therefore appropriate to describe the early work of Brown et al. in some more detail.

The simulations are performed in the staggered fermion formulation. As we have argued above, the staggered formulation represents intrinsically only integer multiples of four continuum flavors. The projection on two or three flavors is enforced by writing l of the effective action in Eq. (3.3.75) as $N_f/4$. Although the local fermionic operator Q describes $N_f = 4 \cdot n$ (n integer) flavors, one allows N_f to take the desired continuum value and uses as a "prescription" for the effective action

$$S_{\text{eff}} = \frac{1}{3}\,\beta\,\sum_p\,\text{Re Tr }U(\partial p) - \frac{1}{4}\,N_{u,d}\ln\det(D + m_{u,d}a)$$

$$- \frac{1}{4}\,N_s\ln\det(D + m_s a) \tag{5.4.20}$$

with $\beta = 6/g_0^2$, $U(\partial p)$ denotes the product of $U's \in SU(3)$ along the boundary ∂p of a plaquette p, $N_{u,d}$ is the number of continuum up- and down flavors, that is 2, and N_s is the number of strange flavors in the continuum, $N_s = 1$. The Dirac operator D is given by Eq. (3.3.69), it describes four flavors of quarks in the continuum limit. Thus, for noninteger n, the representation of the prefactor n as $N_f/4$ cannot be derived from an integration of a local action over fermionic degrees of freedom. The local staggered fermion action leads to $N_f = 4 \cdot n$, hence we call Eq. (5.4.20) a "prescription".

The chiral condensate of staggered fermion fields, defined as

$$\langle \bar{\chi}\chi \rangle = \frac{1}{3}\,\frac{1}{L_s^3 L_0}\sum_x\,\langle \bar{\chi}_x \chi_x \rangle \tag{5.4.21a}$$

with a sum over all lattice sites x is estimated by

$$\langle \bar{\chi}\chi \rangle = \frac{1}{3} \frac{1}{L_s^3 L_0} \left\langle\!\!\left\langle \sum_{xx'} h_x (D+m)_{x,x'}^{-1} h_{x'} \right\rangle\!\!\right\rangle$$

$$= \frac{1}{3} \frac{1}{L_s^3 L_0} \left\langle\!\!\left\langle \sum_{xx'} h_x (m(DD^+m^2)^{-1})_{xx'} h_{x'} \right\rangle\!\!\right\rangle.$$

$$(5.4.21b)$$

For each site x, h_x is an independent, complex three-vector of Gaussian random numbers. This representation of $\langle \bar{\chi}\chi \rangle$ follows from the effective action in terms of pseudofermionic fields h_x. Equation (5.4.21b) shows the nonlocality of the fermionic condensate. The expectation value $<< \ldots >>$ denotes an average over gauge fields and random three-vectors h_x.

The algorithm chosen for evolving the gauge fields with respect to the action Eq. (5.4.20) was the R-algorithm of Gottlieb et al. [236] for which the step size is chosen depending on the number of flavors. The lattice size was fixed to $16^3 \times 4$. As a result, for $m_{u,d} \cdot a = 0.01$ or 0.025 and $m_s \cdot a = \infty$, no transition was found, neither in the three-flavor case for $m_{u,d} \cdot a = 0.025$ and $m_s \cdot a = 0.1$, but a first-order transition occurred for $m_{u,d} \cdot a = 0.025$ and $m_s \cdot a = 0.1$. The results are based on the following type of observations. No transition for the two-flavor case is concluded from the time evolution of the chiral condensate for the ordered and disordered starts. No sign for metastability is seen, as the two starts mix together without clear tunneling events. Also no double peak structure in a histogram of trajectories in phase space is taken as an indication for the absence of a transition. In the case of three light degenerate flavors a two-state signal in the chiral condensate is visible. Over more than 2000 time units the system stays in the ordered (disordered) phase depending on the starting condition. For two light and one heavy flavor - the case that comes closest to realistic mass relations - the order parameter evolves similarly to the two-flavor case. Clear signs of metastability and two-phase coexistence are absent.

Although a translation from lattice units into physical units should be taken with care for the considered β-value of 5.171, it is of interest to estimate bounds on the critical quark masses in units of MeV. Taking a [MeV^{-1}] from Born et al. [237], the set of masses $m_u \cdot a = m_d \cdot a = 0.025$, $m_s \cdot a = 0.1$ corresponds to $m_u = m_d \sim 12$ MeV, $m_s \sim 50$ MeV. These mass values give an upper bound on the critical quark masses at which the chiral transition changes its order.

In the case of three degenerate flavors $N_f = 3$, two state signals are observed for $m_{u,d,s} = 0.025$, while no clear signals of metastability are seen for $m_{u,d,s} = 0.075$ [238], leading to an estimate for the critical lattice quark masses in physical units of 12 MeV $\leq m_{u,d,s}^{crit} \leq$ 38 MeV.

More recent simulations for two degenerate, two plus one, and three degenerate flavors with Kogut-Susskind fermions use an improved action (cf.below) [239]. They come to the same qualitative conclusion that the physical mass point (at vanishing chemical potential) lies in the crossover region. In [239] the investigation is embedded in the location of the so-called chiral critical mass point, representing values of degenerate bare quark masses for which the chiral transition is of second order, while it is of first-order for smaller values. When the bare quark masses are translated into physical units of the pion mass, i.e. m_π [MeV], the value turns out to strongly depend on the used action (improved or not). It is much smaller for an improved action ($m_\pi^c = 67(18) MeV$). Since improved actions produce results closer to the continuum limit, the numbers are more reliable than results for the original staggered fermion action. The small value of the critical mass point confirms earlier calculations in the linear $SU(3) \times SU(3)$ sigma model [240] leading to an estimate of $m_s^c \leq 54 \pm 15.4 MeV$ for the critical strange-quark mass. A location of the physical mass point deeply in the crossover region implies that the chiral critical mass point is of academic interest only, most likely without any influence on the physical mass point. If the physical mass values would be close to the critical mass point, on the other hand, one would expect remnants of a second-order transition in heavy-ion collisions.

At a first glance the absence of first-order signals under realistic quark mass conditions have far-reaching consequences for phenomenological implications in heavy-ion collisions. Many predictions rely on the first-order nature of the chiral transition for three flavors (see section 6.6). From a practical point of view the alternative between a truly first-order transition and a crossover phenomenon may not be distinguishable when the volume is small. The more sensible question to ask is whether the crossover is rapid enough to produce, for example, a sufficiently large gap in entropy densities over a small temperature interval. The jump in entropy density occurs within a β- interval that corresponds to a temperature interval of less than 10 [MeV]. An entropy jump over a finite, but small temperature range is *in principle* sufficient to induce multiplicity fluctuations beyond the statistical noise .

In view of the relevance of the results for the physical mass point, let us briefly summarize their reliability. The considerations are not specific for the work which we have discussed in more detail here, but apply to any study of the phase structure of QCD at the physical mass point. All simulations, reported so far, have been performed on relatively small lattices, typically on $16^3 \times 4$ lattices. A temporal extent of four time slices leads to a transition region in the coupling range, where the lattice is rather coarse grained. For a pure gauge theory $L_0 \geq 10$ is necessary to reach the continuum region, unless one uses an improved action. The effect of fermions is to further lower the effective lattice spacing at which one enters the perturbative scaling region, so that an even larger temporal extent would be necessary to reach the range of asymptotic scaling. Instead of increasing the lattice size one can replace the staggered fermion action as well as the pure gauge part of the action by "improved actions" (see below) with the effect that the continuum limit is reached for smaller lattice extensions. This is actually the improvement in the more recent simulations of the physical mass point, mentioned above, for which the lattice extension is otherwise of comparable size as in the work by Brown et al. [266].

The "distance" from the continuum limit can be visible in results for hadron masses that are obtained in separate $T = 0$-simulations. These calculations are performed to test the scaling properties in the considered coupling regime. One should check several observables: Do the masses of two kaons that are degenerate in their flavor content in the continuum limit still differ by some factor? Is the nucleon over rho mass ratio m_N/m_ρ larger than its physical value of 1.22, and the ratio of m_K/m_ρ smaller than its physical value of 0.64? If they are like that, it suggests that the strange mass entering the lattice K-meson is smaller than its physical value. Hence the first-order transition for three flavors disappears already *before* the strange mass adopts its physical value. Also the pion mass in the simulations is still larger than its physical value, too large for heavier mesons to decay into pions. The unphysical masses of the flavor partners can influence the transition dynamics in a way that is difficult to control.

Other UV-artifacts due to the coarse-grained lattice may be hidden in the results. From the discussion in the previous section it cannot be excluded that the smooth behavior for the two-flavor case is a remnant of the strong coupling $U(1) \times U(1)$ symmetry . Going to larger L_0-values, it may happen that a first-order transition recurs. The same warning applies to the three-flavor case with one heavy and two light flavors, as the $U(1) \times U(1)$ symmetry at strong couplings leads to a second-order transition independent of the number of fermionic flavors.

Another good indicator for the crossover phenomenon is the *baryon-number susceptibility* . It is expected to be small in the low- temperature phase, since the baryon number can only be changed by creating or destroying a baryon, but large in the high-temperature phase, where it is sufficient to create or destroy single quarks to change the baryon number. In [268] the coupling dependence of the baryon-number susceptibility χ has been used to determine the crossover coupling $6/g_{0c}^2$ for two-flavor QCD at given quark masses. The baryon number susceptibility has been also plotted as function of T [MeV] for various lattice sizes. The T-dependence is obtained from $T/m_\rho = 1/(L_0 \cdot a \cdot m_\rho)$ if the zero temperature spectrum calculations of $m_\rho \cdot a$ are extrapolated to the gauge coupling and the quark mass values of the thermodynamic simulations. Once the L_0-dependence is under control, a plot of $\chi(T)$ is of much interest for phenomenological applications. The baryon number susceptibility affects QCD's equation of state, hadronization processes, and heavy-ion collisions.

5.5 Thermodynamics on the lattice

5.5.1 *Thermodynamics for the pure gauge theory*

An important contribution of lattice QCD to phenomenological applications of the QCD transition is a prediction of QCD's equation of state . The behavior of the pressure p, the internal-energy density ϵ and the entropy density s are of most interest for seeking observable effects in heavy-ion collisions. In this section we summarize the lattice expressions of thermodynamic quantities, mainly for an SU(N) pure gauge theory. We outline an approach to derive the equation of state in an entirely nonperturbative framework.

Let us consider the first derivatives of the partition function. The order parameter for the pure gauge theory is the thermal Polyakov loop. Its lattice expression was given in Eq. (5.3.8). From section 2.1 we recall the continuum expressions for the energy density ϵ and the pressure p as $\epsilon = (T^2/V) \cdot \frac{\partial}{\partial T} \ln Z$ and $p = T\frac{\partial}{\partial V} \ln Z$. One way of calculating these quantities on the lattice is to transcribe the derivatives with respect to T and V on the lattice. From lattice QCD at zero temperature we know that we need one bare coupling to tune the lattice spacing in physical units $a[fm]$ by using the lattice QCD parameter $\Lambda_{lat}[MeV]$ as input from outside. When we are interested in simultaneous and independent variations with respect to the physical volume $[fm^3]$ and the physical temperature $[MeV]$, we need two bare couplings g_s and g_τ to tune the physical volume V in units of a_s^3 and the temperature T in units of a_τ^{-1}, so that the derivatives

$$\frac{\partial}{\partial T^{-1}} \quad \longrightarrow \quad \frac{1}{L_0} \frac{\partial}{\partial a_\tau}$$

$$\frac{\partial}{\partial V} \quad \longrightarrow \quad \frac{1}{3a_s^2 L_s^3} \frac{\partial}{\partial a_s} \; . \tag{5.5.1}$$

are replaced accordingly. After performing the appropriate lattice derivatives of $\ln Z$, we set the lattice spacings equal again, $a_s = a_\tau = a$. It is convenient to parameterize the bare gauge couplings in space and time direction, g_s^{-2} and g_τ^{-2}, in terms of a and ξ, where $a = a_s$ and ξ, the asymmetry parameter, is defined as $\xi = a_s/a_\tau$. In terms of these parameters the Wilson action reads

$$S(U) = \beta_m \sum_{i>j, i\neq 0} P_{ij} + \beta_e \sum_x \sum_{i\neq 0} P_{0i} \tag{5.5.2}$$

with couplings

$$\beta_m \equiv \frac{2N_c}{g_s^2(a,\xi)} \xi^{-1}, \qquad \beta_e \equiv \frac{2N_c}{g_\tau^2(a,\xi)} \xi \, , \tag{5.5.3}$$

where N_c stands for the number of colors, the index m stands for "magnetic" or "spacelike" and the index e for "electric" or "timelike" directions, and plaquette terms

$$P_{\mu,\nu} \equiv \text{tr}(1 - U_{x,x+\mu} U_{x+\mu,x+\nu} U_{x+\nu,x+\mu+\nu}^\dagger U_{x,x+\nu}^\dagger) + h.c. \; . \tag{5.5.4}$$

are as introduced in section 3.3, Eqs. (3.3.33)-(3.3.36). The continuum expression for the internal-energy density

$$\epsilon = -\frac{1}{V} \frac{\partial \ln Z}{\partial(1/T)} = \frac{1}{V} \left\langle \frac{\partial S}{\partial(1/T)} \right\rangle \tag{5.5.5}$$

with action S is transcribed to

$$\epsilon = \frac{1}{L_s^3 L_0} \left\langle \frac{\partial S}{\partial a_\tau} \right\rangle \tag{5.5.6}$$

$$= -\frac{1}{L_s^3 L_0 a^4} \xi^2 \left\langle \frac{\partial S}{\partial \xi} \right\rangle$$

$$= -\frac{1}{L_s^3 L_0 a^4} \xi^2 \left((\frac{\partial \beta_m}{\partial \xi}) < \sum_{\mu>\nu, \nu\neq 0} P_{\mu\nu} > + (\frac{\partial \beta_e}{\partial \xi}) < \sum_{\mu\neq 0} P_{0\mu} > \right)$$

for S given by the pure gauge action. Therefore we need the ξ-dependence of g_s, g_τ. This dependence was perturbatively determined in [241] as

$$g_s^{-2}(a,\xi) = g_0^{-2}(a) + c_s(\xi) + O(g_0^2)$$
$$g_\tau^{-2}(a,\xi) = g_0^{-2}(a) + c_\tau(\xi) + O(g_0^2) \tag{5.5.7}$$

with g_0 the bare gauge coupling for the symmetric lattice, and

$$c_s(\xi) \equiv \frac{\partial g_s^{-2}(a,\xi)}{\partial \xi}$$
$$c_\tau(\xi) \equiv \frac{\partial g_\tau^{-2}(a,\xi)}{\partial \xi}, \tag{5.5.8}$$

evaluated in particular for $\xi = 1$ and the case of two and three colors. Now we can express Eq. (5.5.6) in terms of the coefficients c_s, c_τ, the couplings g_s, g_τ and the asymmetry parameter ξ. Alternatively, using instead of ξ the derivatives with respect to a_τ, we obtain

$$\epsilon = \frac{2N_c}{L_s^3 L_0 a^4} (g_0^{-2} \left(< \sum_{\mu<\nu,\mu\neq 0} P_{\mu\nu} > - < \sum_{\nu\neq 0} P_{0\nu} > \right)$$
$$- c_s' < \sum_{\mu<\nu,\mu\neq 0} P_{\mu\nu} > - c_\tau' < \sum_{\nu\neq 0} P_{0\nu} >), \tag{5.5.9}$$

where we have set $g_s = g_\tau = g_0$ after evaluating the derivatives at $\xi = 1$. Here the coefficients c_s', c_τ' are defined via the derivatives with respect to a_τ

$$c_s' = -a \frac{\partial g_s^{-2}}{\partial a_\tau}|_{a_\tau=a_s=a}$$
$$c_\tau' = -a \frac{\partial g_\tau^{-2}}{\partial a_\tau}|_{a_\tau=a_s=a}. \tag{5.5.10}$$

Expression 5.5.9 still contains a contribution from the vacuum energy density ϵ_{vac}, which is a temperature independent constant [269]. While the first two terms of (5.5.9) drop out for temperature zero, the remaining terms lead to

$$\epsilon_{vac} = \frac{2N_c}{L_s^3 L_0 a^4} (-c_s' - c_\tau') < \sum_{\mu<\nu} P_{\mu\nu} > . \tag{5.5.11}$$

This term should be subtracted from the energy density of Eq. (5.5.9). Note that (5.5.9) contains plaquette expectation values that are easily accessible to Monte Carlo simulations. Note also that the energy density scales with $1/a^4$. An uncertainty in $a(g)$ in units of $[MeV]$ is amplified to the fourth power in the results for ϵ. In principle, the coefficients c_σ, c_τ can be determined nonperturbatively. If their perturbative values are used in the strong coupling regime, one must expect inconsistent values for thermodynamic quantities.

In the following we outline an approach as it was proposed by [243] that can be performed entirely in a nonperturbative framework without a direct evaluation of the coefficients c_s and c_τ. The additional quantity now is the β-function $a \cdot dg_0/da$ from QCD, which is known in a nonperturbative region of couplings g_0 from Monte Carlo renormalization-group calculations. Otherwise only plaquette- expectation values have to be calculated. The steps are as follows.

- The free-energy density is calculated from an integration over its derivative with respect to β, since the logarithm of the partition function is not directly accessible within the Monte Carlo approach. With $f = -\frac{T}{V} \ln Z$ one takes the derivative with respect to the gauge coupling $\beta = 2N_c/g_0^2$

$$-\partial \ln \frac{Z}{\partial \beta} = \langle S_G \rangle = 6N_c \, L_s^3 L_0 P_T \,, \qquad (5.5.12)$$

  where S_G is the gauge part of the action (e.g. given by Eq. (3.3.31), (3.3.33)-(3.3.36) and P_T is a short notation of the plaquette-expectation value at temperature T, calculated on a lattice of size $L_s^3 L_0$. If P_0 denotes the corresponding zero temperature expectation value, evaluated on a lattice of size L_s^4, the difference of free energy densities f at coupling β and β_1 is obtained as

$$\left. \frac{f}{T^4} \right|_{\beta_1}^{\beta} = -6L_0^4 \int_{\beta_1}^{\beta} d\beta' \, [P_0 - P_T]. \qquad (5.5.13)$$

- Now an additional assumption enters, which is strictly satisfied only for an infinitely large lattice. It is the relation $\ln Z = V \partial \ln Z/\partial V$, which is valid for homogeneous systems in large volumes. From $p = T \frac{\partial}{\partial V} \ln Z$ and Eq. (5.5.13) one obtains

$$p(\beta) = -[f(\beta) - f(\beta_1)] \,. \qquad (5.5.14)$$

In Eq. (5.5.14) β_1 has been chosen small enough that $p(\beta_1) \sim 0$.

- A second quantity, which is easily calculable on the lattice, is the interaction measure Δ

$$\Delta = \frac{\epsilon - 3p}{T^4} = -12N \, L_0^4 a \frac{dg_0^{-2}}{da}[P_0 - P_T] \, . \tag{5.5.15}$$

As the name suggests, Δ vanishes for an ideal gas (where $\epsilon = 3p$).
- The advantage of calculating first Δ rather than directly ϵ is that Δ depends no longer on c'_σ and c'_τ, but on the QCD-β-function $B(g)$

$$B(g) = -a\frac{dg_0^{-2}}{da} \, . \tag{5.5.16}$$

Here the β-function was named $B(g)$ to avoid confusion with the gauge coupling. To evaluate Δ in a nonperturbative coupling regime requires a knowledge of the β- function for nonperturbative couplings as well. This function has been inferred from Monte Carlo renormalization-group studies [244]. The remaining terms in Δ and f are calculated in a Monte Carlo simulation. Once we have f and Δ, we know ϵ and the entropy density s (with $f = \epsilon - Ts$). Using this approach, the pressure p stays positive and behaves continuously at T_c as it should.

A fully nonperturbative approach alone does not guarantee that the equation of state on the lattice will be relevant for continuum physics. The dominant type of finite-size effects depends on the temperature. At high temperatures there is an intimate relation between finite-size (IR-) and finite-cut-off (UV-)artifacts , although the distinction between them may naively suggest their decoupling. The relation between IR- and UV-artifacts is seen as follows. A reduction of $a_{\tau,\sigma}(g_0)$ [fm] as function of g_0 goes along with an increase in temperature T [MeV], if L_0 is kept fixed. In this way continuum behavior is mixed with high-temperature behavior. For a small lattice spacing, only the high momentum modes are cut off which give the main contribution to the energy density and pressure at high temperatures. For a larger lattice spacing at lower temperatures, lower momentum modes are also cut off, but the energy density and pressure grow approximately with T^4. Thus the finite-cutoff effects, induced by the finite number of time slices, are largest for high temperatures.

Deviations of the interaction measure Δ from ideal-gas behavior is largest in the transition region, going logarithmically to zero with increasing temperature. The interaction measure is less sensitive to finite size/finite cut-off effects than ϵ and p separately. The deviations from the ideal-gas behavior are assumed to be due to the nonperturbative infrared structure of QCD, which is less sensitive to UV-effects. Its behavior indicates that QCD's IR-structure plays an important role even at high temperatures. As mentioned in the introduction, it is an oversimplification if one argues from asymptotic freedom that QCD behaves fully perturbatively at high temperatures.

UV-artifacts induced by the finite temporal extension play a dominant role in the high-temperature region, but a minor role in the transition region. The main contributions to the finite-size effects in the transition region come from the low-momentum modes, in particular if a correlation length becomes large. In case of the pure $SU(2)$ gauge theory, the standard finite-size scaling analysis for second-order phase transitions can be applied to the nonanalytic behavior of the free-energy density. The finite-size effects are controlled by the ratio $L_s/L_0 = TV^{1/3}$ rather than by $1/L_0$ as in the high-temperature region. In particular, the normalized critical energy density $\epsilon^{\text{crit}}/T_c^4$ should scale as function of $T \cdot V^{1/3}$ with critical indices of the three-dimensional Ising model. A finite-size scaling analysis gets much more involved when dynamical fermions are included. For sufficiently large quark masses the singularities of the first-order $SU(3)$ or the second-order $SU(2)$ deconfinement transitions will be rounded anyway. One then has to disentangle which part of the rounding comes from the finite volume and from the finite masses.

5.5.2 *Including dynamical fermions in thermodynamic observables*

When dynamical fermions are taken into account, the internal-energy density ϵ gets an additional contribution from fermions, calculated according to

$$\epsilon_F = \frac{1}{L_s^3 a_\sigma^3 L_0} \left\langle \frac{\partial S_{fermion}}{\partial a_\tau} \right\rangle , \qquad (5.5.17)$$

where $S_{fermion}$ is the action for staggered fermions. Also here the equation of state can be derived in a fully nonperturbative way. The generalization comes from the dependence of observables on the gauge coupling $6/g_0^2$ *and* the bare light quark mass $m_q \cdot a$. Thus the β-function now has two components

$$\beta(6/g_0^2, a \cdot m_q) = (\frac{\partial(6/g_0^2)}{\partial \ln a}, \frac{\partial(a \cdot m_q)}{\partial \ln a}). \qquad (5.5.18)$$

For a number of points in $(6/g_0^2), a \cdot m_q)$-space the β-function can be extracted from data for $(m_\pi \cdot a)$ and $(m_\rho \cdot a)$ measured as functions of $6/g_0^2$ and $a \cdot m_q$. For example, to find the change $\delta(a \cdot m_q)$ for a given change $\delta(6/g_0^2)$ such that the physics remains the same, the mass ratio m_π/m_ρ is kept fixed [245]. An alternative way of finding two equations for the two unknown functions $(a \cdot m_q)(a)$ and $(6/g_0^2)(a)$ is to fit m_π/m_ρ and $m_\rho \cdot a$ as functions of $m_q \cdot a$ and $6/g_0^2$, leading to $a = a(6/g_0^2, m_q \cdot a)$. The inverse function then yields the β-function corresponding to a symmetric change of lattice spacings. The nonperturbative β-function enters the interaction measure according to

$$(\epsilon - 3p) \cdot a^4 = -2\frac{\partial(6/g_0^2)}{\partial \ln a}(P_T - P_0)$$
$$-\frac{\partial(a \cdot m_q)}{\partial \ln a}(\langle \bar{\psi}\psi \rangle_T - \langle \bar{\psi}\psi \rangle_0). \tag{5.5.19}$$

Here $P_{T,0}$ denote the average plaquette expectation values at temperature T or zero, respectively, and $\langle \bar{\psi}\psi \rangle_{T,0}$ are the corresponding light quark condensates.

In general, results for QCD's equation of state, which were obtained so far, must be considered as preliminary and in an exploratory stage in view of an implementation of realistic quark masses at large enough volumes and on sufficiently fine-grained lattices. Nevertheless results for the gluon-energy density in the transition region [270] are already used by experimentalists [271] and are useful for comparing orders of magnitude between the predicted energy density (e.g. of a gas consisting of gluons) and the experimentally available energy densities in heavy-ion collisions.

5.6 Interface tensions

The interface (or surface) tension has been frequently calculated for models of QCD and the electroweak standard model. Like the latent heat the interface tension is an important measure for the strength of a first-order transition. The interface occurs between phases, which coexist at the critical temperature. A large surface tension leads to a strong supercooling effect. The onset of the phase conversion is delayed as the critical temperature is reached. One possible scenario for the phase conversion is droplet formation. Once the system has sufficiently supercooled below the critical temperature, the gain in energy from the conversion to the new phase can compensate for the energy loss in interface-free energy, and the phase conversion sets in.

Another mechanism leading to long, fluffy interfaces is spinodal decomposition, showing a specific time evolution of the long-wavelength modes that is visible in the structure functions (two-point correlations in momentum space) [272].

In applications to QCD different kinds of interfaces must be distinguished. In a first-order chiral transition we have interfaces between phases with broken and restored chiral symmetry. Interfaces between the deconfinement and the confinement phases may occur in a first-order deconfinement transition. They will be considered below for calculations of interface tensions. Furthermore different realizations of the plasma phase are separated by interfaces. Such interfaces occur not only in the region around T_c, but throughout the phase of broken $Z(3)$-symmetry above T_c. One specific phase realization corresponds to a spatial domain characterized by a certain expectation value of the Polyakov loop. The associated interface-free energy density is called an *ordered/ordered interface tension* σ_{oo}, whereas an interface between the plasma and the hadronic phase leads to an *ordered/disordered interface tension* σ_{od}. We write 'plasma phase' as a common name for the deconfinement and/or chiral symmetric phase and 'hadronic phase' for the confinement and/or chiral symmetry-broken phase. Via the relative magnitude of both quantities at T_c, one may gain some insight about the degree of wetting (see below).

In the electroweak standard model interfaces refer to coexisting phases of broken and restored $SU(2)$-symmetry, going along with a nonvanishing (vanishing) Higgs-expectation value, respectively.

Wetting is an alternative or competing mechanism to droplet formation. One phase may spread along a two-dimensional front into the other. In terms of a liquid/gas picture, a droplet of liquid may wet an interface between a gas and a solid when these phases coexist. Whether the wetting is complete or not depends on the relative size of σ_{oo} and σ_{od}.

In connection of QCD the issue of wetting was raised by Frei and Patkós [246]. For some time it was also discussed by other authors in systems which share the global $Z(3)$-symmetry. So far these results are not applicable to QCD transitions under realistic conditions, since fermions have been excluded. Recall that the inclusion of fermion masses may completely wash out the first-order nature of the chiral and the deconfinement transition. A smooth crossover phenomenon prevents the occurrence of different *coexisting* phases and their associated interfaces. If the results of today are corroborated in the future, that both transitions fade away for physical fermion masses, the various interface tensions have no impact on applications in the early universe or relativistic heavy-ion collisions. The same conclusion also applies to the electroweak transition. The physical value of the Higgs mass is certainly much larger than the critical Higgs mass, for which the electroweak phase transition ceases to be of first-order. Nevertheless the measurement of the interface tension was one of the tools to determine the critical endpoint of the Higgs transition. Therefore we summarize different possibilities for its measurement. The first one is the so-called two-coupling method, the second one the histogram method, both methods will be illustrated with examples from QCD. The third method refers to the vacuum tunneling correlation length. It was used for the electroweak part of the standard model.

5.6.1 The two-coupling method

Consider a system in which a domain with volume V_H in the (hadronic) phase H is embedded in a second phase Q (the quark-gluon-plasma phase) with volume V_Q. If we denote by f_h and f_Q the free-energy densities of both phases, the free energy F of the total system is given as

$$F = F_s + (V_H f_H + V_Q f_Q) \, , \qquad (5.6.1)$$

F differs from the sum of the domain free energies just by an amount F_s and F_s equals the free energy associated with the interface separating both domains.

As long as V_H and V_Q are finite, and a homogeneous external field β is applied, which is homogeneous, F_s will actually be zero. Thermal equilibrium forbids the coexistence of phases as stable configurations in a finite volume. The system will tunnel from one phase into the other. A transient coexistence can only result from metastability effects. If one wants to measure the surface tension directly as the excess free energy due to an interface in a finite system, one has to stabilize the interface by an external field gradient $\Delta\beta$. Again the order of limits is essential,

$$\lim_{\Delta\beta \to 0} \lim_{V_H, V_Q \to \infty} F_s/A = \sigma \, , \qquad (5.6.2)$$

where σ is the surface free energy F_s per unit interfacial area A and $\Delta\beta$ stands for a generic field gradient.

For the temperature-driven transitions in QCD an appropriate 'field'-gradient is provided by the temperature. To have a preferred direction, we consider a spacetime lattice with cylindrical spatial geometry, i.e., $L_x \times L_y \times L_z \times N_\tau$ with $L_x = L_y \ll L_z$. Choosing half of the lattice links ($z = 1, 2, \ldots L_z/2$) at a temperature larger than T_c (corresponding to a coupling $\beta \equiv 6/g^2 > \beta_c$) and the other half ($z = L_z/2+1, \ldots, L_z$) below T_c ($\beta < \beta_c$), we introduce into the system an interface between a deconfinement and a confinement phases. It will be located somewhere 'between' $z = L_z/2$ and $L_z/2 + 1$. Early calculations of the surface tension in an $SU(3)$ gauge theory have been performed along these lines [247]. The procedure involved an extrapolation $\Delta\beta \to 0$ in the very end. The result, obtained for a time extension of $N_\tau = 2$, was $\sigma/T_c^3 = 0.24 \pm 0.06$. The value turned out to be too large at least by an order of magnitude, first because of finite-size effects due to the small extension in time direction and second because the extrapolation to $\Delta\beta \to 0$ was not safe. An external field can lead to a suppression of fluctuations of the interface, if it is strong enough, causing the interface to look more rigid than it would be in the $\Delta\beta \to 0$ limit.

It is exactly the extrapolation $\Delta\beta \to 0$ that is subtle and often a weak point of this method. In later simulations for $N_\tau = 4$ (for example by [248]), the value of the interface tension considerably decreased, but the result was neither final.

5.6.2   *The histogram method*

Iwasaki et al. [249] applied the *histogram method* , introduced by Binder [250] for Ising-type systems, to an $SU(3)$ gauge theory. The histogram method is based on the analysis of probability distributions $P(\Omega)$ of order parameters Ω in the vicinity of the phase transition. As we have seen in section 2.6, the probability distribution $P(\Omega)$ develops a characteristic double-peak structure close to T_c. The structure becomes more pronounced as the volume is increased. This behavior was identified as a characteristic signature for a first-order transition in a finite volume. The valley between the peaks corresponds to configurations with interfaces that are more suppressed the larger the volume.

As the suppression comes from the extra costs in interface-free energy, it is plausible that the value of σ may be inferred from the position of the maxima and minimum of $P(\Omega)$ as a function of volume. In the formulas below periodic boundary conditions are assumed with a cylinder geometry satisfying ($L_x, L_y \ll L_z$). Thus two-interfaces will be preferably created in the (x, y)-plane with a total area of $2A = 2L_xL_y \cdot a^2$, a being the lattice constant. The order parameter Ω will be identified with the Polyakov loop

The phenomenological ansatz for the probability distribution $P(\Omega)$ chosen in [249], is then given as

$$P(\Omega) = P_1(\Omega) + P_2(\Omega) + P_m(\Omega)$$

with

$$P_i(\Omega) = c_i \exp\left(-f_i V/T\right) \cdot \exp\left[-\left(\Omega - \Omega_i\right)^2 / d_i^2\right]$$
$$(i = 1, 2) \tag{5.6.3}$$
$$P_m(\Omega) = c_m \exp\left[-\left(f_1 V_1 + f_2 V_2\right)/T - \sigma 2A/T\right],$$

where the following notations are involved. The order parameters Ω_1 and Ω_2 denote the values in the confinement and deconfinement phases, f_1 and f_2 the corresponding free-energy densities, V_1 and V_2 are the associated volumes occupied by each of the coexisting phases. The coefficients c_i and d_i depend on the volume. In Eq. (5.6.3) for $P(\Omega)$ we recognize the superposition of the Gaussians around the characteristic peaks for each phase. The third term $P_m(\Omega)$ gives the probability for finding the mixed phase. Here is the place where the interface tension enters. The probability $P_m(\Omega)$ depends on Ω via the volumes V_1 and V_2, which are occupied by both phases such that $\Omega V = \Omega_1 V_1 + \Omega_2 V_2$.

The weight factors $\exp(-f_i V/T)(i = 1, 2)$ are proportional to the probabilities for the system to reside in the confinement or deconfinement phase. While f_1 and f_2 are degenerate at T_c in the infinite-volume limit, they are in general different in a finite volume. To determine σ in a finite volume V, i.e., σ_V or $\hat{\sigma}_V \equiv \sigma_V/T_c^3$, one has to define $\hat{\sigma}_V$ in such a way that the leading V_i-dependence in the exponent cancels out. The cancellation is achieved if $\hat{\sigma}_V$ is defined according to

$$\hat{\sigma}_V \equiv \frac{-L_\tau^2}{2L_x L_y} \cdot \log \frac{p_{min}}{\left(p_{max,1}\right)^{\gamma_1} \left(p_{max,2}\right)^{\gamma_2}} \cdot \tag{5.6.4}$$

Here $p_{max,1}$ and $p_{max,2}$ are the two maxima of $P(\Omega)$, while p_{min} denotes the minimum between the two peaks. The powers γ_1 and γ_2 denote the weights of the contributions of Ω_1 and Ω_2 to Ω at the minimum, i.e. $\Omega = \gamma_1 \Omega_1 + \gamma_2 \Omega_2$. The infinite-volume limit of $\hat{\sigma}_V$ is the value for the interface tension in units of T_c,

$$\hat{\sigma} \equiv \sigma/T_c^3 = \lim_{V \to \infty} \hat{\sigma}_V \ . \tag{5.6.5}$$

The actual measurement of $\hat{\sigma}_V$ is more subtle. The subtleties concern the choice of the order parameter, the critical coupling in a finite volume and the determination of p_{max} and p_{min}. Candidates for order parameters are the action density or the Polyakov loop. The Polyakov loop is a complex-valued observable. The distribution $P(\Omega)$ develops four peaks near T_c corresponding to the confinement phase with $\Omega_i = 0$ and three realizations of the deconfinement phase in the directions $\exp(i2\pi n/3), n = 0, 1, 2$. To reduce the numerical effort of obtaining high quality data, Iwasaki et al. have projected Ω on the real axis by taking the absolute value Ω_{abs} or by rotating it with $\exp(i2\pi n/3)$ so that $-\pi/3 < \arg \Omega \le \pi/3$ and then taking the real part, leading to Ω_{rot}. The dependence of the results on this choice was checked.

In general, the choice of the critical coupling in a finite volume is a nontrivial issue. It should guarantee the correct infinite volume limit of β_c. Here Iwasaki et al. adjusted β_c so that the peaks of the histograms for Ω_{rot} had equal height.

The actual minimum and maxima of the histograms may be contaminated due to statistical fluctuations. Therefore the extrema were read off from third-order polynomial fits to the histograms in the vicinity of the extrema.

Only the leading volume dependence drops out of $\hat{\sigma}_V$, when it is calculated according to Eq. (5.6.4). Subleading corrections in $1/V$ arise, for example, from fluctuations of Ω in the bulk phases, capillary wave fluctuations of the interfaces, and zero modes corresponding to the translation of interfaces in the direction perpendicular to the interfaces. If the geometry of the lattice deviates from an idealized cylinder in the z- direction, interfaces are no longer restricted to the xy-plane, but sweep out in other directions as well. All of these finite-volume effects have been taken into account by making an appropriate ansatz for the volume-dependent prefactors in the formula for p_{min}/p_{max}.

An important point about the actual measurement of σ concerns the generation of histograms. In case of the $SU(3)$ gauge theory high-statistics histograms were available from the QCDPAX-collaboration [251]. They were used as input in the σ- measurements of [249]. If such histograms are not available for the quantities one is interested in, one can make use of more advanced tools to generate such histograms, so-called multicanonical algorithms [209].

5.6.3   *The vacuum-tunneling correlation length*

The method based on a measurement of the vacuum-tunneling correlation length was used for the Φ^4-theory in four dimensions [273] and for the Ising model in four and three [274] dimensions. Tunneling events between different states create interfaces in finite boxes. For a parameter range, in which tunneling dominates the finite-volume effects, it is possible to extract the interface tension from a finite-size scaling analysis. Roughly speaking, one counts flips of magnetization in space, i.e., one measures the average length ξ_l of magnetization domains in configurations. Choosing an $L^2 \times L_z$-cylinder geometry with $L_z \gg L$, the interfaces build up preferably perpendicular to the cylinder axis. The interface tension σ is then determined according to

$$E_{0a}^{-1} \equiv \xi_L = C(T)\, e^{L^2 \sigma(T)} \,, \tag{5.6.6}$$

where L^2 is the cross-section area spanned in the x, y-directions, ξ_L is the inverse of the vacuum energy splitting E_{0a} in the broken phase. E_{0a} gives the energy splitting between states that become degenerate ground states in the infinite-volume limit, but differ by an energy amount of E_{0a} in the finite volume. This energy determines the tunneling amplitudes between the states with positive and negative magnetization. The tunneling leads to flips in the sign of spin averages and drives the expectation value of the magnetization to zero when a sufficiently large number of Monte Carlo sweeps is made. Equation (5.6.6) was derived in a semi-classical approximation in [275]. The missing L-dependence of the prefactor was taken as an ansatz. Also the L-dependence of σ was neglected as it is expected to be exponentially suppressed for large L. The tunneling correlation length ξ_L itself is determined by the decay properties of appropriate two-point correlations of z-slice spin averages [274].

This third method is applicable as long as finite-size effects and tunneling events are not strongly suppressed. Therefore the temperature should be chosen not too deeply in the broken phase, in which tunneling events are rare and E_{0a} is of the order of the statistical noise, but also not too close to the transition temperature, where the bulk correlation length is no longer small compared to the lattice size and unspecific finite-size effects superimpose on the effects induced by the interface tension.

In the electroweak standard model this method was used by [276] to determine the interface tension between the phases of broken and restored $SU(2)$-symmetry close to the critical endpoint, that is, in a region where the transition is already very weakly first-order and bulk correlation lengths are not small as compared to the lattice size and the tunneling correlation length. Formula (5.6.6) that was *derived* for Φ^4-theories in [275] is now replaced by a phenomenological ansatz $\xi_l = cL^\gamma \exp(-L^2\sigma)$, now with volume-dependent prefactor, and fit-parameters c, γ and σ, where σ corresponds to the interface tension in four dimensions. A result for σ, obtained this way, must be considered as an order of magnitude-estimate at best, since it is well known that measurements of the interface tension are inherently difficult for weak first-order transitions with rough interfaces, independently of the method that is used. The difficulties are expected when one looks closer at the assumptions the derivation of the various formulas is based upon.

5.6.4 *No relics of the type of phase conversion in the early universe?*

When initially the deconfinement transition seemed to be strongly first-order and the interface tension appeared to be large in units of the typical QCD scale, much attention was devoted to the implications of the QCD transition on the phenomenology of the early universe. Unfortunately these early results were contaminated by artifacts of the finite lattice size and lattice constant. Most likely, there are no phenomenological implications of the phase transitions of particle physics for the early universe. For realistic physical quark masses and vanishing or small baryon densities the chiral and the deconfinement transitions of QCD are smooth crossover phenomena. Also the electroweak transition is a crossover phenomenon, since the physical Higgs mass lies above the lower bound obtained in experiments, and this bound lies in the crossover region of the electroweak transition. Nevertheless the phase conversion could have proceeded nonadiabatically: if the crossover were rapid enough and the cooling mechanisms fast enough, there may be remnants from the conversion period, where the universe was out-of-equilibrium. Therefore let us look at the two time-scales that count in the vicinity of the QCD transition. The QCD-time scale is of the order of $1/T_c \sim 1fm/c \sim 0.33 \times 10^{-23}$ s and the Hubble time of the order of $1/\chi = 0.36 \times 10^{-4}s \sim 10^{19}/T_c$ with $\chi = \sqrt{8\pi GB/3}$, where G is the gravitational constant and B the vacuum energy density represented by the bag constant. The time dependence of temperature $T(t)$ follows from Einstein's equations combined with QCD's equation of state that is only known within a number of approximations, to date. The bag-model equation of state in its original form is certainly not the final choice, but for an order of magnitude estimate it leads to an expansion rate that is 10^{-19} times slower than typical time scales in QCD [252], so that the phase conversion in the early universe proceeded in an adiabatic way without producing remnants that could be visible nowadays.

5.7 Other lattice actions for QCD and further reading

When the celebrated principle of local gauge invariance is applied to $SU(3)$-color symmetry, one obtains the classical QCD action in an almost unique way, in the same way as the gauge theory of electroweak interactions. Only a few parameters (the QCD scale parameter Λ and the bare quark masses) have to be taken as experimental input from outside, otherwise the spectrum of mesons and hadrons is predicted. The implementation of this classical action in the continuum into the Euclidean path integral formalism on a space-time lattice is not unique, as we have seen. The difference between different actions should vanish in the continuum limit, but before the continuum limit is taken, it depends on the overall goal which action is favored.

5.7.1 *Simulations with Wilson fermions*

So far we have discussed simulations with Kogut-Susskind or staggered fermions. In the staggered fermion formulation, the chiral limit is obtained by varying the bare mass ma in the Lagrangian to smaller values and extrapolating $ma = 0$ in the end. There remain a chiral $U(1)$-symmetry and a single Goldstone pion on the lattice. The chiral condensate is still a good order parameter because of this symmetry remnant. The rest of flavor symmetry is broken for nonzero lattice spacing. The corresponding pion states are heavier than the Goldstone state because of flavor symmetry breaking. In contrast, the chiral limit in the Wilson fermion formulation must be determined as a one-dimensional submanifold $\kappa_c(\beta)$ in the two-dimensional (β, κ)-plane, where $\beta = 6/g_0^2$ denotes the inverse gauge coupling and κ the hopping parameter. $\kappa_c(\beta)$ is the line of critical hopping parameters, which characterize the chiral limit. The chiral limit can be defined as the vanishing of the pion mass on zero-temperature lattices. Another possibility is to determine $\kappa_c(\beta)$ by the location of zeros in the fermion determinant. The two definitions are in general not equivalent, as the former involves an average over many gauge-field configurations, while the latter definition does depend on the configuration. Other definitions of κ_c on *finite-volume* lattices have been proposed in [253].

One expects a line of finite-temperature phase transitions/crossover phenomena from the confinement/chiral symmetry broken to the deconfinement/chiral symmetric phase will be denoted by $\kappa_T(\beta)$. Both lines should meet in a κ, β-diagram for weak couplings g_0 if the results should have any relevance for continuum physics. In early simulations with small values of N_0 they met (if at all) in the strong-coupling regime. Fukugita et al. [254]were the first to point out such obstacles that hamper the investigation of the confinement phase in the chiral limit with Wilson fermions. Here we do not follow the further development.

5.7.2 *Improved actions*

Nowadays numerical simulations neither start from the Wilson-plaquette action for the pure gauge theory, nor from the Kogut-Susskind or Wilson-fermion action for full QCD, but from a variety of so-called improved actions . It was Kurt Symanzik who initiated the improvement program, later named after him [277]. Improved actions make use of the fact that the lattice action is not unique, but in principle infinitely many actions lead to the same continuum limit if they differ from the "standard" choices by irrelevant terms (cf.section 3.3.) Improved actions are constructed such that their approach of the continuum limit is accelerated. From a practical point of view this means that relatively small lattice sizes, (for example, $(L_s = 12, L_0 = 4, 6)$ may be used, since the $O(a)$-corrections to continuum results are suppressed to some higher power in a as compared to the naive unimproved actions.) For Wilson-type quarks the improved action is called clover action . It was derived in [278]. Naik introduced an improved action for Kogut-Susskind fermions [279]. Later the so-called $P4$-action was introduced as an improved version for staggered fermions, it maintains rotational invariance of the free quark propagator to order p^4. Since improved actions include more terms than the original unimproved actions, the computational effort increases. On the other hand improved actions save computer time as they allow larger lattice spacings and smaller lattices sizes to be used. The saving due to smaller sizes is obvious, but also larger spacings save time, since CPU-time increases with a large power in the inverse lattice spacing $1/a$ [283].

A quantitative measure of acceleration towards the continuum limit is obtained from the difference in (inverse) couplings β, which lead to the same lattice spacing a. For example, the improved action needs only $\beta = 2.0$, while the standard action needs $\beta \sim 5.0$ for $a^{-1} = 1.01[\text{GeV}]$ (if the ρ-mass is calculated on an $8^3 \times 16$ lattice and used as input to set the physical scale). Also in a plot of the energy and pressure for free massless quarks on a lattice with L_0 time slices it is immediately visible that the Naik and P4-actions approach the continuum limit $(L_0 \to \infty)$ more rapidly than either staggered or Wilson fermions [280].

5.7.3 QCD at finite baryon density or nonzero chemical potential

From a physical point of view, QCD transitions which are driven by an increase of density are not less important than temperature- driven transitions on which we focus in this book. Naively, a phase transition should occur when the hadron density rises to the point at which the hadrons begin to overlap. This will be around $0.5 GeV/fm^3$, the energy density in a parton. Applications concern high-density stars in our universe as well as heavy-ion collisions that use nuclei and not anti-nuclei so that there is a quark chemical potential of about $15 MeV$ at RHIC-experiments [281].

The most convenient way of describing quark or nuclear matter at high density is to introduce a chemical potential μ_q for quarks. From asymptotic freedom one would expect that chiral symmetry is restored and quarks are deconfined above a critical value of μ_q, μ_q^{crit}, and above a critical temperature T_c. A critical line $(T_c, \mu_q^{\text{crit}})$ will separate the plasma and the hadron phase in a (T, μ)-diagram.

The naive translation of the continuum expression on the lattice leads to quadratic divergences of the internal-energy density ϵ in the continuum limit. The solution of this earlier difficulty is provided by introducing a chemical potential for quarks $\mu \cdot a$ into the fermion matrix as $e^{\mu a}$ multiplying the forward links in time direction and $e^{-\mu a}$ multiplying the backward links [255].

If the full fermion determinant $\det M$ is included in lattice Monte Carlo simulations, the problem starts for $SU(N)$ gauge theories with $N \geq 3$. In this case the determinant is a complex number for $\mu \neq 0$. Therefore the factor $\det M \exp(-S_g)$ (where S_g denotes the pure gauge part of the action) can no longer be used as a probability for generating configurations in a Monte Carlo simulation as is usually possible when $\det M$ is a positive real number. Nevertheless one can generate gauge-field configurations with some probability $P(U)$, which is frequently chosen as

$P(U) = |\det(M)| \exp(-S_g)$. The expectation value for an observable O is then calculated as

$$\langle O \rangle = \left[\int DU\, O \det(M) e^{-S_g} \right] \Big/ \left[\int DU \det(M) e^{-S_g} \right]$$

$$= \langle O e^{i\phi} \rangle_P \big/ \langle e^{i\phi} \rangle_P \;, \tag{5.7.1}$$

where $\langle \ldots \rangle_P$ denotes the average over configurations with respect to the measure $DU\,P(U)$ and $e^{i\phi}$ is the phase of the determinant. At $\mu = 0$ and $\mu = \infty$, $\langle e^{i\phi} \rangle_P = 1$. Therefore a simulation according to the above prescription is feasible for small and large values of μ. For intermediate values of μ $\langle e^{i\phi} \rangle$ is very small. The phase ϕ fluctuates violently from configuration to configuration. A number of further attempts on the lattice was made to circumvent this problem, but for simulating lattice QCD at finite baryon density and intermediate values of the chemical potential there is no satisfactory solution to date. Conjecture about the phase structure in the T, μ diagram result from simulations in limiting cases of large or small μ, or from calculations based on so-called effective models for QCD in the continuum. [282].

Effective models for QCD in the continuum are not derived from QCD, but share important features like symmetries. In general, they describe specific aspects such as the mesonic behavior in the phase transition region. They are constructed for a limited range of energy scales. As long as they are not derived from QCD (say in a renormalization-group type of approach), their predictions cannot be considered as proven (or derived from first principles), since the very choice of the model may be inappropriate, but they give at least useful hints, in particular on the phase structure of QCD for finite μ [284–287].

Chapter 6

Effective Actions in the Continuum

6.1 Postulating effective actions for QCD

In principle effective actions for QCD should be *derived* from the QCD action in some kind of renormalization-group approach, integrating upon the quark and gluonic degrees of freedom and arriving at new, different, effective degrees of freedom, in terms of which the effective action on the coarse scale gets simplified (for an example of such an attempt see [288]. The effective action could be a linear sigma model for interacting mesons, supposed to describe the phase transition region around T_c of two-flavor QCD. In general there is no need for requiring scale-invariance of the effective action, but symmetry postulates are often imposed and used as guiding principle when a derivation along the renormalization approach is not possible. In contrast to QCD, the applicability range of effective actions is generally limited to a certain scale, for example to low temperatures as in the case of chiral perturbation theory (see below), or to the transition region (linear sigma model, supposed to share the universality with QCD). In addition, effective actions often focus on subsets of the quark and the gluonic degrees of freedom and describe either mesons or baryons, or interacting glueballs, but not in one and the same model.

One may wonder why it is worthwhile considering the approach of effective actions at all, since the fundamental theory of strong interactions is known to be QCD, and lattice simulations from first principles are available. A closer look at the so-called calculations from first principles reveals that often uncontrolled assumptions creep in in unavoidable truncations, or in extrapolations out of the original parameter range, so that it becomes questionable whether the simulations deserve to be called "from first principles". In addition, the very construction of effective models projects on a few degrees of freedom, which are supposed to be the only relevant ones for a certain aspect. The intuition behind such a projection may turn out to be right or wrong. If it turns out to be right, it leads to a physical understanding, since this type of insight can never be gained from purely numerical simulations of full QCD, even not if they would reproduce the experimental results to a high accuracy. Therefore we find it instructive to discuss some of these models in this section.

In the following we illustrate these general remarks with four examples: an $SU(2) \times SU(2)$-invariant scalar model for the phase transition region of two-flavor QCD, an $SU(2) \times SU(2)$-nonlinear sigma model in the framework of chiral perturbation theory, valid at low temperatures, the $SU(3) \times SU(3)$-linear sigma model to study the effect of the strange quark mass on the chiral transition, and a network of gluonic flux tubes to study the deconfinement transition as a percolation phenomenon , that is without symmetry breaking as the driving mechanism. For each of these models we present the action, the guiding principles behind its construction, some of the main results and the limitations.

6.2   QCD and dysprosium

An effective description of QCD that is supposed to be valid just in the vicinity of the chiral transition was considered in [289] and [290]. The approach is based on three hypothesis

- In the limit of two massless flavors, QCD is in the *static* universality class of an $N = 4$-Heisenberg ferromagnet when all other quark masses are sent to infinity and $T \sim T_c$.
- In the limit of two massless flavors, QCD is in the *static* universality class of the ϕ^6-Landau-Ginzburg model when the strange quark mass is near a (tricritical) value m_s^*, the remaining quark masses are infinite, and $T \sim T_c$.
- QCD with two massless flavors is in the *dynamical* universality class of an $O(4)$-antiferromagnet.

These conjectures may be regarded as working hypotheses for lattice simulations of the chiral transition. Before we describe their predictive power, let us comment on the basic assumptions. In spite of the fact that light quark masses have nonvanishing values in reality, their values are small compared to the energy scale of the critical temperature. Thus it makes sense to consider m_u, m_d as perturbations around the chiral limit, where the transition should be of second order if $N_f = 2$. The renormalization-group approach can account for small finite mass values. Their "perturbing" effect on the critical behavior in the chiral limit can be parameterized with critical indices. Thus there is a predictable parametrization of the deviation from the idealized limit. We recall from the second chapter that the mass plays a comparable role to that of the scaling field of the inverse volume in a renormalization-group analysis. Power-law singularities of a second-order transition will be rounded due to the finite volume and due to a finite mass. As we know from the second chapter, the rounding is specific for the second order if the volume is sufficiently large. The deviation from the $L = \infty$-limit can be predicted in a well-known way.

Similarly there is a good chance that the extrapolation from finite masses to the chiral limit is under control. It is under control if the deviations can be parameterized with the critical indices of a second- order transition. For a check of this assumption it would be sufficient to measure certain correlations (e.g. the specific heat) at different small, but finite, quark masses and compare the change in the rounding effects with the predictions of the renormalization-group analysis.

The fact that a theory as intricate as QCD can be reduced to a model as simple as a scalar $O(4)$-model does not necessarily correspond to a crude oversimplification, since the claim is only to describe QCD in the critical region of a second-order phase transition, where the complicated substructure of QCD does not influence the transition dynamics if the second order actually applies.

What, then, is the right choice of the universality class? In the case of two-flavor QCD, Wilczek's proposal is the universality class of an $O(4)$- Heisenberg ferromagnet. This proposal has to be justified by detailed calculations. In view of the $SU(2) \times SU(2)$-linear sigma model , let us first recall the guiding principles for constructing the action of a system that admits a nontrivial phase structure. Landau's free energy was introduced as a framework for discussing the phase structure (see 2.1. It is constructed as a power series in the order-parameter field. Landau's theory corresponds to a mean-field approximation. It leads to wrong predictions of characteristic singularities (e.g. the correlation length at T_c) when large-scale fluctuations in the order-parameter field occur, although the average magnitude of the order parameter is small. The theoretical breakthrough came with Wilson's renormalization-group concepts [291]. It leads to a systematic way of constructing an action in terms of the relevant degrees of freedom at T_c

and answers the question of why the resulting action is representative in some sense. The action is constructed as a limiting (fixed-point) theory after a number of renormalization-group transformations. If the theory is exactly *scale invariant* at T_c, such a limiting theory exits. Scale invariance implies that models at different scales ($|r| \ll \xi$, i.e. scales much smaller than the correlation length ξ but larger than the microscopic scales), share the leading singularity structure.

In case of the chiral transition, these renormalization-group steps are *not* explicitly performed. The limiting theory in terms of pion fields (and their parity partners) is argued to arise out of such a procedure. More precisely only the zero modes of these fields are assumed to survive the iterated renormalization-group steps. The pions are the lightest modes at zero temperature, and the zero modes are the only modes that do not acquire a mass contribution $\propto (2\pi nT)$ from the Matsubara sum. Thus the renormalized mass parameter $\mu^2(T)$ in the resulting action should be understood as an effective mass of the zero modes, which vanishes as T approaches T_c from above.

An essential outcome of the renormalization-group approach is an explanation of *universality*. Universality defines in which sense the limiting theory is representative of a whole class of models, belonging to the same *universality class* . Once the order parameter field is specified, the fixed point theory in terms of these fields depends only on the dimensionality and the symmetry, which is assumed to be broken or restored at the transition. Models with the same underlying symmetry, order-parameter fields, and singularity structure in thermodynamic functions define a universality class.

Thus it is sufficient to find an order-parameter field, construct an action in terms of this field, and restrict the allowed terms by the requirement of chiral symmetry. The order-parameter field for the chiral transition should at least contain the pion multiplet. For two massless flavors the QCD Lagrangian is invariant under the $SU(2)_L \times SU(2)_R \times U(1)_B$-symmetry of independent $SU(2)$-rotations of left- and right-handed fields and the vector baryon number symmetry. At the phase transition, the symmetry is assumed to be broken to $SU(2)_{L+R} \times U(1)_{L+R}$. One choice for an order parameter of the chiral transition is the quark bilinear

$$M_j^i = \left\langle \bar{q}_L^i q_{Rj} \right\rangle \tag{6.2.1}$$

transforming under $SU(2)_L \times SU(2)_R$ according to

$$M \longrightarrow U^+ M V \, , \qquad (6.2.2)$$

where U and V represent independent unitary transformations of the left- and right-handed quark fields. In the following the quark substructure of M_j^i will be disregarded. (This is in general a point that can be questionable and serves further investigation. On the one hand it is an important and natural step to consider composed objects (here condensates) as elementary objects on a coarser scale. On the other hand it is nontrivial to assign the right variables to the new "elementary" objects, assuming that all possible remnants of the substructure can be resumed in the new variables on the coarser scale). As long as M are general complex 2×2-matrices, the Lagrangian has too much symmetry $(U(2) \times U(2))$. A restriction of M to an $SU(2)$-representation removes the additional $U(1)$-symmetry. A possible choice for the $SU(2)$-representation is $O(4) \approx SU(2) \times SU(2)$, where M is parameterized in terms of four real parameters $(\sigma, \vec{\pi}) = \phi^a$ $(a=1,...,4)$

$$M = \sigma + i \vec{\pi} \vec{\tau} \, , \qquad (6.2.3)$$

where $\vec{\tau}$ denote the Pauli matrices. Thus M contains the pion multiplet along with the scalar meson σ. The action in terms of ϕ^a, which is invariant under $SU(2) \times SU(2)$, is given as

$$S = \int d^3x \left\{ \frac{1}{2} \partial^i \phi^a \partial_i \phi_a + \frac{\mu^2}{2} \phi^a \phi_a + \frac{\lambda}{4} (\phi^a \phi_a)^2 \right\} . \qquad (6.2.4)$$

The action takes the same form as the Landau-Ginzburg free energy F in section 2.1. It could be identified with F if the path integral with S of Eq. (6.2.4) were evaluated in the mean- field approximation. Note that S is an action in terms of zero modes. The Euclidean time dependence of ϕ^a has been dropped. Equivalently the $n \neq 0$-Matsubara modes of the original four-dimensional theory are neglected. (The treatment of the $SU(2) \times SU(2)$ sigma-model in connection with chiral perturbation theory will differ in this respect, and time dependence there will be retained.)

Here the action Eq. (6.2.4) should be understood as an effective action for the chiral transition. It coincides with the familiar linear sigma model of Gell-Mann and Levy [292] up to the absence of nucleons and the dimension 3. Nucleon and quark fields are both omitted, as they refer to the "microscopic substructure", which is claimed to be irrelevant for the transition. Moreover the action (6.2.4) agrees with the action of an $N = 4$-Heisenberg ferromagnet , which is believed to model a magnetic transition in dysprosium [293]. This is the reason why Wilczek calls dysprosium an "analog computer" for QCD and explains the title of this section.

For this model one can use results for critical indices from statistical physics and translate their meaning to the mesonic system. The equation of state is then a relation of the form $m = m(<\sigma> = <|\phi|>, T)$ instead of $p = p(V, T)$. It may serve as a working hypothesis for lattice simulations. Its various limits are predictions for the scaling behavior of m_π^2, m_σ^2, or $m_\pi^2 = m_\sigma^2$ in the chiral symmetric phase. In particular, the quantitative predictions for the critical indices β, γ, δ referring to the condensate as a function of m and T are suited for numerical tests.

The role of the strange quark mass

We come now to the second hypothesis concerning the role of the strange-quark mass in the chiral transition . So far we have implicitly assumed that the remaining quark masses m_s, m_c, m_b and m_t are infinite. This is certainly justified for the charm, bottom and top quark masses, which are large compared to the chiral-transition temperature, but the strange-quark mass is just of the order of the transition temperature. Thus it can influence thermodynamic quantities in a nontrivial way. Such an influence is already visible in lattice results. If m_s is infinite, the chiral transition seems to be of second order. If it is zero, renormalization-group arguments predict a first-order transition [294]. Numerical simulations for three light flavors verify this conjecture. Hence a critical value m_s^* should exist, where the second order changes into a first-order transition. Usually such an endpoint, where the order changes from first into second, is called a *tricritical point* . The physical value of m_s is unlikely to coincide with m_s^*, but it may be close by, and if it is so, it is tempting to describe the realistic mass parameters as a perturbation around the idealized tricritical limit. If such an ansatz is justified, the deviations from tricriticality are under control. (Otherwise they are not, when the tricritical point is far apart from the physical mass point.)

According to the second hypothesis a simple model that shares the universality class of QCD with two massless flavors of quarks, $T \sim T_c$ and m_s near m_s^*, is the ϕ^6-Landau-Ginzburg model. Its action reads

$$S = \int d^3x \left\{ \frac{1}{2}(\nabla\vec\phi)^2 + \frac{\mu^2}{2}\vec\phi^2 + \frac{\lambda}{4}(\vec\phi^2)^2 + \frac{\kappa}{6}(\vec\phi^2)^3 - H\sigma \right\} . \quad (6.2.5)$$

The field $\vec{\phi}$ is the same as in Eq. (6.2.4). The explicit symmetry breaking due to $(-H\sigma)$ has been added to account for finite masses $m_{u,d}$. The ϕ^6-term arises as follows. The effect of a finite strange-quark mass is to renormalize the mass and coupling μ^2 and λ in Eq. (6.2.4). For example, one contribution to the renormalization of λ comes from a K-meson exchange between two pions. The "amount" of renormalization depends on m_s. The effect of a mass and coupling renormalization is a shift in T_c as long as λ stays positive. If λ is negative the model gets unstable, and a ϕ^6-term is needed for stabilization. It is easily checked that, for $\lambda < 0$ and fixed, the minimum of the free energy jumps discontinuously from zero to a finite value $|\lambda|/(2\kappa)$ when $\mu^2 = \lambda^2/(4\kappa)$. (For $\lambda > 0$ and fixed, the minimum moves continuously from zero to positive values, when μ^2 goes through zero, as one enters the broken phase.) Hence the value of m_s^* can be defined as the strange-quark mass for which the renormalized coupling λ vanishes. At this point the second order of the $\lambda > 0$ region changes into first order ($\lambda < 0$ region).

Singularities of thermodynamic functions are universal near tricritical points. Thus one may again exploit results from statistical physics. Tricritical exponents of the ϕ^6-model have been calculated [295]. Of particular interest is the result for α, which they find to be $\alpha = \frac{1}{2}$. Note that now $\alpha > 0$ (in contrast to the three-dimensional $O(4)$-model), indicating a true divergence of the specific heat when $T \to T_c$ and $m_s \to m_s^*$. In lattice calculations m_s can be tuned to small values (in principle, in practice this may be just the difficult point). A qualitative change in the shape of the cusp in the specific heat would be a hint of the presence of a tricritical point nearby. This illustrates the predictive power of the second hypothesis.

The third hypothesis concerns the chiral transition and its relation to dynamic universality classes. So far we have dealt with static properties of equilibrium QCD. The third hypothesis may be relevant for off-equilibrium situations in heavy-ion collisions. Here it will not be further elaborated on.

Finally it is a question of size whether corrections due to nonzero modes and heavier mesons are negligible at T_c. In the next section we present an attempt to include heavier mesons, in a four-dimensional $SU(2) \times SU(2)$ nonlinear sigma model. The results cast some doubts on the very ansatz for the three-dimensional action in terms of $\vec{\phi} \equiv (\sigma, \vec{\pi})$.

6.3   The chiral transition in chiral perturbation theory

It is important to recall the physical reason of why chiral perturbation theory is a suitable framework for describing QCD at low temperatures. Chiral perturbation theory is an expansion in small momenta. Its applicability to QCD is based on the fact that strong interactions become *weak* at low energies. This is a consequence of chiral symmetry. The chiral symmetry of QCD in the massless limit implies that the interaction strength is proportional to the square of the energy if the energy is small. At low temperatures the properties of the hadron gas are determined by the lightest excitations. The lightest hadrons are the pions. At low temperatures the average momenta of the pions are small. Thus strong interactions between the pions may be treated *perturbatively* in the framework of chiral perturbation theory.

To be specific, we consider the $SU(2)_R \times SU(2)_L$ chiral symmetry . The particle content is given by the three pion components π^0, π^- and π^+. The fourth component corresponding to the σ-mode is frozen in this description due to the nonlinear realization of the symmetry. The pion field is described by a matrix field $U(x) \in SU(2)$. It transforms under global chiral rotations according to

$$U(x) \longrightarrow V_R U(x) V_L^+ , \tag{6.3.1}$$

where V_R, $V_L \in SU(2)$ and U(x) is parametrized as $U(x) = \exp\left\{ i\vec{\tau}\frac{\vec{\varphi}(x)}{f} \right\}$. Here $\vec{\tau}$ denote the Pauli matrices, f will later be identified with the pion constant, and the components of $\vec{\varphi}$ represent the three components of the pion field.

If one allows only for small four-momenta of the pions, the field $U(x)$ is slowly varying. It is then convenient to expand the Lagrangian in powers of derivatives of the fields $\partial_\mu U$, equivalent to a power series in external momenta p. The ansatz is given as

$$L_{\text{eff}} = L^{(0)} + L^{(2)} + L^{(4)} + \ldots \tag{6.3.2}$$

The upper index counts the number of derivatives ∂_μ, and $L^{(0)}$ can be dropped. Lorentz invariance forbids odd powers in the derivatives. First we consider the chiral limit with a massless pion field. The form of $L^{(2n)}$ is then completely determined by the symmetry requirement, i.e., invariance under $SU(2)_R \times SU(2)_L$ chiral transformations. This leads to

$$L^{(2)} = \frac{f^2}{4} Tr \, \partial_\mu U^+ \partial^\mu U. \tag{6.3.3}$$

Note that this term is just the nonlinear $SU(2) \times SU(2)$ sigma model. The interaction between the pion fields is described in a loop expansion, with $\vec{\varphi}$-fields propagating in the loops. The only place where the finite temperature enters is in the boundary conditions leading to modified propagators. The power-counting rules for graphs associated with the free energy are modified as compared to the rules for chiral perturbation theory in the loop expansion at zero temperature. The only modification consists in having temperature replace the external momenta. There is a one to one-correspondence between the order in the low temperature expansion and the loop expansion of chiral perturbation theory.

The expansion parameter is $\frac{T^2}{8f^2}$ with f an adjustable constant. Since an increasing number in these loops corresponds to a higher order in the low temperature expansion, $L^{(2n)}$, $n = 1, 2 \ldots$ must be expanded in powers of $\vec{\varphi}$ to the order that is needed for the desired accuracy in the low T-expansion. Up to terms of order φ^6 we find, for example, for $L^{(2)}$

$$L^{(2)} = \frac{1}{2} \partial_\mu \vec{\varphi} \partial^\mu \vec{\varphi} + \frac{1}{6f^2} (\vec{\varphi} \partial_\mu \vec{\varphi})(\vec{\varphi} \partial^\mu \vec{\varphi})$$

$$- \frac{1}{6f^2} (\partial_\mu \vec{\varphi} \partial^\mu \vec{\varphi}) \vec{\varphi}^2 + 0(\varphi^6) . \tag{6.3.4}$$

Note that the chiral symmetry fixes $L^{(2)}$ up to one constant f, which may be identified with f_π, the pion-decay constant in the chiral limit. At the order of p^4, two further dimensionless couplings enter the effective Lagrangian. They have to be fixed from experiment. (This is a typical manifestation of a nonrenormalizable action. The number of independent experimental inputs increases with the order of perturbation theory (rather than staying constant).) Away from the chiral limit, the mass terms for the pion field induce an explicit breaking of the chiral symmetry in the effective Lagrangian. In the chiral limit the guiding principle for constructing $\mathcal{L}$ is an expansion in powers of external momenta. It can be maintained in the presence of mass terms, if the quark masses are small compared to the relevant physical scales of the theory. The full effective Lagrangian is Taylor expanded in powers of m. The derivative expansion can be superimposed on the Taylor expansion, term by term. This way additional terms involving powers of m are generated, which break the chiral symmetry explicitly.

The power counting in external momenta of the chiral limit goes through, if m_π is taken to be of the order of p. Since the pion mass square M^2 to lowest order in chiral perturbation theory is given as $M^2 = m_\pi^2 = (m_u + m_d) \cdot B$ (B being the $T = 0$-condensate, m_π being the physical pion mass), the quark masses are counted as quantities of order p^2. Inclusion of mass terms then leads to the following modifications in $L^{(2)}$

$$L^{(2)} = \frac{f^2}{4} \{ Tr(\partial_\mu U^+ \partial_\mu U) - Tr(M^2(U + U^+)) \} , \qquad (6.3.5)$$

$L^{(4)}$ depends now on four input parameters $l_1, \ldots l_4$ [296]. Thus the main effect of finite quark masses is an increase in the number of input parameters. Once the effective action is determined and the input parameters are fixed from experiment, the thermodynamics of the pion gas can be derived in the usual way.

It is tempting to extrapolate results from low temperatures to the transition region. Pions as the lightest hadrons are most easily excited at low temperatures. One may expect that they remain the only relevant degrees of freedom up to T_c. Current quark masses are small compared to the scale of the transition temperature. Thus the naive conclusion would be that their effect, if any, is negligible. Based on these simplifying assumptions one might study the chiral limit of an $O(4)$-model as effective description for the chiral transition region $(T \sim T_c)$. The results of [298] suggest that this line of arguments is too naive.

Chiral perturbation fails to describe the transition region, but leads to reliable predictions for the condensate and other thermodynamic quantities at small temperatures. The influence of finite quark masses on the T-dependence of $\langle \bar{q}q \rangle$, ε and p can be analyzed in the low-temperature region. The effect of heavier mesonic modes may be estimated in a dilute-gas approximation for somewhat higher temperatures. Nonnegligible effects of finite quark masses and heavier mesons show up in the marginal validity range of chiral perturbation theory. Their extrapolations to the transition region can be summarized as follows:

- Below $T = 150$ MeV the nonzero quark masses reduce the temperature dependence of the condensate by roughly a factor of 2.
- The effect of massive states on the energy density $\varepsilon(T)$ is even more significant. In the chiral limit the energy stored in the massive states reaches the order of the energy stored in the pions when $T \sim 130$ MeV. Nevertheless the massive states may be diluted at this temperature.

The main part of the energy is the rest energy of the massive states. At $T \sim 200$ MeV the mean distance between two particles is reduced to $d = 0.9$ fm. This estimate is based on the density formula for a free gas. The approximation of a dilute free gas of massive states is then no longer justified. Below $T \sim 100$ MeV massive states may be neglected.

- At $T \sim 200$ MeV the dilute gas approximation for the massive modes predicts a melting of the condensate, even if the pions are completely ignored. This sheds some light on the idealization of the predominant role of pions in the chiral transition (which we have anticipated in the end of the previous section).

- The extrapolated value of T_c decreases from ~ 190 MeV to ~ 170 MeV when heavier mesons are included in the chiral limit.

- T_c decreases from ~ 240 MeV to ~ 190 MeV when heavier mesons are included at finite quark masses. While heavier mesons accelerate the melting of the condensate, nonvanishing quark masses delay the melting by ~ 20 MeV.

Chiral perturbation theory cannot predict the order of the phase transition, as it loses its validity in the transition region, but it does provide a useful upper bound on the latent heat. This bound should be noticed in particular by phenomenologists, who like a large latent heat for visible effects in heavy-ion collisions. Scenarios based on such an assumption are most likely incompatible with QCD.

6.4 Mass sensitivity of the chiral transition

The aspect of mass sensitivity. In this section we focus on a finite-mass scaling analysis of [299] in a large-N_f approximation. The mass sensitivity of the order of the chiral transition is essential in view of realistic predictions for heavy-ion collisions. To estimate the effect of finite masses on the chiral transition we recall the analogy to a ferromagnet or a liquid/gas system. Quark masses are analogous to the pressure in a liquid/gas system or an external magnetic field in a ferromagnet. If the pressure exceeds a critical value, the first-order transition from a liquid to a gas ceases to occur, it is replaced by a smooth crossover between the liquid and the gas phases. Similarly a first order transition in a $Z(3)$-Potts model becomes a second-order transition for a critical value of the external field and disappears beyond this strength.

When the analogy to a statistical system is translated to an effective model for QCD, the question concerning the effect of finite masses can be posed in the following way. Are the physical meson masses too large for the chiral transition to maintain its first order? In section 4.7.5 we have described an attempt to answer this question within the lattice approach. Here we study the same question on the mesonic level. The bare quark masses $m_s \cdot a$ and $m_{u,d} \cdot a$ should be replaced by meson masses with and without strange quark content.

As a quantitative measure of the distance (in mass parameter space) between the physical and critical meson masses, we consider the ratio of the associated critical to physical light-quark masses $m_{u,d}^{crit}/m_{u,d}$. If this ratio turns out to be much smaller than 1, physical masses lie deeply in the crossover region, and it is difficult to imagine any signatures specific for the phase conversion from the plasma to the hadron phase. The most attractive possibility is a ratio of the order of 1. In that case the physical masses are almost "critical", and effects due to a large correlation length should be visible. Nonuniversal features of the sigma model are then negligible and the reduction of QCD to an effective model in the same universality class is an allowed simplification.

Here we assume that the restoration of the spontaneously broken $SU(3) \times SU(3)$ symmetry is the driving mechanism for the chiral phase transition. Deviations in the spectrum from the idealized octet of pseudoscalar Goldstone bosons are parameterized by terms which break the $SU(3) \times SU(3)$ symmetry explicitly. The assumption in our simplification is that only mesons associated with the $SU(3) \times SU(3)$ multiplets are important for the phase transition. The criterion is chiral symmetry (rather than the size of the meson masses, otherwise one should include ρ-mesons or others as well). The reason why we have chosen $SU(3) \times SU(3)$ rather than $SU(2) \times SU(2)$ is to account for the influence of the strange quark mass on the thermodynamics. With $SU(3) \times SU(3)$ we also include some of the heavier mesons. From the previous section we know that heavier mesons are nonnegligible in the transition region.

We have chosen the sigma model as an effective model for the *low* temperature phase of QCD ($T \leq T_c$). In the low temperature phase quarks are confined to hadrons, and chiral symmetry is spontaneously broken. The meson spectrum reflects some remnants of this symmetry breaking. In the transition region the use of the model becomes questionable; the model certainly fails above T_c as an effective description of the plasma phase.

We use the sigma model in a similar spirit to that in which we used the

$O(4)$-model before. Its action is constructed in terms of QCD's chiral order parameter field ϕ, where ϕ now is a complex 3×3-matrix, parametrized as

$$\phi = 1/\sqrt{2} \sum_{l=0}^{8} (\sigma_l + i\pi_l)\lambda_l. \tag{6.4.1}$$

Here λ_l denote the Gell-Mann matrices, π_l are the pseudoscalar mesons, σ_l are the scalar mesons. The mesonic order-parameter field is a bilinear in the left-handed and right-handed quark fields $\phi_{ij} = \langle \bar{q}_i^L q_j^R \rangle$. In the sigma model the quark structure is ignored by construction for all temperatures $T \geq 0$. In terms of ϕ the Lagrangian reads

$$\begin{aligned}
L = \int d^4 x \Big\{ &\frac{1}{2}\mathrm{Tr}(\partial_\mu \phi \partial_\mu \phi^+)\Big\} - \frac{1}{2}\mu_0^2 \mathrm{Tr}(\phi\phi^+) \\
&+ f_1(\mathrm{Tr}\phi\phi^+)^2 + f_2 \mathrm{Tr}(\phi\phi^+)^2 + g(\det\phi + \det\phi^+) \\
&- \epsilon_0\sigma_0 - \epsilon_8\sigma_8.
\end{aligned} \tag{6.4.2}$$

Note that there are two independent quartic terms with couplings f_1 and f_2. The determinant-terms are cubic in the components of ϕ, g is the "instanton" coupling that takes care on the right η-η'-mass splitting, μ_0^2 is the coupling of the quadratic term. The external field ϵ_0 gives a common mass to the (pseudo)scalar meson octet, while ϵ_8 accounts for the right mass splitting inside the (pseudo)scalar meson octet .

Tree level parametrization at zero temperature
The parameters μ_0^2, f_1, f_2, g, ϵ_0, ϵ_8 of the Lagrangian (6.4.2) should be chosen such that the model reproduces the experimental values of the (pseudo)scalar meson masses. The parametrization of the sigma model is not unique [300–302]. Here we are interested in a tuning of meson masses in terms of a few parameters. Suitable parameters are the external fields ϵ_0, ϵ_8. These induce finite quark masses according to

$$\begin{aligned}
-\epsilon_0 &= \alpha(2\hat{m} + m_s) \\
-\epsilon_8 &= \beta(\hat{m} - m_s) \,,
\end{aligned} \tag{6.4.3}$$

where $\hat{m} \equiv (m_u + m_d)/2$, α and β are constants. Equation (6.4.3) follows from an identification of terms in the Lagrangians for quarks and mesons, which transform identically under $SU(3) \times SU(3)$. The meson masses are determined for given ϵ_0, ϵ_8, once the couplings μ_0^2, f_1, f_2, g are specified and the condensates $\langle\sigma_0\rangle_{T=0}$, $\langle\sigma_8\rangle_{T=0}$ are calculated for given μ_0^2, f_1, f_2, g. Thus we vary the quark and meson masses by varying ϵ_0 and ϵ_8. The chiral limit is obtained for $\epsilon_0 = 0 = \epsilon_8$. The couplings μ_0^2, f_1, f_2, g are then determined from the mass input in the chiral limit , i.e. $m_\pi = m_K = m_\eta = 0$, $m_{\eta'} = 850$, $m_{\sigma_{\eta'}} = 800$, $m_{\sigma_\eta} = 600$, all masses in units of [MeV], $f_\pi = 94$ MeV. Next we keep μ_0^2, f_1, f_2, g fixed to their values in the chiral limit and change ϵ_0, ϵ_8. The choice $\epsilon_8 = 0$, $\epsilon_0 \neq 0$ leads to an $SU(3)$-symmetric case with only one order parameter field σ_0, for which the numerics considerably simplifies. Meson masses with almost experimental values are induced for $\epsilon_0 = 0.0265$ GeV^3, $\epsilon_8 = -0.0345$ GeV^3.

In this way we have constructed a mapping

$$\{m_{u,d}, m_s\} \leftrightarrow (\epsilon_0, \epsilon_8) \leftrightarrow \{m_{\text{Meson}}^2\} \qquad (6.4.4)$$

between quark and meson masses.

It remains to translate the meson condensates at zero temperature to the light and strange quark condensates. In the same way as we obtained the relation (6.4.3), we find here

$$\langle \bar{q}q \rangle = \frac{\epsilon_0}{2\hat{m} + m_s} \langle \sigma_0 \rangle + \frac{\epsilon_8}{2(\hat{m} - m_s)} \langle \sigma_8 \rangle$$

$$\langle \bar{s}s \rangle = \frac{\epsilon_0}{2\hat{m} + m_s} \langle \sigma_0 \rangle - \frac{\epsilon_8}{\hat{m} - m_s} \langle \sigma_8 \rangle \ . \qquad (6.4.5)$$

Eqs. (6.4.5) are derived at zero temperature. We take these relations as temperature independent and use them to determine $\langle \hat{q}q \rangle (T)$, $\langle \hat{s}s \rangle (T)$ from the measured values for $\langle \sigma_0 \rangle (T)$, $\langle \sigma_8 \rangle (T)$.

Critical meson masses in mean-field

Although the method is crude, it is instructive to get a first estimate for critical meson/quark masses. Later the results will be compared with estimates from a large-N-approach. In a mean-field calculation the full effective potential is replaced by the classical part in terms of two constant background fields σ_0, σ_8. For simplicity we consider here only the $SU(3)$-symmetric case, where $\epsilon_8 = 0 = \sigma_8$, $\epsilon_0 \neq 0$, σ_0 denotes a constant background field. The effect of a finite (high) temperature in a mean-field calculation is a renormalization of the quadratic term in the Lagrangian. Thus a finite temperature can be mimicked by tuning μ_0^2 while keeping the other couplings f_1, f_2, g fixed. For a critical field ϵ_0^{crit}, the first- order transition simply disappears, and so does the cubic term in U_{class}. At ϵ_0^{crit}, U_{class} starts with a term proportional to $(\sigma_0 - \sigma_0^{\text{crit}})^4$, where σ_0^{crit} is the minimum of U_{class} for critical values $\mu_0^{2\text{crit}}$, ϵ_0^{crit}. Thus we have

$$U_{\text{class}}(\sigma_0) = -\frac{1}{2}\mu_0^2\sigma_0^2 + \frac{2g}{3\sqrt{3}}\sigma_0^3 + (f_1 + \frac{f_2}{3})\sigma_0^4 - \epsilon_0\sigma_0 \qquad (6.4.6a)$$

$$U_{\text{class}}\big|_{\text{crit}}(\sigma_0) = \frac{1}{4!}\frac{\partial^4 U_{\text{class}}}{\partial\sigma_0^4}\bigg|_{\text{crit}}(\sigma_0 - \sigma_0^{\text{crit}})^4 + o(\sigma_0^5). \qquad (6.4.6b)$$

Here $\big|_{\text{crit}}$ means "evaluated at critical parameters". Note that U_{class} in Eq. (6.4.6a) takes the same form as a free energy functional for a liquid/gas system. It supports the analogy between a liquid/gas system and the chiral transition in QCD as mentioned in the dictionary of correspondent quantities 2.6.3. The vanishing of the first three derivatives in Eq. (6.4.6b) determines σ_0^{crit}, $\mu_0^{2\text{crit}}$ and ϵ_0^{crit} as functions of f_1, f_2, g. In the physical case of $\epsilon_0 = 0.0265$ [GeV^3], $\epsilon_8 = -0.0345$ [GeV^3] we obtain

$$\frac{m_{u,d}^{\text{crit}}}{m_{u,d}} \sim 0.03 \pm 0.02 . \qquad (6.4.7)$$

Such a small ratio of 3 % for the critical to physical light-quark masses would mean that the chiral phase transition is easily washed out by tiny quark masses, and for physical quark masses one is left with a rather smooth crossover phenomenon.

Our interest in the mean-field result is the order of magnitude of this ratio. A first-order transition can have different origins. One such origin is a cubic term in the classical part of the potential. A second one is a ϕ^6-term which may be needed for stabilization of the free energy when the quartic coupling picks up a negative sign due to renormalization effects. For two or more independent relevant couplings, a further first-order transition can be a so-called *fluctuation-induced* transition . Since the linear $SU(3) \times SU(3)$ sigma model contains two such couplings f_1 and f_2, the chiral transition may be mainly fluctuation induced.

If the order of magnitude of the ratio (6.4.7) changes beyond the mean-field level, it casts some doubt on the simplified description of Eq. (6.4.6a) and favors the hypothesis of a fluctuation- induced transition.

Beyond the mean-field level, the temperature dependence of the condensates $< \sigma_0(T) >$, $< \sigma_8(T) >$ is calculated from the minima of the constraint free energy density (under constraints on average values of the order parameter fields σ_0 and σ_8), where the constraint effective potential is evaluated in a large-N_f expansion. The shape of this potential reveals the critical mass parameters, so that the transition is washed out for parameters larger than the critical values. Translated into critical quark masses , one obtains bounds according to

$$m_{u,d}^{\text{crit}} \leq 2.96 \pm 0.85 \text{ MeV}$$

$$\text{(6.4.8a)}$$

$$m_s^{\text{crit}} \leq 54 \pm 15.4 \text{ MeV},$$

or a ratio of

$$m_{u,d}^{\text{crit}}/m_{u,d} \sim 0.26 \pm 0.08. \tag{6.4.8b}$$

In the $SU(3)$-symmetric case with three degenerate flavors, the common critical pseudoscalar mass is only ≤ 51 [MeV], and $m_{u,d}^{\text{crit}} \leq 0.9 \pm 0.14$ MeV. Thus the critical mass values depend on the direction in mass parameter space. More recent lattice results [297] obtain the same order of magnitude for the critical mass point. It should be kept in mind, however, that a direct comparison between lattice quark masses in physical units and current quark masses (as we are using here) is questionable because of an unknown multiplicative renormalization factor. A ratio of 30% is certainly not large enough for predicting visible remnants of a nearby second-order chiral transition. There is some hope that the ratio gets closer to 1, if further fluctuations are included in the effective model and the true nature of the chiral transition is fluctuation induced. From measurements of the energy- and entropy densities one obtains an upper bound on the latent heat in the transition region ($\Delta L \leq 0.2 GeV/fm^3$) that is compatible with estimates from chiral perturbation theory. Contributions to these quantities from from gluonic degrees of freedom were completely neglected.

The choice of the model in form of the parametrization of the $SU(3) \times SU(3)$-sigma model was just one attempt to include a portion of the heavier mesons. This portion is determined by the assumed underlying chiral symmetry. A further assumption is that the deviations from the broken $SU(3) \times SU(3)$ symmetry can be parameterized by two external fields, breaking the symmetry explicitly. Such an ansatz seems to be justified in the sense that the predictions of the pseudoscalar meson masses agree reasonably with experimental values.

An alternative point of view is the following. $SU(2) \times SU(2)$ is the true symmetry with only pions as idealized Goldstone bosons and one external field to account for the finite pion mass, while all other mesons -scalar, pseudoscalar, and vector mesons- are treated on an equal footing. Gerber and Leutwyler [298] take this point of view when they describe all mesons (apart from the pion) in a dilute-gas approximation to study their influence on the chiral-phase transition.

When the quantum virial expansion [303] or the generalized Beth-Uhlenbeck approach [304] are applied to a pion gas, the underlying chiral symmetry of QCD plays a less prominent role in the description than it does in our treatment in the $SU(3) \times SU(3)$-linear sigma model.

6.5 A network of gluonic strings

For flavor symmetries we have just seen that it is difficult to favor one of the two options: explaining the phase transition and the mesonic mass spectrum as result of the spontaneous breaking either of $SU(2) \times SU(2)$ or of $SU(3) \times SU(3)$. In both cases the mesonic mass spectrum can be only reproduced if explicit symmetry breaking is included in the action. Therefore one may look for an explanation of the phase transition that abandons symmetry breaking as the driving mechanism. Such an attempt was made by Patel [305] for the deconfinement transition, and this is the reason why we discuss his proposal of color-flux tube models in this section.

The color flux tube models were developed to describe the deconfinement phase transition, quark degrees of freedom can be included as well, but without an appropriate implementation of chiral symmetry. Let us first consider the case of a pure $SU(2)$- or $SU(3)$ gauge theory. We recall that the spontaneous breaking of the global $Z(2)$- or $Z(3)$ symmetry is associated with the phase transition from the confinement to the deconfinement phase. The order parameter is the expectation value of the Polyakov loop. It vanishes when the free energy to find an isolated test quark in the system grows to infinity. It is different from zero in the deconfinement phase.

Effective $Z(N)$-spin models for the deconfinement transition have been presented in 3.1. They can be derived from the $SU(N)$ lattice gauge theory in the high temperature limit and share the essential extra symmetry of QCD. In the limit of infinitely heavy masses, this is the global $Z(N)$-symmetry. For high temperatures the gauge fields are almost frozen to unity matrices. Lowering the temperature increases the disorder. One big cluster of aligned spins breaks up into several clusters. Below the transition point the system is, if we use "magnetic" language, completed disordered. The formulation in terms of spin systems is natural for $Z(N)$- models. Intuitively it is, however, less clear how this picture should be translated to the original $SU(N)$ degrees of freedom.

Patel's description is complementary to the above model. It starts at $T = 0$ and follows the evolution of the system as the temperature is increased. The symmetry breaking is no longer the driving force for the system to undergo a phase transition. It is replaced by entropy production. It abandons symmetry breaking as a driving force of the phase transition, at least as a primary force. This is desirable as an alternative way of understanding the transition dynamics in mass parameter regions, where the symmetry concept is questionable, because the symmetries are realized only approximately.

Our following description is on a heuristic level. The picture is based on the flux-tube model of the deconfinement transition. Flux tubes connect quarks and antiquarks in strong-coupling expansions on the lattice. Thus one may not automatically expect that they leave some remnant in the continuum limit. On the other hand, they are also the ingredients of phenomenological continuum descriptions like string models for hadrons.

Several properties of the flux tubes or strings have to be specified first. Strings in Patel's models are characterized by three parameters: the string tension σ, the string width w, and the rigidity parameter a. (The notation of the parameter a suggests its actual meaning as an (effective) lattice constant.) We assume that there is a constant energy per unit length along the string, this is σ. The string has a constant width w. It has a certain resistance against bending. It has to go at least a distance a apart, before it can change its orientation. The distance a is of the order of 1 fm. On the lattice a naturally coincides with the lattice constant, the strings bend at right angles. Here the constant a will not be tuned to zero in the end. Its role resembles that of a lattice constant in models of condensed-matter physics (models in continuum spacetime, where the constant a is given by the physical lattice constant). In Patel's flux tube theory the parameters a, σ and w must be fixed from experimental input.

Two further assumptions about the flux tubes must be specified to get a well-defined model. Flux tubes can terminate only on quarks. Their interaction occurs at baryonic vertices. The constants σ, w, a are treated as temperature independent. The only driving force of the phase transition is the increasing entropy of the flux tubes when the temperature is turned on.

The $SU(2)$-case

Based on these ingredients let us see how far heuristic arguments can lead us in a pure $SU(2)$ gauge theory. The only allowed flux-tube structures are closed loops differing only in size and shape, i.e., in the length of the string. Physically these loops may be interpreted as glueballs. The partition function is written as

$$Z = \sum_{loops} N(loops) e^{-E(loop)/T} . \qquad (6.5.1)$$

The sum runs over all loops of fixed length $l \cdot a$ (in physical units). The energy of such a loop is given as $\sigma \cdot l \cdot a$. The combinatorial prefactor N gives the number of loops of length $l \cdot a$. The dependence of N on the rigidity parameter a is essential, as it provides the possibility of a phase transition. How often strings like to bend is a function of temperature and rigidity.

Consider random walks without backstepping of length l (in lattice units) on a lattice of spacing a. Backstepping should be forbidden for a physical string. This number is given as $(2d - 1)^l$ in d dimensions. The constraint that the walks should perform closed loops leads to a power law correction in l to N. It will not be specified further, because it is irrelevant in the large-l-limit. The partition function is then proportional to

$$Z \sim \sum_l \exp\left[-l\left(\frac{\sigma a}{T} - \ln 5\right)\right] \qquad (6.5.2)$$

in three dimensions. The dimension is chosen as 3 since the string model is constructed to provide an alternative description to the three-dimensional $Z(N)$-spin model for the deconfinement transition. As T increases, the average length of the loops becomes larger. A phase transition is signaled if Z diverges. This happens for large l if $(\sigma \cdot a/T = \ln 5)$, that is, at

$$T_c = \sigma a / \ln 5 . \qquad (6.5.3)$$

The effective string tension σ_{eff} defined via $\exp\{-la\sigma_{eff}/T\}$ vanishes continuously at T_c,

$$\sigma_{eff} = \sigma - \frac{T \ln 5}{a} \overset{T \to T_c}{\longrightarrow} 0 , \qquad (6.5.4)$$

while the average length of the loops diverges. Both features suggest a second-order phase transition.

To see its relation to the deconfinement transition, we have to probe the system at T_c with a static $\bar{q}q$-pair. If it costs a finite amount of energy to isolate both quarks at an infinite distance, we have reached the deconfinement phase. Equivalently, deconfinement is manifest if the q and $\bar{q}$ are a finite distance apart, but only rather loosely correlated. Color screening requires that a flux tube connects the quarks of our probe. With increasing temperature the flux reorients itself more and more often and oscillates between its endpoints.

Since one quark knows about its partner only via the connecting string, this information is lost if the string gets infinitely long. This is just what happens at T_c, where it costs no extra free energy to create an infinitely long flux tube. Thus at $T \geq T_c$ both quarks (q and $\bar{q}$) are effectively independent of each other and free. This means deconfinement.

To get a quantitative estimate of the transition temperature, we must fix the string tension and the rigidity parameter in physical units, or the product of both in lattice units. In the strong-coupling approximation of the lattice theory, the lowest-lying O^+-glueball at $T = 0$ is given by a square loop

$$m_{O^+} = 4\,\sigma\,a \ . \tag{6.5.5}$$

The string parameters are assumed to be temperature independent, hence one can take this $T = 0$-result to fix T_c,

$$T_c = m_{O^+}/4\ln 5 \ . \tag{6.5.6}$$

Lattice Monte Carlo calculations give a slightly larger value.

The $SU(3)$-case

Next let us consider the pure $SU(3)$ gauge theory. Two new features must be accounted for. Quarks and antiquarks are no longer in equivalent representations of $SU(3)$. Thus a direction is associated with a string, indicating whether it terminates in a quark or an antiquark. (The quarks and antiquarks are test quarks in the pure gauge theory.) Furthermore, a flux-tube representation of a baryon or antibaryon requires that a string be able to bifurcate at a vertex v, see Fig. 6.5.1. This reflects the previous assumption that flux tubes interact only at baryonic vertices. The diagrammatic rules for allowed string structures in the $SU(3)$ case follow from the allowed vertices, shown in Fig. 6.5.2. This excludes closed loops made up of an odd number of links.

The ansatz for the partition function can be chosen analogously to the $SU(2)$-case. The sum over loops has to be replaced by a sum over more complicated topological structures which resemble nets. A nonvanishing energy v is associated with each vertex to respect the additional bifurcation degree of freedom. At zero temperature the vacuum is filled with closed loops made out of nets of strings. These are virtual "glueballs".

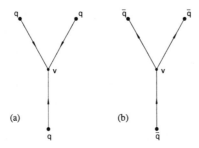

Fig. 6.5.1   String bifurcation for a baryon (a) and an antibaryon (b)

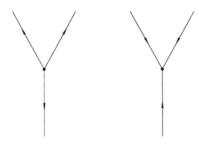

Fig. 6.5.2   Allowed vertices for the SU(3) gauge theory

As the temperature is increased, the size and density of these structures grow. In contrast to the $SU(2)$-case, the phase transition is not induced by a divergence in the length of strings. A new qualitative feature enters, the *connectivity* of the network. The phase transition occurs *if an infinite network is generated*, infinite in the sense that each string is connected with the entire volume of the lattice. Between any given pair of flux-tube segments one can find a flux-tube path on the lattice that connects them.

The definition of "critical temperature" can be made more precise. This type of phase transition is well known in condensed-matter physics under the name of *percolation transition*. Examples for percolation transitions are gelation transitions in the context of polymer chains [306]. The deconfinement transition here resembles a gelation transition, in which the connectivity of the color-flux tube network is determined by the gauge group $SU(3)$. To find a quantitative measure for the connectivity of this network, one has to distinguish between relevant and irrelevant links of the network. If relevant links are removed, the infinite network is destroyed. In the same sense irrelevant links are not essential for the connectivity. The fraction f

of relevant links depends on the temperature and the specific underlying dynamics. It can be estimated that $\frac{2}{3} < f < 1$. The lower bound is realized at T_c, where the network is minimally connected. The upper bound means maximal connectivity.

It is challenging to look at the deconfinement transition as a percolation transition. Again we have to answer, first, what happens to a static test quark, when it is put into the network below or above the percolation transition. Below T_c the free energy of an isolated quark is infinite, because the flux tube originating from this quark may not terminate on any closed loop that is a finite distance apart. It is excluded by the diagrammatically forbidden flux-tube structures. Instead the string attached to the quark has to fuse with the string of an antiquark of the test probe. When both q and $\bar{q}$ can be considered as isolated or free, to separate them that much costs an infinite amount of free energy, since we have assumed a constant energy per unit length of the string.

Above T_c the distance between the q and $\bar{q}$ need no longer be infinite for both to loose the correlation (to behave effectively as free test probes). Already at a finite separation the correlation is weak due to the presence of an infinite network to which they are attached. Now the clusters of closed loops are connected, and the corresponding flux-tube structures for a $\bar{q}q$ pair at finite distance are allowed. The costs in energy due to the long, dense, and connected flux tubes are compensated by the gain in entropy. Thus the free energy of a configuration in which the q and $\bar{q}$ test quarks have lost any information about each other is finite. The q and $\bar{q}$ are effectively free. We have ended in the deconfinement phase.

For a crude estimate of T_c one has to specify the amount of extra energy due to bifurcation points.

In the heuristic way of arguing it remains to explain the first order of the deconfinement phase transition in the $SU(3)$-case. Either one refers to the percolation transition where the transition is known to be of first order, or one proves directly for the colored network of the QCD flux that it costs a finite interface-free energy at T_c for both phases to coexist. Such a coexistence may be visualized as a hole in the network. Roughly speaking, when the loose ends at the boundary of the hole are tied to the network again, it costs interface energy proportional to the area and leads to a reduction in entropy.

Inclusion of matter fields

Dynamical quarks can be introduced into flux-tube models, since the flux-tube picture does not rely on any symmetry argument, and the extra global $Z(N)$-symmetry is explicitly broken in the presence of dynamical quarks. Recall that the limit of a pure gauge theory may be considered as the $m \to \infty$-limit of full QCD. Thus we are interested in the effect when the quark mass is lowered or a $Z(N)$- symmetry breaking field is switched on. Now, where the symmetry is explicitly broken independently of the phase, we have to look for new criteria that tell us the phase of the system at a given temperature. This is the screening property, which is different in the high- and low temperature phases. At low temperatures, color charges are screened due to a breakage of strings. At high temperatures, flux tubes do not break, but get attached to large networks of strings, at least for very heavy masses. For lighter masses, quarks can also break up strings at high temperature. (We see that the clear distinction between low- and high T-phases may get lost depending on the value of the quark mass.)

The probability for string breakage at high temperature is determined by the energy costs of popping a $q\bar{q}$-pair out of the vacuum and the competing gain in entropy by breaking a link. Lighter masses facilitate breaking of links. Less surface energy is necessary to stabilize a hole in an infinitely connected network. Clearly there should be a critical mass at which the surface energy vanishes at T_c and the number of broken flux tubes exceeds the critical number of irrelevant links. Once relevant links are broken, by definition, the network ceases to exist.

In Patel's models one has a heuristic understanding of how the transition temperature should scale as a function of quark masses. As the quark mass decreases, string breakage is facilitated. It becomes more difficult to generate an infinitely connected network. Thus T_c should increase with decreasing m. The opposite tendency is observed in numerical simulations.

The above line of arguments suggests that the transition will always disappear in the $m \to 0$-limit. The energy cost for the creation of a $q\bar{q}$-pair is zero. Recall, however, that it is precisely in the $m \to 0$-limit where the chiral symmetry comes into play and generates dynamical masses at low temperatures. In the chiral limit, the scalar mesons are believed to obtain their finite masses exclusively from chiral symmetry breaking. In the preceding sections we have observed the opposite mass dependence: perturbing around the chiral limit, turning on an external field by increasing m from zero towards realistic values led to the disappearance of the chiral transition. In the chiral limit, the strength of the first order of the chiral transition is said to increase for an increasing number of flavors. On the other hand, in the same chiral limit in Patel's models, there is a gain in entropy due to easy string breakage as a remnant of the deconfinement transition. It is difficult to estimate whether a critical number of flavors exists that will enable one of the competing effects to win. As long as chiral symmetry is not implemented in the massless limit, such competing effects cannot be resolved by Patel's models, even not if matter fields are included. For calculations within these models that go beyond the heuristic level of this presentation, we refer to the original literature [307].

6.6 Further reading

Dual Ginzburg-Landau models explain the confinement mechanism as a dual Meissner effect . A summary and further references can be found in [309]. As an early example for a model that combines mesonic and gluonic degrees of freedom see [308]. Effective models for QCD at finite temperature *and finite baryon density* are discussed for example in [310]. These models predict a region of first-order chiral transitions for large enough baryonic chemical potential μ and finite temperature T.

Chapter 7

Phenomenological Applications to Relativistic Heavy-Ion Collisions

7.1 The QCD transition in the lab

7.1.1 Scales and observables

The experimental possibilities to test the QCD predictions of finite-temperature phase transitions from normal hadronic matter to a quark-gluon plasma are limited if we face the transition temperature of $\approx 10^{12} \, ^\circ K$. Most probably, the QCD transition occurred 10^{-6} sec after the big bang when the universe cooled down to the transition temperature. In contrast to the initial hope there are most likely no phenomenological implications of the phase transitions of particle physics for the early universe, neither from the QCD transition(s), nor from the electroweak transition. For realistic physical parameter values and vanishing or small baryon densities the chiral and the deconfinement transitions of QCD are smooth crossover phenomena. Also the electroweak transition is a crossover for the physical Higgs mass that lies above the lower bound, which was experimentally obtained so far. In the vicinity of the QCD transition (and analogously for the electroweak transition) two very different time scales enter the evolution: the QCD time scale of the order of $1/T_c \sim 1 fm/c \sim 0.33 \times 10^{-23} sec$ and the Hubble time of the order of $1/\chi = 0.36 \times 10^{-4} sec \sim 10^{19}/T_c$ with $\chi = \sqrt{8\pi G B/3}$, where G is the gravitational constant and B the vacuum energy density represented by the bag constant. The time dependence of temperature $T(t)$ follows from Einstein's equations combined with QCD's equation of state that is not (yet) known in its final form, but approximated, for example, by the bag-model equation of state [311], leading to

$$\frac{-dT}{Tdt} = \frac{1}{2t} \propto \frac{T^2}{M_{Pl}} \tag{7.1.1}$$

with M_{PL} denoting the Planck mass, i.e., the expansion rate is 10^{-19} times slower than the typical time scales in QCD, so that the phase conversion in the early universe proceeded in an adiabatic way without producing remnants which could be visible nowadays.

Promising alternatives to reproducing the QCD transition in some kind of little bang in laboratory experiments are heavy-ion collisions at ultra-relativistic energies. In such collisions a large amount of the initial kinetic energy will be concentrated in a short space-time interval and form a *fireball* of matter which could reach thermal equilibrium.

In the last decades, it has become technically feasible to create matter with energy densities up to several hundred times that of ordinary nuclear matter. The available energy for particle production is specified in terms of $\sqrt{s}$, where $\sqrt{s}$ is the total center-of-mass energy. Given the value of $\sqrt{s}$ and the mass number A of colliding ions, one would like to estimate the initial temperatures that can be reached in such a collision. The basic observable is the multiplicity per unit rapidity dN/dy of secondary hadrons which are emitted in the collision. If one extrapolates the known relation between $\sqrt{s}$ and $(dN/dy)_p$ in proton-proton collisions to central nucleus (A)-nucleus (A) collisions, the relation is given as [312]

$$\left(\frac{dN}{dy}\right)_{AA} = A^\alpha \cdot 0.8 \cdot \ln \sqrt{s} \qquad (7.1.2)$$

with $\alpha \geq 1.1$. The multiplicity density of the final state hadrons can be related to the initial state energy density ε either with (approximate) energy conservation (free flow) or entropy conservation. For free flow the relation is given by

$$\varepsilon = \{(dN/dy)_A m_T\}/(\pi R_A^2 \tau). \qquad (7.1.3)$$

Here R_A is the nuclear radius, m_T the transverse mass and τ the equilibration time. With $\tau \approx 1$ fm/c and $m_T \approx 0.5$ $[GeV]$ Eq. (7.1.3) leads to an average initial energy density of $1.5 - 2.5$ $[GeV/fm^3]$ for central $Pb - Pb$ collisions at SPS and $4.6 - 7.8$ $[GeV/fm^3]$ at LHC. An uncertainty is involved in what is called the equilibration time τ. $\tau = 1fm/c$ is just an order of magnitude. (In the context of so-called saturation models τ has been estimated as $0.2fm/c$, leading to correspondingly larger initial energies [313].) Depending on the values used for τ, one obtains for Pb-Pb collisions at the SPS (at CERN in Geneva) for the initial energy density $\epsilon_0^{SPS}(\tau \sim 1fm/c)$ values between $3.5[GeV/fm^3]$ and $16[GeV/fm^3]$, and for a collision energy of $\sqrt{s} = 130A \cdot [GeV]$ for Au-Au collisions at RHIC (at the Brookhaven National Laboratory (BNL) in the US) initial energy density values between $20[GeV/fm^3]$ and $98[GeV/fm^3]$ [314]. In nucleus-nucleus collisions at the Linear Hadron Collider (LHC) at CERN the initial energy density of the equilibrated medium is expected to be in the range of $400 - 1300[GeV/fm^3]$.

From the initial energy density one would like to extrapolate the initial temperature after equilibration. The energy density is translated to a temperature according to an ideal-gas relation $\epsilon \propto T^4$ (although there are strong deviations from ideal-gas relations in the transition region), where the proportionality constant depends on the number of the included degrees of freedom. For example, if we assume that the dominant constituents of the partonic medium produced in heavy-ion collisions are gluons, the proportionality constant stands for the number of effective gluonic degrees of freedom in the transition region that differs from the corresponding number in an ideal gas of gluons. Estimates of this factor are taken from simulations of lattice gauge theory of pure gluon gases. Using for the thermal energy density ϵ the initial energy density ϵ_0, one obtains for the initial temperatures at SPS (CERN) $200[MeV] < T < 330[MeV]$, at RHIC (BNL) $210[MeV] < T < 600[MeV]$, and at LHC (CERN) $1000[MeV] < T1200[MeV]$, respectively. Therefore, in spite of the large quantitative uncertainties, the initial temperatures, at least at LHC, exceed by far T_c, the crossover temperature of $SU(3)$ lattice gauge theory with dynamical fermions. Also the initial energy densities reached at LHC and RHIC lie clearly far above the critical energy density as it is predicted from lattice gauge theory. One should keep in mind, however, that a sufficiently high energy density alone is not sufficient to produce a quark-gluon plasma that can be described by equilibrium thermodynamics. The initial distribution of partons is not at all thermal. It is a very challenging question whether the system has enough time to equilibrate to a quark-gluon plasma before it starts hadronizing. It is another open question whether a thermal nature of the plasma is preserved during hadronization so that the final hadronic particle yields exhibit a thermal equilibrium population. It seems to be extremely hard to answer these questions from so-called first principles, starting from QCD out-of equilibrium.

In applying the formulas which are familiar from equilibrium thermodynamics, we have implicitly made use of the fact that energy, entropy, temperature are well-defined notions in describing the collision. This assumption will be discussed later on.

At this point, let us pause for a comparison with the observation of a phase transition under daily life conditions, the boiling of water. Usually one does not appreciate all the well defined experimental conditions. The fluid container has a fixed volume, the fluid is at rest and in thermal equilibrium. Calibrated thermometers are at hand, whose Hg-column grows linearly in the considered temperature range. The phase transition is easily seen as conversion from the liquid to the vapor phase while the temperature stays constant. Also heating sources are available such that tuning of the temperature does not pose a problem either. The temperature may be tuned adiabatically or as a quench as one likes.

It is not surprising that at temperatures where hadronic matter gets dissolved into its very components, the experimental conditions drastically change. In collider experiments the volume is not fixed in the evolution of the plasma to the hadron gas. At high collision energies the nuclei interpenetrate each other at the collision and recede as Lorentz contracted pancakes, leaving a hot vacuum with secondaries of the collisions between them (in the central rapidity region). After hadronization this hot area is primarily a pion gas.

Length scales

At LHC-energies, for example, the *volume* at the transition is estimated to be 5-8 times larger than the initial volume, for Pb-Pb collisions the critical volume is of the order of 800 - 1200 fm^3. After the transition the system continues expanding until freeze-out, where interactions can be neglected or/and the mean free path of particles gets of the order of the size of the system. (Both definitions of 'freeze-out' lead to different estimates for freeze-out volumes). At LHC freeze-out volumes could be of the order of 10^4 to 10^5 $[fm^3]$ [312].

Time scales

Next we come to various time scales involved in heavy-ion collisions. Note that the total duration of a 'little bang' is only of the order of 10^{-23} sec. (The characteristic length scale of $1[fm]$ is a very short distance for light to pass by.) Time scales are the equilibration time τ_0, the freeze-out time, the delay caused by a possible first-order phase transition, the conversion rate of one phase into the other and the expansion rate. All of them are between say $0.1 - 100[fm/c]$. The largest value refers to an extreme delay of freeze-out due to strong supercooling, the smallest one corresponds to a more recent estimate for the equilibration time (Shuryak, 1992). The equilibration time is the time it takes until the system reaches a stage of local thermodynamic equilibrium, when it has passed an intermediate preequilibrium stage after the bang. 'Local equilibrium' means, that energy, entropy, pressure and temperature can be locally defined. In contrast to our water boiling experiment, the temperature is – after it is defined at all – a function of space and time $T(\vec{x}, t)$ in a volume, whose geometry is not even fixed, but depends on the impact parameter value b.

Volumes of the order of thousands of fm^3 and time scales of some tens of fm$/c$ can be measured with pion-interferometry, especially a time delay due to a transition. For example, the lifetime of the fireball is estimated to be 10-20fm$/c$ for Si+Pb(Au) collisions at 14.6 GeV$/n$-n from pion-interferometry measurements at the AGS-machine [315].

Temperature scales

Temperature scales which should be distinguished are the initial temperature at the onset of local equilibrium T_i, the transition temperature T_c, possible values of superheating/supercooling effects and the decoupling temperature T_d (of the order of the pion mass).

Candidates for thermometers are (among others) thermal photon and dilepton spectra [316, 346] The differential cross section of dilepton production as a function of the invariant lepton pair mass M and the rapidity y, i.e. $(d^2\sigma/dM^2 dy)_{y=0}$, is predicted to scale according to exp $(-M/T)$, T is the temperature of the emitting system. Unfortunately these thermometers like to hide in the background of other dileptons. Thermal dileptons are difficult to be identified in the total dileptonic yield (see below). Also pions have been proposed as thermometers for measuring the freeze-out temperature [317]. Pions from the decay of the $\Delta(33)$-resonance have a characteristic p_T-distribution which is very different from that of primary pions. The ratio of $\Delta(33)$-resonances to nucleons sensitively depends on the temperature via Boltzmann factors. Thus a measurement of the p_T-distribution allows the identification of decay pions and a measurement of the temperature dependent ratio of $\Delta(33)$-resonances to nucleons.

If it is difficult to measure the temperature, let us see, how we can tune it. The initial temperature changes with the initial energy density ($\varepsilon \propto T^4$ for an ideal gas), and that depends on ln $\sqrt{s}$. Thus in principle one could vary s for a given nucleus A. Due to the ln $\sqrt{s}$-dependence this would require a large variation in the incident beam energy, accompanied by a considerable loss in the luminosity, if $\sqrt{s}$ is reduced [312, 318]. For asymmetric collisions between small and big nuclei, the energy can be increased by going from peripheral to central collisions, i.e. by varying the impact parameter. For symmetric collisions the realistic possibility which remains is the variation of A, i.e. the type of nucleus itself. Going from $S - S$ to $U - U$-collisions roughly gives a gain in energy density by a factor of $7 - 8$ at the price that the volume changes as well. Going from fixed target to collider experiments one expects a factor of $2 - 4$ in the increase of the initial temperature T_i. Depending on the experiments, initial temperatures have been estimated between $190[MeV]$ and $900[MeV]$.

Last but not least, to complete our comparison to a phase transition under 'normal' conditions, we have to find observable signatures of the QCD transition. The goal is to identify observables that could reflect almost constant pressure or temperature over an interval where energy and entropy densities (ε and s) rapidly change, where the phase conversion takes place. Typical observables, which are at our disposal in heavy-ion collisions, are multiplicity distributions in rapidity space and average transverse momenta $\langle p_T \rangle$. It turns out that $\langle p_T \rangle$ values can be a measure for the initial pressure and temperature, while (dN/dy) distributions depend on the initial values of ε and s. Thus a $T - \varepsilon$ diagram corresponds to a $\langle p_T \rangle - dN/dy$ plot.

There are a variety of other signatures which are in principle sensitive to a transient plasma and to the transition dynamics as well. Here we restrict our discussion to dilepton production for illustration, but without prejudicing other possible processes and their signatures. The difficulty in general is not to predict some signatures, but signatures which are unique, because in most cases alternative explanations exist as well that are not based on an underlying assumed phase conversion.

We summarize the main complications we have to face in heavy-ion experiments. The basic assumptions that the fireball is large enough and long-lived enough to reach thermal equilibrium, the application of thermodynamic concepts have to be checked and justified. Due to the expansion dynamics, different competing time and length scales are involved. The space-time expansion from the initial thermalization until freeze-out has to be unfolded of the final observables, if one wants to compare the signatures with predictions of static, microscopic equilibrium quantities. Usually the folding is provided by a hydrodynamic description.

Hydrodynamics is a useful computational tool in estimating bulk features like leptonic or hadronic particle yields, multiplicity fluctuations, orders of lifetimes etc. Note also that hydrodynamics describes off-equilibrium situations although it is based on local equilibrium conditions. The system is expanding and cooling and out of global equilibrium. In this sense, it is also a conceptual framework to treat nonequilibrium situations in a way that all the information of equilibrium QCD is not lost, but can be built in. This information contains the equation of state, elementary cross sections, structure functions and other derived quantities of equilibrium thermodynamics which we have partly outlined above.

There are other concepts to treat off-equilibrium aspects in heavy-ion collisions which will not be covered in this book.

The effect of a phase transition on the hydrodynamical flow depends on the type of phase transition dynamics. If the transition is of first order but proceeds smoothly, close to equilibrium, the effect is just to slow down the expansion. Discontinuities in thermodynamic quantities would be reflected in shock-like discontinuities of the fluid. If the transition is of first order but involves metastable states - a supercooled plasma or a superheated hadron gas - deflagrations or detonations may evolve with the possible effect of large multiplicity fluctuations.

Such explosive processes are one source of entropy production during the evolution. Other sources are dissipation effects and the freeze-out transition. All of them are estimated to produce little extra entropy such that the approximation of entropy conservation during the evolution seems to be justified [319].

Although dissipation effects on entropy production may be small, pure glue is rather viscous. It is amusing to estimate the viscosity of a gluon gas at tera degrees in comparison to 'normal' gases. For T close to T_c the shear viscosity is of the order of the Λ-QCD scale. Thus it is 10^{16} times the viscosity of classical gases under 'normal' conditions (where it is 10^{-5} kg$/m \cdot s$) [320].

The main ingredients in a hydrodynamic description are the initial conditions and the equation of state. In order to check whether the basic condition for applying the hydrodynamic description is not violated one has to estimate the mean free path of a particle (here say a quark) in a medium that has to be much smaller than the size L of the medium. The mean free path of a quark at a certain initial energy density for a specified type of heavy-ion collisions is then compared to the diameter of the considered type of nucleus.

Let us anticipate the overall judgment of an experimental manifestation of the phase conversion to a quark-gluon plasma and back to the hadronic phase. It seems to be very hard to find unique signatures for the existence of a thermal quark-gluon plasma. In general, alternative explanations, which completely abandon thermodynamic concepts, exist as well. It is even harder to find signatures that are sensitive to the nature of the phase conversion, in particular if the transition is smooth and proceeding in equilibrium. If there is neither a first-order nor a second-order transition, it is only a rapid crossover and a fast cooling after the "little bang" that offer a chance for observable remnants.

7.1.2   The hydrodynamic framework

The derivation of the hydrodynamic equations can be found in various textbooks [321], and in their adaption to heavy-ion collisions in [322] or [323] or [319]. We sketch the main steps in deriving an appropriate form for heavy-ion collisions. The adaption to heavy-ion collisions amounts to a suitable choice of coordinates and a set of initial conditions which is based on experimental observations.

Hydrodynamic equations describe the evolution of a gas (or fluid) in space and time. The gas is specified by a local temperature, pressure, energy, entropy and velocity. The equations result from constraints of energy and momentum conservation and other conserved quantities in case, such as the baryon number. If we first neglect dissipative effects (viscosity, thermal conductivity), the energy-momentum tensor of a relativistic perfect fluid in motion with velocity u^μ is obtained by a Lorentz boost from its rest frame as

$$T^{\mu\nu} = (\varepsilon + p)u^\mu u^\nu - g^{\mu\nu}p . \tag{7.1.4}$$

The equations for energy-momentum and baryon number conservation are

$$\partial_\mu T^{\mu\nu} = 0 \tag{7.1.5a}$$

and

$$\partial_\mu J^\mu = \partial_\mu(n_B u^\mu) = 0 \tag{7.1.5b}$$

respectively, where $n_B(\vec{x}, t)$ is the local baryon number density. Using Eq. (7.1.4), contraction of Eq. (7.1.5a) with u_ν leads to

$$u^\nu \partial_\nu \varepsilon + (\varepsilon + p)\partial_\nu u^\nu = 0, \tag{7.1.6}$$

where ϵ denotes the energy density, p the pressure. An analogous equation can be derived for the entropy density s, which can be converted to a temperature equation in the baryon free case. Contracting Eq. (7.1.5a) with $(g_{\nu\rho} - u_\nu u_\rho)$ leads to the second hydrodynamic equation

$$(\varepsilon + p)u^\tau \partial_\tau u_\lambda - \partial_\lambda p + u_\lambda u^\tau \partial_\tau p = 0 \ . \tag{7.1.7}$$

The next step is to choose coordinates adapted to a plasma evolution in cylinder geometry, where the z-axis is commonly identified with the beam axis. Now one can express Eqs. (7.1.6) and (7.1.7) in coordinates z and t, the four-velocity u^μ of the matter is written as

$$u^\mu = \frac{1}{\sqrt{1 - v_z^2 - v_r^2}}(1, v_z, v_r, 0) \ . \tag{7.1.8}$$

For vanishing radical velocity v_r (which is frequently used as approximation), the remaining components of u^μ are parameterized according to

$$u^\mu = (\cosh\theta, \sinh\theta, 0, 0) \tag{7.1.9a}$$

where θ is the fluid rapidity, defined via

$$\theta = \arctan v_z \ . \tag{7.1.9b}$$

A more convenient choice of variables are the space-time rapidity η defined as

$$\eta = \frac{1}{2}\ln\frac{t+z}{t-z} \tag{7.1.10}$$

and the proper time τ

$$\tau = \sqrt{t^2 - z^2}$$

with the inverse transformations

$$t = \tau\cosh\eta$$
$$z = \tau\sinh\eta \ . \tag{7.1.11}$$

Note that the fluid rapidity coincides with the space-time rapidity in the case of $v_z = z/t$.

In terms of these new coordinates, the hydrodynamic equations (7.1.6) and (7.1.7) for a longitudinal motion are

$$\tau\frac{\partial\varepsilon}{\partial\tau} + \tanh(\theta - \eta)\frac{\partial\varepsilon}{\partial\eta} + (\varepsilon + p)\left[\frac{\partial\theta}{\partial\eta} + \tanh(\theta - \eta)\tau\frac{\partial\theta}{\partial\tau}\right]$$
$$= 0 \tag{7.1.12a}$$

and

$$\frac{\partial p}{\partial \eta} + \tanh(\theta - \eta)\tau\frac{\partial p}{\partial \tau} + (\varepsilon + p)\left[\tau\frac{\partial \theta}{\partial \tau} + \tanh(\theta - \eta)\frac{\partial \theta}{\partial \eta}\right]$$
$$= 0. \tag{7.1.12b}$$

This set is complemented by a third equation which follows from baryon number conservation (Kajantie et al., 1983)

$$\tau\frac{\partial n_B}{\partial \tau} + n_B\frac{\partial \theta}{\partial \eta} + \tanh(\theta - \eta)\left(\frac{\partial n_B}{\partial \eta} + n_B\tau\frac{\partial \theta}{\partial \tau}\right) = 0 . \tag{7.1.13}$$

Now we have three equations (7.1.12)-(7.1.13) for four unknown functions: the energy density ε, the pressure p, the fluid rapidity θ and the baryon number density n_B, all of them being functions of η and τ in a longitudinal expansion. Thus, in order to find solutions of the set (7.1.12)-(7.1.13), we have to supply one additional equation and to specify the initial conditions. The additional equation is an *equation of state* relating ε and p or T and s. (Other combinations are possible as well.) As an example, we will discuss the bag model equation of state in the next section. Several proposals have been made for the initial conditions. Here we sketch only the Bjorken-Shuryak expansion scenario. We discuss the longitudinal solutions of Eqs. (7.1.12)-(7.1.13) for this choice. The solutions considerably simplify in this special case, which may be one reason why they are frequently used in the hydrodynamic treatments of matter evolution in heavy-ion collisions.

Radial solutions of Eqs. (7.1.12) are rarefaction waves propagating from the boundary into the fluid with the velocity of sound. They differ in an essential way from Bjorken's scaling solution as they are independent of proper time τ. Radial solutions enter measurements of enthalpy and pressure.

Bjorken-Shuryak expansion scenario

In the Bjorken-Shuryak scenario [324, 325] several experimental observations are taken into account. The phenomenon of 'nuclear transparency' leads to a separate treatment of the central rapidity and the fragmentation region. Nuclear transparency means the effect that a large fraction of the incoming energy is carried away by two receding nucleons in a nucleon-nucleon collision at high energy. Similarly, in a nucleus-nucleus collision the baryon contents of the colliding nuclei interpenetrates at the collision and recedes as two Lorentz contracted "pancakes" after the collision. The central rapidity region refers to the fluid of quanta contained in the region between the receding pancakes. In the hadronic phase it mostly consists of pions. Thus it should be a good approximation to neglect the baryon number. Setting $n_B = 0$ leads to a first simplification of the hydrodynamic set of equations. The separate treatment of the central and the fragmentation regions is justified only if both regions are well separated in phase space. Experimental conditions should be checked to guarantee this, otherwise an analysis in this picture is not adequate. At future colliders (RHIC and LHC) these conditions may be satisfied.

Secondly, the pronounced space-time correlations are observed in particle production in the sense that particles with large longitudinal momenta are produced at a late time, those with low momenta promptly, in the center-of-mass system. This is nothing but the twin paradox of special relativity. Particles live longer in case their velocities are higher. In Bjorken and Shuryak's ansatz the effect of time dilatation is incorporated in the boundary conditions. Consider an ensemble of particles which are produced at $z = 0 = t$. If it is only the proper time τ which determines the moment of disintegration, all particles which measure the same τ in their rest frame constitute an initial condition at $\tau = \tau_0$. That is, the initial condition refers to a hyperbola $\sqrt{t^2 - z^2} = \tau_0$ of constant proper time τ_0. The space-time rapidity or light cone variable $\eta = \frac{1}{2} \ln \frac{t+z}{t-z}$ specifies the position on this hyperbola. Two distinct positions are related via a Lorentz boost in z-direction. Since particles in the fluid element are supposed to move as free particles, their velocity component v_z is given by z/t. The physics of a z-slice of a fluid element at time t is equivalent to the physics of a z'-slice at time $t' = z'/v_z$. This is the scaling property in Bjorken's scaling ansatz. For $v_z = z/t$, the space-time rapidity η equals the rapidity y

$$ y = \frac{1}{2} \ln \frac{1 + v_z}{1 - v_z} = \frac{1}{2} \ln \frac{E + p_z}{E - p_z} \,, \qquad (7.1.14) $$

if the four-momentum $\underline{p}$ is parameterized as $\underline{p} = (E, p_z, p_t)$, p_t being the transverse momentum.

A third feature, which is observed in proton-proton collisions is the plateau structure of inclusive cross sections when they are plotted as functions of y. A plateau for central values of y is also expected for nucleus-nucleus collisions. At least the particle multiplicity depends only weakly on y for central rapidities [324]. Accordingly, a further simplifying assumption seems to be justified. The local thermodynamic quantities like ε, p, T, s depend only on τ_0, but not on $\eta(\tau_0)$, when the hydrodynamic expansion commences. Thus the initial condition is invariant under Lorentz boosts in z-direction. The dynamics preserves Lorentz covariance, which is most easily seen from the tensor equation (7.1.5a). Therefore, we will look for solutions $\varepsilon(\tau), s(\tau), T(\tau)$ of Eqs. (7.1.12) depending merely on τ. Inserting $v_z = z/t$ in the collective four-velocity of the fluid leads to

$$
\begin{aligned}
u^\mu &= \frac{1}{\sqrt{1 - v_z^2 - v_r^2}}(1, v_z, v_r, 0) = (t/\tau, z/\tau, 0, 0) \\
&= (\cosh\eta, \sinh\eta, 0, 0) \ .
\end{aligned}
\tag{7.1.15}
$$

A comparison with Eq. (7.1.9a) shows that the fluid rapidity can be identified with the space-time rapidity η, which furthermore coincides with the rapidity y. With $\eta = \theta$, Eqs. (7.1.12) simplify to

$$
\tau\frac{\partial\varepsilon}{\partial\tau} + \varepsilon + p = 0
$$

$$
\frac{\partial p}{\partial y} = 0 \ .
\tag{7.1.16}
$$

The baryon number is set to zero in the following considerations. The entropy equation for $s(\tau)$ simplifies to

$$
\tau\frac{\partial s}{\partial\tau} + s = 0 \ ,
\tag{7.1.17}
$$

where the relations $u^\eta = 0$ and $(u^\tau = u^t \cosh\eta - u^z \sinh\eta)$ have been used. The solution is

$$
\frac{s}{s_0} = \frac{\tau_0}{\tau} \ .
\tag{7.1.18}
$$

As can be seen, upon integration over $d^3x = \tau dy\, d^2x$, the entropy per given rapidity interval remains constant as long as the hydrodynamic equations can be applied. This need not hold throughout all stages of the expansion, especially not close to freeze-out or in the intermediate period, where the plasma converts to the hadronic phase in one or another way. Let us assume that it approximately holds. Then the important feature of Eq. (7.1.18) is that it allows to infer the entropy density in the initial state (more precisely $s_0\tau_0$) from an observation in the final state (the pion multiplicity). Under the same assumptions as above, the temperature equation simplifies in (τ, η)-coordinates to

$$
c_s^2 + \tau\partial_\tau \ln T = 0
\tag{7.1.19}
$$

for $\mu = 0$, where by definition

$$c_s^2 = \frac{\partial p}{\partial \epsilon} \, , \tag{7.1.20}$$

c_s denotes the velocity of sound. Integration of Eq. (7.1.19) gives

$$T = T_0(\tau_0/\tau)^{c_s^2}. \tag{7.1.21}$$

For a massless free gas, the speed of sound is $1/\sqrt{3}$ in units where $c = 1$. Thus the temperature drops more slowly than the entropy density. In fact, the predicted decrease may be too slow, since transverse expansion has been neglected so far.

Finally, we have to solve Eq. (7.1.16) for the energy density as a function of proper time. One possibility is to use an equation of state (in principle, it should be *the* equation of state of QCD) to eliminate the pressure in Eq. (7.1.16). The result is that the energy density does not merely decrease because of the expanding volume in proper time, but also due to the pressure exerted by the gas of the covolume. We come back to the equation of state in the next section. Note that the velocity in transverse direction v_r has been neglected so far. We abbreviate the solutions in 'Bjorken-Shuryak's scenario' as Bjorken's scaling solution.

7.1.3 The bag-model equation of state

Although the bag model leads to a crude description of the equation of state for QCD and is finally not appropriate, and not derived from first principles, it represents a typical phenomenological approach, quite useful for an initial discussion of qualitative implications of first- or second-order phase conversions. Therefore we address a short subsection to the bag model. It is often used in combination with the hydrodynamic equations and leads to quantitative predictions in the end. In the M.I.T.-bag model, the basic features of QCD – confinement and asymptotic freedom – are effectively incorporated via bags [326,327]. In the hadronic phase, quarks and gluons are allowed to move freely or with perturbatively small interactions inside small volumes of space inside the bags. Outside the bags, the quarks are forbidden to move as free particles. The vacuum outside the bags is given a constant energy density B (the bag constant), which keeps the quarks and gluons confined to the bags. During the phase transition, latent heat is necessary to liberate the color degrees of freedom. It turns out to be proportional to B. Its original value $(0.145 \, [GeV]^4)$ [328] was based on fitting the mass spectrum at $T = 0$ and low density in the M.I.T.-bag model. The effective value for the 'vacuum pressure' B which should be used in the quark-gluon plasma phase at a baryonic matter density of $n_B = 1/[fm]^3$ is $0.5 \, [GeV/fm^3]$ [329].

An additive shift B in the energy density of the plasma due to the vacuum energy is obtained, if $\ln Z$ of an otherwise free gas of quarks and gluons is shifted by $-BV/T$, that is

$$T \ln Z \text{ (plasma phase)} = \text{free gas contribution} - BV/T \,. \qquad (7.1.22)$$

The free gas contribution follows from the usual expression for a free gas of particles and antiparticles with mass m, chemical potential μ, and degeneracy factor g. In the large volume limit it is given by

$$\ln Z(T, \mu, V) = \frac{gV}{6\pi^2 T} \cdot \int_0^\infty dK \frac{K^4}{(K^2 + m^2)^{1/2}}$$
$$\cdot \left[\frac{1}{\exp\{[(K^2 + m^2)^{1/2} - \mu]/T\} \pm 1} \right.$$
$$\left. + \frac{1}{\exp\{[(K^2 + m^2)^{1/2} + \mu]/T\} \pm 1} \right], \qquad (7.1.23)$$

where the "+"-sign refers to fermions and the "-"-sign to bosons. Adding up the various contributions from bosons (gluons), fermions (quarks and antiquarks) and the vacuum, one is lead to

$$T \ln Z = \frac{1}{6} N_c N_f V \left(\frac{7}{30} \pi^2 T^4 + \mu_q^2 T^2 + \frac{1}{2\pi^2} \mu_q^4 \right)$$
$$+ \frac{\pi^2}{45} N_g V T^4 - BV. \qquad (7.1.24)$$

Here N_c is the number of colors, N_f the number of flavors, N_g the number of gluons and μ_q the chemical potential due to quarks. The standard thermodynamic relations Eqs. (2.1.1) lead to the following expressions for the energy density, pressure and entropy density in the plasma phase

$$\varepsilon_p = 111aT^4 + B \qquad (7.1.25a)$$
$$p_p = 37aT^4 - B \qquad (7.1.25b)$$
$$s_p = 148aT^3 \,, \qquad (7.1.25c)$$

where $a = \pi^2/90$. These expressions hold for $\mu = 0$, $N_c = 3$, $N_f = 2$,

$N_g = 8$. The general expression for ε_p is

$$\varepsilon_p = \frac{N_c N_f}{\pi^2} \left(\frac{7\pi^4}{60} T^4 \right) + \frac{\pi^2}{15} N_g T^4 + B . \qquad (7.1.26)$$

From Eqs. (7.1.25a and b) we can easily read off the bag model equation of state in the plasma phase as

$$p = \frac{1}{3}(\varepsilon - 4B) , \qquad (7.1.27)$$

which remains valid for $\mu \neq 0$.

Similarly, expressions for ε, p and s are obtained in the hadronic phase, when it is described as a free gas of the lightest mesons and baryons, i.e. pions, nucleons and antinucleons, where the baryonic contribution is sometimes omitted. For pions analytic expressions for ε, p and s can be derived in terms of modified Bessel functions following from Eq. (7.1.23). Contributions of the nucleon-antinucleon gas can be calculated numerically. Heavier mass particles are often omitted for moderate temperatures ($T \leq 250 \, [MeV]$), although the restriction to pions is rather questionable above $T_c \geq 150 \, [MeV]$. Here we state the result for the limit of a gas of *massless* pions

$$\varepsilon_h = 9aT^4 \qquad (7.1.28a)$$

$$p_h = 3aT^4 \qquad (7.1.28b)$$

$$s_h = 12aT^3 \qquad (7.1.28c)$$

such that

$$\varepsilon_h = 3p_h \qquad (7.1.29)$$

is the bag-model equation of state in the hadron phase. The finite T (and finite μ) transition occurs, when the following Gibbs criteria are satisfied

$$p_h = p_p = p_c \qquad (7.1.30a)$$

$$T_h = T_p = T_c \qquad (7.1.30b)$$

$$\mu = 3\mu_q = \mu_c. \qquad (7.1.30c)$$

The indices h, p stand for the hadron and plasma phases, c for the critical value, μ is the chemical potential associated with nucleons. Pressure balance at T_c relates T_c to the bag constant

$$T_c = (B/34a)^{1/4} . \qquad (7.1.31)$$

The latent heat, determined as the gap in the energy densities $\varepsilon_p - \varepsilon_h$ at the transition, is $4B$ in this model, e.g. for the effective bag constant of $0.5 \ [GeV/fm^3]$ the latent heat is $2 \ [GeV/fm^3]$.

The bag model leads to a first-order transition at finite T and vanishing μ. There is a finite gap in energy and entropy densities, while the pressure is continuous. Note that in the mixed phase, the velocity of sound c_s vanishes, since $c_s = dp/d\varepsilon|_{T_c} = 0$. This is true only as long as $\mu = 0$. An exception is a 2nd order transition at $\mu_c \neq 0$, but $T_c = 0$ for a particular value of B, which we will not consider further here [323].

Equations. (7.1.27) and (7.1.29) can be easily combined to a single equation by using θ-functions as projections on the distinct phases above and below T_c. Similarly it is not difficult to formulate an equation of state for a second order transition by smoothly interpolating the step-like behavior of the bag model equation of state. Such an ansatz has been proposed by [319] for $s(T)$. Although the interpolation is ad hoc, it provides a useful check, how sensitively phenomenological implications depend on the order of the transition. The difficulty is to *derive* such an equation for a second-order transition within an effective model. The $O(4)$-model in three dimensions allows a second-order transition, but is supposed to describe only the low-temperature phase of QCD ($T \leq T_c$) or the immediate vicinity above T_c, where pions and sigma mesons have not yet dissolved in their constituents.

The bag model does not provide an adequate description of the transition region. Even at the transition point, the plasma and the hadron phases are treated as noninteracting gases differing only in the degrees of freedom and the vacuum energy. An increasing number of hints warns of a naive counting of the number of modes which are taken to be the same as in the limiting cases of high and low temperatures. In particular the number of degrees of freedom of a hadron gas is a delicate problem, if the change of hadron masses as a function of temperature and density is respected. In this case a counting of pion and nucleon degrees of freedom is certainly insufficient. In the vicinity of T_c the confinement/deconfinement properties should be implemented in a slightly more sophisticated way than with a single parameter B. Nevertheless the bag model is frequently used for temperatures $T \sim T_c$. The reason is probably its very tractable analytic form compared to plots of numerical simulations to date.

At high or low temperatures the bag model is more adequate, e.g the gluonic sector may be treated as a gas of noninteracting glueballs at low T and of gluons at high T.

Various improvements on QCD's equation of state have been proposed within the framework of the bag model. Finite quark masses and perturbative QCD corrections to the partons inside the bag can be taken into account [330,331] leading to corrections of ϵ, p and s. Also strange quarks, finite masses for pions, and higher mass hadrons can be included.

The ultimate goal is an equation of state from lattice QCD with dynamical quarks, which is merely based on nonperturbative ingredients. The numerical data should then be presented in a feasible parametric form to facilitate its handling.

In the next section we will see, what we can directly learn from heavy-ion experiments about QCD's equation of state.

7.2 Signatures sensitive to the nature of the phase transition

Theoretical and experimental tools are described which are sensitive to the nature of the transition, in particular to its order. The sensitivity holds at least in principle. The dependence on the transition dynamics is sometimes hidden in the space-time expansion. The signatures reveal characteristic features like time delays, nucleation of bubbles, strong correlations or large fluctuations, whatever the order of the transition is, in an indirect way.

Let us start with direct experimental tests of the equation of state. We sketch the possibilities to measure two thermodynamic quantities, the entropy and the internal energy.

7.2.1 *Thermodynamic observables*

The basic observables which are at our disposal in heavy-ion collisions are rapidity distributions of final state particles and their transverse momentum distributions. An extraction of the equation of state requires measurable observables which are related to $\varepsilon(T,\mu)$, $s(T,\mu)$ or $p(T,\mu)$. The signals of a first-order transition in a finite volume may be qualitatively very similar to a sharp crossover phenomenon consisting in a rapid rise in the effective number of degrees of freedom over a small range of temperatures, say less than 10 $[MeV]$. These numbers are exposed in s/T^3 or ε/T^4, cf. e.g. Eqs. (7.1.25) of the bag-model equation of state. Thus we have to identify the observables which are related to s, ε, p and T. Roughly speaking, temperature and pressure are measured by the average transverse momentum $\langle p_T \rangle$

(as usual under certain restrictive conditions), energy and entropy by the particle multiplicity distribution in rapidity space.

We consider the baryon-free case, i.e. $\mu = 0$. Four quantities s, ε, p and T have to be determined. An equation of state is a relation between any two of these four variables $(\varepsilon(p), s(T), s(\varepsilon))$. It can be obtained from experiments (and compared with theoretical predictions) if two relations are used as experimental input. For example, if $\varepsilon(T)$ and $p(T)$ were experimentally known, $s(T)$ follows from $dp/dT = s$ and $\varepsilon(p)$ is the equation of state. Alternatively, if the initial condition $s_0(\tau_0)$ is fixed from a measurement of dN/dy, then $s(\tau), T(\tau)$, thus $s(T)$ are known from the hydrodynamic equations, and $p(T)$ follows as integral over s. Suppose that the second (experimentally determined) relation is provided by a measurement of $\langle p_T \rangle$ as a function of $s^{1/3}$, where $\langle p_T \rangle$ is related to s/ε (see below) and $s^{1/3} \propto T$. The knowledge of $(s/\varepsilon)(T)$ yields $\varepsilon(T)$, thus all relations $s(T), p(T), \varepsilon(T)$ are known [319].

In the following we explain relations between dN/dy and s_0 or ε_0, $< p_T >$ and ε/s or T. Similarly one can derive relations between the transverse momentum distributions $\langle p_T \rangle$ and p or $(\varepsilon + p)$ and show why $\langle p_T \rangle$ vs $\langle dN/dy \rangle$ diagrams are roughly equivalent to $(T$ vs $s)$ or $(T$ vs $\varepsilon)$-diagrams. In case of a phase transition a flattening of the $\langle p_T \rangle$-distribution is expected, which was originally proposed by van Hove [347] as a possible signature for a phase transition.

Entropy measurements

The basic step is to identify the entropy density with the particle density in the final state. The final state particles mostly consist of pions. Pions are nearly massless bosons, thus their entropy S is approximately proportional to their number N. For the densities we have

$$s = \alpha \cdot n \tag{7.2.1}$$

with $\alpha = 3.6$ for a free gas of massless pions. The rapidity distribution is obtained from the particle density by an appropriate integration. When the four-volume element is expressed in terms of the space-time rapidity η and the proper time τ, it follows for N

$$N = \int n \, d^4x = \int n\tau \, d\tau \, d\eta \, d^2x \;, \tag{7.2.2}$$

or for the number of particles per unit rapidity at a fixed final proper time τ_f

$$\frac{dN}{d\eta} = \int n(\tau_f, \eta, x) \cdot \tau_f d^2x \; . \tag{7.2.3}$$

From Eqs. (7.2.1) and (7.1.18) we have

$$\frac{dN}{d\eta} = \int d^2x \tau_f \frac{1}{\alpha} s(\tau_f, \eta, x) \int d^2x \tau_0 \frac{1}{\alpha} s_0(\tau_0, \eta, x) \; . \tag{7.2.4}$$

If the initial entropy density is taken to be independent of the transverse coordinates x, we find that the final rapidity distribution of multiplicity is proportional to the initial entropy density

$$\frac{dN}{d\eta} = \tau \frac{1}{\alpha} \cdot s_0(\tau_0, \eta) A_T \; , \tag{7.2.5}$$

where $A_T = \pi R^2$ is the transverse size of the nucleus. For example, the multiplicity per unit rapidity may be about 150 for a central collision of ^{16}O with $R \sim 3$ [fm] on a heavy nucleus [319]. If the initial (= equilibration) time is estimated as 1 fm/c, it implies a value of ~ 20 $[fm]^{-3}$ for the initial entropy density. We will see next what is further needed to extrapolate the initial temperature and the initial energy density.

Energy measurements
To derive a similar expression as Eq. (7.2.5) for the initial energy density ϵ_0, we need a relation between s_0 and ε_0 and ε_0 and T_0 to eliminate T_0. From the thermodynamic relation $(\varepsilon + p = Ts)$ and the bag model equation of state we find

$$(\epsilon_0 - B)(1 + c_s^2) = T_0 s_0 \; . \tag{7.2.6}$$

In Eq. (7.2.6) the factor 1/3 for a massless free gas has been replaced by c_s^2, where c_s denotes the velocity of sound as before. Next we use Eq. (7.1.25a) of the bag model in the plasma phase (assuming that the initial temperature is high enough for the system to be in the plasma phase)

$$\varepsilon_0 - B = gT_0^4 \; . \tag{7.2.7}$$

Again the exponent can be generalized to $1 + 1/c_s^2$, where the value of c_s is left open. The degeneracy factor g in the case of two massless u and d quarks, their antiquarks and gluons is given by

$$g = \frac{\pi^2}{30}(2 \times 2 \times 2 \times 3 \times \frac{7}{8} + 2 \times 8). \tag{7.2.8}$$

The different factors correspond to the spin, quark, antiquark, color, flavor and gluon degrees of freedom, 7/8 arises from the Fermi-Dirac statistics. When T_0 is eliminated in Eq. (7.2.7) via (7.2.6), and (7.2.5) is solved for s_0, we obtain

$$\varepsilon_0 = B + g \left[\frac{\alpha}{A_T(1 + c_s^2)g\tau_0} \frac{dN}{d\eta} \right]^{1+c_s^2}. \qquad (7.2.9)$$

Equation (7.2.9) gives the promised relation between the final state rapidity distribution and the initial energy density ϵ_0.

A second important class of observables comprises average transverse momentum distributions $\langle p_T \rangle$. Their relation to E/S and to the enthalpy can be derived as well as their role of barometers, when the average transverse momentum is related to the values of the initial and final pressure. For these derivations we refer to the original literature [319, 323].

We summarize. Bulk quantities like $\epsilon(T)$, $p(T)$, $s(T)$, derived quantities like the velocity of sound, the enthalpy, the energy per degree of freedom ϵ/s and QCD's equation of state are calculable on the lattice. The equation of state under inclusion of dynamical fermions should finally replace the bag model equation of state which entered the derivation of various relations in the preceding section. Equation (7.2.5) is a remarkable example of a relation between observables which are directly accessible in experiments and on the lattice. Initial energy and entropy are related to the pion multiplicity per unit rapidity. A variation of dN/dy amounts to a variation of s_0 or ϵ_0. On the other hand, $\langle p_T \rangle$ can be a measure for the freeze-out temperature and is sensitive to the pressure. Thus a flattening of $\langle p_T \rangle$ in a $\langle p_T \rangle$ vs dN/dy diagram could be nothing else but a reflection of a slow change of temperature and pressure during a rapid rise in energy and entropy densities, i.e. a rapid crossover or a first-order transition in a finite volume. Unfortunately a sharp crossover or a phase transition are not the only explanations for the shape. The flattening could be just a kinematic effect (lack of available energy). Kinematic constraints like energy conservation require that $\langle p_T \rangle$ goes to zero at the boundaries of the allowed rapidity interval [332]. Rescattering effects or minijet production are further alternatives to explain a first increase and a subsequent flattening effect [312, 333]. This situation is typical for other signatures as well.

7.2.2   Dileptons

We assume that the initial energy densities of a collision between nuclei A of the projectile and nuclei B of the target are high enough so that a quark-gluon plasma is initially generated. Since the size of the system produced in such a collision is not too large and the electromagnetic cross section is small, all dileptons can escape, especially those from the hottest dense state of the plasma. Dileptons will be emitted from all stages of the evolution, from the initial plasma phase, an intermediate possible mixture of plasma and hadron phase till the hadron phase at freeze-out. Thus, differently from hadronic yields, dileptons are sensitive to the whole space-time history of the evolution. The suitable kinematic variables for dileptonic cross sections are the invariant mass M of the dilepton pair and its transverse momentum p_T. A typical observable is the multiplicity of dileptons per invariant mass square M^2, transverse momentum p_T and unit rapidity interval ΔY,

$$R = \frac{dN}{dM^2 d^2 p_T dY} = \int_V d^4 x \frac{dN}{d^4 x d^4 p}. \tag{7.2.10}$$

Here R is a space-time integral over the rate $dN/d^4 x d^4 p$, which is the rate at a given four-momentum p of the dilepton pair at a space-time point $(\vec{x}, t)$. The quantity $dN/d^4 x d^4 p$ depends on the temperature T via $(\vec{x}, t)$, and therefore on the phase of the system. The phase (plasma or hadron) determines the dominant production mechanisms entering the elementary cross sections and the structure functions in $dN/d^4 x d^4 p$. The structure functions reflect the medium where the elementary processes take place (quarks inside a pion or in a heat bath of other partons). Collective effects of the plasma may also influence the rate for a given value of $(\vec{x}, t)$.

In a given phase, usually several sources are responsible for dilepton production. We are mostly interested in thermal collisions reflecting thermodynamic properties of the supposed heat bath environment. Thermal collisions are by far not the only way to produce dileptons. Other mechanisms are Drell-Yan production, preequilibrium production in the plasma phase, and dileptons from pions via ρ-resonance decays in the hadronic phase. Thermal dileptons result from parton collisions in a medium which can be characterized by a local temperature. They are specific for a plasma or a hadron gas which can be described with thermodynamic concepts; they are absent in hadron-hadron collisions.

Finally, we want to find out the mass range ΔM, where R is dominated by thermal dileptons. In this range R should be sensitive to the kinetics of the phase transition and in particular to its order. Let us assume $R(\vec{x}, t)$ has been calculated as a function of $T(\vec{x}, t)$. The space-time integral in Eq. (7.2.10) can be done if the missing relation $T = T(\vec{x}, t)$ is known. Such a relation is provided by the hydrodynamic approach. So far it is the only framework which is detailed enough to predict particle spectra according to Eq. (7.2.10). Further simplifying assumptions are made. To these belong Bjorken's scaling ansatz and the neglection of transverse flow, which lead to entropy conservation as function of time. As we have seen above, the hadronic multiplicity of the final state can be related to the initial conditions under the assumption of entropy conservation. Once the entropy is known as function of time, the time dependence of T follows immediately if $s(T)$ is known. This is the place where the equation of state enters. Further assumptions about the kinetics of the phase transition are necessary to justify the application of hydrodynamic concepts throughout all stages of the evolution.

We will now outline the predictions of Cleymans et al. [334] for thermal dilepton spectra, which are derived for different scenarios of the phase transition. The reason we have selected this reference even though it is not the most recent one in this field is that it explicitly addresses the influence of the order of the phase transition on the dilepton production rate. Although the ingredients do not represent the latest state of the art, the representation is suited for illustrating which features of dileptonic yields are sensitive at all to the dynamics of the phase transition.

Consider the particle rate which is differential in x and p. The general expression for the thermal rate of lepton pairs (here written for $\mu^+\mu^-$ rather than e^+e^-) in the independent particle approximation of kinetic theory [335] is given by

$$
\begin{aligned}
\frac{dN}{d^4x d^4p} = &\int \frac{d^3q_1}{2E_1(2\pi)^3} f(q_1) \int \frac{d^3q_2}{2E_2(2\pi)^3} f(q_2) \\
&\cdot \int \frac{d^3q_+}{2E_+(2\pi)^3} \int \frac{d^3q_-}{2E_-(2\pi)^3} \\
&\cdot |M(p_1\bar{p}_2 \rightarrow \mu^+\mu^-)|^2 \cdot \delta(p - q_1 - q_2) \, .
\end{aligned} \qquad (7.2.11)
$$

Here M is the matrix element for the process of leptoproduction from particles $p_1, \bar{p}_2$ with momenta q_1, q_2, p is the momentum of the lepton pair, q_+ and q_- are the momenta of μ^+ and μ^-. The statistical distributions $f(q_1)$ and $f(q_2)$ measure the probability of finding particles $p_1, \bar{p}_2$ with momenta q_1 and q_2 in the given medium. The dileptons do not receive such

factors, since we assume that rescattering of the electromagnetically interacting particles can be neglected. The cross section is small, and the plasma
volume is assumed not to be too large. In terms of cross sections σ, Eq.
(7.2.11) reads

$$\frac{dN}{d^4x d^4p} = \int \frac{d^3q_1}{(2\pi)^3} \frac{d^3q_2}{(2\pi)^3} v_{q_1 q_2} \sigma(p_1\bar{p}_2 \to \mu^+\mu^-)$$
$$\cdot f_1(q_1)f_2(q_2) \cdot \delta(p - q_1 - q_2) \tag{7.2.12}$$

upon integration over the lepton momenta, and v_{q_1,q_2} denotes the relative
velocity of particles p_1 and $\bar{p}_2$

$$v_{q_1 q_2} = \frac{\sqrt{(q_1 q_2)^2 - m^4_{p_{1,2}}}}{E_1 \cdot E_2}. \tag{7.2.13}$$

In the plasma phase the lowest order process for dilepton production is the
same as in the Drell-Yan production: two quarks annihilate via a virtual
photon to yield a lepton pair

$$q\bar{q} \to \gamma* \to \mu^+\mu^-. \tag{7.2.14}$$

The momentum distribution functions f_1 and f_2 of the quarks q and antiquarks $\bar{q}$ are given by the Fermi-Dirac distributions

$$f_q = \frac{6}{e^{(\underline{u}\underline{q}-\mu)/T} + 1} , \qquad f_{\bar{q}} = \frac{6}{e^{(\underline{u}\underline{q}+\mu)/T} + 1}. \tag{7.2.15}$$

The chemical potential μ is set to zero in the end, $\underline{u}$ is the local four-
velocity of the plasma fluid element in the fixed laboratory frame, $\underline{q}(\underline{\bar{q}})$ is
the four-momentum of the quark (antiquark). As hydrodynamic scenario
a longitudinal expansion will be used with vanishing transverse velocity of
the plasma. In view of that, it is convenient to express the product $\underline{u} \cdot \underline{q}$
in terms of the transverse invariant mass M, the rapidity Y of the lepton
pair, and the plasma space-time rapidity θ according to

$$\underline{u} \cdot \underline{q} = M_T \cosh(\theta - Y) \tag{7.2.16a}$$

with the definitions of the transverse mass M_T

$$M_T = \sqrt{M^2 + p_T^2} \tag{7.2.16b}$$

$$Y = \frac{1}{2}\ln\frac{E + p_z}{E - p_z} \tag{7.2.16c}$$

$$\theta = \operatorname{arctanh} v \tag{7.2.16d}$$

with $v = z/t$ denoting the collective fluid velocity. If the transverse mass is large compared to the temperature, the exponentials are sharply peaked around $\theta = Y$. In this case no final θ-dependence is left. Upon integration over the dilepton momentum, the result for the *dilepton rate in the plasma phase* at a space-time point $\underline{x} = (\tau, x_T, y)$ (y being the local space-time rapidity of a fluid element) and four-momentum $\underline{p} = (M_T \cosh Y, p_T, M_T \sinh Y)$ is given by

$$\frac{dN_P}{d^4x d^4p} = \frac{\alpha^2}{4\pi^4} \left[1 + \frac{2m^2}{M^2}\right] \left[1 - \frac{4m^2}{M^2}\right]^{1/2}$$
$$\cdot e^{-E/T} K_P(\underline{p}, T, \mu) \cdot \sum_i e_i^2. \tag{7.2.17}$$

The index P stands for plasma phase, e_i are the charges of the quarks, α is the electromagnetic coupling, m stands for the lepton mass and K_P is a function, which depends on $\underline{x}$ via T, which we will not specify further now. It is characteristic for the plasma phase.

In the *hadronic phase* it depends on the invariant mass M of the dilepton pair which process makes the leading contribution to the elementary cross section σ in Eq. (7.2.12). For small mass pairs (small compared to the ρ-peak), bremsstrahlung-type emission of soft virtual photons is important. For masses M well above the order of T_c, processes $h\bar{h} \to \gamma* \to \mu^+\mu^-$ play a role, where $h\bar{h}$ are hadrons others than pions. Their thermal production should be suppressed due to the relatively low temperatures, unless the in-medium masses are changed due to effects of chiral symmetry restoration, which may be dramatic! The only process that has been considered so far in [334] is $\pi^+\pi^- \to \rho \to \mu^+\mu^-$. The electromagnetic cross section $\sigma(M)$ is modified by the strong interactions of the pions, leading to

$$\sigma_\pi(M) = F_\pi^2(M) \left(1 - \frac{4m_\pi^2}{M^2}\right)^{1/2} \cdot \sigma(M), \tag{7.2.18a}$$

where

$$\sigma(M) = \frac{4\pi}{3} \frac{\alpha^2}{M^2} \left[1 + \frac{2m_\mu^2}{M^2}\right] \left[1 - \frac{4m_\mu^2}{M^2}\right]^{1/2}. \tag{7.2.18b}$$

The pion form factor F_π is treated in the vector-meson-dominance approximation, only the ρ-pole is kept in the sum over all ρ-like resonances.

Furthermore the functions f_1 and f_2 in Eq. (7.2.12) have to be replaced by Bose-Einstein distributions. The result for the *dilepton rate in the hadron phase* is given by

$$
\frac{dN_H}{d^4x d^4p} = \frac{\alpha^2}{48\pi^4} \left[1 + 2\frac{m_\pi^2}{M^2}\right] \left[1 - 4\frac{m_\pi^2}{M^2}\right]^{3/2} \cdot |F_\pi(M^2)|^2
$$
$$
\cdot \exp\{-E/T\} \cdot K_H(p,T) , \tag{7.2.19}
$$

where K_H is a slightly different function from K_P in Eq. (7.2.17).

Integration over the space-time history

The assumption of Bjorken's scaling solution for the hydrodynamical expansion simplifies the evaluation of the space-time integration. The appropriate representation of the volume element is $d^4x = \tau d\tau dy d^2x_T$, where τ is the proper time in a comoving reference frame with the fluid, y is the space-time rapidity (denoted as η above), x_T are the transverse coordinates. The transverse velocity is neglected, the transverse distributions of thermodynamic quantities are taken as step functions, ε, p, s and T depend then only on the proper time τ. The dileptonic yields per invariant mass, transverse momentum and unit rapidity interval dY are given by

$$
\frac{dN_{Ph}(hydro)}{dM^2 d^2p_T dY} = \frac{1}{2} \int_{\tau_i}^{\tau_f} \tau d\tau \int_{y_{min}}^{y_{max}} dy \frac{dN_{Ph}}{d^4x d^4p}(T(\tau)). \tag{7.2.20}
$$

The index Ph indicates the dependence on the phase, $\tau_f - \tau_i$ is the eigentime duration of a certain phase or a mixture of phases, y_{max} is taken to be the rapidity of the incoming beam. The *equation of state* enters the cooling law $T(\tau)$, the phase duration is sensitive to the dynamics of the transition.

Cleymans et al. [334] have discussed the following scenarios

- a first-order transition described by a bag model equation of state proceeding in equilibrium or with supercooling and subsequent superheating
- a second-order transition with an equation of state taken from lattice Monte Carlo results [336].

As we saw previously, the bag-model equation of state leads to the T^3 dependence of the entropy density and is typical for an ideal gas. In the baryonless plasma phase for two massless flavors, the relation is given by $s_P = 4 \cdot 37\pi^2 T^3/90$. In the hadron phase described as an ideal gas of massless pions we have $s_H = 2\pi^2 T^3/15$. In the case of the second-order transition $s(T)$ has been read off from numerical data of [336], which were obtained in lattice QCD. Using the scaling solution for isentropic flow $s(\tau) \cdot \tau =$

const., we find the relations $T(\tau)$. When the scaled temperature T/T_0 is plotted as a function of the scaled eigen-time τ/τ_0 (where T_0 and τ_0 refer to the initial values chosen as $T_0 = 284$ [MeV] and $\tau_0 = 1$ [fm/c]), the characteristic difference between a first-order transition proceeding in equilibrium and a second-order transition is that the average temperature is higher in the quark-gluon plasma phase and lower in the hadronic phase for a second-order transition. Having $T(\tau)$ at hand, the integration in Eq. (7.2.20) can be done. Depending on the scenario, four rates are distinguished.

- The dilepton rate in the pure plasma phase produced in the time interval between τ_0 and τ_P where $T(\tau)$ is given as $T_0(\tau_0/\tau)^{1/3}$ for the bag model equation of state.
- The dilepton rate in the pure hadronic phase produced during the interval between τ_H and freeze-out time (where the application of hydrodynamic concepts is already questionable).
- If there is a coexistence of phases as it is expected in the first-order case, there is a mixed phase at temperature T_c, where the plasma (hadron) phase contributes a fraction $\frac{dN_{P/H}(\text{mixed})}{dM^2 d^2 p_T dY}$, respectively. Both rates are calculated separately. Their relative weight is described by a factor f, which is the fraction of the entropy in the plasma phase. Its value follows from $s(\tau) \cdot \tau = $ const and Eq. (7.2.21)

$$s(\tau) = f(\tau)s_P + [1 - f(\tau)] \cdot s_H. \qquad (7.2.21)$$

The entropy densities s_P and s_H remain constant for constant $T = T_c$.

The ratio s_P/s_H of entropy densities also gives an estimate for the time scales τ_H and τ_P, as $\tau_H/\tau_P = s_P/s_H = 37/3$ if we assume for a moment that the naive counting of degrees of freedom is appropriate close to T_c. This ratio is frequently quoted in the literature, although the number three in the hadronic phase (resulting from three pions) is most likely incorrect. The large difference in entropy densities between both phases explains the long duration time of the mixed phase. It takes time to rearrange the effective degrees of freedom since they have to be reduced by an order of magnitude for the above counting when the conversion to the hadronic phase sets in.

As a result, the main difference is the extent of the interference region in which both phases contribute to dilepton production. It is considerably smaller for a second-order transition. In this case, the production mechanisms change instantaneously at T_c, while those of both phases would be at work in the coexistence phase of a first-order transition.

The width of the interference region does not yet provide a signal which is suitable for experiments to infer the phase transition dynamics. The sudden change in the production mechanism is reflected in the average transverse momentum $\langle p_T \rangle$ of the dilepton pair or even more clearly in its derivative with respect to M. The average $\langle p_T \rangle$ can be derived from the rate R according to

$$\langle p_T \rangle = \int \frac{dN}{dM^2 dy dp_T} p_T dp_T \Big/ \int \frac{dN}{dM^2 dy dp_T} dp_T. \qquad (7.2.22)$$

A plot of $d \langle p_T \rangle / dM$ as a function of M for both transition scenarios shows a peak in the $d \langle p_T \rangle / dM$ distribution for a second-order transition. Thus a peak in the $d \langle p_T \rangle / dM$ distribution is a possible candidate for a signal of a second-order phase transition (keeping in mind all approximations that have been applied so far). As outlined by Cleymans et al. [334], such a sudden change in the slope of $\langle p_T \rangle$ as a function of M can be also produced by a higher mass resonance like the ρ' (1600). To disentangle the origins of bumps in the $\langle p_T \rangle$ spectra, one should vary the initial temperature. At higher initial temperature the contribution from the plasma phase to the dilepton rate increases. A peak structure in the $\langle p_T \rangle$ spectrum due to a 2nd order transition then shows up already for low initial temperatures, while higher initial temperatures are necessary to see a similar structure in a first-order scenario.

Yet another possibility may lead to a sudden change in $\langle p_T \rangle (M)$. Dilepton rates are calculated in a way that they do not automatically interpolate smoothly at T_c, although they should do so in the case of a second-order transition where no discontinuities should be seen in thermodynamic observables. Cleymans et al. (1987) have investigated the influence of smoothly interpolating the rates between both phases. The bump structure, however, survived the interpolation between $dN_p/d^4x d^4p$ and $dN_H/d^4x d^4p$.

In general, it remains an open problem to calculate the electromagnetic current-current correlation $W(x,p)$ (whose space-time integral is proportional to $dN/dM^2 dp_T dY$) in terms of one effective model which replaces the bag model in the case of a second-order transition.

So far we have been concerned with thermal rates and specific features reflecting the transition dynamics in the rate dependence of the invariant mass M. In heavy-ion experiments it is obviously an objective of higher priority to guarantee that thermal dileptons can be identified from variant background sources. In the order of increasing masses, we have Dalitz pairs, soft gluon bremsstrahlung, hadronic resonance decays, preequilibrium production, and Drell-Yan production in the very early stage of the collision.

Ruuskanen [337] has derived an analytic expression for the integrated thermal rate R. The characteristic feature of this expression concerns the T and M_T-dependence as compared to other production mechanisms ($M_T = (M^2 + p_T^2)^{1/2}$ denotes the transverse mass which is used instead of the transverse momentum p_T). The dependence of the thermal rates on the initial temperature T_i is strong, i.e. proportional to T_i^6 as compared to the T_i^3-dependence for a Drell-Yan rate. Recall that T_i at equilibration time τ_i can be related to the hadronic multiplicity in the final state according to

$$(\tau_i T_i^3)^2 \propto (dN\pi/dy)^2 , \tag{7.2.23}$$

if an isentropic expansion scheme is assumed. This allows an identification of thermal dileptons via their strong (quadratic) dependence on pion multiplicity fluctuations, Drell-Yan rates show only a linear dependence. At fixed multiplicity (i.e., fixed $\tau_i T_i^3$), the thermal rates are expected to show an approximate power law behavior proportional to M_T^{-6} in a range of $M_T/T_i < 5.5 < M_T/T_c$ (Ruuskanen, 1990) with a $\langle p_T^2 \rangle$ dependence proportional to M^2, and an exponential decrease of the dilepton rate for large M_T/T_i, with $\langle p_T^2 \rangle$ showing a linear increase in M. For an initial temperature of 600 MeV, the window of the M_T^{-6} dependence of the transverse mass spectrum lies between 1-3 GeV. Its lower end has to face a background from the hadronic phase, its upper end the Drell-Yan background.

Thus the first goal of a measurement of dilepton rates in a collision of heavy-ion nuclei is to identify a window in M_T of the spectrum which shows the behavior of a thermal rate produced in a plasma phase. It is easier to filter out the M_T^{-6} depending part if high initial temperatures are reached in the experiment.

The overall judgment is that the identification of thermal rates in the different phases is difficult but not unrealistic. Yet it seems to be an order of magnitude more difficult to infer specific features of the transition itself from structures of R or $\langle p_T \rangle$ distributions as functions of M.

To summarize, the approximations that enter predictions of dilepton rates along the lines we have indicated above were the following:

- Uncertainties in the background to thermal dileptons. The Drell-Yan background is the one which is understood best. Corrections due to higher orders in QCD have been assumed to be small, actually they are large in the usual deep-inelastic structure function approach. In the mass range of a few [GeV] which is of special interest for thermal production, the worst perspective is that both the thermal and the Drell-Yan productions – including all their specific signatures – are buried under the yields of plasma oscillations.

- The applicability of hydrodynamic concepts make it possible to evaluate the integral over the space-time history of rates at a given point in phase space. This is marginal in several aspects. The volumes are not so large compared to the mean free paths, at least not close to freeze-out, where by definition the system is so diluted that interactions can be neglected. The scale of equilibration time is estimated in a purely phenomenological way. Thus it cannot really be argued which time τ_0 is to be used for the onset of hydrodynamic expansion. The adiabatic expansion guarantees the conservation of entropy and leads to a simple relation between entropy and time. During the phase conversion this condition may be violated.

- The bag-model equation of state has been used close to T_c, although strong deviations from the ideal gas behavior are expected there. Lattice Monte Carlo calculations should include the effect of dynamical fermions as realistically as possible and derive an equation of state close to T_c.

Given the hydrodynamic approach and the equation of state, further simplifications were made in deriving the results we have indicated. We have neglected effects due to

- a realistic nuclear geometry,
- the impact parameter dependence. In particular, the initial temperature profile depends on the impact parameter $b \neq 0$.
- the transverse coordinate dependence of temperature during all stages of the evolution,
- the transverse flow of the hadron/plasma fluid. When transverse flow is included, hadronic rates can be reduced by an order of magnitude due to the reduced lifetime of mixed and hadronic phases.
- last, but not least, the possible temperature and density dependence of masses that are involved in the dilepton production. Meson masses can be expressed in terms of the chiral condensate $\langle \bar{\psi}\psi \rangle$; they should be sensitive to the order of the transition. Either $\langle \bar{\psi}\psi \rangle$ vanishes smoothly when approaching T_c from below, or two values of $\langle \bar{\psi}\psi \rangle$ coexist at T_c, a large value and a small one.

When dileptons are produced from vector meson decays, the dilepton rates are sensitive to the phase transition dynamics also via vector-meson mass changes in the medium both as a function of T and μ. We have skipped this possible manifestation of the transition dynamics, as we are not aware of any specific prediction to date.

Real photons

Real photons should be understood in contrast to virtual photons in dilepton production. Like dileptons real photons indicate the electromagnetic response to the plasma evolution. They can provide clean signals in the sense that they escape from different stages of the system with rather small interaction cross sections. According to their production conditions, the three major contributions to real photons are *direct photons* from partonic processes in the initial state, *thermal photons* from both phases and *decay photons* produced at a late stage of the evolution. The procedure to calculate the thermal rate of real photons is the same as for dileptons. Once the elementary processes are identified and emission rates are calculated in the stationary plasma or hadronic phase, the result has to be folded with the space-time expansion. In a first approximation one can use Bjorken's similarity flow and correct for transverse flow, nuclear geometry in further steps. For such calculations we refer to [338] and references therein. High multiplicities and short equilibration times favor thermal photon production.

7.2.3 *Strangeness production*

In heavy-ion collisions an enhancement of certain particle ratios involving strange quarks is predicted compared to ratios in hadron hadron or hadron nucleus collisions. Ratios enhanced by a factor of 2 for K^+/π^+ have in fact been observed in experiments at the AGS at Brookhaven and SPS at CERN. A theoretical explanation in terms of a transient quark gluon plasma is not compulsory. Rescattering processes in a hot hadronic gas can also change the ratios in the same direction.

The idea to propose strangeness as a possible signature for a QGP relies on the following arguments. Strangeness production in the plasma should be facilitated by two reasons.

- Independently of the assumed baryon density, the threshold energies for strangeness production in the plasma phase are much lower than the production of strange baryons or mesons in the hadron phase. In the hadron phase, strange mesons or baryons are typically made in a collision of two nonstrange hadrons. The reaction with the lowest threshold energy requires already 671 MeV ($p + n \rightarrow \Lambda^0 + K^+ + n$). In the plasma the threshold for $s\bar{s}$ production is equal to the rest mass of $s\bar{s} \sim 300 \, [MeV]$. For a given temperature the density of noninteracting strange quarks is higher than the density of noninteracting kaons. In a plasma an s-quark has two spin and three color degrees of freedom. If it is bound in a K^- or a $\bar{K}^0$ in the hadron phase it has only 2. Also the kaon mass is $\sim 494 \, [MeV]$ compared to the current s-quark mass of $180 \pm 30 \, [MeV]$.

• The second reason applies to a baryon-rich environment, which can be found in collisions with high baryon stopping or in the rapidity range of fragmentation regions.

  The Pauli principle will prohibit the creation of $u\bar{u}$ and $d\bar{d}$ pairs compared to strange pairs. The light quarks have to supply the large Fermi energy represented by the chemical potential μ_B, while the s-quarks are only suppressed by their finite mass. Moreover $\bar{u}$ and $\bar{d}$ quarks have a high probability to recombine with u and d quarks to form gluons. For strange quarks the recombination to gluons is less likely. Due to the volume expansion and the decreasing temperature of the fireball, the process $gg \underset{\leftarrow}{\overset{\rightarrow}{}} s\bar{s}$ is soon out of thermal equilibrium, that means, it is too slow to proceed in the inverse direction.

For dileptons it has been sufficient (cf. the previous section) to calculate the differential rate per given space-time volume. Due to the very small electromagnetic cross section the dileptons of high invariant mass escape from the plasma without further interaction. Strange quarks are kept in the subsequent evolution. This is the essential difference of strangeness production compared to dilepton production. In various ways strange quarks are incorporated as hadronic constituents during the conversion of the plasma into the hadron phase. After completion of the conversion strangeness is confined to hadrons. The strange hadrons continue to react via strangeness creation, annihilation or exchange reactions until freeze-out. Thus it is not sufficient to calculate the formation rate of strangeness per unit time and volume as if the predicted rate would directly correspond to observed multiplicities (as for dilepton production).

To obtain numbers for particle ratios at the very end, one has to specify the following features (here we add a particular specification as it was used in early calculations [339]):

- the kinematical approach and the geometry. (Bjorken's ansatz of a longitudinal hydrodynamic scaling expansion was used. Possible effects due to viscosity, heat conduction and transverse expansion were neglected. Viscosity and heat conduction would prolong the decrease of temperature as a function of time, transverse expansion will accelerate it. Both effects may or may not approximately cancel.)

- the rapidity range. (The central rapidity range has been chosen, where the baryon number approximately vanishes. Hence the calculations are performed at $\mu_B = 0$. This may not be a favorable choice in view of the different sensitivities of $\bar{u}, \bar{d}$ and $s, \bar{s}$ quarks with respect to μ_B, but it simplifies the discussion a lot.)

- the equation of state. (The bag model equation of state was used, because full QCD's equation of state was not yet available from lattice simulations.)

- the gross features of the phase transition. (Corresponding to the bag model, the phase transition is assumed to be of first order. Two realizations were considered: a smooth transition via the Maxwell construction and a rapid transition with supercooling and subsequent reheating.)

- the observable to measure for the strangeness contents of each phase. (For the hadron phase K^--mesons were considered.)

- the elementary reactions entering the cross sections in the rate formulas.

- an ansatz for the differential rate equation that determines the equilibration time.

Already this list may indicate how challenging it is to obtain a reliable prediction and to judge the validity of the result. This is clearly beyond the scope of this book.

7.2.4   *Pion interferometry*

The overwhelming majority of particles radiated in heavy-ion collisions are pions. We would like to mention pion interferometry as an experimental technique to measure the lifetime and the final-state size of the source that radiates the pions. From the discussion in the previous sections it is evident that we need some measurements delivering information about the equation of state. Production rates of dileptons and strange particles strongly depend on the duration of a certain phase. Three extensive quantities must be measured to obtain an equation of state for a static gas with no conserved charge. In heavy-ion collisions the total energy and entropy can always be estimated from experiment. If we know the volume and the fraction of collective energy at a given time of the collision, we have a single value in an equation of state corresponding to the conditions at that time. The correlation function measured in pion interferometry could give an estimate for the size at freeze-out. The main reason why we mention this experimental tool is the possibility to measure the prolongation of lifetime due to a phase transition.

Interferometry is nowadays a well-known technique. It has been developed by Hanbury-Brown and Twiss in 1954 in astrophysics, as a tool to measure the size of various stellar objects in the visible and radio frequency ranges. Fig. 7.2.1 shows a source which is assumed to emit identical particles from positions P_1 and P_2. The particles are later observed at positions P_3 and P_4. Both emission points may contribute to both observation points, even if the particles are noninteracting, but have small relative momenta. The reason is the symmetrization (antisymmetrization) of the quantummechanical wave function in the case of bosons (fermions). A correlation function is constructed from the number of counts at P_3 and P_4. The particles which are detected in astrophysical interferometry experiments are photons.

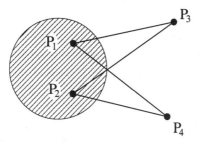

Fig. 7.2.1   Source emitting photons at points P_1 and P_2, which are registered at P_3 and P_4.

Let us see how we can infer the space-time structure of the source from correlations in momentum space. The correlations refer to two-particle correlations in the pion's momenta distributions. They are defined as

$$C(\vec{p}, \vec{q}) = P(\vec{p}, \vec{q})/(P(\vec{p})P(\vec{q})) , \qquad (7.2.24)$$

where $P(\vec{p}, \vec{q})$ denotes the probability to find two pions with three-momenta

$\vec{q}$ and $\vec{p}$ in the same event. In terms of rates, $C(\vec{p}, \vec{q})$ is given as

$$C(\vec{p}, \vec{q}) = \frac{d^6 N}{d^3 p d^3 q} \Big/ \left(\frac{d^3 N}{d^3 p} \frac{d^3 N}{d^3 q} \right) . \qquad (7.2.25)$$

Thus C is measurable as a ratio of two-pion to one-pion inclusive yields. Its width is a measure for the inverse source size. The weaker the correlation the smaller its width, the larger is the source which emits the pions.

Two-pion correlations can arise from different origins. One origin are Bose-Einstein correlations due to the quantum mechanical symmetrization of the outgoing wave functions of identical bosons. In the same way electrons in a metal are anticorrelated in their spatial distribution due to the fact that they occupy only a finite volume in momentum space.

Other causes of correlations due to final-state interactions are hadronic and Coulomb interactions. Coulomb interactions lead to positive correlations between particles of opposite charge if the relative momentum is small. It has to be checked how far their contribution is averaged out due to a comparable number of pairs with equal and opposite charges.

A third source for correlations comes from resonance decays. Pions are detected that are not directly emitted from the hadronic gas at freeze-out, but result from decays of heavier mesons. Such pions may contaminate the information about the source size.

Bose-Einstein correlations are the only origin for correlated emission from the radiating source if the source is completely chaotic. Under this assumption one can calculate the probability $P(\mathbf{p}, \mathbf{q})$ of measuring two pions with momenta $\mathbf{p}$ and $\mathbf{q}$. These probabilities depend on an ansatz for the so-called emission function, that is the probability of emitting a pion of momentum $\mathbf{k} = (p + q)/2$ from a space-time point x. This ansatz depends on the equation of state and the dynamical concept relating the space-time dependence to thermodynamic quantities. The reader, who is interested in pion interferometry, is referred to the literature for concrete calculations and predictions, waiting for experimental verification [340].

7.2.5 *Intermittency analysis*

Experimental manifestations of the QCD transitions are frequently discussed under the assumption that they are of first order. As we have seen above, there are indications that the finite temperature transition resembles more a crossover phenomenon. The realistic set of quark masses may be close to a "critical" set in mass parameter space, which could lead to a second-order transition. A standard physical picture of critical phenomena in spin systems is a diverging correlation between spins leading to clusters of aligned spins of arbitrary length scales. The $3 - d - Z(2)$-spin system is supposed to share the universality class of an $SU(2)$ pure gauge theory, an $O(4)$-ferromagnet is assumed to share the class of 2-flavor QCD.

Thus the question arises how a picture of clusters of aligned spins in color or isospin space can be transferred to typical observables in heavy-ion collisions, e.g. to particle multiplicities. One of the possible answers lies in the *intermittent behavior* in rapidity distributions. The concept of intermittency was originally introduced in studies of turbulent behavior of fluids [341]. In general, intermittency can be defined as the appearance of structure in random media. In the context of particle physics, it has to deal with large fluctuations of charged particle density in small regions of phase space. In heavy-ion collisions intermittency refers to certain moments of rapidity distributions. Other distributions like energy, pseudo-rapidity and azimuthal angles can be considered as well [342].

Let us first consider a toy model that is universal enough for applications in different areas. Consider a set of N balls distributed in a box of total size R into M cells of size $L(M = R/L)$. In the case of rapidity distributions, they correspond to N particles per given unit of some kinematic range. A cell in a box is a rapidity interval in a certain available rapidity range. For an Ising model, the balls are the spins on a space-time lattice R subdivided into cells of size L. We are interested in the ball distributions if the resolution of the lattice is made finer and finer ($M \to \infty$). There are many ways to realize this limit. One possibility is to keep R constant and let L going to zero. This limit is of interest for rapidity distributions, where L stands for the bin size δy of the rapidity interval. Another realization is the thermodynamic limit with $R \to \infty$ for fixed L. This limit is usually taken when critical phenomena are discussed in the infinite volume limit.

The same type of limit is of interest in a second-order finite temperature transition in space-time continuum, which is described by a model of lattice gauge theory. One may expect that intermittency in the limit of ($R \to \infty$, L fixed) has a correspondence in the limit of (R fixed, $L \to 0$).

We define the ℓth normalized moment f_ℓ for a given distribution of N balls as

$$f_\ell(M) = \left[\frac{1}{M} \sum_{m=1}^{M} K_m^\ell \right] \Big/ \left[\left(\frac{1}{M} \right) \sum_{m=1}^{M} K_m \right]^\ell . \tag{7.2.26}$$

Here K_m denotes the number of balls in the mth cell. First, we keep N fixed and vary M. In the extreme case of an equidistribution of N/M balls in each box, it is easily seen that $f_\ell(M)$ is independent of the grain size, i.e.

$$f_\ell(M) = 1 \ \forall \ell \ . \tag{7.2.27}$$

In the other extreme, a strong fluctuation (where all balls are concentrated in one box), we find a logarithmic dependence of $\ln f_\ell$ on the "resolution size" L

$$\ln f_\ell(M) = -(\ell - 1)\ln L + (\ell - 1)\ln R \ . \tag{7.2.28}$$

Such a behavior of a given distribution with N balls is called *intermittent*.

More generally, intermittent behavior is attributed to average values of f_ℓ. The weighted average $\langle \dots \rangle$ of the ℓth moments runs over an ensemble of configurations in a d-dimensional volume R^d divided into $M = (R/L)^d$ cells of equal size L^d. In a multiplicity measurement it could be an average over all events, in the Ising model it is the thermodynamic average with Boltzmann weights.

Intermittency is called the property of fluctuations around some average distribution which lead to a power law behavior of $\langle f_\ell \rangle$ in the number of cells M, or equivalently if

$$\ln < f_\ell(L) > = -\lambda_\ell \ln L + g_\ell(R) \ . \tag{7.2.29}$$

Here $\lambda_\ell > 0$ are constant and $g_\ell(R)$ is independent of L, L being the linear cell size. The constants λ_ℓ are called intermittency indices, they are a measure of the strength of intermittent behavior.

Results for λ_ℓ and g_ℓ in case of a d-dimensional Ising model can be found in [343]. The essential point at which criticality enters the derivation of intermittency in an Ising model is scale invariance at T_c. While the balls of the toy model are the same independently of the partition of the lattice, one has first to argue, why one may choose the same type of variables on all length scales in an Ising model. This is justified by selfsimilarity as $T \to T_c$, if $1 \ll L \ll \xi$ is satisfied. For a given ℓ the intermittency indices can be expressed in terms of the more familiar critical exponents. Like critical exponents they are universal for all models belonging to the same universality class.

The mere divergence of a correlation length ξ is not sufficient for introducing intermittent behavior. In a 1-d-Ising model, where ξ has an *essential* singularity rather than a power-law singularity as in the 2- and 3-dimensional cases, the normalized momenta f_ℓ are bounded from above for all ℓ [344].

We turn now to intermittent behavior in heavy-ion experiments. To identify this behavior in rapidity distributions, we consider moments $\mathcal{C}_\ell$ that have been introduced by Bialas and Peschanski (1986) according to

$$\mathcal{C}_\ell = \frac{1}{M} \sum_{m=1}^{M} (Mp_m)^\ell \,. \tag{7.2.30}$$

Here M is the number of intervals of size δy in a given rapidity interval $\Delta y = M \cdot \delta y$, $p_m (m = 1, \ldots, M)$ denotes the probability for finding particles in any of these rapidity intervals dp_M. The total distribution $P(p_1, \ldots, p_M) dp_1 \ldots dp_M$ of probabilities for finding particles in the intervals $dp_1 \ldots dp_M$ is normalized such that

$$p_1 + \ldots + p_M = 1$$
$$\int dt \int dp_1 \cdots dp_M \; P(p_1, \ldots, p_M) = 1 \,. \tag{7.2.31}$$

The variable t stands for a collection of kinematic cuts like the energy of the collision. The average moment $\langle \mathcal{C}_\ell \rangle$ is obtained as the sum over all configurations in rapidity space weighted by the probability distribution $P(\ )dp$

$$\langle \mathcal{C}_\ell \rangle = \int dt \int dp_1 \ldots dp_M P(p_1, \ldots, p_M; t) \frac{1}{M} \sum_{m=1}^{M} (Mp_m)^\ell \,. \tag{7.2.32}$$

In terms of rapidity variables genuine intermittent behavior is signaled, if

$$\ln \langle \mathcal{C}_\ell \rangle = \phi_\ell \ln(\Delta y / \delta y) = \phi_\ell \ln M \tag{7.2.33}$$

in the limit of $\delta y \to 0$, i.e. a logarithmic dependence on the resolution size in the limit of increasing resolution, which is equivalent to the definition in Eq. (7.2.29). Small values for δy correspond to a fine resolution in rapidity space, Δy is the full considered rapidity interval. The lower bound on δy is given by the experimental resolution. The intermittency indices ϕ_ℓ vary between $O < \phi_\ell \le \ell - 1$. They are a measure for the strength of the intermittent behavior. Intermittency occurs if selfsimilar fluctuations exist on all scales δy. In the case of rapidity distributions intermittency should be seen in contrast to dynamical fluctuations.

In *real experiments* we have to deal with finite size systems. Statistical fluctuations around the probability distribution $dp_1 \ldots dp_M P(p_1, \ldots, p_M; t)$ will always provide a noisy background for fluctuations of intermittent or dynamical origin that we are trying to identify. Unless the multiplicity in the events is very high, the probability p_m for finding particles in the rapidity interval m is different from the *measured fraction* of the total multiplicity N. Bialas and Peschanski [345] have introduced scaled factorial moments to filter out the interesting fluctuations.

At least two physical mechanisms are known which may lead to real intermittent behavior in heavy-ion collisions (real in contrast to fluctuations which are induced by two-particle correlations). One of these mechanisms is *the QCD transition* if its correlation length diverges at T_c, or, adapted to the finite volume, if its correlation length extends the typical volume in the collision. In this case, the anomalous fractal dimension d_ℓ should be approximately independent of ℓ. Other origins are *selfsimilar cascades* which are attempts to model the evolution of the plasma to the hadronic phase. Therefore we have to deal again with the ambiguity of signatures. For further details on how to disentangle the different origins of intermittent behavior we refer to [345].

7.2.6   *Outlook*

Our selection of possible signatures of the phase conversion in heavy-ion collisions corresponds to a subjective choice and is far from being complete. Also the selection of topics should not be considered as representative for the most promising signatures. For example, we have not covered the important case of finite baryon density and the related changes in the phase structure. Zero baryon density may be a reasonable approximation in certain experimental situations, but in general it is not. Also it would be desirable to derive the speed of the cooling after the "little bang" from a detailed knowledge of cooling mechanisms. Probably it is neither a quench nor an adiabatic process. Therefore there is some hope to see effects due to a rapid cooling in spite of the absence of a true finite temperature phase transition. For example, there is at least a chance that a fast cooling amplifies the role of long-wavelength (pionic) modes during the phase conversion of the quark-gluon plasma. Long-wavelength modes relax much more slowly to the equilibrium configuration and can lead to a "misalignment" with respect to the true ground state. In the language of spin models this allows the formation of large clusters of "misaligned" spins, in case of misaligned isospins called "disoriented chiral condensates" [348]. In [348] it was speculated that pionic yield comes in clusters of pions aligned in a single direction in isospin space over a large fraction of the collision volume. When these regions relax coherently to the true ground state, a specific radiation of pions would be emitted.

In analogy to remnants of a quench during the chiral transition, effects of a rapid conversion from the deconfinement to the confinement phase have been discussed in [349] and [350]. In [350] the rapid cooling was implemented via Glauber dynamics. The results are only conclusive if Glauber dynamics can be derived as an effective description of the underlying physical cooling mechanisms involved in the phase conversion. For further reading we recommend the review of Braun-Munziger et al. [314], for example.

References

[1] N.F. Mott, Rev. Mod. Phys. **40** (1968) 677.

[2] P. Braun-Munziger, K. Redlich, and J. Stachel, arXiv:nucl-th/0304013 (2003).

[3] A. Hosoya and K. Kajantie, Nucl. Phys. B **250** (1985) 666.

[4] T. Banks, and A. Casher, Nucl. Phys. B **169** (1980) 103.

[5] B.A. Campbell, J. Ellis, and K.A. Olive, Nucl. Phys. B **345** (1990) 57.

[6] E. Braaten, and R. D. Pisarski, Phys. Rev. Lett. **64**(1990) 1338; Phys. Rev. D **42** (1990) 2156; Nucl. Phys. B **337** (1990) 569; ibid. B **339** (1990) 310.

[7] E. Braaten, and R. D. Pisarski, Phys. Rev. D **45** (1992) 1827; ibid. D **46** (1992) 1829.

[8] T.A. DeGrand, and C. E. DeTar, Phys. Rev. D **34** (1986) 2469.

[9] J. Cleymans, R.V. Gavai, and E. Suhonen, Phys. Rep. **130** (1986) 217.

[10] A. Bialas, and R. Peschanski, Nucl. Phys. B **308** (1988) 857.

[11] F. Wilczek, Int. J. Mod. Phys. A **7** (1992) 3911.

[12] K. Rajagopal, and F. Wilczek, Nucl. Phys. B **399** (1993) 395; ibid. B **404** (1993) 577.

[13] J.D. Bjorken, K. L. Kowalski, and C. C. Taylor, Observing Disoriented Chiral Condensates, Proceedings of the workshop on Physics at Current Accelerators and the Supercollider, Argonne (1993), arXiv:hep-ph/9309235, and SLAC preprint SLAC-PUB-6109 (April 1993).

[14] S. Gavin, A. Gocksch, and R. D. Pisarski, Phys. Rev. D **49** (1994) R3079.

[15] S. Gavin, A. Gocksch, and R. D. Pisarski, Phys. Rev. Lett. **72** (1994) 2143.

[16] C.M. Ko, and M. Asakawa, Nucl. Phys. A **566** (1994) 447c.

[17] C.M. Ko, private communication (1995).

[18] A. Bialas, Phys. Lett. **B442** (1998) 449; Phys. Lett. **B532** (2002) 249.

[19] U.Heinz, *Hunting Down the Quark-Gluon Plasma in Relativistic Heavy-Ion-Collisions*, arXiv: hep-ph/9902424 and CERN-TH/99-45.

[20] B. Banerjee, and R. V. Gavai, Phys. Lett. B **293** (1992) 157.

[21] D.N. Schramm, B. Fields, and D. Thomas, Nucl. Phys. A **544** (1992) 267c.

[22] G.M. Fuller, G.J. Mathews, and C.R. Alcock, Phys. Rev. D **37** (1988) 1380.

[23] J.H. Applegate, C.J. Hogan, and R.J. Sherrer, Phys. Rev. D **35** (1987) 1151.

[24] J. Polchinsky, Nucl. Phys. **B231** (1984) 269.

[25] T. Reisz, Z. Phys. C **53** (1992) 169.

[26] T. Reisz, and B. Petersson, Nucl. Phys. **B353** (1991) 757.

[27] A. D. Linde, Phys.Lett. **70 B** (1977) 306; **100 B** (1981) 37; Nucl.Phys.B **216** (1983) 421.

[28] M.Wortis, *Linked Cluster Expansion*, in: *Phase Transitions and Critical Phenomena*, vol.3, ed. by C.Domb and M.S.Green, Academic Press (London, 1974).

[29] M.Lüscher and P.Weisz, Nucl.Phys.B **300** [FS22] (1988) 325; ibid. B **290** [FS20] (1987) 25; ibid. B **295** [FS21] (1988) 65.

[30] T.Reisz, Nucl.Phys.B **450** (1995) 569; Phys.Lett. **360 B** (1995) 77.

[31] H.Meyer-Ortmanns and T.Reisz, J. Stat. Phys. **87** (1997) 755; Int. J. Mod. Phys. **A14** (1999) 947.

[32] W. Buchmüller, Z. Fodor, T. Helbig, D. Walliser, Ann. Phys. **234** (1994) 260.

[33] L.D. Landau, and E.M. Lifshitz, Statistical Physics (Pergamon, Oxford) (1958).

[34] K. Aizu, Phys. Rev. B **2** (1970) 754.

[35] L. Michel, Rev. Mod. Phys. **52**m (1980) 617.

[36] J.-C. Toledano, Ferroelectrics **35** (1981) 31.

[37] K. Binder, Rep. Prog. Phys. **50** (1987) 783.

[38] F. J. Wegner, Phys. Rev. B **5** (1972) 4529.

[39] K. Wilson, and J. Kogut, Phys. Rep. C **12** (1974) 75.

[40] M.E. Fisher, Rev. Mod. Phys. **46** (1974) 597.

[41] S.-K. Ma, Modern Theory of Phase Transitions, (Reading, MA: Benjamin) (1976).

[42] L.O'Raifeartaigh, A. Wipf and H. Yoneyama, Nuc.Phys.B**271** (1986) 653.

[43] J.-M. Drouffe and J.-B. Zuber, Phys. Rep. 102 (1983) 1 .

[44] H. Meyer-Ortmanns and T. Reisz, Int. J. Mod. Phys. **A14** (1999) 947.

[45] Z. Fodor, J. Hein, K. Jansen, A. Jaster and I. Montvay, Nucl. Phys. **B439** (1995)147.

[46] M. Lüscher and P. Weisz, Nucl. Phys. B300[FS22] (1988) 325.

[47] T. Reisz, Phys. Lett. **360B** (1995) 77.

[48] H. Meyer-Ortmanns and T. Reisz, *Analytical Studies for the critical line and critical end-point of the electroweak phase transition* in: Strong and electroweak matter - 1998, eds. J Ambjorn et al., (World Scientific, Singapore) (1999).

[49] K. Huang, "Statistical Mechanics" John Wiley (New York) (1987).

[50] H. Meyer-Ortmanns and T. Reisz. Int. J. Mod. Phys. **A14** (1999) 947.

[51] K.G. Wilson, J. Kogut, Phys. Rep. C **12** (1974) 75.

[52] F. J. Wegner, Phys. Rev. B **5** (1972) 4529.

[53] I. Cardy, *Scaling and Renormalization in Statistical Physics*, Cambridge Lecture Notes in Physics, Cambridge University Press (Cambridge 1996).

[54] G. Mack, *Multigrid methods in quantum field theory*, in: Proceedings Cargése 1987: Nonperturbative Quantum Field Theory, p. 309-351.

[55] K. Gawdezki, A. Kupiainen, Comm. Math. Phys. **99** (1985) 197.

[56] K. Gawdezki, A. Kupiainen, Comm. Math. Phys. **88** (1983) 77.

[57] V. Rivasseau, *From perturbative to constructive renormalization*, Princeton University Press, (Princeton 1991).

[58] M.N. Barber, *Phase Transitions and Critical Phenomena*, vol. 8, ed. by C. Domb and J.L. Lebowitz (Academic, New York), p. 145 (1983).

[59] J. Cardy, *Scaling and Renormalization in Statistical Physics*, Cambridge Lecture Notes in Physics 5, Cambridge University Press, (Cambridge 1987).

[60] J. L. Alonso et al., Nucl. Phys. B**405** (1993) 574.

[61] N. B. Wilding, J. Phys. :Cond. Mat.9 (1997) 585 and references therein.

[62] K. Rummkainen, M. Tsypin, K. Kajantie, M. Laine and M. Shaposhnikov, Nucl.Phys.B **532** (1998) 283.

[63] C.Borgs, and R. Kotecký, J. Stat. Phys. **61** (1990) 79.

[64] A. Billoire, R. Lacaze, and A. Morel, Nucl. Phys. B **370** (1992) 773.

[65] K. Binder, and D.P. Landau, Phys. Rev. B **30**(1984) 1477.

[66] M. S. Challa, D.P. Landau, and K. Binder, Phys. Rev. B **34** (1986) 1841.

[67] H. Meyer-Ortmanns and T. Reisz, J. Math. Phys.39 (1998) 5316.

[68] H. Meyer-Ortmanns and T. Reisz, J. Stat. Phys. **87** (1997) 755.

[69] H. Meyer-Ortmanns and T. Reisz, Eur. Phys. J. B**27** (2002) 549.

[70] H. Meyer-Ortmanns and T. Reisz, *Analytical Studies of the critical line and critical endpoint of the electroweak phase transition* in: Strong and Electroweak Matter -'98, ed. by J.Ambjorn et al., World Scientific, (Singapore 1999).

[71] T. Reisz, in: *Field theoretical tools for polymer and particle physics*, ed. by H. Meyer-Ortmanns and A. Klümper, Springer (Berlin 1998).

[72] A. Billoire, *A Monte Carlo study of the Potts model in two and three dimensions*, preprint SPHT/91/014 (1991).

[73] A. Billoire, S. Gupta, A. Irbäck, R. Lacaze, A. Morel, and B. Petersson, Phys. Rev. B **42** (1990) 6743.

[74] H. E. Stanley, *An Introduction to Phase Transitions and Critical Phenomena*, Oxford University Press, (Oxford 1971).

[75] C. Callan, R. Dashen, and D.J. Gross, Phys. Lett. B **63** (1976) 334.

[76] E. Witten, Nucl. Phys. B **156** (1979) 269.

[77] R.D. Pisarski, and F. Wilczek, Phys. Rev. D **29** (1984) 338.

[78] A.J. Paterson, Nucl. Phys. B **190** [FS3] (1981) 188.

[79] P. Bak, S. Krinsky, and D. Mukamel, Phys. Rev. Lett. **36** (1976) 52.

[80] H.H. Iacobson, and D.J. Amit, Ann. Phys. (N.Y.) **133** (1981) 57.

[81] D.J. Gross, R.D. Pisarski, and L.G. Yaffe, Rev. Mod. Phys. **53** (1981) 43.

[82] B. Svetitsky, and L.G. Yaffe, Phys. Rev. D **26** (1982) 963.

[83] B. Svetitsky, and L.G. Yaffe, Nucl. Phys. B **210**[FS6] (1982) 423.

[84] R.P. Feynman, Nucl. Phys. B **188** (1981) 479.

[85] P. Pfeuty, and G. Toulouse, *Introduction to the Renormalization Group and Critical Phenomena* (Wiley, New York) (1977).

[86] T. Celik, J. Engels, and H. Satz, Phys. Lett. B **125** (1983) 411.

[87] T. Celik, J. Engels, and H. Satz, Phys. Lett. B **129** (1983) 323.

[88] P. Bacilieri, *et al.*, Phys. Rev. Lett. **61** (1988) 1545.

[89] M. Fukugita, M. Okawa, and A. Ukawa, Phys. Rev. Lett. **63** (1989) 1768.

[90]   T. Banks, and A. Ukawa, Nucl. Phys. B **225** [FS9] (1983) 145.

[91]   T.A. DeGrand, and C.E. DeTar, Nucl. Phys. B **225** (1983) 590.

[92]   F.R. Brown, F.P. Butler, H. Chen, N.H. Christ, Z. Dong, W. Schaffer, L. Unger, and A. Vaccarino, Phys. Rev. Lett. **65** (1990) 2491.

[93]   F.R. Brown, F.P. Butler, H. Chen, N.H. Christ, Z. Dong, W. Schaffer, L. Unger, and A. Vaccarino, Phys. Lett. B **251** (1990) 181.

[94]   F. Wilczek, Int. J. Mod. Phys. A **7** (1992) 3911.

[95]   K. Rajagopal, and F. Wilczek, Nucl. Phys. B **399** (1993) 395; ibid. B **404** (1993) 577.

[96]   S. Gavin, A. Gocksch, and R. D. Pisarski, Phys. Rev. D **49** (1994) R3079.

[97]   ALEPH, DELPHI,L3,OPAL, and the LEP Higgs Working Group, report-no: LHWG Note/2001-03, and arXiv: hep-ex/0107029.

[98]   A.D.Sakharov, JETP Lett.**5** (1967) 24 .

[99]   W. Buchmüller, T. Helbig, and D. Walliser, Nucl.Phys.**B407** (1993) 387.

[100]   W. Buchmüller, Z. Fodor, T. Helbig, D. Walliser, *Ann. Phys.***234** (1994) 260 .

[101]   P. H. Damgaard and U. M. Heller, Nucl.Phys. **B294** (1987) 253 .

[102]   H.G.Evertz, J.Jersak, K.Kanaya, Nucl.Phys. **B285** [FS19] (1987) 229 .

[103]   K. Jansen and W. Seuferling, Nucl. Phys. **B343** (1990) 507 .

[104]   H.Meyer-Ortmanns, Nucl.Phys.**B235**[FS11] (1984) 115 .

[105]   K. Jansen, Nucl.Phys.Proc.Suppl.**47** (1996) 196 .

[106]   B.Bunk,E.-M. Ilgenfritz,J.Kripfganz,and A.Schiller, Phys. Lett. **B284** (1992) 371 ; Nucl.Phys. **B403** (1993) 453 ; E.-M.Ilgenfritz and A.Schiller, Nucl.Phys.B (Proc.Suppl.)**42** (1995) 578 .

[107]   Z.Fodor, J.Hein, K.Jansen, A.Jaster, and I.Montvay, Nucl.Phys.**B439** (1995) 147 ; F.Csikor, Z.Fodor, J.Hein, K.Jansen, A.Jaster, and I.Montvay, Phys.Lett.**B334**(1994) 405 ;Nucl.Phys.B(Proc.Suppl.)**42**(1995) 569 ; F.Csikor, Z.Fodor, J.Hein, J.Heitger, CERN-TH-95-170; Z.Fodor, and K.Jansen, Phys.Lett.**B331**(1994) 119 .

[108]   Y.Aoki, Nucl.Phys.B (Proc.Suppl.)**53** (1997) 609 and arXiv:hep-lat/9612023.

[109]   H.Meyer-Ortmanns and T.Reisz, *Analytical Studies for the critical line and critical end-point of the electroweak phase transition,* in: *Strong and Electroweak Matter-'98,* eds. J.Ambjorn et al., World Scientific, Singapore (1999).

[110]   M. Gürtler, E.-M. Ilgenffritz, J. Kripfganz, H. Perlt, and A. Schiller, Nucl.Phys.**B483** (1997) 383 .

[111]   K. Kajantie, M. Laine, K. Rummukainen, and M. Shaposhnikov, Nucl.Phys.**B466** (1996) 189 .

[112]   K. Kajantie, M. Laine, K. Rummukainen, and M. Shaposhnikov, Phys.Rev.Lett.**77** (1996) 2887 .

[113]   F. Karsch, T. Neuhaus, A. Patkos, and J. Rank, Nucl.Phys.**B53** (Proc.Suppl.) (1997) 623 .

[114]   F.Csikor, Z.Fodor, J.Hein, K.Jansen, A.Jaster, and I.Montvay, Phys.Lett.**B334** (1994) 405 ; J. Hein, and J. Heitger, Phys.Lett.**B385** (1996) 242 .

[115] K. Kajantie, L. Kärkkäinen, and K. Rummukainen, Nucl.Phys.**B333** (1990) 100 .

[116] E.-M. Ilgenfritz and A. Schiller, Nucl.Phys.**B42** (Proc.Suppl.) (1995) 578 .

[117] E. Brezin and J. Zinn-Justin, Nucl.Phys.**B257**[FS14], 867 (1985).

[118] H. Meyer-Ortmanns and T. Trappenberg. J.Stat.Phys.**58** (1990) 185 .

[119] G. Münster, Nucl.Phys.**B324** (1989) 630.

[120] B. Bunk, Nucl.Phys. **B42** Proc.Suppl.(1995) 566 .

[121] M. Gürtler, E.-M. Ilgenffritz, and A. Schiller, Phys.Rev.D**56** (1997) 3888 .

[122] K. Rummukainen, M. Tsypin, K. Kajantie, M. Laine, and M. Shaposhnikov, Nucl.Phys.B **532** (1998) 283 .

[123] E. Marinari, *Optimized Monte Carlo Methods*, Lectures at the 1996 Budapest Summer School on Monte Carlo Methods, cond-mat/9612010.

[124] Positivity of the transfer matrix: C. Borgs and E. Seiler, Nucl. Phys. **B215** (1983) 125; Comm. Math. Phys. **91** (1983) 329.

[125] I. O. Stamatescu and E. Seiler, Phys.Rev.D**25**(1982) 1130.

[126] C. E. DeTar, Phys.Rev.D **32** (1985) 276.

[127] C. E. DeTar, Nucl. Phys. B Proc. Suppl.**42** (1995) 73.

[128] F. Karsch, in *QCD 20 Years Later*, ed. by P. M. Zerwas and H. A. Kastrup (World Scientific Singapore) (1993), Vol.2, p.717, and references therein.

[129] H. Rothe, *Introduction to Lattice Gauge Theories* (World Scientific, Singapore)(1992).

[130] J. B. Kogut, Rev. Mod. Phys. **55** (1983) 775.

[131] F. Karsch, Nucl.Phys.**B205**[FS5](1982) 285.

[132] F. A. Berezin, *The Method of Second Quantization, Pure and Applied Physics*, Vol. 24 (Academic, New York)(1966).

[133] H. B. Nielsen and M. Ninomiya, Nucl. Phys. B **185** (1981) 20; Phys. Lett. **105**B (1981) 219.

[134] M. Creutz, L. Jacobs, and C. Rebbi, Phys. Reports **93** (1983) 201.

[135] P. H. Ginsparg, K.G. Wilson, *Phys. Rev.* **D25** (1982) 2649.

[136] Renormalization of lattice gauge theories with massless Ginsparg-Wilson fermions: T. Reisz and H.J. Rothe, Nucl. Phys. **B575** (2000) 255.

[137] Transfer matrix construction for lattice gauge theories: M. Lüscher, Comm. Math. Phys. **54** (1977) 283.

[138] D. E. Groom et al., Eur. Phys. J. C **15** (2000) 1.

[139] H. N. Brown et al, arXiv:hep-ex/0102017.

[140] G. N. Watson, Philos. Trans. Soc. London, Ser. A **211** (1912) 279; G. H. Hardy, *Divergent Series*, Oxford University Press, (London 1949); F. Nevanlinna, Ann. Acad. Sci. Fenn. Ser. A **12**, no.3 (1918-19); A. D. Sokal, J. Math. Phys. **21(2)** (1980) 261.

[141] C. de Calan and V. Rivasseau, Comm. Math. Phys. **82** (1981) 69.

[142] J. E. Mayer and E. Montroll, J. Chem. Phys. **9** (1941) 2.

[143] J. G. Kirkwood and Z. Salsburg, Faraday Soc.**15** (1953) 28.

[144] J. Glimm and A. Jaffe, *Quantum Physics, A Functional Integral Point of View*, Springer (Berlin, 1981).

[145] Ch. Gruber and A. Kunz, Comm. Math. Phys.**22** (1971) 133.

[146] D. Ruelle, *Statistical Mechanics: rigorous results* (Benjamin, New York 1966).

[147] G. Münster, Nucl. Phys. B **180** [FS2] (1981) 23.

[148] M. Creutz, Phys. Rev. D **21** (1980) 2308.

[149] G. Mack and H. Meyer, Nucl. Phys. B **200** [FS4] (1982) 249.

[150] H. Meyer-Ortmanns, Nucl. Phys. B **235** [FS11] (1984) 115.

[151] R. Marra and S. Miracle Sol'e, Comm. Math. Phys. **67** (1978) 233.

[152] H. Meyer-Ortmanns, Nucl. Phys. B **230** [FS10] (1984) 31.

[153] G. Münster, Nucl. Phys. B **180** [FS2] (1981) 23.

[154] J. B. Kogut, Rev. Mod. Phys. **55** (1983) 775.

[155] M. Creutz, *Quarks, gluons and lattices*, Cambridge University Monographs on Mathematical Physics, Cambridge University Press, (Cambridge, 1983).

[156] K. Osterwalder and E. Seiler, Ann. Phys. **110** (1978) 440.

[157] J. Glimm, A. Jaffe and T. Spencer, Contributions in *Constructive Quantum Field Theory*, ed. by G. Velo and A. Wightman, Springer Lecture Notes in Physics, Vol.25, Springer (Berlin, 1973).

[158] H. Meyer-Ortmanns and T. Reisz, Int. J. Mod. Phys. A**14** (1999) 947.

[159] T. Reisz, Nucl. Phys. B **450** [FS] (1995) 569-602.

[160] T. Reisz, Phys. Lett. **360 B** (1995) 77-82.

[161] I.Montvay and G.Münster, *Quantum Fields on the Lattice*, Cambridge Monographs on Mathematical Physics, Cambridge University Press (Cambridge 1994).

[162] M.Wortis, *Linked Cluster Expansion*, in: *Phase Transitions and Critical Phenomena*, vol.3, ed. by C. Domb and M.S. Green, Academic Press (London, 1974); C. Itzykson, J.-M.Drouffe, *Statistical Field Theory*, vol.2, Cambridge University Press, (Cambridge 1989); A.J. Guttmann, *Asymptotic Analysis of Power-Series Expansions*,in: *Phase Transitions and Critical Phenomena*, vol.13, ed. by C. Domb and J.L. Lebowitz (Academic Press).

[163] M. Lüscher and P. Weisz, Nucl.Phys.B **300** [FS22] (1988) 325; ibid. B **290** [FS20] (1987) 25; ibid. B **295** [FS21] (1988) 65.

[164] T. Reisz, Nucl.Phys.B **450** (1995) 569; Phys.Lett. **360 B** (1995) 77.

[165] H. Meyer-Ortmanns and T. Reisz, J. Stat. Phys. **87** (1997) 755.

[166] H. Meyer-Ortmanns and T. Reisz, Int. J. Mod. Phys. A **14** (1999) 947.

[167] Power-counting theorem in continuum field theories: Y. Hahn, W. Zimmermann, *An elementary proof of Dyson's power-counting theorem*, Comm. Math. Phys. **10** (1968) 330. Cf. also S. Weinberg, *High-energy behavior in quantum field theory*, Phys. Rev **118** (1960) 838.

[168] The forest formula: W. Zimmermann, *Convergence of Bogoliubov's method of renormalization in momentum space*, Comm. Math. Phys. **15** (1969) 208.

[169] The lattice power-counting theorem: T. Reisz, *A power-counting theorem for Feynman integrals on the lattice*, Comm. Math. Phys. **116** (1988) 81. Renormalization of lattice field theories: T. Reisz, *Renormalization of Feynman integrals on the lattice*, Comm. Math. Phys. **117** (1988) 79. M. Lüscher, *Selected topics in lattice field theory*, Lectures given at the summer school on *Fields, Strings and Critical Phenomena*, Les Houches (1988).

[170] IR-power counting and renormalization in continuum field theories: J. H. Lowenstein, W. Zimmermann, *The power-counting theorem for Feyn-*

man integrals with massless propagators, Comm. Math. Phys. **44** (1975) 73; G. Bandelloni, G. Becchi, A. Blasi, R. Collina, *Renormalization of models with radiative mass generation*, Comm. Math. Phys. **67** (1978) 147, and J. H. Lowenstein, *Convergence theorems for renormalized Feynman integrals with zero-mass propagators*, Comm. Math. Phys. **47** (1976) 53.

[171] IR power counting and renormalization of massless field theories on the lattice: T. Reisz, *A convergence theorem for lattice Feynman integrals with massless propagators*, Comm. Math. Phys. **116** (1988) 573, and T. Reisz, *Renormalization of lattice Feynman integrals with massless propagators*, Comm. Math. Phys. **117** (1988) 639.

[172] Renormalization of lattice gauge theories: T. Reisz, *Lattice gauge theory: renormalization to all orders in the loop expansion*, Nucl. Phys.B **318** (1989) 417.

[173] Heavy quark potential for SU(2), both Monte Carlo and weak coupling expansions: A. Irbäck, P. Lacock, D. Miller, B. Petersson, T. Reisz, *The heavy quark potential in SU(2) gauge theory at high temperature*, Nucl. Phys.B **363** (1991) 34.

[174] Faddeev-Popov trick, gauge fixing, derivation of effective action for STALG in the finite volume: T. Reisz, *Functional measure for gauge fields on periodic lattices*, J. Math. Phys. **32** (1991) 515.

[175] Finite T/V perturbation theory, coupled small gauge coupling/large volume expansion: B. Petersson, T. Reisz, *Polyakov loop correlations at finite temperature*, Nucl. Phys.B **353** (1991) 757.

[176] Finite temperature renormalization: C. Kopper, V. Müller and T. Reisz, Annales Henri Poincaré **2** (2001) 387.

[177] IR-problems in finite T QCD: A. D. Linde, *Infrared problems in the thermodynamics of the Yang-Mills gas*, Phys. Lett. **96B** (1980) 289, and *Confinement of monopoles at high temperature*, Phys. Lett. **96B** (1980) 293.

[178] T = 0-Vacuum structure of Yang-Mills theories in the finite volume: M. Lüscher, *Some analytical results concerning the mass spectrum of Yang-Mills gauge theories on a torus*, Nucl. Phys.B **219** (1983) 233, and T = 0 zero modes in perturbation theory: A. Coste, A. Gonzales-Arroyo, J. Jurkiewicz, C. P. Korthals Altes, *Zero momentum contribution to Wilson loops in periodic boxes*, Nucl. Phys.B **262** (1985) 67.

[179] Various analytical techniques for lattice gauge theories: M. Lüscher, P. Weisz, *Efficient numerical techniques for perturbative lattice gauge theory computations*, Nucl. Phys.B **266** (1986) 309.

[180] E. Braaten, R.D. Pisarski, *Soft amplitudes in hot gauge theories: a general analysis*, Nucl. Phys.B **337** (1990) 569.

[181] J.O.Andersen, E.Braaten, M. Strickland, Phys.Rev.D **61** (2000) 014017.

[182] G.Boyd et al., Phys.Rev.Lett. **75** (1995) 4169; Nucl.Phys.B **469** (1996) 419.

[183] W. Buchmüller, Z. Fodor, T. Helbig, and D. Walliser, *The weak electroweak phase transition*, Ann. Phys. **234** (1994) 260.

[184] Appelquist-Carazzone theorem first formulated in: T. Appelquist, J. Carazzone, *Infrared Singularities and Massive Fields*, Phys. Rev.S **11** (1975) 2856.

[185] First proof of the Appelquist-Carazzone theorem in the BPHZ-approach using the alpha-parametric representation of Feynman integrals: J. Ambjørn, *On the Decoupling of Massive Particles in Field Theory*, Comm. Math. Phys. **67** (1979) 109.

[186] The no-go of classical dimensional reduction in renormalizable quantum field theory was studied in: N. P. Landsman, *Limitations to Dimensional Reduction at High Temperature*, Nucl. Phys.B **322** (1989) 498.

[187] A variant of the proof of the Appelquist-Carazzone theorem and its failure at high temperature: N. P. Landsman, *Large-Mass and High-Temperature Behavior in Perturbative Quantum Field Theory*, Comm. Math. Phys. **125** (1989) 643.

[188] The way dimensional reduction is realized in quantum field theory with applications to gauge theories: T. Reisz, *Realization of Dimensional Reduction at high Temperature*, Z. Phys. C (Particles and Fields) **53** (1992) 169.

[189] Dimensional reduction beyond perturbation theory: P. Lacock, D. E. Miller, T. Reisz, *Dimensional reduction of $SU(2)$ gauge theory beyond the perturbative horizon*, Nucl. Phys.B **369** (1992) 501.

[190] Application to $SU(3)$: L. Kärkäinen, P. Lacock, B. Petersson, T. Reisz, *Dimensional reduction and color screening in QCD*, Phys. Lett. **282B** (1992) 121.

Application to QCD: L. Kärkäinen, P. Lacock, B. Petersson, T. Reisz, *Dimensional reduction and color screening in QCD*, Nucl. Phys.B **395** (1993) 733.

[191] Detailed investigation of the heavy quark potential in four dimensions, both by Monte Carlo and perturbation theory, T-dependent flow of the renormalized coupling constant $g(T)$ of the coupled large volume/small gauge coupling expansion: A. Irbäck, P. Lacock, D. Miller, B. Petersson, T. Reisz, *The heavy quark potential in $SU(2)$ gauge theory at high temperature*, Nucl. Phys.B **363** (1991) 34.

[192] Debye mass in resummed perturbation theory to subleading order: A. K. Rebhahn, *Nonabelian Debye screening in one-loop resummed perturbation theory*, Nucl. Phys.B **430** (1994) 319.

[193] Bad convergence of the MS-bar scheme in QCD for use in dimensional reduction: K. Kajantie, M. Laine, K. Rummukainen, M. Shaposhnikov, *3D $SU(N)$+Adjoint Higgs theory and finite-temperature QCD*, Nucl. Phys.B **503** (1997) 357.

[194] Gross-Neveu Model with large but finite number of flavors: T. Reisz, *The Gross-Neveu model and QCDs chiral phase transition*, in *Field theoretical tools for polymer and particle physics*', ed. by H. Meyer-Ortmanns and A. Klümper, Lecture Notes on Physics **508**, (Springer Berlin 1998).

[195] $O(4)$-scenario for QCDs chiral transition: R. D. Pisarski, F. Wilczek, *Remarks on the chiral phase transition in chromodynamics*, Phys. Rev.D **29** (1984) 338; K. Rajagopal, F. Wilczek, *Static and dynamic critical phenomena of a second-order QCD phase transition*, Nucl. Phys.B **399** (1993) 395.

[196] Gaussian scenario for QCDs chiral transition: A. Kocic, J. Kogut, *Can sigma models describe the finite-temperature chiral transition?*, Phys. Rev. Lett. **74** (1994) 3109.

[197] Renormalizability of the three-dimensional Gross-Neveu model in the large-N expansion: B. Rosenstein, S. H. Park, B. J. Warr, *Four-fermi theory is renormalizable in 2+1 dimensions*, Phys. Rev. Lett. **62** (1989) 1433. Renormalizability proof in the framework of constructive field theory: *Constructing the three-dimensional Gross-Neveu model with a large number of flavor components*, Phys. Rev. Lett. **66** (1991) 3233.

[198] Proof that the Polyakov loop is nonvanishing at high temperature, using infrared bounds: C. Borgs, E. Seiler, *Quark deconfinement at high temperature*, Nucl. Phys.B **215** (1983) 125.

[199] K. Kajantie, M. Laine, K. Rummukainen and M. Shaposhnikov, *Generic rules for high-temperature dimensional reduction and their application to the standard model*, Nucl.Phys.B **458** (1996) 90; The electroweak phase transition: A nonperturbative analysis, Nucl.Phys.B **466** (1996) 189; F. Farakos, K. Kajantie, K. Rummukainen and M. Shaposhnikov, *3D physics and the electroweak phase transition: perturbation theory* Nucl.Phys.B **425** (1994) 67.

[200] For Wilson's flow equations for continuous renormalization-group transformations and their use for renormalizability see: J. Polchinski, *Renormalization and Effective Lagrangians*, Nucl. Phys. B **231** (1984) 269.

[201] For strict renormalizability proofs using flow equations see: G. Keller, C. Kopper, M. Salmhofer, *Perturbative Renormalization and Effective Lagrangians in ϕ_4^4*, Helv. Phys. Acta **65** (1992) 33.

[202] G. Keller, C. Kopper, *Perturbative Renormalization of Massless ϕ_4^4 with Flow Equations*, Comm. Math. Phys. **161** (1994) 515.

[203] G. Keller, C. Kopper, *Renormalizability Proof for QED based on Flow Equations*, Comm. Math. Phys. **176** (1996) 193.

[204] C. Kopper, V.F. Müller, *Renormalization proof for spontaneously broken Yang-Mills theory with flow equations*, Comm. Math. Phys. **202** (1999) 89.

[205] As phenomenological application of flow equations see: N. Tetradis, C. Wetterich, *High Temperature Phase Transitions without Infrared Divergencies*, Int. J. Mod. Phys. A **9** (1994) 4029.

[206] N. Tetradis, C. Wetterich, *Critical Exponents from the Effective Average Action*, Nucl. Phys. B **422**[FS] (1994) 541.

[207] As reference which demonstrates the importance of the scale-dependent vacuum energy for renormalization-group transformations see: K. Gawdezki, A. Kupiainen, *Massless Lattice Φ^4 theory in four dimensions: a nonperturbative control of a renormalizable model*, Comm. Math. Phys. **99** (1985) 197.

[208] N. A. Metropolis, M.N. Rosenbluth, A.H. Rosenbluth, E. Teller, and J. Teller, J. Chem. Phys. **21** (1953) 1087.

[209] B. A. Berg, *Markov Chain Monte Carlo Simulations and Their Statistical Analysis*, World Scientific (2000)

[210] D. Toussaint, "Introduction to algorithms for Monte Carlo simulations

and their application to QCD"; lectures presented at the Symposium on New Developments in Hardware and Software for Computational Physics (Buenos Aires) (1988).

[211] H. J. Herrmann and F. Karsch, eds., *Fermion Algorithms, Workshop on Fermion Algorithms* (World Scientific, Singapore)(1991).

[212] D. N. Petcher and D.H. Weingarten, Phys. Lett. **99**B (1981) 333.

[213] S. Duane, A.D. Kennedy, B.J. Pendleton, and D. Roweth, Phys. Lett. B **195** (1987) 216.

[214] G. Parisi and Y.-S. Wu, Sci. Sin. **24** (1981) 483.

[215] M. Fukugita and A. Ukawa, Phys. Rev. Lett. **55** (1985) 1854.

[216] G.G. Batrouni, G.R. Katz, A.S. Kronfeld, G.P. Lepage, B. Svetitsky, and K.G. Wilson, Phys. Rev. D **32** (1985) 2736.

[217] D.J.E. Callaway and A. Rahman, Phys. Rev. Lett. **49** (1982) 613.

[218] D.J.E. Callaway and A. Rahman, Phys. Rev. D **28** (1983) 1506.

[219] S. Duane, Nucl. Phys. B **257** [FS14] (1985) 652.

[220] F. Karsch, private communication (1992).

[221] B. Baumann and B. Berg, Phys. Lett. B **164** (1985) 131.

[222] H.W.J. Blöte and M.P. Nightingale, Physica **112**A (1982) 405; V. Privman, and M.E. Fisher, J. Stat. Phys. **33** (1983) 385.

[223] M. Fukugita, M. Okawa, and A. Ukawa, Phys. Rev. Lett. **63** (1989) 1768.

[224] G. Boyd, J. Fingberg, F. Karsch, L. Kärkkäinen, and B. Petersson, Nucl. Phys. B **376** (1992) 199.

[225] R. D. Pisarski and F. Wilczek, Phys. Rev. D **29**, 338 (1984).

[226] P.H. Damgaard, N. Kawamoto and K. Shigemoto, Nucl. Phys. B **264** (1986) 1.

[227] G. Fäldt and B. Petersson, Nucl. Phys. B **264** (1986) 197.

[228] J.U. Klaetke and K.H. Mütter, Nucl. Phys. B **342** (1990) 764.

[229] K.J.M. Moriarty, Phys. Lett. **106B** (1981) 130.

[230] M. Creutz, Phys. Rev. Lett. **46** (1981) 1441.

[231] G. Bhanot, and M. Creutz, Phys. Rev. D **24** (1981) 3212.

[232] T. Blum, C. DeTar, U. M. Heller, L. Kärkäinen, and D. Toussaint, Nucl. Phys. B Proc.Suppl.**42**(1995) 457 .

[233] F. R. Brown, H. Chen, N.H. Christ, Z. Dong, R.D. Mawhinney, W. Schaffer, and A. Vaccarino, Phys. Rev. D **46** (1992) 5655.

[234] S. Ohta, and S. Kim, Phys. Rev. D **44** (1991) 504.

[235] A. D. Kennedy, J. Kuti, S. Meyer, and B.J. Pendleton, Phys. Rev. Lett. **54** (1985) 87.

[236] S. Gottlieb, W. Liu, R.L. Renken, R.L. Sugar, and D. Touissant, Phys. Rev. D **35** (1987a) 2531; ibid. (1987b) 3972.

[237] K. D. Born, E. Laermann, N. Pirch, T.F. Walsh, and P.M. Zerwas, Phys. Rev. D **40** (1989) 1653.

[238] R. V. Gavai, J. Potvin, and S. Sanielevici, Phys.Rev.Lett.**58**, 2519 (1987).

[239] A. Hasenfratz and F. Knechtli, Phys. Rev. D **63** (2001) 114502.

[240] H. Meyer-Ortmanns and B.J. Schaefer, Phys. Rev. D **53** (1996) 6586.

[241] F. Karsch, Nucl. Phys. B **205**[FS5] (1982) 285.

[242] J. Engels, J. Fingberg, and M. Weber, Nucl. Phys. B **332** (1990a) 737.

[243] J. Engels, J. Fingberg, F. Karsch, D. Miller, and M. Weber, Phys. Lett. B **252** (1990b) 625.

[244] J. Hoek, Nucl. Phys. B **339** (1990) 732; K. Akemi, M. Fujisaki, M. Okuda, Y. Tago, Ph. de Forcrand, T. Hashimoto, S. Hioki, O. Miyamura, T. Takaishi, A. Nakamura, and I. O. Stamatescu, Phys. Rev. Lett., (1993) 3063.

[245] T. Blum, S. Gottlieb, L. Kärkkäinen, and D. Toussaint, Phys. Rev. D **51** (1995a) 5153; T. Blum, S. Gottlieb, L. Kärkkäinen, and D. Toussaint, Nucl. Phys. B (Proc. Suppl.) **42** (1995b) 460.

[246] Z. Frei, and A. Patkós, Phys. Lett. B **229** (1989) 102.

[247] K. Kajantie, L. Kärkkäinen, and K. Rummukainen, Nucl. Phys. B **333** (1990) 100.

[248] B. Grossmann, and M.L. Laursen, Nucl. Phys. B **408** (1993) 637.

[249] Y. Iwasaki, K. Kanaya, L. Kärkkäinen, K. Rummukainen, and T. Yoshié, Phys. Rev. D **49** (1994) 3540.

[250] K. Binder, Z. Phys. B **43** (1981) 119; K. Binder, Phys. Rev. A **25** (1982) 1699.

[251] Y. Iwasaki, K. Kanaya, T. Yoshie, T. Hoshino, T. Shirakawa, Y. Oyanagi, S. Ichii, and T. Kawai, QCDPAX collaboration, Phys. Rev. Lett. **67** (1991) 3343; Y. Iwasaki, K. Kanaya, T. Yoshie, T. Hoshino, T. Shirakawa, Y. Oyanagi, S. Ichii, and T. Kawai, QCDPAX collaboration, Phys. Rev. D **46** (1992a) 4657.

[252] K. Kajantie, and H. Kurki-Suonio, Phys. Rev. D **34** (1986) 1719.

[253] K.M. Bitar, *et al.*, Phys. Rev. D **43** (1991) 2396; M. Bochicchio, L. Maiani, G. Martinelli, G. Rossi, and M. Testa, Nucl. Phys. B **262** (1985) 331; Y. Iwasaki, Y. Tsuboi, and T. Yoshié, Phys. Lett. B **220** (1989) 602.

[254] M. Fukugita, S. Ohta, and A. Ukawa, Phys. Rev. Lett. **57** (1986) 1974.

[255] P. Hasenfratz and F. Karsch, Phys. Lett. **125**B, (1983) 308.

[256] http://wwww.physics.utah.edu/ detar/milc/ .

[257] U.Wolff, Phys. Rev. Lett.**62**(1989) 361 .

[258] J.Goodman and A.D. Sokal, Phys.Rev.D **40**(1989) 2035.

[259] S.Adler, Phys.Rev.D **37**(1988) 458.

[260] U. M. Heller and H.Neuberger, Phys.Rev.D **39** (1989) 616.

[261] C. M. Decker and P.de Forcrand, Nucl. Phys. Proc. Suppl. **17** (1990) 567.

[262] H. Meyer-Ortmanns, Rev. Mod. Phys. **68** (1996) 474.

[263] M. Fukugita, H. Mino, M. Okawa, and A. Ukawa, 1990b, Phys. Rev. Lett. **65** (1990b) 816.

[264] M. Fukugita, H. Mino, M. Okawa, and A. Ukawa, 1990b, Phys. Rev. Lett. **65** (1990b) 816.

[265] S. Gottlieb, W. Liu, R.L. Renken, R.L. Sugar, and D. Touissant, Phys. Rev. D **41** (1991) 622; S. Gottlieb, Nucl. Phys. B (Proc. Suppl.) **20** (1991) 247.

[266] F. R. Brown, F. P. Butler, H. Chen, N.H. Christ, Z. Dong, W. Schaffer, L. Unger, and A. Vaccarino, Phys. Rev. Lett. **65** (1990) 2491; F.R. Brown, F.P. Butler, H. Chen, N.H. Christ, Z. Dong, W. Schaffer, L. Unger, and A. Vaccarino, Phys. Lett. B **251** (1990) 181.

[267] C. Bernard et al., Nucl. Phys. Proc. Suppl.**106** (2002) 429, and arXiv:hep-lat/0110067; C. Schmidt et al., arXiv:hep-lat/0209009; F. Karsch et al., Nucl. Phys. Proc. Suppl. **129** (2004) 614, and arXiv: heplat/0309116.

[268] C. Bernard et al., Nucl. Phys. B Proc. Suppl. **42** (1995) 449; Nucl. Phys. B Proc. Suppl. **42** (1995) 451.

[269] C.Bernard, Phys.Rev.D **9** (1974) 3312 .

[270] F. karsch, E.Laermann, and A. Peikert, Nucl. Phys. B **605** (2001) 579.

[271] P. Braun-Munziger, K. Redlich, and J. Stachel, *Particle Production in Heavy-Ion Collisions*, nucl-th/ 0304013 (2003).

[272] B. Berg, H. Meyer-Ortmanns, and A. Velytsky, Phys. Rev. D **69**; ibid. Phys. Rev. D **70** (2004) 054505.

[273] K. Jansen, J, Jersak, I. Montvay, G. Münster, T. Trappenberg and U. Wolff, Phys. Lett. **213 B** (1998) 203; K. Jansen, I. Montvay, G. Münster, T. Trappenberg and U. Wolff, Nucl. Phys. B **322** (1989) 698.

[274] H. Meyer-Ortmanns and T. Trappenberg, J. Stat. Phys. **58** (1990) 185.

[275] E. Brézin, and J. Zinn-Justin, Nucl. Phys. B **257** [FS14] (1985) 867; J.C. Niel and J. Zinn-Justin, Nucl. Phys. B **280** [FS18] (1986) 355.

[276] M. Guertler, E. M. Ilgenfritz, J. Kripfganz, H. Perlt and A. Schiller, Nucl. Phys. B **483** (1997) 383.

[277] K. Symanzik, Nucl. Phys. B **226** (1983) 187.

[278] B. Sheikhoeslami and R. Wohlert, Nucl. Phys. B **259** (1985) 572.

[279] S. Naik, Nucl. Phys. B **316** (1989) 239.

[280] S. Gottlieb, *Lattice results with three quark flavors*, arXiv:hep-lat/ 0306013.

[281] P. Braun-Munziger, D. Magesho, K. Redlich, and J. Stachel, Phys. Lett. B **518** (2001) 41.

[282] C.R. Alton, S. Ejin, S.J. Hands, O. Kacczmarek, F. Karsch, E. Laevmann, and C. Schmidt, Phys. Rev. D **68** (2003) 014507.

[283] J.F. Lagae, and D.K. Sinclair, Phys. Rev. D **59** (1999) 014511; G.P. Lepage, Phys. Rev. D **59** (1999) 074502; K. Orginos, D. Toussaint, and R.L. Sugar, Phys. Rev. D. **60** (1999) 054503.

[284] J. Berges and K. Rajagopal, Nucl.Phys.B**538** (1999) 215; M. Alford, K. Rajagopal, and F.Wilczek, Nucl.Phys.B **537** (1999) 443.

[285] M.Lutz, M.Klimt and W.Weise, Nucl.Phys.A **542**(1992) 521 .

[286] M.Askawa and K. Yazaki, Nucl. Phys. A **504** (1989) 668 .

[287] T. Hatsuda, in Quark Matter '91: 9th International Conference on Ultra-Relativistic Nucleus-Nucleus Collisions [Nucl.Phys.A **544** (1992) 27c .

[288] U. Ellwanger, Z. Phys. C **58** (1993) 619.

[289] F. Wilczek, Int. J. Mod. Phys. A **7** (1992) 3911.

[290] K. Rajagopal, and F. Wilczek, Nucl. Phys. **B399** (1993) 395; ibid. **B404** 577.

[291] K. Wilson, and J. Kogut, Phys. Rep. **C12** (1974) 75.

[292] M. Gell-Mann, and M. Levy, Nuovo Cimento **16** (1960) 705.

[293] G. Malmström, and D. Geldart, Phys. Rev. B **21** (1980) 1133.

[294] R.D. Pisarski, and F. Wilczek, Phys. Rev.**D29** (1984) 338.

[295] I.D.Lawrie and S.Sarbach, in *Phase Transitions and Critical Phenomena*, ed. by C.Domb and J.Lebowitz (Academic Press, New York, 1984), Vol.9

[296] J. Gasser, and H. Leutwyler, Ann. Phys. **158** (1984) 142.

[297] F. Karsch, C.R. Allton, S. Ejiri, S.J. Hands, O. Kaczmarek, E. Laermann, C. Schmidt, hep-lat/0309116.

[298] P. Gerber, and H. Leutwyler, Nucl. Phys. **B321** (1989) 387.

[299] H. Meyer-Ortmanns, and B.J. Schaefer, Phys. Rev. **D53** (1996) 6586.

[300] L.H. Chan, and R. W. Haymaker, Phys. Rev.**D7** (1973) 415.

[301] S. Gavin, A. Gocksch, and R. D. Pisarski, Phys. Rev. **D49**(1994) R3079.

[302] H. Meyer-Ortmanns, H.J. Pirner, and A. Patkós, Phys. Lett. **B295** (1992) 255.

[303] G.M. Welke, R. Venugopalan, and M. Prakash, Phys. Lett. **B245** (1990) 137.

[304] M. Schmidt, G. Röpke, and H. Schulz, Ann. Phys. **202** (1990) 57.

[305] A. Patel, Nucl. Phys. **B243** (1984a) 411.

[306] P.J. Flory, J. Amer. Chem. Soc. **63** (1941a) 3083; P.J. Flory, J. Amer. Chem. Soc. **63** (1941b) 3091.

[307] A. Patel, Phys. Lett. **B 39** (1984b) 394.

[308] J. Schechter, Phys. Rev.**D21** (1980) 3393.

[309] G.Ripka, *Dual Superconductor Models of Color Confinement*, Lecture Notes in Physics, Springer Berlin, 2004.

[310] K. Rajagopal and F. Wilczek, *The condensed matter physics of QCD*in *Festschrift in honor of B.L. Ioffe, At the Frontier of Particle Physics / Handbook of QCD*, vol.3, ed. by M. Shifman, World Scientific, 2000, p. 2061, see also hep-ph/0011333 .

[311] K. Kajantie, and H. Kurki-Suonio, Phys. Rev. D **34**, (1986) 1719.

[312] H. Satz, in *Proceedings Large Hadron Collider Workshop*, eds. G. Jarlskog and D. Rein, Vol. 1 (1990) p. 188.

[313] K.J. Eskola, K. Kajantie, P. Ruuskanen, and K. Tuominen, Nucl.Phys.**B570**, 379 (2000).

[314] P. Braun-Munziger, K. Redlich, and J. Stachel, arXiv:nucl-th/0304013 (2003).

[315] J. Stachel, E814 collaboration, Nucl. Phys. A **566** (1994) 183c.

[316] E.V. Shuryak, Phys. Lett. **78**B (1978) 150.

[317] G.E. Brown, J. Stachel, and G.M. Welke, Phys. Lett. B **253** (1991) 19.

[318] T. Ludlam, and N.P. Samios, Z. Phys. C **38** (1988) 353.

[319] J.P. Blaizot, and J.Y. Ollitrault, in *Quark-Gluon Plasma*, edited by R.C. Hwa (World Scientific, Singapore) (1990) p. 393.

[320] A. Hosoya, and K. Kajantie, Nucl. Phys. B **250** (1985) 666.

[321] L.D. Landau, and E.M. Lifshitz, *Course of Theoretical Physics Vol. 6, Fluid Mechanics* (Pergamon, London) (1959).

[322] F. Cooper, and G. Frye, Phys. Rev. D **10** (1974) 186.

[323] J. Cleymans, R.V. Gavai, and E. Suhonen, Phys. Rep. **130** (1986) 217.

[324] J.D. Bjorken, Phys. Rev. D **27** (1983) 140.

[325] E.V. Shuryak, Phys. Lett. **78**B (1978) 150.

[326] Chodos, A., R.L. Jaffe, K. Johnson, C.B. Thorn, and V.F. Weisskopf, Phys. Rev. D **9** (1974) 3471.

[327] K. Johnson, Acta Phys. Polonica B **6** (1975) 865.

[328] T.A. DeGrand, R.L. Jaffe, K. Johnson, and J. Kiskis, Phys. Rev. D **12** (1975) 2060.

[329] E.V. Shuryak, *The QCD Vacuum, Hadrons and the Superdense Matter* (World Scientific, Singapore) (1988).

[330] J. Kapusta, Nucl. Phys. **148**B (1979) 461.

[331] E.V. Shuryak, Phys. Rep. **61** (1979) 71.

[332] H.R. Schmidt, and J. Schukraft, J. Phys. G **19** (1993) 1705.

[333] K.J. Eskola, and J. Lindfors, Z. Phys. C **46** (1990) 141.

[334] J. Cleymans, J. Fingberg, and K. Redlich, Phys. Rev. D **35** (1987) 2153.

[335] J. Kapusta, and C. Gale, Phys. Rev. C **35** (1987) 2107.

[336] K. Redlich, and H. Satz, Phys. Rev. D **33** (1986) 3747.

[337] P.V. Ruuskanen, in *Quark-Gluon Plasma*, edited by R. Hwa (World Scientific, Singapore) (1990) p. 519.

[338] P.V. Ruuskanen, Nucl. Phys. A **544** (1992) 169c.

[339] J. Kapusta, and A. Mekjian, Phys. Rev. D **33** (1986) 1304.

[340] see for example U.Heinz, Hanbury-Brown/Twiss interferometry for relativistic heavy-ion collisions: Theoretical aspects. In:Correlations and Clustering Phenomena in Subatomic Physics, eds.M.N.Harakeh, O.Scholten, J.H.Koch, NATO ASI Series **359**, New York, Plenum (1997) 137.

[341] Ya.B Zel'dovich, *et al.*, Usp. Fiz. Nauk **152** (1987) 3 [Sov. Phys. Usp. **30**, 353 (1987)].

[342] W. Ochs, and J. Wosiek, Phys. Lett. B **214** (1988) 617.

[343] H. Satz, Nucl. Phys. B **326** (1989) 613.

[344] D. Hajduković, and H. Satz, *Does the one-dimensional Ising model show intermittency?*, preprint CERN-TH-.6674/92 and BI-TP 92/43 (1992).

[345] A. Bialas, and R. Peschanski, Nucl. Phys. B **273** (1986) 703.

[346] M. Gyulassy, Nucl. Phys. A **418** (1984) 59c; M. Gyulassy, K. Kajantie, H. Kurki-Suonio, and L. McLerran, Nucl. Phys. B **237** (1984) 477.

[347] L. van Hove, Phys. Lett. **118** B (1983) 138; L. van Hove, Z. Phys. C **21** (1983) 93.

[348] K. Rajagopal and F. Wilczek, Nucl.Phys.**B399** (1993) 395; Nucl.Phys.**404** (1993) 577 .

[349] T. R. Miller and M. C. Ogilvie, Nucl.Phys.**B106**(Proc.Suppl.) (2002) 537; Phys.Lett.**B488**(2000)313.

[350] B.A.Berg, H. Meyer-Ortmanns and A. Velytsky, Phys.Rev.**D70** (2004) 054505; Phys.Rev.**D69** (2004) 034501 .

Index